J.C. Loudon

An Encyclopaedia of Trees and Shrubs

Part I

Salzwasser

J.C. Loudon

An Encyclopaedia of Trees and Shrubs
Part I

1. Auflage | ISBN: 978-3-84605-423-9

Erscheinungsort: Frankfurt, Deutschland

Erscheinungsjahr: 2020

Salzwasser Verlag GmbH

Reprint of the original, first published in 1869.

AN

ENCYCLOPÆDIA

OF

TREES AND SHRUBS;

BEING THE

ARBORETUM ET FRUTICETUM BRITANNICUM ABRIDGED:

CONTAINING

THE HARDY TREES AND SHRUBS OF BRITAIN

NATIVE AND FOREIGN,

SCIENTIFICALLY AND POPULARLY DESCRIBED;

WITH THEIR PROPAGATION, CULTURE, AND USES IN THE ARTS;

AND

WITH ENGRAVINGS OF NEARLY ALL THE SPECIES.

ABRIDGED FROM THE LARGE EDITION IN EIGHT VOLUMES, AND

ADAPTED FOR THE USE OF

Nurserymen, Gardeners, and Foresters.

BY J. C. LOUDON, F.L.S. H.S. &c.

LONDON:

FREDERICK WARNE AND CO.,

BEDFORD STREET, COVENT GARDEN.

NEW YORK: SCRIBNER, WELFORD, AND CO.

1869.

PREFACE.

This Abridgement of the *Arboretum et Fruticetum Britannicum* contains.—

1. Characters and short Popular Descriptions of all the species and varieties of hardy trees and shrubs now in British gardens, with directions for their culture; including the soil most suitable for them, their propagation, and their uses in the arts, &c.

2. Engravings of all the species which are described, with the exception of half a dozen. The whole arranged according to the Natural System; all the engravings being to the scale of two inches to a foot, or one sixth of the natural size.

3. The Scientific Names and Scientific Synonymes of all the species; and their Popular Names in the languages of the different countries where they are indigenous or cultivated.

4. An Alphabetical Index to all the species and varieties, with their synonymes.

5. A Tabular Analysis of the Leaves, by which the name of any species of tree or shrub described in the work may, in general, be discovered, from a small portion of a shoot with the leaves on.

6. Specific Characters, Descriptions, and Figures of some species, more particularly of pines, firs, and oaks, which were not in the country in 1838, when the large work was completed.

In a word, though this Abridgement does not include all the interesting and useful information on the natural history of trees which will be found in the larger work, or any of the portraits of entire trees which constitute so distinguished a feature in it, yet it contains all that is necessary to enable the reader to discover the names of the different species, and to ascertain their culture, propagation, and uses in Britain; in short, all that is essential for the nurseryman, gardener, and forester.

The most remarkable circumstance connected with this Abridgement is, that the Author has been able to obtain figures of nearly all the species. For the drawings or specimens from which these additional figures were taken, he is indebted to the kind assistance of various Public Institutions, and of several of the most eminent botanists and possessors of herbariums and living collections in Europe and North America.

The Institutions to which he is under obligations are, the Linnean and Horticultural Societies of London, the British Museum, the Museum of Natural History of Paris, and that of Berlin: and the Botanists who have kindly lent him drawings or specimens include the late A. B. Lambert, Esq.; Sir W. J. Hooker; Dr. Lindley; the late Professor Don; George Don, Esq., who prepared the characters of the Orders and of the Genera; Messrs. Loddiges; the late Professor DeCandolle; M. Alphonse De Candolle; W. Borrer, Esq.; P. B. Webb, Esq.; Baron De Lessert; M. Michaux; Signor G. Manetti; M. Otto; M. Charles Rauch; M. Francis Rauch, who made most of the drawings; and Drs. Torrey and Gray: to all of whom; to the Curators of most of our Botanic and Horticultural Gardens, and those of many foreign ones; and to all Nurserymen and Gardeners, both at home and abroad, who may have rendered him assistance, he begs to return his most sincere thanks.

To the Council of the Horticultural Society of London he is under especial obligations, for their permission to make drawings from the cones and other specimens sent home by their collectors, Douglas and Hartweg, and for authorising him to procure information from their intelligent and experienced superintendent of the arboricultural department of the Garden, Mr. George Gordon, A.L.S.; and to Mr. Gordon he is indebted for the ready and obliging manner in which, at all times, he rendered his assistance.

Bayswater, April, 1842. J. C. L.

CONTENTS.

*** By turning to the pages of the Contents, the whole of the species and varieties, with their synonymes, of any genus or order, may be seen at a glance; and, by turning to the General Index at the end, any particular species, whether known by its general name or its synonyme, may be found at once, both in the Contents and in the body of the work.

ENUMERATION

OF THE

GENERA, SPECIES, AND VARIETIES, WITH THEIR SYNONYMES, SCIENTIFIC AND POPULAR,

IN THE ORDER IN WHICH THEY ARE DESCRIBED.

In the following Table the Synonymes are printed in Italics. English specific names are omitted when they are merely translations of the scientific names. French, German, and Italian generic names are, with few exceptions, not given when they are the same, or nearly the same, as the scientific ones; and the specific names are only given in these languages when they are synonymes, and not mere translations.

Class I. EXO'GENÆ.

Subdiv. I. DICHLAMY'DEÆ.

Subcl. I. THALAMIFLO'RÆ.

Sect. I.

Carpella, that is, the component Parts of compound Capsules or Fruits, numerous; or the Stamens placed opposite the Petals.

Ranunculàceæ.

Tribe I. *Clemati'deæ.*

I. *Cle'matis L.* - 2
Virgin's Bower, Ladies' Bower: Clématite, Fr.; *Waldrebe*, Ger.; *Clematide*, Ital.

§ i. *Flámmula* Dec.

1. Flámmula *L.* - 3
C. *ùrens* Gerard, C. *marítima* All. Ped., C. *suavèolens* Sal. Prod., C. *paniculàta* Thun.: *Sweet-scented Virgin's Bower: Clématite odorante*, Fr.; *Scharfe Waldrebe*, Ger.
2 rotundifòlia *Dec.* 3
C. *fragrans* Ten.
3 marítima *Dec.* - 3
4 rubélla *Dec.* - 3
5 cæspitòsa *Dec.* - 4
C. *cæspitòsa* Scop.
C. *Flámmula* Bert.
6 paniculàta - - 4
C. *paniculàta* Thun.

2. orientàlis *L.* - 4
C. *flàva* Mœnch, C. *glaúca* Willd., C. *ochroleùca* Hort.: *yellow-flowered V. B.*

3. chinénsis *Retz.* - 5
C. *sinénsis* Lour. coch.

4. Vitálba *L.* - - 5
Traveller's Joy.—C. *ùrens* Matth., C. *tértia* Com., *Viórna* Ger. & Lob., *Vitis nigra* Fuch.: *Old Man's Beard, Bindwith, Common V. B., Wild Climber, Great Wild Climber: Clématite brulante*, Fr.; *Gemeine Waldrebe*, Ger.; *Vite bianca*, Ital.

5. virginiàna *L.* - 6
C. *canadénsis* Mill. Dict., C. *cordifòlia* Mœnch supp., C. *triternàta* Hort.: *Broad-lvd Canada V. B.*
2 bracteàta *Dec.* - 6
C. *bracteàta* Mœnch.

6. gràta *Wall.* - 7
C. *odoràta* Hort., C. *triternàta* Hort., C. *nepalénsis* Hort.

7. Viórna *L.* - - 7
C. *purpùrea rèpens* Ray: *Leathery flowered V. B., American Traveller's Joy, Virginian Climber, Purple Climber: Clématite Viorne*, Fr.; *Glockenblüthige Waldrebe*, Ger.
2 cordàta - - 7
C. *cordàta* Sims B. M.
C. *Simsii Swt. H. B.*

8. cylíndrica *Sims* - 8
C. *crispa* Lam., C. *Viórna* Bot. Rep., C. *divaricàta* Jacq.: *Long-flowered V. B.: Clématite à longues Fleurs*, Fr.

9. reticulàta *Walt.* - 9
C. *ròsea* Abbott, C. *Simsii Hook.*

10. Hendersònii Chan. 9

§ ii. *Viticélla* Dec.

11. flórida *Thun.* - 10
Atrágene índica Desf., *Atrágene flórida* Pers.: *Clématite à grandes Fleurs*, Fr.; *Grossblüthige Waldrebe*, Ger.
2 flòre plèno *Hort* - 10
3 fl. pl. violàceo - 10
C. *f. Siebóldii D. Don.*
C. *Siebóldii Paxt.*
C. *bicolor* Hort.

12. cærùlea *Lindl.* - 11
C. *azùrea grandiflòra* Sieb., C. *grandiflòra* Hort.

13. Viticélla *L.* - 11
Viticélla deltöidea Mœnch: *Red-flowered Ladies' Bower*, Gerard: *Italienische Waldrebe*, Ger.
1 cærùlea - - 12
2 purpùrea - - 12
3 múltiplex *G. Don* 12
C. *pulchélla* Pers.
4 tenuifòlia *Dec.* - 12
C. *ten. lusitánica* Tourn.
5 baccàta *Dec.* - 12
C. *campaniflòra* Hort.

14. campaniflòra *Brot.* 12
C. *viornöides* Schrader, C. *parviflòra* Dec.
2 parviflòra *Fis. Göt.* 12

15. críspa *L.* - - 13
C. *flòre crispo* Dill. Elth.

§ iii. *Cheirópsis* Dec.

16. cirrhòsa *L.* - - 13
Atrágene cirrhòsa Pers.: *Traveller's Joy of Candia*, and *Spanish Traveller's Joy*, Gerard; *Spanish Wild Climber*, Parkins.; *Evergreen Clematis: Clématite à Vrilles*, Fr.; *Einfachblättrige Waldrebe*, Ger.
2 pedicellàta *Dec.* - 14
C. *pedicellàta* Swt. H. B.
C. *baleárica* Pers.
C. *cirrhòsa* Sims B. M.
3 angustifòlia - - 14
C. *baleárica* Rich. B. M.
C. *calycìna* Alt.
C. *polymórpha* Hort.
Clématite de Mahon, Fr.

§ iv. *Anemoniflòra.*

17. montàna *Ham.* - 15
C. *anemoniflòra* D. Don.

Other Species.—*C.* holosericea *Pursh*, Angusticifòlia *Nuttall*, Drummóndii Tor. & Gray, parviflòra *Nuttall*, lasiántha *Nutt.*, lineariloba *Dec.*, Pitcheri Tor. & Gray, pubéscens, vitifòlia, Buchaniàna - 15

II. *Atra'gene L.* - 16
Clématis Lam. & Dec.: *Atragene*, Fr. and Ger.

1. alpìna *L.* - - 16
Clématis cærùlea Bauh., *Clématis alpìna* Mill. Dict. & Dec., *A. austrìaca* Scop. & B. M., *A. clematöides* Crantz: *Atragène des Alpes*, Fr.; *Alpen Atragene*, Ger.
2 White flws. *Dec.* 16
3 sibírica ? - - 16
A. *sibírica* L.

Sect. IV.

Fruit gynobasic ; that is, inserted in a fleshy Receptacle, with which the Style is continuous.

Subcl. II. CALYCIFLO'RÆ.

s

Subdiv. II. MONOCHLAMY'DEÆ.

Ulmàceæ

Juglandàceæ.

Betulàceæ.

Empetràceæ.

Class II. ENDO'GENÆ.

Smilàceæ.

SUPPLEMENTARY SPECIES AND VARIETIES,

AND SYNONYMES.

AN ANALYSIS

OF THE

COMMONER TREES AND SHRUBS OF BRITAIN,

WITH

REFERENCE TO THEIR USES IN PLANTATIONS.

*** Where the Generic Names only are given, the reader is left to select the Species and Varieties for himself. This he will readily be able to do by turning to the pages referred to after every Generic Name, and consulting the Popular Descriptions. The insertion of the names of all the Species and Varieties in this Analysis would have swelled it to ten times its present bulk.

TREES

Trees grown chiefly for useful Purposes.

Selected with reference to their Size.

Large-sized timber trees.

Decid. — Æ'sculus Hippocástanum, 124. Cárpinus Bétulus, 916. Cárya álba, 735. Castànea vésca, 911. Fàgus sylvática, 905. Fráxinus excélsior, 639. Júglans règia, 732. Plánera Richárd*i*, 725. Plátanus orientàlis, *P.* occcidentàlis, 927. Pópulus álba, *P.* canéscens, *P.* nìgra, *P.* canadénsis, *P.* monolífera, *P.* fastigiàta, and *P.* angulàta, 819. *Quércus* pedunculàta, *Q.* sessiliflòra, *Q.* *Cérris*, *Q.* rùbra, *Q.* coccínea, and *Q.* palústris, 846. Taxòdium dístichum, 1077. *Tília* europæ`a, 63. *U'lmus* campéstris, *U.* montàna végeta, and other varieties or species of *U'lmus*, 715.

Ev.—*A'*bies excélsa, 1025. Cèdrus Libàni, 1057. *Pìcea* pectinàta, 1036. *Pìnus* sylvéstris, *P.* Larício, *P* Pallasi*àna*, and *P.* *Pinàster*, 950.

Middle-sized timber trees.

Decid. — *A'*cer platanöides, *A.* eriocárpum, and *A.* rùbrum, 79. *A'*lnus glutinòsa, *A.* incàna, and *A.* cordifòlia, 832. *Bétula* papyràcea, 837. *Cérasus* sylvéstris, and *C.* vulgàris, 276. *O'*strya vulgàris, 919. *Pópulus* balsamífera, 819. *Pýrus* commùnis, and *P.* torminàlis, 417. *Quércus* *E'*sculus, 846. Robínia Pseud-*Acàcia*, 233.

Ev. — *A'*bies álba, 1025. *I'*lex *A*quifòlium, 156. *Juníperus* virginiàna, 1080. *Pìcea* balsàmea, 1036. *Quércus* *I'*lex, 846.

Small-sized timber trees.

Decid. — *A'*cer tatáricum, *A.* spicàtum, *A.* striàtum, *A.* O'palus, *A.* monspessulànum, and *A.* campéstre, 79. *Cratæ`*gus, most of the species, 352. *Cydònia* vulgàris, 450. *Cýtisus* Labúrnum, and *C.* alpìnus, 213. *Euónymus* europæ`us, 149. *O'*rnus europæ`a, 651. *Pýrus* *A'*ria, and *P.* aucupària, 417. *Sàlix* càprea, 744.

Ev. — *A'*bies nìgra, 1025. *Juníperus* virginiàna, 1080. *Táxus* baccàta, 939.

Selected as suitable to be grown in Masses. — Social Trees.

Decid.—*A'*cer, 79. *Cýtisus*, 213. *Làrix*, 1053. *Quércus*, 846.

Ev.—*A'*bies, all the species, 1025. *Cèdrus*, 1057. *Pícea*, 1036. *Pìnus*, 950. *U'lmus*, and, in short, most trees, 715.

Selected as suitable for being grown singly, or in scattered Groups. — Solitary Trees.

Decid.—*Æ'*sculus, 124. *Cárpinus*, 916. *Fàgus*, 905. *Sàlix*, 744.

Ev.—*Cupréssus* sempervìrens, 1073.

Selected for the sake of their produce in Timber.

Produced in a short time, with a straight trunk.

Decid.—*Làrix* commùnis, 1053. *Pópulus* monilífera, *P.* fastigiàta, and *P.* álba, 819. *Sálix* álba, and *S.* Russelli*àna*, 744.

Ev.—*A'*bies excélsa, 1025. *Pìcea* pectinàta, 1036. *Pìnus* sylvéstris, and *P.* Larício, 950.

Produced in a short time, with a branchy head.

Decid.—*A'*cer Pseùdo-*Plátanus*, 79. *Æ'*sculus Hippocástanum, 124. *Fàgus* sylvática, 905. *Fráxinus* excélsior, 639. *O'*strya vulgàris, 919. *Plátanus* occidentàlis, 927. *Pópulus* nìgra, *P.* canadénsis, *P.* græ`ca, and *P.* angulàta, 819. *Quércus* *Cérris*, 846. *Sálix* frágilis, 744. *U'lmus* montàna végeta (the Huntingdon elm), and *U.* campéstris, 715.

Ev.—*Pìnus* Pinàster, *P.* sylvéstris, *P.* Larício, and *P.* Pallasi*àna*, 950.

Selected for the Quality of the Timber to be produced.

Soft woods.

Decid.—*A'*lnus, 832. *Sàlix*, 744. *Pópulus*, 819. *Æ'*sculus, 124. Liriodéndron, 36.

Hard woods.

Decid. — *A'*cer, 79. *Ailántus*, 145. *Bétula*, 837. *Cárpinus*, 916. *Cárya*, 735. *Castànea*, 911. *Céltis*, 727. *Fàgus*, 905. *Fráxinus*, 639. Gledítsch*ia*, 249. *Júglans*, 732. *O'strya*, 919. *Plánera*. 725. *Plátanus*, 927. *Quércus*, 846. Robín*ia*, 233. *U'lmus*, 715.

Very hard, compact, and durable woods.
Decid.—*Cérasus*, 276. *Cratæ'gus*, 351. *Cýtisus* Labúrnum, 213. *Euónymus*, 149. *Prùnus*, 270. *Pỳrus*, 417. Robin*ia*, 233.
Ev.—*Búxus*, 703. *Cupréssus*, 1073. *I'lex*, 156. *Juníperus*, 1080. *Phillýrea*, 631. *Thùja*, 1068.
Resinous woods.
Decid.—*Làrix*, 1053.
Ev.—*A'bies*, 1025. *Cèdrus*, 1057. *Picea*, 1036. *Pinus*, 950.

Selected for the character of the Head.
For its general bulk.
Decid.—*A'cer* Pseùdo-*Plátanus*, and *A.* obtusàtum, 79. *Æ'sculus* Hippocástanum, 124. *Cárpinus Bétulus*, 916. *Fàgus* sylvática, 905. *Làrix* europæ'a, 1053. *Pópulus*, most of the species, 819. *Quércus Cérris*, 846. *Sàlix*, several species, 744. *Tília* europæ'a, 63. *U'lmus*, most of the species, 715.
Ev.—*Pinus* sylvéstris, *P.* Laricio, *P.* *Pináster*, and *P.* Pallas*iàna*, 950.
For its upright habit.
Decid.—*A'lnus* glutinòsa laciniàta, 832. *Pópulus* fastigiàta, and *P.* monilífera, 819. *Quércus Cérris*, 846. *U'lmus* campéstris stricta, 715.
Ev.—*A'bies*, 1025. *Picea*, 1036. *Pinus*, 950.
For its spreading habit.
Decid.—*Castànea* vésca, 911. *Fàgus* sylvática, 905. *Fráxinus* excélsior, 639. *Quércus* pedunculàta, 846. *U'lmus* montàna, 715.

Selected with a view to the production of Shelter.
For general purposes.
Decid.—*A'cer* Pseùdo-*Plátanus*, 79. *Cárpinus Bétulus*, 916. *Fàgus* sylvática, 905. *Làrix* europæ'a, 1053. *Pópulus* álba, 819. *Pỳrus A'ria*, 417. *U'lmus* montàna, 715. *Pinus* sylvéstris, 950.
For drawing up young plantations.
Decid.—*Bétula* álba, 837. *Làrix* europæ'a, 1053. *Pỳrus* aucupària, 417. *Pópulus* balsamífera, and *P.* cándicans, 819.
Ev.—*A'bies* excélsa, 1025. *Pinus* sylvéstris, 950.
For protection from the sea breeze.
Decid.—*A'cer* Pseùdo-*Plátanus*, 79. *Hippóphaë* Rhamnöides, 698. *Pópulus* álba, *Pỳrus A'ria*, 417. *Sambùcus* nigra, 513.

Selected for the production of Shade.
For shade in summer only.
Most deciduous species.
For shade in summer and winter.
Most evergreen species.

Selected as adapted for particular Soil.
For dry poor soils.
Decid.—*Ailántus* glandulòsa, 145. *Bétula* álba, 837. *Cárpinus Bétulus*, 916. *Cérasus* vulgàris, and *C.* sylvéstris, 276. *Fàgus* sylvática, 905. Gledítsch*ia* triacánthos, 249. *Hippóphaë* Rhamnöides, 698. *Làrix* europæ'a, 1053. *Pỳrus A'ria*, and *P.* aucupària, 417. Robin*ia*-Pseùdo Acàcia, 233.
Ev.—*Pinus* sylvéstris, *P.* Pináster, and *P.* austrìaca, 950.
For moist soils.
Decid.—*A'cer* eriocárpum, and *A.* rùbrum, 79. *A'lnus* glutinòsa, 832. *Plátanus* occidentàlis, 927. *Pópulus*, 819. *Sàlix*, 744.
Ev.—*A'bies* commùnis, 1025.
For peaty soils.
Decid.—*Bétula* álba, 837. *Pópulus* trémula, and *P.* balsamífera, 819. *Sàlix* càprea, 744.
Ev.—*A'bies* commùnis, 1025.
For good deep soil.
Decid.—*Æ'sculus* Hippocástanum, 124. *Cárya* álba, 735. *Castànea* vésca, 911. *Jùglans* règia, 732. *Pópulus*, most of the species, 819. *Quércus*, most species, 846. *Sàlix*, 744. *U'lmus*, 715.

Selected as adapted for particular Situations.
On elevated surfaces.
Decid.—*Bétula* álba, 837. *Pỳrus A'ria*, and *P.* aucupària, 417. *Sambùcus* nigra, 513.
Ev.—*Pinus* sylvéstris, and *P.* *Cémbra*, 950.
Sheltered by houses.
Decid.—*A'cer* Pseùdo-*Plátanus*, 79. *Plátanus*, 927. *Pópulus* fastigiàta, 819. *Tília* europæ'a, 63. *U'lmus* campéstris, 715.
Exposed to the sea breeze.
Decid.—*A'cer* Pseùdo-*Plátanus*, 79. *Hippóphaë* Rhamnöides, 698. *Pópulus* álba, 819. *Pỳrus A'ria*, 417. *Sambùcus* nigra, 513.

Selected for being grown for special Purposes.
For producing an immediate screen, so dense as to interrupt the view.
Decid.—*A'cer* Pseùdo-*Plátanus*, 79. *Æ'sculus* Hippocástanum, 124. *Cárpinus Bétulus*, 916. *Fàgus* sylvática, 905. *Tília* europæ'a, 63. *U'lmus* campéstris and *U.* montàna, 715.
Ev.—*A'bies* excélsa, 1025. *Pinus* sylvéstris, *P.* Larìcio, and *P.* *Pináster*, 950.
Partially to interrupt the view.
Decid.—*Bétula* álba, 837. *Cérasus* sylvéstris, 276. *Fráxinus* excélsior, 639. Gledítsch*ia* triacánthos, 249. *Pópulus* trémula and *P.* græ'ca, 819. Robin*ia* Pseud-Acàcia, 233.
For producing timber in hedgerows.
Decid.—*A'cer* Pseùdo-*Plátanus*, 79. *Ailántus* glandulòsa, 145. *Cárya* álba, 735. *Castànea* vésca, 911. *Pỳrus* commùnis, 417. *Quércus* pedunculàta, *Q.* sessiliflòra, *Q.* Cérris, *Q.* rùbra, *Q.* coccínea, and *Q.* palústris, 846. *U'lmus* campéstris, 715.
For forming avenues.
Decid.—*A'cer* Pseùdo-*Plátanus*, 79. *Æ'sculus* Hippocástanum, 124. *Cárpinus Bétulus*, 916. *Cárya* álba, 735. *Castànea* vesca, 911. *Cérasus* sylvéstris, and *C.* vulgàris, 276. *Fàgus* sylvática, 905. *Jùglans* règia, 732. *Làrix* europæ'a, 1053. *Plátanus* orientàlis, and *P.* occidentàlis, 927. *Pópulus*, most of the species, 819. *Pỳrus* commùnis, and *P.* Màlus, 417. *Quércus*, many of the species, 846. *Robinia* Pseud-Acàcia, 233. *U'lmus*, most of the species, 715.

Ev.—*A*'bies excélsa, 1025. Cèdrus Libàni, 1057. *Picea* pectinàta, 1036. *Pinus* sylvéstris, *P.* Larício, *P.* Pallas*iàna*, *P.* *Pináster*, &c., 950. Táxus baccàta, 939.

For forming lofty hedges.

Decid. — *A*'cer campéstre, 79. *Bétula* álba, 837. *Cárpinus* *Bétulus*, 916. *Fàgus* sylvática, 905. *Làrix* europæ'a, 1053. *O'strya* vulgàris, 919. *Pópulus* fastigiàta, 819. *Tilia* europæ'a, 63. *U'lmus* campéstris, 715.

Ev. — *A*'bies excélsa, 1025. *I*'lex *A*quifòlium, 156. Juníperus virginiàna, 1060. *Picea* pectinàta, 1036. *Quércus* *I*'lex, 846. Táxus baccàta, 939.

For being periodically cut down as coppice-wood, but not in the shade of other trees.

Decid. — *Bétula* álba, 837. Castànea vésca, 911. *Fráxinus* excélsior, 639. *Quércus* pedunculàta and sessiliflòra, 846.

For being periodically cut down as coppice-wood, in the shade of other trees.

Ev. —*Cérasus* Laurocérasus, 276. *Búxus* sempervirens, 708. *I*'lex *A*quifòlium, 156. *Juniperus*, 1080. *Quércus* *I*'lex, 846. Táxus baccàta, 939

Selected in part for their produce in Fruits or Seeds.

Used in various arts and manufactures.

Decid. — *Æ*'sculus, 124. Fàgus, 905. Quércus, 846.

Used in household economy.

Decid. — Càrya, 735. Castànea, 911. *Cérasus* sylvéstris, and *C.* vulgàris, 276. Jùglans, 732. *Méspilus*, 416. Mòrus, 705. *Prùnus*, 270. *P*ỳrus, 417.

As food for singing-birds, game, or wild animals.

Decid. — *Cratæ'gus*, 352. *P*ỳrus aucupària, 417. *I*'lex *A*quifòlium, 156.

Selected in part for their produce in Bark.

For the tanner.

Decid.—*Bétula* álba, 837. Castànea vésca, 911. *Làrix* europæ'a, 1053. *P*ỳrus aucupària, 417. *Quércus* pedunculàta, and *Q.* sessiliflòra, 846. *Sàlix*, various species, 744.

For other arts.

Ev. — *I* lex *A*quifòlium, 156. Tilia europæ'a, 63.

Selected in part for the use of their Leaves.

As substitutes for spent tan in gardens.

Decid.— Quércus, 846. Castànea, 911.

As producing, by decomposition, leaf-mould.

All trees, especially the broad-leaved kinds.

As winter food for cattle.

Decid.— *Bétula*, 837. *Mòrus*, 715. *Pópulus*, 819. *P*ỳrus, 417. Robin*ia*, 233. Sàlix, 744. Tilia, 63. *U'lmus*, 715.

Trees grown chiefly for Ornamental Purposes.

Selected with reference to their Bulk.

Large-sized ornamental trees.

Decid. — *A*'cer platanöìdes. *A.* eriocárpum, and *A.* rùbrum, 79 *Æ*'sculus Hippocástanum, 124. *Ailántus* glandulòsa, 145. *A*'lnus glutinòsa laciniàta. *A.* incàna, and *A.* cordifòlia, 832. *Betula* populifòlia, and *B.* nigra, 837. Céltis austràlis, 727. Fàgus sylvática purpùrea, 905. Fráxinus americàna, 639. Liriodéndron Tulipifera, 26. *O'strya* vulgàris, 919. Plátanus orientàlis, 927. *Quércus* Cérris fulhaménsis, and *Q.* *C.* Lucombe*àna*, 846. Robin*ia* Pseùd-*A*càcia, 233. Taxòdium distichum, 1077.

Ev. — *A*'bies, 1025. Cèdrus, 1057. *Picea*, 1036. Pinus, 950.

Middle-sized ornamental trees.

Decid. — *A*'cer, 79. *Córylus*, 921. Gleditsch*ia*, 249. Liquidámbar, 932. Magnòl*ia* acuminàta, 21. *Mòrus* álba, 705. *Negúndo*, 122. *O'rnus*, 651. *O'strya*, 919. Pàv*ia*, 128. *Pópulus*, 819. *P*ỳrus, 417. Robin*ia*, 233. Salisbùr*ia*, 944. Sàlix, 744. Tilia, 63.

Ev. — *A*'bies, 1025. Cupréssus, 1073. Juníperus, 1080. Picea, 1036. Pinus, 950. Quércus, 846. Táxus, 939, &c.

Small-sized ornamental trees.

Decid. — *A*'cer, 79. *Amýgdalus*, 261. *Armeniaca*, 267. Broussonèt*ia*, 710. *Catálpa*, 663. *Cérasus*, 276. *Cratæ'gus*, 352. *Cydònia*, 450. *Cý*tisus, 213. *Diospý*ros, 625. *Elæágnus*, 696. *Euónymus*, 149. Gleditsch*ia*, 249. Halès*ia*, 620. *Hippóphaë*, 698. Kölreutèr*ia*, 134. Liquidámbar, 932. Magnòl*ia*, 21. *Méspilus*, 414. *Mòrus*, 705. Pàv*ia*, 128. *Pérsica*, 265. Prùnus, 270. *P*ỳrus, 417. Quércus, 846. *Rhámnus*, 170. Robin*ia*, 233. Sàlix, 744. Virgil*ia*, 197. Xanthóxylum, 142.

Ev. — *Cérasus* lusitánica, 276. *Cupréssus*, 1073. *I*'lex, 156. Juníperus, 1080. *Laúrus*, 681. *Picea*, 1036. Pinus, 950. *Quércus*, 846. Táxus, 939. Thùja, 1068.

Selected with reference to their Form.

With upright oblong or globose heads.

Decid. — *A*'cer obtusàtum, 79. *Æ*'sculus Hippocastanum, 124. *A*'lnus, 832. Cérasus, 276. Fàgus, 905. *Pópulus*, 819. *P*ỳrus *A*'ria, and *P.* aucupària, 417. Quércus *Cérris*, 846. *U'lmus*, 715.

With spreading heads.

Decid. — *Armeniaca*, 267. Cratæ'gus Crús-gálli, 352. *Diospý*ros, 625. *Méspilus*, 414. *Pérsica*, 265.

With conical heads.

Decid. — *Làrix*, 1053. Taxòdium, 1077.

Ev. — *A*'bies, 1025. Cèdrus, 1057. Cupréssus sempervirens, 1073. Juníperus, 1080. *Picea*, 1036. Pinus, 950. Thùja, 1068.

With spiry heads.

Decid. — Cratæ'gus *Oxyacántha* stricta, 352. *Pópulus* fastigiàta, 819. Quércus pedunculàta fastigiàta, 846.

Ev. — Táxus baccàta fastigiàta, 939.

Selected with reference to their Mode of Growth.

With pendent shoots and branches.

Decid. — *Bétula* álba péndula, 837. *Cérasus* semperflòrens, 276. Cratæ'gus *Oxyacántha* reginæ, 352. *Cý*tisus *Labúrnum* péndulum, 213. Fàgus sylvática péndula, 905.

Fráxinus excélsior péndula, and *F. lentiscifòlia* péndula, 639. *Lárix* europæ'a péndula, 1053. *Pàvia* rùbra péndula, 128. *Pópulus* trémula péndula, 819. *Quércus* pedunculàta, 846. *Sàlix* babylónica, 744. *Taxòdium* distichum péndulum, 1077. *U'lmus* montàna péndula, 715

Ev.—*Cúpressus* péndula, 1073. *I'lex Aquifòlium*, T. péndulum (both rare), 156. *Juníperus* recúrva, 1080.

With vertical shoots and branches.

Decid.—*Cárpinus Bétulus*, 916. *Fàgus* sylvática, 905. Gymnócladus canadénsis, 255. *Pỳrus A'ria*, and *P.* aucupària, 417. *U'lmus* montàna fastigiàta, 715; and the other fastigiate or spiry-headed trees enumerated above.

With horizontal shoots and branches.

Decid.—*A'lnus* glutinòsa laciniàta, 832. *Cratæ'gus* Crús-gálli, 352. *Diospỳros Lòtus*, 625. Robin*ia* Pseùd-*Acàcia*, R. umbraculífera, 233. Taxòdium dístichum, 1077. *U'lmus* montàna péndula, 715.

Ev.—*A'bies*, 1025. Cèdrus, 1057. *Picea*, 1036. *Pinus*, 950.

Selected with reference to their Foliage.

Duration.

Deciduous ornamental trees.

Developed early in spring.

Decid.—*A'cer* Pseùdo-*Plátanus*, and *A.* tatáricum, 79. *Cratæ'gus* nigra, and *C.* Oxyacántha præ'cox, 352. *Pópulus* balsamífera, 819. *Sambùcus* nigra, 513. *Tília* europæ'a, 63. *U'lmus* campéstris, 715.

Developed late.

Decid.—*Ailántus* glandulòsa, 145. Cárya, 735. *Catálpa*, 662. *Fráxinus*, 639. Gledítsch*ia*, 249. Gymnócladus, 255. Júglans, 732. *Pỳrus A'ria*, and *P.* vestita, 417. *Quércus Taúzin*, 846. Virgília lùtea, 197.

Dropping early.

Decid.—*A'cer* Pseùdo-*Plátanus*, 79. *Æ'sculus* Hippocástanum, 124. *Fráxinus* excélsior, 639. Pàv*ia*, 128. *Pópulus*, 819. Robin*ia*, 233. *Tília*, 63.

Dropping late.

Decid.—*A'cer* campéstre, and *A.* créticum, 79. *Cárpinus Bétulus*, 916. *Cratæ'gus* Crús-gálli, 352. *Fàgus* sylvática, 905. Liquidámbar imbérbe, 932. Plánera Richárd*i*, 725. *Quércus*, 846.

Persistent, often remaining on in a withered state till spring.

Decid.—*Cárpinus Bétulus*, 916. *Fàgus* sylvática, 905. Liquidámbar imbérbe, 932. *Quércus Cérris*, and *Q. Taúzin*, 846.

Evergreen ornamental trees.

A'bies, 1025. Cèdrus, 1057. *Cérasus* lusitánica, 276. *Cupréssus*, 1073. *I'lex Aquifòlium*, 156. Juníperus, 1080. *Laúrus* nóbilis, 681. Magnòl*ia* grandiflòra, 21. *Picea*, 1036. *Pinus*, 950. *Quércus I'lex*, 846. *Táxus*, 939. *Thùja*, 1068.

Subevergreen ornamental trees.

Cratæ'gus Crús-gálli, and *C.* mexicàna, 352. Magnòl*ia* Thompson*iàna*, 21. *Pỳrus* coronària, 417. *Quércus Cérris* fulhaménsis, *Q. Cérris* Lucombe*àna*, *Q.* hýbrida nàna and *Q.* Túrner*i*, 846.

Form and character.

Simple leaves.

Large.

Decid.—*A'cer* macrophýllum, 79. Castànea vésca, 911. *Catálpa syringæfòlia*, 662. Magnòl*ia* macrophýlla, M. tripétala, and M. acuminàta, 21. *Plátanus*, 927. *Pópulus* heterophýlla, 819. *Pỳrus* vestita, 417. *Quércus* sessiliflòra, and *Q.* nigra, 846. *U'lmus* montana, 715.

Ev.—Magnòl*ia* grandiflòra, 21.

Small.

Decid.—*A'cer* créticum, 79. *Cratæ'gus* Oxyacántha, 352. *Euónymus* europæ'us, 149. *Prùnus* spinòsa, 270. *U'lmus* chinénsis, 715.

Linear, or needle-like.

Decid.—*Lárix*, 1053. Taxòdium, 1077.

Ev.—*A'bies*, 1025. Cèdrus, 1057. *Cupréssus*, 1073. Juníperus, 1080. *Picea*, 1036. *Pinus*, 950. *Táxus*, 939. *Thùja*, 1068.

Compound leaves.

Large.

Decid.—*Ailántus*, 1145. Cárya, 735. Gledítsch*ia*, 249. Gymnócladus, 255. Júglans, 732. Kölreut*èria*, 134. *Fráxinus*, 639. Virgíl*ia*, 197.

Small.

Decid.—*Æ'sculus*, 124. *Cýtisus*, 213. *O'rnus*, 651. *Negúndo*, 122. *Pàvia*, 128. *Ptèlea*, 143.

Colour.

In deciduous trees in early summer.

Light, or yellowish, green.

Decid.—*A'cer* Pseùdo-*Plátanus* lutéscens, and *A.* platanòides, 79. *Catálpa syringæfòlia*, 662. Kölreutèr*ia*, 134. Maclùr*a*, 711. *Negúndo*, 122. *Pópulus* balsamífera, 819.

Ev.—*Quércus* pedunculàta, and *Q.* sessiliflòra, 846. *Sàlix*, 744. *Tília*, 63. Virgíl*ia*, 197.

Dark green.

Decid.—*A'lnus*, 832. Castànea, 911. *Céltis*, 727. *Euónymus*, 149. Liquidámbar, 932. Pàv*ia*, 128. *Prùnus*, 270.

Purple.

Decid.—*A'cer* Pseùdo-*Plátanus* purpuráscens, 79. *Fàgus* sylvática purpùrea, 905. *Quércus* pedunculàta purpuráscens, 846.

Variegated.

Decid.—*A'cer* platanòides variegàtum, and *A.* Pseùdo-*Plátanus* variegàta, 86. Castànea vésca variegàta, 912. *Fráxinus* heterophýlla variegàta, 639. *Quércus* pedunculàta variegàta and *Q. Cérris* variegàta, 846. *U'lmus* campéstris variegàta, 715.

In deciduous trees in autumn before falling off.

Red or purple.

Decid. — A'cer rùbrum, 79. Liquidámbar, 932. Magnòlia auriculàta, 21. Quércus rùbra, Q. nigra, Q. tinctòria, and Q. palústris, 846.

Yellow.

Decid. — A'cer platanöìdes, and A. saccharinum, 79. Bétula álba, 837. Fráxinus excélsior, 639. Kölreutèria, 134. Làrix, 1053. Liriodéndron, 36. Maclùra, 711. Negúndo, 122. Pópulus nìgra, and P. canadénsis, 819. Ptèlea trifoliàta, 143. Pỳrus vestìta, 417. Salisbùria, 844. Virgilia, 197.

Green, or without changing colour.

Decid. — A'lnus, 832. Céltis, 727. Cratæ'gus Crús-gálli, and C. mexicàna, 352. Diospỳros, 625. Pỳrus coronària, 417.

Brown, or without bright colour.

Decid. — A'cer Pseùdo-Plátanus, 79. Ailántus, 145. Càrya, 735. Jùglans, 1080. Pópulus græ'ca, P. trémula, and P. balsamifera, 819

In evergreens.

Light green.

Cèdrus Deodàra, 1057. Cuprèssus thyöìdes, and C. lusitánica, 1073. Juníperus, 1080. Magnòlia grandiflòra, 21. Thùja, 1068.

Dark green.

A'bies excélsa nigra, 1025. Cérasus lusitánica, 276. Cuprèssus, 1073. I'lex, 156. Juníperus, 1080. Laúrus nóbilis, 681. Pìcea, 1036. Pìnus, 950. Quércus I'lex, 846. Táxus, 939.

Variegated.

A'bies excélsa variegàta, 1025. Cuprèssus thyöìdes variegàta, 1073. I'lex Aquifòlium variegàtum, 156. Pìnus Pináster variegàta, 961. Quércus I'lex variegàta, 846. Táxus baccàta variegàta, 939.

Fragrant leaves.

Decid. — Bétula, 837. Jùglans, 732. Liquidámbar, 932. Sàlix pentándra, 744.

Ev. — A'bies, 1025. Juníperus, 1080. Laúrus, 681. Pìcea, 1036. Pìnus, 950. Thùja, 1068.

Selected with reference to their Flowers.

Produced in spring.

Red or purple.

Decid. — A'cer rùbrum, 79. Amýgdalus, 261. Cratæ'gus Oxyacántha ròsea, and C. O. punicea, 352. Pàvia rùbra, 128. Pérsica, 265.

Yellow.

Decid. — A'cer platanöìdes, and A. monspessulànum, 79. Cýtisus Labúrnum, and C. alpìnus, 213. Sàlix, 744. Xanthóxylum, 142.

White.

Decid. — Armeníaca, 267. Cèrasus, 276. Cratæ'gus, 352. Magnòlia conspícua, 21. Prùnus, 270. Pỳrus, 417. Staphylèa, 147.

Variegated.

Decid. — Æ'sculus, 124. Pàvia, 128. Pỳrus Màlus, P. coronària, and P. spectábilis, 417. Robínia viscòsa, 233.

In summer.

Red or purple.

Decid. — Robínia híspida, 233.

Yellow or orange.

Decid. — Kölreutèria, 134. Virgília lùtea, 197.

White or whitish.

Decid. — Cratæ'gus cordàta, and C. mexicàna, 352. Halèsia tetráptera, 620. Liriodéndron, 36. Magnòlia, 21. Méspilus, 414. Robínia Pseùd-Acàcia, 233.

Variegated.

Decid. — Catálpa, 662.

Selected with reference to their Fruit or Seed.

Large and showy.

Decid. — Armeníaca vulgàris, 267. Cydònia vulgàris, 450. Maclùra aurantìaca, 711. Pérsica vulgàris, 265. Prùnus doméstica, 270. Pỳrus commùnis, P. Màlus, and P. M. astracánica, 417.

Small, but conspicuous from colour and quantity.

Decid. — Céltis, 727. Cérasus, 276. Cratæ'gus, 352. Hippóphaë, 698. Pỳrus Màlus prunifòlia, P. M. baccàta, P. A'ria, P. intermèdia, P. torminàlis, P. auriculàta, P. aucupària, and P. americàna, 417. Rhámnus cathárticus, and R. frángula, 170.

Ev. — I'lex Aquifòlium, 156. Juníperus, 1080. Laúrus, 681.

Singular in form or character.

Decid. — Euónymus, 149. Liquidámbar, 932. Magnòlia, 21. Plátanus, 927. Quércus Cérris, and Q. Æ'gilops, 846.

SHRUBS.

Shrubs grown chiefly for useful Purposes.

Selected as adapted for Coppice-Woods.

Decid. — Córnus sanguínea, and C. álba, 501. Córylus Avellàna, 921. Ligústrum vulgàre, 628. Sàlix càprea, 744.

Selected as adapted for producing Shelter for Game.

Decid. — Genísta, 203. Rìbes, 468. Spiræ'a, 299. Symphoricárpos, 541. Vibúrnum, 516.

Ev. — Bùxus, 703. Gaulthèria, 579. Hypéricum, 74. Juníperus, 1080. Mahònia, 80. Rhododéndron, 583. U'lex, 199. Vaccínium, 604.

Selected as adapted for Hedges.

Decid. — Bérberis, 42. Cýtisus scopàrius, 213. Hippóphaë, 698. Ligústrum, 628. Paliùrus, 168. Ròsa, 321. Syrínga, 635.

Ev. — Bùxus, 703. I'lex, 156. Juníperus, 1080. Phillýrea, 631. U'lex, 199.

Shrubs grown chiefly for ornamental Purposes.

Selected with reference to their Bulk.

Large-sized shrubs.

Decid. — *Bácch*aris, 546. *Bérberis*, 42. Bòry*a*, 713. Búddle*a*, 670. *Caragána*, 237. Colùtea, 244. *Córnus*, 501. Cotoneáster, 405. *Euónymus*, 149. Fontanes*ia*, 638. Halimodéndron, 242. *Hamamèlis*, 499. *Hippóphaë*, 698. *Ligústrum*, 628. *Paliùrus*, 168. *Philadélphus*, 460. Piptánthus, 198. *Rhús*, 186. *Ríbes*, 468. *Sambùcus*, 513. Shephérd*ia*, 700. *Spártium*, 202. *Syringa*, 635. *Vibúrnum*, 516.

Ev. — *A'rbutus*, 573. *Búxus*, 703. *I'lex*, 156. *Juniperus*, 1080. *Laúrus*, 681. *Phillýrea*, 631.

Middle-sized shrubs.

Decid. — Amórpha, 230. *Bérberis*, 42. Calóphaca, 243. Calycánthus, 452. *Ceanòthus*, 180. Cephalánthus, 544. Chimonánthus, 454. *Clèthra*, 581. *Cydònia*, 450. *Deùtzia*, 465. *Hibiscus*, 62. Leycestèr*ia*, 543. Potentilla, 319. *Rhododéndron*, 583. *Rhús*, 186. *Ríbes*, 468. *Rósa*, 321. *Spiræ'a*, 299. Staphylèa, 147. Symphoricárpos, 541. *Syringa*, 635.

Ev. — *Aúcuba*, 511. *Bupleùrum*, 495. Gárry*a*, 926. *Juniperus*, 1080. Mahòn*ia*, 50. Prìnos, 163. *Rhododéndron*, 583. *U'lex*, 199. *Vibúrnum Tìnus*, 516.

Small shrubs.

Decid. — *Amýgdalus nàna*, 261. *Androsæ'mum*, 77. *Artemísia*, 549. *Atrapháxis*, 679. *Bétula nàna*, 837. *Cístus*, 54. Comptòn*ia*, 936. *Coriària*, 146. Coronilla, 247. *Dáphne*, 686. Dìrca, 692. Fothergílla, 500. *Genísta*, 203. Hydrángea, 492. *Hypéricum*, 74. *I'tea*, 489. *Jasminum*, 654. *Myrica*, 934. Nitrària, 467. *Onònis*, 229. *Pæònia*, 18. Potentilla, 319. *Rhododéndron*, 583. *Ríbes*, 468. *Spiræ'a*, 299. *Vaccínium*, 604. *Vélla*, 53. Xanthorhìza, 19.

Ev. — Andrómeda, 560. *Bupleùrum*, 495. Cassándr*a*, 562. Cotoneáster microphýlla, 405. *Dáphne*, 686. *Ephèdra*, 937. Kálm*ia*, 598. *Lèdum*, 602. Leiophýllum, 602. Leucóthoë, 568. Lyòn*ia*, 564. Mahòn*ia*, 50. Pernéttya, 578. *Rhododéndron*, 583. *Vaccínium*, 604. *Yúcca*, 1101. Zenòbi*a*, 563.

Under-shrubs.

Decid. — *Arctostáphylos alpina*, 577. *Artemísia*, 549. *Onònis*, 229. Phalerocárpus, 581.

Ev. — *Arctostáphylos U'va úrsi*, 577. Azàlea procúmbens, 601. Callùna, 559. Corèma, 1092. Dabœ'c*ia*, 572. Epigæ'a, 580. *Erica*, 555. Gaulthèr*ia*, 579. Gypsocállis, 557. *Lèdum*, 602. Leiophýllum, 602. *Oxycóccus*, 615. Pernéttya, 578. Phyllódoce, 570. *Rúscus*, 1099. Santolìna, 548.

Selected with reference to their Form as Bushes.

With compact roundish heads.

Decid. — *Artemísia*, 549. *Bérberis* sinénsis, 42. *Bétula* nàna, 837 Calycánthus, 452. Cephalánthus, 544. Coronilla, 247. *Cýtisus*, 213. *Dáphne Mezèreum*, 686. *Genista*, 203. *Hypéricum*, 74. *Ligústrum*, 628. Potentilla, 319. *Rhododéndron*, 583. *Ríbes*, 468. *Spiræ'a*, 299. Symphoricárpos, 541. *Syringa*, 635.

Ev. — *Aúcuba*, 511. *Bupleùrum*, 495. *Búxus*, 703. *Dáphne*, 686. *Hypéricum Kalmiánum*, 74. Mahòn*ia*, 50. *Phillýrea*, 631. *Rhododéndron*, 583. *U'lex*, 199. *Vibúrnum Tìnus*, 604.

With open, rather irregular, heads.

Decid. — Amórpha, 230. *Bácch*aris, 546. Búddle*a*, 670. Colùtea, 244. Cotoneáster, 405. *Philadélphus*, 460. Piptánthus, 198. *Rhús*, 186.

With conical or pointed heads.

Ev. — Cuprèssus, 1073. *I'lex*, 156. Juniperus, 1080. *U'lex stricta*, 199.

Selected with reference to their Mode of Growth.

Bushes with upright shoots and branches.

Decid. — *Caragána*, 237. *Córnus* (*a.*) stricta, 501. *Hibiscus*, 62. *Jasminum frùticans*, 654. Kérr*ia*, 298. Leycestèr*ia*, 543. *Philadélphus*, 460. *Spártium*, 202. *Spiræ'a*, 299. *Támarix*, 458. Myricària, 459.

Ev. — Juniperus c. suècica, 1081. *U'lex* e. stricta, 199. *Táxus* b. stricta, 939. *Thùja* o. stricta, 1068.

Bushes with pendent shoots and branches.

Climbers.

By tendrils, or grasping fibrils.

Decid. — Ampelópsis, 139. *Atrágene*, 16. *Clématis*, 2. *Rhús* radicans, 186. *Técoma*, 661. *Vítis*, 136.

Ev. — Bignòn*ia*, 660. *Hédera*, 497. Smìlax, 1093.

By twining stems.

Decid. — *Aristolòchia*, 701. Berchèm*ia*, 169. *Celástrus*, 154. *Lonícera*, 526. *Menispérmum*, 89. Períploca, 658. Wistàr*ia*, 248.

Ev. — *Lonícera*, 526.

By elongated slender stems.

Decid. — *Jasminum* officinàle, 654. *Lýcium* europæ'um, 665. *Rósa*, numerous sorts, 321. *Rùbus*, 311. *Solànum* Dulcamàra, 663.

Ev. — *Rósa*, several sorts, 321.

Trailers; the branches prostrate, but not generally rooting.

Decid. — *Arctostáphylos* alpina, 577. Decumària bárbara, 466. *Ríbes* prostràtum, 468. *Rùbus*, 311.

Ev. — *Arctostáphylos* U'va úrsi, 577. Azàlea procúmbens, 601. Cotoneáster microphýlla, 405. Gaulthèr*ia*, 519. *Juniperus* Sabina prostràta, 1080. *Vinca*, 657.

Creepers; the branches prostrate and rooting.

Decid. — *Rhús*, 186. *Rùbus*, 311.

Ev. — Epigæ'a rèpens, 580. Mahòn*ia* rèpens, 50. *Oxycóccus*, 615.

Selected with reference to their Foliage.

Duration.

Deciduous.

Developed early in spring.

Decid. — *Cratæ'gus*, 352. *Euónymus*, 149. *Hypéricum*, 74. *Ligústrum*, 628. *Ríbes*, 468. *Rósa*, 321. *Sambùcus*, 513. *Spiræ'a*, 299. *Syringa*, 635.

Developed late.

Decid. — Aràlia, 496. Calycánthus, 452. Cephalánthus, 544. *Hibiscus*, 62. *Paliùrus*, 168. *Técoma*, 661.

Persisting; remaining on in a withered state.
Decid. — Genista, 203. *Hypéricum*, 74. Juniperus recúrva, 1080. *Onònis*, 229. Potentílla, 319.
Evergreen leaves.
Andròmeda, 560. *A'rbutus*, 573. *Arctostáphylos*, 577. *Aúcuba*, 511. Azàlea, 601. *Bérberis*, 42. *Búxus*, 703. Callùna, 559. Cassándra, 562. Cérasus, 276. Corèma, 1092. Cotoneáster, 405. Dabœ'cia, 572. *Dáphne*, 686. *E'mpetrum*, 1091. *E'phedra*, 937. Epigæ'a, 580. *Erica*, 555. Gárrya, 926. Gaulthèria, 579. Gypsocállis, 557. *Hypéricum*, 74. *I'lex*, 156. Juniperus, 1080. Kálmia, 598. *Lèdum*, 602. Leiophýllum, 602. Leucóthoe, 568. Lyònia, 564. Mahònia, 50. Pernéttya, 578. Phillýrea, 631. Phyllódoce, 570. *Prinos*, 163. *Rhámnus*, 170. *Rhododéndron*, 583. *Ròsa*, 321. *Rosmarinus*, 672. *Rúscus*, 1099. *Santolina*, 548. *Táxus*, 939. *Thùja*, 1068. *U'lex*, 199. *Vaccinium*, 604. *Vibúrnum*, 516. *Yúcca*, 1101. Zenòbia, 563.
Evergreen, from the colour of the shoots.
Decid. — Collètia, 178. *Cýtisus* scopàrius, 213. *E'phedra*, 937. *Genista*, 203. Jasminum officinàle, 654. Spártium, 202. *U'lex*, 19.
Subevergreen.
Aristotèlia, 182. *Bérberis* asiática, 42. *Cistus*, 54. *Cýtisus* hirsùtus, and *C.* capitàtus, 213. Leycestèria, 543. *Ligústrum* vulgàre sempervirens, 628. Photinia serrulàta, 403. *Ròsa*, 321. *Vélla*, 53.

Form and character.
Simple.
Large.
Decid. — Calycánthus, 452. Chimonánthus, 454. *Córnus*, 501. Cotoneáster, 405. *Euónymus* latifòlius, 149. *Hamamèlis*, 499. *Hibiscus*, 62. Hydrángea, 492. Leycestèria, 543. *Philadélphus*, 460. Photinia, 403. *Rhododéndron*, 583. *Rhús*, 186. *Ribes*, 468. *Spiræ'a*, 299. *Syringa*, 635. *Vibúrnum*, 516.
Ev. — *Aúcuba*, 311. Cérasus Laurocérasus, and *C.* lusitánica, 276. Gárrya, 926. Gaulthèria *Shállon*, 579. Kálmia latifòlia, 598. *Rhododéndron*, 583. *I'lex*, 156. *Yúcca* gloriòsa, 1101.
Small or narrow.
Decid. — *Amýgdalus*, 261. *Arctostáphylos*, 577. *Atrapháxis*, 679. *Bérberis*, 42. *Bétula* nàna, 837. Bòrya, 713. *Cistus*, 54. Comptònia, 936. Coriària, 146. Cydònia, 450. *Dáphne*, 686. Fontanèsia, 638. Heliánthemum, 58. *Hippóphaë*, 698. *I'tea*, 489. *Ligústrum*, 628. Nitrària, 467. Shephérdia, 700.
Ev. — *A'rbutus*, 573. *Bérberis* dúlcis, 42. *Búxus*, 703. Cotoneáster, 405. *Dáphne*, 686. *Hypéricum* Kalmiànum 74. *Lèdum* latifòlium, 602. Leucóthoe, 568. Lyònia, 564. *Phillýrea*, 631. *Prinos*, 163. *Rhododéndron*, 583. *Rúscus*, 1099. *Vaccinium*, 604. *Yúcca* filamentòsa angustifòlia, 1101.
Linear, heath-like, or needle-like.
Decid. — *Hypéricum* serpyllifòlium, 74. Myricària germánica, 459. Phalerocárpus serpyllifòlia, 581. Phyllódoce empetrifórmis, 570. *Támarix* gállica, 458.
Ev. — Andrómeda, 560. Azalea, 601. Callùna, 559. Corèma, 1092. Dabœ'cia, 572. *E'mpetrum*, 1091. *Erica*, 555. Gypsocállis, 557. Juniperus, 1080. Leiophýllum, 602. *Rosmarinus*, 672.
Compound.
Large.
Decid. — Amórpha, 230. Arália, 496. Colùtea, 244. *Pæònia*, 18. *Rhús* typhina, 186. *Ròsa*, 321. *Sambùcus*, 513. Staphylèa, 147.
Ev. — Mahònia, 50. *Ròsa*, 321.
Small.
Decid. — Calóphaca, 243. *Caragàna*, 237. Coronilla, 247. *Cýtisus*, 213. Genista, 203. Halimodéndron, 242. *Jasminum*, 654. *Onònis*, 229. Potentilla, 319. Xanthorhiza, 19.

Colour.
In deciduous shrubs when first developed, or in early summer.
Light, or yellowish, green.
Decid. — *Caragàna*, 237. *Cýtisus*, 213. Dirca, 692. *Genista*, 203. *Hypéricum*, 74. Xanthorhiza, 19.
Glaucous green.
Decid. — *Artemisia*, 549. *Báccharis*, 546. *Bupleùrum*, 495. Colùtea, 244. *Elæágnus*, 696. Halimodéndron, 242. *Hippóphaë*, 698. Myricària, 459. *Ribes* cèreum, and *R.* triflòrum niveum, 468. Santolina, 548. Shephérdia, 700. *Támarix*, 458.
Ev. — *Yúcca*, 1101. Zenòbia pulverulénta, 563.
Purple.
Decid. — Córylus *Avellàna* purpùrea, 921. *Bérberis* vulgàris purpùrea, 42.
Variegated.
Decid. — Aristotèlia, 182. *Hibiscus*, 62. *Jasminum*, 654. *Ligústrum*, 628. *Philadélphus*, 460. *Rùbus*, 311. *Sambùcus*, 513. *Solànum* Dulcamàra, 663.
Ev. — *Aúcuba*, 511. *Búxus*, 703. *Hédera*, 497. *I'lex*, 156. *Rhododéndron*, 583. *Rhámnus* *Alatérnus*, 170. *Táxus*, 939. *Vibúrnum* *Tinus*, 516. *Vinca*, 657. *Yúcca* gloriòsa variegàta, 1101.
In shrubs in autumn before falling off.
Red.
Decid. — Ampelópsis, 139. *Bérberis*, 42. Cornus, 501. *Euónymus*, 149. Photinia, 403. *Pùnica*, 456. *Rhododéndron*, 583. *Rhús* typhina, 186. *Vaccinium*, 604. *Vitis* vinifera fòliis rubescéntibus, 136.
Yellow.
Decid. — Aristolòchia, 701. *Bérberis*, 42. Bignònia, 660. Búddlea, 670. *Caragàna*, 237. Chimonánthus, 454. *Cýtisus*, 213. *Genista*, 203. *Hypéricum*, 74. *Rhús* Cótinus, 186. *Ribes*, 468. *Técoma*, 661.

Green, or without changing colour.
Decid. — *B*upleùrum, 495. Hydrángea, 492. *Ligústrum vulgàre* semper-virens, 628. *R*ùbus, 311. *Sp*ártium, 202. *U*'lex, 199.
Brown, or without bright colours.
Decid. — *Pæònia*, 18. Potentilla, 319. *R*ùbus, 311. *Spiræ*'a, 299.
Fragrant leaves.
Decid. — *A*rtemísia, 549. *Hyp*éricum, 74. Juniperus, 1080. *Myrica*, 934. *Rh*ús aromática, 186. *Ribes* nigra, 468. *R*ósa rubiginòsa, 321. *Rosma*-rinus, 672. *S*ambùcus, 513. *S*antolina, 548. *Sol*ànum, 663.
Ev. — Juniperus, 1080. *La*úrus, 681.

Selected with reference to their Flowers.
Produced in early spring.
Red, blue, or purple.
Decid. — *Am*ýgdalus nàna, 261. *C*ydònia japónica, 450. *D*áphne *Mezèreum*, 686. *R*hododéndron daùricum, 583. *Ribes* sanguíneum, 468.
Ev. — Gypsocállis, 557 *R*hododéndron daùricum atrovirens, 583.
Yellow or orange.
Decid. — *Caragàna*, 237. Chimonánthus, 454. *C*órnus más, 501. *Kérria* japónica, 298. *Ribes* aúreum, 668.
Ev. — *D*áphne Mahònia, 50. *U*'lex europæ'a, 199.
White or whitish.
Decid. — Cotoneáster, 405. Magnòl*ia* conspícua, 21. *Ribes* cèreum, 468. Staphylèa, 147.
Variegated.
Decid. — Andrómeda polifòlia, 560. *D*áphne *Mezèreum*, 686. *Pæònia Moútan*, 18. *Ribes*, 468.
Produced late in spring, or the beginning of summer.
Red, blue, or purple.
Decid. — Amórpha, 230. *C*istus, 54. *C*ýtisus, 213. Halimodéndron, 242. Heliánthemum, 58. *O*nònis, 229. *R*hododéndron, 583. *Ribes*, 468. *R*òsa, 321. *Spiræ*'a, 299. *Syringa*, 635.
Ev. — *E*rica, 555. Kálm*ia*, 598. *R*hododéndron, 583. *R*òsa, 321.
Yellow or orange.
Decid. — *Bérberis*, 42. Búddle*a*, 670. Calóphaca, 243. *Caragàna*, 237. Cistus, 54. *C*ýtisus, 213. Dírca, 692. *G*enista, 203. Heliánthemum, 58. *Hyp*éricum, 74. *J*asmìnum, 654. Piptánthus, 198. *R*hododéndron, 583. *Ribes*, 468. *R*òsa, 321. *Sp*ártium, 202.
Ev. — *Hyp*éricum, 74. Mahòn*ia*, 50. *R*hododéndron, 583.
White or whitish.
Decid. — Andróme*da*, 560. Bòr*ya*, 713. *C*órnus, 501. Cotoneáster, 405. Deùtz*ia*, 465. *J*asmìnum, 654. *Lig*ústrum, 628. *Ph*iladélphus, 460. Photinia, 403. *Ribes*, 468. *R*òsa, 321. *S*ambùcus, 513. *Spiræ*'a, 299. *Syringa*, 635. *Vib*úrnum, 516.
Ev. — Cassándr*a*, 562. Gárr*ya*, 926. Leucóthoë, 568. Lyòn*ia*, 564. Phillýrea, 631.
Variegated.
Decid. — *A*trapháxis, 679. *C*istus, 54. Colùtea, 244. Coronilla, 247. *O*nònis, 229. *R*hododéndron, 583.
Ev. — Azàlea, 601. *D*áphne, 686. *E*rica, 555. Gypsocállis, 557. Kálm*ia*, 598.
Produced in autumn.
Red, blue, or purple.
Decid. — Ceanòthus, 180. *D*áphne *Mezèreum* autumnàle, 686. *H*ibíscus, 62. *R*òsa, 321. *Ev.* — *E*rica, 555.
Yellow or orange.
Decid. — Chimonánthus, 454. *H*amamèlis, 499. *Hyp*éricum, 74. Paliùrus, 168. *R*òsa, 321. *Sp*ártium, 202. *Ev.* — *U*'lex nàna.
White or whitish.
Decid. — *H*ibíscus, 62. *R*òsa, 321. *S*ambùcus canadénsis, 513. *Sol*ànum *Dulcamàra* álba, 663. *Spiræ*'a, 299.
Ev. — *A*'rbutus, 573. Gárr*ya*, 926. *Vib*úrnum *T*ìnus, 516. *Yúcca*, 1101.
Variegated.
Decid. — Amórpha, 230. Aristotèl*ia*, 182 *H*ibíscus, 62. Leycestèr*ia*, 543. *R*òsa, 321.
Ev. — *A*'rbutus, 573. Gypsocállis, 557.
Fragrant flowers.
Decid. — Calycánthus, 452. *C*érasus *Mahàleb*, 276. Chimonánthus, 454. *C*lématis Flámmula, 2. *C*ýtisus *L*abúrnum frágrans, 213. *D*áphne *Mezèreum*, 686. *Hyp*éricum, 74. *J*asmìnum, 654. Lonícer*a*, 526. *Ph*iladélphus, 460. *R*òsa, 321. *S*ambùcus, 513. *Syringa*, 635. *V*ítis, 136.
Ev. — *J*asmìnum officinàle, 654. *R*òsa sempervìrens, 321. *R*osmarìnus, 672.

Selected with reference to their Fruit or Seed.
Large and showy.
Decid. — *C*ydònia, 450. *Ribes* Grossulària, 468. *R*òsa, 321. *R*ùbus, 311.
Ev. — *A*'rbutus, 573.
Small, but conspicuous from colour and quantity.
Decid. — *Bérberis*, 42. *C*órnus, 501. Cotoneáster, 405. *Cratæ*'gus, 352. *D*áphne, 686. *E*uónymus, 149. *H*ippóphaë, 698. *Lig*ústrum, 628 Lonícer*a*, 526. *L*ýcium, 665. *M*yrica, 934. Periploca, 658. *Rh*ús, 186. *Ribes*, 468. *S*ambùcus, 513. Shepherd*ia*, 700. *Sol*ànum, 663. Symphoricárpus, 541. *V*accínium, 604. *Vib*úrnum, 516. *V*ítis, 136.
Ev. — Cotoneáster, 405. *Cratæ*'gus, 352. Gaulthèr*ia*, 579. *H*édera, 497. *I*'lex, 156. *J*uniperus, 1080. Mahòn*ia*, 50. *R*úscus, 1099. *V*accínium, 604. *Vib*úrnum, 516.
Singular in form or character.
Decid. — Búddle*a*, 670. Calóphaca, 243. Calycánthus, 452. Cephalánthus, 544. Chimonánthus, 454. Colùtea, 244. *E*uónymus, 149. Magnòl*ia*, 21. Paliùrus, 168. Physiánthus, 659. Staphylèa, 147.
Ev. — Magnòl*ia* grandiflòra, 21.

ANALYSIS OF THE GENERA DESCRIBED,

ACCORDING TO THEIR LEAVES.

LEAVES SIMPLE.

Alternate, stipulate, (A) - - Page lxviii.
Alternate, exstipulate, (B) - - - lxix.
Alternate, stipulate, or exstipulate - - lxx.
Opposite, stipulate, (C) - - - lxx.
Opposite, exstipulate, (D) - - - lxx.
Opposite or alternate, stipulate - Page lxxi.
Opposite or alternate, exstipulate - - lxxi.
Opposite or alternate, stipulate or exstipulate - - - - - lxxi.

LEAVES COMPOUND.

Alternate, stipulate, (E) - - lxxi.
Alternate, exstipulate, (F) - - lxxi.
Opposite, stipulate, (G) - - lxxii.
Opposite, exstipulate, (H) - - lxxii.
Alternate or opposite, stipulate - lxxii.

LEAVES SIMPLE OR COMPOUND.—*Alternate, stipulate* - - lxxii.

LEAVES SIMPLE

Alternate, stipulate.

Deciduous.

Entire along the margins.

Three-lobed, fiddle-shaped - - -	LIRIODE'NDRON	Page 36
Linear, caducous, spiny - - -	*U*'LEX - -	- 199
Heart-shaped, reddish - - - -	*CE*'RCIS -	- 256
Coriaceous, roundish, rough - - -	CALYCA'NTHUS	- 452
Coriaceous, oblong-lanceolate, smooth - -	CHIMONA'NTHUS	- 454
Ovate or cuneate - - - -	*HAMAME*'LIS -	- 499
Lanceolate - - - - -	STILLI'NGIA -	- 702

Serrated.

Cordate (one variety laciniate) - - -	*TI*'LIA - -	- 63
Lobed or laciniated - - - -	*VI*'TIS - -	- 136
Oval-lanceolate - - - -	*CELA*'STRUS -	- 154
Ovate or cordate, 3-nerved, with spines in the axils	*PALIU*'RUS -	- 168

Lanceolate, conduplicate when young	*Amy'gdalus*	261
	Pe'rsica	265
Ovate, convolute when young	*Armeni'aca*	267
	Pru'nus	270
Ovate, conduplicate	*Ke'rria*	298
Roundish, downy when young	*Amela'nchier*	411
Lanceolate, often nearly entire	*Me'spilus*	414
Oblong, usually unequal at the base, harsh to the touch	*U'lmus*	715
Oblong, unequal at the base, rough above	*Ce'ltis*	727
Lanceolate or ovate	*Sa'lix*	744
Lobed, with glands in the serratures	Liquida'mbar	932
Deeply cut or lobed, fan-nerved	Salisbu'ria	944
Entire or serrated.		
Ovate, tomentose beneath	*Cydo'nia*	450
Ovate, the disk oblate, and the petiole compressed	*Po'pulus*	819
Ovate, glabrous, aromatic in odour	*Be'tula*	837
Ovate, sometimes cut	*Fa'gus*	905
Oblong-lanceolate	*Casta'nea*	911
Ovate, rough above	*Ca'rpinus*	916
	O'stryа	919
Roundish, rough	*Co'rylus*	921
Dentate, notched, or otherwise cut at the edges.		
Ovate, variously lobed	*Hibi'scus*	62
Denticulate, 3-nerved	*Zi'zyphus*	167
Lobed, angular, or toothed	*Cratæ'gus*	352
Lobed, or deeply cut	*Ri'bes*	468
Ovate or obovate, with soft starry down	Fotherg'illa	500
Palmate and subtrilobate	*Fi'cus*	712
Palmate, sinuated	*Pla'tanus*	927
Entire, serrate or dentate, or otherwise cut at the edges.		
Rough on the upper surface	*A'lnus*	832
Evergreen.		
Entire.		
Full of pellucid dots	Illi'cium	20
Serrated.		
Wedge-shaped, pinnatifid	Cowa'nia	321
Lanceolate, glaucous beneath	Stranvæ'sia	403
Entire or serrated.		
Shining	Photi'nia	403
Deciduous, evergreen, or subevergreen.		
Entire.		
Large, aromatic when bruised	Magno'lia	631
Linear, small, semi-cylindrical	Chenopo'dium	675
Oval or ovate		
Serrated.		
Conduplicate when young	*Ce'rasus*	276
Lobed or sinuated, sometimes lanceolate	*Que'rcus*	846
Besprinkled with resinous dots	*Myri'ca*	934
Entire or serrated.		
Ovate or lanceolate	*Rha'mnus*	170
Ovate or elliptic	*Ceano'thus*	180
Alternate, exstipulate.		
Deciduous.		
Entire.		
Oblong, articulated with the stem	*Asi'mina*	38
Peltate or cordate	Menispe'rmum	39
Cordate, ovate, or lobed	Co'cculus	40
Ovate, serrated at the apex	Nemopa'nthes	154
Lanceolate, soon dropping off	*Spa'rtium*	202
Deeply cut and divided, apparently compound	*Artemi'sia*	549
Membranous, heath-like	Menzie'sia	609
Oblong, coriaceous, shining	*Diospy'ros*	625
Cordate, ovate, or lanceolate	*Syri'nga*	635
Fascicled, lanceolate	*Ly'cium*	665
Fascicled, coriaceous, glaucescent	Crabo'wskia	669
Linear-lanceolate, hoary	Dio'tis	677
Minute, caducous, evergreen bark	Calli'gonum	680
Lanceolate, coriaceous, pale green	Di'rca	692
Oblong or lanceolate	Ny'ssa	693
Lanceolate, beset with scales or stars of hairs	*Elæa'gnus*	696
Linear-lanceolate, scaly and silvery	*Hippo'phae*	698
Cordate, downy beneath	*Aristolo'chia*	701
Ovate, shining	Maclu'ra	711
Linear, in alternate fascicles	*La'rix*	1063
Linear, 2-ranked, flat		
Serrated.	Taxo'dium	1077
Ovate, acuminated	Malachode'ndron	71
Ovate, acute	Stua'rtia	72
Oval, mucronate	Breche'mia	169
Obovate, cuneated, serrated at the tips	Lo'wea	352
Obovate or oval, glabrous on both surfaces	*Cle'thra*	581
Cordate or lobed, and rough	*Mo'rus*	706
Lobed variously, sometimes entire, hairy	Broussone'tia	710
Entire or serrated.		
Oblong-lanceolate, or linear	*Ba'ccharis*	546
Ovate, obovate, or oval-lanceolate	*Sty'rax*	618
Ovate-acuminate	Hale'sia	690

Dentate, notched, or otherwise cut at the edges.		
Lanceolate, acute	I'TEA	489
Oval, mucronate	ZENO'BIA	563
Pinnatifidly dentate, downy, with resinous particles	COMPTO'NIA	936
Evergreen.		
Entire.		
Lobed or peltate, coriaceous, shiny	HE'DERA	497
Linear, chaffy, small, whorled	ERI'CA	555
Acerose, whorled, glabrous	GYPSOCA'LLIS	557
Acerose, trigonal, imbricate in 4 rows	CALLU'NA	559
Linear-lanceolate, mucronulate	ANDRO'MEDA	560
Acerose, very small, imbricate	CASSI'OPE	561
Acerose, on short petioles	CASSA'NDRA	562
Oval, coriaceous	PI'ERIS	570
Linear, spreading, heath-like	PHYLLO'DOCE	570
Acerose, elliptic, flat, downy beneath	DABŒ'CIA	572
Cordate-ovate	EPIGÆ'A	580
Verticilliately ternate, buds naked	KA'LMIA	598
Oval, convex, glabrous, small	LEIOPHY'LLUM	602
Linear or ovate, margins revolute, coriaceous, tomentose beneath	LE'DUM	602
Narrow or linear, crowded	PHLO'MIS	672
	ROSMARI'NUS	672
	LAVA'NDULA	672
	SA'LVIA	673
Linear, sheathed, needle-like	PI'NUS	950
Linear, scattered, needle-like	A'BIES	1025
Linear, 2-rowed, needle-like	PI'CEA	1036
Linear, in alternate fascicles	CE'DRUS	1057
Acerose, imbricate	ARAUCA'RIA	1062
Linear, tongue-shaped, obtuse	E'MPETRUM	1091
Linear, needle-shaped, spreading	CERATI'OLA	1092
	CORE'MA	1092
Alike green on both surfaces, usually floriferous	RU'SCUS	1099
Ensiform, pointed, alike on both surfaces	YU'CCA	1101
Serrated.		
Linear, small, crowded, spreading	BRYA'NTHUS	571
Roundish-oval, small	PHALEROCA'RPUS	561
Entire or serrated.		
Terminating in a tendril, cordate, oblong	MUTI'SIA	551
Oblong-lanceolate, coriaceous, shining	A'RBUTUS	573
Linear, solitary, more or less 2-ranked	CUNNINGHA'MIA	1065
Dentate, notched, or otherwise cut at the edges.		
Ovate-oval, coriaceous, often prickly	I'LEX	156
Oblong or ovate, small, odour of turpentine	DUVAU'A	192
Linear, hoary or silky beneath	STÆHELI'NA	546
Linear, small, rowed, tomentose	SANTOLI'NA	548
Pinnatifid, tomentose, white and mealy beneath	SENE'CIO	550
Spinulose, coriaceous	LEUCO'THOE	568
Ovate, small, approximate, stiff, shining	PERNE'TTYA	578
Ovate, subcordate, ciliate, coriaceous, glabrous, or shining	GAULTHE'RIA	579
Deciduous evergreen.		
Entire.		
Linear, stem-clasping, small	TA'MARIX	458
Linear, broader at the base, sessile, small	MYRICA'RIA	459
Oblong, coriaceous, sessile, glaucous	BUPLEU'RUM	495
Obovate or oval, coriaceous, crowded, glabrous	LYO'NIA	564
Terminated by a sphacelate (withered) apex, or yellow gland	RHODODE'NDRON	583
Ovate or elliptic, smooth, small, with revolute margins	OXYCO'CCUS	615
Lanceolate, bluntish	ARGA'NIA	622
Broad, lanceolate, glabrous	BUME'LIA	628
Lanceolate, smooth on both sides	FONTANE'SIA	638
Ovate-cordate or hastate	SOLA'NUM	663
Spathulate, with a frosty hue	TRAGOPY'RUM	678
Oblong-pointed, sometimes lobed	LAU'RUS	681
Linear-lanceolate, small	OSY'RIS	695
Serrated		
Oblong, coriaceous, smooth, downy beneath	GORDO'NIA	73
Entire or serrated.		
Oval or lanceolate	PRI'NOS	163
Obovate or oblong, full of resinous glands, smooth	ESCALLO'NIA	490
Obovate, coriaceous, shining	ARCTOSTA'PHYLOS	577
Membranous, often beset with resinous dots	VACCI'NIUM	604
Reticulated, alike on both sides	SMI'LAX	1053
Evergreen or subevergreen.		
Dentate, notched, or otherwise cut at the edges.		
Coriaceous, in tufts at the axils	BE'RBERIS	42
Obovate, glaucous	VE'LLA	53
Stipulate or exstipulate.		
Deciduous.		
Dentate.		
At the apex obcuneate	PU'RSHIA	297
Ovate, unequal at the base, rough, dark green	PLA'NERA	725
Deciduous, evergreen, or subevergreen.		
Entire.		
Generally woolly beneath	COTONEA'STER	405

Opposite, stipulate.

- Deciduous.
  - Entire.
    - Minute, caducous, bark evergreen - - COLLE'TIA - - 178
    - Oval-elliptic - - - - SYMPHORICA'RPUS - 541
    - Whorled, oval - - - - CEPHALA'NTHUS - 544
    - Digitate, 7–5-lobed - - - - VI'TEX - - 673
  - Serrate.
    - Ovate-acuminate - - - - DIERVI'LLA - - 525
- Deciduous, evergreen, or subevergreen,
  - Entire, connate - - - - LONI'CERA - - 526
  - Serrate or dentate, petiolate, and shining - - ARISTOTE'LIA - - 182
  - Serrate or entire, petiolate, shining - - - VIBU'RNUM - - 516

Opposite, exstipulate.

- Deciduous.
  - Entire.
    - Ovate-lanceolate, 3-ribbed - - - CORIA'RIA - - 146
    - Lanceolate, glabrous, large - - - CHIONA'NTHUS - 634
    - Ovate or lanceolate, shining - - - PERI'PLOCA - - 652
    - Cordate, large - - - - CATA'LPA - - 670
    - Lanceolate, reticulately veined - - BU'DDLE - - 660
    - Cordate, undivided or 3-lobed, large - - PAULO'WNIA - - 671
    - Lanceolate or ovate, beset with silvery scales - SHEPHE'RDIA - - 700
    - Lanceolate, with very short petioles - - BO'RYA - - 713
  - Serrate.
    - Ovate or cordate, sometimes lobed - - HYDRA'NGEA - - 492
    - Lanceolate, rough, with dots - - - I'VA - - - 548
  - Serrated or entire
    - Oval-acuminate, scabrous - - - { PHILADE'LPHUS - 460; DEU'TZIA - - 465 }
  - Dentate, notched, or otherwise cut at the edges.
    - Variously lobed (one variety deeply lacinaite) - A'CER - - 79
  - Dentate or entire.
    - Glabrous, entire, or toothed at the apex - } DECUMA'RIA - - 466
  - Entire.
    - Ovate-lanceolate, evergreen, pale green with yellow spots - - - - } AU'CUBA - - 511
    - Elliptic, glabrous, small - - - AZA'LEA - - 601
    - Lanceolate, shining - - - - PHILLY'REA - - 631
    - Ovate, ciliate, shining - - - - VI'NCA - - 657
    - Roundish, coriaceous, smooth, shining - - BU'XUS - - 703
  - Serrate.
    - Oblong, with revolute edges, small - - MYGI'NDA - - 156
  - Serrate or entire.
    - Wavy, leathery, dark green, shining - - GA'RRYA - - 926
- Deciduous evergreen, or subevergreen.
  - Entire.
    - Undivided, sessile, or sub-sessile, dotted - - { HYPE'RICUM - - 74; ANDROSÆ'MUM - - 77 }
    - Lanceolate, downy - - - - BENTHA'MIA - - 507
    - Ovate-lanceolate, membranous, glaucous - LEYCESTE'RIA - - 543
    - Lanceolate, glabrous - - - - LIGU'STRUM - - 628
    - Linear and scale-like, caducous, bark evergreen - E'PHEDRA - - 937
  - Serrate.
    - Ovate-lanceolate, sometimes linear, mostly deciduous - - - - - } EUO'NYMUS - - 149
  - Dentate or entire.
    - Petioles embracing the stems - - - CI'STUS - - - 54

Opposite or alternate, stipulate.

- Evergreen.
  - Entire.
    - Rhomb-shaped, bearing a mealy powder - A'TRIPLEX - - 676

Opposite or alternate, exstipulate.

- Deciduous.
  - Entire.
    - Lanceolate, sometimes whorled - - - PU'NICA - - 456
    - Oblong, small-fascicled - - - NITRA'RIA - - 467
- Deciduous, evergreen, or subevergreen.
  - Entire.
    - Smooth, oval-lanceolate - - - - CO'RNUS - - 501
    - Rigid, pale green on both sides - - - VI'SCUM - - 508
    - Glabrous, shining, sometimes small - - DA'PHNE - - 686
    - Linear, more or less 2-rowed - - - { TA'XUS - - 939; TORRE'YA - - 943 }
    - Scale-like, closely imbricated, compressed - THU'JA - - 1068
    - Scales situated under the joints of the branches - CA'LLITRIS - - 1072
    - Linear, adpressedly imbricate - - - CUPRE'SSUS - - 1073
    - Scale-shaped, minute, sometimes linear - - JUNI'PERUS - - 1080

Opposite or alternate, stipulate or exstipulate.

- Subevergreen.
  - Entire.
    - 3-nerved or feather-nerved, hoary or pilose - HELIA'NTHEMUM - 58

LEAVES COMPOUND.

Alternate, stipulate.

- Deciduous.
  - Entire.
    - Biternate or bipinnate, strong-smelled - - PÆO'NIA - - 18
    - Pinnate, 3—5-foliolate - - - - PTE'LEA - - 143
    - Trifoliolate, pilose, or pubescent - - ADENOCA'RPUS - 227

Impari-pinnate	Amo'rpha	230
Impari-pinnate, beset with glands	Eysenha'rdtia	232
Abruptly pinnate, leaflets mucronate	*Caraga'na*	237
Abruptly pinnate, leaflets 2 pairs	Halimode'ndron	242
Impari-pinnate, leaflets orbicular	Calo'phaca	243
Impari-pinnate, stipules small	Colu'tea	246
Pinnate, petioles permanent	*Astra'galus*	246
Abruptly pinnate, and bipinnate, or simple	Gledi'tschia	249
Bipinnate, 4—7 pinnæ	Gymno'cladus	255
Serrate.		
Pinnate, 3—13-foliolate	Xantho'xylum	142
Impari-pinnate, doubly and trebly serrate, very large	Ara'lia	496
Dentate, notched, or otherwise cut at the edges.		
Impari-pinnate, with ovate leaflets	Kölreute'ria	134
Dentate or serrate.		
Trifoliolate, stipules connate	*Ono'nis*	229
Deciduous, evergreen, or subevergreen.		
Entire.		
Trifoliolate, leaflets elliptical-oblong	Pipta'nthus	198
Trifoliolate, often pubescent	Cy'tisus	213
Impari-pinnate, glabrous	Coroni'lla	247
Serrate.		
Impari-pinnate, stipules attached to the petiole	*Ro'sa*	321
Dentate or serrate.		
Digitate, pinnate or lobed, rough	*Ru'bus*	311
Alternate, exstipulate.		
Deciduous.		
Entire.		
Impari-pinnate, with 11—13 leaflets	*So'phora*	195
Impari-pinnate, with 9—11 leaflets	Virgi'lia	197
Impari-pinnate, leaflets petiolate	Robi'nia	233
Impari-pinnate, leaflets oval, pointed	Wista'ria	248
Serrate.		
Trifoliolate or bipinnate, leaflets ovate	*Ci'ssus*	141
Impari-pinnate, 5—19 leaflets	*Ju'glans*	732
Impari-pinnate, 5—15 leaflets	*Ca'rya*	735
Impari-pinnate, 17 leaflets, sessile	Pteroca'rya	743
Dentate, notched, or otherwise cut at the edges.		
Palmate, pinnate, or bipinnate	Ampelo'psis	139
Impari-pinnate, teeth glandulous on the under side	*Aila'ntus*	145
Evergreen.		
Dentate or serrate.		
Pinnate, coriaceous, dark green	Maho'nia	50
Deciduous, evergreen, or subevergreen.		
Entire.		
Impari-pinnate, reddish green	*Pista'cia*	184
Conjugate, trifoliate, tendrilled	Bigno'nia	660
Opposite, stipulate.		
Deciduous.		
Serrate.		
Pinnate, with compound and partial stipules	Staphyle'a	147
Dentate, notched, or otherwise cut at the edges.		
Pinnate, the pinnæ often bi-glandular at the base	*Sambu'cus*	513
Dentate or serrate.		
Pinnately divided, leaflets irregular	Xanthorhi'za	19
Opposite, exstipulate.		
Deciduous.		
Serrate.		
Biternate, leaflets oblong lanceolate	*Atra'gene*	16
Impari-pinnate, leaflets nearly sessile, buds black (One variety has simple leaves, p. 642.)	*Fra'xinus*	639
Impari-pinnate, buds ash-coloured	*O'rnus*	651
Impari-pinnate, leaflets 7—9, petioles marginate	*Te'coma*	661
Dentate, notched, or otherwise cut at the edges.		
Impari-pinnate, 3—5 leaflets	*Negu'ndo*	122
Palmate, leaflets rough	*Æ'sculus*	124
Palmate, leaflets smooth	Pa'via	128
Deciduous, evergreen, or subevergreen.		
Entire.		
Pinnate, in decussating pairs	*Cle'matis*	2
Alternate or opposite, stipulate.		
Subevergreen.		
Entire.		
Pinnately cut, hairy	Potenti'lla	319

LEAVES SIMPLE OR COMPOUND.

Alternate, stipulate.		
Deciduous.		
Serrate or entire.		
Simple, or unequally pinnate	*Rhu's*	186
Simple, but sometimes pinnately divided	*Py'rus*	417
Dentate or serrate.		
Usually simple, but sometimes pinnately cut	*Spiræ'a*	299
Deciduous, evergreen, or subevergreen.		
Entire.		
Lanceolate, linear, or trifoliolate	*Geni'sta*	203
Trifoliolate or pinnate	Jasmi'num	654

EXPLANATION OF SIGNS, &c.

Under the titles of the orders are given signs, intended to show at a glance the general habit of the trees or shrubs described in each order. These signs represent large, small, and middle-sized plants, and are as follows; the first sign in each row indicating a deciduous tree or shrub, the next an evergreen, and so on alternately:—

1. Round-headed trees; such as the oak, ash, elm, beech, chestnut, &c. Deciduous and evergreen.

2. & 3. Spiry-topped or conical trees; such as the spruce fir, silver fir, larch, pine, deciduous cypress, &c. Deciduous and evergreen.

4. Fastigiate trees; such as the Lombardy poplar, evergreen cypress, pyramidal oak, &c. Deciduous and evergreen.

5 Drooping trees; such as the weeping willow, weeping elm, &c.

6 Shrubs of the largest size, and also middle-sized shrubs. Deciduous and evergreen, but exclusive of twiners, climbers, trailers, &c.

7. Under-shrubs, or shrubs of the smallest size. Deciduous and evergreen, but exclusive of twiners, trailers, &c.

8. Twining shrubs; such as the honeysuckle, aristolochia, &c. Deciduous and evergreen.

9. Climbing shrubs; such as the clematis, ampelopsis, vine, &c. Deciduous and evergreen.

10. Trailing shrubs, the branches of which lie prostrate on the ground, but do not root into it; such as many species of willow, *C*ýtisus, &c.

11. Creeping shrubs, or such as send up shoots from their creeping roots; as many species of Spirǽa, &c.

The signs put before each individual species and variety are the same as those used in the *Gardener's Magazine*, and in the *Hortus Britannicus*, viz.

Deciduous tree.
Evergreen tree.
Deciduous shrub.
Evergreen shrub.
Deciduous under-shrub.
Evergreen under-shrub.
Deciduous twiner.
Evergreen twiner.
Deciduous climber.
Evergreen climber.
Deciduous trailer.
Evergreen trailer.
Deciduous creeper.
Evergreen creeper.

Accentuations and Indications.

All the botanic names throughout the Work are accented, and have their origin indicated, as in the *Hortus Britannicus* and the *Gardener's Magazine*. The vowels which are sounded short are marked with an acute accent, thus (′), as A′ceras; and those which are sounded long are marked with a grave accent, thus (`), as A`brus. The origin of each name is indicated thus: if the name has been applied to a plant by the ancients, the first letter is in Italic, as *P*ìnus; if it is commemorative of some individual, the letters additional to the name are in Italic, as Bánks*ia*, Lambert*iàna*, Douglàs*ii*; and if an aboriginal name has been adopted, or if the name is of uncertain derivation, the whole word is in Italic, as, *Ailántus*, *Caragàna*, &c. Where the name would otherwise be in Italic, as in the case of synonymes, headings to paragraphs, &c., these distinctions are, of course, reversed, as P*ìnus*, *Bánks*ia, Ailántus. All the other scientific names, generic or specific, are composed from the Greek or Latin, except a very few which are taken from places: as Araucària, from the country of the Araucanians; Quércus gramúntia, from the estate of Grammont, &c.

The Engraved Figures

Are all to the same scale of 2 in. to 1 ft., or one sixth of the natural size; with the exception of details, which, when given, are generally of the natural size, and indicated by a cross, thus, +.

ABRIDGED

ARBORETUM ET FRUTICETUM BRITANNICUM.

Trees and shrubs, in common with all other flowering plants, are arranged by botanists in two grand divisions; viz. the Exogenous, or Dicotyledonous, Plants, the stems of which increase from without, and the leaves of which have reticulated veins; and the Endogenous, or Monocotyledonous, Plants, the stems of which increase from within, and the leaves of which have parallel veins. The first class includes all the hardy trees and shrubs in Britain, with the exception of shrubs of the genera *Yúcca*, Smìlax, *R*úscus, and one or two others, which belong to the second class. We shall arrange the genera and species under the same subdivisions, subclasses, sections, orders, and tribes, as we have adopted from DeCandolle in our *Hortus Britannicus.*

Class I. EXO'GENÆ.

Stems increasing from without; Leaves with reticulated Veins.

Subdivision I. DICHLAMY'DEÆ.

Calyx and Corolla distinct, by which they are distinguished from Subdivision II., in which the flowers have only a calyx.

It is in consequence of this high developement of the floral envelopes, that the greater part of handsome-flowering trees and shrubs are found in Dichlamýdeæ, it rarely happening that those with a single floral envelope have any brilliant colouring.

Subclass I. THALAMIFLO'RÆ.

Flowers with Petals and Stamens inserted in the Receptacle

This subclass contains all the Polyandrous plants of Linnæus; as the subclass Calycifloræ, in which the stamens are seated on the calyx, contains all the plants of the Linnæan class Icosándria.

Section I.

Carpella, that is, the component Parts of compound Capsules or Fruits, numerous; or the Stamens placed opposite the Petals.

Order I. *R*ANUNCULA'CEÆ.

The Diagnostic, or Distinctive, Character, or, as we shall term it, the Ordinal Character, of this order, is thus given by Dr. Lindley:—" *Polypetalous,* with hypogynous *stamens* [that is, stamens under the pistil]; *anthers* bursting by longitudinal slits; several distinct simple *carpella* [fruits]; exstipulate *leaves*, sheathing at their base; solid *albumen*; and *seeds* without arillus." (*Nat. Syst.*, p. 6.) — Climbing shrubs scarcely woody, and low suffruticose bushes. Natives of Europe, Asia, and North America.

Leaves generally alternate, but sometimes opposite, generally exstipuiate; deciduous, or evergreen; much divided, especially in Clématis, in which the leaves are not articulate with the stem. The petioles often serve as tendrils; and are dilated at the base, forming a sheath half-clasping the stem. Hairs if any, simple. *Inflorescence* small in Xanthorhìza, and some species of Clématis, and large in *Pæònia*. *Seeds* small and pointed, except in *Pæònia*. — The species in British gardens are included in two tribes, Clematídeæ and *Pæoniàceæ*, which contain the genera Clématis, Atrágene, *Pæònia*, and Xanthorhìza.

Tribe I. *Clemati'deæ.*

Tribal Character. Climbers, characterised by having the æstivation of the calyx valvate or induplicate; with no petals, or with the petals flat; the anther opening outwards; the carpels, or seed-vessels, not opening, one-seeded, terminated by a tail, which is the indurated style. Seed pendulous. Leaves opposite. Deciduous and evergreen climbers. — The genera are two, Clématis and Atrágene, which are thus contradistinguished: —

Cle'matis *L.* Petals none. Atra'gene *L.* Petals several.

Genus I.

CLE'MATIS *L.* The Clematis, *or Virgin's Bower.* *Lin. Syst.* Polyándria Polygýnia.

Identification. The word Klematis was, as well as Atragene, used by Theophrastus, to designate the Clématis Vitálba of Linnæus. Clematis was used by Matthiolus, and also by Clusius, who applied it to *C.* Viticélla *L.* and *C.* cirrhòsa *L.*
Synonymes. Ladies' Bower *Gerard*; Clématite, *Fr.*; Waldrebe, *Ger.*; Clematide, *Ital.*
Derivation. The word Clematis, or Klematis, is derived from the Greek word *klēma*, a small branch of a vine; and it is applied to this genus, because most of the plants composing it climb like a vine. The English name of Ladies' Bower was probably adopted from its suitableness for covering bowers; and, as the first kind of clematis brought to England (*C.* Viticélla) was introduced in 1569, during the reign of Elizabeth, the name of Virgin's Bower might be intended to convey a compliment to that sovereign, who, as it is well known, liked to be called the Virgin Queen. Waldrebe is compounded of *wald*, a wood, and *rebe*, the branch of a vine.

Generic Character. *Involucre* none, or situated under the flower, in the form of a calyx. *Calyx* of from four to eight coloured sepals. *Petals* none. *Carpels* numerous, aggregate, terminated by a long, and mostly feathery, tail.— Climbing shrubs, with variously cut opposite leaves. The recent herb of all the species is acrid, and, when applied to the skin, it occasions blisters. (*Don's Mill.*) The seed is pendulous, and the carpels are one seeded; each is terminated by a persistent style, and does not open until ruptured by the germination of the seed.

Leaves compound, opposite in decussating pairs, without stipules, deciduous or evergreen; the petiole possessed of a clasping power for attaching the plant to contiguous bushes, or similar objects; in all the species, more or less persistent after they are decayed. *Flowers* in axillary ramose panicles; small and white in some, and in others larger and highly coloured. *Seed* small, seldom seen divested of its envelope, as that never bursts till after it is committed to the soil. — The species are included in four sections; viz. Flámmula, Viticélla, Cheirópsis, and Anemoniflòra.

Root strong; the fibres rather straight, and not very much branched; extended in the soil rather horizontally than perpendicularly. Stem ligneous, not rigid enough to stand erect. Branches the same, and slender. Well adapted for covering bowers, or for ornamenting verandas or trelliswork. The greater number of the species ripen their seeds in England, and are easily propagated by them, or by layers. The seeds retain their vitality for several years; they are of slow vegetation, and ought to be sown as soon as gathered, in which case they will generally come up the following spring; though, sometimes, not till the second spring. All the species require support by props of some kind; and all, with one or two exceptions, grow freely in any soil that is tolerably dry, but more especially in one that is calcareous. From the acridity of these plants, they are not very liable to be attacked by insects; nevertheless, snails and slugs are occasionally found eating their young herbage.

§ i. *Flámmula* Dec.

Sectional Character. Involucre wanting. Tail of the carpels long, bearded and feathery. Cotyledons distinct (that is, slightly separated) in the seed. (*Don's Mill.*, i. p. 3.) Deciduous.

♣ 1. CLE'MATIS FLAMMULA *L.* The inflammatory-*juiced* Clematis, *or sweet-scented Virgin's Bower.*

Identification. Lin. Sp., 766.; Dec. Prod., 1. p. 2. Don's Mill., 1. p. 4.
Synonymes. C. ùrens *Gerard*; C. marítima *All. Ped.*; C. suavèolens *Salisb. Prod.*; C. paniculàta *Thun.*; Clématite odorante, *Fr.*; scharfe Waldrebe, *Ger.*
Derivation. From *flammare*, to inflame; on account of the blistering qualities of the species.
Engravings. Park. Theat., p. 381. f. 3.; and our *fig.* 1.

Specific Character and abridged Description. Leaves pinnate, smooth; with orbicular, oval, oblong, or linear, entire or three-lobed, acutish leaflets. (*Don's Mill.*) A deciduous climber. South of Europe; in hedges and waste bushy places, not far from the sea, and in soils more or less calcareous. Height 10 ft. to 15 ft. Introduced in 1596. Flowers white, sweet-scented; July to October. Fruit white; ripe in October. Leaves deep green, often remaining on the plants till mid-winter, and dying off black.

1. Clématis Flámmula.

Varieties and their Synonymes. The most distinct is C. F. marítima; the rest are of little importance.

♣ C. *F.* 2 *rotundifòlia* Dec. C. fràgrans *Tenore.* — Leaflets almost orbicular.

♣ C. *F.* 3 *marítima* Dec. — Leaflets linear.

♣ C. *F.* 4 *rubélla* Dec. — Leaflets oval, usually emarginate. Sepals four, reddish on the outside.

⅄ C. *F.* 5 *cæspitòsa* Dec. *C.* cæspitòsa *Scop.*, *C.* Flámmula *Bertol.* — Leaflets minute, entire or cut.

⅄ C. *F.* 6 *paniculàta.* *C.* paniculàta *Thun.* —Flowers with the peduncles simple.

A vigorous-growing plant, the stems of which rapidly attain the length of from 15 ft. to 30 ft. in a state of culture. The leaves are subject to much variation, from soil, situation, and climate. The peduncles of the flowers are sometimes simple, and sometimes branched. The colour of the sepals is white, slightly pubescent on their exterior margins. The whole plant has a dark green hue; and in autumn it is abundantly covered with flowers, the odour of which is of a honied sweetness, exceedingly disagreeable to some persons when near, though at a distance it is not unlike the fragrance of the common hawthorn. From the rapidity of its growth, it will in four or five years cover a very large space of wall, roof, or bower. Its herbage is considered less acrid than that of any other of the European species, notwithstanding its name of Flámmula.

⅄ 2. *C.* ORIENTA'LIS *L.* The Oriental Clematis.

Identification. Lin. Sp., 765.; Dec. Prod., 1. p. 3.; Don's Mill., 1. p. 4.
Synonymes. Flammula scandens apii folio glauco, *Dill. Elth.* 144.; *C.* flàva *Mœnch. Meth.* 296.; the Eastern, or yellow-flowered, Virgin's Bower; *C.* glaúca *Willd.*; *C.* ochroleùca *Hort.*; Clématite orientale, *Fr.*; Morgenländische Waldrebe, *Ger.*
Engravings. Dill. Elth., t. 119. f. 145.; and our *fig.* 2.

Spec. Char., &c. Leaves pinnate; leaflets smooth, wedge-shaped, with three toothed pointed lobes. (*Don's Mill.*) A deciduous climber. Levant and Caucasus. Height 10 ft. to 15 ft. Introduced in 1731. Flowers greenish yellow slightly tinted with russet, sweet-scented; July, August. Fruit white; ripe in October. Leaves somewhat glaucous, dying off black or dark-brown.

Varieties. C. *glaúca* Willd. and C. *ochroleùca* Hort. are, by some, alleged to be varieties of *C.* orientàlis; but we do not consider them sufficiently distinct for varieties, and have, therefore, included these names in our synonymes.

2. Clématis orientàlis.

The general magnitude of this species resembles that of *C.* Flámmula, from which it differs in its ulterior branches being more persistently ligneous, though the main stem in old plants is seldom seen so thick as that of *C.* Flámmula. It is also distinguished from the latter species by throwing up suckers freely, which the other does not. Its leaflets are glaucous, flat, large as compared with those of *C.* Flámmula; and it does not produce flowers so profusely as that species. The flowers are yellowish, and not so strongly scented; and the carpels are dissimilar, though still cottony in appearance when the seed is ripe.

⅃ 3. *C.* CHINE'NSIS *Retz.* The Chinese Clematis.

Identification. Retz. Obs., 2. p. 18.; Dec. Syst., 1. p. 137.; Don's Mill., 1. p. 4.
Synonyme. *C.* sinénsis *Lour. coch.* 1. p. 422.
Engraving. Our *fig.* 3.

Spec. Char., &c. Leaves pinnate; leaflets ovate-lanceolate, quite entire. Peduncles few-flowered, longer than the leaves. Ovaries usually four, with almost naked tails. (*Don's Mill.*) A deciduous climber. China, in the island called Danes. Height 10 ft. to 15 ft. Introduced in 1820. Flowers ?. Leaves purplish green, retained till rendered black by frost.

A plant in the Horticultural Society's Garden, grows vigorously against a wall, producing shoots as long and strong as those of *C.* Flámmula; and retaining its leaves till they are blackened by frost. It has never flowered; but, in its leaves and its general appearance, it seems to resemble *C.* orientàlis, except that the leaves are of a dark purplish green, instead of being glaucous.

3. Clématis chinénsis.

⅃ 4. *C.* VITA'LBA *L.* The White-Vine Clematis, *or Traveller's Joy.*

Identification. Lin. Sp., 766.; Dec. Prod., 1. p. 3.; Smith's Eng. Fl., 3. p. 39.; Don's Mill, 1. p. 4.
Synonymes. Athragenè *Theoph.*; *Vitis* sylvéstris *Dios.*; *C.* latifòlia seu *Atrágene Ray*; *C.* áltera *Matth.*; *C.* tértia *Com.*; Viórna *Ger.* and *Lob.*; *Vitis* nigra *Fuch.*; Vitálba *Dod.*; the Old Man's Beard, Bindwith, the common Virgin's Bower, the wild Climber, the great wild Climber; Clématite brulante, Clématite des Haies, l'Herbe aux Gueux, la Viorne des Pauvres, *Fr.*; gemeine Waldrebe, *Ger.*
Derivation. Because of its "decking and adorning the ways and hedges where people travel," says Gerard, "I have named it the traveller's joy." The name of Old Man's Beard is very appropriate to the white and hairy appearance of the tails of the carpels; and Bindwith, from the shoots being used instead of those of willows for tying up plants. White Vine is supposed to allude to the white appearance of the tails of the carpels in autumn. The French name of Clématite brulante has reference to the acrid properties of the plant; and Clématite des Haies to its growing generally in hedges. The name of l'Herbe aux Gueux refers to the employment of it by the beggars in France, who use it to makes ulcers in their arms and legs, for the purpose of exciting compassion, curing themselves afterwards by the application of the leaves of the beet. La Viorne des Pauvres alludes to the same practice, Viorne being evidently derived from Viórna.
Engravings. Jacq. Austr., 4. t. 308.; Eng. Bot., 612.; and our *figs.* 4. & 5.

4. Clématis Vitálba.

Spec. Char., &c. Leaves pinnate; leaflets ovate-lanceolate, acuminated, cordate at the base, partly cut. Peduncles forked, shorter than the leaves. (*Don's Mill.*) A deciduous climber of vigorous growth. Europe and Britain, in hedges and copses, always indicating a calcareous soil. Height 15 ft. to 30 ft.; in rich soil, and in a sheltered situation, 50 ft. to 100 ft. Flowers white; August, September. Fruit white; ripe from October to February. Leaves long retained, and dying off black or dark brown.

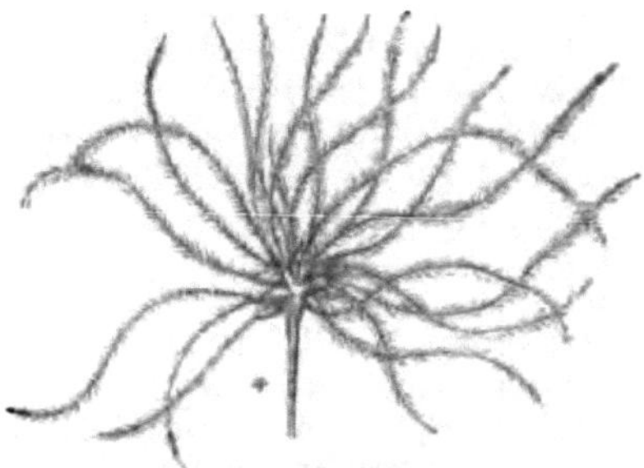

5. Fruit of C. Vitálba.

The stems are woody, more so than those of any other species, angular, climbing to the height of 20 or 30 feet, or upwards, and hanging down from rocky cliffs, ruins, or the branches of trees; or being supported by, and forming tufts on, the upper surface of other shrubs or low trees, which they often so completely cover as to have the appearance of bushes at a distance. The footstalks of the leaves are twined about whatever object they approach, and afterwards become hard and persistent, like the tendrils of a vine. The leaflets are either quite entire, or unequally cut; sometimes very coarsely so. The panicles are axillary and terminal, many-flowered and downy. The flowers are of a greenish white colour, with little show; but they have a sweet almond-like scent. The seeds (*fig. 5.*) have long, wavy, feathery, and silky tails, forming beautiful tufts, most conspicuous in wet weather. The French gardeners use the twigs instead of withs, for tying up their plants; and make very neat baskets of them when peeled, and also beehives. The twigs are in the best state for making these articles in winter; and their flexibility is increased by holding them to the fire before using them. In gardens and plantations the plant is valuable for the rapidity with which it may be made to cover naked walls, unsightly roofs of sheds, or low buildings and arbours; and for a variety of similar purposes.

⅄ 5. *C.* VIRGINIA'NA *L.* The Virginian Clematis.

Identification. Linn. Amœn., p. 275.; Dec. Prod., 1. p. 4.; Don's Mill., 1. p. 5.; Tor. and Gray, 1. p. 8.

Synonymes. *C.* canadénsis trifòlia rèpens *Tourn.*; *C.* canadénsis *Mill. Dict.* No. 5., *Salisb. Prod.* 371.; *C* cordifòlia *Mœnch. Supp.* 104.; *C.* triternàta *Hort.*; the broad-leaved Canada Virgin's Bower; Clématite de Virginie, *Fr.*; Virginische Waldrebe, *Ger.*

Engravings. Den. Brit. (the male plant), t. 74.; and our *fig.* 6.

6. Clématis virginiàna.

Spec. Char., &c. Flowers panicled, diœcious. Leaves ternate; leaflets cordate, acute, grossly toothed, or lobed. (*Don's Mill.*) A deciduous climber. Canada to Florida, in hedges and near rivers. Height 10 ft. to 15 ft. Introduced in 1767. Flowers white; August, September. Fruit white. Decaying foliage brown or black.

Variety.

⅄ C. *v.* 2 *bracteàta* Dec. *C.* bracteàta *Mœnch.* —Leaflets ovate-lanceolate, entire.

The general appearance of this plant is like that of *C.* Vitálba; but it is less robust in all its parts, and less ligneous in its stems and branches; and it is also somewhat more tender. Panicles trichotomously divided, with small leaves at the divisions. Sepals 4, white, obovate, exceeding the stamens. Flowers often diœcious or polygamous. (*Tor. and Gray.*) Miller states that it seldom ripens seeds in England; but, as it is diœcious, it is possible that he possessed only the male plant.

♣ 6. *C.* GRA'TA *Wall.* The grateful-*scented* Clematis.

Identification. Wall. Asiat., 1. t. 98.
Synonymes. *C.* odoràta *Hort.*; *C.* triternàta *Hort.*; *C.* nepalénsis *Hort.*
Engravings. Wall. Asiat., 1. t. 98.; and our *fig.* 7.

Spec. Char., &c. Flowers axillary, panicled; leaves subbiternate, villous; leaflets cordate, acuminated, serrated, 3-lobed; sepals obtuse. (*G. Don.*) A deciduous climber. Nepal, on mountains. Height 10 ft. to 18 ft. Introduced in 1831. Flowers white.

7. Clématis gràta.

Closely resembling *C.* virginiàna, but rather more hoary; and equally hardy, though it has not yet flowered freely in the open air. A shoot introduced into the inside of a stove in the Chelsea Botanic Garden, from a plant on the outside, flowered there in 1833. Frequent in nurseries as *C.* nepalénsis.

♣ 7. *C.* VIO'RNA *L.* The road-ornamenting Clematis, *or leathery-flowered Virgin's Bower.*

Identification. Lin. Sp., 765.; Dec. Prod., 1. p. 7.; Don's Mill., 1. p. 8.; Tor. and Gray, 1. p. 9.
Synonymes. *C.* purpùrea rèpens *Ray*; Flámmula scándens, flore violaceo clauso, *Dill. Elth.*; American Traveller's Joy; the Virginian Climber; the purple Climber; Clématite Viorne, *Fr.*; Glockenblüthige Waldrebe, *Ger.*
Derivation. From *via*, a way, and *ornare*, to ornament. Leather-flowered Virgin's Bower refers to the remarkably thick texture of the sepals; the German name signifies bell-flowered woodvine.
Engravings. Dill. Elth., 118. f. 144; and our *fig.* 9.

Spec. Char., &c. Peduncles 1-flowered. Sepals connivent, thick, acuminated, reflexed at the apex. Leaves smooth, pinnate; leaflets entire, 3-lobed, alternate, ovate, acute, floral ones entire. (*Don's Mill.*) A deciduous climber. Pennsylvania to Georgia. Height 6 ft. to 12 ft. Introduced in 1730. Flowers purple without, whitish within; June to August. Fruit white; ripe in September. Decaying leaves retained long, and dying off black.

Variety. C. *V.* 2 *cordàta.* *C.* cordàta *Sims Bot. Mag.* t. 1816., and our *fig.* 9. from that plate; Clém. Símsii Sweet's Hort. Brit.

8. Clématis Vióma cordàta.

This species is striking in the dissimilarity of its flowers to those of most other species. It is of vigorous growth, and, exclusive of its flowers, assimilates to *C.* Viticélla; but its stems and branches are less decidedly ligneous. The stems are numerous, slender, and round; the peduncles of the flower are long, deflexed towards the tip, rendering the flowers pendulous; the sepals never open, except at their extreme ends, which are bent back, giving the whole flower a bell shape, but with the mouth of the bell narrower than the body. The sepals are of a greenish purple or reddish lilac on the outside, and of a very pale green within. The stamens scarcely emerge from the sepals. The carpels are broad and flat; as they ripen, the tail becomes bent in and plumose, and of a brownish green colour. It is most ornamental as a single plant, trained to a rod or to a wire frame. As its branches are not very decidedly ligneous or persistent, but consist mostly of annual shoots from a suffruticose base, and are not much branched, the plant does not exhibit a bushy head. It thrives best in bog earth, kept somewhat moist, in which circumstance it differs from most of the woody species of *Clématis*. It may be increased by layers, though not so readily as from seeds, which it produces in abundance. This species, *C.* cylíndrica, and *C.* reticulàta, being neither very woody, nor very luxuriant in growth, may all be treated as herbaceous plants, to be supported during the flowering season by temporary props. Few border plants, indeed, will be found more elegant or more ornamental when so managed.

9. Clématis Viórna.

Ⅎ 8. *C.* CYLI'NDRICA *Sims.* The cylindrical-*flowered* Clematis.

Identification. Sims, in Bot. Mag., t. 1160.; Dec. Prod., 1. p. 7.; Don's Mill., 1. p. 8.; Tor. and Gray, 1. p. 10.
Synonymes. *C.* críspa *Lam.*, but not of Linn.; *C.* Viórna *Andr.* in *Bot. Rep.*; *C.* divaricàta *Jacq.*; the long-flowered Virgin's Bower; Clématite à longues Fleurs, *Fr.*
Engravings. Bot. Mag., t. 1160.; Bot. Rep., t. 71.; and our *fig.* 10.

Spec. Char., &c. Peduncles 1-flowered. Sepals thin, acuminated, reflexed at the apex, with wavy margins. Leaves slender, pinnate; leaflets stalked, ovate or oblong, middle one sometimes trifid, floral ones entire. (*Don's Mill.*) A deciduous climber. North Carolina and Florida. Height 3 ft. to 5 ft. Introduced in 1802. Flowers large, purplish blue, nodding; June, August. Fruit white; ripe in September.

DeCandolle describes this species as related to *C.* Viórna, reticulàta, and críspa, and discriminated it from these. *C.* cylíndrica, he says, differs from *C.* Viórna, in the segments of its leaves being entire and not trifid; in the flowers being blue, not reddish lilac and pale within, and twice the size of those of *C.* Viórna; in the sepals

10. Clématis cylíndrica.

being not leathery, but somewhat of the consistence of paper, with the margin waved; the ovaries 12-15, not 25-30. *C.* cylíndrica differs from *C.* reticulàta in its leaves being in consistence papery, not leathery; scarcely veined, not reticulately veined; and in other points. *C.* cylíndrica closely resembles *C.* críspa in habit and mode of flowering; but differs from it in its sepals being waved in the margin, not rolled backwards, in its larger flowers, and especially in its carpels having long bearded tails, and not naked ones. *C.* Viórna and *C.* cylíndrica, seen together in a living state, are very dissimilar in appearance. *C.* Viórna has vigorous long branches and reddish flowers, which are acorn-like in figure, except that they have a spreading mouth; there is also obvious dissimilarity in the foliage and shoots, *C.* cylíndrica being almost herbaceous.

10a. Fruit of Clématis cylíndrica.

‡ 9. *C.* RETICULA'TA *Walt.* The net-veined-*leaved* Clematis.

Identification. Walt. Fl. Car., 156.; Dec. Prod., 1. p. 7.; Don's Mill., 1. p. 8.; Tor. and Gray, 1. p. 10.
Synonymes. *C. ròsea Abbott*; *C. Simsii* Hook.; the netted Virgin's Bower; the reticulated Clematis.
Engravings. Dend. Brit., t. 72.; and our *fig.* 11.

Spec. Char., &c. Peduncles 1-flowered. Sepal connivent. Leaves coriaceous, netted with nerves, smooth, pinnate; leaflets stalked, 3-lobed or entire, ovate. (*Don's Mill.*) A deciduous climber. S. Carolina and Georgia. Height 6 ft. to 8 ft. Introduced in 1812. Flowers pale purplish red; June to August. Fruit white; ripe in September.

Leaflets all petiolulate, 1 in. to 1½ in. long, undivided or variously lobed, the lowest pair 3-parted, sometimes rather acute and mucronate. Peduncles longer than the leaves. Flowers as large as in *C.* críspa. Sepals dull purple, ovate-lanceolate, velvety externally. Tails of the carpels long. (*Tor. and Gray.*) In *C.* Viórna the sepals do not divaricate, except in their recurved tips; while in *C.* reticulàta the sepals expand in the mode of those of *C.* Viticélla. A side view of a flower less expanded resembles more the flower of *C.* cylíndrica, but the cylindrical portion is shorter. The flowers (sepals) of the two are different in colour. The leaves of *C.* reticulàta are veined, as is implied in the specific name. The stems are scarcely ligneous.

11. Clématis reticulàta.

‡ 10. *C.* HENDERSO'NII Chandler. Henderson's Clematis.

Engravings. Our *figs.* 12. and 13.

Spec. Char., &c. Peduncles 1-flowered, much longer than the petioles of the leaves. Sepals long, wrinkled, reflexed. Leaves bipinnate, leaflets ovate-acuminate. A deciduous climber. Hybrid. Height 8 ft. to 10 ft. Cult. 1835. Flowers bluish purple; June to September. Fruit white.

The stems and foliage bear a general resemblance to those of *C.* Viticélla, while the flowers, in magnitude and colour, and the leaflets in shape and veining, resemble those of *C.* integrifòlia; but the sepals expand much wider, in the manner of those of *C.* Viticélla. This plant is apparently a hybrid between *C.* Viticélla and *C.* integrifòlia, having the flowers of the latter, and the leaves and stems of the former. It was raised by Mr. Henderson, nurseryman, of Pine-apple Place, and first flowered in the nursery of Mr. Chandler, by whom it was named. It may fairly be described as one of the most ornamental species of this section, from the largeness of its flowers, their long footstalks, which make them stand out distinctly from the foliage, the great profusion with which they are produced, and the long time the plant continues to produce them. Layers.

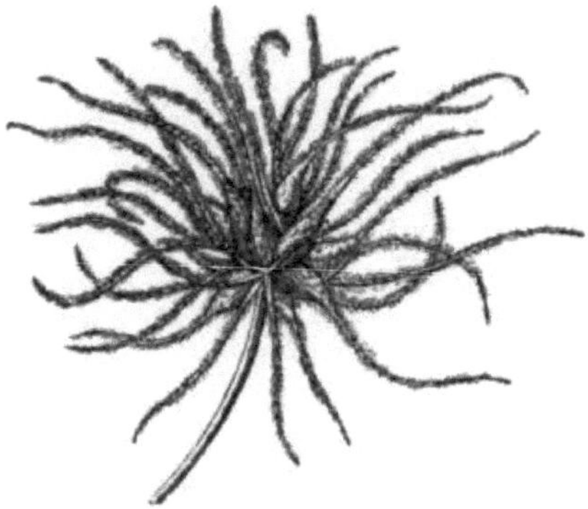

12. Fruit of *Clématis Hendersonii.*

13 *Clématis Hendersonii.*

§ ii. *Viticélla* Dec.

Derivation. From *viticula*, a small vine; on account of the plants climbing like *Vitis vinifera L.*

Sect. Char. Involucre wanting. Tail cf the pericarp (that is, of the carpel) short, beardless. Leaves ternate, or decompound. Stems climbing. (*Don's Mill.*, i. p. 9.) Deciduous.

⅄ 11. *C.* FLO'RIDA *Thun.* The florid, *or showy-flowered*, Clematis.

Identification. Thunb. Fl. Jap., 240.; Dec. Prod., 1. 8.; Don's Mill., 1. p. 9.
Synonymes. *Atrágene* indica *Desf.*; *Atrágene flórida* *Pers.*; Clématite à grandes Fleurs, *Fr.*; grossblüthige Waldrebe, *Ger.*
Engravings. Sims Bot. Mag., t. 834.; Andr. Bot. Rep., t. 402.; and our *fig.* 14.

Spec. Char., &c. Peduncles 1-flowered, longer than the leaves. Leaves ternately decompound; leaflets ovate, acute, quite entire. Sepals oval-lanceolate, much pointed. (*Don's Mill.*) A deciduous climber. Japan. Height 10 ft. to 12 ft. Introduced in 1776. Flowers white; June to September. Fruit ?.

Varieties.

⅄ C. *f.* 2. *flòre plèno* Hort. has the stamens changed into floral leaves, which may be denominated petals. It is very handsome, but the petals have frequently a tinge of green, which renders it less ornamental than the single species, in which the centre of the flower is comparatively inconspicuous, while the sepals are large, and of a pure white.

⅄ C. *f.* 3 *flòre plèno violàceo*; *C.* f. Siebóldii D. Don in *Sweet's Brit. Flow.-Gard.* 2d ser. p. 396.; *C.* Siebóldti Paxton's Mag. Bot.

p. 147.; *C.* bícolor *Hort.*; and our *fig.* 14.; is in all respects the same as the last, except that the petals, or centre of the flower, are of a rich violet colour, approaching to purple. It was sent from Japan to Europe by Dr. Sieboldt; and introduced into England in 1836. It is a most ornamental plant, and as hardy and easily propagated as the other variety or the species.

14. C. flórida fl. pl. violàceo.

15. Clématis flórida.

The stem is slender and striated; climbing to the height of 15 ft. or upwards when it is trained to a wall with a favourable exposure, though never becoming very woody. The flowers are large and handsome, either in a single or double state; and these, with the neatness of its foliage, and the slenderness of its stems and branches, give it such an air of elegance, that no lover of plants should be without it, who has a garden in which it will thrive. North of London it requires a wall; and in Scotland, as well as in France and Germany, it is generally kept in the green-house. A mode of pruning plants of this species, by cutting them down to the ground annually, though not generally practised, is said to produce vigorous shoots and fine flowers. This species never ripens seeds in England, and is therefore only propagated by layers.

⅃ 12. *C.* CÆRU'LEA *Lindl.* The blue, *or violet-flowered*, Clematis.

Identification. Lindl., in Bot. Reg., t. 1955.
Synonymes. C. azùrea grandiflòra *Sieb.*; *C* grandiflòra *Hort.*
Engravings. Bot. Reg., t. 1955.; and our *fig.* 16.

Spec. Char., &c. Leaves spreading, hairy, ternate. Segments ovate-acute, entire. Peduncles 1-flowered. Sepals 6 to 8, oblong, lanceolate, acute, membranaceous. Margin distended. (*Lindl.*) A deciduous climber. Japan. Height 10 ft. to 12 ft. Introduced in 1836. Flowers blue; June and July. Fruit ?.

16. Clématis cærùlea

A free-growing and profuse-blooming plant, with the habit of *C.* flórida. Flowers large, violet-coloured, with deep purple stamens. It differs from *C.* flórida in the colour, delicacy, and transparency of its blossoms, and also in its leaves being only once ternate, and in the sepals not touching and overlapping each other at the edges. Culture and propagation as in *C.* flórida.

⅃ 13. *C.* VITICE'LLA *L.* The Vine-Bower Clematis.

Identification. Lin. Sp., 762.; Dec. Prod., 1. p. 9.; Don's Mill., 1. p. 9.
Synonymes. Viticélla deltòidea *Mœnch*; the red-flowered Lady's Bower, *Gerard*; Italienische Waldrebe, *Ger.*
Engravings. Flor. Græc., t. 516.; Bot. Mag., t. 565.; and our *fig.* 17.

Spec. Char., &c. Peduncles 1-flowered, longer than the leaves. Leaves ternately decompound, lobes or leaflets entire. Sepals obovate, spreading. (*Don's Mill.*) A deciduous climber. South of Europe, in hedges, on calcareous soil. Height 10 ft. to 15 ft. Introduced in 1569 Flowers blue or purple; July to September. Fruit white; ripe in October. Decaying foliage black or brown.

Varieties.

- ♣ C. *V.* 1 *cærùlea.* — Flowers blue.
- ♣ C. *V.* 2 *purpùrea.* — Flowers purple.
- ♣ C. *V.* 3 *múltiplex* G. Don. *C.* pulchélla *Pers.* — Flowers double, blue. This variety produces more robust, more extended, and fewer shoots, than the single-flowered blue or purple varieties.
- ♣ C. *V.* 4 *tenuifòlia* Dec., *C.* tenuifòlia lusitánica *Tourn.*; and
- ♣ C. *V.* 5 *baccàta* Dec., *C.* campaniflòra *Hort.*; are varieties which we have not seen.

C. Viticélla, and all its varieties, are tolerably robust and vigorous in their growth, and decidedly ligneous; though plants, individually, do not endure many years. They are, perhaps, the most beautiful and most estimable of all the kinds of clematis, for the purposes of floral decoration. For the mere covering of bowers and other objects, they are, however, less suited than *C.* Vitálba and *C.* Flámmula; as these grow faster, extend farther, and each yields a greater aggregate of herbage, and so covers better: but none of them can vie with C. Viticélla and its varieties in beauty; more especially with the single purple and the single blue.

17. Clématis Viticélla.

♣ 14. *C.* CAMPANIFLO'RA *Brot.* The bell-flowered Clematis.

Identification. Brot. Flor. Lus., 3. p. 352.; Dec. Prod., 1. p. 9.; Don's Mill., 1. p. 9.
Synonymes. C. viornöides, received at the Chelsea Botanic Garden by this name, from the Berlin Botanic Garden (*D. Don*, in *Sw. Fl.-Gard.*, 2d ser., t. 217.); *C.* viornöides *Schrader*, *Hort. Brit.* No. 28757.; *C.* parviflòra *Dec.*, according to Sweet.
Engravings. Lod. Bot. Cab., 987.; Sw. Br. Fl.-Gard., 2d series, t. 217.; and our *fig.* 18.

Spec. Char., &c. Peduncles 1-flowered, somewhat longer than the leaves. Leaves biternately decompound; leaflets entire, or 3-lobed. Sepals half-spreading, dilated at the apex, wavy. (*Don's Mill.*) A deciduous suffrutescent climber of vigorous growth. Portugal, in hedges. Height 11 ft. to 15 ft. Introd. 1810. Flowers white, slightly tinged with purple; July and August. Fruit white. Decaying foliage dark brown.

18. Clématis campaniflòra.

Variety.

- ♣ C. *c.* 2 *parviflòra.* *C.* parviflòra *Fisch. of Göttingen.* — Flowers rather smaller than in the species, sepals crisped at the edges. H. S.

The habit of growth of this plant is that of *C.* Viticélla, to which it also comes nearest in affinity, but, though less woody, its shoots are much more robust; the much smaller and white flowers, and pointed sepals connivent (that is, lying close together) below, will readily distinguish it. It seldom ripens wood in England, but is readily propagated by layers.

♣ 15. *C.* CRI'SPA *L.* The curled-*sepaled* Clematis.

Identification. Lin. Sp., 765.; Dec. Prod., 1. p. 9.; Don's Mill., 1. p. 9.
Synonyme. *C.* flòre crìspo *Dill. Elth.*
Engravings. Dill. Elth., 1. t. 73. fig. 84.; Bot. Mag., 1982.; and our *fig.* 19.

19. Clematis crispa.

Spec. Char., &c. Peduncles 1-flowered, shorter than the leaves. Leaves entire, 3-lobed, or ternate, very acute. Sepals connivent at the base, but reflexed, and spreading at the apex. (*Don's Mill.*) A deciduous climber. Virginia to Florida. Height 3 ft to 5 ft. Introduced in 1726. Flowers purple; July to September. Fruit brownish; ripe in October.

Leaves glabrous, or slightly hairy. Flowers one third smaller than in *C.* Viórna, bright purple. Tail of the carpels thick and rigid, about half an inch long. (*Tor. and Gray.*) The flowers of this species are pretty, but perhaps never produced in sufficient quantity to render it highly decorative. The sepals have their tips reflexed, and waved with transverse wrinkles. The stems are weak, and do not generally rise higher than 3 or 4 feet. The plants frequently die down to the ground, so that this species requires to be treated more as herbaceous than ligneous. It ripens seeds plentifully.

§ iii. *Cheirópsis* Dec.

Derivation. From *cheir*, the hand, and *opsis*, resemblance; in allusion to the form of the bracteas.

Sect. Char. Involucre in the form of a calyx, from two joined bracteas situated at the top of the peduncle just under the flower. Tails of pericarps bearded. Climbing or rambling shrubs, with simple or ternate leaves. The old petioles persistent, and the new leaves and the peduncles produced in clusters from the axils of these. (*Dec. Syst.*, i. 162.) Evergreen.

♣ 16. *C.* CIRRHO'SA *L.* The tendriled-*petioled* Clematis.

Identification. Lin. Sp., 766.; Dec. Prod., 1. p. 9.; Don's Mill., 1. p. 9.
Synonymes. *Atrágene* cirrhòsa *Pers. Syn.* 2. p. 98.; Traveller's Joy of Candia, and Spanish Traveller's Joy, *Johnson's Gerard*; Spanish wild Climber, *Parkinson*; the evergreen Clematis; Clématite à Vrilles, Clématite toujours verte (*Bon Jard.*), *Fr.*; einfachblättrige (simple-leaved) Waldrebe, *Ger.*
Derivation. The word cirrhòsa, which means cirrhose, or tendriled, is applied to this species from the peculiarly grasping and tendril-like action of its petioles, which retain their hold even after the leaflets have fallen. The French word Vrilles signifies tendrils; and the German word einfach alludes to its comparatively simple leaves.
Engravings. See the Varieties.

Spec. Char., &c. Peduncle 1-flowered, with an involucre. Leaves simple, or variously divided; evergreen. An evergreen climber. Spain and the Balearic Isles. Height in British gardens, in the climate of London, in a warm situation, against a wall, 5 ft to 10 ft. Introduced in 1596. Flowers greenish or yellowish white; March and April. Fruit ?. Foliage of the broad-leaved varieties forming a fine dark green mass.

Varieties.

20. Clématis cirrhòsa pedicellàta.

21. C. cirrhòsa angustifòlia

Ł C. *c.* 2 *pedicellàta* Dec.; *C.* pedicellàta *Sweet's Hort. Brit.* p. 2., and *Don's Mill.* i. p. 9.; *C.* baleárica *Pers.*; *C.* cirrhòsa *Sims B. Mag.* t. 1070.; and our *fig.* 20.; has the pedicel between the involucre and the flower considerably longer than in the species.

Ł C. *c.* 3 *angustifòlia. C.* baleárica *Rich. Bot. Mag.* t. 959., and our *fig.* 21.; *C.* calycìna *Ait.*; *C.* polymórpha *Hort.* Clématite de Mahon, *Fr.* — The leaves of this variety vary exceedingly, from those shown in *fig.* 23. of the natural size, taken from a plant in the Horticultural Society's Garden, to those shown in *fig.* 22., reduced from Smith's *Flora Græca.* Introduced into England by M. Thouin, in 1783.

The varieties are all elegant evergreen climbing shrubs, rising to the height of from 6 ft. to 10 ft., and branching freely, so as to become, in two or three years, very thick bushy plants.

22. Clématis cirrhòsa angustifòlia.

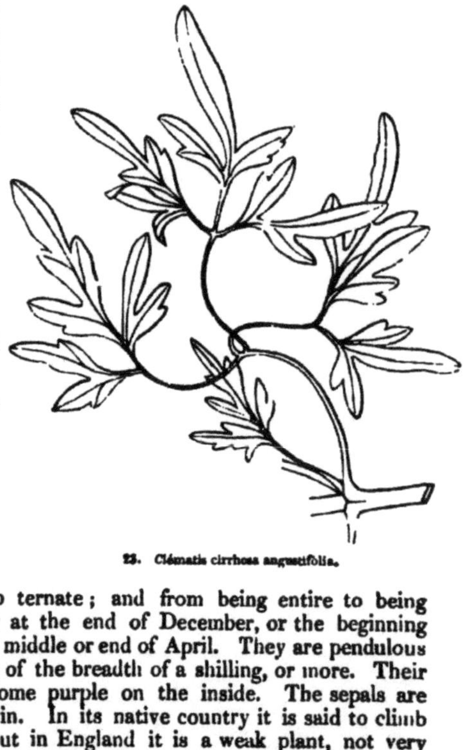

23. Clématis cirrhosa angustifòlia.

The leaves vary from simple to ternate; and from being entire to being deeply cut. The flowers appear at the end of December, or the beginning of January, and continue till the middle or end of April. They are pendulous and bell-shaped, the mouth being of the breadth of a shilling, or more. Their colour is greenish white, with some purple on the inside. The sepals are downy without, and smooth within. In its native country it is said to climb up and overwhelm the trees; but in England it is a weak plant, not very readily kept. In nurseries it is generally cultivated in pots, and kept in a green-house, or in a cold-frame. The principal beauties of this species consist in its bright evergreen verdure, and in the earliness of its flowering in spring; and these properties may be best obtained by training it against a wall with a southern aspect. Layers and cuttings.

§ iv. *Anemoniflòra.*

Derivation. From the flowers being like those of the *Anemòne sylvéstris.*

Sect. Char. Peduncles axillary, 1-flowered, aggregate, not bearing an involucre. Carpel with a feathery tail. Leaves deciduous.

⅄ 17. *C.* MONTA'NA *Ham.* The Mountain Clematis.

Identification. C. montàna *Ham. MSS.*; Dec. Prod., 1. p. 9.; Royle, p. 51.
Synonymes. C. anemoniflòra *D. Don Prod. Fl. Nepal.* p. 192., *Don's Mill.* 1. p. 9.
Engravings. Wall. Pl. Asiat. Rar., 3. p. 12. t. 217.; Swt. Br. Fl.-Gard., 2. s. t. 253.; and our *figs.* 24. and 25. *Fig.* 25. is from the plant in the Hort. Soc. Garden, and *fig.* 24. from a specimen of that at Montreal, Kent.

Spec. Char., &c. Peduncles 1-flowered, not bracteated, several together. Leaves ternately parted, the segments ovate-oblong, acuminate, toothed, the teeth in the mode of incisions. Sepals elliptic-oblong, mucronulate, spreading. (*D. Don.*) A deciduous climber. Himalayan Mountains at 5000 ft. to 7000 ft. elevation. Height 10 ft. to 15 ft., or in sheltered situations 30 ft. to 50 ft. Introduced in 1831. Flowers white; April in Nepal, May in England. Fruit white; ripe in August. Decaying foliage brown, and dropping more freely than in most of the other kinds.

A highly ornamental species. The plant is large and branching; the bark thick, ash-coloured, and deciduous. Leaves numerous, pale green. Flowers numerous, about the size and form of those of *Anemòne sylvéstris L.*, borne several together, each upon a separate, upright, slender peduncle, about 3 in. long. Sepals 4, 1 in. long, pure white, faintly stained with pink outside at the base. Styles clothed with long white silky hairs; from which it may be inferred that this species will have its fruits terminated with feathery tails, in a state of maturity. In the climate of England it proves to be quite hardy, and seems to flourish as well as on its native mountains. It grows with great vigour in a loamy soil, flowers profusely early in the season, and is readily increased by layers. A very desirable species.

24. Clématis montàna.

25. Clématis montàna.

Other Species of Clématis.—There are several other species of Clématis described in books, some of them as introduced, and others as not yet in cultivation in Britain; but we have refrained from describing any species of which we have not seen living plants. In Torrey and Gray's *Flora of North America*, C. *holosericea* Pursh, C. *ligusticifòlia* Nutt., C. *Drummóndii Tor. & Gray*, C. *parviflòra* Nutt., C. *lasiántha* Nutt., C. *lineariloba* Dec., and C. *Pitcheri Tor. & Gray*, are described as woody species, none of which, even by name, are yet in British gardens. C. *pubéscens*, *vitifòlia*, *Buchaniàna*, and some others, mentioned by Drs. Wallich and Royle, are yet to introduce from the Himalayas; and there are several names in DeCandolle's *Prodromus* of which living plants are not in our gardens.

Genus II.

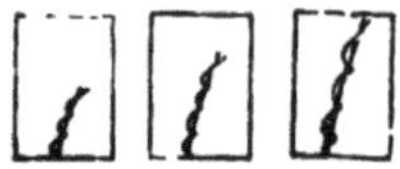

ATRA'GENE *L.* The Atragene. *Lin. Syst.* Polyándria Polygýnia.

Identification. Lin Gen., p. 615.; Don's Mill., 1. p. 10.
Synonymes. Clématis *Lam.* and *Dec.*; Atragene, *Fr.* and *Ger.*
Derivation. The name of *Atrágene* appears to be taken from two Greek words; *athros*, pressed, and *genos*, birth; alluding, as it is supposed, to the manner in which the branches press against or clasp the trees that support them. It was first used by Theophrastus, and was by him applied to Clématis Vitálba *L.*

Gen. Char. *Involucre* none. *Sepals* 4, somewhat induplicate in the bud. *Petals* numerous, shorter than the sepals. *Cariopsides* (carpels) terminated by a bearded tail. *Cotyledons* approximate in the seed. (*Don's Mill.*)

Leaves compound, opposite, generally exstipulate, deciduous; leaflets variously cut. *Flowers* axillary, pedunculate; purple, blue, or white. *Climbing shrubs*, natives of Europe and North America.

The atragenes differ from the clematises in producing leaves and one flower from the same bud contemporaneously; whereas in most clematises the flowers are produced upon wood developed previously to their appearance, and during the same season. Hence the winter buds of *Atrágene* are larger than those of *Clématis*, from their including the flower as well as the leaves of the succeeding year. In atragenes the leaves are less divided than in many of the species of *Clématis*, and they are always divided ternately. All the species of *Atrágene* described in this work have petioles, which not only clasp objects, like those of *Clématis*, but maintain the hold for more than the season, like the vine. All extremely interesting from the beauty of their blossoms. The culture is the same as in *Clématis*, and the propagation generally by layers.

⅄ 1. *Atra'gene alpi'na L.* The Alpine Atragene.

Identification. Lin. Sp., 764.; Don's Mill., 1. p. 10.
Synonymes. Clématis cærùlea *Bauh.*; Atrágene austrìaca *Scop.* and *Bot. Mag.*; Atrágene clematìdes *Crantz*; Clématis alpìna *Mill. Dict.* No. 9.; *C.* alpìna *Dec. Prod.* 1. p. 10.; Atragène des Alpes, *Fr.*; Alpen Atragene, *Ger.*
Engravings. Bot. Rep., t. 180.; Bot. Mag., t. 530.; and our *fig.* 26.

Spec. Char., &c. Peduncles 1-flowered, longer than the leaves. Leaves biternate; leaflets ovate-lanceolate, acuminated, serrate. Petals somewhat spathulate, blunt. (*Don's Mill.*) A deciduous climber. South of Europe, on mountains, in calcareous soils. Height 8 ft. Introduced in 1792. Flowers blue; May to July. Fruit white; ripe in August. Decaying leaves brownish, and in general parting more freely from the stems than in *Clématis*.

Varieties. DeCandolle mentions its varying with white flowers; and *A.* sibírica *Lin.*, described below as a species with yellowish white flowers, appears to us nothing more than a variety of *A.* alpìna.

The stems are numerous, branching, weak, forming knots at the joints where the leaves and flowers are protruded. One flower, on a longish scape, springs from between the leaves. The sepals are twice the length of the petals, and are blue on both sides. The petals are small, of a dirty white, and in general 12 in number. Very ornamental. Layers.

26. Atrágene alpìna.

⅄ 2. *A.* (A.) SIBI'RICA *L.* The Siberian Atragene.

Identification. Sims, Bot. Mag., t. 1951.; Don's Mill., 1. p. 10.
Synonymes. Atrágene alpina *Gmel. Sib.* 4. p. 194., *Pall. Flor. Ross.* 2. p. 69.; Clématis sibírica *Mill. Dict.* No. 12., and *Dec. Prod.* 1. p. 10.
Engravings. Sims, Bot. Mag., t. 1951.; and our *fig.* 27.

Spec. Char., &c. Peduncles 1-flowered, almost equal in length with the leaves. Leaves biternate; leaflets oblong-lanceolate, acuminated, serrated. Petals emarginate at the apex. (*Don's Mill.*) A deciduous climber. Siberia, on mountains. Height 6 ft. to 12 ft. Introduced in 1753. Flowers white; June and July. Fruit white: ripe in August. Decaying leaves brownish.

Variety. A blue-flowered variety of this species is mentioned in *Bot. Mag.*, t. 1591., which is probably the *A.* ochoténsis of Pallas, or possibly nothing more than *A.* alpina *L.*

There is a considerable similarity in this to the last, in foliage and habit of growth; but it is less robust and less branchy; its branches are more ligneous-looking, and the segments of the leaves longer. The calyxes of the flower are white, longer, and with the tips rather connivent than spreading. The bark and foliage are of a lighter colour, and the flowers longer than those of *A.* alpina; and the latter are perhaps less numerous.

27. Atrágene (a.) sibírica.

⅄ 3. *A.* AMERICA'NA *Sims.* The American Atragene.

Identification. Sims, in Bot. Mag., t. 887.; Don's Mill., 1. p. 10.
Synonyme. Clématis verticillàris *Dec. Prod.* 1. p. 10., *Tor. & Gray*, 1. p. 10.
Engravings. Bot. Mag., 887.; and our *fig.* 28.

Spec. Char., &c. Peduncles 1-flowered; leaves whorled, in fours, ternate; leaflets stalked, cordate lanceolate, acuminated, entire or somewhat lobed or serrated. Petals acute. (*Don's Mill.*) A deciduous climber. Vermont to Carolina, on mountains and rocky places. Height 10 ft. to 15 ft. Introduced in 1797. Flowers purplish blue; May to July. Fruit white; ripe ?. Decaying leaves dark brown.

Variety.

⅄ A. *a.* 2 *obliqua* Dou. MS.—Leaflets bluntly serrated.

This species is distinguishable from all the other Clematídeæ described in this work, by the peculiarity of its leaves being disposed, not oppositely in alternately decussating pairs, but in whorls of four. This is an anomalous characteristic, which DeCandolle has expressed by his specific epithet verticillàris. The flowers are very large, and campanulate. Sepals oblong-lanceolate, bright purplish blue. (*Tor. and Gray.*) Layers.

28. Atrágene americàna.

Other Species and Varieties of Atrágene. — A. *ochoténsis* Pall. we consider as a variety of *A.* sibírica *L.* A. *columbiàna* Nutt., *C.* columbiàna *Tor. & Gray*, i. p. 11., has ternate leaves, and pale blue flowers smaller than those of *A.* americàna. It is a native of the Rocky Mountains, but has not yet been introduced.

Tribe II. *PÆONIA'CEÆ* Dec.

Trib. Char. At once distinguishable from Clematídeæ, by the character of the anthers opening to admit the escape of the pollen on the side next the ovaries. In Clematídeæ, the anthers open on the side outward to the ovaries. The æstivation is also imbricate, and the carpels from one-seeded to many-seeded Suffruticose deciduous shrubs, of low growth, natives of temperate climates.

Leaves compound, alternate or opposite, stipulate, deciduous, but without possessing a clasping power. *Flowers* very large in Pæonia, very small in Xanthorhiza; and the following are the distinctive characteristics of these genera: —

PÆO'NIA L. Sepals persistent. Petals orbicular, sessile.
XANTHORHI'ZA *L.* Sepals deciduous. Petals 2-lobed, unguiculate.

GENUS I.

PÆO'NIA L. THE PÆONY. *Lin. Syst.* Polyándria Di-Pentagýnia.

Identification. The term Pæonia was applied by the Greeks to these plants, which have continued to bear that name ever since.

Synonymes. Peony, Piony; Pivoine, *Fr.*; Gichtterrose, and Päonie, *Ger.*; Rosa del Monte, *Span.*; Peonia, *Ital.*

Derivation. The term Pæonia is generally said to have been given by Hippocrates and Dioscorides, in commemoration of Pæon, the physician who first used it in medicine; but Professor Don thinks it more probable that it is derived from Pæonia, a mountainous country of Macedonia, where some of the species grow wild. Gichtterrose, *Ger.*, signifies the gouty rose, from the knobby or gouty appearance of the roots of the herbaceous species.

Gen. Char. *Calyx* of 5 leafy, unequal, permanent sepals. *Petals* from 5 to 10, somewhat orbicular. *Stamens* numerous. *Disk* fleshy, girding the ovaries. *Carpels* follicular, from 2 to 5, large, many-seeded, terminated with thick bilamellate stigmas. *Seeds* rather globose, shining.

Leaves compound, alternate, biternate or bipinnate. *Flowers* large, rosy, or rosy and white, usually with a strong disagreeable smell. A suffruticose shrub. Height from 3 ft. to 10 ft. Native of China and Japan.

There is but one ligneous species, *P. Moútan;* but there are several varieties; all undershrubs, which never attain a great height, and the wood of which always retains a herbaceous character, with a large pith. The roots are ramose rather than tuberous. The whole plant is narcotic and poisonous. The varieties are all beautiful, and hardy in most parts of Great Britain; though, from vegetating early, they commonly suffer from spring frosts.

▲ 1. *P. MOU'TAN* Sims. The Moutan, *or Tree*, Pæony.

Identification. Sims, Bot. Mag., t. 1154.; Dec. Prod., 1. p. 65.; Don's Mill., 1. p. 65.

Synonymes. Pæònia arbòrea *Donn Hort. Can.*; *P. suffruticòsa Bot. Rep.*; Pivoine Moutan, and Pivoine en Arbre, *Fr.*; baumartige Gichtterrose, *Ger.*; Hoa-Ouang, and Pé-Leang-Kin, *Chinese.*

Derivation. The word Moutan has been applied to this species of pæony, in China, for above 1400 years. *P.* arbòrea and *P.* suffruticòsa signify the tree and the sub-shrubby pæony. The German name signifies the tree-like gouty rose. The Chinese name Hoa-Ouang signifies the king of flowers, alluding to the beauty of the plant; and Pé-Leang-Kin, a hundred ounces of gold, in allusion to the high price which some of the varieties bear in China.

Spec. Char., &c. Segments of leaves oval-oblong, glaucous underneath. Carpels 5, villose. (*Don's Mill.*) A deciduous suffrutescent bush. China. Height 3 ft. to 6 ft. Introduced in 1787. Flowers pink; May. Fruit brownish green; ripe in September.

Decaying leaves brown or black.

Varieties.

▲ P. M. 1 *papaveràcea* Andrews. Bot. Rep., t. 463.; Lod. Bot. Cab., 547.; Bot. Mag., 2175.; and our *fig.* 29.—Petals from 8 to 13, white, with a purple spot at the base of each. Capsules altogether enclosed in the urceolus, or disk. Introduced in 1805. Professor Don remarks (*Sw. Br. Fl.*-

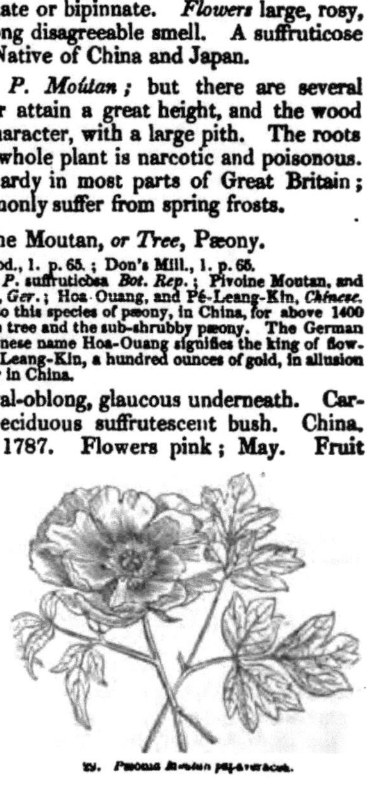

29. *Pæonia Moutan papaveracea.*

Gar., 2d ser., 238.) that *P. M.* papaveràcea appears to be really the normal form of the species, as the late Mr. George Anderson suggested in his paper on the subject in the *Linnæan Transactions*, vol. xii.

■ P. M. 2 *Bánksii Andrews*. Bot. Rep., t. 448.; Bot. Reg., 379.; Bot. Mag., t. 1154.; and our *fig.* 30.—Flowers double. Petals slightly tinged with blush, becoming nearly white at the edges, marked at the base with purplish red. In the centre of the flower are some elongated petals, which sometimes appear to rise from amongst the germens. Cultivated in 1787.

30. Pæònia Moùtan Bánksii.

Other Varieties. Upwards of twelve are described in the first edition of this work, and the number is continually increasing, in consequence of cross fecundation with one another, and with the herbaceous species. They are all very beautiful, and well deserving of cultivation.

The *Pæònia Moùtan*, in a sheltered situation, will attain the height of from 6 ft. to 10 ft. in ten years: and no plant can be a more gorgeous ornament of the garden than such a bush, abounding as it does in leaves striking from their branched character and numerous segments, and in very magnificent flowers of extraordinary size; both leaves and flowers being produced early in the spring. On its first importation, this plant was grown in sandy peat; but it has since been found to thrive best in deep rich loamy soil. An open situation is preferable, both on account of maturing the wood and leaves, and for displaying the flowers to advantage; but the plant must be sheltered from the cold spring winds, unless it is intended to cover it, when it is in flower, with a movable glass or canvass case. The protection given to this plant is necessary, not so much to prevent it from being injured during winter (for it will bear the winters of Paris without any protection, if the wood has been properly ripened), as to protect the tender leaves and flowers when they first appear, in April and May, from being blackened by the frost. Seeds are frequently produced from which new varieties are raised, and any variety may be increased by division of the root; by grafting on the tubers of herbaceous pæonies, any time from the middle of September to the middle of March; by budding, a mode said to be practised by the Chinese; by layers, which is the most general mode; by ringing a branch beneath each bud, and then pegging down the branch, and covering it with soil; and by cuttings. The details of these modes of propagation will be found in the first edition of this work.

GENUS II.

XANTHORHI`ZA *L.* THE YELLOW-ROOT. *Lin. Syst.* Polyándria Mono-Tri-gýnia.

Gen. Char. Calyx of 5 deciduous sepals. *Petals* 5, of two roundish lobes raised on a pedicel. *Stamens* 5-10. *Ovaries* 5-10. *Carpels* 2—3-seeded, but usually solitary from abortion. (*Don's Mill.*, i. p. 65.) — There is only one species known.

Leaves compound, opposite, stipulate, deciduous; pinnately divided, toothed, and serrated. *Flowers* in racemes, axillary, compound, appearing with the leaves.

♠ 1. X. APIIFO'LIA *L'Hérit.* The Parsley-leaved Yellow-Root.

Identification. L'Hér. Stirp. nov., p. 79. t. 38.; Dec. Prod., 1. p. 65.; Don's Mill., 1. p. 65.; Tor. and Gray, 1. p. 40.
Synonymes. Xanthorhiza apiifòlia; Zanthorise à Feuilles de Persil, *Fr.*; Sellerie-blättrige Gelbwurz, *Ger.*
Derivation. From the Greek words *xanthos*, yellow, and *rhiza*, a root; applied from the deep yellow colour of the roots. The French name needs no explanation; and the German is a literal translation of the English one.
Engravings. Lam. Ill., t. 854.; Bot. Mag., 1736.; and our *fig.* 31.

Spec. Char., &c. Flowers minute, dark purple, often by abortion polygamous. A low, suffrutescent, deciduous shrub. Flowers dark purple; May. Height 2 ft. to 3 ft. Pennsylvania. Introduced in 1776. Decaying leaves yellowish or brown, dropping in September.

A small shrub with yellow creeping roots, which attain a large size, and throw up numerous suckers; with irregularly pinnate leaves, branched racemes, and small purplish flowers (which are usually unisexual from abortion) rising from the scaly buds. The flowers appear early in May, and continue a month or upwards before they drop off. We have never heard of its ripening seeds in Europe; nevertheless, this may have occurred, and been overlooked, from the inconspicuousness of the shrub, and the smallness of its fruit. Suckers, or division of the root.

31. Xanthorhiza apiifòlia.

ORDER II. WINTERA'CEÆ.

ORDINAL CHARACTER. *Calyx* of 2—6 deciduous sepals, and 2 to many petals; the sepals and petals, when more than two, disposed ternately. *Carpels* whorled, very rarely solitary from abortion. — Evergreen shrubs, or low trees, chiefly natives of warm climates.

Leaves simple, alternate, stipulate, evergreen; full of pellucid dots, and coriaceous. Properties aromatic and stimulant. Illícium is the only genus of this order which contains species that will stand out in the open air in Britain.

GENUS I.

ILLI'CIUM *L.* THE ILLICIUM, *or* ANISEED TREE. *Lin. Syst.* Polyándria Polygýnia.

Identification. Lin. Gen., 611.; Tor. and Gray, 1. p. 42.; Dec. Prod., 1. p. 77.; Don's Mill., 1. p. 78.
Synonymes. Badiane, or Anis-étoilé, *Fr.*; Sternanis, *Ger.*
Derivation. The generic name Illicium is formed from the Latin word *illicio*, to allure, on account of the agreeable aromatic smell of all the species. It is called the Aniseed Tree, from its smell bearing a strong resemblance to that of aniseed. Badiane appears to be an aboriginal French word; Anis-étoilé, and Sternanis, signify literally the starry anise, and may allude to the starry disposition of the parts of the flower and of the capsules.

Gen. Char. *Calyx* of 3-6 petal-like sepals. *Carpels* stellately disposed, capsular, opening on the upper side, 1-seeded. (*Don's Mill.*, i. p. 79.)

♠ 1. ILLI'CIUM FLORIDA'NUM *Ellis.* The Florida Illicium.

Identification. Lin. Mant., 395.; Dec. Prod., 1. p. 77.; Don's Mill., 1. p. 79.
Synonymes. The Florida Aniseed Tree, red-flowered Anise-seed Tree, *Mor. Hist.*; Badiane de la Floride, *Fr.*; unächter (spurious) Sternanis, *Ger.*
Engravings. Bot. Mag., 439.; Lod. Bot. Cab., t. 209.; and our *fig.* 32.

Spec. Char., &c. Petals 27—30, dark purple, outer ones oblong, inner ones lanceolate. (*Don's Mill.*) An evergreen glabrous shrub. West

Florida to Louisiana, in swamps. Height 4 ft. to 6 ft. Introduced in 1766. Flowers dark reddish purple, with the odour of anise; April to June. Fruit none in England. Decaying leaves reddish brown, dropping in June.

32. Illicium floridanum.

A compact, many-stemmed, bushy, evergreen, slow-growing shrub, attaining, in the neighbourhood of London, the height of 6 or 8 feet or upwards, and flowering every year. The leaves are oblong-lanceolate, quite entire, pointed at both extremities, smooth, shining, and, in common with the whole plant, have a rich reddish hue. The flowers are numerous, solitary, and terminal; and bear some general resemblance to those of Calycánthus flóridus. The manner in which the plant is propagated in the London nurseries is, generally, by forming stools of it in a cold-pit, and laying down the shoots, which require two years to root sufficiently to admit of their being separated from the parent plant; but it is sometimes propagated by cuttings both of the young and of the old wood. This very handsome evergreen shrub is sufficiently hardy to have resisted the winter of 1837-8, in several situations in the climate of London.

ORDER III. MAGNOLIA'CEÆ.

ORD. CHAR. *Calyx* of 3 deciduous sepals. *Corolla* of 3—12 petals, disposed in threes. *Anthers* adnate, elongated. *Carpels* numerous, disposed along a spiked axis. *Leaves* destitute of pellucid dots, stipulate when young. *Stipules* convolute, and enclosing the unexpanded leaves. — Evergreen and deciduous trees and shrubs, chiefly natives of warm climates.

Leaves simple, alternate, stipulate, evergreen or deciduous; oblong, not dotted, more or less coriaceous, articulated distinctly with the stem, and, when expanding, rolled together like those of Ficus. *Flowers* large, mostly white or yellowish. *Seeds* roundish, large, red or brown. — The species hardy in British gardens are included under the genera Magnòlia and Liriodéndron, the differential characters of which are as follows: —

MAGNO'LIA L. Carpel dehiscent; that is, opening to admit the escape of the seed.

LIRIODE'NDRON *L.* Carpel indehiscent; that is, not opening to admit the escape of the seed.

GENUS I.

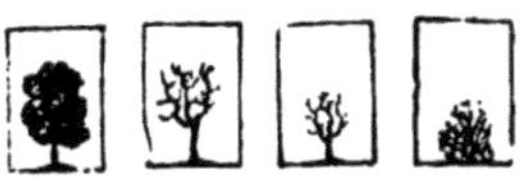

MAGNO'LIA L. THE MAGNOLIA. *Lin. Syst.* Polyándria Polygýnia.

Identification. Lin. Gen., 690.; Dec. Prod., 1. p. 79.; Don's Mill., 1. p. 82.
Synonymes. Magnolie, *Fr.* and *Ital.*; Bieberbaum *Hart.*, and Magnolie *Willd.*, *Ger.*
Derivation. The name Magnòlia was given to this genus by Linnæus, in honour of *Pierre Magnol*, professor of medicine, and prefect of the botanic garden at Montpellier. The German name Bieberbaum, beaver wood, is applied generically by Hartweg in the *Hortus Carlsruhensis*; but, in America, Beaver-wood appears to be applied only to M. glauca.

Gen. Char. *Calyx* of 3 deciduous sepals, that resemble petals. *Corolla* of from 6—9 petals. *Stamens* numerous. *Pistils* numerous. *Carpels* disposed compactly in spikes, opening by the external angle, 1—2-seeded, permanent. *Seeds* baccate, somewhat cordate, pendulous, hanging out beyond the carpels by a very long umbilical white thread.

Leaves simple, alternate, stipulate, deciduous or evergreen; entire, large, oblong or oval, stipulate. *Flowers* terminal, solitary, large, odoriferous. *Seed* large, roundish, produced in conical strobiles. Trees and shrubs, natives of North America and Asia.

One of the species is a lofty evergreen tree; but the others are deciduous, and partly trees and partly shrubs. The seeds are mostly of a scarlet colour. The roots are branched, and yet but sparingly supplied with fibres. Magnolias may be cultivated in most parts of Britain, and of the middle and southern states of Europe; but, north of London and Paris, some of the species require protection during winter, or to be kept in the greenhouse. A deep sandy soil, and a situation sheltered from the north and east, will suit most of the species; though some, as M. glaúca, for example, thrive best in a moist peaty soil. Few of the species ripen seeds in England, but most of them do so in France. From these seeds, or from such as are imported, all the American species, except M. grandiflòra, are most frequently raised; but the species from Asia are increased by layers, as are occasionally some of the more rare of the American species. In no case whatever would we recommend purchasing any species of magnolia not grown in a pot; because plants so grown may be sent to any distance without injury to the roots, which are few and succulent, and easily damaged by exposure to the air and light. The hardy species of this genus are included in two sections, Magnoli*ástrum* and Gwillími*a*.

§ i. *Magnoli*ástrum.

Derivation. *Magnòlia*; and *astrum*, from *ad instar*, an affixed particle, signifying likeness.

Sect. Char. American species, with one spathe-like bractea enclosing the flower-bud; ovaries approximate; anthers bursting outwards. (*Don's Mill.*, i. p. 83.)

♀ 1. MAGNO'L*IA* GRANDIFLO'RA *L.* The large-flowered Magnolia.

Identification. Lin. Sp., 755.; Dec. Prod., 1. p. 80.; Don's Mill., 1. p. 82.; Tor. and Gray, 1. p. 42.
Synonymes. Laurel-leaved Magnolia, the large-flowered evergreen Magnolia, the Laurel Bay, big Laurel, the large Magnolia; Laurier tulipier, *Fr.* in Louisiana; Magnolie à grandes Fleurs, *Fr.*; grossblumiger Magnolie, or Bieberbaum (Beaver-wood Tree), *Ger.*
Engravings. Mill. Ic., 2. t. 172.; the plate in vol. v. of Arb. Brit. 1st edit.; and our *fig.* 34.

Spec. Char., &c. Evergreen. Leaves oval-oblong, coriaceous, upper surface shining, under surface rusty. Flowers erect, 9—12 petals, expanding. (*Don's Mill.*) A large evergreen tree. North Carolina. Height in North America 60 ft. to 70 ft.; in England 20 ft. to 30 ft. Introduced in 1737. Flowers white, fragrant; June to September. Strobile brown, with scarlet seeds; ripe in October. Decaying leaves yellow and orange, dropping in June. Young wood green.

Varieties.

♀ *M. g.* 2 *obovàta* Ait. — Leaves obovate-oblong. Flowers expanded. (*Hort. Kew.*, iii. p. 329.) This seems to be the only variety found in a wild state. In British gardens it is a magnificent plant, the broad ends of its leaves forming a conspicuous feature, and distinguishing it readily from the original species, the leaves of which are pointed; but it does not flower freely.

? *M. g.* 3 *exoniénsis* Hort. M. g. lanceolàta *Ait.*; M. g. stricta *Hort.*; M. g. ferruginea *Hort.* *The Exmouth Magnolia.* (Bot. Mag., t. 1952.; Bot. Cab., t. 1814.; the plate in Arb. Brit., 1st edition, vol. v.; and our *fig.* 33.) — The leaves are oblong-elliptical, generally rusty

33. *Magnòlia grandiflòra exoniénsis.*

underneath. Flowers somewhat contracted. This is the most distinct of all the varieties of the species, and, on account of its flowering early and freely, the one best deserving of general culture. It forms a tall, fastigiate, elegant bush, or tree, and has attained the height of 30 ft., as a standard, at various places in the South of England.

? *M. g.* 4 *angustifòlia* Hort. — Leaves lanceolate, pointed at both extremities, wavy. A very distinct variety, introduced from Paris about 1825, which has not yet flowered in England.

? *M. g.* 5 *præ'cox* Andry.—Leaves oval-oblong. Flowers fully expanded. This is an early variety, introduced from Paris about 1830. The flowers are as large as those of any of the varieties, and they are produced from the end of May till the approach of winter.

Other Varieties. In consequence of the great demand for this species in the nurseries, many slight variations have been noticed by cultivators, and named as distinct. In the garden of the London Horticultural Society, in 1834, there were plants with the following names: — *M. g. vèra*, *M. g. latifòlia*, *M. g. exoniénsis var.*, and *M. g. rubiginòsa.* In the London nurseries are — *M. g. rotundifòlia* Swt., *M. g. ellíptica* Ait., and various others. In the nursery of M. Roy, at Angers, are 18 varieties, among which are included *M. g. longifòlia undulàta*, *M. g. exoniénsis à fleur demi-double*, *M. g. canaliculé*, *M. g. floribúnda*, *M. g. fòliis variegàtis*, &c. At Desio, near Monza, there is a variety called *M. g. magordénsis.*

Selection of Varieties. *M. g. obovàta* deserves the preference for the magnificence of its foliage; and *M. g. exoniénsis*, because it flowers early and freely; and because, from the fastigiate form of the tree, it is less liable to be injured by a heavy fall of snow; it seems also to grow faster than any of the other varieties. Where the tree is to be trained against a wall, *M. g. præ'cox* deserves the preference, on account of the largeness of its flowers, and because they appear early, and continue during the whole summer. *M. g. angustifòlia* deserves culture on account of its foliage, which is quite distinct from that of all the other varieties. The species sold in the nurseries as the common broad-leaved Magnòl*ia* grandiflòra is frequently raised from American, French, or Italian seeds; and, hence, the plants, though they grow freely, do not flower for 20 or 30 years after being planted out. For this reason, when it is desired to have plants of the Magnòl*ia* grandiflòra which will flower early, those plants which have been raised by layers from flowering trees ought to have the preference; or the Exmouth, or some other variety, should be made choice of, because the varieties are always raised from layers.

In its native country, M. grandiflòra is a tree varying from 60 ft. to 100 ft., or upwards, in height; but in Europe, except in some situations in Spain and Italy, and a few in the South of England, it is chiefly to be considered as a wall tree.

34. Magnòlia grandiflòra.

A deep sandy loam, dry at bottom, and enriched with vegetable mould or heath soil, seems to suit all the varieties of this species. When these are to be trained against a wall, any aspect may be chosen, except, perhaps, the north-east. To display the flowers to the greatest advantage, to a spectator walking in a direction nearly parallel to the wall, the ground plan of the latter should be curvilinear, by which means a direct or front view of a considerable portion would be brought before him. In the London nurseries, propagation is generally effected by forming stools either in warm situations in the open air, to be protected during winter, or in cold-pits. The shoots are laid down in autumn, and require two years to become sufficiently rooted for separation; they are then potted, and kept in pits or under glass during winter, and set in the open air, in a shady place, during summer, till wanted for final planting. M. grandiflòra is also occasionally raised from American seeds. In planting, the ball should be carefully broken by the hand, and the roots spread out in every direction, and covered with heath mould, or a mixture of leaf mould and sandy loam. The soil ought to be made firm to the fibrous roots, not by treading, but by abundant watering, and, if the plant be large, by fixing with water; that is, while the earth is being carefully put about the roots by one man, another should pour water from a pot held 6 ft. or 8 ft. above it, so that the weight of the water may wash the soil into every crevice formed by the roots, and consolidate it there. Shading will be advisable for some weeks after planting. If the Exmouth variety be chosen, layers will produce flowers in a year or two after being separated from the parent plant, if kept in pots; but, when they are planted out, and grow freely, so as to make shoots of 2 or 3 feet every season, they will probably not flower for three or four years. Whether the tree be against a wall or trellis, or treated as a standard, all the pruning it will require, after it has begun to grow freely, will be, to cut out the stumps from which the flowers or the strobiles have dropped off, and any dead or decaying wood, and any branches which cross and rub on each other. Magnolias against a wall require very little protection, even when young; and this can easily be given by mulching the ground at the roots, and covering their branches with a mat, or with the fronds of the spruce fir.

፣ 2. M. GLAU'CA *L.* The glaucous-*leaved* Magnolia.

Identification. Lin. Sp., 2. p. 755.; Tor. and Gray, 1. p. 42.; Dec. Prod., 1. p. 80.; Don's Mill., 1. p. 82.
Synonymes. M. fràgrans *Salisb.*; Swamp Sassafras, Beaver-wood, white Bay, small Magnolia, Swamp Magnolia; Magnolie glauque, Arbre de Castor, *Fr.*; grauer Bieberbaum, *Ger.*
Derivation. It is named Swamp Sassafras on account of its growing in boggy places, and resembling in qualities the *Laúrus Sássafras*; and Beaver-wood, because the root is eaten as a great dainty by the beavers, and these animals are caught by means of it. It also grows in the swamps, which they inhabit; and Michaux tells us that it is felled by them for constructing their dens and houses, in preference to any other tree, on account of the softness of the wood.
Engravings. Lodd. Bot. Cab., t. 215.; Sims Bot. Mag., 2164.; the plate of this species in Arb. Brit., 1st edit. vol. v.; and our *fig.* 35.

Spec. Char., &c. Almost deciduous. Leaves elliptical, obtuse, under surface glaucous. Flower 9—12-petaled, contracted. Petals ovate, concave. (*Don's Mill.*) A shrub, or low tree, sometimes sub-evergreen. Massachusetts to Missouri in swamps. Height in America 3 ft. to 10 ft.; 6 ft. to 20 ft. in England. Introduced in 1688. Flowers white, 2 in. to 3 in. broad, very fragrant; June and September. Strobile brownish. Seeds deep scarlet; ripe in October. Decaying leaves yellow, brown, or black. Naked young wood green.

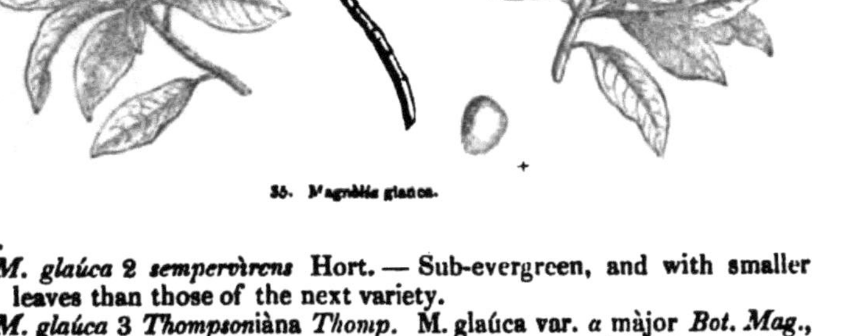

35. Magnòlia glauca.

Varieties.

- ፣ *M. glaúca* 2 *sempervìrens* Hort. — Sub-evergreen, and with smaller leaves than those of the next variety.
- ፣ *M. glaúca* 3 *Thompsoniàna* *Thomp.* M. glaúca var. *α* màjor *Bot. Mag.*, new edition, p. 36. The plate of this in the Arb. Brit., first edition, vol. v.; and our *fig.* 36.— It was noticed about 1820, in a pot of seedlings, by Mr. Thompson, in his nursery at Mile-end; and by him kept distinct, and propagated under the above name.

36. Magnòlia glaùca Thompsoniàna.

Other Varieties. *M. glaùca Gordon*iàna and *M. glaùca Burchell*iàna are names found in nurserymen's catalogues, of varieties said to have double or semi-double flowers. *M. g. longifòlia* Pursh is supposed to be an aboriginal variety, and sub-evergreen; but we think it probably the same variety as M. g. Thompson*iàna*, which may have come up wild in America, as well as in Mr. Thompson's nursery. *M. g. Cardòn*ii, M. Cárdon *J. Knight*, is a variety imported from Belgium, where it was found by Mr. Knight of the Exotic Nursery, in the nursery of M. Cardon, after whom he has named it.

A low tree, nearly evergreen in moist soils, with a slender stem, covered with a smooth whitish bark. The wood is white and spongy; the young shoots of a fine green. The leaves are smooth, of a bluish green on their upper surface, and whitish or glaucous and a little hairy underneath. The flowers are produced in May or June, at the extremity of the last year's shoots. They have six concave white petals, and have an agreeable odour. The spike or strobile of fruits is an inch or more in length, conical, an inch in diameter in the widest part, and of a reddish brown colour when ripe. When the plant is in a soil supplied with moisture during the summer, it continues to produce flowers till the autumn, and retains part of its leaves all the winter: in dry situations the leaves drop off. Seeds are frequently ripened in England: they are of a bright scarlet, and they hang down by slender white threads, as in all the other American species. The young shoots are from 1 ft. to 18 in. in length, and the plant, in ordinary circumstances, will attain the height of 12 ft. in ten years. Plants are generally raised from seeds imported from America, which should be sown in pots of bog earth about the beginning of March, and placed in gentle heat under glass. In a year they will be fit to transplant into small pots; and every year they should be shifted into others of a larger size, till wanted for final planting out. M. glaùca Thompson*iàna*, and the other varieties, are propagated by layers, which require two years to root properly.

♈ 3. MAGNOLIA TRIPETALA *L.* The three-petaled Magnolia.

Identification. Lin. Sp., 2. p. 756.; Michx., 3. p. 90.
Synonymes. M. umbrélla *Lam., Nouv. Duh., Dec. Prod., Don's Mill., Tor. & Gray*; M. frondòsa *Salisb.*; the Umbrella Tree; Umbrella Magnolia; Elkwood; Magnolie Parasol, and Arbre Parasol *Fr.*; dreyblättriger Bieberbaum, dreiblättrige Magnolie, *Ger.*
Derivation. This species is called the Umbrella Tree, according to Michaux, because its leaves, which are thin, oval, entire, and acuminate at both extremities, 18 in. or 20 in. long, and 7 in. or 8 in. broad, are often disposed in rays at the extremity of vigorous shoots; and these display a surface of 2½ ft. in diameter, in the form of an umbrella. The tree is called Elkwood in the mountains of Virginia, probably from the resemblance which the points of the shoots bear to the horns of the elk. The French names merely signify umbrella tree, and the German ones the three-petaled beaver tree, or magnolia.
Engravings. Michx. Arb., 3. t. 5.; Lodd. Bot. Cab., t. 418.; the plate of this species in Arb. Brit. 1st edit. vol. v.; and our *fig.* 37.

Spec. Char., &c. Deciduous. Leaves lanceolate, spreading, adult ones smooth, younger ones pubescent underneath. Petals 9—12, exterior ones pendent. (*Don's Mill.*, i. p. 83.) A deciduous tree of the middle size. Pennsylvania to Georgia, in moist soil. Height 30 ft. to 40 ft. in America; 15 ft. to 30 ft. in England. Introduced in 1752. Flowers white, 7 in. to 8 in. in diameter, with an unpleasant odour; May to July. Strobiles rose-coloured, 4 in. to 5 in. long; ripe in October. Decaying leaves dark brown or black. Naked young wood of a fine mahogany brown.

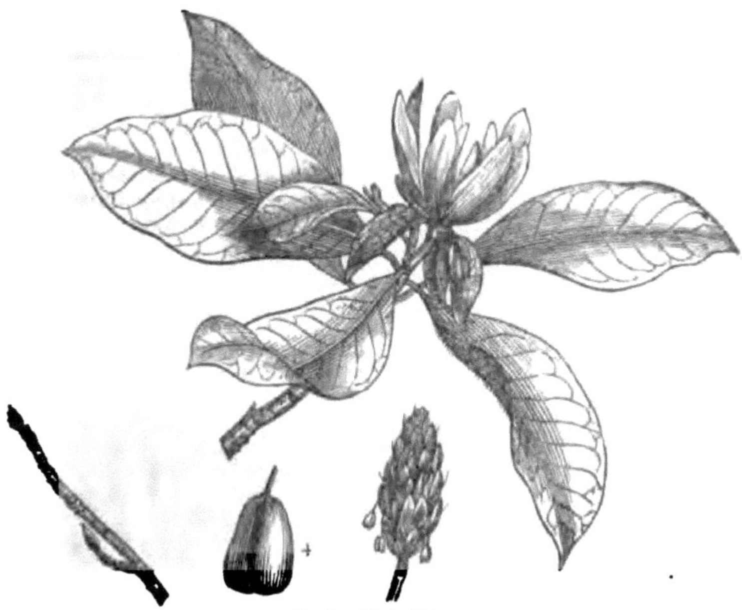

37. *Magnòlia tripétala.*

This tree, both in America and Europe, is remarkable for the largeness of its leaves and its flowers. The wood is spongy, brittle, with a large pith, soft, porous, and of very little use. The bark upon the trunk is grey, smooth, and polished; and, if cut while green, it exhales a disagreeable odour. In Britain the tree sends up various shoots from the root, to replace the stems, which are seldom of long duration; so that a plant that has stood thirty or forty years in one spot has had its stems several times renewed during that period. The leaves are 18 or 20 inches long, and 7 or 8 inches broad. The flowers are 7 or 8 inches in diameter, with large white flaccid petals; they are borne on the extremities of the last year's shoots, have a languid luxurious appearance, and a sweet but heavy odour. The fruit, which is conical, is 5 or

6 inches long, and about 2 in. in diameter; it is of a beautiful rose colour, and contains usually from 50 to 60 seeds. This species is very hardy, and can withstand the most rigorous winters, when the summer has been sufficiently hot to ripen the wood thoroughly. As it is a short-lived tree, and consequently flowers early, there is not the same objection to raising plants of it from seed, as there is to raising plants in that manner of M. grandiflòra, which is a long-lived species. The soil should be a deep, rich, sandy loam, and the situation sheltered and shaded. Exposure to the sun is injurious; and, trained against a south wall, the plant suffers extremely. A sheltered glade in a shrubbery or wood, where the tree is sufficiently distant from others not to be injured by their roots, is the most desirable site. In the nurseries it is almost always propagated by seeds, which should be sown immediately after they are gathered, as when they are left exposed they become rancid and lose their vital qualities; though, if enveloped in moist moss or earth, they may be preserved for several months. The plants should be kept in pots until required for final transplanting.

♆ 4. M. MACROPHY'LLA *Mx.* The long-leaved Magnolia.

Identification. Mich. Bor. Amer., 1. p. 327.; Dec. Prod., 1. p. 80.; Don's Mill., 1. p. 83.; Tor. and Gray, 1. p. 43.
Synonymes. Large-leaved Umbrella Tree, *Amer.*; Magnòl*ia* Michaúx*ii* Hort.; Magnolier à grandes Feuilles, Magnolier bannanier, *Fr.*; grosseblättrige Bieberbaum, *Ger.*
Engravings. Bot. Mag., 2189.; the plate in Arb. Brit., 1st edit. vol. v.; and our *fig.* 38.

Spec. Char., &c. Deciduous. Leaves very large, oblong-obovate, somewhat panduriform, cordate at the base, under surface whitish, glaucous. Petals 6—9, ovate. (*Don's Mill.*) A deciduous tree of the middle size. North Carolina and Georgia. Height 30 ft. to 40 ft. in America; 15 ft. to 30 ft. in England. Introduced in 1800. Flowers white, with a purple spot near the base of each petal; 8 in. to 10 in. in diameter, fragrant; June and July. Strobile rose-coloured; ripe in October. Decaying leaves yellow, brown, or black. Naked young wood of a whitish brown.

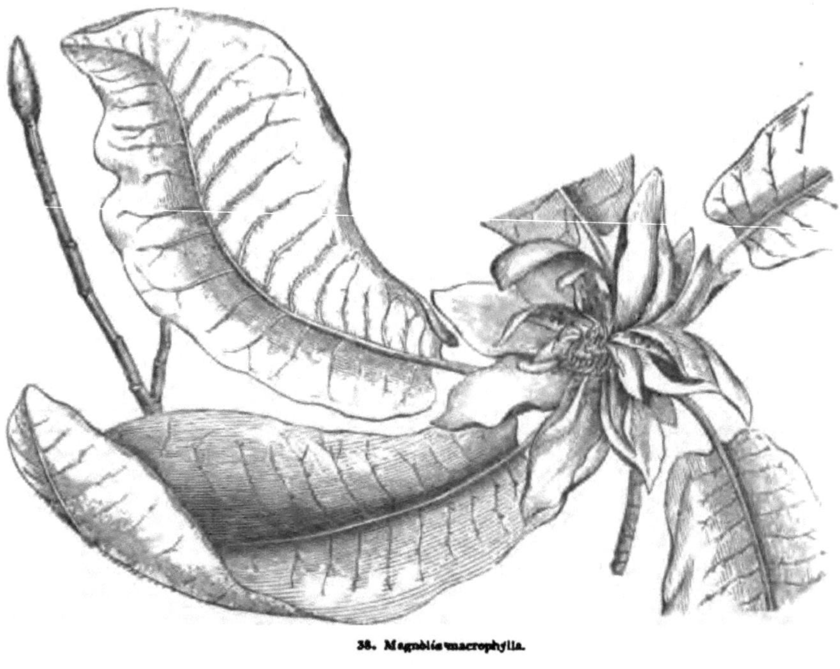

38. Magnòlia macrophýlla.

The general appearance of this tree greatly resembles that of Magnòlia tripétala. The terminal arrangement of the leaves is the same, and it is remarkable that in America the two trees are almost always found together. In point of size, it exceeds the M. tripétala, both in its leaves and general height; but it is seldom found higher than 35 ft., which exceeds the height of the other by a sixth part only. The body of the tree is covered with a smooth and very white bark, by which, in the winter, when stripped of its leaves, it is readily distinguished from M. tripétala. At this season, also, it may be distinguished by its buds, which are compressed, and covered with a soft and silvery down; whereas in M. tripétala they are prominent and rounded at the end. The leaves, in its native country, are 35 in. long, and 9 or 10 inches broad; and in vigorous plants, in England, they sometimes even exceed these dimensions. They are borne on petioles short in comparison with the size of the leaves, and are of an oblong oval shape, pointed at the extremity, and cordiform at the base; their colour is light green above, and glaucous beneath. The fruit is about 4 in. long, nearly cylindrical, and of a vivid rose-colour when arrived at maturity. Young plants of this species grow very slowly till they are thoroughly established, which will require, in general, two years. The year's shoots may then be from 1 ft. to 2 ft.; so that in ten years a plant may attain the height of 12 or 15 feet. It may be considered a short-lived tree, and, like all such, it comes into flower when young. It has rarely, if ever, been propagated in this country by inarching or layers, and very seldom from seeds; and, hence, the plant is very sparingly distributed. Soil, propagation, &c., as in M. tripétala. Seeds are ripened in France, and young plants imported from that country, or from North America.

⚘ 5. M. ACUMINA'TA *L.* The pointed-*leaved* Magnolia.

Identification. Lin. Sp., 756.; Dec. Prod., 1. p. 80.; Don's Mill., 1. p. 83.; Tor. and Gray, 1. p. 43.

Synonymes. M. rústica, and M. pennsylvánica, of some; the blue Magnolia, *Eng.*; the Cucumber Tree, *U. S.*; Magnolier acuminé, Magnolier à Feuilles pointées, *Fr.*; zugespitzer Bieberbaum, *Ger.*

Derivation. This species is called the Cucumber Tree, in America, from its fruit resembling a small cucumber. The other names are translations of the botanic one.

Engravings. Mich. Arb., 3. p. 82. t. 3.; Bot. Mag., 2427.; and the plate in Arb. Brit., 1st edit. vol. v.; and our *fig.* 39.

Spec. Char., &c. Deciduous. Leaves oval, acuminate, under surface pubescent. Flowers 6—9-petaled. (*Don's Mill.*) A deciduous tree of large size. New York to Georgia. Height in America 60 ft. to 80 ft., with the trunk 4 ft. to 5 ft. in diameter at the base; in England 30 ft. to 50 ft. Introduced in 1736. Flowers yellowish within, glaucous without, slightly fragrant; May to July. Strobile cylindrical, brownish red, 3 in. long; ripe in October. Decaying leaves dark brown or black. Wood of a mahogany brown.

Varieties.

⚘ *M. a.* 2 *Candóllii Savi.*—Leaves ovate oblong, acute. Flowers greenish. Figured in Savi's *Bibl. Ital.*, p. 224.

⚘ *M. a.* 3 *máxima* Lodd.—Leaves much larger than those of the original species. Introduced by Messrs. Loddiges, and cultivated in different nurseries.

Other Varieties. The Magnòlia acuminàta being frequently raised from seed, and the seedlings varying much in the size of their leaves, and in the presence or absence of pubescence, both on the leaves and wood, it would be easy to select several varieties apparently as distinct as those above mentioned, such as M. striàta, latifòlia, &c. In the Goldworth Nursery, Woking, Surrey, are some which appear remarkably distinct.

Trunk straight, branches numerous, shoots regularly distributed. The leaves are from 6 in. to 7 in. long, and ftom 3 in. to 4 in. broad, upon old trees, but double that size upon young vigorous-growing plants. Michaux describes them as oval, entire, and very acuminate; but, in the seedlings raised in British nurseries, they are found sometimes ovate, nearly orbiculate,

39. Magnòlia acuminàta.

and cordate acuminate. The flowers are 4 or 5 inches in diameter, bluish, and sometimes white, with a tint of yellow. They have but a feeble odour, and the petals are never fully expanded, though, as they are large and numerous, they have a fine effect in the midst of the superb foliage. Plants raised from seeds do not usually produce flowers till they are eight or ten years old, when the tree will probably be from 15 ft. to 20 ft. in height; but plants raised from layers produce flowers in two or three years. The fruit is about 3 in. long, and nearly 1 in. in diameter. It is nearly cylindrical, and often a little larger at the summit than at the base: it is convex on one side, and concave on the other; and, when green, it nearly resembles a young cucumber: it becomes rose-coloured when ripe; and, as in the case of the other species, the seeds, before they drop, remain suspended for some time by long white threads. The wood of this tree is of a fine grain, and of an orange colour. A free, deep, and rather moist soil answers best for this species; but, as it is much hardier than any of the others in this section, it will grow in almost any soil that is moderately free, and not overcharged with moisture. It is generally propagated in the London nurseries by layers, the plants so produced flowering much sooner than seedlings; but the latter, as they make far more durable plants, should always be preferred when this species is used as a stock to graft or inarch others on. It is so used very generally, not only for M. auriculàta and cordàta, but for M. conspícua and Soulangeàna. The plants are, in some nurseries, grown in the free soil; but it is always preferable to rear them in pots; because, in that case, they are not checked by transplanting, and at least a year is gained in their growth.

? 6. M. (A.) CORDA'TA *Mx.* The heart-*leaved* Magnolia.

Identification. Mich. Bor. Amer., 1. p. 328.; Dec. Prod., 1. p. 80.; Don's Mill., 1. p. 83.; Tor. and Gray, 1. p. 43.

Synonymes. The heart-leaved Cucumber Tree, *Amer.*; Magnolier à Feuilles en Cœur, *Fr.*; herzblättriger Bieberbaum, *Ger.*

Engravings. Bot. Mag., t. 325.; Bot. Cab., 474.; the plate in Arb. Brit., 1st edit. vol. v.: and our *fig.* 40.

Spec. Char., *&c.* Deciduous. Leaves broadly ovate, subcordate, acute, under surface tomentose, upper surface smooth. Petals 6—9, oblong. (*Don's Mill.*) A deciduous tree of the middle size. Carolina to Georgia, on mountains. Height 20 ft. to 40 ft. and 50 ft. in America, and 20 ft. to 30 ft. in England. Introduced in 1800. Flowers yellow slightly streaked with red, with a disagreeable odour, seldom expanding fully; June and July. Strobile like that of M. acuminàta, but smaller; ripe in October. Decaying leaves dark brown or black. Naked young wood hoary brown.

40. Magnòlia cordàta.

This tree, in its native country, has a trunk 12 or 15 inches in diameter, straight, and covered with a rough and deeply furrowed bark. Its leaves are from 4 in. to 6 in. in length, and from 3 in. to 5 in. wide, smooth and entire. The flowers are from 3 in. to 4 in. in diameter, and are succeeded by fruit about 3 in. long, and nearly 1 in. in thickness, of a similar form to those of the preceding species. The soil, situation, propagation, &c., may be considered the same as for M. acuminàta; but, as M. (a.) cordàta seems, in its native country, to inhabit higher and drier localities than M. acuminàta, it may probably be placed in still more exposed situations than that species in Britain.

♀ 7. M. AURICULA'TA *Lam.* The auricled-*leaved* Magnolia.

Identification. Willd. Sp., 2. p. 1258.; Dec. Prod., 1. p. 80.; Don's Mill., 1. p. 83.
Synonymes. M. Fràseri Walt., Tor. & Gray; M. auriculàris *Salisb.*; Indian Physic, and long-leaved Cucumber Tree, *Amer.*; Magnolier auriculé, *Fr.*; geöhrter (eared) Bieberbaum, *Ger.*
Engravings. Bot. Mag., 1206.; the plate in Arb. Brit., 1st edit. vol. v.; and our *fig.* 42.

Spec. Char., *&c.* Deciduous. Leaves smooth, under surface somewhat glaucous, spathulately obovate, cordate at the base, with blunt approximate auricles. Sepals 3, spreading. Petals 9, oblong, attenuate at the base. (*Don's Mill.*) A smooth deciduous tree of the middle size. Carolina to Florida, and on the Alleghany Mountains. Height 30 ft. to 40 ft. in America, and 20 ft. to 30 ft. in England. Introduced in 1786. Flowers white; April and May. Strobile oval oblong, rose-coloured; ripe in October. Decaying leaves of a rich yellowish brown. Naked young wood smooth, and of a purplish mahogany colour, with small white dots.

Variety.

M. a. 2 *pyramidàta.* M. pyramidàta *Bartr.*; M. Fràseri pyramidàta *Nutt.*, *Tor. & Gray.* The plate in Arb. Brit. 1st edit. vol. v.; and

41. Magnòlia auriculàta pyramidàta.

our *fig.* 41. — Leaves shorter than those of the species, and the plant altogether weaker. It is found in the western parts of Carolina and Georgia, but only in two or three localities. Propagated by inarching on M. auriculàta, but it requires two years to adhere, and seldom makes a vigorous plant.

This tree has a straight trunk 12 or 15 inches in diameter, often without branches for half its height; the branches spread widely, and ramify but sparingly; and this circumstance, Michaux observes, gives the tree a very peculiar air, so that it may readily be known at a distance, even in winter.

42. Magnòlia auriculàta.

The leaves are of a light green colour, of a fine texture, 8 or 12 inches long, and from 4 in. to 6 in. broad : on young and vigorous trees they are often one third, or even one half, larger. The flowers are 3 or 4 inches in diameter, of a milky white, and of an agreeable odour, and are situated at the extremity of the young shoots. The fruit is oval, 3 or 4 inches long, and, like that of

Magnòlia tripétala, of a beautiful rose colour when ripe; it differs from those of the other species by a little inferiority of size, and by a small appendage which terminates the carpels. Each carpel contains one or two seeds. The wood is soft, spongy, very light, and unfit for use. The bark is grey, and always smooth, even on the oldest trees. When the epidermis is removed, the cellular integument, by contact with the air, instantly changes from white to yellow. In England, annual shoots of young plants are from 1 ft. to 2 ft. or more in length; and the height which the tree usually attains in 10 years is from 10 ft. to 15 ft. The soil for this species ought to be free and deep; and the situation low, sheltered, and moist, rather than dry. As seeds are not very easily procured, the common mode of propagation is by layers, or by inarching on M. acuminàta. Two years are required before the plants can be separated from the parent stock.

§ ii. *Gwillímia Rott. in Dec. Syst.*

Derivation. Named in honour of *General Gwillim*, some time governor of Madras. (*Don's Mill.*, 1. p. 83.)

Sect. Char. Asiatic species, generally with two opposite spathe-like bracteas enclosing the flower-bud. Anthers bursting inwards. Ovaries somewhat distant. (*Don's Mill.*) Trees or shrubs; natives of Asia.

♀ 8. M. CONSPI'CUA *Salisb.* The *Yulan, or* conspicuous-*flowered* Magnolia.

Identification. Dec. Prod., 1. p. 81.; Don's Mill. 1., p. 83.
Synonymes. M. prècia *Correa*; M. Yùlan Desf.; Yu lan, *Chinese*; the Lily-flowered Magnolia; Magnolier Yulans, *Fr.* Yulans Bieberbaum, *Ger.* Magnolia dai Fiori di Giglio, *Ital.*
Derivation. The epithet prècia was given to this magnolia by M. Correa, because it produces its flowers before its leaves. Yu lan signifies the lily tree. Giglio is the Italian for a lily.
Engravings. Bot. Mag., 1621.; the plate of this species in Arb. Brit. 1st edit. vol. v.; and our *fig.* 44.

Spec. Char., &c. Deciduous. Leaves obovate, abruptly acuminated; younger ones pubescent, expanding after the flowers. Flowers erect, 6—9-petaled. Styles erect. (*Don's Mill.*) A deciduous tree of the middle size. China. Height in China 40 ft. to 50 ft.; in England 20 ft. to 30 ft. Introduced in 1789. Flowers white, fragrant; February and April. Strobiles brownish; ripe in September. Decaying leaves dark brown or black. Naked young wood ash-coloured or greyish brown.

Varieties, or Hybrids.

♀ *M. c.* 2 *Soulangeàna.* M. Soulangeàna An. Hort. Soc. Par.; Magnolier de Soulange, *Fr.* (M. Soulangeàna Swt. Brit. Fl.-Gard., t. 260.; and our *fig.* 43.) — The leaves, wood, and general habit of this tree bear a close resemblance to those of M. conspícua. The flowers resemble in form those of M. purpùrea var. grácilis or of M. purpùrea, and the petals are slightly tinged with purple. It was raised at Fromont, near Paris, from the seeds of a plant of M. conspícua, which stood near one of M. purpùrea, in front of the château of M. Soulange-Bodin; the flowers of the former of which had been accidentally fecundated by the pollen of the latter.

43. Magnòlia conspícua Soulangeàna.

Other Varieties, or Hybrids. M. conspícua has ripened seeds in various places; and, as it fertilises readily with M. purpùrea and M. grácilis, many new varieties may be expected when the attention of cultivators is more especially directed to the subject. *M. c. S. speciòsa* and *M. c. S. Alexandrina* are in British gardens, but they are not worth keeping distinct from M. c. Soulangeàna.

44. Magnòlia conspícua.

This is a very showy tree, distinguishable from all the other magnolias of both sections, by its flowers expanding before any of the leaves. The tree assumes a regular conical shape, with a grey bark and numerous branches and twigs, which generally have a vertical, rather than a horizontal, direction. The young shoots are from 1 ft. to 18 in. in length, and the tree, in ten years, will attain the height of from 10 ft. to 15 ft., flowering the second or third year after grafting. It is nearly as hardy as the American species; flowering freely every year, as a standard, in the neighbourhood of London, when the wood has been properly ripened during the preceding summer. A rich sandy loam seems to suit this species best; but it will grow in any deep free soil, properly drained, and moderately enriched. The situation, when it is to be treated as a standard, ought to be sufficiently open to admit of ripening the wood in autumn, and yet not so warm as to urge forward the flower-buds prematurely in spring, as they are very liable to be injured by frost; from which, however, they may be protected by a very slight covering (during nights and frosty days) of gauze or bunting, stretched over the tree horizontally, and supported by posts. Against a wall, the tree shows itself in its greatest beauty; and there it can easily be protected, by a projecting coping, from the severest weather ever experienced in the neighbourhood of London. In warm situations, sloping to the south or south-east, the tree has a fine effect planted in front of a bank of evergreens; and, indeed, wherever it is planted, evergreens should be placed near it, and, if possible, so as to form a background, on account of the flowers expanding before the tree is furnished with any leaves. The species and all the varieties are propagated by layers, or by inarching on M. purpùrea or on M. acuminàta. When inarched on M. purpùrea, the tree is comparatively dwarfed, by which it is rendered very convenient for use as a shrub, or for growing in pots, and forcing; but, when it is intended to form a tree, it should either be inarched on M. acuminàta, or raised from layers or seeds. It generally requires two years before the plants can be separated from the parent stock. Some plants of this species have been raised from seed ripened in Europe; and we have no doubt that, when this magnificent tree becomes better known and more generally in demand, it will be raised in this way extensively in France and Italy, and supplied to the British nurseries from these countries.

■ 9. M. PURPU'REA *Sims*. The purple-*flowered* Magnolia.

Identification. Sims, in Bot. Mag.
Synonymes. M. obovàta *Thun.*, and *Don's Mill.* 1. p. 84.; M. discolor *Vent.*; M. denudàta *Lam.*; the obovate-leaved Magnolia; Magnolier discoloré *Bon. Jard.*, and Magnolie bicoloré *Dum., Fr.*; rothe Bieberbaum, *Ger.*
Engravings. Sal. Par., t. 87.; Bot. Mag., t. 390; and our *fig.* 45.

Spec. Char., &c. Deciduous. Leaves obovate, acute, reticulately veined; almost smooth. Flowers erect, of 3 sepals and 6 obovate petals; styles very short. (*Don's Mill.*) A deciduous shrub, with large dark green foliage. Japan. Height 3 ft. to 5 ft. Introduced in 1790. Flowers purple outside, white within; March to May. Strobile brownish; ripe in September. Decaying leaves black. Naked young wood greenish brown.

Varieties.

■ *M. p.* 2 *grácilis*. M. *Kóbus* Dec. and G. Don; M. tomentòsa *Thun.* in *Lin. Trans.* Kæmpf. Icon., t. 42.; Par. Lon., t. 87. — The two main points of difference between it and M. purpùrea are, the paler green, and somewhat narrower shape, of the leaves; and the longer and more slender form of the flower, the points of the petals of which are slightly turned back; while the flower of M. purpùrea is more cup-shaped, and the petals at the points are rather turned inwards. The petals of M. grácilis are on the exterior entirely of a dark purple, whereas those of M. purpùrea melt off into white at their upper extremities. A number of plants of this variety, which stood in the Hammersmith Nursery as border shrubs, and flowered freely every year, were killed down to the ground in the winter of 1837–8.

Other Varieties. In DeCandolle's *Prodromus*, and in Don's *Miller*, three varieties are described: *M. p. denudàta* Lam., distinguished by the flowering branches being without leaves; *M. p. díscolor* Vent., which is said to be rather more tender than the species; and *M. p. liliflòra* Lam., the petals of which are white on both sides. These varieties were originally described by Kæmpfer; but, as far as we know, none of them are in British gardens. Several plants of this species having been raised from seed ripened in this country, the plants may exhibit slight shades of difference, as has been the case with certain seedlings raised in the Brentford Nursery; but, as far as we have observed, none of these are worth keeping distinct. The only variety which we consider truly distinct is *M. p. grácilis*, considered as a species by Salisbury and other botanists, but which, we are convinced, is nothing more than a race, or a variety. At Desio, a variety has been raised which grows only 1½ ft. high, and which Signor Cassoretti, the garden director there, calls *M. obovàta pùmila*.

45. Magnòlia purpùrea.

A deciduous shrub, attaining, in the gardens about London, the height of from 4 ft. to 8 ft. in as many years, and seldom growing much higher as a bush. The stems are numerous, but not much branched; the leaves are large, of a very dark green; and the plant produces a profusion of flowers, which do not expand fully till a day or two before they drop off; and which, unless the weather is warm, do not expand at all, but wither on the plant, and disfigure it. The flowers are large, more or less purple (according to the season, but never wholly dark purple) without, and always white within. The bark, when bruised, has an aromatic odour. A very ornamental species, which no garden ought to be without. This species is generally considered as requiring a mixture of heath soil, or sandy peat, with loam; but in many gardens about London it succeeds perfectly both in sand

and clay; the latter soil being rendered free by sand, leaf mould, or manure, and drainage. The situation, when the plant is treated as a bush, ought to be open, in order that the wood may be ripened; and the plant should be detached, in order that it may be covered with foliage and blossoms on every side. North of London, in most situations, it requires a wall, and few plants are more deserving of one. Against a wall, it will reach the height of 15 ft. or 20 ft. In the London nurseries, it is generally propagated by layers; but it will also strike by cuttings, both of the ripened and the herbaceous wood. The stools are generally formed in pits; or, if in the open ground, they are covered with mats during winter. Seeds have been ripened both in England and France; and from these plants have been raised in some few nurseries. The plants, whether raised from layers, cuttings, or seed, should always be kept in pots till wanted for final planting. This species often serves as a stock for grafting the other kinds on, which belong to this section.

Genus II.

LIRIODE'NDRON *L.* The Tulip Tree.

Gen. Char. *Carpels* 1—2-seeded, disposed in spikes, indehiscent, deciduous, drawn out into a wing at the apex. *Calyx* of 3 deciduous sepals. *Corolla* of 6 petals, conniving into a bell-shaped flower. (*Don's Mill.*, i. p. 86.) — There is only one species; a deciduous tree of the first rank, native of North America.

Leaves simple, alternate, stipulate, deciduous; 3-lobed, the terminal lobe emarginately truncate, the lateral ones with two sinuses. *Stipules* flat. *Flowers* terminal, solitary, greenish yellow, orange within. — The only species in British gardens is the Liriodéndron Tulipífera.

♈ 1. Liriode'ndron Tulipi'fera *L.* The Tulip-bearing Liriodendron, *or* Tulip Tree.

Identification. Lin. Sp., 755.; Dec. Prod., 1. p. 82.; Don's Mill., 1. p. 86.
Synonymes. The Poplar, White Wood, Canoe Wood, the Tulip Tree, *Amer.*; Virginian Poplar, Tulip-bearing Lily Tree, Saddle Tree, *Eng.*; Tulipier de Virginie, *Fr.*; Virginischer Tulipeerbaum, *Ger.*
Derivation. This tree is called Liriodéndron, from *leirion*, a lily, and *dendron*, a tree; from the flowers resembling those of a lily, though more correctly those of a tulip, as the specific name implies. It is called Poplar, from its general resemblance to trees of that genus; White Wood, from the colour of its timber; Canoe Wood, from the use to which it is applied by the native Indians: Tulip Tree, from its tulip-like flowers; and Saddle Tree, from the form of its leaves. The French and German names are literal translations of the words Virginian tulip tree.
Engravings. Bot. Mag., 275.; Duh., tom. 3. t. 18.; the plate in Arb. Brit., 1st edit. vol. v.; and our *fig.* 46.

Spec. Char., &c. Leaves smooth, truncate at the top; 4-lobed, resembling a saddle in shape. Flowers large, solitary, terminal; variegated with green, yellow, and orange colour; furnished with two deciduous bracteas under flowers. (*Don's Mill.*) A smooth deciduous tree of large size. Canada to Florida. Height 70 ft. to 140 ft., and trunk 8 ft. to 9 ft. in diameter, in America; 50 ft. to 90 ft. in England. Introduced in 1688. Flowers greenish yellow without, orange within; June and July. Strobile brown; ripe in October. Decaying leaves rich yellow and brown. Naked young wood smooth, and of a mahogany brown.

Variety.

♈ *L. T.* 2 *obtusíloba* Michx., integrifòlia *Hort.*, Yellow Wood, or Yellow Poplar, has the leaves with blunter lobes than the species, but is in no other respect different from it.

Other Varieties. *L. T. acutifòlia* Michx. has never, we believe, been introduced. *L. T. flàva* Hort. has yellow flowers. As the tulip tree is almost always raised from seeds, it is probable that the flowers of seedlings will vary in their shades of colour, and any desirable variation may be perpetuated by propagating the plant possessing it by layers or inarching.

46. Liriodéndron Tulipífera.

In the developement of its leaves, the tulip tree differs from most other trees. The leaf-buds, in general, are composed of scales closely imbricated, which, in the spring, are distended by the growth of the minute bundle of leaves that they enclose, till they finally fall off. The flowers, which are large, brilliant, and on detached trees very numerous, are variegated with different colours, among which yellow predominates; they have an agreeable odour, and, surrounded by the luxuriant foliage, they produce a fine effect The fruit is composed of a great number of thin narrow scales, attached to a common axis, and forming a conical spike 2 or 3 inches in length. Each fruit contains 60 or 70 carpels; of which never more than a third, and, in some seasons, not more than seven or eight in the whole number, are matured. It is also observed, that, during ten years after it begins to yield fruit, almost all the seeds are unproductive; and that, on large trees, the seeds from the highest branches are the best. The heart, or perfect, wood of the tulip tree is yellow, approaching to a lemon colour; and its sap, or alburnum, is white. The annual shoots of young plants, in the neighbourhood of London, are from 18 in. to 2 ft. in length; and the tree will, in favourable circumstances, attain the height of from 15 ft. to 20 ft. in ten years; seldom, however, flowering till it is upwards of twenty years old. The height, in England, frequently exceeds 70 ft.; and it has ripened seeds here, occasionally, from which young plants have been raised. It ripens its fruit very generally in France; though it is observed, in the *Nouveau Du Hamel*, that these seeds do not vegetate so freely as those which are imported from America. Deep, loamy, good soil best suits the tulip tree; and the situation most favourable is one which, while it is sheltered from high winds, is, at the same time, sufficiently exposed to the light and air to admit of the maturation of its leaves on every side, and the perfect ripening of its wood, without which it can neither resist the severe frosts of winter, nor form blossom buds. The species is seldom, if ever, propagated otherwise than by seeds, which come up best in heath soil, very fine mould, or sandy loam, in a shady situation, kept rather moist; but the varieties are multiplied by layers or inarching. When the seeds are sown in autumn, they generally come up the following spring; but, sown in spring or the beginning of summer, they generally remain a year in the ground. The tulip tree, like the magnolias, having roots furnished with but few fibres, does not transplant readily; and, therefore, the plants ought either to be kept in pots, or, if in the free ground, transplanted in the nursery every year; or, if neither of these modes be practicable, removed to their final situation, when not more than two, or at most three, years old. The tree is, like the magnolias, not very patient of the knife, either in a young or in an old state; and, from the bitter qualities of the leaves, it does not seem to be much attacked by insects. As tulip trees raised from seed seldom flower before they are twenty or thirty years old, it is much to be wished, that nurserymen would propagate them by grafting or inarching from flowering trees, in consequence of which the plants would probably flower the second or third year.

ORDER IV. *ANONA'CEÆ.*

ORD. CHAR. The distinctive characteristics of this order from that of Magnoliàceæ are: *Anthers* with an enlarged four-cornered connectivum, which is sometimes nectariferous; *albumen* pierced by the substance of the seed-coat; *leaves* without stipules, conduplicate in the bud; *properties* aromatic. — Trees or shrubs mostly natives of warm climates

Leaves simple, alternate, exstipulate, deciduous; distinctly articulated with the stem, entire; leaves and branches pubescent when young, the leaves commonly minutely punctate, with pellucid dots. *Flowers* axillary. — The hardy species, in British gardens, are included in the genus *Asímina* Adans., formerly *Anòna* L., and are natives of North America.

GENUS I.

ASI'MINA Adans. THE ASIMINA. *Lin. Syst.* Polyándria Polygýnia.

Identification. Adans. Fam., 2. p. 365.; Dec. Prod., 1. p. 87.; Don's Mill., 1. p. 91.
Synonymes. *Annòna* L.; Orchidocárpum *Mx.*; Porcèliæ sp. *Pers.*; Uvària *Tor. & Gray*; Custard Apple; Asiminier, and Anone, *Fr.*; Flaschenbaum, *Ger.*
Derivation. *Asimina* is Latinised from a word of Canadian origin, the meaning of which is not known. Orchidocárpum was, it is probable, intended to express a likeness between the figure of the fruit, and that of some species of *O'rchis.* Porcèlia is a name given by Ruiz, in honour of *Antonio Porcel*, a Spanish promoter of botany. *Anòna* is a South American word that signifies a mess, or dish of food, to be eaten with a spoon. Uvària is from *uva*, a grape, to which, however, the fruit has little resemblance. The German name, Flaschenbaum, flask tree, is given from the shape of the fruit.

Gen. Char. *Calyx* 3-parted. *Petals* 6, spreading, ovate-oblong, inner ones smallest. *Anthers* numerous, nearly sessile. *Ovaries* many, but for the most part only 3, ovate or oblong. *Carpels* the same number as the ovaries, baccate, sessile. *Seeds* many, disposed in a single or double row. (*Don's Mill.*) — Low trees or shrubs, deciduous, with white or purplish flowers, and fruit about the size of small plums. Rather tender, and difficult of culture. Only one species is truly hardy in the climate of London.

● 1. *A.* TRI'LOBA *Dun.* The three-lobed-*calyxed* Asimina.

Identification. Dun. Monog.; Dec. Prod., 1. p. 87.; Don's Mill., 1. p. 91.
Synonymes. *Annòna* triloba *L.*, and *Mx.* in *Arb.*; Porcèlia triloba *Pers.*; Orchidocárpum arietinum *Mx. Bor. Am.*; Uvària triloba *Tor. & Gray*; the Papaw, *Amer.*; Asiminier de Virginie, and Annone à trois Lobes, *Fr.*; dreylappiger (three-lobed) Flaschenbaum, *Ger.*
Engravings. Mill. Icon., 1. t. 35.; Tor. and Gray, 1. p. 45.; Mx. Arb., 3. t. 9.; and our *fig.* 47.

Spec. Char., &c. Leaves oblong-cuneated, acuminated, and, as well as the branches, smoothish. Flowers on short peduncles; outer petals roundish-ovate, four times longer than the calyx. (*Don's Mill.*) A low deciduous tree. Middle, southern, and western states of North America. Height 15 ft. to 20 ft. in North America; 7 ft. to 10 ft. in England. Introduced in 1736. Flowers dark purple and yellow. Fruit yellowish, esculent; ripe in August in America, rarely seen in England. Decaying leaves rich yellowish brown. Naked young wood dark brown.

A small tree, densely clothed with long leaves, lying over one another in such a manner as to give a peculiarly imbricated appearance to the entire plant. The flowers are campanulate and drooping, and appear before the leaves; the outer petals are purple, and vary in colour in different plants; in some being very dark, and in others light, inclining to yellow. All parts of the tree have a rank, if not a fetid, smell; and the fruit is relished by few persons except the negroes, who call it papaw. The fruit ripens in America in the beginning of August, and is about 3 in. long and 1½ in. thick, oval, irregular, and swelling into inequalities. In British gardens, the plant is always raised from American seeds; and, to thrive, it requires to be planted in sandy peat or deep sand, and kept moist. In England it may be considered as a curious, slow-growing, deciduous shrub, or low tree, well deserving a place in gardens, but which ought always to be isolated, and at some distance from rapid-growing plants. Relatively to growth, it may be placed near Dírca palústris, some of the daphnes, or Illícium.

4?. *Asimina* triloba.

Other Species of Asímina. — A. *parviflòra* and A. *grandiflòra* are North American shrubs, seldom growing higher in their native habitats than 1 ft. to 2 ft., and rather too tender for the climate of London.

ORDER V. MENISPERMA'CEÆ.

ORD. CHAR. *Flowers* unisexual. *Sepals* and *petals* similar. *Stamens* monadelphous, or rarely free. *Ovaries* somewha connected at the base; with one or many *styles*; many-celled. *Fruit*, in most, baccate or drupaceous, one-seeded or many-seeded, oblique or lunulate, compressed, with the *seeds* of the same form. *Embryo* curved or peripheric. *Albumen* none, or very sparing and fleshy. (*Don's Mill.*) — Climbing or twining flexible shrubs, natives of North America and Asia.

Leaves simple, alternate, exstipulate, deciduous; stalked, usually cordate or peltate, palmately veined, and always with the middle nerve terminating in an awn or point. *Flowers* in axillary racemes in most species, small. — The species in British gardens are included in the genera Menispérmum and Cócculus, which are thus contradistinguished: —

MENISPE'RMUM *L.* Sepals and petals quaternary. Male flowers with 15—20 stamens.

CO'CCULUS *Bauh.* Sepals and petals ternary. Male flowers with 6 stamens.

GENUS I.

MENISPE'RMUM *L.* THE MOONSEED. *Lin. Syst.* Diœ'cia Dodecándria.

Identification. Tourn.; Dec. Prod., 1. p. 102.; Don's Mill, 1. p. 112.; Tor. and Gray, 1. p. 48.
Synonymes. Ménisperme, *Fr.*; Mondsaame, *Ger.*
Derivation. From *mēnē*, the moon, and *sperma*, a seed; from the seeds being crescent-shaped.

Gen. Char. *Sepals* and *petals* disposed in a quaternary order, in two or three series. *Male flowers* with 16 to 20 stamens; *female flowers* with 2 to 4 ovaries. *Drupe* baccate, roundish-kidney-shaped, 1-seeded. — Climbing shrubs natives of North America and Dauria.

Leaves simple, alternate, peltate or cordate, entire, smooth. *Peduncles* axillary, or supra-axillary. Male and female peduncles rather dissimilar. *Flowers* small, greenish white. — The species are all of the easiest culture in common soil, and are propagated by dividing the root, or by cuttings.

♄ 1. M. canade'nse. The Canadian Moonseed.

Identification. Lin. Sp.; Dec. Prod., 1. p. 102.; Don's Mill., 1. p. 112.; Tor. and Gray, 1. p. 48.
Synonymes. M. canadénse var. α *Lamarck*; M. angulàtum *Moench*; Ménisperme du Canada, *Bon. Jard.*; Canadischer Mondsame, *Ger.*
Engravings. Schkuhr. H., 3. t. 337.; Lam. Dict., t. 824.; and our *fig.* 48.

Spec. Char., *&c.* Leaves peltate, smoothish, somewhat cordate, roundish-angular; angles bluntish, terminal one abruptly awned, mucronate. Racemes solitary, compound. Petals 8. (*Don's Mill.*) A deciduous, suffruticose, long, slender twiner. Canada to Carolina. Height 8 ft. to 12 ft. Introduced in 1713. Flowers small, greenish yellow; June and July. Berry black; ripe in September. Decaying leaves greenish brown.

48. Menispér. canadénse.

Varieties.

♄ *M. c.* 2 *lobàtum* Dec. M. virgínicum *L.* (Dill. Elth., t. 178. fig. 219.) — This variety is distinguished by the angles of the leaves being acutish, and the flowers of a greenish white.

♄ *M. c.* 3 *smilácinum.* M. smilácinum *Dec.* (Jac. Icon., t. 269.; and our *fig.* 49) — Leaves smoother, and racemes more simple than in the species.

49. Menispérmum canadénse smilácinum.

Roots thick and woody, with numerous very slender shoots, which, though somewhat ligneous, never attain any considerable diameter, and are not of many years' duration. The stem twines in a direction contrary to the sun's apparent motion, and is smooth and even, having more the appearance of a herbaceous plant, than of a shrub.

♄ 2. Menispe'rmum dau'ricum *Dec.* The Daûrian Moonseed.

Identification. Dec. Prod., 1. p. 102.; Don's Mill., 1. p. 112.
Synonymes. Trilophus Ampelisàgria *Fisch.*; M. canadénse var. β *Lam.*
Engravings. Deless. Icon., 1. t. 100.; and our *fig.* 50.

Spec. Char., *&c.* Leaves peltate, smooth, cordate, angular; angles acute, terminal one acuminated hardly mucronate. Racemes in pairs, capitulate. (*Don's Mill.*) A twining, deciduous, suffruticose shrub. Dauria, on rocky hills, near the river Chilca. Height 5 ft. to 10 ft. Introduced in 1818. Flowers yellowish; June and July. Berries black; ripe in September.

50. Menispérmum daûricum.

Resembles the preceding species, and probably only a variety of it.

Genus II.

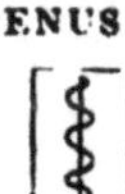

♄ CO'CCULUS *Bauh.* The Cocculus. *Lin. Syst.* Diœ'cia Hexándria.

Identification. Bauh. Pin., 511.; Dec. Prod., 1. p. 96.; Don's Mill., 1. p. 104.

Synonymes. Menispérmum *L.*; Wendlándia Willd.; Andróphilax *Wendl.*
Derivation. From *coccus*, the systematic name of cochineal; which is applied to this genus on account of the greater number of the species bearing scarlet berries.

Gen. Char. *Sepals* and *petals* disposed in a ternary order, in 2, very rarely in 3, series. *Male flowers* with 6 free stamens opposite the petals; *female* ones with 3 or 6 carpels. *Drupes* baccate, 1 to 6, usually obliquely reniform, somewhat flattened, 1-seeded. *Cotyledons* distant. (*Don's Mill.*)

Leaves simple, alternate, exstipulate, deciduous; cordate or ovate, entire or lobed. *Flowers* small. — The only hardy species is C. carolìnus, a native of Carolina, of the same culture as Menispérmum.

♃ 1. Co'cculus caroli'nus *Dec.* The Carolina Cocculus.

Identification. Dec. Prod., 1. p. 98.; Don's Mill., 1. p. 107.
Synonymes. Menispérmum carolìnum *Lin.*; Wendlándia populifòlia *Willd.*, *Pursh*, and *Dill.*; Andróphilax scándens *Wendl.*; Baumgártia scándens *Moench*; Ménisperme de la Caroline, *Fr.*; Carolinischer Mondsaame, *Ger.*
Engravings. Dil. Elth., 223. t. 178. f. 219.; Wendl. Obs., 3. t. 16.; and our *fig.* 51.

Spec. Char., &c. Leaves cordate or ovate, entire, obtuse, and somewhat 3-lobed; under surface velvety pubescent. Male racemes floriferous from the base, female ones 3-flowered. (*Don's Mill.*) A twining, deciduous, suffruticose shrub. North Carolina and Georgia. Height 6 ft. to 10 ft. Introduced in 1759. Flowers small, greenish; June and July. Fruit red; ripe in September. Decaying leaves yellowish or brownish.

51. Cócculus carolìnus.

Leaves extremely variable in form, 2 in. to 4 in. long, often quite entire, but usually with several sinuated, obtuse lobes. Drupe red, as large as a small pea.

Order VI. *BERBERA'CEÆ.*

Ord. Char. *Sepals* usually 6, in two whorls, deciduous, and furnished with petal-like scales on the outside. The *petals* are equal in number with the sepals, and the *stamens* equal in number with the petals, and opposite to them. The *anthers* "open by reflexed valves; that is to say, the face of each cell of the anther peels off except at the point, where it adheres as if it were hinged there;" a structure so remarkable, Dr. Lindley observes, as to be "found in no European plants except *Berberàceæ* and the laurel tribe." (*Penny Cyc.*, vol. iv. p. 259.) — Bushy shrubs, which throw up numerous suckers; natives of the temperate climates of Europe, Asia, and North America.

Leaves simple or compound, alternate, generally exstipulate, deciduous or persistent; shoots generally furnished with prickles; the sap, and the colour of the leaves and bark, more or less yellow. *Flowers* generally yellow. — The genera containing species hardy in British gardens are two, *Bérberis* and Mahònia, which are thus contradistinguished: —

Be'rberis L. Petals with 2 glands on the inside of each. Stamens toothless. Leaves undivided.

Maho'nia Nutt. Petals without glands. Stamens furnished with a tooth on each side. Leaves pinnate.

Genus I.

BE'RBERIS L. The Berberry. *Lin. Syst.* Hexándria Monogýnia.

Identification. Lin. Gen., 442.; Dec. Prod., 1. p. 105; Don's Mill., 1. p. 114.
Synonymes. Pipperidge Bush; E'pine vinette, *Fr.*; Berberitze, *Ger.*
Derivation. *Berberys* is the Arabic word used for this plant by Averrhoes and other writers on medicine; but some persons derive the name from the Greek word *berberi*, signifying a shell, from the leaves of the common species having a hollow surface. Bochart says that the word *Bérberis* is derived from the Phœnician word *barar*, which signifies shining like a shell, from their shining leaves. Gerard says that the word Berbery is a corruption of *amyrberis*, the name given to the plant by Avicenna. Du Hamel says that *Bérberis* is derived from an Indian word signifying mother of pearl. Pipperidge bush, or piprage tree, Gerard says, is Dr. Turner's name for the plant, and it is still given to it in Cambridgeshire. E'pine vinette signifies the acid, or sorrel, thorn, from the taste of the fruit and leaves.

Gen. Char. *Sepals* 6, guarded on the outside by 3 scales. *Petals* 6, with 2 glands on the inside of each. *Stamens* toothless. *Berries* 2—3-seeded. *Seeds* 2, rarely 3, laterally inserted at the base of the berries, erect, oblong, with a crustaceous coat and fleshy albumen. *Cotyledons* leafy, elliptical. *Radicle* long, capitellate at the tip. (*Don's Mill.*) *B.* heterophýlla *Juss.* has toothed stamens.

Leaves simple, alternate, exstipulate, deciduous or evergreen; toothed or serrated, coriaceous, with numerous small leaves produced at the axils of the larger ones, often abortive in the form of prickles. *Flowers* yellow. *Fruit* red, in some kinds black, purple or white in others.—Shrubs natives of Europe, North America, and Asia; characterised in a general view by being crowded with suckers, and having axillary tufts of leaves and spines.

The species are all readily propagated by seeds which most of them ripen in England; and also by side suckers and root suckers, which almost all of them throw up in abundance.

A. *Leaves thin, deciduous. Flowers solitary.*

⁂ 1. *B.* sibi'rica *Pall.* The Siberian Berberry.

Identification. Pall. Fl. Ross., 2. p. 42.; Dec. Prod., 1. p. 106.; Don's Mill., 1. p. 117.; Pen. Cyc., 4. p. 260.
Synonymes. *B.* altàica *Pal.*; Vinettier de Sibérie, *Fr.*
Engravings. Pall. Fl. Ross., 2. t. 67.; and our *fig.* 52. after Hayne, and *fig.* 53. after Pallas.

Spec. Char., &c. Spines 3—7-parted. Leaves lanceolate-obovate, ciliately serrated. Peduncles 1-flowered, shorter than the leaves. (*Don's Miller.*) An erect deciduous shrub. Siberia, on hills and the lower mountains. Height 2 ft. to 3 ft. Introduced in 1790. Flowers yellow; May and June. Berry red; ripe in September.

52. *Bérberis sibírica.*

53. *Bérberis sibírica.*

B. *Leaves thin, mostly deciduous. Flowers in Racemes.*

⁂ 2. *B.* vulga'ris *L.* The common Berberry.

Identification. Lin. Sp., 472.; Dec. Prod., 1. p. 105.; Don's Mill., 1. p. 115.
Synonymes. *B.* ætnénsis *Presl*; *B.* macrocárpa of some; Pipperidge Tree, *Dr. Turner*; E'pine vinette, *Fr.*; gemeine Berberitze, *Ger.*
Engravings. Eng. Bot., t. 49.; Willd. Baum., t. 89.; and our *fig.* 54., in which *a* is a specimen in flower, *b* a specimen in fruit, *c* a flower of the natural size, and *d* a fruit of the natural size.

Spec. Char., &c. Spines 3-parted. Leaves somewhat obovate, ciliately serrated. Racemes many-flowered, pendulous. Petals entire. (*Don's Mill.*) A spreading, many-stemmed, deciduous shrub. Europe, and Britain in hedges and copses, and naturalised in many parts of Asia and America. Height 6 ft. to 10 ft. Flowers yellow; May and June. Berries red; ripe in September. Decaying leaves reddish yellow. Naked wood yellowish white.

54 Bérberis vulgàris.

Varieties.

- B. v. 2 *lùtea.* — Fruit yellow, sometimes stoneless.
- B. v. 3 *álba.*—Fruit white.
- B. v. 4 *violàcea.* — Fruit violaceous
- B. v. 5 *purpùrea. B.* innominàta *Kalm.* — Fruit purple; leaves narrow, hardly ciliated.
- B. v. 6 *nìgra.* — Fruit black; leaves oblong, ciliately serrated, serratures few. The fruit of this plant is said by Tournefort, who found it on the banks of the Euphrates, to be of delicious flavour.
- B. v. 7 *dúlcis.*—Fruit red, somewhat less acid than that of the common berberry. Leaves of a bright shining green. Native of Austria, where it was first considered to be a distinct species, till the fruit of plants raised from its seed was found to be as acid as that of the common berberry. It is now, however, propagated by layers; the leaves and fruit are considerably larger than those of the species, and the fruit is found perfectly sweet and agreeable to eat. In short, this variety is to the common berberry, what the apple is to the crab.
- B. v. 8 *aspérma. The seedless Berberry.* — Fruit destitute of seeds. Miller, and also Du Hamel, both say that suckers taken from this variety commonly produce fruit with seeds; that, as the tree grows older, the seeds become fewer, and that it is the age of the plant that at last causes the fruit to be seedless; in that case this plant must be considered more a variation than a variety. *B.* v. aspérma is said by Du Hamel to produce the best fruit for preserving; and it is from it that the delicious *Confitures d'E'pine vinette,* for which Rouen is so celebrated, are made. (*Nouv. Duh.*, iv. p. 13.)
- B. v. 9 *longifòlia* Booth. — Leaves longer than those of the species.
- B. v. 10 *glaúca. B.* glaúca *Booth.* — Leaves glaucous. Mr. Gordon considers this plant as related to *B.* sibírica; but, as it has not yet flowered in the Horticultural Society's Garden this point cannot be determined. (See *Gard. Mag.*, vol. xvi. p. 2.)
- B. v. 11 *mìtis.* — Shoots without spines. Leaves glaucous, rather broader.
- B. v. 12 *provinciàlis* Schrad. — Young shoots brown. Leaves and fruit as in the common berberry.

All these varieties are in the London Horticultural Society's Garden.

Other Varieties. In the Horticultural Society's Garden a number of alleged species of berberries have been raised from seed, which have all proved varieties of *B.* vulgàris, and most of them so slight, as to be scarcely worth keeping distinct. (See *Gard. Mag.*, vol. xvi. p. 2.)

The common berberry will live for two or three centuries, without increasing much in size. The wood is hard and brittle, of a yellow colour, and but little used except for dyeing. The rate of growth, when the plant is young, is rapid; and, in consequence, in five or six years it will attain the height of

7 or 8 feet; but it grows slowly afterwards, unless the suckers are removed from it as they are produced. It is seldom seen above 10 ft. high; but there are examples of trees of it 30 ft. high, probably of 30 years' growth. The inner bark, both of the stems and roots, affords a yellow dye. The leaves are agreeably acid, and, according to Gerard, were used in his time "to season meat with, and instead of a salad, like sorrel." The berries are so acid, that birds seldom touch them. They are not eaten raw, but are excellent when preserved with sugar in syrup, or candied. They are also made into jelly and rob, both of which are not only delicious to the taste, but extremely wholesome; and they are pickled in vinegar, when green, as a substitute for capers. The plant is cultivated in gardens as a fruit tree or fruit shrub; and the variety, or rather variation, in which the seeds are said to be wanting, and that in which the fruit is sweet, are recommended in preference. The plant makes an excellent hedge; but there exists a prejudice against it among agriculturists, from its supposed influence in producing blight, or mildew, on the corn adjoining it. This opinion is of unknown antiquity; but it is now generally considered to be an erroneous prejudice.

3. *B.* (v.) EMARGINA'TA *Willd.* The emarginate-*petaled* Berberry.

Identification. Willd. Enum., 1. p. 395.; Dec. Prod., 1. p. 105.; Don's Mill, 1. p. 115.
Synonyme. Ausgerandete (serrated) Berberitze, *Ger.*
Engravings. Hayne Abbild., t. 62.; and our *figs.* 55. and 56.

Spec. Char., &c. Spines 3-parted. Leaves lanceolate-obovate, ciliately serrated. Racemes scarcely pendulous, shorter than the leaves; petals emarginate. (*Don's Mill.*) A deciduous shrub. Siberia. Height 5 ft. to 7 ft. Introduced in 1820. Flowers yellow; May and June. Berries red; ripe in September. Decaying leaves yellow. Naked young wood whitish yellow.

55. *Bérberis emarginàta.*

56. *Bérberis emarginàta.*

Closely resembling *B.* vulgàris, of which it is, doubtless, only a variety; but it is one half smaller in all its parts, and has the petals emarginate, and the leaves decidedly glaucous.

4. *B.* (v.) CRE'TICA *L.* The Cretan Berberry.

Identification. Lin. Sp., 472.; Dec. Prod. 1. p.106.; Don's Mill., 1. p. 115.
Synonymes. *B.* crética buxifòlia *Tourn*; Vinettier de Crète, *Fr.*; Cretische Berberitze, *Ger.*; Candian Berberry.
Engravings. Fl. Græc., t. 342.; and our *figs.* 57. and 58.

Spec. Char., &c. Spines 3—5-parted. Leaves oval-oblong, entire, or somewhat serrated. Racemes 3—8-flowrd., rather shorter than the leaves.

57. *Bérberis crética.*

58. *Bérberis crética, nat. size.*

(*Don's Mill.*) A deciduous shrub, crowded with shoots. Crete, Candia, and, perhaps, Japan. Height 3 ft. to 4 ft. Introduced in 1759. Flowers yellow; May and June. Berries ovate, black; ripe in September. Decaying leaves whitish yellow. Naked young wood also whitish.

The leaves are produced without any obvious order; they are small, and in their shape they resemble those of the narrow-leaved variety of the common box. The berries are ovate, black, 2-seeded, more astringent than acid; stigma on a very short style.

■ 5. *B.* (v.) CRATÆ'GINA *Dec.* The Cratægus-like Berberry.

Identification. Dec. Syst., 2. p. 9.; Don's Mill., 1. p. 116.
Engraving. Our *fig.* 59. from a specimen in the Hort. Soc. Garden.

Spec. Char., &c. Spines simple. Leaves oblong, reticulated, hardly serrated. Racemes many-flowered, crowded, spreading, scarcely longer than the leaves. (*Don's Mill.*) A deciduous glaucous-leaved shrub. Asia Minor. Height 4 ft. to 6 ft. Introduced in 1823. Flowers yellow; May and June. Berries red; ripe in September.

59. *Bérberis cratæ'gina*, nat. size.

Distinguished from all the other species, by the leaves being long, flaccid, entirely glaucous, or whitish Young shoots brown. A plant bearing this name in the Horticultural Society's Garden is 5 ft. high, with the leaves much longer than those of *B.* vulgàris; serrated, as in that species, and decidedly glaucous. In other respects we can see no difference.

■ 6. *B.* IBE'RICA *Stev.* The Iberian Berberry.

Identification. Don's Mill., 1. p. 115.; and Lindl., Pen. Cyc., 4. p. 61.
Synonymes. *B.* vulgàris ? v. ibérica *Dec. Syst.* 2. p. 6.; *B.* sinénsis *Wal.*
Engravings. Dend. Brit., t. 96., as *B.* sinénsis; and our *fig.* 60.

Spec. Char., &c. Spines simple, and 3-parted; leaves obovate-oblong, quite entire. Racemes many-flowered; petals entire. (*Don's Mill.*) A deciduous shrub. Iberia. Height 3 ft. to 5 ft. Introduced in 1790. Flowers yellow; May and June. Berries dark purple; ripe in September. Decaying leaves yellowish red. Naked young wood reddish yellow.

Readily distinguished from the common berberry by its smaller and smoother leaves, its red shoots, and its almost upright racemes; and from *B.* sinénsis by the leaves being comparatively entire.

60. *B.* ibérica.

■ 7. *B.* CANADE'NSIS *Mill.* The Canadian Berberry.

Identification. Pursh's Fl. Amer. Sept., 1. p. 219.; Dec. Prod., 1. p. 106.; Don's Mill., 1. p. 115. Tor. and Gray, 1. p. 50.
Synonymes. *B.* vulgàris *Mx. Fl. Bor. Amer.*, 1. p. 205.; *B.* vulgàris var. canadénsis *Martyn's Mill.*, No. 1.
Engravings. Hayne Abbild., t. 63.; and our *fig.* 61. after that author.

Spec. Char., &c. Branches verrucose, dotted, with short triple spines; leaves spatulate, oblong, remotely serrate, with somewhat bristly teeth; racemes sub-corymbose, few-flowered; petals emarginate; berries subglobose, or oval. (*Tor. and Gray.*) A deciduous shrub. Canada to Georgia. Height 2 ft. to 3 ft., in England 5 ft. Introduced in 1759. Flowers yellow;

May and June. Berries red; ripe in September. Decaying leaves yellowish green and reddish. Naked young wood whitish yellow.

61. B. (v.) canadénsis.

Leaves much smaller and narrower than in *B.* vulgàris, attenuate at the base, but nearly sessile; the margins serrulate, with 6–8 distant, often inconspicuous, mucronate teeth. Raceme 5–8-flowered, nodding; flowers smaller than in *B.* vulgàris; fruit smaller and much shorter. Stem and roots yellow; the former rarely exceeding 3 ft. in height. Found in the Alleghany Mountains, Virginia and Carolina, Tenessee, and Georgia. (*Tor. and Gray.*) Introduced into England in 1759, but probably lost, as we have seen no plant answering this description in British gardens.

♠ 8. *B.* SINE'NSIS *Desf.* The Chinese Berberry.

Identification. Desf. Catal. Hort. P., 150.; Dec. Prod., 1. p. 106.; Don's Mill., 1. p. 115.
Synonyme. *B.* vulgàris *Thunb. Jap.* 1. p. 146.
Engravings. Our *figs.* 62. and 63. from a specimen in the Hort. Soc. Garden.

Spec. Char., &c. Spines 3-parted. Leaves oblong, obtuse, entire, or the lower ones a little toothed. Racemes many-flowered, nodding. (*Don's Mill.*) A deciduous shrub with slender shoots. China. Height 3 ft. to 5 ft. Introduced in 1800. Flowers yellow; May and June. Berries oval, dark red; ripe in September. Leaves smooth, sharply serrated. Decaying leaves of a fine yellowish red. Naked young wood reddish yellow.

The plant at the Horticultural Society's Garden, and at Messrs. Loddiges's, has smooth leaves, red shoots, and closely resembles *Bérberis* ibérica.

62. Bérberis sinénsis.

63. Bérberis sinénsis, nat. size.

C. *Leaves leathery, evergreen, or sub-evergreen. Flowers solitary, or in Clusters.*

9. *B.* DU'LCIS *D. Don.* The sweet-*fruited* Berberry.

Identification. Swt. Brit. Fl.-Gard.; Lindl. Pen. Cyc., 4. p. 261.
Engravings. Swt. Brit. Fl. Gard., 2d ser., t. 100.; and our *fig.* 64.

Spec. Char., &c. Spines long, slender, simple, or 3-parted. Leaves obovate obtuse, with or without a bristly point, quite entire, glaucous on the under side. Flowers solitary, on slender stalks, twice as long as the leaves. (*Lindl., Pen. Cyc.,*) An evergreen shrub, with shining leaves. Straits of Magellan to Valdivia. Height 2 ft. to 5 ft. Introduced in 1830. Flowers yellow; March to June. Berries round, black, about the size of a black currant; ripe in August. Decaying leaves bright yellow; dropping in May and June.

64. *Bérberis dúlcis.*

An elegant evergreen bush, which, in some places, has attained the height of 5 ft. The flowers are large, of a fine bright yellow, more expanded than they are in many species, and, from their long slender stalks, they hang down in a very graceful manner. In its native country, the fruit is used, both green and ripe, as we use gooseberries, for making pies and tarts and preserves, for which it is most excellent. It is quite hardy and evergreen.

10. *B.* BUXIFO'LIA *Lam.* The Box-leaved Berberry.

Identification. Lam. Ill., t. 253. fig. 3.; Don's Mill., 1. p. 117.
Engravings. Lam. Ill. t. 253. fig. 3.; and our *fig.* 65.

Spec. Char., &c. Spines 3-parted. Leaves ovate, or ovate-lanceolate, smooth, quite entire. Pedicels longer than the leaves, either solitary, 1-flowered, or in threes, rising from a short peduncle. (*Don's Mill.*) A small twisted sub-evergreen shrub. Native of the Straits of Magellan. Height 2 ft. to 3 ft. Introduced ?. Flowers yellow. December to March. Berries bluish purple, 4-seeded.

65. *Bérberis buxifòlia.*

Said to be nearly allied to *B.* dúlcis. A very valuable addition to our hardy evergreens; though, at present, rare in British gardens.

11. *B.* ACTINACA'NTHA *Mart.* The ray-spined Berberry.

Identification. Rœmer and Schultes, 7. p. 12.; Hook. Bot. Mis., 3. p. 135.
Engraving. Our *fig.* 66.

Spec. Char., &c. Spines palmate, 3—5 divisions. Leaves ovate elliptic, rigid, coriaceous, toothed, mucronate. Peduncles 4—5, sub-umbellate, shorter than the leaves. An evergreen shrub, with numerous spreading branches, and long white spines, generally in threes, but sometimes more numerous. Straits of Magellan. Height 3 ft. to 4 ft. Introduced ? 1830. Flowers deep yellow, a little larger than those of *B.* vulgàris; May and June. Fruit ?.

66. *Bérberis actinacántha.*

A very desirable species, nearly allied to *B.* heterophýlla. Our engraving is of a specimen taken from a very handsome plant in the rich collection of evergreens at Elvaston Castle. Dr. Hooker mentions a variety with smaller leaves than the species, but it does not appear to be introduced. Quite hardy.

■ 12. *B.* HETEROPHY'LLA *Jus.* The various-leaved Berberry.

Identification. Juss. in Poir. Dict., 8. p. 622.; Dec. Prod., 1. p. 108.; Don's Mill., 1. p. 117.; Lindl., Pen. Cyc., 4. p. 261.
Synonymes. *B.* ilicifòlia *Forst.*; *B* tricuspidàta *Smith.*
Engravings. Hook. Exot. Fl., 1. t. 14.; and our *fig.* 67.

67. B. heterophylla.

Spec. Char., &c. Spines 3-parted. Leaves ovate-lanceolate, glabrous, some of them entire, others furnished with 3 pungent teeth. Pedicels solitary, 1-flowered, hardly longer than the leaves. Filaments toothed. (*Don's Mill.*) An evergreen shrub. Straits of Magellan. Height 3 ft. to 4 ft. Introduced in 1823. Flowers orange yellow; May and June. Berries red; ripe in September. Decaying leaves greenish yellow; June and July.

Much branched, and the older branches covered with dark wrinkled bark. The leaves clustered, and of two kinds; the old ones terminated with a sharp spinose point, and having a lateral spinule on each side above the middle, and the younger ones being pale green, unarmed, and having their margins entire and softish. The old leaves are also quite rigid, dark green, and shining.

■ 13. *B.* EMPETRIFO'LIA *Lam.* The Empetrum-leaved Berberry.

Identification. Lam. Ill., t. 253.; Dec. Prod., 1. p. 107.; Don's Mill., 1. p. 117.; Pen. Cyc., 4. p. 261.
Engravings. Lam. Ill., t. 253. fig. 4.; Sw. Brit. Fl.-Gar., 2. s. t. 350.; and our *fig.* 68.

68. B. empetrifòlia.

Spec. Char., &c. Spines 3-parted. Leaves linear, quite entire, with revolute margins. Pedicels 1—2, 1-flowered. (*Don's Mill.*) An elegant, decumbent, evergreen bush. Cordilleras of Chili in subalpine woods. Height 1 ft. to 2 ft. Introduced in 1830. Flowers yellow; December to March. Berries ? yellow; ripe in July. Wood reddish brown.

Branches slender, twiggy, angular, covered with a chestnut-coloured bark. Leaves fasciculate, linear, mucronate, revolute, and entire at the margins, glaucous; about half an inch long, and nearly a line in breadth. Flowers large, spreading. A very curious and pretty plant, in general aspect much more like a heath than a berberry. It is perfectly hardy, and deserves a place in every collection. It is readily increased by layers, in heath soil.

D. *Leaves leathery, evergreen or sub-evergreen. Flowers in Racemes.*

■ 14. *B.* DEALBA'TA *Lindl.* The whitened-*leaved* Berberry.

Identification. Bot. Reg., t. 1750.; Pen. Cyc., 4. p. 261.
Synonyme. *B.* glaùca *Hort.*
Engravings. Bot. Reg., t. 1750.; and our *fig.* 69.

69. Bérberis dealbàta.

Spec. Char., &c. Spines scarcely any. Leaves roundish, coarsely toothed, rather glaucous, white beneath. Racemes very short and compact, pendulous. (*Pen. Cyc.*) An upright evergreen bush, with white or glaucous leaves. Mexico. Height 3 ft. to 5 ft. Introduced in 1830. Flowers yellow, sweet-scented; December to March. Berries yellowish purple; ripe in August.

A tall, slender, evergreen bush, with deep brown branches, and scarcely any spines. The leaves are sometimes wedge-shaped and 3-toothed, but more frequently nearly round, with two or three spiny teeth on every side. A curious and beautiful species, well deserving of cultivation. It is quite hardy, and readily increased by layers, which root the same season that they are made.

■ 15. *B.* ASIA'TICA *Roxb.* The Asiatic Berberry.

Identification. Roxb. in Dec. Syst., 2. p. 13.; Dec. Prod., 1. p. 107.; Don's Mill., 1. p. 116; Pen. Cyc., 4. p. 261.
Synonymes. The Lycium of Dioscorides, *Royle* in *Linn. Trans.*, xvii. p. 83.; *B.* tinctòria *Lech.*; the Raisin Berberry, *Pen. Cyc.*
Engravings. Deless. Icon. sel., 2. t. 1.; and our *fig.* 70.

Spec. Char., &c. Spines trifid, or simple. Leaves oval, cuneated or elliptical, mucronate, smooth, under surface glaucous, entire or spinulosely toothed. Racemes short, many-flowered, corymbose, shorter than the leaves. Pedicels elongated, one-flowered. Berries oval. (*Don's Mill.*) A vigorously growing sub-evergreen shrub, crowded with numerous luxuriant suckers. Nepal, on mountains. Height 6 ft. to 8 ft. Introduced in 1820. Flowers yellow; May and June. Berries purplish, with a fine bloom; ripe in July. Decaying leaves yellow and red.

70. *Bérberis asiática.*

Leaves somewhat resembling those of *B.* heterophýlla, but not glaucous. The plant is easily distinguished from that species by the very short racemes of its flowers, by their being produced much earlier, and by the smoothness of its shoots. On July 20. 1837, a fine plant of *B.* asiática, in the grounds at Syon, was covered with fruit, while one of *B.* aristàta, standing close beside it, was covered with its beautiful rich yellow blossoms, many of which were not fully expanded. The fruit is oblong, pinkish or purplish, wrinkled, and covered with a fine thick bloom like that of the best raisins. The plants in 5 or 6 years attain the height of 6 or 8 feet.

■ 16. *B.* ARISTA'TA *Dec.* The bristled-*tooth-leaved* Berberry.

Identification. Dec. Prod., 1. p. 106.; Don's Mill., 1. p. 115.
Synonymes. *B. Chitria* Buch.; *B.* angustifòlia *Roxb.*; *B.* sinénsis *Desf.*
Engravings. Hook. Exot. Flor., t. 98.; Bot. Reg., t. 729.; and our *fig.* 71

Spec. Char., &c. Lower spines 3-parted, simple; leaves obovate-acute, tapering much to the base, ending in a mucro (prickly point) at the apex, membranous, smooth on both sides, serrated, with 4 or 5 bristly teeth. Racemes nodding, many-flowered, longer than the leaves. Berries oblong. (*Don's Mill.*) A vigorous-growing sub-evergreen shrub, crowded with suckers which sometimes grow 8 ft. to 9 ft. long in a season. Nepal, on mountains 5000 ft. to 8000 ft. of elevation. Height 6 ft. to 10 ft. Introduced in 1820. Flowers yellow; June and July. Berries purplish, with a fine bloom; ripe in September. Decaying leaves yellow and scarlet. Naked young wood yellowish brown.

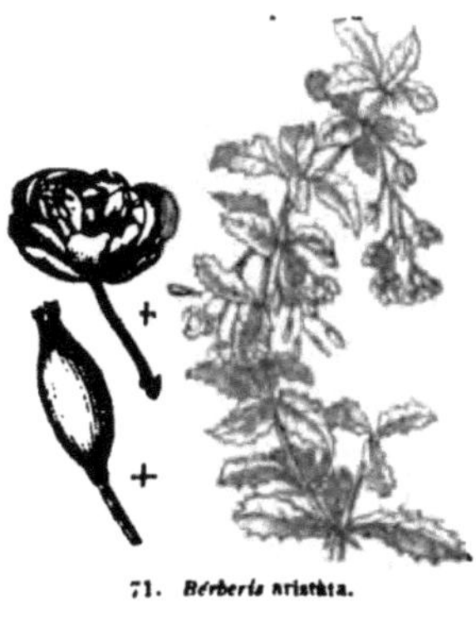

71. *Bérberis aristàta.*

Very distinct from any of the preceding species or varieties, growing with extraordinary vigour, and capable of being formed into a very handsome small tree. The root and wood are of a dark yellow colour, and form the yellow wood of Persian authors; they are used as a dye, and, being bitter and a little astringent, they, as well as the bark, are employed in medicine. (*Royle's Illust.*, p. 63.) In Nepal, the fruit of this species is dried, like grapes for forming raisins, in the sun. A most desirable plant, calculated to produce a splendid effect, both when in flower and when in fruit, upon an open lawn. As a rapid grower, it ought not to be planted near slow-growing shrubs or trees.

Other Species of Bérberis. — B. *Coriària* Royle, a species having the same general appearance as *B.* aristàta, has been raised in the Horticultural Society's Garden, and there are plants 3 ft. high, but they have not yet flowered. Plants have been raised in the Horticultural Society's Garden, and in some nurseries, from seeds received from Mexico and Nepal; but, though these have new names, it is not certain that they will all prove new species, and therefore we consider it better not to record them till they have flowered. In *Hook. Bot. Mis.* vol. iii., B. *chilénsis* Gill., B. *ruscifòlia* Lam., B. *corymbòsa* Hook. et Arn., B. *glomeràta* Hook. et Arn., and B. *Grevilleàna* *Gill.*, are described, or mentioned, as having been found in South America, and Dr. Hooker has specimens of them in his herbarium. Numerous varieties of *Bérberis* vulgàris are raised in the London gardens, under continental names, as if they were species, but very few of them are worth keeping distinct. See in *Gard. Mag.* for 1840, p. 1., Mr. Gordon's Report on those raised in the Horticultural Society's Garden in 1839.

Genus II.

MAHO'N*IA* Nutt. The Mahonia, *or Ash Berberry.* *Lin. Syst.* Hexándria Monogýnia.

Identification. Nutt. Gen. Amer., 1. p. 307.; Dec. Prod., 1. p. 108.; Don's Mill., p. 117.
Synonymes. *Bérberis* of authors; Odostèmon *Raf.*; Ash Berberry *Pen. Cycl.*
Derivation. Named by Nuttall in honour of *Bernard M'Mahon*, a seedsman at Philadelphia, the author of the *American Gardener's Calendar*, and an ardent lover of botanical science.

Gen. Char. *Sepals* 6, guarded on the outside by three scales. *Petals* 6, without glands on the inside. *Stamens* furnished with a tooth on each side at top of the filament. *Berries* 3—9-seeded. (*Don's Mill.*)

Leaves compound, pinnate, alternate, exstipulate, evergreen; the leaflets coriaceous, with the margins toothed or serrated. *Flowers* yellow. *Fruit* mostly black. — Natives of the north-west coast of America, and also of Nepal, and perhaps Japan.

Though some botanists think that the characters ascribed to this genus, and those ascribed to *Bérberis*, as exhibited in p. 41., are not sufficient to keep them separate as genera; yet the habits of the species of one, as to the mode of growth, foliage, and inflorescence, are so distinct from those of the other, as to induce us to adopt the genus Mahònia. The species in British gardens are all of comparatively slow growth, and admit but of slow multiplication by layers, which require to remain on two years, and scarcely at all by cuttings. Some of them, however, seed freely, and are readily propagated in this way. The seeds of all the species of Mahònia, and also of those of *Bérberis*, if sown immediately after they are ripe, and protected through the winter from frost, will come up the following spring.

■ 1. M. fascicula'ris *Dec.* The crowded-*racemed* Mahonia, *or Ash Berberry.*

Identification. Dec. Prod., 1. p. 108.; Don's Mill., 1. p. 118.
Synonymes. *Bérberis* pinnàta *Lag.*, *Bot. Reg.*, *Bot. Mag.*, and *Tor. & Gray*; *B.* fasciculàris *Pen. Cyc.* In the same work it is stated that Mahònia diversifòlia is the same as this species; though it is figured and described by Sweet, as a species from Monte Video: see *Swt. Br. Fl.-Gar.*, 2d series, t. 56.
Engravings. Bot. Reg., t. 702.; Bot. Mag., t. 2396.; and our *fig.* 72.

72. Mahònia fasciculàris.

Spec. Char., &c. Leaves of 3—6 pairs with an odd one, the lowest pair near the base of the petiole. Leaflets ovate-lanceolate, rather distant, one-nerved, spiny-toothed, with 4 or 5 teeth on each side. Racemes nearly erect, much crowded. Filaments bidentate. (*Don's Mil.*) An upright evergreen shrub. California and Mexico, on mountains. Height 5 ft. to 8 ft. Introduced in 1819. Flowers yellow; Mar. to May. Berries purple; ripe in September. Decaying leaves rich yellow; drop in June.

Very handsome. "Perhaps the most showy of all the family." (*Hook.*) It is readily distinguished, even at a distance, from the other mahonias, by the glaucous green and subdued tone of colour of its leaves; those of all the others being of a darker green, and more or less shining. The plant is rather too tender to be treated as a detached bush, unless some slight protection be given to it during very severe frosts; but it will grow freely against a wall with scarcely any protection. Layers and seeds.

■ 2. M. *A*QUIFO'LIUM *Nutt.* The Holly-leaved Mahonia, *or Ash Berberry.*

Identification. Nutt. Gen. Amer., 1. p. 212.; Dec. Prod., 1. p. 108.; Don's Mill., 1. p. 118.
Synonyme. *Bérberis Aquifòlium Ph.*, *Pen. Cyc.*, and *Tor. & Gray.*
Engravings. Pursh. Fl. Amer. Sept. 1. t. 4.; Bot. Reg., t. 1425.; and our *fig.* 73.

73. Mahònia Aquifòlium

Spec. Char., &c. Leaves of 4 pairs of leaflets with an odd one, the lower pair distant from the base of the petiole; leaflets ovate, approximate, cordate at the base, one-nerved, spiny-toothed, with 9 or 6 teeth on each side. Racemes erect, and much crowded. Filaments bidentate. (*D.'s Mill.*) A shining evergreen shrub. New Albion to Nootka Sound. Height 5 ft. to 7 ft. in its native country, probably 10 ft. in England. Introduced in 1823. Flowers yellow; April and May. Berries purple; ripe in September.

Varieties. One variety, M. A. *nutkànum* Dec., is mentioned by De Candolle; and another, found at the junction of the Portage river with the Columbia, by G. Don. Torrey and Gray consider Mahònia rèpens and M. pinnàta *Menzies* as only varieties of this species; an inadvertence excusable in those who have not seen the plants in a living state.

One of the handsomest of evergreen shrubs, attaining the height of 6 ft. in 6 years, quite hardy, producing a profusion of bunches of yellow flowers during April and May. In its native country it grows in rich vegetable soil, among rocks, or in woods, where it forms a thick and rich undergrowth. According to Dr. Lindley, it is "perhaps the handsomest hardy evergreen we yet possess. Its foliage is of a rich, deep, shining green, becoming purple in the winter; it bears fruit in some abundance, which consists of clusters of roundish black berries, having their surface covered with a rich violet bloom. It most resembles M. fasciculàris, from which its large shining leaves at once distinguish it." (*Penny Cyc.*, iv. p. 262.) Layers and seeds.

⁂ 3. M. NERVO'SA *Nutt.* The nerved-*leaved* Mahonia, *or Ash Berberry.*

Identification. Nutt. Gen. Amer., 1. p. 212.; Don's Mill., 1. p. 118.
Synonymes. *Bérberis* nervòsa *Ph.*, and *Tor. & Gray*; Mahònia glumàcea *Dec.*; *Bérberis* glumàcea *Pen. Cyc.*
Engravings. Pursh Fl. Amer., 1. t. 5.; Bot. Reg., t. 1426.; and our *fig.* 74.

74. Mahònia nervòsa.

Spec. Char., &c. Leaves of 5—6 pairs, with an odd one, the lower pair distant from the petiole; leaflets ovate, acuminated, and remotely spiny-toothed, somewhat 3—5-nerved, with 12 or 14 teeth on each side. Racemes elongated. Filaments bidentate. (*Don's Mill.*) An evergreen undershrub. North-west of N. America, on the river Columbia, in shady pine woods. Height 2 ft. to 3 ft. Introd. in 1822. Flowers yellow; October to March. Berries roundish, glaucous purple, or deep blue; ripe in July.

According to Torrey and Gray, the stem is so low, that it often scarcely rises from the ground, and, indeed, is much shorter than the leaves, which are 1 ft. to 2 ft. in length. Racemes spicate, often 6 in. to 8 in. long. Flowers larger than in M. *A*quifòlium. The petioles of the leaves, Dr. Lindley says, "are jointed at every pair of leaflets, in the manner of a bamboo stem." The plant is hardy, and will thrive in a shady border of peat soil. One of the handsomest of undershrubs.

⁂ 4. M. RE'PENS *G. Don.* The creeping-*rooted* Mahonia, *or Ash Berberry.*

Identification. G. Don, in Loud. Hort. Brit., No. 28182.; and in Don's Mill., 1. p. 118.
Synonymes. *Bérberis A*quifòlium *Lindl. Bot. Reg.*, t. 1176.; *Bérberis* rèpens *Pen. Cyc.* iv. p. 262.; *B. A*quifòlium var. rèpens *Tor. & Gray*., 1. p. 50.
Engravings. Bot. Reg., t. 1176.; and our *fig.* 75.

75. Mahònia rèpens.

Spec. Char., &c. Leaflets 2—3 pairs, with an odd one, roundish ovate, opaque, spiny-toothed. Racemes diffuse. Root creeping. Filaments bidentate. (*Don's Mill.*) An evergreen undershrub. West coast of N. America, on the Rocky Mountains. Height 1 ft. to 2 ft. Introduced in 1822. Flowers yellow; April and May. Berries purplish black; ripe in September.

Variety.

■ *M. r.* 2 *rèpens-fasciculàris.* — Habit of M. fasciculàris, with larger and more robust foliage, resembling that of M. rèpens. A sport, or a hybrid, produced accidentally in the Sawbridgeworth Nursery.

The shoots consist chiefly of short unbranched suckers, with the leaves somewhat glaucous on both surfaces. The racemes of flowers are terminal, numerous, fascicled, diffuse, rising from scaly buds. The plant, in British gardens, is perfectly hardy, and produces a profusion of rich yellow flowers in April and May. Layers or suckers; but it does not root readily. Seeds are sometimes produced.

Other Species of Mahònia are no doubt in British gardens; but as they have been only raised lately from Nepal or Mexican seeds, nothing can be recorded of them with that degree of certainty and detail which is suitable for this work. *Mahònia tenuifòlia*, a Mexican species with pinnate leaves, and entire quite smooth leaflets, on very long slender footstalks, has been raised in the Horticultural Society's Garden, and may probably be as hardy as M. fasciculàris; but this is uncertain. *M. nepalénsis*, *M. acanthifòlia*, *M. tragacanthöìdes*, and *M. caraganæfòlia*, are said to be very desirable species.

SECTION II.

Carpella solitary, or connate; Placenta parietal (that Part of the Capsule which the Seeds are attached to adhering to the Sides or Walls of the Ovary or Germen), attached to the Walls or Cells of the Ovary.

ORDER VII. CRUCIA'CEÆ.

ORD. CHAR. The order Cruciàceæ is readily recognised by the cruciform arrangement of the *petals*, which are always four, in conjunction with tetradynamous *stamens*, and the *fruit* a silique or silicle. — Though there are several species which, technically considered, are ligneous plants, such as *Alýssum* saxátile, *Ibèris* sempervìrens, Cheiránthus *Chèiri*, and some others; yet, in a popular point of view, the only shrub included in the order is the *Véll*a Pseùdo-*Cýtisus*.

GENUS I.

VE'LLA *L.* THE VELLA. *Lin. Syst.* Tetradynàmia Siliculòsa.

Derivation. The word *Vèlla* is Latinised from the word *velar*, the Celtic name of the cress.

Gen. Char. *Stamens* the 4 longer in 2 pairs, the 2 of each pair grown together. *Style* ovate, flat, tongue-shaped, at the tip of the silicle. *Silicle* ovate, compressed, its valves concave. *Partition* elliptic. *Cotyledons* folded, the embryo root disposed in the sinus of the fold. (*Dec. Syst.*)

Leaves simple, alternate, exstipulate, sub-evergreen; toothed or serrated, glaucous. *Flowers* in axillary spikes, yellow, seldom succeeded by seed pods in the climate of London. — Shrub low, suffruticose, native of Spain.

a 1. *VE'LLA* PSEU'DO-*CY'*TISUS *L.* False-Cytisus, *or shrubby*, Cress-Rocket.

Identification. Lin. Sp. 895.; Dec. Prod., 1. p. 223.; Don's Mill., 1. p. 254.
Synonymes. *Vèlla* integrifòlia *Sal.*; Faux-cytise, *Fr.*; strauchartige (shrubby) Velle, *Ger.*
Engravings. Cav. Ic., 1. 42.; and our *fig.* 76.

Spec. Char., &c. Petals yellow, with long dark purple claws. Larger stamens perfectly connate by pairs (*Don's Mill.*) A low subevergreen shrub. Spain, on calcareous hills. Height 2 ft. to 4 ft. Introd. in 1759. Flowers yellow; April and May. Silique greenish yellow; ripe in July. Decaying leaves yellow.

76. *Vèlla* Pseùdo-Cytisus.

Branches arched, spreading, somewhat decumbent. Leaves glaucous green. Somewhat tender, but requires no protection in the climate of London, when planted on dry soil. It is a short-lived plant, like all the suffruticose Crucíferæ, but it may readily be renewed by cuttings or seeds.

ORDER VIII. *C*ISTA'CEÆ.

ORD. CHAR. *Sepals* 5, two of them being exterior. *Petals* 5, very fugitive. *Stamens* numerous. *Fruit* capsular, 3—5-valved, 5—10-celled, with parietal placentæ. *Embryo* inverted. Properties balsamic. (*Lindl.*)

Leaves simple, opposite or alternate (the lowest leaves always opposite), stipulate or exstipulate, deciduous or sub-evergreen; generally pubescent, pubescence simple or stellate. *Flowers* large, showy, white, red, or purple.—Shrubs low, suffrutescent, many subherbaceous; natives of Europe and Africa.

The *C*istàceæ have no medical properties; but the resinous balsamic substance called ladanum or labdanum is produced from *C.* créticus, *C.* ladaníferus, *C. l*aurifòlius, and one or two other species. Their use in gardens is for ornamenting rockwork, or for keeping in pits during the winter, and planting out in flower-borders in spring; as, from the tenderness of the finer species, they are unfit for a permanent place in a shrubbery or arboretum. Most of even the larger-growing kinds require some protection during winter: but they will all grow freely in any soil that is dry; and they are readily propagated by seeds, which, in fine seasons, they produce in abundance, or by cuttings; the plants, in both cases, flowering the second year. Though easily propagated, the *C*istàceæ do not readily bear transplanting, having very few fibres, and these rambling to a great distance from the main root. Plants for sale ought, therefore, to be always kept in pots; and, in the winter season, they should be protected by some slight covering during severe weather. The hardy ligneous species are included in two genera; which are thus contradistinguished by DeCandolle and G. Don:—

*C*I'STUS *L.* Capsule 10—5-celled.
HELIA'NTHEMUM *Tourn.* Capsule 1-celled, 3-valved.

GENUS I.

***C*I'STUS *L.* THE CISTUS, *or* ROCK ROSE. *Lin. Syst.* Polyándria Monogýnia.**

Derivation. From the Greek word *kistē*, a box or capsule, or the Anglo-Saxon, *cist*, a hollow

vessel; on account of the shape of its capsules. In Martyn's *Miller*, the name is said to be derived from that of the youth Cistus, whose story is to be found in Cassianus Bassus. Others derive it from *kis*, a worm or weevil.
Identification. Tourn., Lin., Dec., G. Don.
Synonymes. Holly Rose, *Gerard*; Gum Cistus; Ciste, *Fr.*; Cisten Rose, *Ger.*; Cisto *Ital.*

Gen. Char. *Calyx* of 5 sepals. *Sepals* disposed in a double series; 2 outer ones unequal, sometimes wanting. *Petals* 5, equal, somewhat cuneated, caducous. *Stamens* numerous, usually exserted from the glandular disk. *Style* filiform. *Stigma* capitate. *Capsule* covered by the calyx, 5- or 10-valved, with a seminiferous partition in the middle of each valve, therefore 5- or 10-celled. *Seeds* ovate, angular. *Embryo* filiform, spiral. *Leaves* simple, opposite, exstipulate, sub-evergreen, entire or toothed; the petioles embracing the stem. *Flowers* axillary, or many flowered peduncles; large, beautiful, resembling a rose, red or white.—Shrubs or subshrubs, natives of the South of Europe and North of Africa. Only two or three species or subspecies, and their varieties, are hardy in the climate of London.

1. CI'STUS PURPU'REUS *Lam.* The purple-*flowered* Cistus, *or Rock Rose.*

Identification. Lam. Dict., 2. p. 14.; Dec. Prod., 1. p. 264.; Don's Mill., 1. p. 298.
Synonymes. C. créticus *Hort. Kew.*; the purple Gum Cistus. the purple shrubby Cistus; Ciste pourpre, *Fr.*; purpurrothe Cisten Rose, *Ger.*
Engravings. Bot. Reg., t. 408.; Swt. Cist., t. 17.; and our *fig.* 77.

77. Cistus purpureus.

Spec. Char., &c. Leaves oblong-lanceolate, obtuse or acute, and more or less rugose; reticulately veined, with undulated margins. Petioles short, hairy, connected at the base, and sheathing the stem. Flowers terminal, from 1 to 6, on short peduncles. Bracteas sessile, leaf-like, pubescent, broad and concave at the base, where they are connected, and terminating in acute points. Pedicels short, and with the calyx hairy. Calyx of 5 sepals. Petals 5 or 6, obovate or wedge-shaped; very much imbricate, more or less crumpled. Stamens numerous, filaments smooth. Style very short; and stigma large, capitate, 5-lobed, papillose. (*Swt. Cist.*) A sub-evergreen low bush. Levant. Height 3 ft. to 4 ft. Introduced in 1659. Flowers large, bright reddish purple, with a yellow spot at the base. June and July. Capsule brown; ripe in Sept. Decaying leaves brown.

Branches numerous, erect, and clothed with a brownish pubescence. The flowers are very large and handsome, of a bright reddish purple, with a yellow spot at the base, above which is a large dark purple velvet mark, surrounded with red, and slightly branched. The petals are imbricate, and much crumpled. It flowers abundantly in June and July; grows very fast, is easily propagated by cuttings, and is very ornamental.

2. C. INCA'NUS *L.* The hoary Cistus, *or Rock Rose.*

Identification. Lin. Sp., 737.; Smith's Fl. Græc., 494.; Don's Mill., 1. p. 298.
Synonymes. C. álbidus *Hort.*; C. cymòsus *Dec.*; Ciste cotonneux, *Fr.*; bestaubte Cisten Rose, *Ger.*
Engravings. Bot. Mag., t. 43.; Swt. Cist., t. 44.; and our *fig.* 78.

78. Cistus incànus.

Spec. Char., &c. Leaves spathulate, tomentose, wrinkled, somewhat 3-nerved, sessile, somewhat connate at the base, upper ones narrower. Peduncles 1—3-flowered. (*Don's Mill.*) A hoary evergreen shrub. Spain and France. Height 2 ft. to 3 ft. Introduced in 1597. Flowers reddish purple, with the petioles emarginate. June and July. Capsule brown; ripe in September. Decaying leaves brown.

Quite hardy in dry soil.

■ 3. *C.* CORBARIE'NSIS *Pourr.* The Corbières Cistus, *or Rock Rose.*

Identification. Dec. Prod., 1. p. 265.; Don's Mill., 1. p. 299.
Synonymes. *C. salviæfòlius β Dec. Prod. Fl. Fr.*, 4. p. 813.; *C.* populifòlius minor, in some nurseries; *C. hybridus Pourr.*, not of Vahl.
Engravings. Swt. Cist., t. 8.; and our *fig.* 79.

Spec. Char. &c. Leaves stalked, somewhat cordate, ovate, acuminated, with fringed margins, wrinkled on both surfaces, and very glutinous. Peduncles long, 1—3-flowered. (*Don's Mill.*) A sub-evergreen shrub. South of France, on the mountains of Corbières; and also in Spain. Height 2 ft. to 3 ft. Introduced in 1656. Flowers white; May and June. Capsule brown; ripe in August.

79. C. corbariénsis.

A handsome plant, and, according to Sweet, one of the hardiest species of the genus, thriving well in common garden soil, and in any situation where it is not too moist. It continues in bloom for about two months; and every day during that period the plant is covered with a profusion of handsome white flowers, the margins of which are tinged with rose colour. The rose-coloured buds are also very pretty before the flowers expand.

■ 4. *C.* POPULIFOLIUS *L.* The Poplar-leaved Cistus, *or Rock Rose.*

Identification. Lin. Sp. 736.; Don's Mill., 1. p. 300.
Synonymes. Cistus populifòlius *Cav. Icon.*; Ciste à Feuilles de Peuplier, *Fr.*; Pappel-blättrige Cisten Rose, *Ger.*
Engravings. Swt. Cist. 23.; and our *fig.* 80.

Spec. Char., &c. Leaves stalked, cordate, acuminate, wrinkled, smooth. Flowers cymose. Peduncles bracteate. Bracteas oblong. Sepals acuminate, clammy. (*Don's Mill.*) A sub-evergreen shrub of vigorous growth. France and Spain. Height 5 ft. to 7 ft. Introduced in 1656. Flowers white, with distinct petals; May to July. Capsule brown; ripe in September.

Leaves dark green, cordate, clammy, with undulate margins. One of the most robust species of the genus, and also one of the hardiest. A plant 7 ft. high, in the grounds at Syon, stood through the winter of 1837–8 uninjured, without the slightest protection.

80. Cistus populifòlius.

■ 5. *C.* LAURIFO'LIUS *L.* The Laurel-leaved Cistus, *or Rock Rose.*

Identification. Lin. Sp. 736.; Clus. Hist. 1. p. 78. f. 1.; Don's Mill., 1. p. 300.
Synonymes. Ciste à Feuilles de Laurier. *Fr.*; Lorbeer-blättrige Cisten Rose, *Ger.*
Engravings. Clus. Hist., 1. p. 78. f. 1.; Swt. Cist., t. 52.; and our *fig.* 81.

Spec. Char., &c. Leaves stalked, ovate-lanceolate, 3-nerved, upper surface glabrous, under surface tomentose. Footstalks dilated, and connate at the base. Capsules 5-celled. (*Don's Mill.*) A sub-evergreen bush. South of France, and Spain. Height 4 ft. to 5 ft. Introduced in 1771. Flowers large, white, with light red bracteas; July and Aug. Capsule brown; ripe in October.

81. Cistus laurifòlius.

A very robust species, with large green laurel-like leaves. It produces an abundance of flowers, which, with their light red bracteas, are very ornamental before they expand, resembling, at a distance, the bursting buds of roses. It requires no protection; and may be raised from seeds, which it ripens in abundance; and also by cuttings, which, however, do not strike so freely as in some of the other species.

■ 6. *C.* LADANI'FERUS *L.* The Ladanum-bearing *Gum* Cistus, *or Rock Rose.*

Identification. Lin. Sp. 737.; Don's Mill., 1 p. 300.
Synonymes. Ciste ladanifère, *Fr.*; Ladanum Cisten Rose, *Ger.*; Ladano, *Ital.*

Spec. Char., &c. Leaves almost sessile, connate at the base, linear-lanceolate, 3-nerved, upper surface glabrous, under surface tomentose. Capsule 10-celled. Petals imbricate. (*Don's Mill.*) A sub-evergreen shrub. Spain and Portugal, on hills. Height 4 ft. to 5 ft. Introduced in 1629. Flowers large, white, 1 in. to 2 in. broad; June and July. Capsule brown; ripe in September.

82. C. ladanif. maculàtus.

Varieties. C. *l.* 1 *albiflòrus* Dec. Prod. i. p. 266., Swt. Cist. t. 94.; Lèdon, i., *Clus. Hist.* i. p. 78. ic.; and C. *l.* 2 *maculàtus* Dec. Prod. l. c. (Swt. Cist.; and our *fig.* 82.); *C.* l. 3 plenifòlius *Ait. Hort. Kew.* iii. p. 305.; are varieties of this species.

The leaves are lanceolate, and nearly sessile, of a deep green; the flowers terminating the branches, solitary, white, and large. The variety *C.* ladaníferus maculàtus, which our *fig.* 82. represents, is a plant of very great beauty, and no collection should be without it.

■ 7. *C.* (v.) CY'PRIUS *Lam.* The *Gum* Cistus, *or* Cyprus *Rock Rose.*

Identification. Lam. Dict., 2. p. 16.; Don's Mill., 1. p. 300.
Synonymes. C. ladaniferus *Bot. Mag.* t. 112.; Cistus stenophýllus *Link Enum.* 2. p. 74.?; *C.* salicifòlius of some.
Engravings. Swt. Cist., t. 39.; and our *fig.* 83.

Spec. Char., &c. Leaves stalked, oblong-lanceolate, upper surface glabrous, under surface clothed with hoary tomentum. Peduncles generally many-flowered. Petals spotted. Capsules 5-celled. (*Don's Mill.*) A splendid sub-evergreen shrub. Island of Cyprus. Height 5 ft. to 6 ft. Introduced in 1800. Flowers white, 2 in. to 3 in. across, imbricated, each petal having a dark rich brownish crimson spot at the base; June and July. Capsule brown; ripe in September.

83. Cistus cýprius.

One of the handsomest species of the genus, and so closely resembling *C.* ladaníferus, as, in our opinion, to be nothing more than a variety of that species. Young cuttings, Sweet observes, planted under hand-glasses in autumn, will strike root; but the best way is to raise them from layers or from seed. There was, in 1834, a plant of this species at Minard, in Argyllshire, 7 ft. 9 in. high, with a head 12 ft. in diameter, which is clothed with flowers every year.

Other Species of Cistus are described in Sweet's *Cistineæ*, as nearly equally hardy with the above; but the experience of the winter of 1837-8 has induced us to omit them. Those who intend to treat them as garden plants, and can afford them a little protection during winter, will find 36 species, besides varieties, described in the first edition of this work, and several of them figured. Those who intend only to have a collection of showy species, without much regard to their names, will have recourse to the mode recommended in the concluding paragraph on the Heliánthemum. (p. 61.) The following species or subspecies were found tolerably hardy in the Canterbury Nursery: *C.* heterophýllus, *C.* créticus, *C.* críspus, *C.* Cupaniànus, *C.* hirsùtus, *C.* láxus, *C.* villòsus, *C.* oblongifòlius, *C.* undulàtus, *C.* salviæfòlius, *C.* longifòlius, *C.* psilosépalus.

Genus II.

HELIA'NTHEMUM. The Helianthemum, *or Sun Rose*. *Lin. Syst.* Polyándria Monogýnia.

Identification. Tourn. Inst., 248. t. 128.; Dec. Prod., 1. p. 266.; Don's Mill., 1. p. 301.
Synonymes. Cisti species of Lin.; Helianthème Sonnen Gurtel, *Ger.*; Eliantemo, *Ital.*
Derivation. From *hēlios*, the sun, and *anthemon*, a flower; because the flowers open with the rising of the sun in the morning, and the petals fall off with the setting of the sun in the evening. The flowers of Heliánthemum, as well as of Cistus, only last for a few hours when the sun shines; and if the weather is dull, and the sun does not make its appearance, the flowers do not open, but remain unexpanded. Should this continue for several days together, they will decay in the bud.

Gen. Char. *Calyx* of 3—5 sepals; the two outer usually smaller than the inner ones, rarely larger. *Petals* 5, usually regularly denticulated at the top. *Stigma* capitate. *Style* sometimes almost wanting, sometimes straight, sometimes oblique, and sometimes bent at the base. *Ovary* triquetrous. *Capsule* 3-valved. *Seeds* angular, smooth.

Leaves simple, opposite or alternate, stipulate or exstipulate, sub-evergreen; 3-nerved or feather-nerved. *Flowers* axillary or terminal. *Pedicels* usually furnished with bracteas at the base. — Shrubs or subshrubs, low, prostrate, resembling herbaceous plants; natives of Europe; and of the easiest culture in any common soil.

♄ 1. H. vulga're *Gært.* The common Helianthemum, *or Sun Rose.*

Identification. Gært Fruct., 1. p. 371. t. 76.; Don's Mill., 1. p. 311.
Synonyme. Cistus Heliánthemum *Lin. Sp.* 1. p. 744., *Sm. Engl. Bot.* 1321., *Curt. Fl. Lond.* 5. t. 36.
Engravings Eng. Bot., t. 1321.; and our *fig.* 84.

Spec. Char., &c. Stem suffruticose, procumbent, branched, branches elongated. Leaves scarcely revolute at the margins; under surface cinereously hoary; upper surface green, pilose, somewhat ciliated. Lower leaves somewhat orbicular, middle ones ovate-elliptical, upper ones oblong. Stipules oblong-linear, ciliated, longer than the footstalks of the leaves. Racemes loose. Pedicels and calyxes pilose. (*Don's Mill.*) A procumbent sub-evergreen undershrub. Europe and Britain, in dry and hilly pastures. Height 6 in. to 1 ft. Flowers yellow; May to September.

84. H. vulgàre.

Varieties. There is a very handsome double-flowered variety, with pale yellow flowers; and another, called Lee's new double yellow, with dark yellow flowers; both of which are in general cultivation in the nurseries. De Candolle also notices two forms of the species: one with tomentose pubescent branches, and stipules scarcely longer than the footstalks of the leaves; and another with branches glabrous at the base, but pubescent upwards, and the stipules twice or thrice the length of the petioles of the leaves.

The stamens, if touched during sunshine, spread slowly, and lie down upon the petals. (*Smith.*) "A very variable species," even in a wild state; and, in gardens, many beautiful varieties, single and double, have been originated from it by cross fecundation with Continental helianthemums, and perhaps with cistuses.

♄ 2. H. (v.) surreja'num *Mill.* The Surrey Helianthemum, *or Sun Rose.*

Identification. Mill. Dict., No. 15.; Swt. Cist., t. 28.; Don's Mill., 1. p. 311.
Synonyme. Cistus surrejànus *Lin. Sp.* 743., *Smith's Eng. Bot.*
Engravings. Dill. Elth., 177. t. 145. f. 174.; Swt. Cist., t. 28.; Eng. Bot., t. 2207.; and our *fig.* 85.

Spec. Char., &c. Stem suffruticose, procumbent. Leaves ovate-oblong, rather pilose. Racemes many-flowered, terminal. Pe-

85. Heliánthemum surrejànum.

tals narrow, lanceolate, jagged. (*Don's Mill.*) A procumbent, sub-evergreen, suffruticose, undershrub. Britain, in Surrey, near Croydon. Height 6 in. to 1 ft. Flowers yellow, with the petals distinct, and the calyxes pilose; July to October.

♄ 3. H. (v.) SERPYLLIFO'LIUM *Mill.* The Wild-Thyme-leaved Helianthemum, *or Sun Rose.*

Identification. Mill. Dict. No 8.; Don's Mill., 1. p. 311.
Synonyme. Cistus serpyllifòlius *Lin. Sp.* 743.
Engravings. Swt. Cist., t. 60.; and our *fig.* 86.

86. H. serpyllifolium.

Spec. Char., &c. Stem suffruticose. Branches ascendent, glabrous at the base, and pilose at the apex. Leaves oblong-elliptical, with revolute margins; under surface hoary-tomentose; upper surface intensely green, shining, at first rather pilose, afterwards almost smooth. Stipules and bracteas green, ciliated. Calyxes canescent, with inconspicuous down, and with the nerves sparingly pilose. (*Don's Mill.*) A trailing, sub-evergreen, suffruticose undershrub. Alps of Styria and Austria, and in Britain, in Somersetshire. Height 6 in. to 1 ft. Introduced in 1731; discovered in England in 1826. Flowers large, yellow, with the petals distinct; May to September.

♄ 4. H. (v.) GRANDIFLO'RUM *Dec.* The large-flowered Helianthemum, *or Sun Rose.*

Identification. Dec. Fl. Fr., 4. p. 821.; Don's Mill., 1. p. 311.
Synonyme. Cistus grandiflòrus *Scop. Carn.* ed. 2. No. 648. t. 26.
Engravings. Swt. Cist., t. 69.; Scop. Carn., ed. 2. t. 25.; and our *fig.* 87.

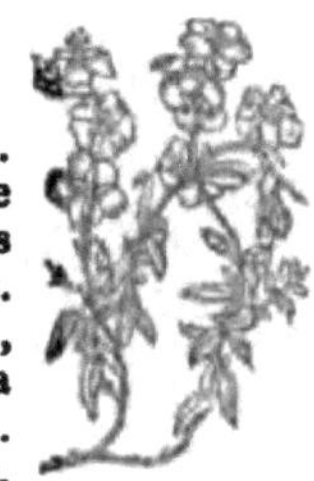
87. H. grandiflòrum.

Spec. Char., &c. Stem suffruticose, ascending. Branches hairy. Upper leaves flattish, oblong, rather pilose; upper surface green, under surface sometimes pale cinereous. Stipules ciliated, rather longer than the footstalks of the leaves. Flowers large. Calyxes rather hairy. (*Don's Mill.*) A neat, little, bushy, sub-evergreen, suffruticose undershrub, bearing a close resemblance to H. vulgàre, but larger in all its parts. Pyrenees. Height 6 in. to 1 ft. Introduced in 1800. Flowers large, pale yellow; June to August.

♄ 5. H. (v.) TAU'RICUM *Fisch.* The Taurian Helianthemum, *or Sun Rose.*

Identification. Fisch. MSS.; Don's Mill., 1. p. 312.
Engravings. Swt. Cist., 105.; and our *fig.* 88.

88. H. táuricum.

Spec. Char., &c. Stem suffruticose, much branched, procumbent. Branches procumbent, beset with long hairs. Leaves oblong-lanceolate, with rather revolute margins, pilose on both surfaces, green above, and paler beneath. Stipules lanceolate-linear, ciliated, longer than the petiole. Flowers large. Calyx shining, rather hairy. Petals imbricate. (*Don's Mill.*) A procumbent, sub-evergreen, suffruticose undershrub. Tauria. Height 6 in. to 1 ft. Introduced in 1820. Flowers large, pale yellow; May to October. Resembles H. grandiflòrum; but differs from it in the branches spreading flat on the ground, and extending to a great distance.

♄ 6. H. (v.) APENNI'NUM *Dec.* The Apennine Helianthemum, *or Sun Rose.*

Identification. Dec. Fl. Fr., 4. p. 824; Don's Mill., 1. p. 313.
Synonymes. Cistus apennìnus *Lin. Sp.* 744.?, *Dill. Elth.* 170.; Cistus híspidus β *Lam. Dict.* 2. p. 26.; Erba bottoncina, *Ital.*
Engravings. Swt. Cist., t. 62.; and our *fig.* 89.

89. H. apennìnum.

Spec. Char., &c. Stem suffruticose, branched. Branches spreading, hoary tomentose. Leaves stalked, oblong linear, with the margins scarcely revolute; under surface tomentose, upper surface glaucescent, but at length becoming smooth. Stipules awl-shaped, longer than the footstalks of the leaves. Calyxes covered with very short hairs, striated, cinereously glaucous, bluntish. (*Don's Mill.*) A procumbent, sub-evergreen, suffruticose, undershrub. Germany and Italy, on dry hills and places exposed to the sun. Height 6 in. to 1 ft. Introduced in 1731. Flowers white, with the petals distinct, and marked with yellow at their base; May to August.

ꝉ 7. H. (v.) MACRA'NTHUM *Swt.* The large-flowered Helianthemum, *or Sun Rose.*

Identification. Don's Mill., 1. p. 313.
Engravings. Swt. Cist., t. 103.; and our *fig.* 90.

90. H. macránthum.

Spec. Char., &c. Stem suffruticose. Branches procumbent, rather tomentose. Leaves flat, ovate oblong, acutish; smooth above, and densely tomentose beneath; pale cinereous. Stipules rather pilose; about equal to, or longer than, the petioles. Calyx striated, pilose. Petals distinct. (*Don's Mill.*) A procumbent, sub-evergreen, suffruticose undershrub. Originated in gardens. Height 6 in. to 1 ft. Cultivated in 1828. Flowers larger than in any other species; cream-coloured white, with the petals distinct, spreading, and very slender at the base, where they are marked with yellow; May to August.

91. H. m. múltiplex.

Variety.

ꝉ *H. m.* 2 *múltiplex* (Swt. Cist., t. 104.; and our *fig.* 91.) is a beautiful plant, not only on account of its fine double flowers, but of its habit of growth. It ought to be in every cistacetum.

ꝉ 8. H. (v.) CANE'SCENS *Swt.* The canescent-*leaved* Helianthemum, *or Sun Rose.*

Identification. Swt. Cist., t. 51.; Don's Mill., 1. p. 313.
Engravings. Swt. Cist., t. 51.; and our *fig.* 92.

Spec. Char., &c. Stem suffruticose, branched, diffuse. Branches ascending, rather tomentose, canescent. Leaves flat, or hardly revolute at the margins; under surface tomentosely hoary, upper surface greenish glaucous. Lower leaves ovate oblong, obtuse; upper ones lanceolate, acute. Stipules linear, ciliated, somewhat longer than the footstalks. Calyxes smoothish, but with the nerves pubescent. Petals imbricated. (*Don's Mill.*) A procumbent, sub-evergreen, suffruticose undershrub. Originated in gardens. Height 6 in. to 1 foot. Cultivated in 1826. Flowers reddish crimson; May to August.

A splendid plant, with reddish crimson flowers; the petals imbricated, and having a small orange spot at the base of each. Sweet considers it as having the darkest-coloured, if not the handsomest, flowers of the genus. The flowers are also, he says, very large for the size of the plant. It is nearly related to H. rhodánthum, but is readily distinguished from it by its canescent leaves, and stronger habit of growth.

92. H. canéscens.

ꝉ 9. H. (v.) HYSSOPIFO'LIUM *Tenore.* The Hyssop-leaved Helianthemum, *or Sun Rose.*

Identification. Tenor. Syn. Fl. Neap., p. 48.; Don's Mill., 1. p. 314.

Spec. Char., &c. Stem suffruticose, ascending. Branches hairy-tomentose. Lower leaves oval; upper ones oblong-lanceolate, green on both surfaces, flat, hairy. Calyxes hairy. Petals imbricate. (*Don's Mill.*) A sub-evergreen suffruticose undershrub, with the branches ascending. Originated in gardens. Height 1 ft. to 1½ ft. Cultivated in 1827. Flowers variable from saffron to red; May to August. Capsule brown; ripe in September. Varying considerably by culture.

Varieties.

- ♄ *H.* h. 1 *crocàtum* (Swt. Cist., t. 92.) has flowers saffron-coloured, with more or less of a ferrugineous tint, and may represent the species.
- ♄ *H.* h. 2 *cùpreum* (Swt. Cist., t. 58.) has flowers of a reddish copper-colour.
- ♄ *H.* h. 3 *múltiplex* (Swt. Cist., t. 72.; and our *fig.* 93.) has double flowers, of a reddish copper-colour.

93. H. (v.) hyssopifòlium.

All the three forms of this variety are splendid plants; they are hardy, of luxuriant growth, flowering freely, and of the easiest culture, either in pots or on banks of light sandy soil, covered with flints or stones. The flowers of the copper-coloured variety, and also the leaves, are larger than those of the two other kinds. The double-flowered variety appears to be of a more upright habit of growth, and not quite so robust as the others.

♄ 10. H. SCABRO'SUM *Pers.* The rough Helianthemum, *or Sun Rose.*

Identification. Pers. Ench., 2. p. 76.; Brot. Fl. Lus., 2. p. 265.; Don's Mill., 1. p. 302.
Synonyme. Cistus scabrosus *Ait. Hort. Kew.* 2. p. 236.
Engravings. Swt. Cist., t. 81.; and our *fig.* 94.

94. H. scabròsum.

Spec. Char., &c. Erectish. Branches pilosely tomentose, scabrous, canescent. Leaves sessile, tapering to the base, oblong-ovate, acutish, roughish, 3-nerved, with waved revolute margins; upper surface green, under surface clothed with grey tomentum. Peduncles terminal, 1—2-flowered, shorter than the leaves. Calyx 3-sepaled, hairy. Petals distinct. (*Don's Mill.*) A sub-evergreen erect undershrub. Portugal. Height 2 ft. to 3 ft. Introduced in 1775. Flowers rich yellow; June to August. Capsule brown; ripe in Sept.

Differs from most other species of Heliánthemum, in growing erect. It forms a very handsome little bush, but it requires a sheltered situation, and a dry soil.

Many other Kinds of Helianthemums described in Sweet's *Cistineæ* are, perhaps, as hardy as those we have selected; but it would be of little use giving them here, the greater part having been lost during the winter of 1837–8. In the first edition of this *Arboretum* 99 species are described, besides varieties. Supposing a cultivator about to form a collection of Cistàceæ, we should attach much less importance to his being able to procure all the sorts of Cistus and Heliánthemum described in Sweet's *Cistineæ*, than to his obtaining all the sorts easily procurable, whatever names they might pass under, and cross-fecundating them so as to produce new forms. There can be no doubt whatever that the sorts of both the genera Cistus and Heliánthemum might, by cross-fecundation, be increased *ad infinitum*; and, considering their very great beauty as border and rockwork shrubs, we think they merit the attention of cultivators at least as much as many florist's flowers.

SECTION III.

Ovarium solitary; Placenta central. (The Column in the Fruit to which the Seeds are attached central, and not adhering to the Side as in Section II.)

ORDER IX. *MALVA'CEÆ.*

ORD. CHAR. *Calyx* with a valvate æstivation, mostly with an involucre. *Stamens* with the filaments monadelphous, and the anthers 1-celled. *Pubescence* starry. (*Lindl.*)—Trees or shrubs, deciduous, natives of warm climates.

Leaves simple, alternate, stipulate, deciduous; more or less divided. *Hairs* stellate, axillary. *Flowers* on peduncles, large, showy.—The only genus containing hardy species is *Hibíscus*. The genus Lavátera contains some species which have an arborescent appearance, but which are in fact only suffrutescent biennials or triennials.

GENUS I.

HIBI'SCUS *L.* THE HIBISCUS. *Lin. Syst.* Monadélphia Polyándria.

Identification. Lin. Gen., 846.; Dec. Prod., 1. p. 446.; Don's Mill., 1. p. 476.
Synonymes. Ketmie, *Fr.*; Eibisch, *Ger.*; Ibisco, *Ital.*
Derivation. The word *hibiskos* is one of the names given by the Greeks to the mallow. The *Hibíscus* of Pliny appears to be an umbelliferous plant; while that of Virgil is a plant with pliant branches, which was made into baskets. The word *Hibiscus* is supposed by some to be derived from *ibis*, a stork, which is said to feed on some of the species. Ketmie (*Fr.*) is derived from *Kétmia*, the name given to the genus by Tournefort. Eibisch is the German aboriginal word for the mallow.

Gen. Char. *Calyx* encompassed by a many-leaved, rarely by a few-leaved, involucel, or one with its leaves connate. *Petals* not auricled. *Stigmas* 5. *Carpels* joined into a 5-celled 5-valved capsule, with a dissepiment in the middle of each valve on the inside. *Cells* many-seeded, rarely 1-seeded.

Leaves simple, alternate, stipulate, deciduous; variously lobed and toothed, generally ovate-wedge-shaped. *Flowers* pedunculate, large, showy. —The only hardy ligneous species is *H.* syrìacus.

☗ 1. *H.* SYRI'ACUS *L.* The Syrian Hibiscus, *or Althæa Frutex.*

Identification. Lin. Sp., 978.; Dec. Prod., 1. p. 448.; Don's Mill., 1. 478.
Synonymes. Ketmie des Jardins, *Fr.*; Syrischer Eibisch, *Ger.*
Derivation. It is called Althæa from the resemblance of its flowers to those of the *Althæ'a rósea*.
Engravings. Cav. Diss., 3. t. 69. f. 1.; Bot. Mag., t. 83.; and our *fig.* 95.

95. *Hibiscus syriacus.*

Spec. Char., &c. Stem unarmed, arboreous. Leaves ovate, wedge-shaped, 3-lobed, toothed. Pedicels hardly longer than the leaves. Involucel 6—7-leaved. (*Don's Mill.*) A deciduous shrub. Syria and Carniola. Height 6 ft. Introduced in 1596. Flowers large, single or double, white, red, purple, or variegated; August and September. Capsule greenish brown; ripe in October. Decaying leaves yellowish green. Naked wood white.

One of our most ornamental hardy shrubs, of which there are the following varieties:—

- ☗ H. *s.* 2 *fòliis variegàtis.*—Leaves variegated.
- ☗ H. *s.* 3 *flòre variegàto.*—Flowers variegated.
- ☗ H. *s.* 4 *flòre purpùreo.*—Flowers purple.
- ☗ H. *s.* 5 *flòre purpùreo plèno.*—Flowers double, purple.
- ☗ H. *s.* 6 *flòre rùbro.*—Flowers red.
- ☗ H. *s.* 7 *flòre álbo.*—Flowers white.
- ☗ H. *s.* 8 *flòre álbo plèno.*—Flowers double, white.

Branches numerous, upright, white-barked; their general character being rather fastigiate than spreading. Leaves variously lobed; flowers axillary, large, and bell-shaped. Conspicuously ornamental; and the more valuable, because it produces its flowers at a time of the year when few shrubs are in bloom. It forms beautiful garden hedges; more especially when the different sorts are planted in a harmonious order of succession, according to their colours; and when the plants are not clipped, but carefully pruned with the knife. In the colder parts of Britain, and in the north of Germany, few ornamental shrubs better deserve being planted against a wall. It will grow in almost any soil not too wet; but, like all the *Malvàceæ*, seems to prefer one which is sandy, deep, and rich, rather than poor. An open airy situation, where it will ripen its wood, is essential. The single-flowered varieties are propagated by seed, which come up true to their respective colours; and the double-flowered varieties are propagated by layers, by grafting on the common sorts, and sometimes by cuttings of the ripened wood, planted in sand in autumn, and covered with a hand-glass during the winter.

ORDER X. TILIA'CEÆ.

ORD. CHAR. *Sepals* 4 or 5, with a valvate æstivation, mostly without an involucre. *Petals* 4 or 5, or rarely not any. *Stamens* hypogynous, generally numerous, with filaments separate, and anthers 2-celled. Mostly *glands* between the petals and ovarium. *Ovary* and *fruit* single, of 4—10 carpels grown together; cells in the fruit, at least in some, not so many as the carpels. (*Lindley.*) — Trees and shrubs chiefly from warm climates.

Leaves simple, alternate, stipulate, deciduous; cordate. *Flowers* panicled, yellowish, fragrant, with an oblong bractea united to the common stalk. *Capsule* downy. — The only genus which is perfectly hardy is *Tilia*; native of Europe and North America.

GENUS I.

TI'LIA L. THE LIME TREE. *Lin. Syst.* Polyándria Monogýnia.

Identification. Lin. Gen., 660.; Dec. Prod., 1. p. 512.; Don's Mill., 1. p. 552.
Synonymes. Line Tree, *Gerard*; Lind, *Anglo-Sax.*; Tilleul, *Fr.*; Bast-holts, *Ger.*; Linde, *Ger.* and *Dutch*; Tiglia, *Ital.*; Tilo, *Span.*; Lipa, *Russ.*
Derivation. In London and Wise's *Retired Gardener* the name of *Tilia* is derived from the Greek word *ptilon*, a feather, from the feathery appearance of the bracteas; but others derive it from the Greek word *tilai*, light bodies floating in the air like wool or feathers. Tilleul is from *tailler*, either because the tree bears pruning well, or the wood may be easily carved. Bast-holts is literally bark wood, in allusion to the use of the bark, in forming mats.

Gen. Char. *Calyx* 5-parted. *Petals* 5. *Stamens* numerous, free, or somewhat polyadelphous. *Ovary* globose, villous, 1-styled, 5-celled; cells 2-ovuled. *Nut* coriaceous, 1-celled, 1—2-seeded, from abortion. (*Don's Mill.*, i. p. 540.) — Timber trees, with a tough fibrous bark, large deciduous leaves, mellifluous flowers, and a remarkable bractea attached to the peduncle of each of the cymes of flowers. Natives of Europe and North America.

Leaves simple, alternate, stipulate, deciduous; cordate. — The species are three, according to some; and more than twice that number, according to others. Our opinion is, that they may be all included under two, *T.* europæ'a and *T.* americàna.

¶ 1. *T.* EUROPÆ'A *L.* The European, *or common*, Lime Tree.

Identification. Lin. Sp. 733.; Don's Mill., 1. p. 552.; Smith's Eng. Fl., 3. p. 16.
Synonymes. *T.* intermèdia *Dec. Prod.* 1. 513.; or *T.* vulgàris *Hayne Dend.*; *T.* europæ'a borèalis *Wall.*

96. Tília europæ'a.

Spec. Char., &c. Petals without scales. Leaves cordate, acuminated, serrated, smooth, except a tuft of hair at the origin of the veins beneath, twice the length of the petioles. Cymes many-flowered. Fruit coriaceous, downy. (*Don's Mill.*) A large deciduous tree. Europe, and Britain in some aboriginal woods. Height 60 ft. to 90 ft. Flowers yellowish white; August and September. Fruit yellow; ripe in October. Decaying leaves yellow, or yellowish brown. Naked young wood reddish, or yellowish brown.

Varieties. The extensive distribution and long cultivation of this tree in Europe have given rise to the following varieties, or races, described by most botanists as species:—

A. *Varieties differing in respect to Foliage.*

♈ T. *e.* 1 *parvifòlia.* *T.* microphýlla *Vent., Willd., Dec.,* and *G. Don*; *T.* e. var. γ *L.*; *T.* ulmifòlia *Scop.*; *T.* sylvéstris *Desf.*; *T.* parvifòlia *Ehrh., Hayne Dend.*; *T.* cordàta *Mill.*; Tilleul à petites Feuilles *Fr.*; kleinblättrige Linde, or Winterlinde, *Ger.* (Willd. Holzart, t. 106.; Engl. Bot., t. 1705.; and our *fig.* 97.) — Leaves cordate, roundish, acuminated, sharply serrated; smooth above, glaucous and bearded beneath on the axils of the veins, as well as in hairy blotches. Fruit rather globose, hardly ribbed, very thin and brittle. Native of Europe, in sub-mountainous woods; in England, frequent in Essex and Sussex. This variety is distinguishable, at first sight, from all the others, by the smallness of its leaves, which are only about 2 in. broad, and sometimes scarcely longer than their slender footstalks. The flowers are also much smaller than in any of the other varieties; they expand later; and they are very fragrant, having a scent like those of the honeysuckle There was, in 1834, a subvariety of this in the garden of the Hort Soc., under the name of *T.* parvifòlia glaúca.

97. T. microphýlla.

♈ T. *e.* 2 *grandifòlia.* *T.* platyphýlla *Scop.*; *T.* cordifòlia *Bess.*; *T* europæ'a *Desf.*; *T.* grandifòlia *Ehrh.* and *Smith*; broad-leaved downy Lime Tree; Tilleul à grandes Feuilles, or Tilleul de Hol-

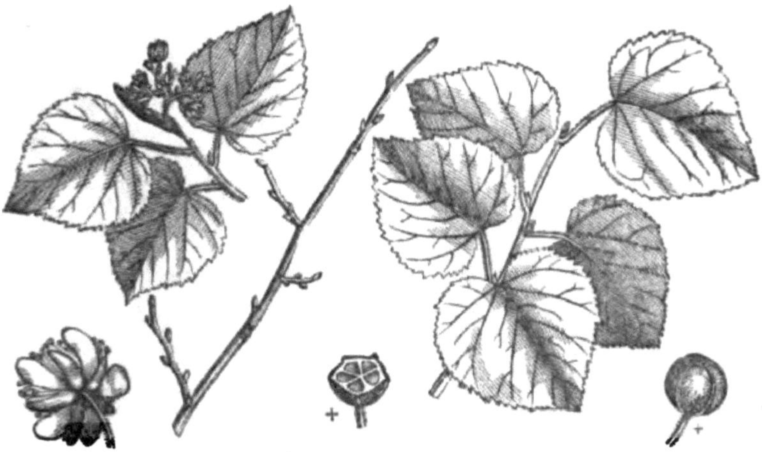

98. Tilia europæ'a platyphylla.

lande, *Fr.* (Vent. Diss., p. 6. t. 1. f. 2.; the plate in Arb. Brit., 1st edit., vol. v.; and our *fig.* 98.) — Leaves cordate, roundish, acuminated, sharply serrated, downy beneath; origin of their veins woolly. Branches hairy. Cymes 3-flowered. Fruit woody, downy, turbinate, with 5 prominent angles. This tree is readily distinguished from *T.* e. parvifòlia by its much larger and rougher leaves, and, also, by its rougher bark and hispid branches.

Ɨ T. *e.* 3 *intermèdia.* *T.* intermèdia *Hayne*; *T.* platyphýlla minor *Hort.* (The plate of this variety in Arb. Brit., 1st edit., vol. v.; and our *fig.* 99.) — Leaves intermediate between *T.* e. grandifòlia

99. Tilia europæ'a intermèdia.

and *T.* e. parvifòlia. This variety is the most common in Britain; *T.* e. grandifòlia in the South of Europe; and *T.* e. parvifòlia in the North of Europe, and especially in Sweden.

Ɨ T. *e.* 4 *laciniàta.* *T.* platyphýlla laciniàta *Hort.*; *T.* asplenifòlia nòva *Hort.* (The plate in Arb. Brit., 1st edition, vol. v.; and our *fig.* 100.) — Leaves deeply and irregularly cut and twisted, scarcely two on the tree being alike. Apparently a subvariety of *T.* e. parvifòlia. Height 20 or 30 feet.

100. Tilia europæ'a laciniàta.

B. *Varieties differing in the Colour of the young Shoots.*

Each of the varieties included in Division A may have subvarieties differing in the colour of the young wood; but we shall only notice those in general cultivation.

♀ T. *e.* 5 *rùbra.* *T.* corinthìaca *Bosc*; *T.* coràllina *Hort. Kew.*; *T.* europæ'a β rùbra *Sibthorp*; *T.* europæ'a γ *Smith's Flor. Brit.* 571.; *T.* grandifòlia β *Smith's Eng. Flora*, 3. 19. — Young shoots of a bright red; apparently a variety of *T.* e. intermèdia.

♀ T. *e.* 6 *parvifòlia aùrea.* (The plate in Arb. Brit., 1st. edit. vol. v.; and our *fig.* 101.) — Young shoots of a rich yellow. Height 15 ft. to 20 ft.

101. Tilia parvifòlia aurea.

♀ T. *e.* 7 *grandifòlia aùrea.* — Twigs of a fine yellow. Leaves large. Tree 60 ft. to 80 ft.

Other Varieties. There is a variety with variegated leaves, but it is such a ragged ill-looking plant that we deem it altogether unworthy of culture. Host, in his *Flora Austriaca*, has the following names, which he considers as species: *T. vitifòlia*; corylifòlia; grandifòlia *Sm.*; coràllina, syn. europæ'a *Hook. Lond.*; mutàbilis; làte bracteàta; præ'cox; pyramidàlis; intermèdia; tenuifòlia; oblìqua; europæ'a *Sm.*; parvifòlia *Sm.*; argéntea, syn. álba *Waldst. et Kit. Icon.* t. 3. (*Fl. Aus.*, vol. ii. p. 59—63.)

The wood of the lime tree is of a pale yellow or white, close-grained, soft, light, and smooth, and not attacked by insects. It weighs, per cubic foot, when green, 55 lb.; half-dry, 45 lb.; and dry, 37 lb.; and it loses a third part of its weight, and a fourth part of its bulk, by drying. (*Baudrill.*) It is used by pianoforte-makers for sounding-boards, and by cabinet-makers for a variety of purposes. It is carved into toys, and turned into domestic utensils of various kinds, and into small boxes for the apothecaries. The most elegant use to which it is applied is for ornamental carving, for which it is superior to every other wood. This wood is said to make excellent charcoal for gunpowder; even better than alder, and nearly as good as hazel. Baskets and cradles were formerly made from the twigs; and shoemakers and glovers are said to prefer planks of lime tree for cutting the finer kinds of leather upon. The leaves of the lime tree, in common with those of the elm and the poplar, were used, both in a dried and in a green state, for feeding cattle, by the Romans; and they are still collected for the same purpose in Sweden, Norway, Carniola, and Switzerland. One of the most important uses of the lime tree, in the North of Europe, is that of supplying material for ropes and bast mats; the latter of which enter extensively into European commerce. The Russian peasants weave the bark of the young shoots for the upper parts of their shoes, the outer bark serves for the soles; and they also make of it, tied together with strips of the inner bark, baskets and boxes for domestic purposes. The outer bark of old trees supplies them, like that of the birch, with tiles for covering their cottages. Ropes are still made from the bark of the tree in Cornwall, and in some parts of Devonshire. The fishermen of Sweden make nets for catching fish of the fibres of the inner bark, separated, by maceration, so as to form a kind of flax; and the shepherds of Carniola weave a coarse cloth of it, which serves them for their ordinary clothing. The sap of the lime tree, drawn off in spring, and evaporated, affords a considerable quantity of sugar. The honey produced by the flowers is considered superior to all other kinds for its delicacy. London and Wise recommended the lime tree, as preferable to the elm, for sheltering gardens or orchards; because the roots do not, like those of the elm, spread out and impoverish all around them. A deep and rather light soil is recommended; but the largest trees are generally found in a good loamy soil. In dry situations, the lime never attains a large size, and it loses its leaves earlier than any other tree. Being a tree of the plains, rather than of the mountains, it does not appear suitable for exposed surfaces: but it requires a pure air rather than otherwise; for, though, it is found in towns on the Continent, and sparingly so in Britain, the smoke of mineral coal seems more injurious to it than it is to the platanus, the elm or some other trees. It is seldom propagated otherwise than by layers, which are made in the nurseries in autumn and winter; and which become rooted, so as to admit of being taken off, in a year. The tree in Britain appears seldom to ripen its seeds.

♆ 2. *T.* (EUR.) A'LBA *Waldst. & Kit.* The white-*leaved* European Lime Tree.

Identification. Waldst. and Kit. Pl. Hung.; Don's Mill. 1. p. 553.
Synonymes. *T.* americàna *Du Roi*; *T.* argéntea *Desf.*, *Dec. Cat. Hort. Monsp.*, and *Dec. Prod.* 1. p. 513.; *T.* rotundifòlia *Vent.* and *N. Du Ham.*; *T.* tomentòsa *Mœnch.*
Engravings. Waldst. and Kit. Pl. Hung., 1. t. 3.; Wats. Dend., t. 71.; the plate in Arb. Brit., 1st edit., vol. v.; and our *fig.* 102.

Spec. Char., &c. Petals each with a scale at the base inside. Leaves cordate, somewhat acuminated, and rather unequal at the base, serrated, clothed with white down beneath, but smooth above, 4 times longer than the petioles. Fruit ovate, with 5 obscure ribs. (*Don's Mill.*) Fruit evidently ribbed.

102. *Tilia álba.*

(*Steven.*) Host says that he has always found the calyx 6-sepaled, and the corolla 12-petaled. A large tree. Hungary. Height 30 ft. to 50 ft. Introduced in 1767. Flowers yellowish white, very fragrant; June to August. Fruit yellow; ripe in October.

Our own opinion is, that this is nothing more than a very distinct race of the common lime; notwithstanding the circumstance of its having scales to its petals, which no one of the other varieties of *T.* europæ'a is said to possess. Even allowing this structure to be permanent in the Hungarian lime, the tree bears such a general resemblance to *T.* europæ'a in all its main features, that it seems to us impossible to doubt the identity of their origin. We are strengthened in this opinion by the circumstance of its being found only in isolated stations in the Hungarian forests. We have, however, placed this lime by itself, rather than among the other varieties; because, from the whiteness of its foliage, it is far more obviously distinct than *T.* e. grandifòlia or *T.* e. parvifòlia. The tree is at once distinguishable from all the other species and varieties by this white appearance, even at a considerable distance, and by the strikingly snowy hue of its leaves when they are ruffled by the wind. Its wood and shoots resemble those of the common lime; but it does not attain the same height as that tree.

♀ 3. *T.* AMERICA'NA *L.* The American Lime Tree.

Identification. Lin. Sp., 733.; Hort. Kew.; Willd. Spec.; Tor. and Gray.
Synonymes. *T.* glàbra *Vent.*; *T.* caroliniàna *Wangenh.*; *T.* canadénsis *Michaux*; *T.* glàbra *Dec.*, *Hayne's Dendr.*, and *Don's Mill.*; the smooth-leaved, or black, Lime Tree, and Bass Wood, *Amer.*
Engravings. Vent. Diss., t. 2; Wats. Dendr. Brit., t. 134.; the plate in Arb. Brit., 1st edit., vol. v.; and our *fig.* 105.

Spec. Char., &c. Petals each with a scale at the base, inside. Leaves profoundly cordate, abruptly acuminate, sharply serrated, somewhat coriaceous, smooth. Petals truncate and crenate at the apex, equal in length to the style. Fruit ovate, somewhat ribbed. (*Don's Mill.*) A large tree. Canada, Virginia, and Georgia. Height 70 ft. to 80 ft. in America; in England 60 ft. to 70 ft. Introduced in 1752. Flowers yellowish white; July and August. Fruit the size of a large pea, yellow; ripe in October. Decaying leaves yellowish brown. Naked young wood dark brown.

Varieties. Those which we shall give as such are described in the *Flora* of Torrey and Gray as species; but with the following remark, which we think fully justifies us in not considering them more specifically distinct than the different alleged species of the European lime. "There is great uncertainty respecting the synonymes. Indeed, nearly all the characters which have been employed for distinguishing them are either inconstant, or are common to them all. A careful examination of the flowers in the living plants may afford more certain marks of discrimination."

♆ T. *a.* 2 *heterophýlla.* *T.* heterophýlla *Vent.*, *Pursh*, *Dec. Prod.*, *Tor. & Gray.* — A tree of 30 ft. to 50 ft. high, found on the banks of the Ohio and Mississippi, and introduced in 1811. Leaves glabrous and deep green above, very white and velvety tomentose beneath; the veins dark-coloured, and nearly glabrous, with coarse mucronate serratures. Petals obtuse, crenulate. Staminodia spatulate, entire. Style hairy at the base. (*Tor. and Gray*, i. p. 240.) Leaves 4 in. to 8 in. in diameter, very oblique, and more or less cordate, with a short abrupt acumination; somewhat shining above: the veins on the under surface very conspicuous, in contrast with the white pubescence. Cyme few-flowered, loose. Style longer than the petals. (*Ibid.*)

♆ T. *a.* 3 *álba.* *T.* álba *Michx.*, *Tor. & Gray*; *T.* laxiflòra *Pursh.* (The plate of this tree in Arb. Brit., 1st ed., vol. v.; and our *fig.* 103.) — A

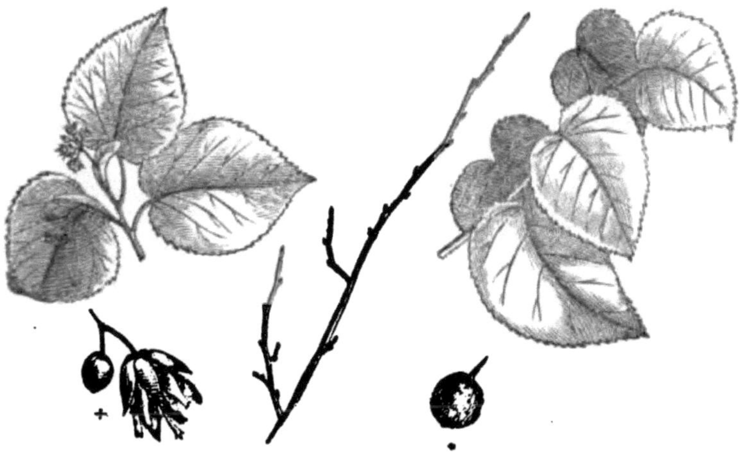

103. Tilia americàna álba.

large tree in Pennsylvania and Maryland. Height 30 ft. to 50 ft. Introduced in 1820. Flowers yellowish white, very pale; June. Leaves 3 in. to 4 in. in diameter, with a short abrupt acumination, cordate, somewhat unequal at the base; the under surface rather thinly pubescent, very pale, but scarcely white. Staminodia (scales) two thirds the length of the petals. Filaments slightly pentadelphous. (*Tor. and Gray*, i. p. 240.) We have only seen this tree in the H. S., where, 10 years planted, it was about 10 ft. high in 1837.

♆ T. *a.* 4 *pubéscens.* *T.* pubéscens *Ait.*, *Vent.*, *Tor. & Gray.* (The plate of this tree in Arb. Brit., 1st ed., vol. v.; and our *fig.* 104.) — A large tree found in Carolina and Florida, along the sea coast. Height 50 ft. to 70 ft. Introduced in 1726. Flowers pale yellow; June. Leaves 3 in. to 4 in. in diameter; the under surface, when young, rather paler than the upper, but at length nearly the same colour; serratures broad and short. (*Tor. and Gray*, i. p. 240.)

104. Tília americàna pubéscens.

On a general view of the trees, the most obvious external differential characteristics of the European and American limes appear to us to be, that the former have regularly cordate, and the latter obliquely cordate, leaves. The other American limes we consider to be nothing more than varieties of this species. Layers. These trees only thrive in warm sheltered situations. The American lime is readily distinguished from the European limes by the largeness of its leaves, which are 3 in. to 4 in. wide, heart-shaped, acutely

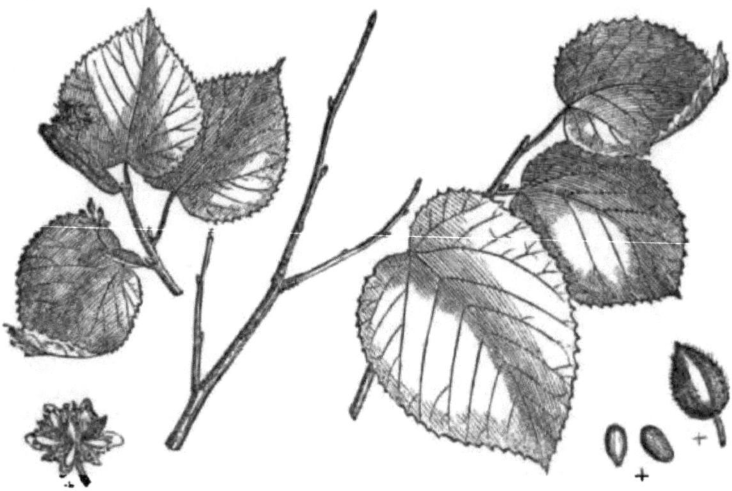

105. Tília americàna.

pointed, coarsely and mucronately serrated, deep green and glabrous on their upper sides, and pale green beneath. Some of them have a tendency to be slightly pubescent; but they are generally smooth and shining. In winter, this species is readily recognised by the robust appearance of the trunk and branches, and by the dark-brown colour of the bark on the young shoots. This circumstance alone is a very marked distinction; and has, no doubt, procured for the species the name of the black lime tree.

Other American Limes. — T. *p. leptophýlla* Vent. is mentioned by Torrey and Gray as having the leaves very thin and papyraceous. There were in 1834 other varieties in the H. S. G., but the plants had not attained sufficient size to enable us to state any thing with certainty respecting them. All the species and varieties of American limes are delicate in this country; they are readily injured by spring frosts, but, where they thrive, are readily distinguished from other limes by their very large cordate leaves, and rough bark, even on the young wood. Some of the oldest and largest American limes in England are in Gatton Park, Surrey, near the lake; and at Croome, near Upton upon Severn. There is an old tree, also, at Purser's Cross, Fulham; and a remarkably handsome young tree at White Knights.

Order XI. TERNSTRÖMIA'CEÆ.

ORD. CHAR. *Calyx* with an imbricate æstivation. *Stamens* with filaments monadelphous or polyadelphous, and anthers 2-celled to 4-celled. — Trees and shrubs of warm climates.

Leaves simple, alternate, exstipulate, deciduous or evergreen; mostly coriaceous, now and then with pellucid dots. *Flowers* axillary, large, showy. — The hardy genera in British gardens are Malachodéndron, Stuártia, and Gordònia; which are thus contradistinguished: —

MALACHODE'NDRON *Cav.* Calyx unibracteate. Styles free. Carpels capsular, 5, connected.

STUA'RTIA Cav. Sepals connected, bibracteate. Style 1. Capsule 5-celled, 5-valved. Seeds wingless.

GORDO'NIA Ellis. Sepals distinct. Style 1. Capsule 5-celled, 5-valved. Seeds winged.

Genus I.

MALACHODE'NDRON *Cav.* The Malachodendron. *Lin. Syst.* Monadélphia Polyándria.

Identification. Cav. Diss., 5. p. 502.; Dec. Prod., 1. p. 528.; Don's Mill., 1. p. 572.
Synonymes. Stuártia L'Hérit.; Stewártia L.
Derivation. From *malakos*, soft, and *dendron*, a tree; in allusion, perhaps, to the quality of the timber; or, possibly, from the flowers resembling those of the mallow, the Greek name for which is *malachē*.

Gen. Char. *Calyx* 5-cleft, furnished with two bracteas at the base. *Petals* 5, with a crenulate limb. *Ovary* 5-furrowed. *Styles* 5, unconnected. *Stigmas* capitate. *Carpels* 5, capsular, connected, 1-seeded. *Seeds* unknown. (*Don's Mill.*) — A deciduous shrub or low tree; native of North America.

Leaves simple, alternate, exstipulate, deciduous. *Flowers* axillary, solitary, large, showy.

1. MALACHODE'NDRON OVA'TUM *Cav.* The ovate-*leaved* Malachodendron.

Identification. Cav., l. c.; Lindl. Bot. Reg.; Don's Mill., 1. p. 272.
Synonymes. Stuártia pentagýnia *L'Hérit.*; Stewártia Malachodéndron *Mill.*; Stewartia à cinq Styles, *Fr.*
Engravings. Smith's Exot. Bot., t. 101.; N. Du Ham. vol. vi. t. 44.; Bot. Reg., t. 1104.; and our *fig.* 106. of a plant in flower, and *fig.* 107. of a shoot from a stool.

Spec. Char., &c. Leaves ovate, acuminated. Flowers axillary, solitary, almost sessile. Petals waved, cut, of a pale cream-colour. (*Don's Mill.*) A deciduous tree; in England, a large shrub. North Carolina to Georgia. Height

106. M. ovàtum.

107. M. ovàtum.

7 ft. to 12 ft. Introduced in 1795. Flowers arge, white; July and August. Capsules brownish; ripe in September. Decaying leaves purplish brown. Naked young wood dark brown.

The soil in which it is generally grown is a mixture of loam and peat, in which the latter prevails; but, in the Mile End Nursery, it shoots vigorously, and flowers freely, in deep sandy loam. The situation should be sheltered; and shaded rather than otherwise. The usual mode of propagation is by layers; and the stools are sometimes protected, during winter, by mats.

Genus II.

STUA'RT*IA* Cav. The Stuartia. *Lin. Syst.* Monadélphia Pentándria.

Identification. Cav. Diss., 5. p. 393.; Dec. Prod., 1. p. 528.; Don's Mill., 1. p. 573.
Derivation. Named in honour of *John Stuart*, Marquess of Bute, the patron of Sir John Hill, and a distinguished promoter of botanical science.

Gen. Char. *Calyx* permanent, 5-cleft, rarely 5-parted, furnished with two bracteas at the base. *Petals* 5. *Ovary* roundish. *Style* 1, filiform, crowned by a capitate 5-lobed stigma. *Capsule* woody, 5-celled, 5-valved; cells 1—2-seeded. *Seeds* wingless, ovate, even. (*Don's Mill.*)

Leaves simple, alternate, exstipulate, deciduous; ovate, acute. *Flowers* axillary, solitary, large. — A deciduous shrub, or low tree, native of North America.

▲ 1. Stua'rt*ia* virgi'nica *Cav.* The Virginian Stuartia.

Identification. Dec. Prod., 1. p. 528.; Don's Mill., 1. p. 573.
Synonymes. Stewártia Malachodéndron *Lin. Sp.*, 982.; Stuártia marilándica *Bot. Rep.*; Stewartia à un Style, *Fr.*; eingriffliche (one-styled) Stuartie, *Ger.*
Engravings. Lam. Ill., t. 593.; Bot. Rep., t. 397.; and our *fig.* 108. of a plant in flower, and *fig.* 109. of a shoot from a stool.

108. Stuártia virgínica.

109. Stuártia virgínica.

Spec. Char., &c. Flowers large, white, with purple filaments and blue anthers, usually in pairs. Leaves ovate, acute. Petals entire. (*Don's Mill.*) A deciduous shrub. Virginia to Carolina, in swamps. Height 6 ft. to 8 ft.

Introduced in 1742. Flowers large, white; July to September. Capsules brownish; ripe in October. Decaying leaves reddish brown. Naked young wood dark brown.

The general appearance of the plant is the same as that of the preceding genus; but it forms a smaller bush, and the foliage has a redder hue. The flowers are of the same size, white, with crisped petals, purple filaments, and blue anthers. This plant is not so extensively cultivated as the other, from its being more tender, and of somewhat slower growth; but its beauty, and the circumstance of its flowering from July to September, when but few trees or shrubs are in blossom, render it desirable for every collection. It thrives best in a peat soil, kept moist; but it will also grow in deep moist sand. Layers.

GENUS III.

GORDO'NIA Ellis. THE GORDONIA. *Lin. Syst.* Monadélphia Polyándria.

Identification. Ellis, in Phil. Trans., 1770.; Dec. Prod. 1. p. 528.; Don's Mill., 1. p. 573.
Derivation. Named in honour of *James Gordon*, a celebrated nurseryman at Mile End, near London, who corresponded with Linnæus.

Gen. Char. *Calyx* of 5 rounded coriaceous sepals. *Petals* 5, somewhat adnate to the urceolus of the stamens. *Style* crowned by a peltate 5-lobed stigma. *Capsules* 5-celled, 5-valved; cells 2—4-seeded. *Seeds* ending in a leafy wing fixed to the central column, filiform. (*Don's Mill.*)

Leaves simple, alternate, exstipulate, sub-evergreen or deciduous; serrated or nearly entire. *Flowers* axillary (or terminal), solitary, large. — Trees or shrubs, sub-evergreen or deciduous; natives of North America.

■ ■ 1. GORDO'NIA LASIA'NTHUS *L.* The woolly-flowered Gordonia, *or Loblolly Bay.*

Identification. Lin. Mant., 570.; Dec. Prod., 1. p. 528.; Don's Mill., 1. p. 573.; Tor. and Gray, 1. p. 223.
Synonymes. Hypéricum Lasiánthus *Lin. Sp.* 1101., *Catesb. Carol.* 1. t. 44., *Pluk. Amalth.* t. 352.; Gordonia à Feuilles glabres, and Alcée de la Floride, *Fr.*; langstielige Gordonie, *Ger.*
Engravings. Cav. Diss., 6. t. 171.; Bot. Mag., t. 668.; and our *fig.* 110.

Spec. Char., &c. Pedicels axillary, usually shorter than the leaves. Leaves oblong, coriaceous, smooth, serrated. Calyx silky. Capsules conoid, acuminated. (*Don's Mill.*) A sub-evergreen tree; in England a shrub, deciduous in dry soils, but retaining its leaves in warm moist situations. Virginia to Florida, in swamps. Height 50 ft. to 80 ft. in America; 8 ft. to 10 ft. in England. Introduced in 1739. Flowers white, 4 in. across, scented; July and August. Capsule oval, brownish; ripe in September.

110. Gordònia Lasiánthus.

Trunk straight. This most beautifully flowering plant well deserves to have a suitable soil prepared for it, and to be treated with more care after it is planted than it appears to have hitherto received in England. The soil ought to be peat, or leaf-mould and sand; and it should be so circumstanced as always to be kept moist. Layers or American seeds.

■ 2. G. PUBE'SCENS *L'Her.* The pubescent Gordonia.

Identification. L'Her. Stirp., p. 156.; Dec. Prod., 1. p. 528.; Don's Mill., 1. p. 573.; Tor. and Gray, 1. p. 223.
Synonymes. *Lacathèa flórida Sal. Par. Lond.* t. 56.; Franklinía americàna *Marsh.*; the Franklinia, *Amer.*; behaarte Gordonie, *Ger.*
Engravings. Sal. Flor. Lond., t. 56.; Michx., t. 59.; and our *fig.* 111.

111. Gordònia pubéscens.

Spec. Char., &c. Flowers almost sessile. Leaves obovate-lanceolate, pubescent beneath, somewhat serrated, membranaceous. Petals and sepals rather silky on the outside. (*Don's Mill.*) A deciduous tree in America, in Britain a shrub. Georgia and Florida. Height in America 30 ft. to 50 ft.; in England 4 ft. to 6 ft. Introduced in 1774. Flowers large, white, fragrant, 3 in. across, with yellow filaments; May to August. Capsule globose, brownish; ripe in October.

Somewhat hardier than the preceding species, but requiring the same general treatment.

Order XII. *H*YPERICA'CEÆ.

Ord. Char. *Sepals* 4 or 5, unequal, with an imbricate æstivation. *Stamens*, in nearly all, numerous, and in 3 or more parcels. *Fruit*, a capsule or berry of many valves and many cells; the edges of the valves curved inwards. *Seeds* attached to a placenta in the axis, or on the inner edge of the dissepiments.

Leaves simple, opposite, exstipulate, deciduous or evergreen; entire, copiously dotted with immersed, pellucid, resinous glands. *Flowers* terminal or axillary, generally yellow. *Sap* yellow, resinous.— Shrubs, natives of Europe, North America, and Asia. The genera in British gardens are two, which are thus contradistinguished: —

Hype'ricum L. Capsule membranous. Stamens polyadelphous.
Androsæ'mum Chois. Capsule baccate. Stamens monadelphous.

Genus I.

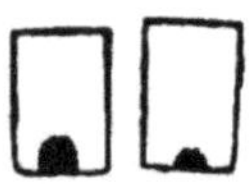

*H*YPE'RICUM *L.* The St. John's Wort. *Lin. Syst.* Polyadélph. Polyánd.

Identification. Lin. Gen., 892.; Juss., 255.; Dec. Prod., 1. p. 543.; Don's Mill., 1. p. 601.
Synonymes. Fuga Dæmonum; Mille Pertuis, *Fr.*; Johanniskraut, *Ger.*; Iperico, *Ital.*
Derivations. The name of *Hypéricum* is as old as the time of Dioscorides; but its origin and meaning are uncertain. Some derive it from the Greek words *huper*, under, and *eikōn*, an image; and suppose it to signify that the upper part of the flower represents a figure. Others state that *huper* signifies through, and that the name alludes to the pellucid dots in the leaves, which form small lenses, through which, when held up to the light, images might be seen. The French name of the plant, Mille Pertuis, a thousand pores, is evidently derived from the same source. The English name, St. John's Wort, and the German one, Johanniskraut, are taken from the country people formerly, both in England and Germany, being in the habit of gathering this plant on St. John's day, to use it to protect themselves from evil spirits. This plant, with some others, was employed to make what was called John's fire, which was supposed to be a security, for those who kindled it, against witchcraft and all attacks of demons. For this reason, also, the *Hypéricum* received the name of Fuga Dæmonum.

Gen. Char. *Capsules* membranous. *Stamens* numerous, free or joined at the bases into 3 or 5 bundles. *Petals* 5. *Sepals* 5, more or less connected at the base, unequal, rarely equal. *Styles* 3 to 5, rarely connate in one, permanent. *Capsule* 1- or many-celled, many-seeded, 3—5-valved. Integument of *seed* double. *Embryo* with the radicle situated at the umbilicus, and with semicylindrical cotyledons. (*Don's Mill.*)

Leaves simple, opposite, exstipulate, sub-evergreen or deciduous; ovate-oblong or lanceolate, sessile or subsessile, usually full of pellucid dots on their disks, and some dark ones on their edges, lodging an essential oil. *Flowers* terminal, racemose, yellow. — Low sub-evergreen shrubs; natives of Europe, North America, and Asia; of easy culture in common soil; and propagated by division, suckers, cuttings, or seeds.

§ i. Ascyreia Chois. Prod. Hyp.

Derivation, From *a*, not, and *skuros*, hard; that is to say, plants soft to the touch.

Sect. Char. Sepals connected at the base, and unequal. Stamens numerous. Styles 3 to 5. Flowers terminal, large, few, sub-corymbose. (*Don's Mill.*, i. p. 601.)

A. *Styles commonly* 3.

1. *H.* ELATUM *Ait.* The tall St. John's Wort.

Identification. Ait. Hort. Kew., ed. 2. vol. 3. p. 104.; Don's Mill., 1. p. 601.; Tor. and Gray, 1. p. 166.
Engravings. Wats. Dend. Brit., t. 85.; and our *fig.* 112.

Spec. Char., &c. Younger stems reddish. Leaves ovate-oblong, acute, dilated at the base, somewhat emarginate, with the margins somewhat revolute. Flowers corymbose. Peduncles bibracteate. Sepals ovate-oblong. (*Don's Mill.*) A sub-evergreen shrub. North America. Height 3 ft. to 5 ft. Introduced in 1762. Flowers yellow; July and August. Capsule reddish brown; ripe in October. Decaying leaves yellowish.

112. Hypéricum elàtum.

2. *H.* HIRCINUM *L.* The Goat-scented St. John's Wort.

Identification. Lin. Sp., 1103.; Don's Mill., 1. p. 602.
Synonymes. Trágium *Clus.*; Androsæ'mum fœ'tidum *Bauh.*, *Park.*, and *Ray*; Mille Pertuis à Odeur de Bouc, *Fr.*
Engravings. Schkuhr Handb. 3. t. 213. f. 3.; Wats. Dend. Brit., t. 6.; and our *fig.* 113.

Spec. Char., &c. Branches winged. Leaves somewhat emarginate at the base, dilated, sessile, acute at the apex, ovate-lanceolate, with glandular margins. Peduncles bibracteate. Stamens exceeding the corolla in length. Seeds 2, appendiculated. (*Don's Mill.*) A deciduous or sub-evergreen undershrub. Sea coast of Spain. Height 3 ft. to 4 ft. Introduced in 1640. Flowers yellow; July to September. Capsule reddish brown; ripe in October. The leaves of this species, when bruised, have a very disagreeable smell, resembling that of a goat, whence its name.

113. H. hircinum.

Varieties.

- H. *h.* 2 *obtusifòlium* Dec. — Leaves blunter than the species. Found on the mountains of Corsica, on humid rocks.
- H. *h.* 3 *minus* Dec. is a smaller plant than the others, figured in *Dend. Brit.* t. 87.

B. *Styles commonly* 5.

H. KALMIANUM L. Kalm's St. John's Wort.

Identification. Willd. Sp., 3. p. 438.; Don's Mill., 1. p. 603.; Tor. and Gray, 1. p. 188.
Synonymes. *H.* Bartràmium Mill.; Virginia St. John's Wort.
Engraving. Our *fig.* 114.

Spec. Char., &c. Branches tetragonal. Leaves linear-lanceolate. Flowers 3 to 7, in a terminal corymb. Sepals lanceolate, bluntish. (*Don's Mill.*) A sub-evergreen undershrub. Canada to Virginia. Height 1½ ft. in America; 2 ft. to 3 ft. in England. Introduced in 1759. Flowers yellow; June and July. Capsule reddish brown; ripe in October.

114. H. Kalmianum.

A neat compact bush, one of the most ornamental of the hardy species of the genus. The general hue of the entire plant is yellow, and the calyxes and the capsule, before they are ripe, particularly so. Flowers very numerous, in upright raceme-like corymbs.

⸙ ⸙ 4. *H. Ura'lum* Ham. The Urala St. John's Wort.

Identification. D. Don Prod. Nep., p. 218.; Don's Mill., 1. p. 603.
Derivation. From its name, *Urala swa*, in the Newar language.
Engravings. Bot. Mag., t. 2375.; and our *fig.* 115.

115. *Hypéricum Urálum.*

Spec. Char., &c. Branches compressed, 2-edged. Leaves elliptical, mucronulate, smooth, shining. Flowers terminal, somewhat corymbose. Sepals oval, very blunt. Petals orbicular. Styles shorter than the stamens. (*Don's Mill.*) A neat sub-evergreen undershrub. Nepal, on the tops of mountains. Height 2 ft. Introduced in 1823. Flowers yellow; July to September. Capsule reddish brown; ripe in October.

In mild situations, and on a dry soil, it may safely be left through the winter without any protection; but this should not be the case where the situation is cold, and the soil tenacious or humid.

⸙ 5. *H.* calyci'num *L.* The *large*-calyxed St. John's Wort.

Identification. Lin. Mant., 106.; Willd., 3. p. 1442.; Hook. Scot., 221.; Dec. Prod., 1. p. 546.; Smith's Eng. Fl., 3. p. 323.; and Don's Mill., 1. p. 603.
Synonymes. *Androsæ'mum* constantinopolitànum flòre máximo, *Wheeler's Journey*, 205.; the large-flowered St. John's Wort; the large-flowering Tutsan; the terrestrial Sun; Aaron's Beard; Mille Pertuis à grandes Fleurs, *Fr.*; grossblumiger Johanniskraut, *Ger.*; Asciro *Ital.*
Derivation. This species was called *Androsæ'mum* by the old writers on botany, on account of the tinge of red in different places on the stems, and the redness of the anthers, which were supposed to give it the appearance of being spotted with blood. It was called Constantinopolitan from its having been found near that city, in 1676, by Sir George Wheeler, Bart. The large size of its flowers is remarkable, and has given rise to most of its other names. The name of the Terrestrial Sun is very appropriate to the large golden flowers, with their long ray-like stamens, lying glittering on a bed of dark green shining leaves, which spread over the surface of the ground. The number and length of the stamens are, doubtless, also the origin of the name of Aaron's Beard.
Engravings. Eng. Bot., v. 29. t. 2017.; Bot. Mag., t. 146.; and our *fig.* 116.

Spec. Char., &c. Stem tetragonal, dwarf. Leaves ovate, coriaceous, broad, full of pellucid dots. Flowers large, terminal, solitary. Sepals large, obovate, spreading; capsule nodding. (*Don's Mill.*) A beautiful little evergreen undershrub, with dark green shining leaves. Levant, Olympus, Britain, on the western coast of Scotland, and in Ireland near Cork, in woods. Height 1 ft. to 1½ ft. Flowers of a bright golden yellow, with innumerable reddish tremulous anthers; June to September. Capsule reddish brown; ripe in October.

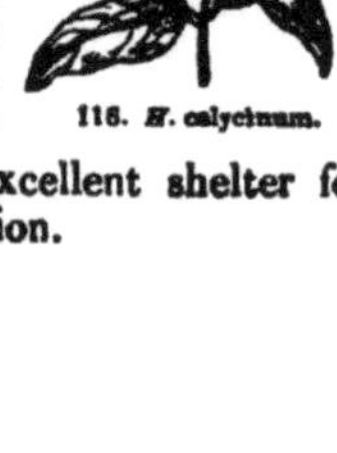

116. *H. calycinum.*

Valuable for covering banks, rockwork, or the surface of the ground in old shrubberies or picturesque woods, especially for the latter purpose, as it thrives perfectly well under the drip and shade of trees. The root creeps, and a small plant will soon extend itself in every direction, especially if the soil be light, so as to cover a great many square yards in a very short space of time. It is an excellent shelter for game. It may be readily increased to any extent by division.

§ ii. *Perforària* Chois.

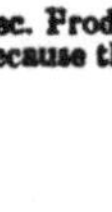

Identification. Chois. Prod. Hyp., p. 44.; Dec. Prod., 1. p. 546.; Don's Mill., 1. p. 603.
Derivation. From *perforatus*, perforated; because the leaves are full of pellucid dots, which gives them the appearance of being perforated.

Sect. Char. Calyx of 5 equal sepals, toothed in some with glandular teeth, but entire in others, connected at the base. Stamens numerous, free or disposed in 5 sets. Styles commonly 3. Herbs or undershrubs. Flowers axillary, or in terminal panicled corymbs. Leaves rarely linear. (*Don's Mill.*) Undershrubs, from 1 ft. to 3 ft. in height.

A. *Sepals entire.*

ꭤ ꭤ 6. *H.* PROLI'FICUM *L.* The prolific St. John's Wort.

Identification. Lin. Mant., 106.; Don's Mill., 1. p. 605.; Tor. and Gray, 1. p. 159.
Synonymes. *H. foliòsum Jacq., Hort. Schönbr.* 3. p. 27.; *H. Kalmiànum* Du Roi Harbk. 1. p. 310.
Engravings. Wats. Dend. Brit., t. 88.; Jacq. Hort. Schönb., t. 299.; and our *fig.* 117.

117. *H. prolíficum.*

Spec. Char., &c. Stem round. Branches angular. Leaves linear-lanceolate, with revolute edges, full of pellucid dots. Corymbs few-flowered. Sepals ovate-lanceolate, stamens very numerous. Styles usually connected together. (*Don's Mill.*) A sub-evergreen shrub. New Jersey to Florida, in swamps. Height 1 ft. to 4 ft. Introduced in 1758. Flowers yellow; June to August. Capsule reddish brown; ripe in October.

Frequent in gardens, and forming a dense leafy bush, covered with flowers great part of the summer, and with seed-pods in the autumn. Readily distinguished from *H.* Kalmiànum, by the leaves, bracts, and sepals being much smoother and shining.

B. *Sepals toothed, usually with the Teeth glandular.*

ꭤ 7. *H.* EMPETRIFO'LIUM *Willd.* The Empetrum-leaved St. John's Wort.

Identification. Willd. Spec., 3. p. 1452.; Don's Mill., 1. p. 610.
Engravings. Dend. Brit., t. 141.; and our *fig.* 118.

118. *H. empetrifòlium.*

Spec. Char., &c. Stems suffruticose, round, with subulate branchlets. Leaves linear, ternary, with revolute margins. Calyx small, obtuse. Petals without glands. (*Don's Mill.*) A neat little evergreen shrub. South of Europe, near the Mediterranean; and in Greece. Height 1 ft. to 2 ft. Introduced in 1820. Flowers yellow; May to August.

One of the neatest species of the genus, but somewhat tender.

Other Species of Hypéricum.—The only truly hardy shrubby species of Hypéricum are, *H.* elàtum, *H.* hircìnum, *H.* calycìnum, *H.* Kalmiànum, and *H.* prolíficum. The other hardy species are of such low growth, that they may be considered, for all practical purposes, as herbaceous plants. H. *nepalénse* Royle appeared to be hardy in the Hort. Soc. Garden, but it was destroyed by the winter of 1837–8. H. *adpréssum* Bartr., H. *rosmarinifòlium* Lam., H. *galiöides* Lam., H. *fasciculàtum* Lam., and some other shrubby or frutescent species, are described by Torrey and Gray, but we are not aware of their having been yet introduced.

GENUS II.

ANDROSÆ'MUM *Chois.* THE ANDROSÆMUM, *or* TUTSAN. *Lin. Syst.* Polyadélphia Polyándria.

Identification. Chois. Prod. Hyp., 37.; Dec. Prod., 1. p. 543.; Don's Mill., 1. p. 601.
Synonymes. Hypéricum, *L.*; Androsème, *Fr.*; Johanniskraut, *Ger.*; Androsemo, *Ital.*

Derivation. From *anēr*, *andros*, a man, and *haima*, blood; the capsules, when crushed between the fingers, giving out a blood-coloured juice. Tutsan is a corruption of *toute saine*, all heal; and it was applied to the plant formerly from its supposed vulnerary properties.

Gen. Char. *Capsule* baccate; usually 1-celled. *Calyx* 5-parted, with unequal lobes. *Petals* 5. *Styles* 3. *Stamens* numerous, disposed in 3 sets. (*Don's Mill.*)

Leaves simple, opposite, exstipulate, sub-evergreen. *Flowers* terminal. The whole plant closely resembling a *Hypéricum*. — Suffruticose. Indigenous in Britain.

1. *A.* OFFICINA'LE *Allioni.* The officinal Androsæmum, *or common Tutsan.*

Identification. All. Ped., No. 1440.; Dec. Prod., 1. p. 543.; Don's Mill., 1. p. 601.

Synonymes. Clymenon Italòrum *L'Obel*; *Hypéricum Androsæ'mum Lin.*, *Willd.*, *Smith*, and *Hooker*; Park Leaves (because it is frequently found wild in parks); Androsème officinale, *Fr.*; breitblättriges (broad-leaved) Johanniskraut, *Ger.*; Cicilliana *Ital.*

Engravings. Blackw., t. 94.; Eng. Bot., t. 1225.; and our *fig.* 119. in flower, and *fig.* 120. showing the fruit.

119. *Androsæ'mum officinàle.*

Spec. Char., &c. Leaves ovate, and somewhat heart-shaped, sessile, widely spreading. A sub-evergreen, suffruticose shrub, forming a dense bush, with many stems. Europe; and in England in moist shady woods. Height 2 ft. to 3 ft. Flowers yellow, 1 in. across; July to September. Capsule brownish purple, and lastly, almost black; ripe in October.

The fruit is an ovate capsule, assuming the appearance of a berry: it is at first yellowish green, then red or brownish purple; and, lastly, almost black when ripe. The juice of the capsules, and also that of the leaves, is claret-coloured. The latter, when bruised, have an aromatic scent, and were formerly applied to fresh wounds; and hence the French name of *la toute saine*. In gardening, the plant is valuable as growing under the drip of trees, and thriving and flowering freely in almost any soil or situation. It is readily propagated by division of the root.

120. *Androsæ'mum officinàle.*

ORDER XIII. ACERA'CEÆ.

ORD. CHAR. *Flowers* either unisexual or bisexual. *Calyx* and *corolla* equal in the number of their parts, with an imbricated æstivation; the corolla sometimes absent. *Petals* without appendages. *Stamens* inserted upon a disk, which arises from below the pistillum, not agreeing in number with the divisions of the calyx and corolla. *Pistillum* 2-lobed, each lobe having a wing at its back. *Style* 1. *Stigmas* 2. *Fruit* formed of two samaræ, or keys, each containing 1 cell and 1 erect seed. *Embryo* curved, with leafy shriveled cotyledons, and no albumen. (*Lindl.*) — Deciduous trees or shrubs, natives of the temperate climates of Europe, North America, and Asia.

Leaves simple, opposite, exstipulate, deciduous, rarely evergreen; variously lobed, rarely pinnate. — The species in British gardens are included in the genera *A'cer* and *Negúndo*, which are thus contradistinguished: —

A'CER *L.* Flowers polygamous. Leaves lobed.
NEGU'NDO Mœnch. Flowers diœcious. Leaves pinnate.

Genus I.

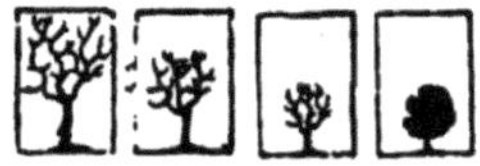

A'CER *L.* The Maple. *Lin. Syst.* Polygàmia Monœ'cia.

Identification. Lin. Gen., No. 1115.; Mœnch Meth., 334.; Dec. Prod., 1. p. 593.; Don's Mill., 1. p. 648.
Synonymes. E'rable, *Fr.*; Ahorn, *Ger.*; Acero, *Ital.*; and Arce, *Spanish.*
Derivation. From *acer*, hard or sharp, derived from *ac*, Celtic, a point. The name is supposed to be applied to this genus because the wood of some species is extremely hard, and was formerly much sought after for the purpose of making pikes and lances.

Gen. Char. *Sexes* hermaphrodite, or monœciously polygamous. *Flowers* with a calyx and corolla. *Calyx* divided into 5 parts, or some number between 4 and 9. *Petals* the same in number. *Stamens* 8, or some number between 5 and 12. *Anthers* 2-lobed. *Carpels* 2, very rarely 3, each a samara; that is, a fruit which is called in Britain, vernacularly, a key.— Deciduous trees, natives of Europe, North America, and Asia.

Leaves simple, opposite, exstipulate, deciduous, rarely evergreen; variously lobed, toothed. *Flowers* axillary, corymbose. *Fruit* a samara. Decaying leaves rich yellow in some, and red or brown in others.

Several of the species produce useful timber; and sugar is one of the constituent parts of the sap of all of them. They all prefer a situation sheltered rather than exposed; a free, deep, loamy soil, rich rather than sterile, and neither very wet, nor very dry. They are propagated by seeds and layers, or by grafting. The maturity of the seed may be proved by opening the key, and observing if the cotyledons are green, succulent, and fresh; if the green colour of the cotyledons is wanting, the seeds are good for nothing. The seeds of all the species may either be sown in autumn, after they are gathered, or in spring: and the latter method is preferable where moles abound, as they are very fond of the seeds. Sown in spring, they come up in five or six weeks afterwards; with the exception of those of the *A.* campéstre, which never come up till the second or third year. The seeds should not be covered with more than a quarter to half an inch of soil. The surface of the ground in which they are sown may be advantageously shaded with leaves, fronds of ferns, heath, or straw.

A. *Leaves simple, or only slightly or occasionally lobed.*

1. *A.* oblo'ngum *Wall.* The oblong-*leaved* Maple.

Identification. Wall. in Litt.; Dec. Prod., 1. p. 593.; Don's Mill., 1. p. 648.
Synonymes. *A. laurifòlium D. Don, Prod. Fl. Nep.* p. 249.; *A. Buxinpàla* Hamilt.
Engravings. Our *figs.* 121, 122., reduced to our usual scale; and the figures of the leaves, of the natural size, as given in the plate, p. 95.

121. *A. oblóngum.*

Spec. Char., &c. Leaves oblong-lanceolate, acuminated, quite entire, coriaceous, smooth, rounded at the base. Racemes compound; wings of fruit parallel, smooth, separated. (*Don's Mill.*) A low, deciduous tree; in Britain, a rather tender shrub. Nepal. Height 20 ft. Introduced in 1824. Flowers greenish yellow; February. Keys?; ripe?.

122. *A. oblóngum.*

This species is rather tender, and somewhat difficult to keep in the open ground, unless when planted against a wall. Though the leaves are generally entire, yet they are sometimes lobed, or show a tendency to become so.

♀ 2. *A.* TATA'RICUM *L.* The Tartarian Maple.

Identification. Lin. Sp., 1495.; Dec. Prod., 1. p. 593.; Hayne Dend., p. 209.; Don's Mill., 1. p. 648.
Synonymes. E'rable de Tartarie, *Fr.*; Tartarische Ahorn, *Ger.*; Zarza-modon, or Locust Tree, *Russ.*
Engravings. Pall. Fl. Ros., t. 3.; Tratt. Arch., 1. No. 1.; Wats. Dend. Brit., t. 160.; the plate of this species in Arb. Brit., 1st edit., vol. v.; our *fig.* 123.; and the figure of the leaves, of the natural size, in the plate forming p. 96.

Spec. Char., &c. Leaves cordate, undivided, serrated, with obsolete lobes. Racemes compound, crowded, erect; wings of fruit parallel, young ones puberulous. (*Don's Mill.*) A low deciduous tree, native of Tartary. Height 20 ft. to 30 ft. Introduced in 1759. Flowers pale greenish yellow, sometimes slightly tinged with red; May and June. Keys brown; ripe in August. Decaying leaves reddish yellow, or brown. Naked young wood brown.

When raised from seed, the plant will come into flower in 5 or 6 years; and, in good soil, it will attain the height of 15 ft. in 10 years. According to

123. *A'cer tatáricum.*

some, it will thrive in a moister soil than most others. In ornamental plantations, it is valuable on account of the early expansion of its leaves, which appear before those of almost every other kind of *A'cer.*

B. *Leaves 3-lobed, or trifid; rarely 5-lobed.*

♀ 3. *A.* SPICA'TUM *Lam.* The spiked-*flowered* Maple.

Identification. Lam. Dict., 2. p. 381.; Dec. Prod., 1. p. 593.; Don's Mill., 1. p. 648.; Tor. and Gray, 1. p. 246.
Synonymes. *A.* montánum *Ait. Hort. Kew.*, 3. p. 435.; *A.* pennsylvánicum *Du Roi Harbk.* t. 2.; *A.* parviflòrum *Ehrh.*; Mountain Maple, E'rable de Montagne, *Fr.*; Berg Ahorn, *Ger.*; Acero di Montagna, *Ital.*
Engravings. Trat. Arch., No. 13.; the plate of this species in Arb. Brit., 1st. edit., vol. v.; our *fig.* 124.; and the figure of the leaves, of the natural size, in the plate forming p. 97.

Spec. Char., &c. Leaves cordate, 3- or slightly 5-lobed, acuminated, pubescent beneath, unequally and coarsely serrated. Racemes compound, erect. Petals linear. Fruit smooth, with the wings rather diverging. (*Don's Mill.*) A deciduous shrub, or small tree. Canada to Georgia. Height 6 ft. to 10 ft. in America; 18 ft. to 20 ft. in England. Introduced in 1750. Flowers small, greenish, raceme many-flowered; May and June. Keys often reddish; ripe in August. Decaying leaves yellowish red. Naked young wood brown.

124. A'cer spicàtum

Very ornamental in autumn, from its small keys, which are fixed upon slender pendulous spikes, and have their membranous wings beautifully tinged with red when ripe. Michaux states that this species, grafted upon the sycamore, is, like the A'cer striàtum, augmented to twice its natural dimensions; a fact which we have never had an opportunity of seeing verified.

Ỵ 4. *A.* STRIA'TUM *L.* The striped-*barked* Maple.

125. A'cer striàtum.

Identification. Lam. Dict., 2. p. 381.; Dec. Prod., 1. p. 593.; Don's Mill., 1. p. 648.
Synonymes. A. pennsylvánicum *Lin. Sp.* 1496., and *Tor. & Gray*, 1. p. 246.; A. canadénse *Marsh.*

and *Duh. Arb.* 1. t. 12.; Snake-barked Maple, Moose Wood, striped Maple. Dog Wood; E'rable jaspé, *Fr.*; gestreifter Ahorn, *Ger.*; Acero screziato, *Ital.*

Engravings. Mill., t. 7.; Tratt. Arch., No. 11.; Mich. Fel. Arb., 2. t. 17.; the plate of this species in Arb. Brit., 1st edit., vol. v.; our *fig.* 125.; and *fig.* 146. of the leaves, of the natural size, forming p. 98, 99.

Spec. Char., &c. Leaves cordate, 3-lobed, acuminated, finely and acutely serrated. Racemes pendulous, simple. Petals oval. Fruit smooth, with the wings rather diverging. (*Don's Mill.*) A deciduous tree, with green bark, striped with white. Canada to Georgia. Height 8 ft. to 10 ft. in America; 10 ft. to 20 ft. in England. Introduced in 1755. Flowers yellowish green, on long peduncles; May. Keys brown, and remarkable for a cavity on one side of the capsules; ripe in August. Decaying leaves yellowish green. Naked young wood green, striped with white and black.

The buds and leaves, when beginning to unfold, are rose-coloured; and the leaves, when fully expanded, are of a thick texture, and finely serrated. From the great beauty of its bark, this tree deserves a place in every collection. It is propagated by seeds, which are received from America; or by grafting on *A.* Pseùdo-*Plátanus.*

C. *Leaves 5-lobed.*

♀ 5. *A.* MACROPHY'LLUM *Pursh.* The long, *or large*, leaved Maple.

Identification. Pursh, Fl. Amer. Sept., 1. p. 267.; Dec. Prod., 1. p. 594.; Don's Mill., 1. p. 649.; Tor. and Gray, 1. p. 246.

Engravings. Hook. Fl. Bor. Amer., 1. t. 38.; the plate of this species in Arb. Brit., 1st edit., vol. v.; our *fig.* 126.; and *figs.* 147. and 148. of the leaves, of the natural size, forming p. 100, 101. and 102, 103.

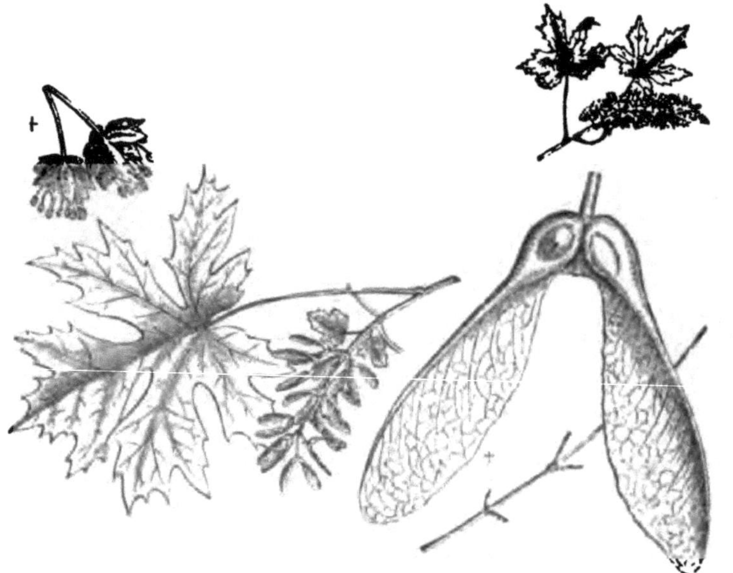

126 *A'cer macrophýllum.*

Spec. Char. &c. Leaves digitately 5-lobed, with roundish recesses. Lobes somewhat 3-lobed, repandly toothed, pubescent beneath. Racemes compound, erect. Stamens 9, with hairy filaments. Ovaries very hairy. (*Don's Mill.*) A deciduous tree of large size. North-west coast of North America, on the alluvial banks of rivers. Height 40 ft. to 90 ft. Introduced in 1826. Flowers yellow, fragrant; April and May. Keys brown; ripe in September. Decaying leaves yellowish brown. Naked young wood brown. Buds green.

Leaves nearly 1 ft. broad. Carpels sometimes 3. Sap as abundant as in any species, except in *A.* sacchárinum; the wood soft, whitish, but beautifully veined. (*Tor. and Gray.*) This species is quite hardy in the climate of London, and promises to form a most valuable addition to our ornamental, and, possibly, to our timber, trees. The tree in the Hort. Soc. Garden is between 40 ft. and 50 ft. high, after having been thirteen years planted; and it has flowered, and ripened some seeds.

ꝑ 6. *A.* PLATANÖI'DES *L.* The Platanus-like, *or Norway*, Maple.

Identification. Lin. Sp., 1496.; Dec. Prod., 1. p. 649.; Don's Mill., 1. p. 649.
Synonymes. E'rable plane, or E'rable de Norvège, *Fr.*; spitz Ahorn, or spitzblättriger Ahorn, *Ger.*; Acero riccio, *Ital.*
Engravings. Duh. Arb., 1. t. 10. f. 1.; the plate of this species in Arb. Brit., 1st edit., vol. v.; our *fig.* 129.; and *fig.* 149. of the leaves, of the natural size, forming p. 104, 105.

Spec. Char., &c. Leaves cordate, smooth, 5-lobed. Lobes acuminated, with a few coarse acute teeth. Corymbs stalked, erectish, and, as well as the fruit, smooth. Fruit with divaricated wings (*Don's Mill.*) A deciduous tree, above the middle size. Norway to Switzerland, but not in Britain. Height 30 ft. to 60 ft. Introduced in 1683. Flowers rich yellow; April and May. Keys brown; ripe in September and October. Opening foliage and flowers, in spring, of a bright yellow; when decaying, also, of a fine yellow. Naked young wood smooth, brown.

Varieties.

ꝑ A. p. 2 *Lobèlii.* *A.* Lobèlii Tenore; *A.* platanöides *Don's Mill.* i. p. 649. (Our *fig.* 127.; and *fig.* 150. of the leaves, of the natural size, in the plate forming p. 106.) — The leaves are very slightly heart-shaped, irregularly toothed, 5-lobed, with the lobes more or less abruptly pointed. The bark of the young wood striped, somewhat in the manner of that of *A.* striàtum; by which circumstance the plant, in a young state, is readily distinguished from *A.* platanöides. A large tree, native of the kingdom of Naples, and found on mountains. One of the most beautiful acers in cultivation; but very little known, though it was introduced about 1683. There is a tree of it at Croome, above 20 ft. high, which has ripened seeds.

127. *A'cer platanöides Lobèlii.*

ꝑ A. p. 3 *variegàtum* Hort. *A. p.* álbo variegàtum Hayne. — Leaves variegated with white.

ꝑ A. p. 4 *aúreo variegàtum* Hort. — Leaves variegated with yellow.

ꝑ A. p. 5 *laciniàtum* Dec. *A. p.* crispum *Lauth*; Eagle's Claw, or Hawk's Foot Maple. (The plate of this variety in Arb. Brit., 1st edit., vol. v.; our *fig.* 128.; and *fig.* 151. of the leaves, of the natural size, in the plate forming p. 107.) — Leaves deeply and variously cut.

128. *A'cer platanöides laciniàtum.*

It is frequently produced from seed, being found by nurserymen among seedlings of the species.

The tree, in general appearance, at a distance, is like the common sycamore; but, on a nearer approach, the leaves are found of a smoother and finer texture. The roots extend considerably, both downwards and laterally. The bark is green on the young shoots, but it afterwards becomes of a reddish brown, dotted with white points: that of the trunk is brown, and rather cracked. The buds are large and red in autumn, becoming of a still darker red in the course of the winter: those on the points of the shoots are always the largest. The leaves

129. *A'cer platanöides.*

are thin, green on both sides, and shining. When the petiole is broken, an acrid milky sap issues from it, which coagulates with the air. The leaves are about 5 in. long, and nearly the same in width. The petioles are longer than the leaves. About the end of October, the leaves become either of a clear, or a yellowish, red, and then drop off. The flowers appear just before the leaves, near the end of April: they form a short raceme, somewhat corymbose. The fruits, or keys, have their wings yellow. It is not till the tree has attained the age of nearly 40 years that it produces fertile seeds, though it will flower many years before that period. The rate of growth of this species, when once established, is from 18 in. to 3 ft. long every year, till it attains the height of 20 or 30 feet; which, in favourable situations, it does in ten years. The wood weighs, when dry, 43 lb. 4 oz. per cubic foot; is easily worked, takes a fine polish, and absorbs and retains all kinds of colours. It may be used for all the various purposes of the wood of the common sycamore. Sugar is made from the sap in Norway, Sweden, and Lithuania. Seeds are ripened in England in abundance.

♀ 7. *A.* SACCHA'RINUM *L.* The Sugar Maple.

Identification. Lin. Sp., 1496.; Dec. Prod., l. p. 595.; Don's Mill., 1. p. 650.; Tor. and Gray, 1. p. 248.
Synonymes. Rock Maple, Hard Maple, Bird's-eye Maple, *Amer.*; Acero del Canadà, *Ital.*
Engravings. Michx. Fl. Arb., 2. t. 15.; the plate of this species in Arb. Brit., 1st edit., vol. v.; our *fig.* 130.; and *fig.* 152. of the leaves, of the natural size, in the plate forming p. 108, 109.

Spec. Char., &c. Leaves cordate, smooth, glaucous beneath, palmately 5-lobed; lobes acuminated, serrately toothed. Corymbs drooping, on short peduncles. Pedicels pilose. Fruit smooth, with the wings diverging. (*Don's Mill.*) A deciduous tree. Canada to Georgia. Height in America 50 ft. to 80 ft.; in England 30 ft. to 40 ft. Introduced in 1735. Flowers small, yellowish, and suspended on long, slender, drooping peduncles; April and May. Keys brown; ripe in September. Decaying leaves rich yellow. Naked young wood smooth, whitish brown.

130. *A'cer sacchárinum.*

Variety.

♀ A. s. 2. *nigrum.* *A. s. β* nigrum *Tor. & Gray*; *A.* nigrum *Michx.*; the black Sugar Tree, or Rock Maple, *Michx. Arb.* 2. t. 16. — Leaves pale green beneath, the veins of the lower surface and petioles minutely villous, pubescent; wings of the fruit a little more diverging. (*Tor. and Gray*, i. p. 248.) Michaux, who considered this variety a species, says the leaves resemble those of the species in every respect, except that they are of a darker green, and of a thicker texture, and somewhat more bluntly lobed. The tree is indiscriminately mixed with the common sugar maple, through extensive ranges of country in New Hampshire, Vermont, and Connecticut; but is readily distinguished from it by the smaller size which it attains, and the darker colour of its leaves. The soil in which it flourishes best is a rich, strong, sandy loam; and there it usually grows to the height of 40 or 50 feet.

Closely resembling *A.* platanöides in foliage, except in being somewhat glaucous beneath, and in the fruit being much more divergent. Bark of the trunk white. Leaves 3 in. to 5 in. in length, generally wider than long. The buds have a fine ruddy tint, especially in spring before they expand. The tree in England is rather tender, and never attains a large size; but in America the timber is valuable, and the sap produces sugar. American seeds.

♀ 8. *A.* Pseu'do-*Pla'*tanus *L.* The Mock Plane Tree, *the Sycamore, or Great Maple.*

Identification. Lin. Sp., 1469.; Don's Mill., 1. p. 648.
Synonymes. Plane Tree, *Scotch*; E'rable Sycamore, E'rable blanc de Montagne, fausse Platane, grand E'rable, *Fr.*; Ehrenbaum, weisser Ahorn, gemeine Ahorn, *Ger.*; Acero Fico, *Ital.*
Engravings. Duh. Arb., 1. t. 36.; Schmidt Arb., 1. p. 34.; the plate of this species in Arb. Brit., 1st edit., vol. v.; our *fig.* 132.; and *fig.* 153. of the leaves, of the natural size, in the plate forming p. 110, 111.

Spec. Char., &c. Leaves cordate, smooth, with 5 acuminated, unequally toothed lobes. Racemes pendulous, rather compound; with the rachis, as well as the filaments of stamens, hairy. Fruit smooth, with the wings rather diverging. (*Don's Mill.*) A deciduous tree of the first rank. Europe and Britain, in wooded mountainous situations. Height 30 ft. to 80 ft. Flowers greenish yellow, mostly hermaphrodite; May and June. Keys reddish brown; ripe in October. Decaying leaves brown, with dark blotches. Naked young wood reddish brown. Buds large, green, or tinged with red.

131. *A. P. opulifolium.*

Varieties.

♀ A. *P.* 2 *opulifòlia.* *A.* opulifòlium *Hort.*; *A.* trilobàtum *Hort.*; *A.* barbàtum *Hort.* (Our *fig.* 132.) — Leaves and fruit smaller than in the species, as shown in *fig.* 155. of the leaves, in the plate forming p. 114. We have no doubt of this being the *A.* opulifòlium of Thuillier and Villars, L'E'rable duret, and L'E'rable ayart, *Fr.*, which is said by these authors to resemble *A.* Pseùdo-*Plátanus*, but to be much smaller. It is a native of the Alps and Pyrenees. Introduced in 1812. Height 15 ft. to 20 ft.

♀ A. *P.* 3 *longifòlia.* *A.* longifòlium *Booth.*—Leaves more deeply cut, and the petioles much longer than in the species. Altogether a tree of very remarkable aspect.

♀ A. *P.* 4 *flàvo variegàta.* — *A.* P. lutéscens *Hort.*; the Corstorphine Plane. — Leaves variegated with yellow. The original tree stands near an old pigeon-house in the grounds of Sir Thomas Dick Lauder, Bart., in the parish of Corstorphine, near Edinburgh. Seeds of this variety, sown, have produced plants with the character of the parent to a certain extent.

♀ A. *P.* 5 *álbo variegàta* Hayne.—Of all the variegated varieties of *A*'cer, it must be acknowledged that this is the most ornamental; especially in spring, when the leaves first expand.

♀ A. *P.* 6 *purpùrea* Hort. — Leaves of a fine purple underneath. This variety was found in a bed of seedlings, in Saunders's Nursery, Jersey, about 1828, and is now to be met with in all the principal nurseries. The tree has a very singular effect when the leaves are slightly ruffled by the wind, alternately appearing clothed in purple and in pale green. In spring, when the leaves first expand, the purple bloom is less obvious than when they become matured, at which time it is very distinct.

Other Varieties. In the garden of the Hort. Soc. there is a variety called *Hodgkins's Seedling*, with yellow-blotched leaves; and another, called *Leslie's*

Seedling. In Hayne's *Dendrologische Flora* there are, also, the following varieties: A. *P. stenóptera*, A. *P. macróptera*, and A. *P. micróptera*, which differ in the proportions of the wings of the keys, and do not appear worth farther notice. In all seed beds and young plantations some of the plants will be found with the petioles and the buds red, and others with the petioles and the buds greenish yellow: such trees, when of considerable size, are very distinct in their general aspect, when in bud, and when they have newly come into leaf; but after midsummer, when the leaves are fully matured, and begin to get rusty, the trees are scarcely distinguishable. Different plants also differ much in the time of their coming into leaf, and of dropping their leaves; and some of the more remarkable of these it might be worth while to propagate by extension.

132. *A'cer Pseùdo-Plátanus.*

The growth of the common sycamore is very rapid compared with that of most other species of *A'cer*, particularly when it is in a deep, free, rich soil, and in a mild climate. It arrives at its full growth in 50 or 60 years; but it requires to be 80 or 100 years old before its wood arrives at perfection. In marshy soil, or in dry sand, and even on chalk, the tree never attains any size. It produces fertile seeds at the age of 20 years, but flowers several years sooner; sometimes even perfecting its seeds sooner also. The longevity of the tree is from 140 to 200 years, though it has been known of a much greater age. The wood weighs per cubic foot, newly cut, 64 lb.; half-dry, 56 lb.; dry, 48 lb. It loses, in drying, about a twelfth part of its bulk. When the tree is young, it is white; but, as the tree gets older, the wood becomes a little yellow, and often brown, especially towards the heart. It is compact and firm, without being very hard; of a fine grain, sometimes veined, susceptible of a high polish, and easily worked, either on the bench, or in the turning-lathe. It does not warp, and is not likely to be attacked by worms. It is used in joinery and turnery, and cabinet-making; by musical instrument makers; for cider-presses; and, sometimes, for gun-stocks. Formerly, when wooden dishes and spoons were more used than they are at present, it was much in demand, especially in Scotland, by the manufacturers of these articles. As underwood, the sycamore shoots freely from the stool, to the age of 80 or 100 years. As a timber tree, it is most advantageously cut down at the age of 80 years, or from that age to 100. As an ornamental tree, it produces the best effect, either singly, or in groups of two or three, placed sufficiently near to form a whole, but not so as to touch each other; and in rows or avenues. The varieties with variegated leaves are very ornamental in the beginning of summer; but their leaves are almost always more or less imperfect, especially on the edges, and fall off much sooner in the autumn than those of the species. The leaves of the purple variety are not liable to the same objection as those of the variegated sorts. Seeds; and the varieties by grafting on the species.

♀ 9. *A.* OBTUSA'TUM *Kit.* The obtuse-*lobed-leaved* Maple.

Identification. Kit. in Willd. Spec., 4. p 948.; Dec. Prod., 1. p. 594.; Don's Mill., 1. p. 649.
Synonymes. *A.* neapolitànum *Tenore*; *A.* hýbridum in the Lond. Hort. Soc. Gard. in 1834; the Neapolitan Maple.
Engravings. Tratt. Arch., 1. No. 14.; the plate of this species in Arb. Brit., 1st edit., vol. v.; our *fig.* 133.; and *fig.* 154. of the leaves, of the natural size, forming p. 112, 113.

Spec. Char., &c. Leaves cordate, roundish, 5-lobed; lobes bluntish (or pointed), repandly toothed, velvety beneath. Corymbs pendulous. Pedicels hairy. Fruit rather hairy, with the wings somewhat diverging. (*Don's Mill.*) A deciduous tree of the first rank, of as rapid growth as *A.* Pseùdo-*Plátanus.* Hungary, Croatia, and many parts of Italy, on hills and mountains. Height 40 ft. to 60 ft. Introduced in 1825. Flowers greenish yellow, few in a panicle; May and June. Keys brown; ripe in September. Decaying leaves dark brown. Naked young wood smooth and brown. Buds prominent, green.

133. *A'cer* obtusàtum.

Varieties.— In the Neapolitan territory, this tree is probably somewhat different in its habit and aspect from what it is in Hungary; and hence, the A. *neapolitànum* of Tenore may be considered a variety. The following also appear to belong to this species:—

♀ A. *o.* 2 *coriàceum.* *A.* coriàceum *Bosc.* (Don's Mill., 1. p. 649.; and our *fig.* 134.)—Leaves coriaceous, the same length as breadth, 3-5-lobed, denticulated, smooth. Corymbs loose. Wings of fruit erectly divergent. Native of ?. (*Don's Miller.*) There are small plants of this *A'cer* in the collection of Messrs. Loddiges, which appear to us to belong to this species, though it is with considerable doubt that we have placed it here. Possibly it may belong to *A.* platanöides, as we once thought, or to *A.* O'palus.

134. *A. o.* coriàceum.

♀ A. *o.* 3 *ibéricum.* *A.* ibéricum *Bieb. Fl. Taur.* p. 247.—Leaves shining, glaucous beneath, bluntly three-lobed; lobes furnished with one or two teeth; lateral ones marked with the middle nerve to the insertion of the petiole. Petioles a little shorter than the leaves. Tree 20 ft. in height. A native of Georgia. (*Don's Mill.*, i. p. 649.) As we have only seen plants a few inches high, we may be mistaken in considering *A.* ibéricum as a variety of *A.* obtusàtum.

‡ A. o. 4. *lobàtum*, *A.* lobàtum *Fisch.*, has the leaves 7-lobed, according to *Don's Miller*, but the young plants bearing this name in the Hort. Soc. Garden, which was raised from seeds received from Dr. Fischer of Petersburg, appears obviously to belong to *A.* obtusàtum.

D. *Leaves 5-, rarely 7-lobed.*

‡ 10. *A.* O'PALUS *Ait.* The Opal, *or Italian*, Maple.

Identification. Ait. Hort. Kew., 3. p. 436.; Dec. Prod., 1. p. 594.; Don's Mill., 1. p. 649.; Webb Iter Hispan., 60.
Synonymes. *A.* rotundifòlium *Lam. Dict.* 3. p. 382.; *A.* Italum *Lauth Ac.* No. 8.; *A.* villòsum *Presl*; l'E'rable Opale, E'rable à Feuilles rondes, or E'rable d'Italie, *Fr.*; Loppo, *Ital.*
Derivation. The specific appellation of O'palus has been given to this species, probably from the thick opal-like aspect of the leaves.
Engravings. Baudril. Traité, &c., vol. 5. p. 13.; the plate of this species in Arb. Brit., 1st edit., vol. v.; our *fig.* 135.; and *fig.* 156., of the leaves, of the natural size, in the plate forming p. 115.

Spec. Char., &c. Leaves more or less heart-shaped, roundish, 5-lobed, smooth beneath; the lobes generally obtuse, and coarsely serrated. Flowers in drooping corymbs. Keys smooth. (*Pen. Cyc.*) A low deciduous tree. Corsica. Height 8 ft. to 12 ft. Introd. 1752. Flowers whitish; May to June. Keys small, brown; ripe Sept. Decaying leaves yellowish brown.

A branchy tufted tree, covered with smooth leaves, somewhat coriaceous, roundish, indented, with five blunt lobes, deep green on the upper surface, and

135. A'cer O'palus.

somewhat glaucous underneath, with long red petioles. Its flowers are whitish, in short racemes; and the small fruits, or keys, which succeed them, are almost round. It is found in forests and on mountains in Corsica; in Spain, on the Sierra Nevada; and in Italy, where, from the denseness of its shade, it is sometimes planted by road sides, and in gardens near houses. The red colour of the petioles of the leaves, of the fruits, and even the red tinge of the leaves themselves, more especially in autumn, give it rather a morbid appearance. It pushes later in the spring than most of the other species.

‡ 11. *A.* CIRCINA'TUM *Pursh.* The round-*leaved* Maple.

Identification. Pursh Fl. Amer. Sept., 1. p. 267.; Dec. Prod., 1. p. 595.; Don's Mill., 1. p. 651. Tor. and Gray, 1. p. 247.
Engravings. Hook. Amer., t. 39.; our *fig.* 136.; and *fig.* 157. of the leaves, of the natural size, in the plate forming p. 116.

Spec. Char., &c. Leaves orbicular, rather cordate at the base, 7-lobed, smooth on both surfaces; lobes acutely toothed; nerves and veins hairy at their origin. (*Don's Mill.*) A deciduous tree of the middle size. N. W. coast of North America, between lat. 43° and 49°. Height 20 ft. to 40 ft. Introd. 1826. Flowers with the sepals purple, and the petals white; April and May. Keys purplish brown, with thin straight wings, which are so divaricate as to form right angles with the peduncle; the lower margin scarcely thickened. (*Tor. and Gray*). Decaying leaves of a fine reddish yellow. Naked young wood reddish brown.

156. *Acer circinàtum.*

Branches slender, pendulous, and crooked; often taking root, in the manner of those of many species of *Ficus*. Bark smooth; green when young, white when fully grown. Leaf the length of the finger, upon rather a short footstalk, membranaceous, heart-shaped, with 7—9 lobes, and 7—9 nerves; smooth above, except hairs in the axils of the nerves; downy beneath, and in the axils of the nerves woolly: lobes ovate, acute, and acutely serrated; the sinuses acute: the nerves radiate from the tip of the petiole, and one extends to the tip of each lobe. Flowers of a middling size, in nodding corymbs, that are on long peduncles. (*Hook. Fl. Bor. Amer.*) This is a very marked and beautiful species; distinguishable, at sight, by the regular form of its leaves, and their pale reddish green colour. Though this fine tree has been in the country since 1826, it seems to have been comparatively neglected, for there is no good specimen that we know of in the neighbourhood of London. At High Clere, a thriving tree has ripened seeds for some years past; so that there can be no doubt of its hardiness.

▲ 12. *A.* PALMA'TUM *Thunb.* The palmate-*leaved* Maple.

Identification. Thunb. Fl. Jap., p. 161.; Dec. Prod., 1. p. 595.; Don's Mill., 1. p. 650.
Engravings. Tratt. Arch., 1. No. 17.; and *fig.* 158. of the leaves, of the natural size, in the plate forming p. 117.

Spec. Char., &c. Leaves smooth, palmately divided into 5—7 lobes down beyond the middle; lobes acuminated, oblong, serrated. Umbels 5—7-flowered. (*Don's Mill.*) A low deciduous tree; in England a rather delicate shrub. Japan. Height in Japan, 20 ft. Introd. in 1820. Flowers greenish yellow and purple; May. Keys ?. Decaying leaves reddish yellow.

This species requires the protection of a wall; having been, like *A.* oblongum, killed to the ground in the open air, in the Hort. Soc. Garden, in the winter of 1837–8.

¥ 13. *A.* ERIOCA'RPUM *Michx.* The hairy-fruited, *or white*, Maple.

Identification. Michx. Fl. Amer. Bor., 2. p. 213.; Don's Mill., 1. p. 650.
Synonymes. *A.* dasycárpum *Willd. Spec.* 4. p. 985., *Tor. & Gray*, 1. p. 248.; *A.* tomentòsum *Hort. Par.*; *A.* glaúcum *Marsh.*; *A.* virginiànum *Duh.*; *A.* rùbrum *Wangenh.*; white, silver-leaved, or soft, Maple, *United States*; Sir Charles Wager's Maple; E'rable à Fruits cotonneux, or E'rable blanc, *Fr.*; rauher Ahorn, *Ger.*; Acero cotonoso, or Acero bianco, *Ital.*
Engravings. Desf. Ann. Mus., 7. t. 25.; the plate of this species in Arb. Brit., 1st edit., vol. v.; our *fig.* 137.; and *fig.* 159. of the leaves, of the natural size, in the plate forming p. 118.

Spec. Char., &c. Leaves truncate at the base, smooth and glaucous beneath, palmately 5-lobed, with blunt recesses, and unequally and deeply toothed lobes. Flowers conglomerate, on short pedicels, apetalous, pentandrous. Ovaries downy. (*Don's Mill.*) A middle-sized tree. North America, from lat. 43° to Georgia. Height in America 10 ft. to 40 ft.; in England 30 ft. to 50 ft. Introduced in 1725. Flowers small, pale yellowish purple; March and April. Keys brown; ripe in July.

Varieties. There are several names in nurserymen's catalogues, such as *A.* coccíneum, *A.* macrocárpum, *A.* flóridum, *A.* Pàvia which are only very slight varieties of *A.* eriocárpum. The last-named variety, introduced by Messrs. Booth, has received the absurd name of *Pàvia*, from the upper surface of the leaves being slightly wrinkled, somewhat in the manner of those of the horsechestnut. As the species seeds freely, endless varieties may be obtained from seed beds.

157. Acer eriocárpum.

Distinguished from *A.* rùbrum by the leaves being more decidedly 5-lobed, the lobes deeply cut, and the whole leaf more tomentose. A very desirable species, from the rapidity of its growth, the graceful divergent direction of its branches, the beauty of its leaves, and the profusion of its early flowers. In mild seasons, these flowers begin to burst from their buds in the first week in January; and they are often fully expanded by the end of February or beginning of March. It requires a deep free soil, and more moisture than most of the other species. It ripens its seeds, both in America and Britain, by midsummer, or earlier; and, if these are immediately sown, they come up, and produce plants which are 8 or 10 inches high by the succeeding autumn.

¶ 14. *A.* RU'BRUM *L.* The red-*flowering, or scarlet,* Maple.

Identification. Lin. Spec., 1496.; Dec. Prod., 1. p. 595.; Don's Mill., 1. p. 650.; Tor. and Gray, 1. p. 249.

Synonymes. *A.* virginiànum *Herm.*; *A.* coccíneum *Ait. & Mich.*; *A.* glaûca *Marsh. Arbust.*; *A.* caroliniànum *Walt.*; *A.* sanguíneum *Spach*; soft Maple, Swamp Maple, red Maple; Erable rouge, *Fr.*; rother Ahorn, *Ger.*

Engravings. Mich. Arb., 2. t. 14.; Schmidt Arb., 1. t. 6.; the plate of the tree in Arb. Brit., 1st edit., vol. v.; our *fig.* 138.; and *fig.* 160. of the leaves, of the natural size, in the plate forming p. 119.

138. Acer rùbrum.

Spec. Char., &c. Leaves cordate at the base, glaucous beneath, deeply and unequally toothed, palmately 5-lobed, with acute recesses. Flowers conglomerate, 5-petaled, pentandrous. Ovaries smooth. (*Don's Mill.*) A large tree with numerous divergent slender branches. Canada to Florida. Height in America 30 ft. to 80 ft.; in England 30 ft. to 60 ft. Introduced in 1656. Flowers small, dark red, appearing a fortnight before the leaves; March and April. Keys brown; ripe in September.

Variety.

♀ A. *r.* 2 *intermèdium* Lodd. seems intermediate between this species and *A.* eriocárpum.

In England distinguished at sight from *A.* eriocárpum by the leaves being much less cut, and less white beneath, and by the tree being generally less vigorous. The red-flowered maple, whether we regard the beauty of its flowers and opening leaves in early spring, its red fruits in the beginning of summer, or its red fol age in autumn, deserves to be considered one of the most ornamental of hardy trees. Contrary to the general character of the maples, this species is said to thrive best in moist soil, which must, however, at the same time, be rich; and, for the tree to attain a large size, the situation ought to be sheltered. In Britain it is chiefly propagated by layers; but, on the Continent, almost always by seeds, which ripen before midsummer, even sooner than those of *A.* eriocárpum, and, if sown immediately, come up the same season. The seeds, even when mixed with soil, do not keep well; and, in general, but a small proportion of those sent home from America vegetate in Europe

♀ 15. *A.* MONSPESSULA'NUM *L.* The Montpelier Maple.

Identification. Lin. Spec., 1497.; Dec. Prod., 1. p. 595.; Don's Mill., 1. p. 649.
Synonymes. *A.* trilobum *Mœnch*; *A.* trifòlium *Duh.*; *A.* trilobàtum *Lam.*; E'rable de Montpellier, *Fr*; Französischer Ahorn, *Ger.*; Acero minore, *Ital.*
Engravings. Schmidt Arb., 1. t. 14.; and Krause, t. 101.; the plate of this species in Arb. Brit., 1st edit., vol. v.; our *fig.* 139.; and *fig.* 161. of the leaves, of the natural size, in the plate forming p. 120.

Spec. Char., &c. Leaves cordate, 3-lobed; lobes almost entire, and equal. Corymbs few-flowered, pendulous. Fruit smooth, with the wings hardly diverging. A low tree. South of Europe. Height 15 ft. to 40 ft. Introduced in 1739 Flowers pale yellow; May. Keys brown; August.

139. *A'cer monspessulanum.*

In general aspect the tree resembles *A.* créticum, which has much shorter footstalks, and coriaceous leaves. It also resembles *A.* campéstre, which, however, has the leaves 5-lobed, while in *A.* monspessulànum they are only 3-lobed. See the figures of leaves in p. 120. and 121. The leaves, in mild seasons, remain on through the greater part of the winter. Seeds; which it ripens in great abundance.

♈ 16. *A.* CAMPE'STRE *L.* The *common, or* Field, Maple.

Identification. Lin. Spec., 1497.; Hayne Dend., p. 211.; Dec. Prod., 1. p. 595.; Don's Mill., 1. p. 649.
Synonymes. E'rable champêtre, *Fr.*; kleiner Ahorn, Feld Ahorn, *Ger.*; Galluzzi, or Pioppo, *Ital.*
Engravings. Engl. Bot., t. 304.; Willd. Abbild., t. 213.; our *fig.* 141.; and *fig.* 162. of the leaves, of the natural size, in the plate forming p. 120.

Spec. Char., &c. Leaves cordate, with 5 toothed lobes. Racemes erect. Wings of fruit much divaricated. (*Don's Mill.*) A low tree or shrub. Britain. Height 15 ft. to 30 ft., sometimes 40 ft. and upwards. Flowers yellowish green; May and June. Keys brown; ripe in September. Decaying leaves yellow. Naked young wood pale brown.

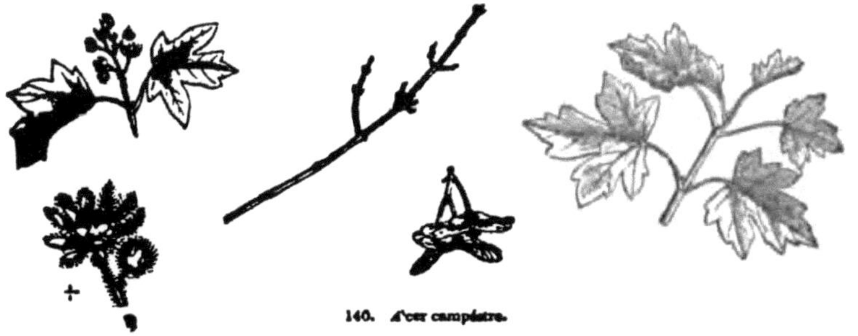

140. A'cer campéstre.

Varieties.

♈ A. *c.* 2 *fòliis variegàtis.*— Next to the variegated-leaved variety of *A.* Pseùdo-*Plátanus*, this seems the handsomest of all the variegated-leaved maples; the leaves preserving, with their variegation, the appearance of health, and the blotches and stripes of white, or whitish yellow, being distinctly marked.

♈ A. *c.* 3 *hebecárpum* Dec. Prod. i. p. 598. *A.* campéstre *Wallr.* in *Litt. Trat. Arch.* i. No. 7; *A.* mólle *Opiz.*— Fruit clothed with velvety pubescence.

♈ A. *c.* 4 *collìnum* Wallr. in *Litt. Dec. Prod.* i. p. 594. *A.* affìne and *A.* macrocárpum *Opiz.*— Fruit smooth. Lobes of leaves obtuse. Flower smaller. Native of France.

♈ A. *c.* 5 *austrìacum* Tratt. Arch. i. No. 6. (The plate of this tree in Arb. Brit., 1st. edit., vol. v.)—Fruit smooth. Lobes of leaves somewhat acuminated. Flowers larger than those of the species. Native of Austria, Podolia, and Tauria. (*Don's Mill.*) This variety is larger in all its parts than the original species, and is of much freer growth; the main stem rises erect and straight, and sends out its branches regularly on every side, so as to form a sort of cone, almost like a fir. A subvariety of this sort, with variegated leaves, is propagated in the Bollwyller Nursery.

141. A. c. hyrcànum.

Other Varieties. A. *c. lævigàtum*, leaves very smooth and shining; A. *c. nànum*, habit dwarf; and, perhaps, some others, are in the collection of Messrs. Loddiges. A. *taùricum*, leaves larger and less divided than in the species; and A. *hyrcànum* (*fig.* 141.) with the leaves variously cut, are also in some collections.

Differing from *A.* monspessulànum in having the flowers produced upon the young shoots; as well as in the racemes of flowers being erect. The wood weighs 61 lb. 9 oz. a cubic foot in a green state, and 51 lb. 15 oz. when perfectly dry. It makes excellent fuel, and the very best charcoal. It is compact, of a fine grain, sometimes beautifully veined, and takes a high polish. It was celebrated among the ancient Romans for tables. The wood of the roots is frequently knotted; and, when that is the case, it is used for the manufacture of snuffboxes, pipes, and other fanciful productions. A dry soil suits this species best, and an open situation. Seeds; which often remain eighteen months in the ground before they vegetate, though a few come up the first spring. The varieties are propagated by layers.

♆ 17. *A.* CRE′TICUM *L.* The Cretan Maple.

Identification. Lin. Spec., 1497.; Dec. Prod., 1. p. 594.; Don's Mill., 1. p. 649.
Synonymes. *A.* heterophýllum *Willd. En.*; *A.* sempervirens *L. Mant.*; *A.* obtusifòlium *Sibthorp*; E′rable de Crète, *Fr.*; Cretischer Ahorn, *Ger.*
Engravings. Flor. Græc., t. 361.; Schmidt Arb., t. 15.; the plate of this species in Arb. Brit., 1st edit., vol. v.; our *fig.* 142., from the Flora Græca; and *fig.* 163. of the leaves, of the natural size, in the plate forming p. 121.

Spec. Char., &c. Leaves permanent, cuneated at the base, acutely 3-lobed at the top. Lobes entire, or toothleted; lateral ones shortest. Corymbs few-flowered, erect. Fruit smooth, with the wings hardly diverging. (*Don's Mill.*) A diminutive, slow-growing, sub-evergreen tree. Candia, and other islands in the Grecian Archipelago. Height 10 ft. to 30 ft. Introd. 1752. Flowers greenish yellow; May and June. Keys brown; ripe in September.

There is a general resemblance between *A.* créticum, *A.* monspessulànum, and *A.* campéstre; but the first is readily known from both, by its being evergreen, or sub-evergreen, and by its leaves having shorter footstalks, and being less deeply lobed. In a young state, the leaves are often entire or nearly so. It is oftener seen as a shrub than as a tree; and it seems to thrive better in the shade than any other *A*cer. Seeds, layers, or grafting on *A.* campéstre.

142. *A′cer créticum.*

Other Species of A′cer.—*A.* barbàtum *Michx.*, given in our first edition, has been omitted, because the plant in the Hort. Soc. Garden has always appeared to us nothing more than *A.* platanöìdes, and because Torrey and Gray consider it a doubtful species, and probably described by Michaux from "specimens of *A.* sacchárinum; the only species, so far as we know, which has the sepals bearded inside." (*Tor. and Gray*, i. p. 249.) *A.* opulifòlium given in our first edition as a species, we have now satisfied ourselves, from having been able to examine larger plants, is nothing more than a variety of *A.* Pseùdo-*Plátanus* diminished in all its parts. There are several names of species of *A*′cer in the works of European botanists, the plants of which would require to be procured and studied in a living state: such as A. *granaténse* Bois., a native of Spain; A. *parvifòlium* Tausch; also some natives of the Himalayas; and the following in North America as given by Torrey and Gray; A. *glàbrum* Torr., a shrub of the Rocky Mountains; A. *tripartìtum* Nutt. MSS., a shrub of the Rocky Mountains allied to *A.* glàbrum; A. *grandidentàtum* Nutt. MSS., a shrub or low tree from the Rocky Mountains, supposed to be the same as *A.* barbàtum *Douglas*, mentioned in Hooker's *Flor. Bor. Amer.*, i. p. 112. The names of several other species, not yet introduced, will be found in the first edition of this work.

A'cer oblóngum. The oblong-*leaved* Maple.

Leaves of the natural size.

143

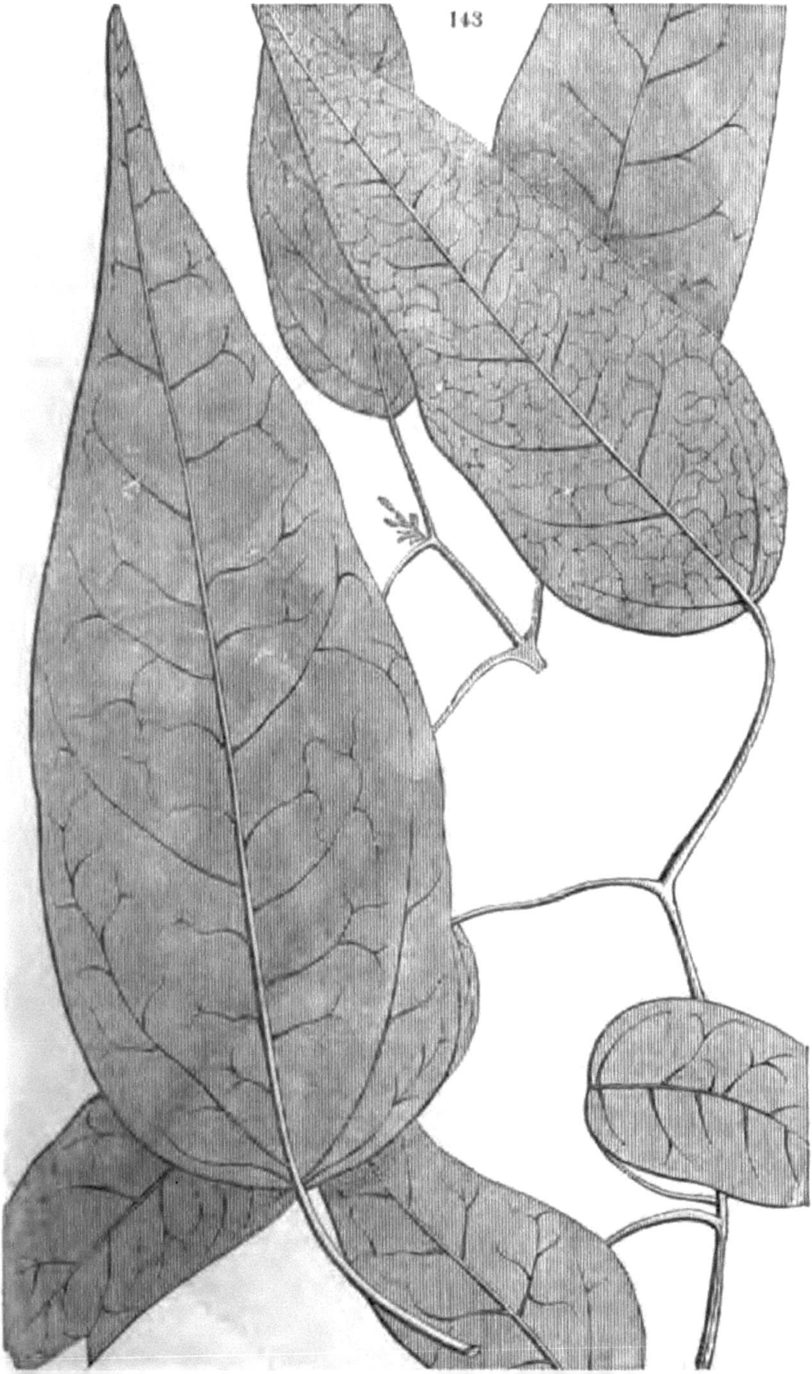

A`cer tatáricum. The Tartarian, *or entire-leaved,* Maple.
Leaves and fruit of the natural size.

144

A'cer *spicàtum*. The spike-*flowered, or mountain*, Maple.

Leaves and fruit of natural size.

145

A'cer striàtum. The striped-
Leaves of

146

bark, or Pennsylvanian, **Maple.**
the natural size.

146

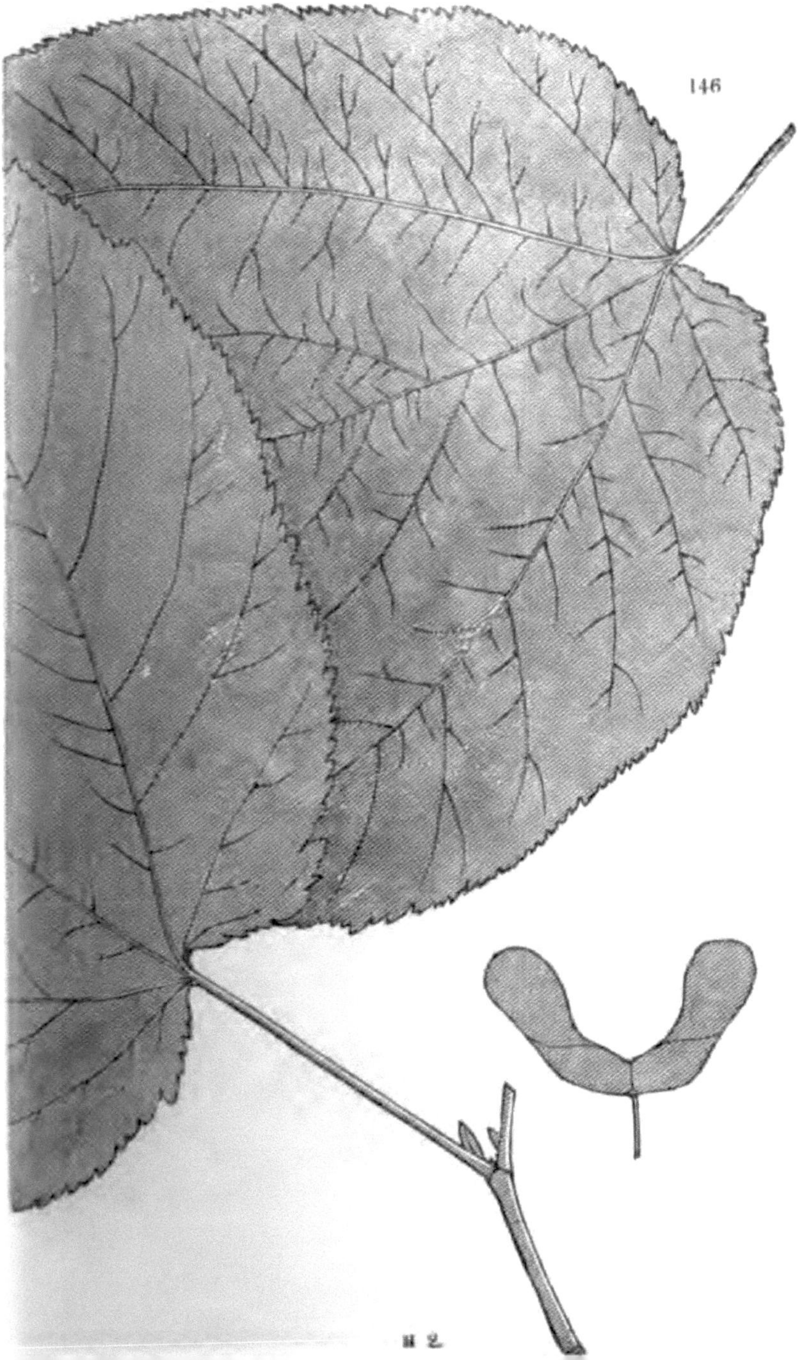

A'cer macrophýllum. The
Part of leaf, and

147

large-leaved Maple. Plate I.
fruit, of the natural size.

147

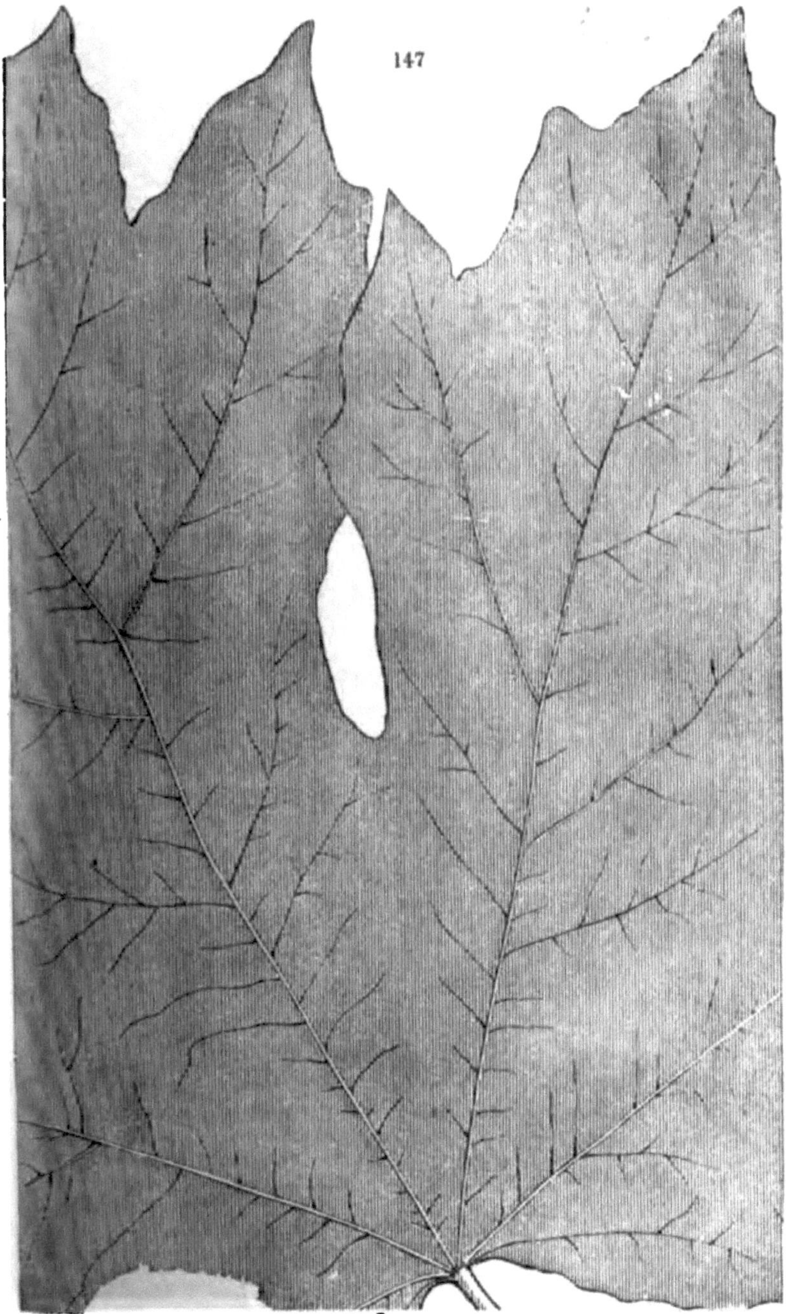

A`cer *macrophýllum.* The

Smaller leaves, also of the natural size,

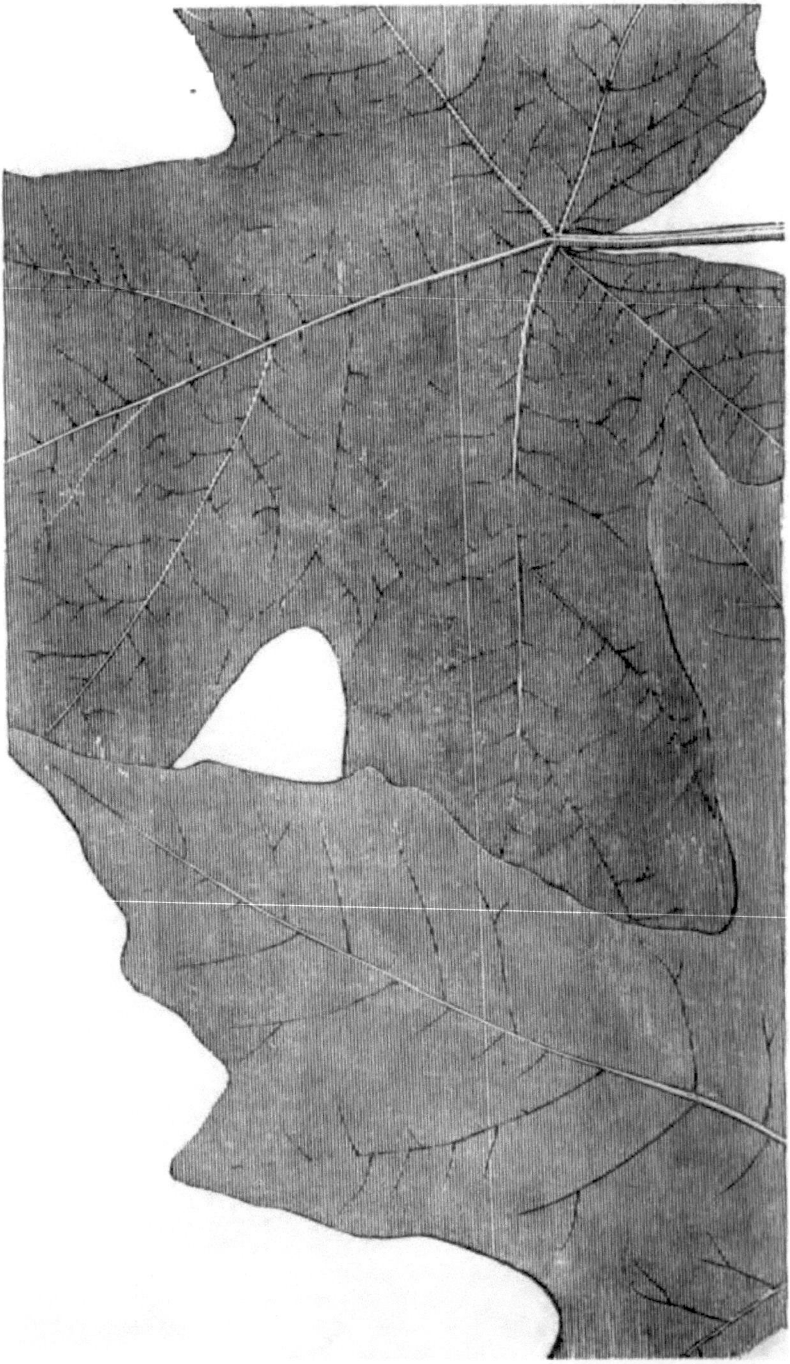

large-leaved Maple. Plate II.

to show how much they vary on the same tree.

148

A'*cer platanoïdes*. The Platanus-

Leaves and fruit of

149

ıike, *or Norway*, Maple.
the natural size.

149

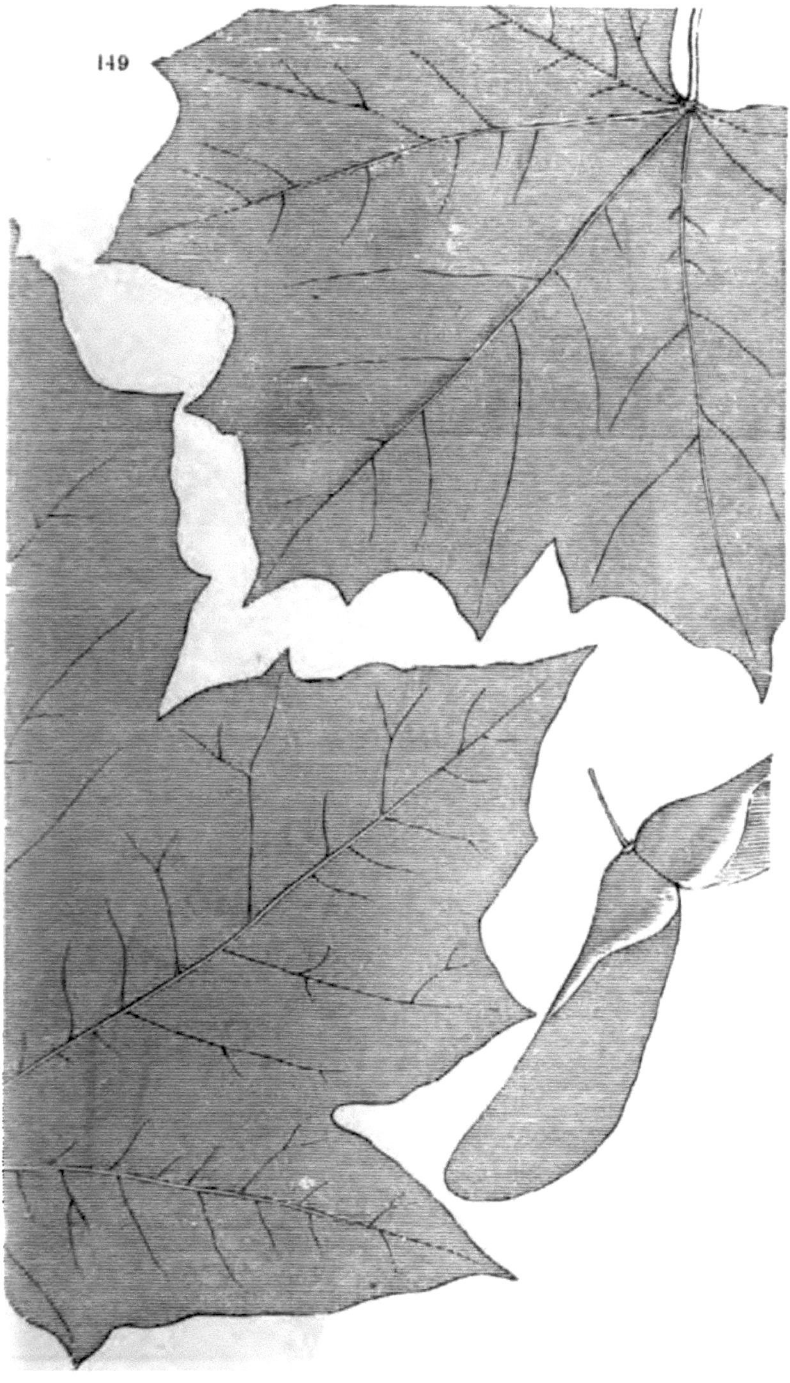

A`cer (*platanoïdes*) *Lobèlii*. L'Obel's Maple.

Leaves of the natural size.

150

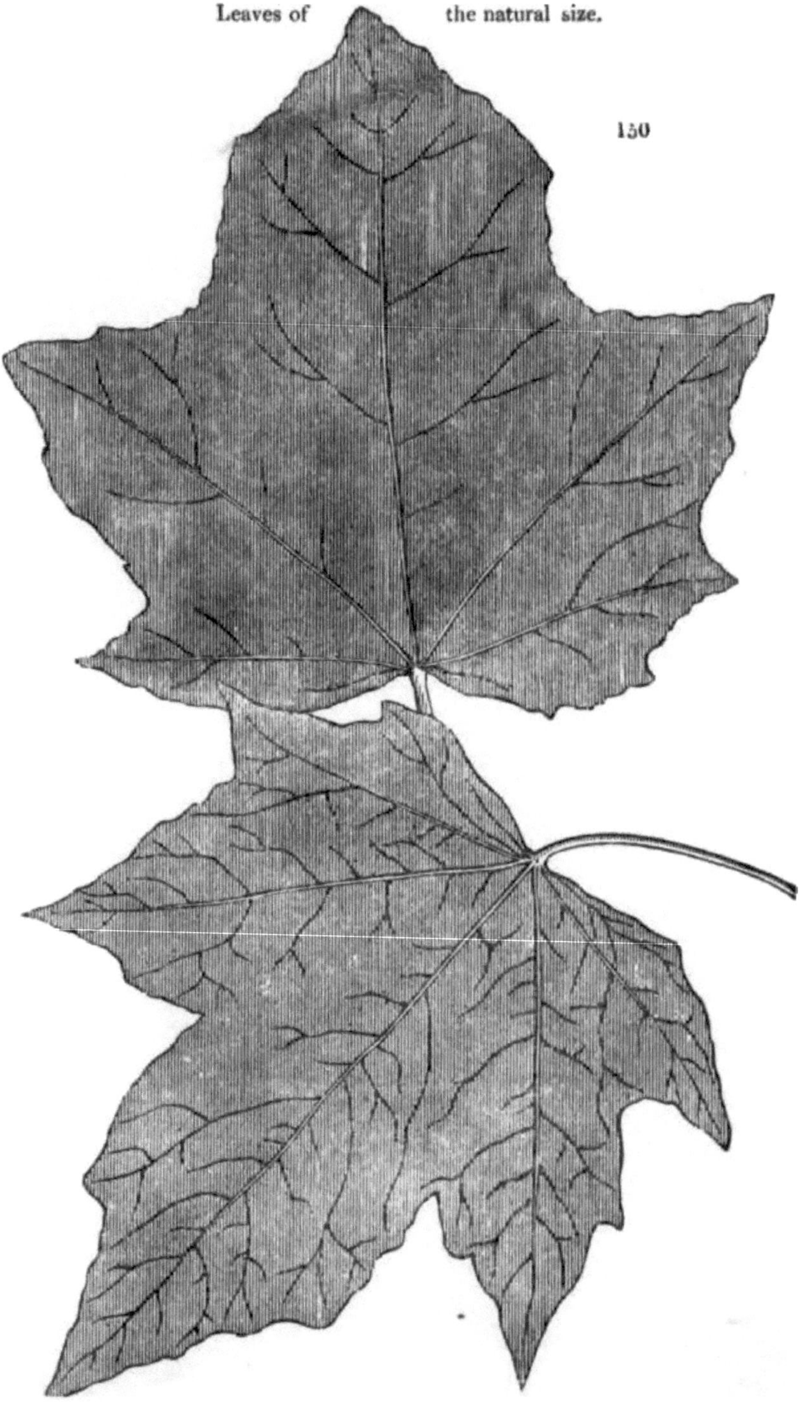

A'*cer platanöìdes laciniàtum.*
The cut-*leaved* Platanus-like, *or Eagle's claw*, Maple.

Leaves of natural size.

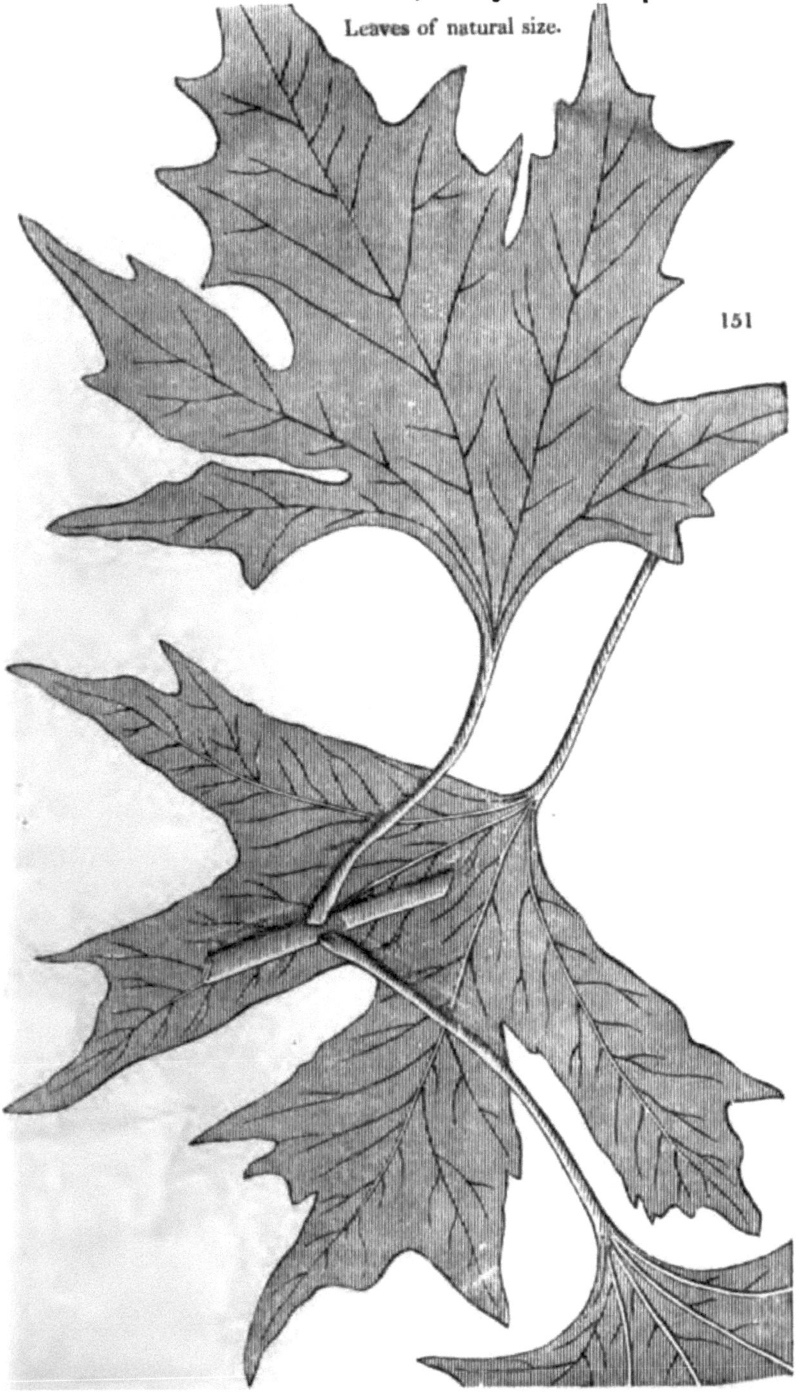

151

A`cer sacchárinum.
The leaves and fruit

152

The Sugar Maple.
of the natural size.

152

A`cer *Pseùdo*-*Plátanus*. The

Leaves and fruit

153

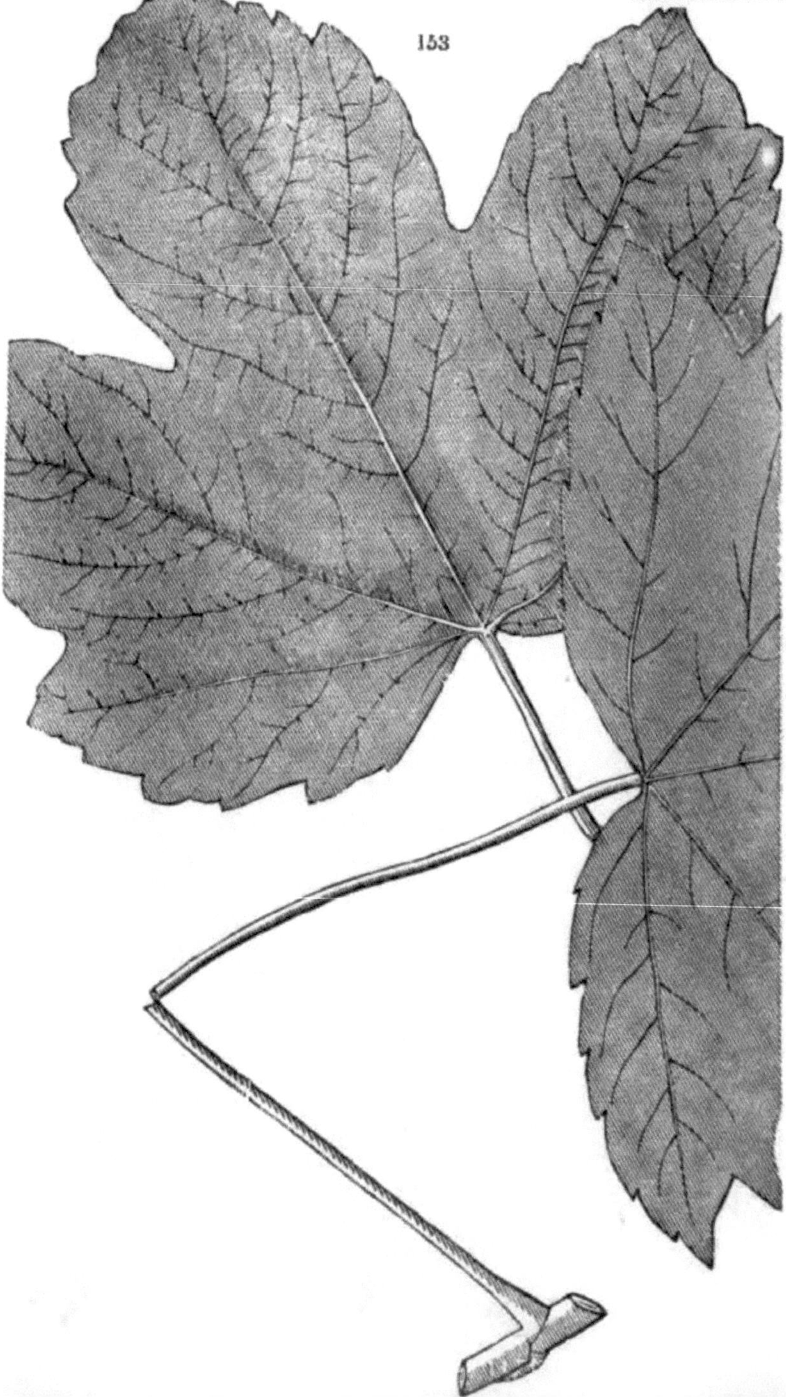

False Plane, *or Sycamore,* Maple.
of the natural size.

153

A ***cer obtusàtum.*** **The obtuse-**
Leaves of the

154

lobed-leaved, or Neapolitan, Maple.
natural size.

154

1

A'cer *Pseùdo-Plátanus opulifòlia.*
The Opulus-leaved False Plane, *or Sýcamore.*
Leaves and fruit of the natural size.

155

A`cer O'palus. The Opal, *or Italian*, Maple.

Leaves of the natural size.

156

A'cer circinàtum. The round-*leaved* Maple.

Leaves of the natural size.

157

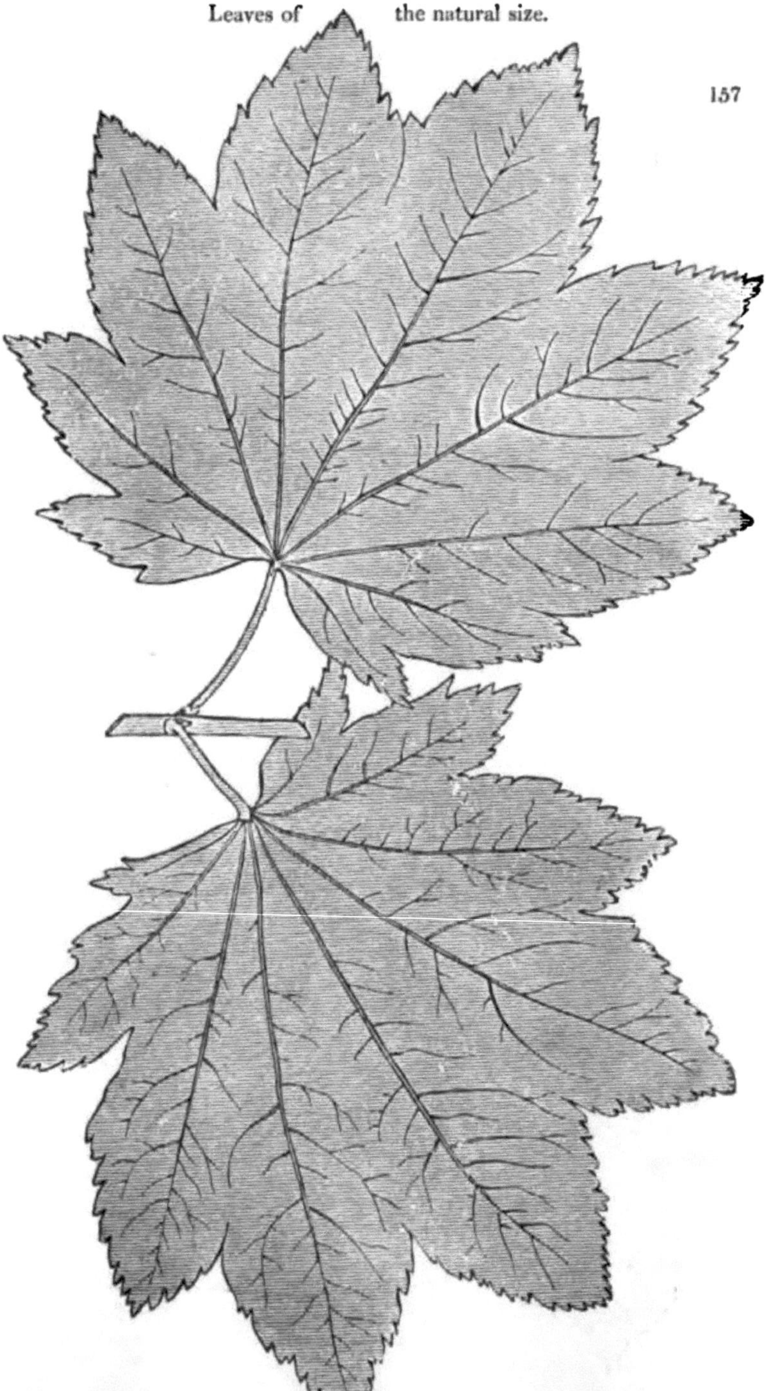

A'*cer palmàtum*. The palmate-*leaved* Maple.

Leaves of the natural size.

158

A'cer eriocárpum. The woolly-fruited Maple.

Leaves and fruit of the natural size.

159

A'*cer rùbrum.* The red-*flowered* Maple.

Leaves and fruit of the natural size.

160

A'*cer monspessulànum, and* A. *campéstre.* The Montpelier Maple, and the *common*, or field, Maple.

Leaves and fruit of the natural size.

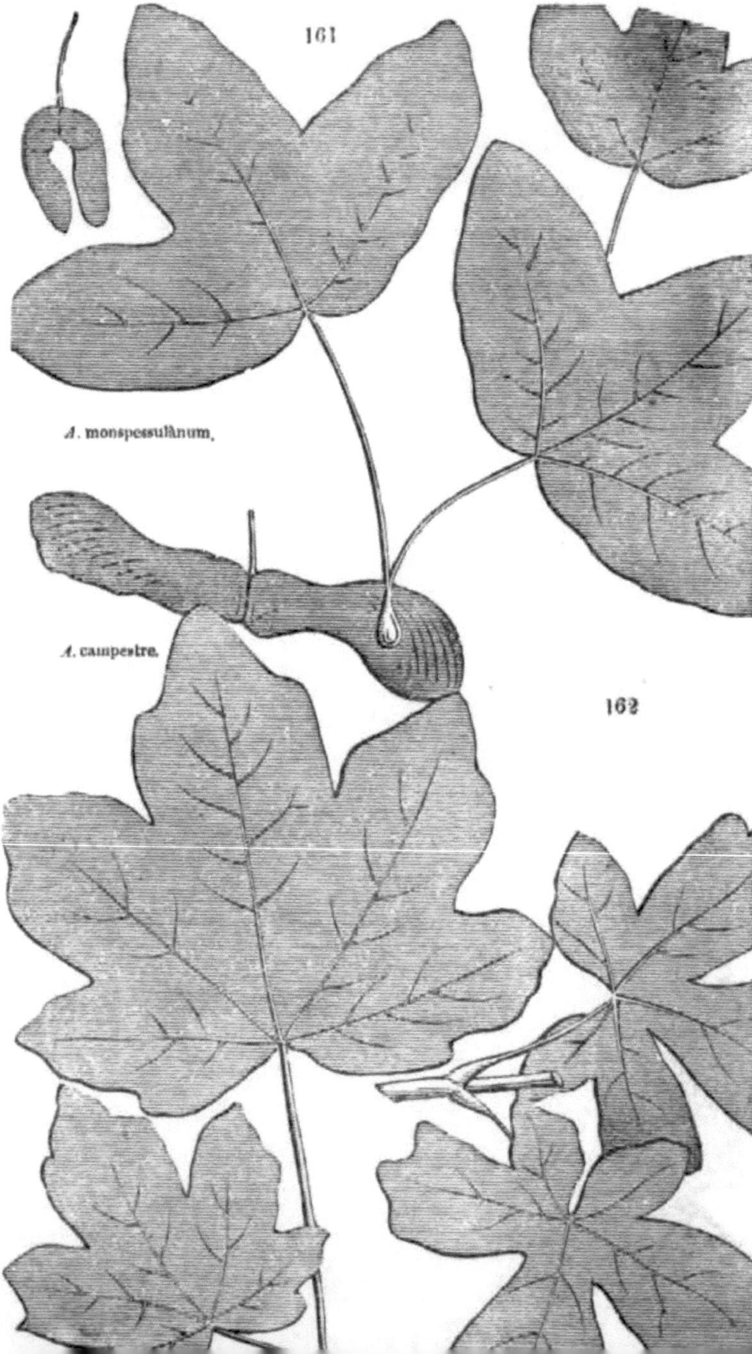

A'*cer créticum.* The Cretan, *or various-leaved,* Maple.

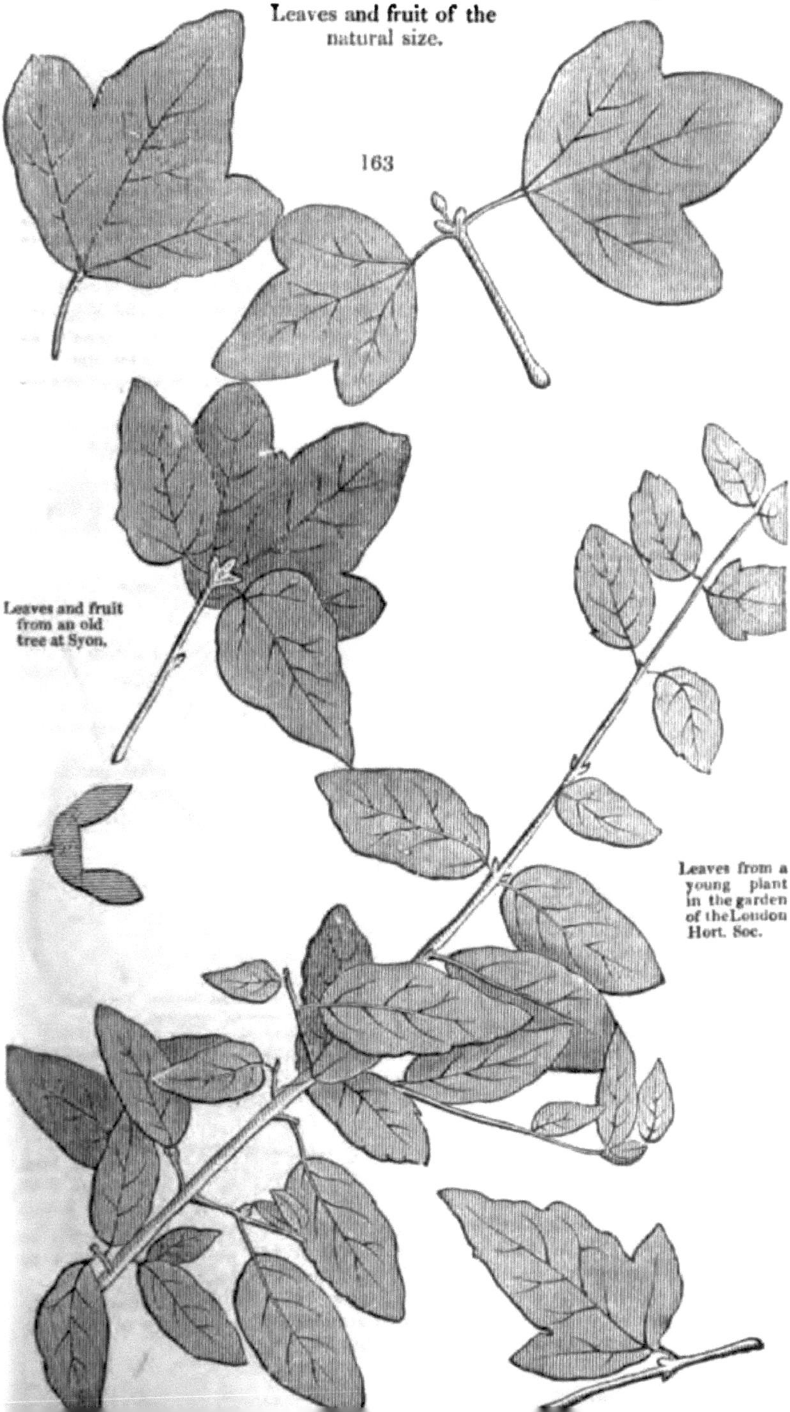

Genus II.

NEGU'NDO Mœnch. The Negundo, *or* Box Elder. *Lin. Syst.* Diœ'cia Pentándria.

Identification. Mœnch Meth., 334.; Dec. Prod., 1. p. 596.; Don's Mill., 1. p. 647. and 651.
Synonymes. A'cer *Lin.*; *Negúndium* Rafinesque.
Derivation. This genus was constituted from A'cer Negúndo L.; but the meaning of the latter word is unknown. Probably, it may be merely the Illinois name of Gigueres (from *giguer*, to romp, alluding to the tremulous and playful motion of the long pinnated leaves) Latinised.

Gen. Char. *Sexes* diœcious. *Flowers* without a corolla. *Calyx* with 4—5 unequal teeth. *Male flowers* upon thread-shaped pedicels, and disposed in fascicles; *anthers* 4—5, linear, sessile. *Female flowers* disposed in racemes. (*Dec. Prod.*) — Deciduous trees, natives of North America.

Leaves compound, opposite, exstipulate, deciduous; impari-pinnate. — There is only one species in British gardens.

↑ 1. *N.* fraxinifo'lium *Nutt.* The Ash-leaved Negundo.

Identification. Nutt. Gen. Amer., 1. p. 253.; Dec. Prod., 1. p. 596.; Don's Mill., 1. p. 651.
Synonymes. A'cer *Negúndo* L., *Mich. Arb.*; *N. aceröides Mœnch* and *Torr. & Gray*; *Negúndium* americànum *Rafin.*; the Ash-leaved Maple, the Black Ash; E'rable à Feuilles de Frêne, *Fr.*; E'rable à Giguières, *Illinois*; Eschenblättriger Ahorn, *Ger.*; Acero americano, *Ital.*
Engravings. Mich. Arb., 2. t. 16.; Schmidt Arb., 1. t. 12.; Wats. Dend., t. 172.; the plate of this species in Arb. Brit., 1st edit., vol. v.; and our *fig.* 164. from Schmidt.

164. *Negúndo fraxinifòlium.*

Spec. Char., &c. Leaves of from 3 to 5 leaflets, the opposite ones coarsely and sparingly toothed, the odd one oftener 3-lobed than simple. (*Dec. Prod.*) A deciduous tree, of the middle size. Canada to Carolina. Height 15 ft. to 30 ft.; in England 30 ft. to 40 ft. Introduced in 1688. Flowers yellowish green, appearing with the leaves; April. Keys brown; ripe in August. Decaying leaves of a rich yellow. Naked young wood smooth, and of a fine pea green. The tree in the Hort. Soc. Garden is a male; but there is a female plant in the collection of W. Borrer, Esq., Henfield.

Varieties.

↑ N. f. 2 *críspum* G. Don. (The plate in Arb. Brit., 1st edit., vol. v.; and our *fig.* 165.) — Leaves variously cut and curled. The plant of this variety in the arboretum of the Hort. Soc. is a male: the inflorescence consists of pendulous panicles of flowers, that are green, with some redness from the colour of the anthers; and each is placed upon a slender peduncle of about 1 in. long.

↑ N. f. 3 *violàceum* Booth. — Young shoots covered with a violet bloom. This appearance is not uncommon in the young shoots of different species of A'cer as well as in *Negúndo.*

105. *Negúndo fraxinifòlium críspum.*

A rapid-growing tree; very ornamental, from its compound leaves, and the fine pea-green of its young shoots; arriving at maturity in 15—20 years. American seeds, which ought to be sown as soon as possible, or layers, in any common soil.

Other Species of Negúndo.—*N* califórnicum *Tor. & Gray*, found by Douglas, is supposed to be a new species; but neither fruit, nor full-grown leaves, are described.

Order XIV. ÆSCULA'CEÆ.

Synonymes. Castanàceæ *Link*; Hippocastàneæ *Dec.*

Ord. Char. *Calyx* campanulate, 5-lobed. *Ovary* roundish, trigonal. *Seeds* large and globose; albumen wanting. *Embryo* curved, inverted; with fleshy, thick, gibbous cotyledons, not produced above ground in germination. *Plumule* large, 2-leaved. — Deciduous trees, natives of North America and Asia.

Leaves compound, opposite, exstipulate, deciduous; leaflets 5—7, serrated. *Flowers* terminal, in racemes, somewhat panicled. — All the known plants of this order cross-fecundate freely, and by most botanists they are included in one genus; but so numerous are the garden varieties, that we have thought it more convenient to follow those authors who separate the species into two genera. These are *Æ'sculus* and Pàv*ia*, which are thus contradistinguished: —

Æ'sculus *L.* Capsule echinated.
Pa'via Boerh. Capsule smooth.

Genus I.

Æ'SCULUS *L.* The Horsechestnut. *Lin. Syst.* Heptándria Monogýnia.

Identification. Lin. Gen., No. 462.; Dec. Prod., 1. p. 597.; Don's Mill., I. p. 562.
Synonymes. Hippocástanum *Tourn.*; Marronier d'Inde, *Fr.*; Rosskastanie, *Ger.*
Derivation. The word Æ'sculus, derived from *esca*, nourishment, is applied by Pliny to a species of oak, which had an eatable acorn. The word Hippocástanum, from *hippos* a horse, and *castanea*, a chestnut, is said by some to have been given to this tree ironically, the nuts, though they have the appearance of sweet chestnuts, being only fit for horses; and by others, because it is said the nuts are used in Turkey, for curing horses of pulmonary diseases.

Gen. Char. *Calyx* campanulate. *Petals* 4—5, expanded, with an ovate border. *Stamens* with the filaments recurved inwardly. *Capsules* echinated. *Leaflets* sessile, or almost sessile. (*Don's Mill.*) — Deciduous trees, natives of Asia and North America.

Leaves palmately divided, with stalked leaflets, generally rough. *Capsule* rough. *Buds* generally covered with resin. — Two species and several varieties are in British gardens.

The common horsechestnut is invariably propagated by the nuts, which are sown when newly gathered, or in the following spring; and in either case they will come up the succeeding summer. All the other sorts, as being varieties of the species, are propagated by budding or grafting. Soil deep sandy loam. Only the first three sorts described below can be considered as true horsechestnuts; the remainder, to which some other names might be added, we consider as hybrids between Æ'sculus and some kind of Pàvia, most probably P. flàva.

♀ 1. Æ. Hippoca'stanum *L.* The *common* Horsechestnut.

Identification. Lin. Sp., 488.; Dec. Prod., 1. p. 597.; and Don's Mill., 1. p. 652.
Synonymes. Hippocástanum vulgàre *Tourn.*; Marronier d'Inde, *Fr.*; gemeine Rosskastanie, *Ger.*; Marrone d'India, *Ital.*
Engravings. Woodv. Med. Bot., t. 128.; the plate of this species in the Arb. Brit., 1st edit., vol. v.; and our *fig.* 166.

Spec. Char., &c. Leaflets 7, obovately cuneated, acute, and toothed. A large deciduous tree. Asia and North America. Height 50 ft. to 60 ft. Introduced in 1629. Flowers white, tinged with red; May. Fruit brown; ripe in October. Decaying leaves dark brown. Naked young wood brown. Buds long, large, greenish brown, covered with resin.

Varieties.

- ♀ Æ. *H.* 2 *flòre plèno.*—Recorded in nurserymen's catalogues, but not common.
- ♀ Æ. *H.* 3 *aúreo-variegàtum.* — The leaves are blotched with yellow, but they have a ragged and unhealthy appearance, and are by no means ornamental.
- ♀ Æ. *H.* 4 *argénteo-variegàtum.* — Leaves blotched with white.
- ♀ Æ. *H.* 5 *incìsum* Booth. Æ. asplenifòlia *Hort.* — Leaflets cut into shreds.

Other Varieties. In Booth's *Catalogue* are the names Æ. H. críspum, nìgrum, præ'cox, striàtum, tortuòsum, &c., but none of these, nor any other variety which we have seen, is worth culture.

A tree of the largest size, with an erect trunk, and a pyramidal head. The leaves are large, of a deep green colour, and singularly interesting and beautiful when they are first developed. When enfolded in the bud, they are covered with pubescence, which falls off as the leaves expand. The growth, both of the tree and of the leaves, is very rapid; both shoots and leaves being sometimes perfected in three weeks from the time of foliation. The wood weighs, when newly cut, 60 lb. 4 oz. per cubic foot; and, when dry, 35 lb. 7 oz.; losing, by drying, a sixteenth part of its bulk. It is soft, and

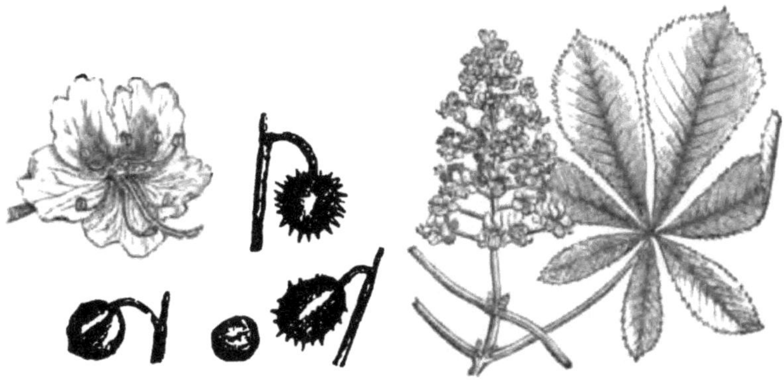

166. Æ'sculus Hippocástanum.

unfit for use where great strength, and durability in the open air, are required; nevertheless, there are many purposes for which it is applicable when sawn up into boards; such as for flooring, linings to carts, packing-cases, &c. The nuts may be used when burned as a kind of ley, or substitute for soap. The nuts, if wanted for seed, should be gathered up as soon as they drop, and either sown or mixed with earth; because, if they are left exposed to the air, they will lose their germinating properties in a month. Some nurserymen cause the nuts to germinate before sowing them, in order to have an opportunity of pinching off the extremity of the radicle; by which means the plants are prevented from forming a taproot; or, at least, if a taproot be formed, it is of a much weaker description than it otherwise would be, and the number of lateral fibres is increased; all which is favourable for transplanting. When the tree is intended to attain the largest size, in the shortest time. the nut ought to be sown where the tree is finally to remain; because the use of the taproot is mainly to descend deep into the soil, to procure a supply of water, which, in dry soils and seasons, can never be obtained in sufficient quantities by the lateral roots, which extend themselves near the surface in search of nourishment and air.

¥ 2. Æ. (H.) OHIOE'NSIS *Michx*. The Ohio Æsculus, *or Horsechestnut*.

Identification. Mich. Arb., 3. p. 242.; Dec. Prod., 1. p. 597.; Don's Mill., 1. p. 652.
Synonymes. Æ. ohioénsis *Lindl.*; ? Æ. pállida *Willd.*; Æ. echinàta *Muhl.*; Æ. glàbra *Tor. & Gray*; Pàvia ohioénsis *Michx.*; Pàvia glàbra *Spach*; Ohio Buckeye, fetid Buckeye, *Amer.* All these synonymes appear to us very doubtful, when we compare them with the tree bearing this name in the Hort. Soc. Gardens, and with Dr. Lindley's description of it in Bot. Reg. for 1838, t. 51.
Engravings. Michx. Arb., 2. t. 92.; Bot. Reg., 1838, t. 51.; and our *fig*. 167. from Michaux.

Spec. Char., &c. Stamens nearly twice the length of the (yellowish white) corolla; petals 4, spreading, a little unequal, the claw scarcely the length of the campanulate calyx; thyrsus racemose, loosely flowered; leaflets 5, oval or oblong, acuminate, fine and unequally serrate, glabrous. (*Tor. and Gray*, i. p. 251.) A deciduous tree of the middle size. Pennsylvania and Virginia. Height in America 10 ft. to 30 ft.; in the climate of London apparently the same as the common horsechestnut. Introduced in ? 1820. Flowers white, yellow, and red; May and June. Fruit brown; ripe in October. Bark rough, fetid. Branches of the thyrsus of flowers short, 4—6-flowered; the flowers mostly unilateral, small (not half the size of those of the common horsechestnut). Fruit prickly, resembling that of the cultivated horsechestnut. but scarcely half the size. (*Tor. and Gray*, i. p. 251.)

167. Æ. (H.) ohioénsis.

According to Michaux, the American horsechestnut is commonly a bush or low tree, from 10 ft. to 12 ft. in height; but it is sometimes 30 or 35 feet high, trunk 12 or 15 inches in diameter. He found it only on the banks of the Ohio; but Torrey and Gray give as its habitats the western parts of Pennsylvania, Virginia, Ohio, Kentucky. The tree in the Hort. Soc. Garden is of equally vigorous growth with the common horsechestnut; the leaves are larger, and of a bright green: on the supposition that this is the Æ. ohioénsis of *Michx.* and *Tor. & Gray*, we have no doubt in our own mind that it is only a variety of the common horsechestnut, Dr. Lindley, however, is of a different opinion, considering it as a distinct species. (See *Bot. Reg.*, 1838, t. 51.)

♀ 3. Æ. (H.) RUBICU'NDA *Lois.* The reddish-*flowered* Æsculus, *or Horse chestnut.*

Identification. Loiseleur Herb. Amat.; Dec. Prod., 1. p. 597.; Don's Mill., 1. p. 652.
Synonymes. Æ. cárnea *Hort.*, and *Lindl. Bot. Reg.*; Æ. ròsea *Hort.*; Æ. coccinea *Hort.*; Æ. Hippocástanum var. rubicúndum *Schubert*; Æ. Watsonidna Spach; Marronier rubicund, *Fr.*; scharlachrothe Rosskastanie, *Ger.*; Whitley's fine scarlet.
Engravings. Herb. Amat, t. 367.; Hayne Abbild., t. 22.; Bot. Reg., t. 1056., as Æ. cárnea; Wats. Dendr., t. 121., as Æ. cárnea; the plate of this species in Arb. Brit., 1st edit., vol. v.; and our *fig.* 168.

Spec. Char., &c. Petals 4, with the claws shorter than the calyx. The flowers are scarlet, and very ornamental; the leaves of a deeper green than those of any other sort, and they have a red spot at the base of the petioles of the leaflets on the under side. The flowers come out of a dark red, and die off still darker. Fruit prickly. A deciduous tree, below the middle size. ? Hybrid from North America, Height 20 ft. to 30 ft. Cultivated in 1820. Flowers red; May and June. Fruit brown; ripe in October.

It is doubtful whether this tree is a native of North America, or originated in British gardens. It passes under different names in different nurseries,

168. Æ'sculus (Hippocástanum) rubicúnda.

as will be seen by our list of synonymes, and may be considered as differing little, if at all, from Æ. cárnea *Lindl.* It is distinguished from Pàvia rùbra by its larger and rougher leaves; and from Æ. Hippocástanum by the leaves being fuller and more uneven on the surface, and of a deeper green. The tree is also smaller, and of much less vigorous growth than the common horsechestnut. It is, without doubt, the most ornamental sort of the genus.

Variety.

Ɏ Æ. (*H.*) 2 *r. ròsea.* Æ'sculus ròsea *Hort.*—This variety differs from Æ. (H.) rubicúnda, in having the leaflets without a red spot at the base of the petioles. The flowers come out of a pale red, and die off about the same shade as the flowers of Æ. (H.) rubicúnda are when they first appear.

Other Varieties. There are several names in gardens, and in nurserymen's catalogues, which appear to belong to Æ. (H.) rubicúnda, but how far they are worth keeping distinct, we are very doubtful. *Whitley's new scarlet,* of which there is an imported tree in the Fulham Nursery, is said to have flowers of a darker scarlet than any of the above-named varieties; and, if so, it may be recorded as Æ. (H.) r. 3 Whitlèii. Æ. (*H.*) *americàna* of the same nursery belongs also to Æ. rubicúnda.

Ɏ 4. Æ. GLA'BRA *Willd.* The smooth-*leaved* Æsculus, *or Horsechestnut.*

Identification. Willd. Enum., p. 405.; Dec. Prod., 1. p. 597.; Don's Mill., 1. p. 652.
Engravings. Hayne Abbild., t. 24.; and our *figs.* 169, 170.

Spec. Char., &c. Claws of the petals of about the length of the calyx. Leaflets of a pale green, very smooth. Flowers of a greenish yellow. A deciduous low tree. North America. Height 20 ft. to 30 ft. Introduced in 1812. Flowers yellow; June. Fruit brown; ripe in October.

This sort is very distinct; but it is evidently not the Æ'sculus glàbra of Torrey and Gray, but probably a variety of the Æ'sculus flàva of these authors, with rough fruit. The whole plant is comparatively glabrous, and even the fruit partakes of that quality. The tree is of less vigorous growth than Æ. rubicúnda; and the shoots take a more upright direction. It appears to lose its leaves sooner than most of the other sorts.

169. Æ'sculus glàbra.

170. Æ'sculus glàbra.

Ɏ 5. Æ. (G.) PA'LLIDA *Willd.* The pale-*flowered* Æsculus, *or Horsechestnut.*

Identification. Willd. Enum., p. 406.; Hayne Dend., p. 44.; Dec. Prod., 1. p. 597.; Don's Mill., 1. p. 652.
Synonyme. Gelbliche Rosskastanie, *Ger.*
Engravings. Hayne Abbild., t. 25.; and our *fig.* 171.

Spec. Char., &c. Petals with the claws shorter than the calyx. Stamens twice as long as the corolla. A deciduous low tree. Originated in gardens. Height 20 ft. to 30 ft. Cultivated in 1812. Flowers pale yellow; May and June. Fruit brown; ripe in October.

This sort so closely resembles Æ. glàbra, as to leave no doubt in our mind of its being a variety of that species. It is of somewhat more robust growth, and the leaves are, perhaps, not quite so smooth.

171. *Æ'sculus glàbra pállida.*

Other Kinds of Æ'sculus are enumerated in some nurserymen's catalogues; but we know of none worth notice, except those already recorded.

GENUS II.

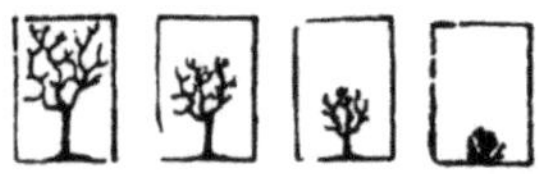

PA'VIA Boerh. THE PAVIA, *BUCKEYE, or SMOOTH-FRUITED HORSECHESTNUT TREE.* *Lin. Syst.* Heptándria Monogýnia.

Identification. Boerh. Lugd., t. 260.; Dec. Prod., 1. p. 598.; Don's Mill., 1. p. 652.
Derivation. In honour of *Peter Paw*, a Dutch botanist, once professor of botany at Leyden. Buckeye has reference to the conspicuousness of the hilum of the seed, when taken out of the husk.

Gen. Char. *Calyx* tubular. *Petals* 4, erect, narrow. *Stamens* straight. *Capsules* unarmed. (*Don's Mill.*) — Middle-sized deciduous trees or shrubs, natives of North America; distinguishable from the horsechestnuts by the smoothness of their fruit, and the comparative smallness of their flowers, which have their petals erect and narrower.

Leaves palmate, with 5—7 leaflets, smooth. *Flowers* small, with erect and narrow petals. *Buds* blunt, not covered with resin. — There are three species, and several varieties or hybrids, in British gardens.

Distinguished from the common horsechestnuts, by being smaller and smoother in all their parts. There are probably only three aboriginal species; but there are several beautiful garden varieties, or hybrids. Culture the same as for the common horsechestnut.

♆ 1. P. RU'BRA *Lam.* The red-*flowered* Pavia.

Identification. Lam. Illust.; Dec. Prod., 1. p. 598.; Don's Mill., 1. p. 653.
Synonymes. *Æ'sculus* Pàvia Lin and Tor. & Gray; *Æ.* Pàvia var. *α* rùbra *Hayne Dend.* p. 44.; Pàvia parviflòra *Hort.*; small Buckeye, *Amer.*; Marronier Pavie, or Pavie à Fleurs rouges, *Fr.*; rothe Rosskastanie, *Ger.*; Marrone di Paw, *Ital.*
Engravings. Lam. Illust., t. 273., Hayne Abbild., t. 21.; Wats. Dend., t. 120.; Krause, t. 55.; the plate of the tree in Arb. Brit., 1st edit., vol. v.; and our *fig.* 173.

Spec. Char., &c. Corolla of 4 petals, that are longer than the stamens. Leaflets 5, elliptic-oblong, tapered to both ends, and smooth, as is the petiole; axils of the nerves hairy on the under surface of the leaf. (*Dec. Prod.*) A slender-growing tree. Virginia and Carolina, on mountains. Height 10 ft. to 20 ft. Introduced in 1711. Flowers brownish scarlet; May and June. Fruit brown; ripe in October. Decaying leaves brown. Naked young wood reddish brown.

Varieties.

♀ *P. r.* 2 *argùta* G. Don. (B. Reg., t. 993.; our *fig.* 172.) — A handsome small tree, with dark brownish red flowers, differing little from those of P. rùbra. Introduced in ? 1820.

172. Pàvia rùbra argùta.

♀ *P. r.* 3 *sublaciniata* Wats. Dend. Brit. t. 120. Æ. P. serràta *Hort.* — Leaflets acutely serrated: in other respects it differs little from the species.

■ *P. r.* 4 *hùmilis*. P. hùmilis *G. Don*; and Æ'sculus hùmilis *Lodd.* (Bot. Reg., t. 1018; and our *fig.* 173.) — A diminutive, weak, straggling form of the species, probably obtained from some sport, and which, on its own root, is only a re-

173. Pàvia rùbra hùmilis.

cumbent bush, from 2 ft. to 3 ft. in height; but which, when grafted on the common horsechestnut, forms the very beautiful pendulous low tree of which there is a plate in the *Arb. Brit.*, 1st edit., vol. v.

In addition to these varieties, there are the three forms which are enumerated below.

In its native country this species varies in magnitude from a low rambling shrub to a tree of 20 ft. or more in height. In England P. rùbra is in cultivation in various forms: as a tree, in which character it has, at Syon (see our plate in the *Arb. Brit.*, 1st edit., vol. v.), attained the height of 26 ft.; as a pendulous tree of 12 or 14 feet in height (see our plate in the *Arb. Brit.*,

174. Pàvia rùbra.

1st edit., vol. v., under the name of P. r. péndula); and as a trailing shrub, under the name of P. hùmilis, in the London Hort. Soc. Garden, and in the arboretum of Messrs. Loddiges.

♆ 2. P. FLA'VA *Dec.* The yellow-*flowered* Pavia.

Identification. Dec. Prod., 1. p. 598.; Don's Mill., 1. p. 653.
Synonymes. Æ'sculus flàva *Ait.*, *Hayne*, and *Tor. & Gray*; Æ. lùtea *Wangh.*; Pàvia lùtea *Poir.*; the sweet Buckeye, big Buckeye, *Amer.*; the yellow Pavia; the yellow Horsechestnut.
Engravings. Wats. Dend. Brit., t. 163.; the plate in Arb. Brit., 1st edit., vol. v.; and our *fig.* 175.

Spec. Char., &c. Petioles pubescent, flattish towards the tip. Leaflets 5—7, pubescent beneath, and above upon the nerves. (*Dec. Prod.*) A deciduous tree of the middle size. Virginia and Georgia, in fertile valleys. Height 30 ft. to 60 ft. in America; 30 ft. to 40 ft. in England. Introduced in 1764. Flowers yellow; April and May. Fruit brown; ripe in October. Decaying leaves yellow, tinged with brown. Naked young wood yellowish brown.

A more vigorous and rigid-growing tree than P. rùbra, with the branches

175. Pàvia flàva.

upright; whereas in P. rùbra they are spreading, slender, and pendulous. Leaves paler than in P. rùbra. To thrive, it, like all the other Æsculàceæ, requires a deep rich soil. Propagated by budding, because the colour of the flowers is found to vary much in plants raised from seed.

♀ 3. P. (F.) NEGLE'CTA *G. Don.* The neglected Pavia.

Identification. Loud. Hort. Brit., p. 143.; Don's Mill., 1. p. 653.; Swt. Hort. Brit., p. 83.
Synonyme. Æ'sculus neglécta *Lindl.* in *Bot. Reg.*
Engravings. Bot. Reg., t. 1009.; and our *fig.* 176.

176. Pàvia (flàva) neglécta.

Spec. Char., &c. Leaflets 5, lanceolate, serrulated, tapering to the base, flat, rather plicate, smooth beneath, but pilose in the axis of the veins. Calyx campanulate, obtusely 5-toothed, about the length of the pedicel. Stamens rather longer than the corolla. Superior petal veined. (*Don's Mill.*) A deciduous tree. North America. Height 20 ft. to 30 ft. Introduced in 1823. Flowers yellow and red; May and June, a week earlier than P. flàva. Fruit brown; ripe in October. Leaves with rufous down on the veins on the upper side. Flowers pale yellow, veined with red, disposed in thyrsoid racemes. Capsules unarmed, but the ovary tomentose.

A tree resembling Pàv*ia* flàva but smaller. The plant in the Hòrt. Soc. Garden was purchased from M. Catros of Bordeaux, under the name of Æ. ohioénsis. In the *Bot Reg.* it is said to be most nearly related to Æ. (Pàv*ia*) flàva, but to differ from it in the flowers appearing a week or 10 days earlier, and in the leaflets being more glabrous, with rufous down on the veins on the upper side, and with hairs in the axils of the veins on the under surface.

♀ 4. P. MACROCA'RPA *Hort.* The long-fruited Pavia.

Synonyme. Æ'sculus Pàv*ia* macrocárpa *Lodd. Cat.*, 1830.
Engravings. The plate in Arb. Brit., vol. v.; and our *fig.* 177.

Spec. Char., &c. Leaves glabrous on both sides. Leaflets ovate-lanceolate. A deciduous tree under the middle size. A garden hybrid between

177. Pàv*ia* macrocárpa.

Æ'sculus and Pàv*ia* ? rùbra. Height 20 ft. to 30 ft. Cultivated in 1820. Flowers pale red and yellow, nearly as large as the common horsechestnut; May and June. Fruit brown; ripe in October.

The leaves are large, glabrous on the upper surface, and shining. The branches are spreading and loose; and the whole tree has an open graceful appearance, quite different from that compactness of form and rigidity of branches which belong to most of the tree species and varieties both of Æ'sculus and Pàvia.

■ 5. P. DI'SCOLOR *Swt.* The two-coloured-*flowered* Pavia.

Identification. Swt. Hort. Brit., p. 83.; Don's Mill., p. 653.
Synonymes. Æ'sculus discolor *Ph.* and *Bot. Reg.*; Æ'sculus Pàvia β discolor *Tor. & Gray.*
Engravings Bot. Reg., t. 310.; and our *fig.* 178.

Spec. Char., &c. Leaflets 5, acuminate at both ends, tomentose beneath, unequally serrulated. Raceme thyrsoid, many-flowered. Corolla of four conniving petals, with their claws the length of the calyx. Stamens 7, shorter than the corolla. (*Don's Mill.*) A deciduous tree-like shrub. Virginia and Georgia, in fertile valleys and on mountains. Height 3 ft. to 10 ft. Introduced in 1812. Flowers variegated with white, yellow, and purple; May and June. Fruit brown; ripe in October. Decaying leaves yellowish brown. Naked young wood of a brownish stone-colour.

The whole plant, including the young wood, is covered with pubescence. The flowers are large, showy, continuing a long time expanding, and numerous though they are but sparingly succeeded by fruit. When the plant is raised from seed, it is remarkable for its thick, fleshy, carrot-like roots, which, in free soil, penetrate perpendicularly to the depth of 8 or 10 feet before they branch. Unless when grafted on Æ. Hippocástanum, it is seldom seen above 4 or 5 feet in height; but it is a very free flowerer, and, considered as a shrub, is in May, when it is in flower, one of the most ornamental that the British arboretum affords.

178. Pàvia discolor.

■ 6. P. MACROSTA'CHYA *Lois.* The long-racemed Pavia.

Identification. Lois. Herb. Amat.; Dec. Prod., 1. p. 598.; Don's Mill., 1. p. 652.
Synonymes. Æ'sculus parviflòra *Walt.*, and *Torr. & Gray*; Æ macrostáchya *Mx.* and *Hayne*; Pàvia álba *Poir.*; Pávia edùlis *Poit. Arb. Fr.* t. 88.; Macrothýrsus discolor *Spach*; Pavier à longs E'pis, Pavier nain, *Fr.*; langährige Rosskastanie, *Ger.*
Engravings. Lois. Herb. Amat., t. 212.; Hayne Abbild., t. 26.; and our *fig.* 179.

Spec. Char., &c. Stamens much longer than the corolla. Racemes very long. Root stoloniferous. Flowers white. (*Dec. Prod.*) A deciduous shrub, with numerous radicled shoots. South Carolina and Georgia. Height in America 2 ft. to 4 ft.; in the climate of London 10 ft. to 15 ft. Introduced in 1820. Flowers white, with long projecting stamens, which give the spike a fine fringed appearance; July and August. Fruit brown; ripe in October.

The shoots are slender, spreading, and rooting at the joints where they happen to rest on the soil, with ascendent extremities. The tree comes into flower about a month or six weeks later than the other Æsculàceæ, and continues flowering,

179. Pàvia macrostachya

in the case of large plants on moist soil, for three months or longer, forming one of the greatest floral ornaments of the shrubbery, at a season when very few trees or shrubs are in flower. The fruit, which is small, seldom ripens in England: but in America it is said to be eaten, boiled or roasted; and M. Poiteau, accordingly, has included this species of Pàvia in his list of fruit trees. Layers; or seeds, when they can be procured, and which ought to be sown as soon as possible after they are ripe.

Other Kinds of Pàvia.—*Pàvia califórnica* (*Æ.* califórnica *Nutt.*) has been described by Torrey and Gray, but is not yet introduced. *P. Lyònii* is in the Hort. Soc. Garden, but has not yet flowered there. We have omitted in this edition P. hýbrida, described by DeCandolle as a truly intermediate plant between P. rùbra and P. flàva, with yellow, white, and purple flowers; because the only plant which we have seen bearing this name, that in the Hort. Soc. Garden, has the flowers yellow, and appears merely a very slight variety of P. flàva. In nurserymen's catalogues there are several names which we have not noticed; for the truth is, that the different kinds of *Æ'sculus* and P*àvia* cross-fecundate so freely, and seedlings vary so much, that there is no limit to the number of varieties that might be produced. The great error (because it creates so much confusion in the nomenclature) consists in giving these varieties to the world as species.

It is almost unnecessary to observe, that all the most valuable varieties, of both *Æ'*sculus and Pàv*ia*, are best perpetuated by budding or grafting, and that collectors ought always to see that the plants they purchase have been worked. Pàv*ia* rùbra as a tree, P. díscolor either as a shrub or grafted standard high, and P. macrostàchya as a shrub, ought to be in every collection, whether small or large. Pàv*ia* hùmilis, when grafted standard high on the common horsechestnut, forms an ornament at once singular and beautiful. As the horsechestnut is to be found in most plantations, those who are curious in the species and varieties might graft them on the upper branches of old trees; or young trees might be headed down, and one kind grafted on each.

Order XV. SAPINDA'CEÆ.

Ord. Char. *Flowers* polygamous. — *Males* with the calyx more or less deeply 4—5-parted. *Petals* 4—5, or occasionally absent, alternate with the sepals. *Disk* fleshy. *Stamens* 8—10, inserted into the disk. — *Hermaphrodite flowers* with the calyx, petals, disk, and stamens as in the males. *Ovary* 3-celled. *Cotyledons* incumbent. *Plumule* 2-leaved. (*Lindl.*) — A tree, a native of China.

Leaves compound, alternate, stipulate, deciduous. *Flowers* terminal, in racemose panicles, small, white or yellow. — There is only one hardy species belonging to this order in British gardens, namely, Kölreutèr*ia* paniculàta *Laxm.*

Genus I.

KÖLREUTE'R*IA* Laxm. The Kolreuteria. *Lin. Syst.* Octándria Monogýnia.

Identification. Laxm. Acad. Petr., 16. p. 561.; L'Hérit. Sert., 18. t. 19.; Willd. Spec. Pl., 380.; Dec. Prod., 1. p. 616.; Don's Mill., 1. p. 672.
Synonymes. Sapíndus sp. *Lin. fil.*; Cölreuteria, *Ital.*
Derivation. In honour of *John Theophilus Kölreuter*, once professor of natural history at Carlsruhe, and celebrated for his researches on the pollen of plants.

Gen. Char., &c. *Calyx* of 5 sepals. *Petals* 4, each with 2 scales at the base. *Capsule* 3-celled, inflated. *Seeds* ovate-globose, the seed-coat penetrating into the seed, and occupying in the place of an axis the centre of the embryo, which is spirally convoluted. (*Dec. Prod.*)

Leaves impari-pinnate, of many pairs of leaflets, that are ovate and coarsely toothed. *Flowers* yellow, in panicles. — A deciduous tree.

¶ 1. K. PANICULA'TA *Laxm.* The panicled-*flowering* Kolreuteria.

Identification. Laxm. Acad. Petr., 16. p. 561. ; Dec. Prod., 1. p. 616. ; Don's Mill., 1. p. 672.
Synonymes. Sapindus chinénsis *Lin. fil. Supp.* p. 221. ; K. paullinioìdes L'Hérit. Sert. ; Savonnier paniculé, *Fr.* ; rispentragende Kölreuterie, *Ger.*
Engravings. N. Du Ham., 1. t. 36. ; Bot. Reg., t. 320. ; and the plate of the tree in Arb. Brit., 1st edit., vol. v. ; and our *fig.* 180.

Spec. Char., &c. Leaves impari-pinnate, with ovate leaflets, coarsely toothed. Flowers polygamous. A deciduous tree of the middle size. North of China Height 20 ft. to 40 ft. in the climate of London. Introduced in 1763. Flowers yellow, in terminal, racemose, spreading panicles ; July and August. Fruit a bladdery capsule, whitish brown ; ripe in October. Decaying leaves deep yellow. Naked young wood brown.

It is very hardy; the hermaphrodite plants not unfrequently ripening seeds in the neighbourhood of London. It has not only a very fine appearance

180. Kölreutèria paniculàta.

when in flower, but also in autumn, when the tree is covered with its large bladdery capsules, and the leaves change to a deep yellow, which they do before they fall off. It is of the easiest culture in any common soil, and is readily propagated either by seeds or cuttings of the root or branches. In the London nurseries it is generally propagated by seed.

ORDER XVI. VITA'CEÆ.

ORD. CHAR. *Calyx* small. *Petals* 4 or 5. *Stamens* equal in number to the petals ; filaments distinct, or slightly cohering at the base. *Anthers* versatile. *Ovarium* 2-celled. *Fruit* a pulpy berry. *Seeds* 4 or 5, fewer by abortion; embryo erect; albumen hard. — Climbing shrubs, with tumid separable joints.

Leaves simple or compound, opposite or alternate, stipulate, deciduous; the lower ones opposite, the upper alternate. *Flowers* axillary, racemose, sometimes by abortion changing to tendrils, which are generally opposite to the leaves; small, green. — Shrubs, trailing and climbing, deciduous, and including the grape vine, which may be considered as the type of the order. The genera which contain hardy species are three, which are thus contradistinguished: —

Vi'tis. Style wanting. Petals 5.
AMPELO'PSIS. Style 1. Petals 5.
Ci'ssus. Style 1. Petals 4.

GENUS I.

VI'TIS L. THE GRAPE VINE. *Lin. Syst.* Pentándria Monogýnia.

Identification. Lin. Gen., 284.; Dec. Prod., 1. p. 633.; Don's Mill., 1. p. 695.
Synonymes. Giud, *Celtic*; Vid, *Span.*; Vigne, *Fr.*; Vite, *Ital.*; Wein, *Ger.*

Gen. Char. *Flowers* hermaphrodite, diœcious or triœcious. *Calyx* commonly 5-toothed. *Petals* 5, cohering at the top, separating at the base, and deciduous. *Stamens* 5. (*Dec. Prod.*) — Climbing tendriled shrubs, deciduous; natives of Asia and North America.

Leaves simple, alternate, stipulate, lobed or serrated. *Flowers* in thyrsoid racemes, small, and of a greenish yellow. — There are several species in British gardens, the principal of which is the grape vine.

⊥ 1. *V.* VINI'FERA *L.* The wine-bearing Vine.

Identification. Lin. Spec., 293; Dec. Prod., 1. p. 633.; Don's Mill., 1. p. 695.
Synonymes. Vigne, *Fr.*; gemeiner Weinstock, *Ger.*; Vite da Vino, *Ital.*
Engravings. Duh. Arb. Fr., 2. t. 16.; Jacq. Ic., 1. p. 53.; and our *fig.* 181.

Spec. Char., &c. Leaves lobed, toothed, sinuated, or serrated, naked or downy. (*Dec. Prod.*) A deciduous tendriled climber. Syria. Stem 20 ft. to 50 ft. Cultivated in 1648, or probably from the time of the Romans. Flowers greenish yellow, scented; June and July. Fruit green, red, or black; ripe in October. Decaying leaves yellow or red. Naked young wood yellowish brown.

181. *Vitis vinifera.*

Varieties. The grape vine has been in cultivation from the remotest period of history, in the warmest parts of the temperate zones of the Old World. The varieties have been described at length by Du Hamel in France, Don Roxas de Clementi in Spain, and Sickler in Germany. The varieties of the vine as a fruit shrub, and all that relates to their propagation and culture, will be found treated of in our *Encyclopædia of Gardening;* and we shall here only notice those which we think deserving of cultivation, as ornamental and fragrant-flowered climbers.

⅄ V. v. 2 *fòliis incànis.* Miller's Grape, or Miller's black Cluster Grape. — Leaves almost entire, small, woolly, and whitish. Fruit round, small, in compact bunches, black. This variety is selected on account of the whiteness of its leaves.

⅄ V. v. 3 *fòliis rubescéntibus.* The Claret Grape; Tenturier, *Fr.* (*N. Du Ham.*, var. 75., not Clairette *Du Ham.*, var. 12.) — The leaves are larger than those of the preceding variety, and more lobed and notched: in the autumn, before they die off, they change to a deep claret colour, in which state they are highly ornamental.

182. *Vitis vinífera apiifòlia.*

⅄ V. v. 4 *apiifòlia laciniòsa* L. The Parsley-leaved Grape Vine; Ciotat, *Fr.*; Vite d'Egitto, *Ital.* (*fig.* 182.) — The leaves are beautifully laciniated, middle-sized, and the fruit black. A very handsome climbing shrub which has been in cultivation for its fruit since 1648.

⅄ 2. *V.* LABRU'SCA *L.* The wild Vine, *or Fox Grape.*

Identification. Lin. Spec., 293.; Dec. Prod., 1. p. 634.; Don's Mill., 1. p. 711.; Tor. and Gray, 1. p. 244.
Synonymes. *V. taurina Walt.*; filziger Wein, *Ger.*; Abrostine, *Ital.*
Engravings. Plum. Icon., t. 259. fig. 1.; Jacq. Schœn., t. 426.; and our *fig.* 183.

Spec. Char., &c. Sexes diœcious or polygamous. Leaves heart-shaped, rather 3-lobed, acutely toothed beneath, and the peduncles tomentose and rather rusty. (*Dec. Prod.*) Canada to Georgia. Climbing stem 10 ft. to 30 ft. Introduced in 1656. Flowers greenish yellow; June and July. Fruit red or black; ripe in October. Decaying leaves brown or black.

183. *Vitis Labrùsca.*

Varieties. Several varieties, with red, white, or black fruit, are known in the gardens of North America, from which wine is made; such as the Isabella, Schuylkill or Alexander's, the Catawba, and Bland's Grape, which have doubtless been produced from seeds of this species. (*Tor. and Gray.*)

Leaves 4 in. to 6 in. or more in diameter, often distinctly 3-lobed, short, mucronate, and densely tomentose beneath. Berries 6—7 lines in diameter, globose, usually very dark purple when ripe, but sometimes amber-coloured or greenish white, of a strong musky flavour, and filled with a tough pulp. (*Ibid.*)

⅄ 3. *V.* ÆSTIVA'LIS *Michx.* The Summer Vine, *or Grape Vine.*

Identification. Michx. Fl. Bor. Amer., 2. p. 230.; Dec. Prod., 1. p. 634.; Don's Mill., 1. p. 711.; Tor. and Gray, 1. p. 244.
Synonymes. *V. vinífera americàna Marsh.*; *V. intermèdia Muhl.*; and *V. palmàta Vahl.*
Engravings. Jac. Hort. Schœn., t. 425.; and our *fig.* 184.

Spec. Char., &c. Sexes diœcious or polygamous. Leaves broadly heart-shaped, with from 3 to 5 lobes; the under surface of the young ones invested with a cottony down; of the adult ones, smooth. Racemes fertile, oblong. Berries small. (*Dec. Prod.*) A tendriled climber. Connecticut to Florida. Stem 20 ft. to 30 ft. Introduced in 1656. Flowers greenish yellow; June. Fruit dark blue; ripe in October.

184. *Vitis æstivàlis.*

Leaves 4 in. to 7 in. wide, often deeply lobed, with the sinuses rounded;

the lower surface, particularly in the young state, clothed with a reddish cobweb-like pubescence; when old, somewhat glabrous. Berries 3—4 lines in diameter, deep blue, of a pleasant flavour; ripe in October. (*Tor. and Gray.*) Perhaps only a variety of the preceding species. *V. labruscöides Muhl.* is also probably a synonyme or a variety of that species.

ꝉ 4. *V.* CORDIFO'LIA *Michx.* The heart-*shape*-leaved Vine, *or Chicken Grape.*

Identification. Michx. Fl. Bor. Amer., 2. p. 231.; Dec. Prod., 1. p. 654.; Don's Mill., 1. p. 711.; Tor. and Gray, 1. p. 244.
Synonymes. *V.* incisa *Jacq. Schœn.* t. 427.; *V.* vulpina *Lin. Spec.* p. 293., *Walt. Flor. Car.* 243.; Winter Grape; Frost Grape.
Engravings. Jacq. Schœn., t. 247.; and our *fig.* 185.

185. *Vitis cordifòlia.*

Spec. Char., &c. Sexes diœcious or polygamous. Leaves heart-shaped, acuminate, toothed in the mode of incisions, smooth on both surfaces. Racemes loosely many-flowered. Berries small, greenish, ripened late. (*Dec. Prod.*) A tendriled climber. Canada to Florida, in thickets along rivers. Stem 10 ft. to 20 ft. Introduced in 1806. Flowers greenish yellow; June. Fruit greenish; ripe in November.

Leaves thin, 3 in. to 6 in. in diameter, often slightly 3-lobed, and rarely sinuated. Berries nearly black when mature, about a quarter of an inch in diameter, ripening late in autumn; acid, but tolerably well flavoured after having been touched by frost. (*Tor. and Gray.*)

ꝉ 5. *V.* RIPA'RIA *Michx.* The river-side, *or sweet-scented,* Vine.

Identification. Michx. Flor. Bor. Amer., 2. p. 821.; Dec. Prod., 1. p. 635.; Don's Mill., 1. p. 711.
Synonymes. *V.* odoratissima *Donn Hort. Cant.*, and *Lodd. Cat.*; Vigne de Battures, *Amer.*
Engravings. Bot. Mag., t. 2429.; Dend. Brit., 1. p. 13.; and our *fig.* 186.

186. *Vitis ripària.*

Spec. Char., &c. Leaves heart-shaped, shallowly 3-cleft, toothed in the mode of incisions and unequally. Footstalk, and the margin of the nerves, pubescent. Racemes loose. Fruit small. (*Dec. Prod.*) A tendriled climber. Canada to Virginia. Stem 20 ft. to 30 ft. Introduced in 1806. Flowers greenish yellow, with the fragrance of mignonette; June and July. Fruit dark purple, or amber; ripe in October.

Leaves 4 in. to 6 in. in diameter, thin; teeth very coarse, acuminate. Berry 3—4 lines in diameter, dark purple, or amber colour, when ripe. (*Tor. and Gray.*)

ꝉ 6. *V.* VULPI'NA *L.* The Fox Grape, *or Bullet Grape.*

Identification. Willd. Sp., 1. 1181.; Walt. Car., p. 243.; Tor. and Gray, 1. p. 245.
Synonymes. *V.* rotundifòlia *Michx.*, *Pursh*, *Dec.*, *Don's Mill.*, and *Arb. Brit.* 1st edit.; Muscadine Grape.
Engraving. Our *fig.* 187.

Spec. Char., &c. Branches minutely verrucose. Leaves cordate, shining on both surfaces, somewhat 3-lobed, coarsely toothed, the teeth not acuminate. Racemes composed of numerous capitate umbels. Berries large. (*Tor. and Gray.*) A tendriled climber. Virginia to Florida. Stem 20 ft. to 30 ft. Introduced in 1806. Flowers greenish yellow; June and July. Fruit deep blue; ripe in October.

The stem of this species has a smooth bark, and climbs to the summit of some of the highest trees. Leaves 2—3 in. in diameter; the lower surface more shining than the upper; sinus deep, but rather acute. Fruit 7—8 lines in diameter, covered with a coriaceous integument, the flower not unpleasant. This, according to Torrey and Gray, appears to be the original *V.* vulpina of Linnæus.

Other Species of Vìtis. — The American species have been considerably reduced in number by Messrs. Torrey and Gray; but it appears to us, that the reduction might have been carried still farther. Some species are described as natives of the Himalayas, and 130 varieties are noticed by Rafinesque in his *Monograph of American Vines* (see *Gard. Mag.*, vol. viii. p. 248.); but they are not yet known in this country. Indeed, from the appearance of the above-described species in the Hort. Soc. Garden, we are much inclined to think they are only varieties of the same species. They certainly do not differ more from each other than the known varieties of the common cultivated vine.

187. *Vitis vulpina.*

Genus II.

AMPELO'PSIS *Michx.* The Ampelopsis. *Lin. Syst.* Pentándria Monogýnia

Identification. Michx. Fl. Bor. Amer., 1. p. 159.; Dec. Prod., 1. p. 632.; Don's Mill., 1. p. 694.
Synonymes. *Vitis* sp. and *Cissus* sp.; Ampelosside, *Ital.*
Derivation. *Ampelos*, a vine, and *opsis*, resemblance; similarity in the habits of the species.

Gen. Char. *Calyx* almost entire. *Petals* 5, falling off separately. *Stigma* capitate. *Ovary* not immersed in a disk, including 2—4 ovules. (*Dec. Prod.*)

Leaves compound, alternate, exstipulate, deciduous; palmate, pinnate, or bipinnate. *Flowers* small. — Tendriled climbers, natives of North America. The species in British gardens are two, of the easiest culture in any common soil; and one of them, A. *h*ederàcea, is among the most ornamental of hardy climbers.

♣ 1. A. hedera'cea *Michx.* The Ivy-like Ampelopsis, *or Five-leaved Ivy.*

Identification. Michx. Fl. Bor. Amer., 1. p. 160.; Dec. Prod., 1. p. 633.; Don's Mill., 1. p. 694.
Synonymes. *Hédera* quinquefòlia *Linn. Spec.* 292.; *Vitis* quinquefòlia *Lam. Ill.* No. 2815.; *Cissus* hederàcea *Ph. Fl. Amer. Sept.* 1. p. 170.; *Cissus* quinquefòlia *Hort. Par.*; *Vitis* hederàcea *Willd. Spec.* p. 1182.; Ampelópsis quinquefòlia *Hook. Fl. Bor. Amer.* 1. 114., and *Tor. & Gray*; Vigne Vierge, *Fr.*; Jungfern Reben, or wilder Wein, *Ger.*; Vite del Canadà, *Ital.*
Engravings. Cornut. Canad., t. 100.; and our *fig.* 188.

Spec. Char., &c. Leaves digitate, of from 3 to 5 leaflets, that are stalked oblong, toothed with mucronated teeth. Racemes dichotomously corymbose. (*Dec. Prod.*) A tendriled climber. Canada to Georgia. Stem 30 ft. to 50 ft., or more. Introduced in 1629. Flowers yellowish green; June and July. Berry black or dark blue; ripe in October. Decaying leaves deep purple or red, or yellowish red.

Variety.

♃ *A.* h. 2 *hirsùta* Tor. & Gray. A. hirsùta *Donn;* *C*íssus *h*ederàcea β hirsùta *Pursh.*—Leaves pubescent on both sides, leaflets ovate. The plants of this variety in British gardens do not die off of so intense a crimson as the species.

188 Ampelópsis hederàcea.

Stem attaching itself to trees and walls by expansions of the extremities of the tendrils. Panicle many-flowered. Petals at first somewhat cohering, at length spreading. Berry about as large as a small pea, the peduncles and pedicels bright crimson; and the foliage in autumn, before it dies off, of a deep crimson. The most vigorous-growing and generally ornamental climber in Europe. It thrives in almost every soil and situation from Warsaw to Naples, and in town, as well as in the country.

♃ 2. A. BIPINNA'TA *Michx.* The bipinnate-*leaved* Ampelopsis.

Identification. Michx. Fl. Bor. Amer., 1. p. 160.; Dec. Prod., 1. p. 633.; Don's Mill., 1. p. 694.
Synonymes. *V*ìtis arbòrea *Willd. Spec.* 1. p. 1183.; *V*ìtis bipinnàta *Tor. & Gray*; *C*íssus stáns *Pers. Syn.* 1. p. 183., *Ph. Fl. Amer. Sept.* 1. p. 170.; Vite del Carolina, *Ital.*
Engravings. Pluk. Mant., p. 412. fig. 2.; and our *fig.* 189.

Spec. Char., &c. Leaves bipinnate, smooth; leaflets cut in a lobed manner. Racemes pedunculate, almost doubly bifid. Berries globose and cream-coloured. (*Dec. Prod.*) A tendriled climber. Virginia to Georgia. Stem 10 ft. to 20 ft. Introduced in 1700. Flowers greenish white; June to August. Berry black; ripe in October. Decaying leaves purplish red, sometimes yellowish red.

189. Ampelópsis bipinnàta.

Stem upright, or somewhat twining, glabrous. Panicle short, spreading, and without tendrils. Berry globose, depressed, as large as a small pea, blackish when ripe, slightly hairy. (*Tor. and Gray.*) A very handsome climber, of easy culture, and much admired for the beauty of its foliage. Compared with A. *h*ederàcea, it is of slow growth, the shoots in the climate of London being seldom more than 18 in. or 2 ft. in a season.

Other Species of Ampelópsis. — *A. incìsa*, *V*ìtis incìsa *Nutt.*, is described in Torrey and Gray's *Flora;* but we are not aware of its having been introduced. *A.* cordàta *Michx.* (the *C*íssus Ampelópsis of Persoon, and *V*ìtis indivìsa of Willdenow) is described in the first edition of this work; but, as we consider it a very doubtful species, we have omitted it in this abridgement. A. capreolàta *G. Don*, *V*ìtis capreolàta *D. Don*, and A. bótrya *Dec.*, are also omitted, as not having been yet introduced.

Genus III.

CI'SSUS *L.* The Cissus. *Lin. Syst.* Tetrándria Monogýnia.

Identification. Lin. Gen., No. 147.; Dec. Prod., 1. p. 627.; Don's Mill., 1. p. 689.
Synonymes. Ampelópsis, and Vitis in part.
Derivation. *Kissos* is the Greek name of the ivy, which these plants in some manner resemble.

Gen. Char. *Calyx* almost entire. *Petals* 4, falling off separately. *Ovary* 4-celled. *Berry* 1—4-seeded. (*Dec. Prod.*)

Leaves compound, alternate, exstipulate, deciduous; trifoliate. *Flowers* axillary, small, greenish. *Fruit* a berry. —Climbing shrubs, only one of which is hardy in British gardens.

1. *C.* orienta'lis. The oriental Cissus, *or Ivy Vine.*

Identification. Lam. Ill. p. 332. Don's Mill., 1. p. 694.
Engravings. Lam. Ill., t. 84. fig. 2.; and our fig. 190.

Spec. Char., &c. Leaves bipinnate, smooth; leaflets ovate, serrated. (*Don's Mill.*) A deciduous climber. Levant. Stem 5 ft. to 10 ft. Introd. in 1818. Flowers yellowish green; June and July. Berry, ?.

We have only seen the plant bearing this name in the collection of Messrs. Loddiges, from which our figure is taken, and which, as it does not agree very well with the specific character, is perhaps not the true plant. At all events, the plant figured is handsome, and as vigorous and hardy as Ampelópsis bipinnàta.

190. Cissus orientàlis.

Order XVII. XANTHOXYLA'CEÆ.

Ord. Char. *Flowers* unisexual, regular. *Calyx* in 3—5 divisions. *Petals* the same number, longer than the calyx; æstivation generally twisted. *Stamens* equal in number to the petals; in the female flowers wanting or imperfect. *Ovary* with as many carpels as there are petals. *Fruit* either berried or membranous.—Trees or shrubs, chiefly natives of warm climates. (*Lindl.*)

Leaves compound, alternate or opposite, without stipules; abruptly or unequally pinnate; with pellucid dots. *Flowers* axillary or terminal; grey, green, or pink.—The species in British gardens are comprised in three genera, which are thus contradistinguished :—

Xantho'xylum *L.* Flowers bisexual. Carpels 1—5, 2-valved. Leaves abruptly and impari-pinnate.

Pte'lea *L.* Flowers bisexual. Fruit compressed, 2—3-celled; cells winged. Leaves of 3 leaflets, rarely of 5 leaflets.

Aila'ntus Desf. Flowers polygamous. Carpels 3—5, membranous. Leaves abruptly or impari-pinnate.

GENUS I.

XANTHO'XYLUM *L.*, and *H. B. et Kth.* THE XANTHOXYLUM, *or* *TOOTHACHE TREE.* *Lin. Syst.* Diœ'cia Tri-Pentándria.

Identification. Lin. Gen., No. 150. and 1109.; Dec. Prod., 1. p. 725.; Don's Mill., 1. p. 801.
Synonymes. Zanthóxylum (it is thus spelled in many botanical works); Kampmánnia Rafin.; Clavalier, *Fr.*; Zahnwehholz, *Ger.*; Santossilo, *Ital.*
Derivation. From *xanthos*, yellow, and *xulon*, wood; from the yellowness of the wood, more especially of the roots. The French name means club tree, and the German name, the toothache tree.

Gen. Char. *Calyx* short, 3—4-parted. *Petals* equal in number to the lobes of the calyx, but longer, very rarely wanting.—*Male flowers.* *Stamens* equal in number with the petals.—*Female flowers.* *Stamens* sometimes wanting, or very short. *Ovaries* 5—1, sometimes equal in number to the petals. *Capsules* 1—5, 1—2-seeded. *Seeds* globose, dark, shining. (*Don's Mill.*)

Leaves compound, alternate, stipulate, deciduous; pinnately 3—13-foliolate. *Flowers* axillary, small, greenish or whitish. *Inflorescence* various.—Deciduous low trees or shrubs, natives of North America, with prickles on the branches, petioles, and midrib of the leaflets. The species in British gardens are of easy culture in any common soil, and are easily propagated by seeds, layers, or cuttings of the roots.

♀ 1. X. FRAXI'NEUM *Willd.* The Ash-*leaved* Xanthoxylum, *or common Toothache Tree.*

191. Xanthóxylum *fraxineum*

Identification. Willd. Sp., 4. p. 757.; Dec. Prod., 1. p. 726.; Don's Mill., 1. p. 802.
Synonymes. Zanthóxylum ramiflòrum *Mich. Fl. B. A.* 2. p. 235.; Z. Clàva Hérculis var. *Lin. Sp.* 1455., *Lam. Dict.* 2. p. 38.; Z. americànum *Mill. Dict.* No. 2., and *Tor. & Gray*, 1. p. 214.; Z. mite *Willd. Enum.*; Z. caribæ`um *Gært. Fruct.*, but not of Lam.; Z. tricàrpum *Hook.* not of Michx.; Clavalier à Feuilles de Frêne, *Fr.*; Eschen-blättriges Zahnwehholz, *Ger.*; Prickly Ash, *Amer.*; Frassino spinoso, *Ital.*
Engravings. Du Ham. Arb., 1. t. 97.; the plate of this species in Arb. Brit., 1st edit., vol. v.; and our *fig.* 191.

Spec. Char., &c. Leaves pinnate, of 4 to 5 pairs of leaflets, and an odd one; the leaflets ovate, obscurely sawed, equal at the base; the petiole round, and devoid of prickles; prickles in the situation of stipules. Flowers in axillary umbels, without petals. (*Dec. Prod.*, i. p. 726, 727.) A low deciduous tree or shrub. Canada to Virginia. Height 10 ft. to 15 ft. Introd. 1740. Flowers yellowish, with red anthers; April and May. Seeds large, black; ripe in September. Decaying leaves yellowish green. Naked young wood ash-coloured and greenish.

Variety.

Ŧ ▲ X. f. 2 *virgínicum*, the X. virgínicum of Lodd. Cat., of which there is a plant in the garden of the London Horticultural Society, and several in the arboretum of Messrs. Loddiges, appears to us only a variety of X. *fraxíneum*; probably the same as X. (*f.*) tricárpum.

Ŧ ▲ 2. X. (F.) TRICA'RPUM *Michx.* The three-fruited Xanthoxylum, *or Toothache Tree.*

Identification. Michx. Fl. Bor. Amer., 2. p. 335.; Dec. Prod., 1. p. 726.; Don's Mill., 1. p. 803.
Synonymes. Z. caroliniànum *Lam.*, *Tor. & Gray*, 1. p. 214.; *Fagàra fraxinifòlia Lam. Ill.* 1. t. 334.
Engravings. Lam. Ill., 1. t. 334.; and our *fig.* 192.

Spec. Char., &c. Leaves pinnate; the leaflets 3 to 5 pairs, and an odd one, all on short stalks, oblong oval, acuminate, finely sawed, oblique at the base. Petioles and branches prickly. Panicles terminal. Petals 5. (*Dec. Prod.*) A low tree or shrub. North Carolina to Florida. Height 10 ft. to 15 ft. Introd. 1806. Flowers greenish; June. Seeds large, black; ripe Oct.

192. Xanthóxylum tricárpum.

Leaves and bark very aromatic and pungent. Prickles very sharp. The bark of this and the preceding species is imported from New York, and sold in Covent Garden Market as a cure for the rheumatism. Probably a variety of the preceding species.

Other Species of Xanthóxylum.—X. mìte *Willd.*, treated as a species by some authors, is made a synonyme of X. *fráxineum* by Torrey and Gray, and it probably bears the same relation to that species that Gledítschia inérmis does to G. triacánthos. Our opinion is, that there is only one species of the genus in British gardens.

Genus II.

**PTE`LEA *L.*** THE PTELEA, *or SHRUBBY TREFOIL.* *Lin. Syst.* Monœ`cia Tetra-Pentándria.

Identification. Lin. Gen., No. 152.; Dec. Prod., 2. p. 82.; Don's Mill., 1. p. 806.
Synonymes. *Bellùcia* Adans.; Orme de Samarie, *Fr.*; Lederblume, *Ger.*
Derivation. From *ptelea*, the Greek name of the elm, adopted by Linnæus.

Gen. Char. *Calyx* short, 4—5 parted. *Petals* 4—5, longer than the calyx. —*Male flowers.* *Stamens* 4—5, longer than the petals.—*Female flowers.*

Stamens 4—5, very short. *Style* short. *Fruit* compressed, indehiscent, samara-like, turgid, 2—3-celled. *Seeds* oblong. (*Don's Mill.*)

Leaves compound, alternate, stipulate, deciduous; pinnate, 3- rarely 5-foliolate, with pellucid dots, the lateral leaflets inequilateral. *Flowers* whitish, cymose: cymes corymbed or panicled.—Deciduous shrubs or low trees, natives of North America and Asia. There is only one species in British gardens, which is of the easiest culture, and is propagated by seeds and cuttings, put in in autumn, and covered with a hand glass.

⚘ ■ 1. *P.* TRIFOLIA'TA *L.* The three-*leafleted* Ptelea, *or Shrubby Trefoil.*

Identification. Lin. Sp., 173.; Willd. Sp. Pl., 1. 670.; Dec. Prod., 2. p. 82.; Don's Mill., 1. p. 806.; Tor. and Gray, 1. p. 215.
Synonymes. Orme de Samarie à trois Feuilles, *Fr.*; dreyblättrige Lederblume, *Ger.*
Engravings. Dill. Elth., t. 122.; Schmidt Arb., 2. t. 76.; the plate in Arb. Brit., 1st. edit., vol. v.; and our *fig.* 193.

Spec. Char., &c. Leaf of three leaflets that are ovate acute, the middle one much tapered towards its base. Flowers in corymbs, usually tetrandrous. (*Dec. Prod.*) A low tree or shrub. Lake Erie to Florida and Texas. Height 6 ft. to 10 ft. Introd. 1704. Flowers whitish; June and July; Capsules greenish; ripe in October. Decaying leaves of a remarkably clear rich yellow. Naked young wood dark purplish brown.

Varieties.

⚘ ■ P. *t.* 2 *pentaphýlla* Munchh. has 5 leaflets, H. S.

⚘ ■ P. *t.* 3 *pubéscens* Pursh has the leaflets pubescent.

When this plant is pruned up with a single stem, it forms a handsome low tree with a hemispherical head; but in British gardens it is more frequently

193. *Ptèlea trifoliàta.*

found as a large shrub, with numerous stems proceeding from the rootstock. The shoots and leaves pubescent when young. Ovary of the staminate flowers abortive. Odour of the flowers disagreeable. Capsules with flattened wings, somewhat resembling those of the elm.

Other Species of Ptèlea. — P. *Baldwínii* is described by Torrey and Gray as a shrub not more than a foot high, but it has not yet been introduced

Genus III.

AILA'NTUS **Desf. The Ailanto.** *Lin. Syst.* Polygàmia Monœ'cia.

Identification. Desf. Act. Acad. Par., 1786, p. 263 ; Dec. Prod., 2. p. 88. ; Don's Mill., 1. p. 807.
Synonymes. Rhùs *Ehrh.*, *Ellis*, and *Mœnch* ; Verne du Japon, *Fr.* ; Götterbaum, *Ger.*
Derivation. Ailanto is the name of *Ailántus* glandulòsa *Desf.* in the Moluccas. It was long considered as a species of Rhùs, whence the French name ; and the meaning of the aboriginal word being, it is said, tree of heaven, hence the German name, Götterbaum, tree of the gods.

Gen. Char. *Male flowers.* *Calyx* 5 cleft. *Petals* 5, longer than the calyx. *Stamens* 10, the 5 opposite the petals shortest. *Disk* central.—*Hermaphrodite*, or *female, flowers.* *Calyx, petals*, and *disk* as in the male, but with fewer stamens. *Ovaries* 3—5, distinct. *Samaræ* 3—5, oblong ; 1-celled, 1-seeded. (*Don's Mill.*)

Leaves compound, alternate, exstipulate, deciduous; impari-pinnate. *Flowers* terminal, small, greenish. — One species, a deciduous tree from China.

♈ 1. *A.* glandulo'sa *Desf.* The glandulous-*leaved* Ailanto.

Identification. Desf. Act. Acad. Par. 1786, p. 263. ; Dec. Prod., 1. p. 89. ; Don's Mill., 1. p. 807.
Synonymes. *A.* procèra *Sal. Prod.* p. 271. ; Rhùs hypselodéndron *Mœnch* ; *R.* cacodéndron *Ehrh.* ; *R.* sinénse *Ellis* ; Aylanthe glanduleux, *Fr.* ; drüsiger Götterbaum, *Ger.* ; Albero di Paradiso, *Ital.*
Engravings. Wats. Dend. Brit., t. 108. ; the plate of the tree in Arb. Brit., 1st edit., vol. v. ; and our *fig.* 194.

Spec. Char., &c. Leaves impari-pinnate ; the leaflets coarsely toothed at the base ; the teeth glandulous on the under side. (*Dec. Prod.*) A large tree. North of China. Height 50 ft. to 60 ft. Introd. 1751. Flowers whitish green, exhaling a disagreeable odour; August. Capsules like the keys of the ash, but smaller ; ripe in October. Decaying leaves brownish, but dropping with the first frost, without any great change of colour. The leaflets often separating from the petiole of the leaf, and leaving it for some weeks attached to the tree. Naked young wood rusty brown, without buds.

The leaves on vigorous young trees are sometimes 6 ft. in length. The fruit, which has been ripened at White Knights, resembles the keys of the ash, but is smaller. The tree grows with great rapidity for the first 10 or 12 years, producing shoots from 3 ft. to 6 ft. in length at first, and attaining the height of 15 or 20 feet in 5 or 6 years, in favourable situations. Afterwards its growth is much slower. It grows in any soil, though one that is light and somewhat humid, and a sheltered situation, suit it best. In France, it is said to thrive on chalky soils, and attain a large size where scarcely any other tree will grow. It is readily propagated by cuttings of the roots.

194. *Ailántus* glandulòsa.

Section IV.

Fruit gynobasic ; that is, inserted into a fleshy Receptacle, with which the Style is continuous.

Order XVIII. CORIA'CEÆ.

Ord. Char. *Flowers* either hermaphrodite, monœcious, or diœcious. *Calyx* campanulate, 5-parted. *Petals* 5. *Stamens* 10. *Carpels* 5.—Low shrubs, natives of temperate and warm climates.

Leaves simple, opposite or alternate, exstipulate, deciduous; entire. *Buds* scaly. *Flowers* in terminal and axillary racemes. *Fruit* in some poisonous, in others edible. — There is only one hardy genus, Coriària; the species of which are low shrubs, natives of Europe and Asia.

Genus I.

CORIA'RIA *Niss.* The Coriaria. *Lin. Syst.* Diœ'cia Decándria.

Identification. Niss. in Act. Par. 1711, t. 12.; Dec. Prod., 1. p. 739.; Don's Mill., 1. p. 818.
Synonymes. Redoul, *Fr.*; Gerberstrauch, *Ger.*
Derivation. From *corium*, a hide; *C. myrtifòlia* being used both in tanning leather and in dyeing it black.

Gen. Char. *Flowers* either hermaphrodite, monœcious, or diœcious. *Calyx* 5-parted. *Petals* 5, sepaloid, smaller than the lobes of the calyx. *Stamens* 10, hypogynous, 5 between the lobes of the calyx and the angles of the ovarium, 5 between the petals and the furrows of the ovarium. *Anthers* bursting by longitudinal slits. *Style* none. *Stigmas* 5, long, awl-shaped. *Carpels* 5, surrounding a fleshy axis; when ripe, close together, but separate, not opening, 1-seeded, surrounded with glandular lobes. (*Lindl.*)

Leaves simple, opposite, exstipulate, deciduous; 3-ribbed. *Branches* square, opposite. — Low suffruticose shrubs, of easy culture in common soil, and propagated by division of the root.

1. C. myrtifo'lia *L.* The Myrtle-leaved Coriaria.

Identification. Lin. Sp., 1467.; Dec. Prod., 1. p. 739.; Don's Mill., 1. p. 818.
Synonymes. Fustet des Corroyeurs, or Redoul à Feuilles de Myrte, *Fr.*; Myrtenblättriger Gerberstrauch, *Ger.*
Engravings. Lam. Ill., t. 822.; Wats. Dend. Brit., t. 103.; and our *fig.* 195.

Spec. Char., &c. Leaves ovate-lanceolate, acute, three-nerved, on short footstalks, glabrous. Flowers in rather upright racemes. (*Dec. Prod.*) A low, deciduous, suffruticose shrub, consisting of numerous suckers. South of Europe, and the North of Africa. Height 2 ft. to 3 ft. Introduced 1629. Flowers greenish; May to August. Carpels in the form of a berry, black; ripe in October. Leaves drop off of a brownish green.

195. Coriària myrtifòlia.

Found in hedges and waste places, throwing up numerous suckers. An ornamental undershrub, chiefly remarkable for its myrtle-like leaves, and the handsome frond-like form of its branches. Suckers in any common soil.

196. C. nepalénsis.

Other Species of Coriària. — C. *nepalénsis* Wall. Pl. As. Rar. t. 289., and our *fig.* 196., from a specimen gathered in the Hort. Soc. Gardens, a native of Nepal, at heights of from 5000 ft. to 7000 ft., appears to be quite hardy, and of robust growth. C. *sarmentòsa* Forst., from New Zealand, is probably hardy also, but has not yet been introduced.

Subclass II. CALYCIFLO'RÆ.

Petals separate, inserted in the Calyx.

Order XIX. STAPHYLEA'CEÆ.

Ord. Char. *Sepals* 5, connected at the base, coloured, with an imbricated æstivation. *Petals* 5, alternate. *Stamens* 5. *Disk* large. *Ovary* 2—3-celled. *Fruit* membranous or fleshy. — Shrubs, natives chiefly of warm climates. (*Lindl.*)

Leaves compound, opposite, stipulate, deciduous. *Flowers* terminal. *Fruit* a bladdery capsule. — The only hardy ligneous plants belonging to this order are contained in the genus Staphylèa.

Genus I.

STAPHYLE'A *L.* The Staphylea, *or Bladder-nut Tree.* *Lin. Syst.* Pentándia Di-Trigýnia.

Identification. Lin. Gen., No. 374.; Dec. Prod., 2. p. 2.; Don's Mill., 2. p. 2.
Synonymes. Staphylodéndron *Tourn.*; Staphilier, faux Pistachier, *Fr.*; Pimpernuss, *Ger.*; Stafilea, *Ital.*
Derivation. Abridged from Staphylodéndron, its name before the days of Linnæus, derived from *staphulē*, a bunch or cluster, and *dendron*, a tree; the flowers and fruits being disposed in clusters, and the plant being ligneous.

Gen. Char. *Calyx* of 5 coloured sepals, connected at the base, in æstivation imbricate. *Petals* 5, in æstivation imbricate. *Stamens* 5, perigynous, alternate with the petals, and opposite the sepals. A large urceolate disk, or nectary, within the corolla. *Ovarium* 2- or 3-celled, superior. *Fruit* membraneous. *Seeds* with a bony testa, and a large truncate hilum. (*Lindl.*)

Leaves compound, opposite, stipulate, deciduous; pinnate, with both common and partial stipules. *Flowers* in terminal stalked racemes. — Two hardy species, low trees or shrubs; natives of Europe and North America, of easy culture in any common soil, and propagated by seeds, which ought to be sown as soon as they are ripe, or by cuttings.

■ ‡ 1. S. trifo'lia *L.* The three-leaved Staphylea, *or Bladder-nut Tree.*

Identification. Lin. Sp., 386.; Dec. Prod., 2. p. 2.; Don's Mill., 2. p. 2.; Tor. and Gray, 1. p. 256.
Synonymes. Staphilier à Feuilles ternées, *Fr.*; Virginische Pimpernuss, *Ger.*
Engravings. Schmidt Baum., t. 81.; our *fig.* 197. in flower, and *fig.* 198. in fruit.

Spec. Char., &c. The leaf of 3 leaflets, which are ovate, acuminate, regularly sawed, and, when young, pubescent; the style smooth; the capsule bladdery. (*Dec. Prod.*) A deciduous shrub or low tree. Canada to South Carolina, and west to Arkansas, in moist places. Height 6 ft. to 12 ft. Introduced in 1640. Flowers whitish; May and June. Nuts globose, in a bladdery capsule, white; ripe in October. Decaying leaves greenish yellow.

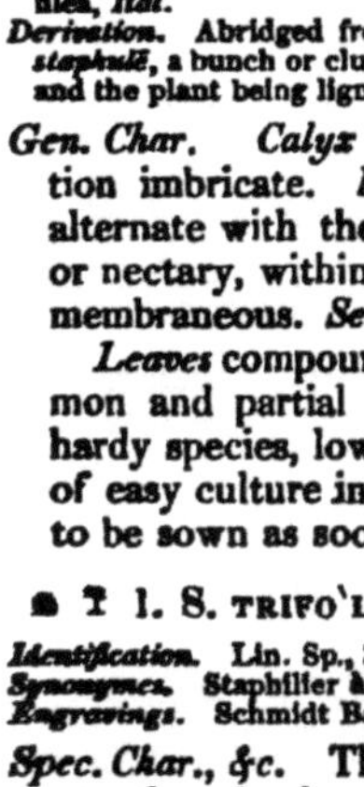

197. S. trifolia.

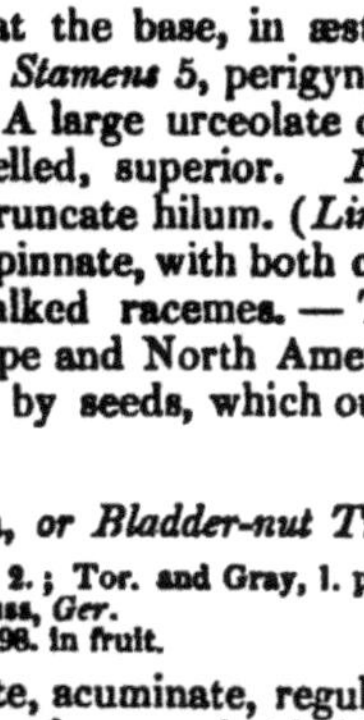

198. S. trifolia.

Branches slender, smooth, and dotted. Petioles pubescent above. Partial stipules mostly none. Petals obovate-spatulate, ciliate at the base. Stamens rather exserted; filaments hairy below; anthers cordate; the lobes somewhat united at the tip. Capsule 2 in. long; the carpels (sometimes 4) distinct at the summit, tipped with the persistent styles, and opening by the inner suture; seeds smooth and polished, all but one often abortive. (*Torrey and Gray.*) When not trained to a single stem, this shrub throws out abundance of shoots resembling suckers from the collar; but, if these be removed as they are produced, it will form a very handsome low tree. Seeds, suckers, layers, or cuttings, in any common soil, kept moist. The largest plants of this species, in the neighbourhood of London, are at Syon.

■ ♆ 2. S. PINNA'TA *L.* The pinnated-*leaved* Staphylea, *or Bladder-nut Tree.*

Identification. Lin. Sp., 386.; Dec. Prod., 2. p. 3.; Don's Mill., 2. p. 3.
Synonymes. Staphylodéndron pinnàtum *Ray*; Staphilier à Feuilles allées, *Fr.*; gemeine Pimpernuss, *Ger.*; Lacrime di Giobbe, or Pistacchio falso, *Ital.*; Job's Tears.
Engravings. Eng. Bot., t. 1560.; Hayne Abbild., t. 36.; and our *fig.* 199.

Spec. Char., &c. Leaves pinnate, of 5—7 oblong, perfectly glabrous, serrate leaflets; the flowers in racemes; the capsules membranous and bladdery. (*Dec. Prod.*) Shrub or low tree. South of Europe, and ? England in hedges. Height 6 ft. to 12 ft. Flowers whitish; May and June. Nuts globose white, in a bladdery capsule; ripe in October. Decaying leaves yellowish green. Naked young wood greenish, with green buds.

199. Staphylea pinnata.

A smooth branching shrub, throwing up many side suckers, in gardens often from 6 ft. to 12 ft. high, and exhibiting a much more luxuriant growth than the preceding species. The nuts, in some parts of Europe, are strung for beads by the Roman Catholics. The kernels taste like those of the pistacia, and are eaten in Germany by children. The flowers contain a great deal of honey, and are very attractive to bees. In the London nurseries, the plant is generally cultivated by side suckers, by cuttings put in during the month of September, or by seeds, which are ripened in abundance. The seeds ought to be sown as soon as they are ripe; because, as they contain an oil, they very soon become rancid. They will come up the following June, with two large, lance-shaped, seminal leaves; though sometimes they do not come up for two years.

ORDER XX. CELASTRA'CEÆ.

ORD. CHAR. *Sepals* 4—6: æstivation imbricate. *Petals* 4—6. *Stamens* 4—6, alternate with the petals, opposite the sepals, indistinctly perigynous. *Ovary* superior, free, girded with a fleshy disk, with 2—4 cells. *Ovules* erect, rarely pendulous. *Fruit* capsular, baccate, drupaceous, or samarideous. *Seeds*, in most, attended with an aril. (*Lindl.*)

Leaves simple, alternate or opposite, generally stipulate, deciduous, or evergreen. *Flowers* whitish or greenish, in axillary cymes.— Shrubs or low trees, generally deciduous; natives of both hemispheres.

The species are chiefly remarkable for the form and colours of their fruits; their flowers being neither large nor showy, nor their properties valuable in medicine, or general economy. All the species are readily increased by layers, by cuttings struck in sand, or by seeds in any common soil. The genera containing hardy species are Euónymus, Celástrus, and Nemopánthes, which are thus contradistinguished:—

EUO'NYMUS *Tourn.* Sexes mostly hermaphrodite. Fruit a dehiscent capsule, of 3—5 cells. Seed with an aril. Leaves mostly opposite.

CELA'STRUS *L.* Sexes mostly hermaphrodite. Fruit a dehiscent capsule of 2—3-cells. Seed with an aril. Leaves alternate.

NEMOPA'NTHES *Rafin.* Sexes polygamous or diœcious. Fruit an indehiscent berry

Genus I.

EUO'NYMUS *Tourn.* The Euonymus, *or* *Spindle Tree.* *Lin. Syst.* Tetra-Hex-ándria Monogýnia.

Identification. Tourn. Inst., t. 388.; Lin. Gen., 271.; Dec. Prod., 2. p. 3.; Don's Mill., 2. p. 3.
Synonymes. Fusain, Bonnet de Prêtre, or Bois à Lardoire, *Fr.*; Spindelbaum, *Ger.*; Evonimo, *Ital.*
Derivation. The word *Euónymus* is formed from the Greek, and signifies of good repute. The French word Fusain means a spindle, alluding to the use of the wood for making spindles. Bonnet de Prêtre alludes to the form of the capsules, which, when opened, bear some resemblance to a priest's cap; and it is called Bois à Lardoire from the use made of the wood for skewers or larding pins. The German name is literally spindle tree.

Gen. Char. *Calyx* 4—5-lobed, flat, covered by the peltate disk at the base. *Petals* 4—6, spreading, inserted in the disk. *Stamens* 4—6, inserted above the disk in rather prominent glands. *Capsule* 3—5-celled, 3—5-angled. *Seeds* 1—4 in each cell, and wrapped in pulp or aril. (*Don's Mill.*)

Leaves simple, opposite, serrate. *Stipules* mostly none. *Peduncles* axillary, 1—many-flowered. *Inflorescence* cymose. — Deciduous shrubs or low trees; sometimes trailing, or climbing by rootlets.

♈ ♠ 1. *E.* europæ'us *L.* The European Euonymus, *or Spindle Tree.*

Identification. Lin. Sp., 286.; Dec Prod., 2. p. 4.; Don's Mill., 2. p. 3.
Synonymes. *E.* vulgàris *Mill. Dict.*; Prick-timber *Gerard*; Louse Berry, Dogwood, Gatteridge Tree; Fusain d'Europe, *Fr.*; Bonnet de Prêtre commun, *Fr.*; gemeine Spindelbaum, *Ger.*; Berette di Prete, *Ital.*
Derivation. The English name Prick-timber, or Prick-wood, alludes to the employment of the wood in making toothpicks and skewers, which were formerly called pricks; and it is called Dogwood, because the wood of *Córnus sanguínea* and that of the *Euónymus europæ'us* are used indiscriminately for the same purposes, both being called Gatteridge Tree; the meaning of which we do not know. It is called Louse-wood, because the powdered leaves were formerly put on the heads of children to chase away lice.
Engravings. Smith's Eng. Bot., t. 362.; the plate of the tree in Arb. Brit., 1st edit., vol. v.; and our *fig.* 200.

200. *Euónymus europæ'us.*

Spec. Char., &c. Branches smooth. Leaves lanceolate-ovate, very finely sawed. Flowers about 3 upon one peduncle; the petals oblong, rather acute. Lobes of the capsule obtuse. (*Dec. Prod.*) A deciduous shrub, or low tree. Europe and Britain, in hedges and copse woods. Height 6 ft. to 12 ft. Flowers greenish white; May. Fruit scarlet, produced in great abundance, and very showy; ripe in September. Decaying leaves reddish. Naked young wood green or reddish green.

Varieties.

- ♀ ▲ E. *e.* 2 *latifòlius* Lodd. Cat. has rather broader leaves than the species.
- ♀ ▲ E. *e.* 3 *fòliis variegàtis* Lodd. Cat. has variegated leaves, but never looks healthy.
- ♀ ▲ E. *e.* 4 *frúctu álbo* Lodd. Cat. has white capsules.
- ▲ E. *e.* 5 *nànus* Lodd. Cat. is a dwarf-growing plant.

Nos. 2. and 4. of these varieties are, in our opinion, alone worth cultivating.

Roots numerous and whitish, forming a dense mass of network, and not extending to a great distance from the stem. The branches are numerous and opposite; and the wood hard and fine-grained. The leaves and bark are acrid, poisonous, and fetid when bruised. The capsules are of a fine rose colour, except in the white-capsuled variety, and the seeds are each invested with an aril of a fine orange colour. In a state of cultivation the tree attains the height of 30 ft. or upwards, and, though almost entirely neglected in pleasure-grounds, it forms a singularly handsome object in autumn, when covered with its ripe fruit. Seeds; in any common soil not over moist.

♀ ▲ 2. *E.* VERRUCO'SUS *Scop.* The warted-*barked* Euonymus, *or Spindle Tree.*

Identification. Scop. Carn., ed. 2. No. 268.; Dec. Prod., 2. p. 4.; Don's Mill., 2. p. 4.
Synonymes. *E.* europæ'us lepròsus *Lin.*; Fusain galeux, ou verruqueux, *Fr.*; warziger Spindelbaum, *Ger.*
Engravings. Nouv. Du Ham., 3. t. 8.; Schmidt Arb., t. 72.; and our *fig.* 201.

Spec. Char., &c. Branches warted with prominent lenticular glands. Leaves ovate, slightly serrate. Flowers three on a peduncle. Petals ovate. Capsule bluntly 4-cornered. (*Dec. Prod.*) A deciduous shrub or low tree. Austria, Hungary, and Carniola. Height 6 ft. to 12 ft. Introduced in 1763. Flowers purplish brown; May and June. Fruit reddish purple; ripe in September. Decaying leaves reddish green. Naked young wood green, with brown and white spots.

201. *Euónymus verrucòsus.*

A shrub of somewhat fastigiate habit of growth, with rough warty branches. This species is cultivated in collections chiefly for the singularity of its appearance, being among spindle trees what the warted ash is among ash trees. It ripens seeds, and is readily increased by cuttings.

♀ ▲ 3. *E.* LATIFO'LIUS *C. Bauh.* The broad-leaved Euonymus, *or Spindle Tree.*

Identification. C. Bauh. Pin., 428.; Dec. Prod., 2. p. 4.; Don's Mill., 2. p. 4.
Synonymes. *E.* europæ'us var. 2. *Lin.*; Fusain à larges Feuilles, *Fr.*; breitblättriger Spindelbaum, *Ger.*
Engravings. Jacq. Fl. Austr., t. 289.; Bot. Mag., 2384.; the plate of the species in Arb. Brit., 1st edit., vol. v.; and our *fig.* 202.

Spec. Char., &c. Branches smooth. Leaves broad-ovate, toothleted. Peduncles trichotomous, many-flowered. Petals oval, obtuse. Lobes of capsule acutely angled, wing-formed. (*Don's Mill.*) A deciduous shrub or low tree. South of France to Tauria, in groves. Height 10 ft. to 20 ft. Introduced in 1730. Flowers white, becoming purplish; June and July. Fruit deep red, and very showy; ripe in September. Decaying leaves purplish red. Naked young wood reddish green, with long pointed green buds, tinged with red.

In British gardens, this forms much the handsomest species of the genus,

202. Euónymus latifòlius.

from its broad shining leaves, and its large red pendulous fruits, with orange-coloured seeds, which, when the capsules open, are suspended from the cells somewhat in the manner that the seeds of the magnolias hang from their strobiles. Even the wood of this species, during winter, is much handsomer than that of any other, the branches being regularly divaricate, with a clean bark, of a reddish green, and with long pointed dark brown buds; by which alone this species may be distinguished from all the others. Unfortunately for this species, it is generally treated as a shrub, and crowded among other shrubs or trees; so that it is never allowed a chance of attaining either its full size or its proper shape.

■ 4. *E.* NA'NUS *Bieb.* The dwarf Euonymus, *or Spindle Tree.*

Identification. C. Bieb. Fl. Taur. Suppl., p. 160; Dec. Prod., 2. p. 4.; Don's Mill., 2. p. 4.
Synonyme. *E. caucásicum Lodd.*
Engraving. Our *fig.* 203. from Messrs. Loddiges's plant.

Spec. Char., &c. Branches smooth, somewhat herbaceous. Leaves lanceolate, entire, nearly opposite. Flowers 4-cleft, from 1 to 3 on a peduncle. (*Dec. Prod.*) A trailing undershrub, with the aspect of the widow wail (*Cneòrum tricóccum*). Northern Caucasus. Height 1 ft. Introduced in 1830. Flowers greenish white; July and August. Fruit?; ripe?. Shoots slender, recumbent, and with the leaves of a deep green.

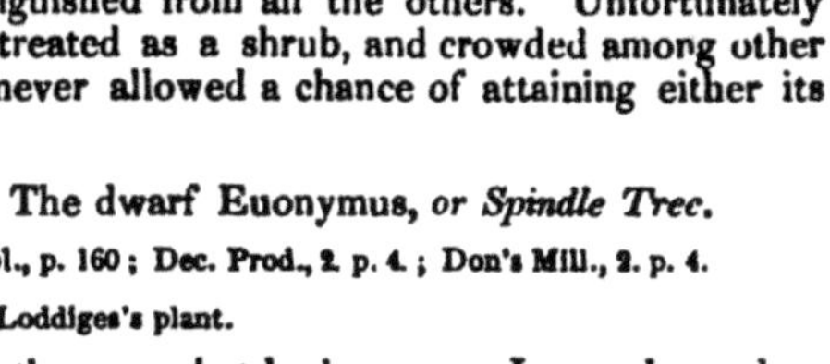

203. Euónymus nànus.

A very neat little plant, apparently quite hardy, and well adapted for rock-work.

? ■ 5. *E.* ATROPURPU'REUS *Jacq.* The dark-purple-*flowered* Euonymus, *or Spindle Tree.*

Identification. Jacq. Hort. Vind., 2.; Dec. Prod., 2. p. 4.; Don's Mill., 2. p. 5.; Tor. and Gray, 1. p. 257.
Synonymes. *E. caroliniénsis Marsh. Arb. Amer.* No. 1.; and, probably, *E. latifòlius Marsh. Arb. Amer.* No. 2.; Burning Bush, *Amer.*
Engravings. Jacq. Hort. Vind., 2. t. 120.; Schmidt Arb., t. 78.; and our *fig.* 204.

Spec. Char., &c. Branches smooth. Leaves stalked, lanceolate, serrated.

Flowers many upon a peduncle; the peduncle compressed. Petals orbiculate. Capsules angulately furrowed, smooth. (*Dec. Prod.*) A shrub or low tree. Canada to Florida. Height 4 ft. to 12 ft. Introduced in 1756. Flowers dark purple; June and July. Capsule crimson. Seeds white, with a red aril; ripe in October. Decaying leaves purplish red. Naked young wood purplish green

Branches slightly 4-sided. Leaves 2 in. to 5 in. long. Parts of the flower usually in fours; petals roundish obovate. Capsules smooth, deeply lobed. This and the other American species of *Euónymus* are rarely found in a thriving state in Britain: as it appears to us, from not being planted in moist shady situations, and in peat or sandy soil.

204. Euónymus atropurpùreus.

6. *E.* AMERICA'NUS *L.* The American Euonymus, *or Spindle Tree.*

Identification. Lin. Sp., 286.; Dec. Prod., 2. p. 4.; Don's Mill., 2. p. 5.; Tor. and Gray, 1. p. 258
Synonymes. *E.* sempervirens *Marsh.*; *E.* alternifolius *Mœnch*; the Burning Bush, Strawberry Tree, *Amer.*
Engravings. Nouv. Du Ham., 3. t. 9.; Schmidt Arb. t. 75.; our *fig.* 205., representing the plant in flower; and *fig.* 206., representing it in seed, with the warty capsule.

Spec. Char., &c. Branches smooth. Leaves almost sessile, elliptic-lanceolate, sawed. Flowers 1 to 3 on a peduncle. Petals sub-orbiculate. Capsule echinately warty. (*Dec. Prod.*) A sub-evergreen recumbent shrub. Canada to Florida, among rocks, and in moist woodlands. Height 2 ft. to 6 ft. Introduced in 1686. Flowers greenish yellow, tinged with purple; May and June. Capsule deep crimson. Seeds white, with a scarlet aril; ripe in October. Decaying leaves and naked shoots green.

205. Euónymus americànus

206. Euónymus americànus.

207. E. a. angustifòlius.

Varieties.

E. *a.* 2 *angustifòlius.* Var. β *Tor. & Gray.* (Our *fig.* 207.)—Leaves narrowly elliptical or oblong, slightly falcate, the margin minutely serrated. Possibly the *E.* angustifòlius of Pursh, which Torrey and Gray had only seen in a herbarium.

E. *a.* 3 *sarmentòsus* Nutt. Var. γ *Tor. & Gray.*—Shoots trailing and often rooting; leaves ovate-lanceolate.

E. *a.* 4 *obovàtus* Nutt. Var. δ *Tor. & Gray; E.* obovàtus *Dec. Prod.* 2. p. 4., *Don's Mill.* 2. p. 5. (Our *fig.* 208.)—Trailing and rooting; leaves obovate, or oval-obovate, obtuse or slightly acuminate, acute at the base.

208. E. a. obovàtus.

Erect. Leaves oval or elliptical lanceolate, the uppermost often slightly falcate, mostly acuminate, acute or obtuse (rarely subcordate) at the base. (*Tor. and Gray*, var. *a*.) Branches slender, green. Leaves 1 in. to 2 in. long, coriaceous, nearly evergreen in the southern states. Seeds smaller than in *E.* atropurpùreus. The scarlet fruits, according to Pursh, resemble, at a distance, those of *A'*rbutus *U'*nedo. They form a great ornament, he says, to this almost evergreen shrub, and have given rise, in America, to its common name, the burning bush. Of easy culture in moist soil, and a shady situation. Cuttings or seeds.

? ■ 7. *E.* HAMILTONIA'NUS Wall. Hamilton's Euonymus, *or Spindle Tree.*

Identification. Wall. Fl. Ind., 2. p. 403.; Don's Mill., 2. p. 4.
Synonyme. *E.* atropurpùreus *Wall. Fl. Ind.* 2. p. 402.
Engraving. Our *fig.* 209., from a vigorous plant in the Hort. Soc. Garden.

Spec. Char., &c. Branches smooth, terete. Leaves lanceolate, finely serrated. Peduncles dichotomous, 6-flowered. Flowers tetrandrous. Petals 4, lanceolate cordate. Ovary 4-lobed, 4-celled, each cell containing 2 ovules. (*Don's Mill.*) A low tree or shrub. Nepal. Height 10 ft. to 20 ft. Introduced in 1825. Flowers yellowish green; June and July. Fruit ? purple; ripe in ? October. Decaying leaves and naked young wood green.

209. Euónymus Hamiltoniànus.

A free-growing species, with an erect stem; the young shoots green; the leaves large; bark of the older shoots white. Left to itself, as a standard, it forms a dense fastigiate bush, with numerous suckers; but, trained to a single stem, it would doubtless form a handsome small tree. A plant against the wall, in the Horticultural Society's Garden, flowers freely every year; but has not yet ripened fruit. The plant in the open garden was killed to the ground by the winter of 1837-8, but sprang up again with vigour. In the Liverpool Botanic Garden it was not injured.

210. E. japónicus.

Other Species of Euónymus. — E. *japónicus* Thunb. (our *fig.* 210.), and E. *japónicus fòliis variegàtis*, E. *garciniæfòlius Roxb.*, and some other species, are in London gardens; but they can only be considered as half-hardy. In the Canterbury Nursery, *E.* j. fòliis variegàtis has been found hardier than the species. In the Horticultural Society's Garden, *E.* japónicus, trained against a wall, was but little injured by the winter of 1837-8. The following species, shortly described in our first edition, Mr. Don considers as likely to prove "truly hardy;" some of them are introduced, and are in green-houses: *E.* gróssus *Wall.*, *E.* micránthus *D. Don*, *E.* lùcidus *D. Don*, *E.* echinàtus *Wall.*, *E.* tíngens *Wall.*, *E.* glàber *Roxb.*, *E.* fimbriàtus *Wall.*, *E.* índicus *Heyne*, *E.* vàgans *Wall.*, *E.* subtriflòrus *Blume*, *E.* Thunbergiànus Blume, *E.* péndulus *Wall.*, and *E.* frígidus *Wall.*

GENUS II.

CELA'STRUS *L.* THE CELASTRUS, *or STAFF TREE. Lin. Syst.* Pentándria Monogýnia.

Identification. Lin. Gen., 270.; Dec. Prod., 2. p. 5.; Don's Mill., 2. p. 6.
Synonymes. Euonymöides *Mœnch*; Célastre, *Fr.*; Celaster, *Ger.*
Derivation. From *kēlas*, the latter season: the fruit remaining on the tree all the winter. The *kēlastros* of the Greeks is supposed to be the *Euónymus.*

Gen. Char. Calyx small, 5-lobed. *Petals* 5, unguiculate. *Ovary* small, immersed in a 10-striped disk. *Stigmas* 2—3. *Capsule* 2—3 valved. *Seed* 1, in a large fleshy aril. (*Don's Mill.*)

Leaves simple, alternate, stipulate, deciduous; stipules minute. — One hardy species; a climbing shrub, a native of North America.

⊥ 1. *C.* SCA'NDENS *L.* The climbing-*stemmed* Celastrus, *or Staff Tree.*

Identification. Lin. Sp., 285.; Dec. Prod., 2. p. 6.; Don's Mill., 2. p. 6.; Tor. and Gray, 1. p. 257.
Synonymes. Bourreau des Arbres, *Fr.*; Baummörder, *Ger.*; Bittersweet, Waxwork, *Amer.*
Engravings. Nouv. Du Ham., 1. t. 95.; and our *fig.* 211.

Spec. Char., &c. Thornless, climbing, smooth. Leaves oval, acuminate serrate. Flowers diœcious. (*Dec. Prod.*) A deciduous twining shrub. Canada to Virginia. Height 5 ft. to 20 ft. Introduced in 1736. Flowers small, pale yellowish green; June. Capsules orange; ripe in September. Seed reddish brown, coated with a bright orange aril, changing at last to scarlet.

211. Celástrus scándens.

The stems are woody and flexible, and twist themselves round trees and shrubs, or round each other, to the height of 12 or 15 feet, or upwards, girding trees so closely as, in a few years, to destroy them; whence the French and German names, which signify "tree strangler." The leaves are about 3 in. long, and nearly 2 in. broad, serrated, of a lively green above, but paler on the under side. The plant prefers a strong loamy soil, rather moist than dry; and is readily propagated by seeds, layers, or cuttings.

Other Species of Celástrus. — C. *bullàtus*, described from a figure of Plukenet, is, according to Torrey and Gray, a doubtful plant. C. *nepalénsis* and C. *pyracanthifòlius* are in Messrs. Loddiges's collection, but rather tender.

GENUS III.

NEMOPA'NTHES *Rafin.* THE NEMOPANTHES. *Lin. Syst.* Polygàmia Diœ'cia.

Identification. Rafin. Journ. Phys., 1819, p. 96.; Dec. Prod., 2. p. 17.; Don's Mill., 2. p. 13.
Synonyme. Ilicöides *Dum. Cours.* 1. vol. 4. p. 27.
Derivation. From *nemos*, a grove, and *anthos*, a flower; it being generally found in groves.

Gen. Char. Calyx small, scarcely conspicuous. *Petals* 5, distinct, oblong, linear, deciduous. *Stamens* 5, alternating with the petals. *Ovary* hemispherical, covered with clammy juice. *Style* wanting. *Stigmas* 3—4, sessile; in the male flowers hardly manifest. *Berry* globose, 3—4-celled. (*Don's Mill.*)

Leaves simple, alternate, ovate, exstipulate, deciduous; quite entire.—One hardy species.

■ 1. N. CANADENSIS Dec. The Canadian Nemopanthes.

Identification. Dec. Mém. Soc. Gen., 1. p. 44.; Pl. Rar. Hort. Gen. t. 3.; Don's Mill., 2. p. 13.
Synonymes. Ilex canadénsis *Michx. Flor. Bor. Amer.* 2. p. 299.; N. fasciculàris *Rafin.*; Ilex delicátula *Bart. Fl. Vir.* p. 67.; ? Prinos lùcidus *Ait. Hort. Kew.* 2. p. 478.; Houx du Canade, *Fr.*
Engravings. Michx. Fl. Bor. Amer. 2. t. 49., as Ilex canadénsis; and our *fig.* 212.

212. N. canadénsis.

Spec. Char., &c. Leaves ovate, quite entire, or serrated at the apex, smooth Pedicels usually solitary, 1-flowered, very long. Flowers white. Berries large, beautiful crimson, very ornamental. (*Don's Mill.*) A deciduous shrub. Canada to Carolina, on mountains. Height 3 ft. to 5 ft. Introduced in 1802. Flowers small, white; April and May. Berry large, beautiful crimson; ripe in October. Very ornamental.

Plants of this species in the Hort. Soc. Garden, and at Messrs. Loddiges's, under the name of Prìnos lùcidus, form neat fastigiate shrubs, rather of slow growth. A few years ago there were some very handsome plants of this species at White Knights. Suckers or layers in loamy soil; or cuttings of the young wood in sand under a glass.

213. Màytenus chiléns̀is

Other Species of Celastràceæ.—Màytenus *chiléns̀is* Dec. (our *fig.* 213.), a handsome evergreen branchy shrub, with twiggy branchlets. The flowers are in axillary clusters, with the corolla of a yellowish green colour, not showy. It is a native of Chile, and stood eight or ten winters against a south wall in the Hort. Soc. Garden, and was thought to be tolerably hardy, but was killed to the ground in the winter of 1837–8, and did not spring up again. It also stood several winters in the open garden, as a standard, and promised to be a valuable addition to our hardy evergreen shrubs, which it may possibly yet prove.

ORDER XXI. AQUIFOLIACEÆ.

Identification. Lindley's Key, p. 63.
Synonymes. Celastrineæ, tribe Aquifoliàceæ, in part, *Dec. Prod.* 2. p. 11.; Ilicineæ, in part, *Lindl. Introd. to N. S.* p. 178., *Don's Mill.* 2. p. 14.

ORD. CHAR. *Calyx* and *corolla* with an imbricate æstivation. *Sepals* 4—6. *Corolla* hypogynous, with 4—6 lobes, and as many stamens inserted into it alternately to its lobes. *Ovary* 2—6-celled; a pendulous ovule in each cell. *Fruit* fleshy, indehiscent, with 2—6 stones, each containing a pendulous seed, which has large fleshy albumen. (*Lindl.*)—Low trees or shrubs, chiefly evergreen. Natives of Europe and North America.

Leaves simple, alternate or opposite, exstipulate, evergreen or deciduous; coriaceous. *Flowers* small, axillary, solitary or fascicled.—The genera containing hardy species are three, and are thus contradistinguished: —

MYGINDA Jacq. Sexes hermaphrodite. Stamens 4. Fruit 1-celled, 1-seeded. Shrubs with square branchlets; leaves opposite, subcoriaceous, and flowers upon trifidly or trichotomously branched peduncles.

ILEX *L.* Sexes hermaphrodite, diœcious, or polygamous. Stamens 4—5. Fruit including 4 or 5 nuts. Evergreen shrubs with, mostly, coriaceous leaves. Flowers many on a peduncle.

PRINOS *L.* Sexes mostly diœcious or polygamous. Stamens 6. Fruit including 6 nuts. Shrubs with leaves deciduous or persistent, and flowers 1 upon a peduncle.

Genus I.

MYGI'NDA Jacq. The Myginda. *Lin. Syst.* Tetrándria Monogýnia.

Identification. Jacq. Amer., p. 24.; Dec. Prod., 2. p. 12.; Don's Mill., 2. p. 16.
Synonymes. I'lex *Pursh*; Oreóphila *Nutt.* in *Tor. and Gray.*
Derivation. So named by Jacquin in honour of *Francis von Mygind*, a German botanist.

Gen. Char. *Calyx* small, 4-cleft. *Corolla* deeply 4-cleft, subrotate. *Stamens* 4, alternating with the segments of the corolla, and shorter than them, and inserted in its throat. *Ovary* roundish. *Stigmas* 4. *Drupe* ovate, 1-celled. (*Don's Mill.*)

Leaves simple, mostly opposite, exstipulate, evergreen; entire or remotely serrulated. *Flowers* axillary, subsolitary, minute. — One species is hardy.

1. M. myrtifo'lia *Nutt.* The Myrtle-leaved Myginda.

Identification. Nutt. Gen., 1. p. 109., Don's Mill., 2. p. 16.; Dec. Prod., 2. p. 13.
Synonymes. I'lex Myrsinites *Pursh*; Oreóphila myrtifòlia *Nutt.*
Engravings. Hook Fl. Bor. Amer., t. 41.; and our *fig.* 214.

214. M. myrtifòlia.

Spec. Char., &c. Leaves oblong, blunt, serrated, smooth, with revolute edges. Peduncles very short usually solitary, 1-flowered. Style short, club-shaped 4-lobed at the apex. (*Don's Mill.*) A low evergreen shrub. N. W. coast of North America, and the Rocky Mountains. Height 2 ft. to 4 ft. Introduced in 1818. Flowers small, white; May to August. Drupe dark purple; ripe in Oct.

Plants of this species are in the arboretum of Messrs. Loddiges; where it is increased by cuttings, and grows in common soil.

Genus II.

I'LEX *L.* The Holly. *Lin. Syst.* Tetrándria Tetragýnia.

Identification. Lin. Gen., No. 172.; Dec. Prod., 2. p. 13.; Don's Mill., 2. p. 16.
Synonymes. Aquifòlium *Tourn. Inst.* t. 371., *Gært. Fruct.* 2. t. 92.; Houx, *Fr.*; Stechpalme, or Hülse, *Ger.*; Ilice, *Ital.*
Derivation. Theophrastus, and other Greek authors, named the holly Agria; that is, wild, or of the fields; and the Romans formed from this the word Agrifolium; and called it, also, Aquifolium, from *acutum*, sharp, and *folium*, a leaf. C. Bauhin and Loureiro first named it I'lex, on account of the resemblance of its leaves to those of the Quércus I'lex, the true Ilex of Virgil. Linnæus adopted the name of I'lex for the genus, and preserved the name of Aquifòlium for the most anciently known species. The name of Holly is, probably, a corruption of the word holy, as Turner in his *Herbal* calls it Holy, and Holy Tree; probably from its being used to commemorate the holy time of Christmas, not only in houses, but in churches. The German name Christdorn, the Danish name Christorn, and the Swedish name Christtorn, seem to justify this conjecture.

Gen. Char. *Calyx* 4—5-toothed, permanent. *Corolla* 4—5-cleft, subrotate. *Stamens* 4—5, alternating with the segments of the corolla. *Ovary* sessile. *Stigmas* 4. *Berry* containing 4—5 1-seeded nuts. (*Don's Mill.*)

Leaves simple, alternate, exstipulate, mostly evergreen; ovate, oval, or ovate-lanceolate, coriaceous, serrated, toothed, or quite entire. *Flowers* axillary, aggregate, small, generally white. *Fruit* a drupe, mostly red.—Low trees and shrubs, chiefly evergreen, natives of Europe, North America, and the Himalayas, generally of slow growth, and of long duration. Loamy soil, rather dry than moist.

A. *Leaves spiny-toothed.*

‡ 1. *I.* AQUIFO`LIUM *L.* The prickly-leaved, *or common,* Holly.

Identification. Lin. Sp., 181.; Fl. Dan., 508.; Dec. Prod., 2. p. 14.; Don's Mill., 2. p. 16.
Synonymes. The holly, being a native of most parts of Europe, and being every where much admired, has several names in most living European languages: Hulver, Hulfere, and Holme, *Eng.*; Le Houx, *Fr.*; Stechpalme, Stechlaub, Hulse, Christdorn, Mausdorn, Kleesebusch, *Ger.*; Schubbig hardkelk, *Dutch*; Stikpalme, *Danish*; Jernek, Christtorn, *Swedish*; Waefoseheld, Ostrokol, Padub, *Russ.*; Agrifolio, *Ital.*; Acebo, *Span.*; Azevinho, *Port.*
Engravings. Smith Eng. Bot., t. 496.; the plate of the species in Arb. Brit, 1st edit., vol. v.; and our *fig.* 215.

Spec. Char., &c. Leaves oblong, shining, wavy, spiny-toothed. Peduncles axillary. Flowers nearly umbellate. A handsome, conical, evergreen tree. Europe and Britain. Height 20 ft. to 30 ft. in a wild state; and

215. *Ilex Aquifolium.*

twice that height, or upwards, in a state of cultivation. Flowers white; May. Fruit red; ripe in September, and remaining on the tree all the winter. The lower leaves are very spinous; while the upper ones, especially on old trees, are entire. Decaying leaves yellow, dropping in June or July.

Varieties. In general the variegation of plants, more especially of trees and shrubs, is accompanied by a ragged, or otherwise unhealthy, appearance in the leaves; but the holly is one of the very few exceptions to this rule. The variegations of the holly are chiefly confined to the modification of white and yellow in the leaves: but there are some sorts in which the variation results from the state of the leaves with reference to prickles, to magnitude, and to form; and others consist of differences in the colour of the fruit, which is red, yellow, or white, and black. These varieties are, for the most part, without names, and those in the following groups appear to us to be all that are truly distinct; but the shades of difference under each name in these groups are almost innumerable.

a. *Varieties designated from the Form, Magnitude, Thickness, Surface or Margin of the Leaf.*

‡ I. A. 2 *heterophýllum* Hort. — Leaves variously shaped.
‡ I. A. 3 *angustifòlium* Hort. — Leaves narrow.
‡ I. A. 4 *latifòlium* Hort. — Leaves broad. Shoots dark purple. There is a fine specimen at Elvaston Castle, where it is called the Water Holly.
‡ I. A. 5 *altaclerénse* Hort. The High Clere Holly. — Leaves broad, thin, and flat.
‡ I. A. 6 *marginàtum* Hort. (*fig.* 216.) — Leaves without prickles, coriaceous, nearly as broad as long, and with a thickened margin.

216. *I. A.* marginàtum.

217. *I. A.* laurifòlium.

218. *I. A.* ciliàtum.

‡ I. A. 7 *laurifòlium* Hort. (*fig.* 217.) — Leaves small, oval-lanceolate without prickles, about the size and shape of those of *Laúrus nóbilis.*
‡ I. A. 8 *ciliàtum* Hort. (*fig.* 218.) — Leaves oval-acuminate, small; the prickles along the margins like hairs.
A. 9 *ciliàtum mìnus* Hort. — Leaves thinner and smaller than in the preceding variety.
A. 10 *recúrvum* Hort. (*fig.* 219.) — Leaves recurved.

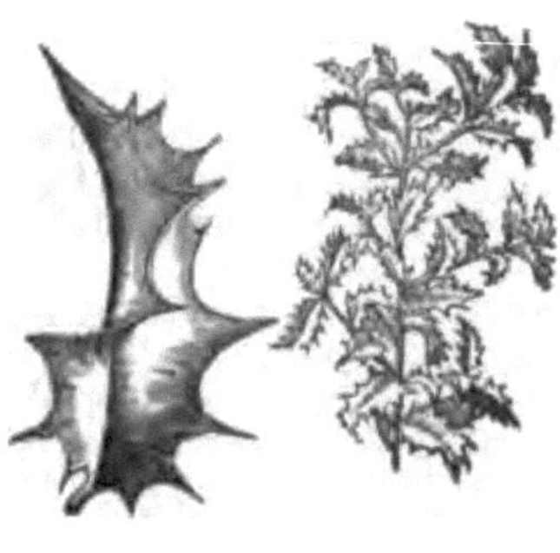
219. *Ilex Aquifòlium recúrvum*

220. *Ilex Aquifolium serratifolium.*

‡ I. A. 11 *serratifòlium* Hort. (*fig.* 220.) — Leaves serrated.
‡ I. A. 12 *críspum* Hort. — Leaves curled.
‡ I. A. 13 *fèrox* Hort. Hedgehog Holly; Houx-hérisson, *Fr.* (*fig.* 221.) — The disk of the leaf has its edges rolled back ; and a somewhat

221. *I. A. ferox.*

cylindrical figure is hence given to it; and, as the surface abounds in prominences and prickles, it has a curious appearance, not unaptly compared to that of a hedgehog.

? I. A. 14 *crassifòlium* Hort. (*fig.* 222.) — Leaves thick and fleshy.

? I. A. 15 *senéscens* Sweet. — Leaves spineless, thin.

All these varieties may be seen in Messrs. Loddiges's collection, and it is no small proof of their value for town gardens, that they thrive in the smoky atmosphere of that magnificent establishment.

222. *I. A. crassifòlium.*

b. *Varieties designated from the Colours of the Leaf.*

Under the general name of variegated hollies, twenty or thirty varieties, some of them with, and some of them without, popular names, are obtainable in the principal London nurseries. Having examined and compared the different shades of variegation in the plants in the very complete collection of Messrs. Loddiges, we think they may be all included in the following groups: —

? I. A. 16 *álbo-marginàtum* Hort. — Leaves with white edges. Of this variety the subvarieties in Loddiges's arboretum are marked 5, 15, 18, and 24, which have all long and narrow leaves, with edgings of white or pale yellow along their margins; and 4, 6, 7, 12, 17, 22, 23, and 28, which have larger leaves, and a greater breadth of margin variegated; the white or pale yellow forming in some cases one third, or even one half, of the surface of the leaf.

? I. A. 17 *aúreo-marginàtum* Hort. — Leaves with yellow edges. The following subvarieties are in Messrs. Loddiges's arboretum, Nos. 19 and 20 with dark yellow margins; and Nos. 1, 2, 8, 9, 10, 13, and 29, with margins of dark and light yellow. Another subdivision of this group consists of plants with broad leaves, in what may be called a transition state from green to variegated, viz., with greenish yellow or very pale green blotches or margins. When such plants become old, they are generally very distinctly variegated with yellow. Examples in the Hackney arboretum are Nos. 3, 20, and 21.

? I. A. 18 *álbo-pictum* Hort. — Leaves spotted with white. This variety has a considerable portion of the centre of the disk of the leaf white, and of a somewhat transparent appearance; the edges of the disk of the leaf being green.

? I. A. 19 *aúreo-pictum* Hort. — Leaves spotted with yellow. The following subvarieties are in Messrs. Loddiges's arboretum. Nos. 11, 14, 16, 26, 27, and 30.

? I. A. 20 *fèrox argénteum* Hort. — The hedgehog holly with leaves blotched with white.

? I. A. 21 *fèrox-aúreum* Hort. — The hedgehog holly with leaves blotched with yellow.

c. *Varieties designated from the Colour of the Fruit.*

? I. A. 22 *frúctu lùteo* Hort. — Fruit yellow.

? I. A. 23 *frúctu álbo* Hort. — Fruit white.

? I. A. 24 *frúctu nìgro* Hort. — Fruit black.

The holly makes the most impenetrable and the most durable of all vegetable fences; and it has this great advantage over deciduous-leaved trees and shrubs, that it is seldom liable to be attacked by insects; and, if shorn, the outer surface becomes impenetrable even to birds, who cannot build their nests in it. The wood is almost as white as ivory, except in the centre of very old trunks, where it is somewhat brown. It is very hard, with a fine grain, susceptible of a high degree of polish, and is readily stained with black, green, blue, or red. It weighs, when dry, at the rate of 47 lb. 7 oz. per cubic foot. The veins of the wood, and its annual layers, are so small as scarcely to be perceptible. It is applied to a great many purposes, in joinery, cabinet-making, and turnery; in engineering, in mathematical-instrument-making; and it is even used for wood-engraving. The bark affords birdlime, which is prepared by boiling and evaporation. The holly attains the largest size in a rich sandy loam; but it will grow, and even thrive, on almost any soil, provided it is not overcharged with moisture. As its seeds, like those of the hawthorn, do not come up the first year, to save ground and the expense of weeding, the berries are commonly buried in the soil, or kept mixed up in a heap of earth for one year; after which, if sown in autumn, they will come up the following June. The varieties are propagated by budding and grafting. These operations are performed at the usual times, and in the usual manner; but it has been observed by Tschoudi, that cleft-grafting does not succeed nearly so well with the holly as whip-grafting or budding. In England, the stocks budded or grafted on are generally of four or five years' growth; and the grafting is effected in March, and the budding in July. The variegated kinds are also propagated by cuttings, which are made in autumn, of the ripened summer shoots. They are planted in sandy soil, in a shady border, and covered with hand-glasses; and they generally put forth roots the following spring. Holly hedges should never be clipped, because, when the leaves are cut through the middle, they are rendered unsightly; and the shoots should therefore be cut with a knife close to a leaf. The proper season for cutting would appear to be just after the leaves have attained maturity; because at that season, in the holly, as in the box, the wound is comparatively soon obliterated by the healing over produced by the still abundant sap.

‡ 2. *I.* (*A.*) BALEA'RICA *Desf.* The Minorca Holly.

Identification. Desf. Arb., 2. p. 362.; Don's Mill., 2. p. 17.
Synonymes. *I. Aquifolium* var. ? *Lam. Dict.* 3. p. 145.; *I. maderénsis Willd. Enum. Suppl.* 8. according to Link.
Engraving. Our *fig.* 223.

Spec. Char., &c. Leaves ovate, acute, flat, shining, entire or spiny-toothed. Umbels axillary, few-flowered, short. (*Don's Mill.*) An evergreen low tree. Minorca and Madeira. Height 10 ft. to 20 ft. in England; in Madeira 60 ft. to 80 ft. Introduced in 1744. Flowers white; May and June. Drupe scarlet; ripe in December. Decaying leaves yellow, dropping in June and July.

223. *I.* (*A.*) balearica

A very distinct variety of the common holly, readily distinguished at sight, by its yellowish green leaves, which are sharply acuminated, but very slightly waved at the edges, and with few prickles. It is propagated by budding or grafting on the common holly.

‡ 3. *I.* OPA'CA *Ait.* The opaque-*leaved, or American,* Holly.

Identification. Ait. Hort. Kew, 1. p. 177.; Dec. Prod., 2. p. 14.; Don's Mill., 2. p. 17.
Synonymes. *Agrifolium vulgàre Clayt. Flor. Virgin.*; *I'lex Aquifolium Gronov.* and *Walt. Fl. Car.* 241.; dunkelblättrige Stechpalme, *Ger.*; Agrifolio a Foglie di Quercia, *Ital.*
Engravings. The plate of the species in Arb. Brit., 1st edit., vol. v.; and our *fig.* 224.

Spec. Char., &c. Leaves ovate, flat, coriaceous, acute, toothed in a scalloped manner, spiny, and glabrous, but not glossy. Flowers scattered, at the base of only those branches that are a year old. Teeth of the calyx acute. Sexes diœcious. (*Dec. Prod.*) A beautiful evergreen low tree. Canada to Carolina. Height in England 10 ft. to 20 ft.; in Carolina 60 ft. to 80 ft. Introduced in 1744. Flowers white; May and June. Drupe scarlet; ripe in December, remaining on the trees all the winter.

Variety.

¥ I. *o.* 2 *laxiflòra*, *I.* laxiflòra *Lam.*, *I.* opàca var. *Nutt.*, has the flowers on loosely branched peduncles, and the drupe yellow. Introduced in 1811. Several other varieties are mentioned by Rafinesque.

224. *I'lex opàca.*

In America, this species is applied to all the uses which the common holly is in Europe. It forms hedges; is an ornamental tree or shrub in gardens; is employed for making birdlime; and the wood is used in turnery and cabinetmaking. Propagation as in the common holly.

225. *I. magellánica.*

I'lex magellánica (*fig.* 225.), of which there is a small plant in the Hort. Soc. Garden, promises to be a very handsome species, and tolerably hardy, as it stood the winter of 1837–8 against a wall.

226. *I. Peràdo.*

B. *Leaves toothed, serrated, or crenate, but not spiny.*

■ 4. *I. Peràdo* Ait. The Perado Holly.

Identification. Ait. Hort. Kew., 1. p. 169.; Don's Mill., 2. p. 17.
Synonyme. *I.* maderénsis *Lam. Dict.* 3. p. 146.
Engravings. N. Du Ham., v. t. 35.; Bot. Cab., t. 549.,; and our *fig.* 226.

Spec. Char., &c. Leaves ovate, with an entire acumen, or having very few teeth, shining. Umbels short, axillary, few-flowered. (*Don's Mill.*) An evergreen shrub. Madeira. Height in England 5 ft. to 10 ft. Introduced in 1760. Flowers white or reddish; May and June. Drupe large, red; ripe in October.

Commonly treated as a green-house plant, but quite hardy in the Hort. Soc. Garden; where, and also in several other places, it stood the winter of 1837–8, without any protection, uninjured.

■ 5. *I. Cassi'ne* Ait. The Cassine-*like, or broad-leaved Dahoon*, Holly.

Identification. Ait. Hort. Kew., 1. p. 170.; Dec. Prod., 2. p. 14.; Pursh Fl. Am. Sep. 1. p. 117.; Don's Mill., 2. p. 17.
Synonymes. Aquifòlium carolinénse *Catesb. Car.* 1. t. 31.; *I.* caroliniàna *Mill. Dict.* No. 3.; *I.* cassinöìdes *Link Enum.* 1. p. 148.; the Cassena of the American Indians. *Rafin.*
Engravings. Catesb. Car., 1. t. 31.; and our *fig.* 227.

Spec. Char., &c. Leaves ovate-lanceolate, sharply serrated, flat; the midribs, petioles, and branchlets glabrous. The flowers upon lateral corymbosely branched peduncles. (*Dec. Prod.*) An evergreen shrub. Lower Carolina to Florida, in shady swamps. Height 6 ft. to 10 ft. Introduced in 1700. Flowers white; August. Drupe red; ripe in December.

227 Ilex Cassine.

The fruit is rather smaller than that of the common holly; it continues on the trees the most part of the winter, untouched by birds; and, being of a bright red, and large in proportion to the leaves, which are about the size of those of the common arbutus, the plant makes a fine appearance, both in its native country and in England. Commonly propagated by seeds; but it will also strike by cuttings, or it may be grafted on the common holly.

■ 6. *I.* ANGUSTIFO'LIA *Willd.* The narrow-leaved Holly.

Identification. Willd. Enum., 1. p. 172.; Dec. Prod., 2. p. 14.; Don's Mill., 2. p. 17.
Synonymes. *I.* myrtifòlia *Walt. Carol.* 241., *N. Duh.*, and *Lodd. Cat.*; *I.* rosmarinifòlia *Lam. Ill.* 1. p. 356.
Engravings. N. Du Ham., 1. t. 4.; and our *fig.* 228.

Spec. Char., &c. Leaves linear-lanceolate, sawed at the tip, rather revolute in the margin; the midrib, petiole, and branchlets glabrous. Flowers in stalked lateral cymes. (*Dec. Prod.*) An evergreen shrub. Height 6 ft. to 10 ft. Virginia to Georgia, in swamps. Introduced in 1806. Flowers white; June. Drupe globular and red; ripe in December.

228. *I.* angustifòlia.

A very handsome species, but not very common. There are plants of it at Messrs. Loddiges, and in the H. S. Garden, under the name of *I.* myrtifòlia.

♀ 7. *I.* VOMITO'RIA *Ait.* The emetic Holly, *or South Sea Tea.*

Identification. Ait. Hort. Kew., 1. p. 278.; Dec. Prod., 2. p. 14.; Don's Mill., 2. p. 17.
Synonymes. *I. Cassine* vèra *Walt. Carol.* 241.; *I.* ligústrina *Jacq. Coll.* 4. p. 105., *Icon. Rar.* t. 310., *Wendl. Hort.* t. 31.; *Cassine Peràgua* Mill. Icon. t. 83. f. 2.; *I. Cassèna* Michx. Fl. 1. p. 229.; *I.* religiòsa *Bart. Fl. Virg.* 69.; *I.* floridàna *Lam. Ill.* No. 1731.; Houx apalachine, *Fr.*; true Cassena, Cassena, *Florida*; the Yapon, *Virginia*; the evergreen Cassena, or Cassioberry Bush, *Eng.*
Engravings. Jacq. Icon. Rar., t. 310.; Wendl. Hort., t. 31.; Mill. Icon., t. 83. f. 2.; and our *fig.* 229.

Spec. Char., &c. Leaves oblong or elliptic, obtuse at both ends, crenately serrated, and, with the branchlets, glabrous. Flowers in subsessile lateral umbels. (*Dec. Prod.*) An evergreen low tree. Carolina to Florida, along the sea coast. Height 10 ft. to 12 ft. Introduced in 1700. Flower white; June and July. Drupe red, like that of the common holly; ripe in December.

229. Ilex vomitòria.

Not very common in British collections; but there are plants of it in Loddiges's arboretum, and in the garden of the Hort. Soc.

C. *Leaves quite entire, or nearly so.*

♀ ■ 8. *I.* DAHOO'N Wall. The Dahoon Holly.

Identification. Walt. Fl. Carol., 241.; Dec. Prod., 2. p. 14.; Don's Mill., 2. p. 17.
Synonyme. *I. Cassine* Willd. Hort. Berol. 1. t. 31

Engravings. Willd. Hort. Berol., t. 31., under the name of *I. Cassine*; our *fig.* 230. from a plant in Loddiges's arboretum, and *fig.* 231. from the Hort. Berolin.

Spec. Char., &c. Leaves lanceolately elliptical, nearly entire, almost revolute in the margin; the midrib, petiole, and branchlets villous. Flowers disposed in corymbose panicles, that are upon lateral and terminal peduncles. (*Dec. Prod.*) A beautiful evergreen shrub or low tree. Carolina to Florida, in swamps. Height 8 ft. to 10 ft. Introduced in 1726. Flowers white; May and June. Drupe red; ripe in December.

The leaves of this species are very numerous, and resemble those of *Laúrus Borbònia*. The plant is rare in British gardens, and seldom ripens fruit. It is most commonly kept in green-houses or pits: but there was a plant in 1836 in the open air, in the Mile End Nursery, which was 20 ft. high, with a head 30 ft. in diameter; and which had stood there many years, without the slightest protection.

230. *I'lex Dahoòn.*

231. *I'lex Dahoòn.*

Other Species of I'lex are described by authors as natives of North America and Nepal, some of which are introduced, and may probably be found hardy, but we can state nothing with certainty respecting them. In this genus, as in most others containing numerous species which are not in general cultivation, there is a great deal of uncertainty. *I. Cassìne*, *I.* vomitòria and *I. Dahoòn* are probably the same species.

GENUS III.

PRI'NOS L. THE PRINOS, *or WINTER BERRY*. *Lin. Syst.* Hexándria Monogýnia, or Polygàmia Diœ'cia.

Identification. Lin. Gen., No. 461.; Dec. Prod., 2. p. 16.; Don's Mill., 2. p. 20.
Synonymes. Agèria *Adans. Fam.* 2. p. 166.; Apalanche, *Fr.*; Winterbeere, *Ger.*
Derivation. From *prinos*, the Greek name for the holly, which the present genus much resembles; or, according to others, from *prion*, a saw, on account of the serrated leaves of the species.

Gen. Char. *Flowers* 6-cleft, hexandrous; usually diœcious, or polygamous from abortion. *Fruit* with 6 nuts. In other respects the character is the same as that of I'lex. (*Don's Mill.*)

Leaves simple, axillary, exstipulate, deciduous or evergreen; oval or lanceolate, entire or serrated; dying off of a greenish yellow. *Flowers* on axillary pedicels, usually single, small, mostly white. — Shrubs, evergreen and deciduous; natives of North America.

In habit of growth the species are all more or less fastigiate, and send up numerous suckers from the collar; but, if these were removed, the plants would form neat miniature trees. Propagated by suckers, or by cuttings of the young wood in sand under a glass, or by seeds. Common soil, kept moist. There is a close general resemblance among all the deciduous species, which leads us to doubt whether they are any thing more than varieties.

§ i. *Prinöides* Dec.

Sect. Char. Flowers usually 4—5- sometimes 6-cleft. (*Dec. Prod.*, ii. p. 16.) Leaves deciduous.

1. *P.* DECI'DUUS *Dec.* The deciduous Winter Berry.

Identification. Dec. Prod., 2. p. 16.; Don's Mill., 2. p. 20.
Synonymes. I'lex prinöides *Ait. Hort. Kew.* 2. p. 278.; I'lex decidua *Walt. Fl. Carol.* 241.
Engraving. Our *fig.* 232. from a plant in the Hort. Soc. Garden.

232. Prinos deciduus.

Spec. Char., &c. Leaves deciduous, elliptic-lanceolate, tapered to the petiole, shallowly serrated; the midrib villous beneath. Peduncles axillary; those of the male flowers several together; of the female ones, singly. Berries red. (*Dec. Prod.*) A deciduous shrub. Virginia to Georgia, on rocky shady banks of rivers. Height 3 ft. to 5 ft. Introduced in 1736. Flowers white; June and July. Berries large, crimson; ripe in December.

Plants of this species are in Loddiges's nursery, under the name of I'lex prinöides.

2. *P.* AMBI'GUUS *Michx.* The ambiguous Winter Berry.

Identification. Michx. Fl. Bor. Amer., 2. p. 236.; Dec. Prod., 2. p. 17.; Don's Mill., 2. p. 20.
Synonyme. Cassine caroliniàna *Walt. Fl. Carol.* p. 242.
Engravings. Wats. Dend. Brit., t. 29.; and our *fig.* 233.

233. P. ambiguus.

Spec. Char., &c. Leaves deciduous, oval, acuminate at both ends; both adult ones and young ones glabrous in every part. Peduncles of the male flowers crowded together in the lower parts of the branchlets; of the female ones, singly. (*Dec. Prod.*) A deciduous shrub. New Jersey to Carolina, in wet sandy woods. Height 4 ft. to 5 ft. Introduced in 1812. Flowers white; June to August. Berries red; ripe in November.

The leaves are subimbricate, serrated, acute at the apex, and the berries small, round, smooth, and red. There is a handsome plant of this species in the arboretum of Messrs. Loddiges, which, in 1835, was 5 ft. high. It is of easy culture in any free soil; and is propagated by seeds, cuttings, or layers.

§ ii. *Agèria* Dec.

Sect. Char. Flowers usually 6-cleft. Leaves deciduous. (*Dec. Prod.*, ii. p. 17.)

3. *P.* VERTICILLA'TUS *L.* The whorled Winter Berry.

Identification. Lin. Spec., 471.; Dec. Prod., 2. p. 17.; Don's Mill., 2. p. 20.
Synonymes. *P. padifòlius Willd. Enum.* p. 394.; *P. Gronòvii* Michx. Fl. Bor. Amer. 2. p. 236.; *P. confértus Mœnch*; *P. prunifòlius Lodd. Cat.*

Engravings. Wats. Dend. Brit., t. 30.; Dun. Arb., l. t. 23.; our *fig.* 234. in flower, and *fig.* 235. from the Hort. Soc. Garden.

234. *Prinos verticillàtus.*

Spec. Char., &c. Leaves deciduous, oval, acuminate, serrated, pubescent beneath. Male flowers in axillary umbel-shaped fascicles; the female ones aggregate, the flowers of both sexes 6-parted. (*Dec. Prod.*) A deciduous shrub. Canada to Virginia, in wet woods. Height 6 ft. to 8 ft. Introduced in 1736. Flowers white; June to August. Berries red or crimson, turning purple; ripe in November.

235.

There are two handsome plants of this species in Loddiges's arboretum, 7 ft. high, one of which is under the name of *P. prunifòlius.*

▲ 4. *P.* LÆVIGA'TUS *Pursh.* The smooth-*leaved* Winter Berry.

Identification. Pursh Fl. Sept. Amer., 1. p. 220.; Dec. Prod., 2. p. 17.; Don's Mill., 2. p. 20.
Engravings. Wats. Dend. Brit., t. 29.; and our *fig.* 236.

Spec. Char., &c. Leaves deciduous, lanceolate, serrated, the teeth directed forwards, acuminate; glabrous on both surfaces, except on the nerves beneath, where they are slightly pubescent; upper surface glossy. Flowers 6-cleft; the male ones scattered; the female ones axillary, solitary, almost sessile. (*Dec. Prod.*) A deciduous shrub. New York to Virginia, on the Alleghany Mountains. Height 6 ft. to 8 ft. Introduced in 1812. Flowers white; July. Berries large, dark red; ripe in November.

236. *Prinos lævigàtus.*

The plant of this species in Loddiges's arboretum was 4 ft. high in 1835. Readily distinguished by its somewhat more succulent leaves and shoots, the latter, when young, tinged with dark purple.

237. *Prinos lanceolàtus.*

▲ 5. *P.* LANCEOLA'TUS *Pursh.* The lanceolate-*leaved* Winter Berry.

Identification. Pursh Fl. Sept. Amer., 2. p. 27.; Dec. Prod., 2. p. 17.; Don's Mill., 2. p. 20.
Synonymes. *P. canadénsis Lyon*; *P. lùcidus Hort.*
Engraving. Our *fig.* 235. from a plant in the Hort. Soc. Garden.

Spec. Char., &c. Leaves deciduous, lanceolate, remotely and very slightly serrulate, smooth on both surfaces. Male flowers aggregate, triandrous; female ones mostly in pairs, peduncled, and 6-cleft. (*Dec. Prod.*) A deciduous shrub. Carolina to Georgia, in low grounds. Height 6 ft. to 8 ft. Introduced in 1811. Flowers white; June and July. Berries small, scarlet; ripe in November.

The plant in Loddiges's arboretum is 8 ft. high.

§ iii. *Wintérlia* Mœnch.

Sect. Char. Flowers, for the most part, 6-cleft. Leaves permanent. (*Dec. Prod.*, ii. p. 17.)

6. *P.* GLA'BER *L.* The glabrous Winter Berry.

Identification. Lin. Spec., 471.; Dec. Prod., 2. p. 17.; Don's Mill., 2. p. 20.
Synonyme. Ink berry, *Amer.*
Engraving. Our *fig.* 238. from nature. The figure under this name in Wats. Dend., t. 27., is that of *P.* coriàceus *Pursh.*

Spec. Char., &c. Evergreen. Leaves lanceolate, with wedge-shaped bases, coriaceous, glabrous, glossy, somewhat toothed at the tip. Flowers mostly three on an axillary peduncle that is usually solitary. Fruit black. (*Dec. Prod.*) An evergreen shrub. Canada to Florida, in shady woods. Height 3 ft. to 5 ft. Introduced in 1759. Flowers white; July and August. Berries black; ripe in November.

238. Prìnos glàber.

A low but very handsome evergreen shrub, which, in its native country, makes a fine show, when covered with its black berries. In Loddiges's arboretum it has attained the height of 4 ft., with a regular ovate shape, densely clothed with shining foliage.

7. *P.* CORIA'CEUS *Pursh.* The coriaceous-*leaved* Winter Berry.

Identification. Pursh Fl. Sept. Amer., 1. p. 221.; Dec. Prod., 2. p. 17.; Don's Mill., 2. p. 21.
Synonyme. *P.* glàber *Wats.*
Engravings. Wats. Dend. Brit., t. 27., and Bot. Cab., 450., under the name of *P.* glàber; and our *fig.* 239.

Spec. Char., &c. Evergreen. Leaf lanceolate, with a wedge-shaped base, coriaceous, glabrous, glossy, entire. Flowers in short, sessile, axillary corymbs, many in a corymb. (*Dec. Prod.*) A handsome, tall, evergreen shrub. Georgia, in sandy woods near the banks of rivers. Height 8 ft. to 10 ft. Introduced in 1820. Flowers white; June and July. Berry black; ripe in November.

Varieties. This species varies, with leaves broader, obovate-lanceolate, and acuminate; and narrower, lanceolate, and acute. (*Dec. Prod.*) The broad-leaved variety appears to be that figured in *Lod. Bot. Cab.*, t. 450.

239. Prìnos coriàceus.

The general aspect of this plant is that of *Ilex Dahoòn*. It is a handsome shrub, well deserving a place in collections.

Other Species of Prìnos. — P. *dùbius* G. Don. and P. *atomàrius* Nutt. have been introduced, and, probably, some others; but we have only noticed those of which we have seen living plants, and which we consider to be tolerably distinct, though there are probably only three species of *Prìnos*; two of which are deciduous, and the other evergreen.

ORDER XXII. *RHAMNA'CEÆ.*

ORD. CHAR. *Calyx* 4—5-cleft; æstivation valvate. *Corolla* of 4—5 petals; in some absent. *Petals* cucullate, or convolute, inserted into the orifice of the calyx. *Stamens* 4—5, opposite the petals, perigynous. *Ovary* superior, or half-superior, 2-, 3-, or 4-celled, surrounded by a fleshy disk. *Ovules* one in a cell, erect, as are the seeds. *Fruit* fleshy and indehiscent, or dry and separating into 3 divisions.—Trees or shrubs, often spiny, and generally deciduous. (*Lindl.*)

Leaves simple, alternate, very seldom opposite, with minute stipules, deciduous or evergreen. *Flowers* axillary or terminal. — Chiefly natives of Europe or North America.

They are ornamental in British gardens and shrubberies, chiefly from the variety of their foliage, and from their berries; but some of them, as *Ceanòthus*, from their flowers. They are all of easy culture; and they are propagated by seeds, cuttings, or layers. The hardy genera in British gardens are six; which are characterised as follows: —

ZI'ZYPHUS *Tourn.* Petals 5. Styles 2—3. Fruit an ovoid mucilaginous drupe. Nuts 1—3-celled. Seed compressed. A deciduous low tree or shrub.

PALIU'RUS *Tourn.* Petals 5. Styles 3. Fruit dry, indehiscent, girded with a broad membranaceous wing, 3-celled. Seed ovate. Spiny shrubs.

BERCHE'MIA Necker. Petals 5. Style 1. Stigmas 2. Fruit an oblong dry drupe; the nut 2-celled. A twining deciduous shrub from Carolina.

RHA'MNUS *Lam.* Petals in some absent. Style 2—4-cleft. Fruit nearly dry, or berried, 2—4-celled. Seed oblong. Shrubs or small trees, deciduous or evergreen; chiefly natives of Europe, but some of N. America and Asia.

COLLE'TIA Comm. Corolla none. Style ending in 3 teeth. Fruit a 3-celled capsule. Spiny shrubs; natives of Peru or Chile.

CEANO'THUS *L.* Petals 5. Styles 2—3, united. Fruit a dry berry, 3-celled, rarely 2—4-celled. Seed ovate. Shrubs, evergreen or deciduous, from North America.

GENUS I.

ZI'ZYPHUS *Tourn.* THE JUJUBE. *Lin. Syst.* Pentándria Di-Trigýnia.

Identification. Tourn. Inst., t. 403.; Gært. Fruct., 1. p. 43; Lam. Ill., t. 185.; Dec. Prod., 2. p. 198.; Brongn. Mém. Rham., p. 47.
Synonymes. Jujubier, *Fr.*; Judendorn, *Ger.*; Giuggiolo, *Ital.*
Derivation. From *zizouf*, the Arabic name of the lotus.

Gen. Char. *Calyx* spreading, 5-cleft. *Petals* 5, obovate, unguiculate, convolute. *Stamens* 5, exserted. *Anthers* ovate, 2-celled. *Disk* flat, pentagonal, expanded, adhering to the tube of the calyx. *Ovary* 2—3-celled, immersed in the disk. *Styles* 2—3. *Fruit* fleshy, containing a 1—2-celled nut. (*Don's Mill.*, 2. p. 23.)

Leaves simple, alternate, stipulate, deciduous; 3-nerved. *Flowers* axillary. — Only one hardy species. A low tree or shrub from Syria.

❦ 1. *Z.* VULGA'RIS *Lam.* The common, *or cultivated*, Jujube.

Identification. Lam. Ill., 185. f. 1.; Dec. Prod., 2. p. 19.; Don's Mill., 2. p. 23.
Synonymes. Rhámnus Zizyphus *Lin. Spec.* 282., *Pall. Fl. Ross.* 2. t. 59.; *Z. sativa* *Desf. Arb.* 2. p. 373., *N. Du Ham.* t. 16., but not of Gært.; *Z. Jujuba* Mill. Dict. No. 1., but not of Lam.; Jujubier cultivé, *Fr.*; Brustbeeren, *Ger.*; Giuggiolo, *Ital.*
Engravings. Lam. Ill., 185. f. 1.; N. Du Ham., 3. t. 16.; and our *fig.* 240.

Spec. Char., &c. Branchlets glabrous. Leaves ovate, retuse, denticulate, glabrous; or, beneath, pubescent along the nerves. Prickles not any, or twin, one of them recurved. Drupe ovate-oblong. (*Dec. Prod.*) A deciduous tree. The South of Europe and Syria. Height in the South of Europe 20 ft. to 30 ft.; and in England 5 ft. to 10 ft. Introduced in 1640. Flowers greenish yellow; August and September. Fruit blood-red or saffron; rarely seen in England.

Stem thick, cylindrical, somewhat twisted. The bark is brown, and rather chapped. The branches are numerous, pliant, armed with prickles, zigzag in their direction; the prickles at the joints being two of unequal size, of which one is almost straight, and the other shorter and quite straight. The leaves are alternate and oval-oblong, somewhat hard and coriaceous. The flowers are small, axillary, of a pale yellow colour, with short peduncles. The fruit is oval-oblong, resembling that of the olive; at first green, afterwards yellow, and entirely red when ripe. The juice of the fruit is used for making the jujube lozenges. The plant is tolerably hardy; having stood the winter of 1837-8 in the Hort. Soc. Garden. It is easily increased by cuttings of the roots, whether of young or old trees; or by suckers, which it throws up in the greatest abundance. Seeds of it may also be procured from Italy.

Other Species of Zízyphus. — Z. *sinénsis* Lam. has been cultivated in the Hort. Soc. Garden but it is only half-hardy; and the same may be said of Z. *spìna Christi*, Z. *flexuòsa*, and Z. *incúrva*, which are marked in some catalogues as hardy.

240. Zizyphus vulgaris.

Genus II.

PALIU'RUS L. The Paliurus, *or Christ's Thorn.* *Lin. Syst.* Pentándria Trigýnia.

Identification. Tourn. Inst., t. 386.; D. Don Prod. Fl. Nep., p. 189.; Dec. Prod., 1. p. 22.; Don's Mill., 2. p. 23.; Brongn. Mém. Rham., p. 46.
Synonymes. Paliûre, Porte-chapeau, *Fr.*; Judendorn, *Ger.*; Paliuro, *Ital.*
Derivation. From *pallō*, to move, and *ouron*, urine; in allusion to its diuretic qualities; or from Paliurus, the name of a town in Africa, now called Nabil.

Gen. Char. *Calyx* spreading, 5-cleft. *Petals* 5, obovate, convolute. *Stamens* 5, protruding. *Anthers* ovate, 2-celled. *Disk* flat, pentagonal. *Ovary* 3-celled. *Styles* 3. *Fruit* dry, indehiscent, expanding into a membrane round the disk, containing a 3-celled nut. (*Don's Mill.*)

Leaves simple, alternate, stipulate, deciduous; nerved with spines in the axils. *Flowers* axillary, greenish yellow.

Two species are hardy, and very ornamental from their shining leaves, and abundance of rich greenish yellow flowers, which are succeeded by fruit of rather a singular form. Propagated by seeds, which they produce in England, in abundance.

ꝑ ♣ 1. *P.* aculea'tus *Lam.* The prickly Paliurus, *or Christ's Thorn.*

Identification. Lam. Ill., t. 210.; Fl. Fr., ed. 3., No. 4081.; N. Du Ham., 3. t. 17.; Don's Mill., 2. p. 23.
Synonymes. *P.* pétasus *Dum. Cours.* 6. p. 266.; *P.* austràlis *Gært. Fruct.* 1. t. 43. f. 5.; *P.* vulgàris *D. Don Prod. Fl. Nep.* 189.; *Rhámnus Paliùrus Lin. Spec.* 281.; *Zízyphus Paliùrus Willd. Spec.* 1. p. 1183., *Sims Bot. Mag.* t. 1893.; Christ's Thorn, or Ram of Libya, *Gerard*; E'pine de Christ, Argalon, Porte-chapeau, *Fr.*; geflügelter Judendorn, *Ger.*; Giuggolo salvatico, *Ital.*; Xlin, in the herb-shops of Constantinople, where the seeds are sold as a medicine, and as a yellow dye.
Engravings. Lam. Ill., t. 210.; N. Du Ham., 3. t. 17.; Bot. Mag., t. 1893.; the plates of this species, both in a young and an old state, in Arb. Brit., 1st edit., vol. v.; and our *fig.* 241.

Spec. Char., &c. Branchlets pubescent. Leaves ovate, serrulated, quite smooth, 3-nerved, with two spines at the base, one straight, the other recurved. Flowers in axillary crowded umbellules; few in an umbellule. Wing of capsule crenated. (*Don's Mill.*) A branching deciduous shrub, or low tree. South of Europe, and North and West of Asia. Height 15 ft. to 30 ft. Introduced in 1596. Flowers greenish yellow; June and July. Fruit yellow; ripe in September.

The fruit is buckler-shaped, flat and thin, but coriaceous. From the singular appearance of this fruit, which has the footstalk attached to the middle, which is raised like the crown of a hat; and the flattened disk, which re-

241. Paliùrus aculeàtus.

sembles its brim; the French have given this tree the name of *porte-chapeau.* On both shores of the Mediterranean, it grows to about the same height as the common hawthorn, on rocky sterile places. In many parts of Italy the hedges are formed of this plant, as they are of the hawthorn in Britain; it is also the common hedge plant in Asia. Any common soil; seeds, or cuttings of the root.

♣ 2. *P.* (A.) VIRGA'TUS *D. Don.* The twiggy Christ's Thorn.

Identification. D. Don in Bot. Mag.; and Fl. Nep., 189.; Don's Mill., 2. p. 23.
Engravings. Bot. Mag., t. 2535.; and our *fig.* 242.

Spec. Char., &c. Branches smooth. Leaves obliquely cordate, or elliptical, 3-nerved, shining; wing of fruit entire. (*G. Don.*) A deciduous shrub. Nepal, on mountains. Height 10 ft. to 15 ft. Introduced in 1819. Flowers greenish yellow, in axillary corymbs; July and August. Fruit yellow; ripe in September.

The only tree which we have seen of this species is in the Chelsea Botanic Garden, where in general aspect it bears a close resemblance to *P.* aculeàtus, of which it is in all probability only a variety.

242. Paliùrus virgàtus.

GENUS III.

BERCHE'MIA Neck. THE BERCHEMIA. *Lin. Syst.* Pentándria Monogýnia.

Identification. Neck. Elem., 2. p. 122.; Dec. Prod., 2. p. 22.; Brongn. Mém. Rham., 49.; Don's Mill., 2. p. 27.
Synonyme. Œnóplia *Hedw. F. Gen.* 1. p. 151., and *Schult. Syst.* 5. p. 962.
Derivation. From *Berchem*, probably the name of some botanist.

Gen. Char. *Calyx* with a hemispherical tube, and 5 erect segments. *Petals* 5, convolute. *Stamens* included within the petals. *Anthers* ovate, 2-celled.

Disk annular, rather flat. *Ovary* half-immersed in the disk, 2-celled. *Style* short, bifid at the apex. *Fruit* dry, indehiscent, 2-celled. (*Don's Mill.*)

Leaves simple, alternate, exstipulate, deciduous; nerved. *Flowers* terminal, diœcious by defect; small, greenish yellow. — A twining deciduous shrub; a native of Carolina; of easy culture in any common soil, and propagated by seeds, or cuttings of the root.

⚘ 1. B. VOLU'BILIS *Dec.* The twining Berchemia.

Identification. Dec. Prod., 2. p. 22.; Don's Mill., 2. p. 27.
Synonymes. Rhámnus volùbilis *Lin. fil. Suppl.* 132., *Jacq. Icon. Rar.* t. 336.; Zizyphus volùbilis *Willd. Spec.* 1. p. 1102.; Œnóplia volùbilis *Schult. Syst.* 5. p. 332.; Supple Jack, *Virginian.*
Engravings. Jacq. Icon. Rar., t. 336.; our *fig.* 243. in flower, and *fig.* 244. in fruit, from nature.

Spec. Char., &c. Branches glabrous, rather twining. Leaves oval, mucronate, somewhat waved. Flowers diœcious. Drupes oblong. (*Dec. Prod.*) A deciduous twining shrub. Carolina and Virginia, in deep swamps. Height, in America, 20 ft. to 50 ft.; in British gardens, 8 ft. to 10 ft. Introduced in 1714. Flowers greenish yellow; June and July. Fruit violet-coloured; ripe in October.

According to Pursh, this species, in Virginia, ascends the highest trees, and is known by the name of Supple Jack. The stems twine round one another,

243. Berchèmia volùbilis.

244. Berchèmia volùbilis.

or any object which they may be near. In British gardens, they are seldom seen above 8 or 10 feet high; probably from little attention being paid to place the plant in a deep sandy or peaty soil, and to supply it with abundance of moisture in the growing season. In fine seasons it ripens fruit.

GENUS IV.

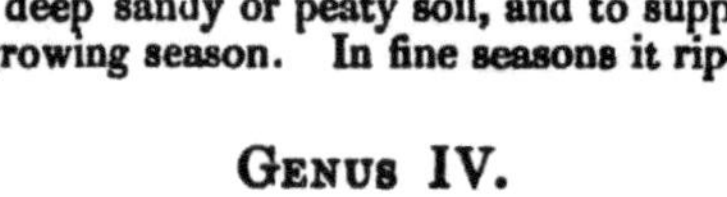

RHA'MNUS *Lam.* THE BUCKTHORN. *Lin. Syst.* Pentándria Monogýnia.

Identification. Lam. Dict., 4. p. 461.; Lam. Ill., t. 128.; Gært. Fruct., 2. p. 106.; Dec. Prod., 2. p. 23.; Don's Mill., 2. p. 29.; Brongn. Mém. Rham., p. 53.
Synonymes. Nerprun, *Fr.*; Wegdorn, *Ger.*; Ramno, *Ital.*; the Ram, or Hart's,Thorne, *Gerard*; Box Thorn.
Derivation. From the Celtic word, *ram*, signifying a tuft of branches; which the Greeks have changed to *rhamnos*, and the Latins to *ramus*.

Gen. Char. *Calyx* urceolate, 4—5-cleft. *Stamens* bearing ovate 2-celled anthers. *Disk* thin, covering the tube of the calyx. *Ovary* free, 3—4-celled. *Styles* 3—4, connected or free. *Fruit* baccate, containing 3—4 indehiscent nuts. (*Don's Mill.*)

Leaves simple, alternate, stipulate, deciduous, sub-evergreen, or evergreen; feather-nerved; the stipules never converted into prickles. *Flowers* axillary, aggregate, often unisexual. *Fruit* not eatable, generally black, rarely red or yellow.

Deciduous or evergreen shrubs, with the tips of their branches often becoming spines. One or two species have the habit of low trees, and some of them are sub-procumbent or procumbent; all of them, except the latter, being distinguished by an upright stiff mode of growth, and numerous strong thorns in their wild state; whence the name of ram, or buck, thorn. The flowers in all the species are inconspicuous; but *R. Alatérnus* and its varieties are most valuable evergreen shrubs, and several of the other species are ornamental, both from their foliage and their fruit; the latter of which is also useful in dyeing. All the species are easily propagated by seeds or layers, and most of them by cuttings; and they will all grow in any soil that is dry. They all vary much in magnitude by culture, in common with most plants which, in a wild state, grow in arid soils.

§ i. *Marcorélla* Neck.

Synonymes. Rhámnus and *A*latérnus of Tourn.

Sect. Char. Flowers usually diœcious, and 5-cleft. Fruit a berry, with 3 seeds, or, from abortion, 2 seeds. Seeds deeply furrowed, with the raphe in the bottom of the furrow. Leaves usually permanent; coriaceous, and glabrous. (*Dec. Prod.*, ii. p. 23.)

A. ALATE'RNUS Tourn. *Flowers racemose, 5-cleft. Evergreen Shrubs.*

■ 1. *R. A*LATE'RNUS *L.* The Alaternus.

245. *Rhámnus Alatérnus*

Identification. Lin. Spec., 281.; Dec. Prod., 2. p. 23.; Don's Mill., 2. p. 30.
Synonymes. *A*latérnus Phillýrea *Mill. Dict.* No. 1.; Alaterna, *Ital.*
Derivation. From *alternus*, a generic name adopted from Dioscorides, designating the alternate position of the leaves.
Engravings. Mill. Dict., t. 16. f. 1.; N. Du Ham., 3. p. 42. t. 14.; and our *fig.* 245.

Spec. Char., &c. Leaves ovate-elliptical, or lanceolate, coriaceous, quite smooth, serrated. Flowers diœcious, disposed in short racemes. (*Don's Mill.*) An evergreen shrub. South of Europe and North of Africa. Height 10 ft. to 20 ft. Introduced in 1629. Flowers green, without any corolla; April to June. Berry black; ripe in October.

Varieties.

■ R. A. 2 *baleárica* Hort. Par. The *R*hámnus rotundifòlius of Dumont. — Leaves roundish. We take this as the first variety, assuming the species to be what is called *R. A.* latifòlia, which is the commonest variety in British nurseries.

247. *R. A.* angustifòlia.

246. *R. A.* hispánica.

■ R. A. 3 *hispánica* Hort. Par. (Our *fig.* 246.) — Leaves ovate, a little toothed.

■ R. A. 4 *angustifòlia*. *R.* Clùsii Willd. (Mill. Icon., t. 16. fig. 2.; and our *fig.* 247.) Leaves long and narrow.— This variety is so distinct, that it is by many authors considered as a species. It is of remarkably rapid growth. There are two subvarieties of it, the gold-striped, and the silver-striped; both of remarkably free growth.

■ R. A. 5 *fòliis maculàtis*.—Leaves blotched with yellow.

■ R. A. 6 *fòliis aùreis*.—Leaves edged with yellow.

R. A. 7 *fòliis argénteis.* — This variety, which is very conspicuous from the large proportion of the leaves which is white, is more tender than some of the others. It generally does best against a wall, and is well worth a place there, on account of its splendid appearance, especially in winter.

In British gardens, this shrub is particularly valuable for the rapidity of its growth in almost any soil and situation, more especially the narrow-leaved variety. It is less injured by the smoke of coal than most other evergreens. The species, and all the varieties, are readily propagated by cuttings, which are taken off in autumn, and planted in sandy soil, in a shady border, and covered with a hand-glass. As the roots are not very productive of fibres, when large plants are chosen, they should be such as have been reared in pots, in order that they may receive no check from removal.

248. Rhámnus hýbridus.

2. *R.* HY'BRIDUS *L'Hérit.* The hybrid Alaternus.

Identification. L'Hérit. Sert., t. 5.; Dec. Prod., 2. p. 23.; Don's Mill., 2. p. 33.
Synonyme. *R.* burgundiacus *Hort. Par.*; *R.* sempervirens *Hortulan.*
Engravings. L'Hérit. Sert., t. 5.; and our *fig.* 248.

Spec. Char., &c. Leaves oblong, acuminated, serrated, smooth, shining, hardly permanent, rather coriaceous. Flowers androgynous. (*Don's Mill.*) A garden hybrid, a sub-evergreen shrub, raised from *R.* alpinus, fecundated by *R. A*latérnus, and forming a very distinct and desirable kind, which, in British gardens, grows to the height of 10 or 12 feet. The flowers are green, and appear in May or June.

B. *R*HA'MNUS Dec. *Flowers 4-cleft, in Fascicles.*

a. *Branchlets terminating in a Thorn.*

3. *R.* CATHA'RTICUS *L.* The purging Buckthorn.

Identification. Lin. Spec., 280.; Dec. Prod., 2. p. 24. Don's Mill., 2. p. 30.
Synonyme. The White Thorn of the modern Greeks.
Engravings. Eng. Bot., t. 1629.; N. Du Ham., 2. t. 10.; the plate of this species in Arb. Brit., 1st edit., vol. v.; and our *fig.* 249.

Spec. Char., &c. Erect. Leaves ovate, toothed. Flowers in fascicles, polygamo-diœcious. Berries 4-seeded, rather globose. (*Don's Mill.*) A deciduous shrub or low tree. Europe and Britain, in woods and thickets, on calcareous loamy soil. Height 10 ft. to 12 ft.: in cultivation, 12 ft. to 15 ft. Flowers yellowish green, with very narrow petals; May. Berry black; ripe in September. Decaying leaves yellowish green. Naked young wood whitish.

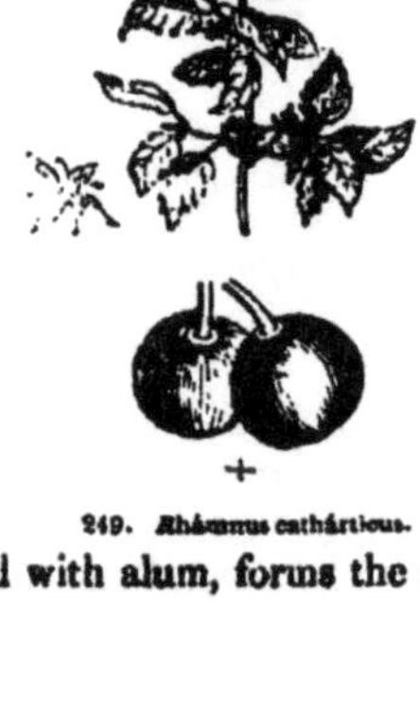
249. Rhámnus cathárticus.

The flowers are, for the most part, hermaphrodite, and in a wild state abundant and clustered; but in a state of cultivation they are fewer, and nearly solitary. The juice of the unripe berries is of the colour of saffron, and it is used for staining maps or paper: they are sold under the name of French berries. The juice of the ripe berries, mixed with alum, forms the sap green of painters; but, if the berries be gathered late in the autumn, the juice is purple. Plants of this species attain the height of 9 ft. in 10 years.

♠ 4. *R.* TINCTO'RIUS *Waldst.* The Dyer's Buckthorn.

Identification. Waldst. et Kit. Pl. Rar. Hung., 3. p. 255.; Dec. Prod., 2. p. 24.; Don's Mill., 2. p. 31.
Synonyme. *R. cardiospérmus Willd. Herb.*
Engravings. Hayne Abbild., t. 97. and our *fig.* 250.

Spec. Char., &c. Erect. Leaves ovate, crenate-serrated. Petioles villous. Flowers crowded, diœcious. Berries obcordate, 3—4-seeded. (*Don's Mill.*) A deciduous shrub. Hungary, in hedges. Height 8 ft. Introduced in 1820. Flowers greenish yellow; May and June. Berries yellow; ripe in September.

A plant of this species, in the garden of the London Horticultural Society, was, in 1834, 3 ft. high, after being 7 years planted.

250. Rhámnus tinctòrius.

♠ 5. *R.* INFECTO'RIUS *L.* The staining Buckthorn, *or Avignon Berry.*

Identification. Lin. Mant., 49.; Dec. Prod., 2. p. 24.; Don's Mill., 2. p. 31.
Synonymes. Rhámnus Lýcium *Scop. Carn.* ed. 2. n. 260.; dwarf, or yellow-berried, Buckthorn; Nerprun des Teinturiers, Graine d'Avignon, Nerprun teignant, *Fr.*; färbender Wegdorn, *Ger.*
Engravings. Ard. Mém., 78. t. 14.; N. Du Ham., vol. v. t. 73.; and our *fig.* 251.

Spec. Char., &c. Leaves ovate-lanceolate, serrulated, smoothish. Flowers diœcious, bearing petals in both sexes. (*Don's Mill.*) A deciduous, subprocumbent shrub. South of Europe, in rocky places; common about Avignon and the Vaucluse. Height 2 ft. Introduced in 1683. Flowers greenish yellow; June and July. Berry 3-celled, black; ripe in September.

The root fixes itself so firmly in the fissures of the rocks, that the plant can scarcely be pulled up. The stem divides immediately into branches, that are very much subdivided, and form a very close head, the shoots having numerous spines, both terminating and lateral. The berries are used for dyeing leather yellow; and the Turkey leather, or yellow morocco, is generally supposed to be coloured by them.

251. Rhámnus infectòrius.

♠ 6. *R.* SAXA'TILIS *L.* The Stone Buckthorn.

Identification. Lin. Sp., 1671.; Dec. Prod., 2. p. 24.; Don's Mill., 2. p. 31.
Synonymes. *R.* longifòlius *Mill. Dict.*; Stein Wegdorn, *Ger.*; Lycio Italiano, *Ital.*
Engravings. Jacq. Austr., t. 43.; Hayne Abbild., t. 98.; Schmidt, 3. t. 157.; and our *fig.* 252.

Spec. Char., &c. Procumbent, or erectish. Leaves ovate-lanceolate, serrulated, smoothish. Flowers diœcious, female ones destitute of petals. (*Don's Mill.*) A procumbent deciduous shrub. South of Europe, among rocks, in Austria, Switzerland, Italy, and Greece. Height 1 ft. Introduced in 1752. Flowers greenish yellow; June and July. Berries black, containing three whitish seeds, each enclosed in a dry whitish membrane, which separates, when ripe, into two parts with elastic force; ripe in September.

252. Rhámnus saxátilis.

♠♠ 7. *R.* BUXIFO'LIUS *Poir.* The Box-leaved Buckthorn.

Identification. Poir. Dict., 4. p. 463.; Dec. Prod., 2. p. 24.; Don's Mill., 2. p. 31.
Synonymes. ? *R.* buxifòlius *Brot. Fl. Lus.* 1. p. 301.; Lýcium buxifòlium *Bauh.*
Engravings. Du Ham., 3. t. 3. No. 12.; and our *fig.* 253.

Spec. Char., &c. Diffuse. Leaves ovate, quite entire, mucronate, smooth,

253. R. buxifolius.

coriaceous, green on both surfaces. (*Don's Mill.*) A sub-evergreen shrub. Spain, Italy, and the Levant, on hills. Height 2 ft. to 4 ft. Introd. in 1820. Flowers greenish yellow; June and July. Berry ? black; ripe ?.

A very neat shrub, of which there is a good specimen in the Chelsea Botanic Garden, which is quite hardy.

254. R. lyciöides.

■ 8. *R.* LYCIÖI'DES *Lin.* The Lycium-like Buckthorn.

Identification. Lin. Spec., 279.; Dec. Prod., 2. p. 25.; Don's Mill., 2. p. 31.
Engravings. Cav. Icon., 2. t. 182.; and our *fig.* 254.

Spec. Char., &c. Erect. Leaves linear, quite entire, obtuse, smooth. Flowers hermaphrodite. (*Don's Mill.*) A deciduous shrub. Spain, on the limestone hills of Valencia. Height 3 ft. to 4 ft. Introduced in 1752. Flowers greenish yellow; May and June. Fruit ?.

■ 9. *R.* ERYTHRO'XYLON *Pall.* The red-wooded Buckthorn.

Identification. Pall. Fl. Ross., 2.; Dec. Prod., 2. p. 25.; Don's Mill., 2. p. 31.
Engravings. Pall. Fl. Ross., 2. t. 62.; Itin., French edit., t. 90.; and our *fig.* 255.

Spec. Char., &c. Erect. Leaves linear, lanceolate, quite entire or serrated, smooth. Flowers hermaphrodite. Berries oblong. (*Don's Mill.*) A deciduous shrub. Mongolia and Siberia, near rivers. Height 6 ft. Introduced in 1823. Flowers greenish yellow; July and August. Berries black ripe in September.

Variety.

255. R. Erythróxylon.

256. R. E. angustissimum.

■ R. *E.* 2 *angustissimum* Dec. Prod. 2. p. 25., *R. lyciöides Pall. Fl. Ross.* t. 63., and our *fig.* 256., has the leaves narrow, smaller, and very finely serrulated. Native of Caucasus.

Delights in a warm situation; and in cold and humid places, Pallas observes, it is never met with. The wood, on account of its hardness and red colour, is used by the Mongols for making their images; and the berries, when macerated in water, afford them a deep yellow colour. The plant, in its wild state, is a prickly bush; but, when cultivated, the spines no longer appear.

b. *Branchlets not terminated by Spines.*

■ 10. *R.* DAHU'RICUS *Pall.* The Dahurian Buckthorn.

Identification. Pall. Fl. Ross., 2. t. 61.; Dec. Prod., 2. p. 25.; Don's Mill., 2. p. 31.
Engravings. Pal.. Fl. Ross., 2. t. 61.; and our *fig.* 257.

257. Rhámnus dahùricus.

Spec. Char., &c. Erect. Leaves oblong-ovate, serrated, smooth, veiny. Flowers diœcious, female ones with bifid stigmas. (*Don's Mill.*) A deciduous shrub. Dahuria, near the river Arguinus. Height 5 ft. Introd. in 1817. Flowers greenish yellow; May and June. Berry black; ripe Sept.

The general appearance of the plant is that of *R.* cathárticus, but it is without thorns. The wood is red, and is called sandal wood by the Russians.

11. *R.* ALNIFO`LIUS *L'Hérit.* The Alder leaved Buckthorn.

Identification. L'Hérit. Sert., t. 5.; Dec. Prod., 2. p. 25.: Don's Mill., 2. p. 32.
Engravings. Hayne Abbild., t. 61.; and our *fig.* 258.

Spec. Char., &c. Erect. Leaves obovate or ovate, serrulated, obliquely lineated with lateral nerves, acuminated or obtuse, smoothish beneath, except the nerves. Flowers hermaphrodite or diœcious. Pedicels 1-flowered, aggregate. Calyxes acute. Fruit turbinate. (*Don's Mill.*) A deciduous shrub. Hudson's Bay to Pennsylvania, in sphagnous swamps. Height 2 ft. to 4 ft. in America; 6 ft. to 8 ft. in England. Introd. in 1778. Flowers greenish yellow; June and July. Berry black, fleshy, 3-seeded; ripe in September.

258. Rhámnus alnifòlius.

Variety.

R. *a.* 2 *frangulöides* Dec. *R.* frangulöides *Michx.* (N.Du Ham.3. t. 15.; our *fig.* 259.)—Leaves oval, serrated, pubescent on the nerves beneath. Canada to Virginia. Introd. 1810.

259. R. a. frangulöides.

12. *R.* ALPI`NUS *Lin.* The Alpine Buckthorn.

Identification. Lin. Spec., 213.; Don's Mill., 2. p. 32.
Engravings. N. Du Ham., 3. t. 13.; Bot. Cab., t. 1077.; Schmidt, 3. t. 157.; the plate of the tree in Arb. Brit., 1st edit., vol. v.; and our *figs.* 260. and 261.

260. Rhámnus alpìnus.

261. Rhámnus alpìnus.

Spec. Char., &c. Erect, twisted. Leaves oval lanceolate, crenate-serrated, smooth, lineated with many parallel nerves. Flowers diœcious, female ones with 4-cleft stigmas. (*Don's Mill.*) A deciduous shrub. Alps of Switzerland, Dauphiné, and Carniola. Height 5 ft. to 10 ft. Introd. 1752. Flowers greenish; May and June. Berries black; ripe in September.

Variety.

■ R. a. 2 *grandifòlius* (*fig.* 262.) has much larger leaves than the species. It forms, when well grown, a very striking and handsome object, from the large size of its leaves and buds.

This is a very distinct species, and remarkable for its twisted leaves. There are strong plants of both the species and the variety in the arboretum of Messrs. Loddiges, and of the variety in the garden of the Hort. Soc., which, in 10 years, have attained the height of 8 ft., with numerous suberect branches, clothed with a purplish bark.

262. Rhámnus alpìnus grandifòlius.

263. Rhámnus pùmilus.

✱ 13. *R.* PU'MILUS *Lin.* The dwarf Buckthorn.

Identification. Lin. Mant., 49.; Don's Mill., 2. p. 32.
Synonymes. *R.* rupéstris *Scop. Carn.* 1. t. 5.; Ranno spaccasassi, *Ital.*
Engravings. Scop. Carn., 1. t. 5.; Schmidt Arb., 3. t. 155.; and our *fig.* 263.

Spec. Char., *&c.* Plant procumbent, much branched. Leaves ovate, serrated, smooth. Flowers hermaphrodite. (*Don's Mill.*) A deciduous procumbent shrub. Mount Baldo in the Alps, and Carniola, in the fissures of rocks. Height 2 ft. Introd. 1752. The flowers greenish yellow within and red without, the stamens white; June and July. Berries black; ripe in Sept.

The plant bearing this name in the London gardens we have never seen in such a thriving state as to enable us to decide whether or not it is truly distinct.

§ ii. *Frángula* Tourn.

Identification. Tourn. Inst., t. 383.; Dec. Prod., 2. p. 26.; Don's Mill., II. p. 32.

Sect. Char. Flowers hermaphrodite, rarely diœcious, 5-cleft, sometimes 4-cleft. Seeds smooth, compressed, with the hilum white and exserted, and with the raphe lateral, on the surface of the inner testa. Embryo flat. Leaves membranous, caducous, quite entire, lined with approximate parallel nerves. (*Don's Mill.*)

† ■ 14. *R.* CAROLINIA'NUS *Walt.* The Carolina Buckthorn.

Identification. Walt. Car., p. 101.; Dec. Prod., 2. p. 27.; Don's Mill., 2. p. 32.; Tor. and Gray, 1. p. 262.
Engraving. Our *fig.* 264.

264. Rhámnus caroliniànus.

Spec. Char., &c. Erect. Leaves oval-oblong, almost entire, smooth. Umbels stalked. Flowers hermaphrodite. Berries globose. (*Don's Mill.*) A deciduous shrub or tree. North Carolina to Florida, on the banks of rivers. Height 6 ft. to 8 ft., sometimes a tree 30 ft. to 40 ft. Introduced in 1819. Flowers greenish; May and June. Berries black; ripe in September.

In America this species, though usually a shrub in North Carolina, is in Georgia a considerable tree. Leaves 3 in. to 6 in. long, and 1 in. to 2 in. wide; sometimes acuminate, irregularly serrated; sometimes the margin is waved. Fruit as large as a small pea, mostly 3-seeded. (*Tor. and Gray.*)

■ ♆ 15. *R.* FRA'NGULA *L.* The breaking Buckthorn, *or Berry-bearing Alder.*

Identification. Lin. Spec., 280.; Don's Mill., 2. p. 32.
Synonymes. Nerprun Bourgène, Aune noir, *Fr.*; glatter Wegdorn, *Ger.*; Alno nero, *Ital.*
Derivation. The name of Frángula, breaking, is applied to this species, from the brittleness of its branches.
Engravings. Eng. Bot., t. 250.; Œd. Fl. Dan., t. 278.; the plate of the species in Arb. Brit. 1st edit., vol. v.; and our *fig.* 265.

Spec. Char., &c. Leaves oval, quite entire, lineated with 10 or 12 lateral nerves, and, as well as the calyx, smooth. Flowers hermaphrodite. (*Don's Mill.*) A deciduous shrub, or low tree. Europe and part of Siberia, in woods and thickets; not uncommon in England, but rare in Scotland. Height 3 ft. to 6 ft., wild; 8 ft. to 10 ft. in British gardens. Flowers whitish, with purple anthers; May. Berries dark purple; ripe in September. Decaying leaves reddish green. Naked young wood dark brown.

Variety.

■ ♆ R. *F.* 2 *angustifòlia* Hort. has narrower leaves. The plant of this species in the Horticultural Society's Garden is very distinct; and, in 1835, was 6 ft. high, after being 10 years planted.

265. *R.* Frángula.

Branches numerous, alternate, leafy, round, smooth, and blackish. From a quarter to half an ounce of the inner bark, boiled in small beer, is a sharp purge. The bark dyes yellow, and, with a preparation of iron, black. The flowers are particularly grateful to bees. The charcoal prepared from the wood is preferred by the makers of gunpowder to any other.

■ ♆ 16. *R.* LATIFO'LIUS *L'Hérit.* The broad-leaved Buckthorn.

Identification. L'Hérit. Sert., 5. t. 8.; Dec. Prod., 2. p. 26.; Don's Mill., 2. p. 32.
Engravings. L'Hérit. Sert., 5. t. 8.; Dend. Brit., t. 11.; Willd. Abbild., t. 100.; the plate of this tree in Arb. Brit., 1st edit., vol. v.; and our *fig.* 266.

Spec. Char., &c. Leaves elliptical, acuminate, quite entire, lineated with 12 or 15 lateral nerves; younger leaves and calyxes villous. Flowers hermaphrodite. (*Don's Mill.*) A deciduous shrub, with the habit of a low tree. Azores, on the mountains of St. Michael. Height 10 ft. to 15 ft. Introduced in 1778. Flowers greenish; July. Berries black or red, both colours appearing on the same plant at once; ripe in September.

266. Rhámnus latifòlius.

The leaves are larger than those of any other species, except *R.* alpìnus grandifòlius; and the whole plant is remarkable for its robust appearance, and the conspicuous opposite nerves which proceed from the middle of the leaves. It deserves a place in every collection.

Other Species of Rhámnus. — There are various kinds described by authors, several of which are said to have been introduced, but they are either lost or synonymous with kinds already described. Among those which remain to be introduced are some which promise to be useful additions to this genus ; such as : —

■ R. *persicifòlius* Bert. (Moris. Stirp. Sard. fig. 2.) — An erect shrub, with lanceolate, minutely crenated leaves, pubescent on the under side, and on long petioles. Calyx free. Sardinia. Height 10 ft. to 12 ft. March and April.

■ R. *amygdálinus* Desf. Atl. 1. p. 198. — A native of the North of Africa, in the fissures of rocks, where it grows to the height of 3 ft., and produces berries used for dyeing yellow, like those of *R.* saxátilis.

■ R. *prunifòlius* Smith Prod. Fl. Græc. 1. p. 157. — A native of Crete, on the highest mountains, and probably only a variety of one of the preceding sorts.

■ R. *Sibthorpiànus Schult. Syst.* 6. p. 286. ; *R.* pubéscens *Sibth. Fl. Græc.* t. 239. — A native of Mount Parnassus, and nearly allied to *R.* alpìnus and *R.* Frángula.

■ R. *Purshiànus Dec. Prod.* 2. p. 25. *R. a*lnifòlius of Pursh, but not of L'Héritier. (Hook. Flor. Bor. Am. 1. p. 123. t. 43., and our *fig.* 267.) — A shrub growing to the height of 6 ft., native of North America, on the banks of the Koorkoosky.

267. Rhámnus Purshiànus.

268. Rhámnus oleifòlius.

■ R. *oleifòlius* Hook. Fl. Bor. Am. 1. p. 123. t. 44., Tor. & Gray, 1. p. 200., and *fig.* 268. from Hooker, is an evergreen shrub, with coriaceous leaves, unarmed shoots, and the flowers small, in axillary crowded panicles. It is a native of the north-west coast of California, where it forms a handsome shrub from 6 ft. to 12 ft. high.

■ R. *umbellàtus* Cav. Icon. 6. p. 2. t. 504. — A shrub, growing 6 ft. high in Mexico ; was raised in 1839 in the Hort. Soc. Garden, from seeds sent home by M. Hartweg ; but it is probably only half-hardy.

*R. l*aurifòlius *Nutt.*, *R.* cròceus *Nutt.*, *R.* lanceolàtus *Pursh*, *R.* parvifòlius *Tor. & Gray*, *R.* ferrugíneus *Nutt.*, *R.* califórnicus *Esch* , and *R.* texénsis *Tor. & Gray*, are described in Tor. and Gray's *Flora of North America ; R.* pubéscens *Fl. Græc.*, and several others, are described in Don's *Miller*, and in the first edition of this work.

Genus V.

COLLE'T*IA* Com. The Colletia. *Lin. Syst.* Pentándria Monogýnia.

Identification. Kunth Nov. Gen. Amer., 7. p. 58. ; Dec. Prod., 2. p. 28. Don's Mill., 2. p. 34.
Synonyme. *Rhámnus* in part.
Derivation. Named by Commerson, in honour of *Collet*, his friend and countryman, who wrote upon the plants of Brest.

Gen. Char. Calyx campanulate, membraneous, coloured. *Petals* wanting, or very minute, linear. *Stamens* with ovate 2-celled, or reniform 1-celled, anthers. *Disk* short, cup-shaped, adnate to the bottom of the calyx. *Ovary* free, 3-celled. *Style* simple, elongated. *Stigma* 3-lobed. *Fruit* guarded at the base by the permanent tube of the calyx, tricoccous, dehiscent. (*Don's Mill.*)

Leaves, when present, simple, opposite, stipulate, deciduous; very minute, and quite entire. *Flowers* axillary, fascicled, or racemose; and, when the leaves are absent, rising from beneath the base of the spines. — Much-branched shrubs, with divaricating, decussately opposite branches, and spiny branchlets.

1. C. HO'RRIDA *Lindl.* The bristly Colletia.

Identification. Lindl. in Bot. Reg.? Ad. Brong.
Synonyme. C. fèrox *Gill. et Hook.* in *Bot. Mis.* 1. 154. t. 44. β.
Engravings. Bot. Reg., t. 1776.; and our *fig.* 269.

Spec. Char., &c. Spines rigid, simple, or much branched. Peduncles mostly in pairs. Calyx ovate-oblong. Stamens sessile. (*Lindl.*) A spiny shrub, evergreen, from the colour of its branches and branchlets. Chili and Mendoza, on mountains. Height 3 ft. to 4 ft. Introduced in 1823. Flowers greenish white, stained with dull purple; May to July. Berry whitish, about the size of a small pea; ripe in Sept.

The young branches are furnished with "bright green sawed scales" as leaves; they are placed opposite, and at the base of each is a small stipule. The leaves and stipules speedily fall off, "leaving the branches to act as leaves, by the aid of their soft parenchyma, with which they are clothed in the form of bark." Hence, plants of any size appear one mass of naked spiny green branches in winter; and, in summer, having leaves and flowers all over the points of the branchlets. It is a most desirable addition to our evergreen shrubs; and, as it escaped the winter of 1837–8, it may be safely recommended as hardy, for climates not much colder than that of London. It grows in common garden soil, in a dry situation, fully exposed to the sun. It has not yet been propagated otherwise than by Chilian seeds, which are frequently received under the name of Retanilla. We have no doubt, however, that it might be increased by layers, or by cuttings in sand under a glass.

269. Collètia hórrida.

Other Species of Collètia. — There are plants in the London gardens, under the name of C. spinòsa and C. ulícina, which we consider merely as varieties of C. hórrida; though the former has white flowers, and the latter broader leaves than those which we have described. They are considered more tender than C. hórrida; but this may possibly be owing to the plants having been

270. Collètia ulícina. C. E'phedra. Collètia spinòsa.

younger. C. E'*phedra* Vent. Choix, t. 16. (the *Rhámnus E'*phedra *Domb.*, and the *Retanílla E'*phedra *Brong.*) is said to have survived the criterion winter at Liverpool. In *fig.* 270., *a* represents C. ulícina; *b*, C. *E'*phedra (*Retanílla E'*phedra *Brong.*); and *c*, C. spinòsa.

Genus VI.

CEANO'THUS *L.* The Ceanothus, *or Red Root.* *Lin. Syst.* Pentándria Monogýnia.

Identification. Brongn. Mém. Rham., p. 62.; Hook. Fl. Bor. Amer., 1. p. 124.; Dec. Prod., 2. p. 31.; Don's Mill., 2. p. 37.
Synonymes. Rhámnus species *L., Juss., Lam.*; Céanothe, *Fr.*; Säkebbaum, *Ger.*
Derivation. From *keanothus*, a name employed by Theophrastus to designate a spiny plant, derived from *keō*, to cleave: the modern genus has, however, nothing to do with the plant of Theophrastus. The English name, Red Root, is given to the plant in America, from the red colour of the roots, which are of a large size in proportion to the branches.

Gen. Char. *Calyx* with a subhemispherical tube, and 5 connivent segments. *Petals* 5, unguiculate, cucullate, deflexed. *Stamens* with ovate 2-celled anthers. *Disk* spongy, annular. *Ovary* spherical, girded by the disk, 3-celled. *Styles* 3, diverging, terminated by small papilliform stigmas. *Fruit* tricoccous, girded by the circumcised tube of the calyx. (*Don's Mill.*)

Leaves simple, alternate, stipulate, persistent or deciduous; ovate or elliptical, serrated or entire. *Flowers* terminal or axillary, in elongated racemes. — Shrubs, natives of North America, very ornamental in British gardens, and easily propagated by cuttings of the young wood, planted in sand, and covered with a hand-glass. Most of the species produce seeds freely in British gardens, and they all grow in any common garden soil.

1. *C.* americа'nus *L.* The American Ceanothus, *or Red Root; or New Jersey Tea.*

Identification. Lin. Spec., 281.; Dec. Prod., 2. p. 31.; Don's Mill., 2. p. 37.; Tor. and Gray, 1. p. 264.
Engravings. Bot. Mag., t. 1497.; and our *fig.* 271.

Spec. Char., &c. Leaves ovate, acuminate, serrated, pubescent beneath. Thyrse elongated, axillary, with a pubescent rachis. (*Don's Mill.*) A deciduous suffrutescent low shrub. Canada to Florida, in woods and copses. Height 1 ft. to 3 ft; in British gardens, 2 ft. to 4 ft. Introduced in 1713. Flowers white; June to August. Fruit black; ripe in September.

Varieties. Torrey and Gray describe three varieties, *C.* a. 2 Pítcheri, *C.* a. 3 herbàceus (*C.* perénnis *Pursh*, *C.* ovàtus *Desf.*), and *C.* a. 4 intermèdius (*C.* intermèdius *Pursh*, *fig.* 271.); but we scarcely think they are worth keeping apart in collections.

271. C. a. intermèdius.

272. Ceanòthus americànus.

Stems shrubby, or suffruticose. Leaves 2 in. to 3 in. long. The leaves and stems of the plant are pubescent; and the flowers, being produced in great numbers together, are very ornamental. They are succeeded by bluntly triangular capsules; and, about London, in fine seasons, the seeds ripen. Any soil that is tolerably dry. Seeds or cuttings. The leaves of this plant, dried, were used by the Americans as a substitute for Chinese tea, during the war of independence.

■ ■ 2. *C.* azu'reus *Desf.* The azure-*flowered* Ceanothus, *or Red Root.*

Identification. Desf. Cat., 1815, p. 232.; Dec. Prod., 2. p. 31.; Don's Mill., 2. p. 37.
Synonymes. *C.* cærùleus *Lag. Gen. et Spec.* 1816, p. 11.; *C.* bicolor *Willd.* in *Schult. Syst.* 7. p. 65.
Engravings. Bot. Reg., t. 291.; Lodd. Bot. Cab., t. 110.; and our *fig.* 273.

Spec. Char., &c. Leaves ovate-oblong, obtuse, acutely serrated, smooth above, hoary and downy beneath. Thyrse elongated, axillary, with a downy rachis. Pedicels smooth. (*Don's Mill.*) A sub-evergreen shrub. Mexico. Height 6 ft. to 10 ft. Introduced in 1818. Flowers bright blue; May to September. Fruit black, enclosing 3 seeds; ripe in October.

Variety.

♣ ♣ C. *a.* 2 *intermèdius*, *C.* intermèdius *Hort.*, has the habit of *C.* azùreus, with pale flowers, like those of *C.* americànus, varying with different shades of blue. It was raised by Mr. Masters of Canterbury, from seeds of *C.* azùreus fecundated by *C.* americànus.

273. C. azureus.

A very handsome shrub, profusely covered with brilliant celestial blue flowers in large panicles. In Mexico its bark is considered as a febrifuge. It is the most robust-growing species of the genus, attaining, in 3 or 4 years from seed, the height of 5 or 6 feet, or more, against a wall. It was at first treated as a green-house plant, but lately it has been found to be nearly as hardy as the North American species. In the winter of 1837-8 these plants were greatly injured, but none of them killed. North of London the plant is less vigorous.

♣ ♣ 3. *C.* (A.) THYRSIFLO'RUS *Esch.* The Thyrse-flowered Ceanothus.

Identification. Esch. in Mem. Acad. St. Petersburg (1826); Hook. Fl. Bor. Am., 1. p. 125.; Don's Mill., 2. p. 37.; Tor. and Gray, 1. p. 266.
Synonyme. C. ovàtus cyàneus *Booth, Baumann, &c.*
Engraving. Our *fig.* , in p. .

Spec. Char., &c. Leaves oval, 3-nerved, serrulated, smooth. Stem many-angled; panicle thyrsoid in the axillary branches. (*Don's Mill.*) A sub-evergreen shrub, or small tree. Monterey, Upper California, and north-west coast of North America. Height in America 5 ft. to 20 ft. Introduced in ?1830. Flowers bright blue; May to November. Fruit black; ripe about a month after flowering.

In its native country, in favourable situations, this species becomes a small tree, with a stem sometimes as thick as a man's arm, and strongly angular branches. In British gardens it forms a free-flowering highly ornamental shrub, with much of the habit of *C.* azùreus; from which it chiefly differs in having the flowers in a close, instead of in an elongated, thyrse. Notwithstanding this difference, we consider it as only a variety of that species.

♣ ♣ 4. *C.* VELUTI'NUS *Doug.* The velvety-*leaved* Ceanothus.

Identification. Hook. Flor. Bor. Amer., 1. p. 125.; and Tor. and Gray, 1. p. 265.
Engravings. Hook. Flor. Bor. Amer., 1. t. 45.; and our *fig.* 274.

Spec. Char., &c. Branches somewhat pendulous. Leaves orbicular, elliptical or elliptical ovate, obtuse, subcordate, glandularly crenate, serrulate, coriaceous, glabrous, and shining (as if varnished) above, velvety, canescent, and strongly 3-ribbed beneath. Panicles axillary, elongated, on rather long peduncles. (*Tor. and Gray.*) A shrub, probably sub-evergreen. North-west coast of North America, on subalpine hills. Height 3 ft. to 8 ft. Not introduced. Flowers white. Fruit dry, 2—3-seeded.

274. C. velutinus.

This is apparently a very desirable species; and, as it is so abundant as to cover the whole declivities of hills, forming thickets very difficult to penetrate, we have no doubt that it will soon be introduced. Branches nearly glabrous. The leaves abound with an aromatic resin.

⸗ 5. *C.* COLLI'NUS *Doug.* The Hill-side Ceanothus.

Identification. Doug. in MSS.; Fl. Cab., t. 13.
Engravings. Fl. Cab., t. 13.; and our *fig.* 275.

Spec. Char., &c. Branches decumbent, round, and smoothish. Leaves ovate or elliptic, somewhat clammy, glandular, serrated, upper surface shining, under surface covered with adpressed hairs, 3-nerved. Stipules awl-shaped. Panicles axillary. (*Knowles and Westcott.*) A hardy, evergreen, low, decumbent shrub. North America. Height 1 ft. Introduced in 1827. Flowers white, produced in great abundance; June and July. Fruit brown; ripe in September.

Layers, which root readily, or seeds.

275. Ceanothus collinus.

Other Species of Ceanòthus. — *C.* ovàtus and *C.* intermèdius, we have seen, on the authority of Torrey and Gray, are only varieties of *C.* americànus; and we have no doubt that this will be the case with *C.* ovàlis, *C.* sanguíneus, *C.* oregànus, and other species described by authors. In short, there appears to us no assignable limits to the sports and hybrids that may be produced in this genus.

ORDER XXIII. HOMALINA'CEÆ.

ORD. CHAR. *Calyx* funnel-shaped, its tube usually adnate to the ovary, its limb with 5—15 lobes. *Petals* inserted into the calyx, as many as its lobes, alternate with them, smaller than they, and deemed by some an inner whorl of lobes of the calyx. *Glands* present in front of the segments of the calyx. *Stamens* arising from the base of the petals, either singly, or in threes or sixes. *Anthers* 2-celled, opening longitudinally. *Ovary* 1-celled, with numerous ovules. *Styles* 3—5, simple. *Fruit* berried or capsular. *Seeds* small. (*Lindl.*) — Trees or shrubs; natives of South America.

Leaves simple, alternate, with deciduous stipules, sub-evergreen; toothed or entire. *Flowers* axillary, in spikes, racemes, or panicles. — The species in British gardens belong to the genera Aristotèl*ia* and Azàr*a* (the latter rather tender), which are thus contradistinguished: —

ARISTOTE'L*IA* L'Hérit. Corolla of 5 petals. Stamens 15—18, polyadelphous. Fruit a globose, free, 3-celled berry. Cells 1—2-seeded.

AZA'R*A* R. et P. Corolla none. Stamens numerous. Fruit a globose 1-celled, 5-seeded berry.

GENUS I.

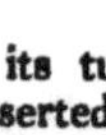

ARISTOTE'L*IA* L'Hérit. THE ARISTOTELIA. *Lin. Syst.* Polyadélphia Polyándria.

Identification. L'Hérit. Stirp., p. 31.; Dec. Prod., 2. p. 56.; Don's Mill., 2. p. 56.
Derivation. Named in commemoration of *Aristotle*, the celebrated philosopher and naturalist.

Gen. Char. *Calyx* campanulate, profoundly 5-cleft. *Petals* 5, inserted in the base of the calyx, and alternating with its lobes. *Stamens* 15—18, generally 3 or 4 in each bundle, placed in front of the lobes of the calyx. *Anthers* opening by two pores at the apex. *Ovary* free. *Styles* 3, somewhat connected at the base. *Berry* globose, 3-celled. *Seeds* angular. (*Don's Mill.*)

Leaves simple, opposite, stipulate, sub-evergreen; stalked and shining. *Flowers* in axillary racemes.

■ ■ ‡ ‡ 1. A. *MA'CQUI* L'Hérit. The Macqui Aristotelia.

Identification. L'Hérit. Stirp., p. 31.; Dec. Prod., 2. p. 56.; Don's Mill., 2. p. 58.
Synonymes. A. glandulòsa *R. et P. Fl. Per. Syst.* p. 126.; A. *Màqui* in Dec. Prod., 2. p. 56.
Engravings. L'Hérit. Stirp., t. 16.; Wats. Dend. Brit., t. 44.; N. Du Ham., t. 33.; the plate of this tree in Arb. Brit., 1st edit., vol. v.; and our *fig.* 276.

Spec. Char., &c. Calyx deeply 5-cleft. Styles 3, somewhat connected at the base. A sub-evergreen shrub, or low tree. Chili. Height in British gardens 7—18 ft. Introduced in 1733. Flowers small, green, purplish, and yellow; May and June. Berry very dark purple; ripe in September.

Variety.

■ ■ *A.* M. 2 *fòliis variegàtis.*—The variegated-leaved Macqui Aristotelia.

276. *Aristotèlia Mácqui.*

In Chili this plant forms an evergreen shrub, with diffuse branches, growing to the height of 6 ft. The flowers are not very showy; but they are succeeded by berries about the size of a pea, very dark purple, and at length becoming black, which are acid and eatable. In British gardens, it forms a sub-evergreen shrub or low tree, of very vigorous growth; so much so, in a young state, that, from the shoots not being matured, they are frequently killed down to the ground, and the foliage more or less injured. Notwithstanding this, the aristotelia frequently flowers, and even ripens fruit; and, in all probability, if the tree were planted in dry and rather poor soil, so as to grow slowly, and not make more wood every year than it could ripen properly, it would attain a large size, and form a very handsome hardy evergreen shrub or tree. The plant grows vigorously in any common garden soil, producing shoots 3 ft., 4 ft., or 5 ft. in length when young; and it is readily propagated by cuttings or by layers.

2.7. Azàra dentàta.

Other Species of Aristotèlia. — One has been raised in the Chelsea Botanic Garden, from South American seeds, which Mr. Dillwyn found to stand the winter of 1837–8 better than A. *Mácqui*.

■ *Azàra dentàta* R. & P., Don's Mill. i. p. 257. (Bot. Reg. t. 1788., and our *fig.* 277.) is an evergreen shrub or low tree, growing to the height of 12 ft. in Chili. It stood 8 years in the Hort. Soc. Garden, against a wall, and, though killed by the winter of 1837–8, it may yet ultimately prove tolerably hardy. *A. integrifòlia*, if a different species, may possibly be found hardy also.

ORDER XXIV. ANACARDIA'CEÆ.

Identification. Lindley, in Introd. to N. S.
Synonymes. Terebinthàceæ, tribe 1. Anacardièæ *R. Br.*, and tribe 2. *Sumachìneæ* Dec. Prod. 2. 66.

ORD. CHAR. *Flowers* generally unisexual. *Calyx* usually 5-parted. *Petals* equal in number to the divisions of the calyx, cohering at the base when the disk is absent. *Stamens* same number, or twice that number. *Disk*, when present, annual. *Ovarium* usually solitary. *Styles* 1—3, sometimes wanting. *Fruit* indehiscent. (*Lindl.*) — Low deciduous or evergreen trees, natives of Asia and Africa.

Leaves simple or compound, alternate, exstipulate, deciduous or evergreen; without pellucid dots. *Flowers* terminal or axillary, in panicles, with bracts. — The hardy species belong to the genera *Pistàcia*, *Rhús*, and Duvaúa, which are thus contradistinguished: —

PISTA'CIA *L.* Flowers diœcious, apetalous, amentaceous. Stigmas 3. Drupe dry, containing a 1-celled, 1-seeded nut.

RHU'S *L.* Flowers polygamous. Styles or stigmas 3. Drupe nearly dry, containing a 1-celled, 1—3-seeded nut.

DUVAU'A Kth. Flowers monœcious or diœcious. Styles 3—4, short. Drupe containing a coriaceous 1-seeded nut.

GENUS I.

PISTA'CIA *L.* THE PISTACHIA TREE. *Lin. Syst.* Diœ'cia Pentándria.

Identification. Lin. Gen., 1108.; Dec. Prod., 2. p. 64.; Don's Mill., 2. p. 61. and 65.
Synonyme. Terebinthus *Juss.*
Derivation. From the Greek word *Pistakia*, derived from *Psittakion*, the name of a city; or from the Arabic word *Foustaq*, the Arabian name of *Pistàcia vèra*.

Gen. Char. *Flowers* diœcious, and without petals; disposed in amentaceous racemes, each scale with one flower. *Calyx* 3—5-cleft. *Stamens* 5, inserted into a calycine disk, or into the calyx; with 4-cornered, almost sessile, anthers. *Ovary* 1—3-celled. *Stigmas* 3, and thickish. *Fruit* a dry ovate drupe; nut bony, and usually 1-celled, with a single seed affixed to the bottom. *Cotyledons* thick, fleshy, oily, and bent back upon the radicle.—Small trees, natives of the South of Europe and Asia.

Leaves compound, impari-pinnate, deciduous or evergreen; dying off of a beautiful reddish purple; young shoots tinged with purple.

♀ 1. *P.* VE'RA *L.* The true Pistachia Tree.

Identification. Lin. Spec., 1454.; Dec. Prod., 2. p. 64.; Don's Mill., 2. p. 65.
Synonymes. Pistacia officinārum *Hort. Kew.*; Pistachier, *Fr.*; Pistazienbaum, *Ger.*; Pistacchio, *Ital.*; Alfocigo, *Span.*
Engravings. Blackw. Icon., t. 461.; N. Du Ham., 4. t. 17.; and our *fig.* 278.

Spec. Char., &c. Leaves deciduous, impari-pinnate, of 3—5 leaflets, rarely of 1; the leaflets ovate, a little tapered at the base, indistinctly mucronate at the tip. (*Dec. Prod.*) A deciduous tree. Syria. Height 20 ft. Introduced in 1770. Flowers small, brownish green; April and May. Fruit reddish, an inch long, ovate; ripe in Syria in September, rarely seen in England.

Varieties. The following are considered by some authors as species:—

♀ P. *v.* 2 *trifòlia* Lin. Spec. 1454., Bocc. Mus. ii. t. 93., has leaves usually of 3 leaflets.

♀ P. *v.* 3 *narbonénsis* Bocc. Mus. t. ii. 693., *P.* reticulàta *Willd.*, has pinnate leaves, the leaflets having prominent veins. H. S.

278. Pistacia vèra.

Cultivated in the South of France, and in Italy, for its fruit; the nut of which is sometimes eaten raw, but more frequently in a dried state, like almonds. In British gardens, the tree is not much planted, from its being generally supposed to require a wall; but, in favourable situations, it will grow as a standard or a bush in any common garden soil, and may be propagated either by nuts procured from abroad, or by cuttings.

♀ 2. *P.* TEREBI'NTHUS *Lin.* The Turpentine Pistachia, *or Venetian or Chian Turpentine Tree.*

Identification. Lin. Spec., 1455.; Dec. Prod., 2. p. 64.; Don's Mill., 2. p. 65.
Synonymes. *T.* vulgàris *Tourn. Inst.* 579.; *P.* vèra *Mill. Dict.* No. 4.; Pistachier Térébinthe, *Fr.*; Terpentin Pistacie, *Ger.*; Terebinto, *Ital.*
Engravings. Woodv. Med. Bot., 415. t. 153.; and our *fig.* 279.

Spec. Char., &c. Leaves deciduous, impari-pinnate, of about 7 leaflets, that are ovate-lanceolate, rounded at the base, and at the tip acute and mucronate. (*Dec. Prod.*) A deciduous tree. South of Europe and North of Africa. Height 30 ft. Introduced in 1656. Flowers dull yellow and crimson; June and July. Fruit dark blue, hardly bigger than a large pea.

Variety.

♀ P. T. 2 *sphærocárpa* Dec.—Fruit larger and rounder than that of the species.

The general appearance of the tree is that of *P.* vèra, but the leaves are larger, and the fruit only a third of the size; the leaflets are, also, lanceolate, instead of being subovate. The red hue of the branches, especially when young, is very beautiful; and the leaves are

279. P. Terebinthus.

also more or less tinged with red. The fruit is round, not succulent, and somewhat furrowed; at first green, and afterwards reddish; but black, or of a very dark blue, when ripe. The leaves and flowers emit a very resinous odour, which spreads to a considerable distance, more especially at sunset, when the dew is falling, after a very warm day. The substance called Venice or Chian turpentine is the resin which exudes from this tree. In British gardens, the tree is not very common, though it is generally considered as the hardiest of the genus; and, with *P.* vèra, may be planted in warm sheltered situations in the open border.

♀ 3. *P.* Lenti'scus *L.* The Mastich Tree.

Identification. Lin. Spec., 1455.; Dec. Prod., 2. p. 65.; Don's Mill., 2. p. 66.
Synonyme. Corno capra, *Ital.*
Engravings. Woodv. Med. Bot., t. 152.; and our *fig.* 280.

Spec. Char., &c. Evergreen. Leaves abruptly pinnate; leaflets 8, lanceolate; petiole winged. (*Dec. Prod.*) An evergreen tree. Southern Europe, Northern Africa, and the Levant. Height 20 ft. Introduced in 1664. Flowers green; April and May. Fruit brownish; ripe in October.

Varieties.

♀ P. L. 2 *angustifòlia* Dec., *P.* massiliénsis *Mill. Dict.*, *P.* angustifòlia massiliénsis *Tourn.*, has leaflets almost linear, and the tree seldom exceeds 10 ft. in height.

♀ P. L. 3 *chìa* N. Du Ham. iv. p. 72.; *P.* chìa *Desf. Cat. Hort. Par.* — A native of Scio, where it produces the mastich.

280. Pistàcia Lentiscus.

The species bears a general resemblance to the two preceding ones, in summer, when they are clothed with foliage; but it differs from them in being evergreen, and in having the leaves much smaller. The leaves have sometimes 5 leaflets on each side; and the petioles are so much winged as to appear like pinnæ. The tree in the South of Europe, and the North of Africa, is cultivated in gardens, as well as found in a wild state; but in British gardens it is not so hardy as *P. Te*rebínthus, and north of London should always be planted against a wall.

Other Species of Pistàcia. — P. *atlántica* Desf., a deciduous tree from Mount Atlas, is said to have been introduced in 1790, but it requires the protection of a frame or green-house.

Genus II.

RHU'S *L.* The Rhus, *or Sumach.* *Lin. Syst.* Pentándria Trigýnia and Diœ'cia Pentándria.

Identification. Lin. Gen., 369.; Lam. Ill., t. 207.; Kunth Gen. Tereb., p. 5.; Dec. Prod., 2. p. 66.; Don's Mill., 2. p. 61. and p. 69.
Synonymes. Sumach, *Fr.* and *Ger.*; Ru, *Ital.*
Derivation. From *rhoos*, or *rhous*, Greek, or from *rhudd*, or *rud*, Celtic, red; in allusion to the colour of the fruit and leaves of some of the species in autumn. Others derive *Rhùs* from the Greek verb *rheō*, I run, from the habit of the roots running and spreading under ground to a considerable distance from the tree. Sumach is derived from *Simaq*, the Arabic name of the plant.

Gen. Char. *Sexes* hermaphrodite, diœcious, or polygamous. *Calyx* small, 5-parted, persistent. *Petals* ovate, and inserted into a calycine disk, or into the calyx. *Stamens* 5, inserted into a calycine disk. *Ovary* single, subglobular, of 1 cell. *Styles* 3, short, or wanting. *Stigmas* 3. *Fruit* an almost dry drupe of 1 cell, with a bony nut, which includes a single seed; and, in some instances, 2—3 seeds. (*Dec. Prod.*)—Deciduous shrubs. Natives of Europe, Asia, and North and South America.

Leaves simple or unequally pinnate, alternate, stipulate, deciduous. *Flowers* in terminal racemes, or panicles. — The leaves vary much, both in form and magnitude; and they generally die off, in autumn, of a dark red, or a bright scarlet, or yellow, when they are very ornamental. Most of the species are poisonous, some highly so; and they all may be used in tanning, and dyeing yellow or black. They are all easily propagated by cuttings of the root, and some of them by cuttings of the branches.

Some of the hardy species are rambling climbers, and others tree-like bushes.

§ i. *Cótinus* Tourn.

Sect. Char. Leaves undivided. Flowers hermaphrodite.

♣ 1. *R.* Co'TINUS *L.* The Cotinus Rhus, *or Venetian Sumach.*

Identification. Lin. Spec., 383.; Dec. Prod., 2. p. 67.; Don's Mill., 2. p. 69.
Synonymes. Cótinus Coggýgria *Scop. Carn.* ed. 2. No. 368., *Mœnch Meth.* 73.; Cótinus coriàcea *Duh. Arb.* 1. t. 78.; Venus Sumach, Venice Sumach, wild Olive; Sumach Fustet, or Arbre aux Péruques, *Fr.*; Perücken Sumach, *Ger.*; Scotano, *Ital.*
Derivation. The term Cótinus is derived from *cotinos*, a name under which Pliny speaks of a tree with red wood, which is supposed to grow in the Apennines.
Engravings. Jacq. Aust., t. 210.; and our *fig.* 281.

Spec. Char., &c. Leaves obovate. (*Dec. Prod.*) A deciduous rambling shrub. Spain to Caucasus; and, according to Torrey and Gray, probably of North America. Height 4 ft. to 6 ft. wild; 6 ft. to 8 ft. in a state of culture. Introduced in 1656. Flowers pale purplish, or flesh colour; June and July. Fruit white; ripe in September. Decaying leaves of a fine reddish yellow. Naked young wood smooth brown.

281. *R. Cótinus.*

The flowers are disposed in loose panicles, and are hermaphrodite. The drupe is half-heart-shaped, smooth, and veiny; and its nut is triangular. Many of the flowers are abortive; and their pedicels, after flowering, lengthen, and become hairy. A highly ornamental shrub, more especially when covered with its large loose panicles of elongated hairy pedicels. It is easily known from all the other species by its simple, obovate, smooth, stiff, lucid green leaves, rounded at the points, and supported by long footstalks, which remain on till they are killed by frost, so that the plant is almost a sub-evergreen. A dry loam suits it best; and it is propagated by pegging down the branches flat to the ground, and strewing earth over them, through which young shoots rise up, which root at the base, and may be removed in autumn.

§ ii. Sùmach *Dec.*

Sect. Char. Leaves impari-pinnate; leaflets more than 3 in the leaves of each of the first 6 species of this section. Flowers in panicles, polygamous, diœcious, or hermaphrodite.

♣ ⚲ 2. *R.* TYPHI'NA *L.* The Fever Rhus, *or Stag's Horn Sumach.*

Identification. Lin. Spec., 380.; Dec. Prod., 2. p. 67.; Don's Mill., 2. p. 70.; Tor. and Gray, 1. p. 217.
Synonymes. *R.* virginiàna *Bauh. Pin.* p. 517.; Virginian Sumach.
Engravings. N. Du H., 2. t. 47.; Wats. Dend. Brit., t. 17. and t. 18.; and our *fig.* 282., the male.

Spec. Char., &c. Leaf of 8—10 pairs of leaflets, and the odd one, that are lanceolate, acuminate, serrated, hairy beneath. Petiole and branches hairy.

(*Dec. Prod.*) A shrub, with the habit of a low deciduous tree. Canada to Carolina, in rocky dry situations. Height 20 ft. Introduced in 1629. Flowers, female dark purple, male greenish yellow and purple; July and August. Fruit hairy, purple; ripe in October. Decaying leaves dark purple or red, sometimes mixed with yellow. Naked young wood dark brown, hairy. DeCandolle has characterised two forms of this species as follows: —

‡ R. *t.* 1 *arboréscens.* — A tree between 10 ft. and 25 ft. high; leaf slightly downy beneath.

■ R. *t.* 2 *frutéscens.* — Shrubby, between 2 ft. and 10 ft. high; and its leaf downy and whitish beneath.

‡ R. *t.* 3 *viridiflòra.* *R.* viridiflòra *Poir.* — Flowers green. Possibly nothing more than the male plant.

*R*hús typhìna, in British gardens, is either a large shrub, or a low tree with a woody stem and a head composed of many irregular branches, generally crooked and deformed. The young shoots are covered with a soft velvet-like down, resembling that of a young stag's horn, both in colour and texture; whence, and probably also from the crookedness of the branches, the common name. The cellular tissue of the wood is of an orange colour, with a strong aromatic odour, and a copious resinous juice. The leaves are 2 ft. to 3 ft. long, and they are very conspicuous in autumn, before they drop off, when they change to a purplish or yellowish red. The flowers are produced in close spikes at the ends of the branches; they are often polygamous or diœcious by abortion, and the female ones are followed by seeds enclosed in woolly, simple, succulent covers. As the plant is of open irregular growth, and not of long duration, it should never be placed where it is intended to act as a screen. Like all objects the chief beauty of which consists in their singularity, it produces the most striking effect when standing alone on a lawn.

282. *Rhús typhìna.*

■ ‡ 3. *R.* (? T.) GLA'BRA *Lin.* The glabrous Rhus, *or Scarlet Sumach.*

Identification. Lin. Spec., 380.; Dec. Prod., 2. p. 67.; Don's Mill., 2. p. 70.; Tor. and Gray, i. p. 217.
Engravings. Wats. Dend. Brit., t. 15.; and our *fig.* 283., the female.

Spec. Char., &c. Leaf glabrous, of 8—10 pairs of leaflets, and an odd one; leaflets lanceolate-oblong, serrate, whitish beneath. Branches glabrous. (*Dec. Prod.*) A deciduous shrub or low tree. Canada to Georgia. Height 5 ft. to 18 ft. Introduced in 1726. Flowers, male greenish yellow, female greenish red. Fruit red; ripe in October. DeCandolle has distinguished three forms of this species; namely: —

■ R. *g.* 1 *hermaphrodìta.* *R.* glàbra *Willd. Spec.* i. p. 1478., *Dill. Elth.* t. 243. — Flowers hermaphrodite, greenish.

■ R. *g.* 2 *dioica* Lam. Ill. t. 207. f. 1. — Flowers diœcious, greenish.

■ R. *g.* ? 3 *coccìnea.* *R.* caroliniànum *Mill. Dict.*; *R.* élegans *Ait.*, *Lodd. Cat.*, *Dend. Brit.* t. 16. — Flowers diœcious, red. This variety is dis-

283. *Rhús glàbra.*

tinguished by a more upright habit of growth, and smoother branches and leaves, than *R.* glàbra. The leaves are glaucous underneath; and the fruit is of a rich velvety crimson.

The general appearance of the species is similar to that of *R.* typhìna; but the leaves and the entire plant are smaller, the branches more spreading and smooth, and the leaflets wider, less serrated, and of a deeper green.

4. *R.* VENENA'TA *Dec.* The poisonous Rhus, *Poison Wood, or Swamp Sumach.*

Identification. Dec. Prod., 2. p. 68.; Don's Mill., 2. p. 71.; Tor. and Gray, 1. p. 218.
Synonymes. *R.* vèrnix *Lin. Spec.* 380., *Big. Med. Bot.* 1. p. 96. t. 10.; Toxicodéndron pinnàtum *Mill. Dict.* No. 5.; Poison Sumach, Poison Elder.
Engravings. Wats. Dend. Brit., t. 19.; and our *fig.* 284

Spec. Char., &c. Leaf rather glabrous than pubescent, of 5—6 pairs of leaflets, and the odd one, which are ovate-lanceolate, acuminate, entire, and beneath reticulately veined. (*Dec. Prod.*) A deciduous shrub. Canada to Georgia, and west to Louisiana, in swamps. Height 15 ft. to 20 ft. Introd. 1713. Flowers green; July. Berry smooth, greenish white; ripe in ? October. Decaying leaves intense red, or purple. Naked young wood purplish green.

284. Rhus venenata.

The leaves are divided like those of *R.* typhìna and *R.* glàbra; but they are quite different from those of both kinds, in being smooth, shining, and having the leaflets very entire, narrow, and pointed, and the veins of a purplish red colour. The whole shrub is in a high degree poisonous; and the poison is communicated by touching or smelling any part of it. In British gardens it is not very common; but it well deserves culture, on account of the beauty of its smooth shining foliage at all seasons, and of its almost unparalleled splendour in the autumn, from the time that the leaves begin to change colour, till they ultimately drop off, of an intense purple or scarlet, with the first frost.

? 5. *R.* CORIA'RIA *Lin.* The hide-tanning Rhus, *or the Elm-leaved Sumach.*

Identification. Lin. Spec., 379.; Dec. Prod., 2. p. 67.; Don's Mill., 2. p. 70.
Derivation. Coriària alludes to the use made of this plant by the Romans, and also by the Turks, in tanning leather.
Engravings. N. Du Ham., 2. t. 46.; Wats. Dend. Brit., t. 136.; and our *figs.* 285. and 286.

285. Rhus Coriaria.

Spec. Char., &c. Leaf villose, of 5—7 pairs of leaflets, and the odd one; leaflets elliptical, and toothed with large and blunt teeth. The petiole smooth at the tip, a little margined. (*Dec. Prod.*) A deciduous shrub or low tree. Portugal to Tauria, on rocks in exposed situations. Height 15 ft. to 20 ft. Introduced in 1629 Flowers whitish green, in large loose panicles; July and August. Fruit red; ripe in October, rare in England. Decaying leaves purplish red.

286. R. Coriaria.

The general habit of this plant resembles that of *R.* typhìna; but it is much smaller in all its parts. The leaflets are about 2 in. long, and ½ in. wide, of a pale green, serrated, and in general appearance resembling the leaves of the common elm. Culture as in *R.* typhìna.

■ 6. *R.* COPALLI'NA *Lin.* The Gum Copal Rhus, *or Mastich-tree-leaved Sumach.*

Identification. Lin. Spec., 380.; Dec. Prod., 5. p. 68.; Don's Mill., 2. p. 72.; Tor. and Gray, 1. p. 217.
Engravings. Jacq. Hort. Schön., t. 341.; Pluk. Alm., p. 56. f. 1.; and our *fig.* 287.

Spec. Char., &c. Leaf glabrous above, a little pilose beneath, of 5—7 pairs of leaflets, and the odd one; leaflets lanceolate and entire. Petiole winged and jointed. Root stoloniferous. Flowers yellow green. Sexes diœcious. (*Dec. Prod.*) A deciduous shrub. Canada to Florida. Height 3 ft. to 8 ft. Introduced in 1688. Flowers yellowish green; July and August Berries red; ripe in September. Decaying leaves purplish red.

Varieties. Three forms are given by Torrey and Gray: — α, Leaflets entire, usually acuminate, which may be considered as the species: β, leaflets coarsely and unequally serrate: and γ, leaflets (about 21) small, oblong, acute at the base; obtuse and slightly mucronate at the apex; petiole nar rowly winged. Jacquin has

■ *R. c.* 2 *leucántha* Jac. Hort. Schön., t. 342. — Root not stoloniferous. Panicles more contracted than in the species.

287. *Rhús copallìna.*

The leaves and general habit of the plant are those of *R.* typhìna, but it seldom grows to the height of more than 4 or 5 feet in British gardens. The branches are smooth, and the leaflets entire with acute points; they are light green on both sides, and in autumn change to a fine purple. The petiole, as in *R.* Coriària, is somewhat winged towards its tip, which, with other circumstances, induces us to think that they may both be varieties of the same species. The leaves are used as tobacco by the Indians of the Missouri and the Mississippi.

⁎ ♣ 7. *R.* TOXICODE'NDRON *L.* The Poison-Tree Rhus, *or Sumach.*

Identification. Tor. and Gray, 1. p. 218.
Synonymes. *R.* Toxicodéndron, and *R.* radicans *L.*, *Dec.*, *Don's Mill.*, *&c.*
Engraving. Our *fig.* 000. in p. 000.

Spec. Char., &c. Stem erect, decumbent, or climbing by radicles. Leaves 3-foliolate, somewhat pubescent; leaflets (membranaceous) broadly oval or rhomboid, acuminate, entire or toothed, the lateral ones inequilateral. Panicles racemed, axillary, subsessile. Drupe subglobose, smooth. (*Torrey and Gray.*) A low rambling or climbing shrub. Canada to Georgia, in shady damp places. Stems 10 ft. to 20 ft. as a climber; or 3 ft. to 5 ft. high as a bush. Introduced in 1640. Flowers greenish, mostly diœcious; June and July. Berry pale chestnut; ripe in September. Decaying leaves purplish red.

Varieties. The following forms are given by Torrey and Gray: —

⁎ R. T. 1 *quercifòlium* Tor. & Gray. *R.* T. β quercifòlium *Michx.* — Not climbing; leaves entire, or variously and irregularly sinuatedly toothed, or lobed. The *R.* Toxicodéndron of the London gardens, readily distinguished from the two following varieties, by its deeply sinuated, or almost pinnatifid, leaflets. It grows to the height of

3 ft. to 4 ft. with several upright stems, forming a small bush, from the base of which proceed many prostrate runners.

R. *T.* 2 *radìcans* Tor. & Gray. *R. T. α* vulgàre *Michx.*; *R.* T. *β* radìcans *Tor.* (Bot. Mag. t. 1806. and N. Du Ham. 2. t. 48., and our *figs.* 288. and 289.) — Climbing; leaves more commonly entire, or nearly so. The *R*hús radìcans of the London gardens, readily known from the preceding variety by its trailing or climbing stem, and by its entire leaflets.

288. Rhús Toxicodéndron radìcans.

289. Rhús Toxicodéndron radìcans.

R. *T.* 3 *microcárpon* Tor. & Gray *R.* Toxicodéndron *γ* microcárpon *Michx.* — Leaves oval-oblong; fruit smaller.

These varieties, which have been hitherto, for the most part, treated as belonging to two species, *R.* radìcans and *R.* Toxicodéndron, are common in many parts of North America; sometimes covering the surface of the ground to a great extent, and at other times climbing to the top of the highest trees, and penetrating the bark with their fibrous roots. The terrible effects of their poison are frequent, and well authenticated.

§ iii. *Lobàdium* Dec.

Sect. Char. Leaf of 3 leaflets, and palmately disposed on the tip of the common petiole, cut in a serrate manner; the teeth large. Flowers in a dense catkin. Sexes polygamous. There are two-lobed glands under the ovary, alternate with the stamens. Styles 3, short, distinct. Drupe rather compress ed, villose. Nut smooth. Aromatic shrubs. (*Dec. Prod.*, ii. p. 72.)

8. *R.* AROMA'TICA *Ait.* The aromatic Rhus, *or Sumach.*

Identification. Ait. Hort. Kew., 1. p. 367.; Dec. Prod., 2. p 73.; Don's Mill., 2. p. 75.; Tor. and Gray, 1. p. 219.

Synonymes. *R.* suavèolens *Ait.*; *R.* trifoliàta *Lodd. Cat.*; *R.* canadénsis *Marsh.*; Lobàdium aromáticum *Raf.*; Turpínia Raf.; Schmálzia Desv.; *Myrica* trifoliàta *Hort.*; Toxicodéndron crenàtum *Mill. Dict.* No. 5.

Engravings. Turp. in An. du Mus., 5. p. 446. t. 30.; and our fig. 290.

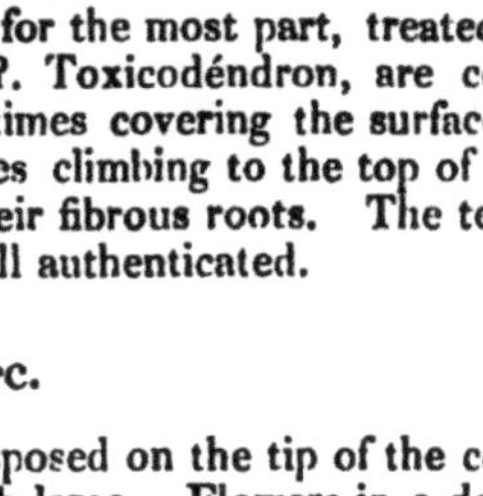

290. Rhús aromàtica.

Spec. Char., &c. Leaves pubescent when young (at length coriaceous, and often glabrous); leaflets sessile, rhomboid-ovate, unequally and incisively toothed, the terminal one narrowed at the base. (*Tor. and Gray.*) A small aromatic shrub. Pennsylvania to Carolina and Georgia. Height 1 ft. to 4 ft. Introd. in 1772. Flowers small, yellow; April and May. Fruit small, light red; ripe in September.

Drupes the size of a small pea, light red, more

or less hispid, slightly compressed, agreeably acid. This species varies greatly in the degree of pubescence of the leaves. *R.* suavèolens *Ait.* only differs in having the leaves almost glabrous. H. S.

Other Species of Rhús.—Several names are in the London catalogues, which are synonymes of kinds which have been lost, or are not distinctly known by us. *R.* pùmila *Michx*, *R.* diversíloba *Tor. & Gray* (*R.* lobàta *Hook.*), *R.* trilobàta *Nutt.*, *R.* laúrina *Nutt.*, are described in Torrey and Gray's *Floras*, but they have not yet been introduced; or, if they have, they exist only as small plants. Some plants of *Rhús* have been raised in the Hort. Soc. Garden, from seeds sent from the snowy mountains of Nepal, which will doubtless prove hardy.

GENUS III.

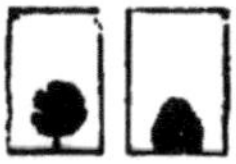

DUVAU'*A* Kth. THE DUVAUA. *Lin. Syst.* Polygàmia Monœ'cia.

Identification. Kth. Gen. Tereb., p. 8.; Dec. Prod., 2. p. 74.; Don's Mill., 2. p. 76.
Synonymes. Schìnus sp. *Andr.*; Amỳris sp. *Cav.*
Derivation. Called Duvaúa, after *M. Duvau*, a French botanist, known as the editor of the original edition of Richard's *Analyse du Fruit*; and for some observations on *Verónica*." (*Lindley*, in *Bot. Reg.*, t. 1568.)

Gen. Char. *Calyx* persistent, with 4—5 segments. *Corolla* of 4—5 concave petals. *Sexes* monœciously polygamous. *Stamens* 8—10, inserted under a pitcher-shaped calycine disk, which has as many sinuses and as many teeth as there are stamens: these are opposite the sinuses, half of them opposite the petals, and half alternate with them. *Anthers* in the fruit-bearing flowers barren. *Ovary* conical. *Styles* 3—4. *Stigmas* capitate. *Fruit* a globose drupe, with a leathery nut.—Chilian trees and shrubs, becoming spiny as they advance in growth. (*Dec. Prod.*)

Leaves simple, alternate, exstipulate, evergreen; generally oblong or ovate, toothed, small. *Flowers* in axillary racemes, greenish yellow.—There are four species in cultivation, which are all very handsome evergreen bushes, with bright shining foliage.

The foliage emits, when bruised, a strong but not unpleasant odour, of the nature of turpentine. The leaves of D. ovàta, and, doubtless, those of every species of Duvaúa, when thrown upon water, move about in a manner which may be compared to a fleet of ships employed in manœuvring, or to persons engaged in dancing. Seeds have been produced plentifully in the Hort. Soc. Garden by D. depéndens, trained to a south wall; and seeds of D. latifòlia are often imported from Chili. Cuttings of the ripe wood root in sand, under a bell-glass, in a gentle heat. D. depéndens was but little injured at Kew, in the Chelsea Botanic Garden, and in the Hort. Soc. Garden, by the winter of 1837–8; and D. ovàta was not injured at all, and may be considered as an evergreen shrub, as hardy in the climate of London as Aristotèlia *Mácqui.*

1. D. DEPE'NDENS *Dec.* The drooping-*branched* Duvaua.

Identification. Dec. Prod., 2. p. 74.; Don's Mill., 2. p. 76.
Synonymes. Amỳris polỳgama *Cav. Icon.* 3. p. 20. t. 239.; Schìnus depéndens *Ort. Decad.* 8. p. 102.; Duvaúa depéndens *a* *Hook. Bot. Misc.* 2. p. 176.
Engravings. Cav. Ic., t. 239.; Bot. Reg., t. 1573.; and our *fig.* 291.

Spec. Char., &c. Leaves mostly, especially upon the flower-bearing branches, obovate, and very obtuse, or even emarginate, with scarcely any denticulations. Racemes scarcely exceeding the leaves in length. Stamens mostly 10. Flowers smaller than those of D. ovàta. (*Lindl.*) An evergreen tree; in British gardens a wall shrub. Chili. Height in England 10 ft. to 12 ft.

291. Duvaúa depéndens.

Introduced in 1790. Flowers yellowish white; June and July. Berries black; ripe n Sept.

There is an old plant in the Botanic Garden at Kew, and a tree in the Chelsea Botanic Garden, which is 12 ft. high, with a trunk 7 in. in circumference. The plant in the Hort. Soc. Garden passed seven winters against a wall with a southern exposure, till the winter of 1837–8, when it was killed down to the ground; but it has sprung up again vigorously.

❧ ⚘ 2. D. OVA`TA *Lindl.* The ovate-*leaved* Duvaua.

292. Duvaúa ovàta.

Identification. Lindl. in Bot. Reg., t. 1568.
Engravings. Bot. Reg., t. 1568.; and our *fig.* 292

Spec. Char., &c. Leaves ovate, toothed, in most acute at the tip, in some obtuse. Racemes a little longer than the leaves. Stamens mostly 8. (*Lindl.*) An evergreen tree; in British gardens a shrub. Chili, on mountains. Height in the climate of London 6 ft. to 10 ft. against a wall. Introduced in 1825. Flowers yellowish white; June and July. Berries black; ripe in September.

Probably a variety of the preceding species. It was wholly uninjured by the winter of 1837–8, in the Horticultural Society's Garden.

❧ ⚘ 3. D. LATIFO'LIA *Gill.* The broad-leaved Duvaua.

Identification. Gillies MSS.; Lindl. in Bot. Reg., t. 1580.
Synonymes. D. depéndens γ *Hook. Bot. Misc.*; Huing han, *Chili.*
Engravings. Bot. Reg., t. 1580.; and our *fig.* 293.

293. Duvaúa latifòlia.

Spec. Char., &c. Leaves oblong, acute, coarsely toothed, so waved as to seem in some measure plicate. Racemes dense, the length of the leaves. Stamens 8. (*Lindl.*) An evergreen tree; in British gardens a shrub. Chili, on mountains. Height 6 ft. to 12 ft. against a wall. Introduced in 1826. Flowers greenish white; June and July. Berries black; ripe in October.

"Whatever," observes Dr. Lindley, "may be thought" of the distinctness, as species, "of D. ovàta and D. depéndens, there can be no doubt that D. latifòlia is a totally distinct species; for not only are the leaves, in their outline, surface, and colour, and the whole plant in its habit, very different, but we find it maintain all its peculiarities unchanged when raised from seeds."

Other Species of Duvaúa.—D. dentàta Dec., Schìnus dentàta *Bot. Rep.*, was introduced in 1795, and is doubtless as hardy as any of the above kinds; since all of them are safest when planted against a wall. *Duvaúa sinuàta* Lindl. appears equally hardy with D. depéndens in the Hort. Soc. Garden. It differs from the others in producing the flowers before the leaves, and in being deciduous. All the species well deserve culture as evergreen bushes, in shrubberies where the soil is dry and sandy, the situation sheltered, and the surface sloping to the south. A concurrence of circumstances of this kind is not unfrequent in country residences, both in England and Scotland; and two examples which occur to us at the moment we are writing are, Bury Hill in Surrey, and Blair Drummond in Stirlingshire.

ORDER XXV. LEGUMINA'CEÆ.

ORD. CHAR. *Calyx* with 5 divisions, either partitions, teeth, or clefts, the odd one anterior to the axis of inflorescence. *Fruit* a legume. *Seed* with the radicle next the hilum. (*Lindl.*)—Trees and shrubs natives of every climate.

Leaves alternate, stipulate, generally compound; deciduous, or sometimes evergreen; petiole tumid at the base. *Stipules* 2 at the base of the petiole, and 2 at the base of each leaflet. *Pedicels* usually articulated, with 2 bractlets under the flower.

The name of Leguminàceæ is applied to this extensive and truly natural order, on account of the seeds of all the species being produced in leguminous pods, bearing more or less resemblance to those of the common pea or bean; and quite different from the siliquose pods of cruciferous plants.

The ligneous species are trees and shrubs, for the most part deciduous; and they are disposed through almost every part of the world. The order contains some of our finest ornamental shrubs and low trees, such as Robínia, Cýtisus, Wistària, Genísta, U'lex, Amórpha, Halimodéndron, Acàcia, Gledítschia, Cércis, and various others. It also contains some considerable trees, which belong to the genera Robínia, Gledítschia, Sophòra, &c. The genera containing hardy ligneous plants are in number twenty-three, which, after De Candolle and G. Don, we place in characterised sections, and ascribe to them short characters, that are more or less contradistinctive.

Sect. I. *SOPHO'REÆ.*

Sect. Char. Corolla, in most, papilionaceous. Stamens 10, with the filaments distinct. Legume not jointed. Cotyledons flat, leafy. Embryo with the radicle beside the edges of the cotyledons. Leaves simply pinnate, or simple.

SOPHO'RA R. Br. Legume necklace-shaped, including many seeds. Leaf with more than three leaflets.

VIRGI'LIA Lam. Legume compressed, including many seeds. Leaf with more than three leaflets.

PIPTA'NTHUS *Swt.* Legume compressed, including 6 seeds. Leaf with its leaflets 3.

Sect. II. *LO'TEÆ.*

Sect. Char. Corolla papilionaceous. Stamens 10, the filaments of all connate, or those of 9 connate, and that of one distinct. Legume not jointed. Embryo with the radicle beside the edges of the cotyledons. The cotyledons flattish; in germination, converted into leaves furnished with stomata. Leaves simply pinnate, or simple.

U'LEX L. Calyx 2-parted, 5-toothed. Legume oval-oblong, turgid, scarcely longer than the calyx, containing but few seeds, though the ovules are many. Habit spiny.

SPA'RTIUM *Dec.* Standard roundish. Keel acuminate. Branches rush-like. Leaf simple.

GENI'STA *Lam.* Standard oblong-oval. Keel oblong, not wholly including the stamens and pistils. Leaves with 3 leaflets, or, in some, simple.

CY'TISUS *Dec.* Standard ovate. Keel very obtuse, including the stamens and pistil. Leaves, in all, with three leaflets.

ADENOCA'RPUS *Dec.* Stamens with the filaments connate. Legume bearing stalked glands all over it.

ONO'NIS *L.* Calyx with 5 linear segments. Standard striate. Legume containing few seeds; in most, turgid.

AMO'RPHA *L.* Corolla consisting of the standard only.

EYSENHA'RDTIA *H. & B.* Corolla with the standard, and 2 keel petals distinct.

ROBI'NIA Dec. Legume flat; that edge to which the seeds are attached margined. Leaf impari-pinnate.

CARAGA'NA Lam. Legume rather cylindrical. Leaf abruptly pinnate.
HALIMODE'NDRON *Fisch.* Legume stipitate, inflated, bladdery. Leaf abruptly pinnate.
CALO'PHACA *Fisch.* Stamens with the filaments of 9 connate, that of one distinct. Legume sessile, with concave valves bearing hairs, some soft, some rigid and glanded.
COLU'TEA *R. Br.* Legume stipitate, much inflated, glabrous.
ASTRA'GALUS Dec. Legume with its lower suture so bent in towards the opposite one as to cause the legume to seem, more or less, 2-celled.

Sect. III. *HEDYSA'REÆ.*

Sect. Char. Corolla papilionaceous. Stamens usually with the filaments connate in one of three modes; the 10 connate; 9 connate and one distinct; or connate by fives: in a few cases all are distinct. Legume dividing transversely into 1-seeded joints, called loments. Embryo with the radicle beside the edges of the cotyledons, which are flattish, and, in germination, are converted into leaves furnished with stomata. Leaves simply pinnate, or simple.

CORONI'LLA *Neck.* Calyx campanulate, usually shorter than the claws of the petals. Carina acute. Stamens diadelphous. Seeds ovate, or cylindrical.

Sect. IV. *PHASEO'LEÆ.*

Sect. Char. Corolla papilionaceous. Stamens usually with 9 filaments connate, and one distinct. Legume not jointed, including many seeds, that are separated from one another with a cellular, transverse, membraneous partition, that is in some cases not complete. Embryo with the radicle beside the edge of the cotyledons, which are thick, and, in germination, either remain under ground, or are changed into thick leaves that scarcely have stomata. Leaves simply pinnate, or simple.

WISTA'*RIA* Nutt. Leaf impari-pinnate.

Sect. V. *CASSIE'Æ.*

Sect. Char. Corolla, in most of the species, of equal petals; in some sub-papilionaceous. Stamens with the filaments distinct. Leaves doubly or triply pinnate; in some simple.

GLEDI'TSCHI*A* L. Sexes diœciously polygamous. Corolla of 3—5 equal petals. Legume in most long and narrow. Seeds compressed. Leaves compoundly divided. Bearing prickles in most.
GYMNO'CLADUS *Lam.* Sexes, by defect, diœcious. Corolla of 5 equal petals. Legume compressed and broad. Seeds scarcely compressed. Leaves compoundly divided.
CE'RCIS L. Sexes hermaphrodite. Corolla sub-papilionaceous, of 5 unequal petals the side ones, or wings, longer than the others. Leaves simple.

Sect. I. *SOPHO'REÆ.*

GENUS I.

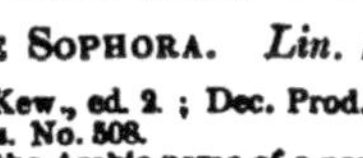

SOPHO'RA R. Br. THE SOPHORA. *Lin. Syst.* Decándria Monogýnia.

Identification. R. Brown in Hort. Kew., ed. 2.; Dec. Prod., 2. p. 95.; Don's Mill., 2. p. 109.
Synonyme. *Sophòræ* spec. *Lin. Gen.* No. 508.
Derivation. Altered from *sophero*, the Arabic name of a papilionaceous flowering tree.

Gen. Char. *Calyx* 5-toothed, campanulate at the base, or somewhat attenuated. *Petals* of the keel usually concrete at the apex. *Legume* somewhat moniliform, wingless, many-seeded. (*Don's Mill.*)

Leaves impari-pinnate, with 11—13 leaflets, generally exstipulate. *Flowers* yellow, white, or blue, in simple racemes, or panicles.—The only hardy species are deciduous trees, natives of Japan or China.

Ⓣ 1. *S.* JAPO'NICA *L.* The Japan Sophora.

Identification. Lin. Mant., 78.; Dec. Prod., 2. p. 98.; Don's Mill., 2. p. 109.
Synonyme. *S.* sinica *Rosier Journ. Phys.* 14. p. 248., *Dec. Légum.* t. 4. f. 1.
Engravings. Red. in N. Du Ham., 3. t. 21.; Dec. Légum., t. 4. f. 1.; the plate of this species in Arb. Brit. 1st edit., vol. v.; and our *fig.* 294.

Spec. Char., &c. Leaves pinnate, with 11—13 leaflets, which are oblong-ovate, acute, and smooth; panicle loose, terminal; pods smooth. A deciduous tree of the middle size. Japan. Height 40 ft. to 50 ft. Introduced in 1763. Flowers cream-coloured; August and September. Pods rarely produced in England. Decaying leaves yellow and green. Naked young wood dark green, like that of *Jasmìnum officinàle.*

Varieties.

Ⓣ S. *j.* 2 *variegàta* Hort. has the leaves variegated, but is not worth cultivating as an ornamental plant.

Ⓣ S. *j.* 3 *péndula* Hort., and the plate of this tree in our 1st edit. vol. v., has pendulous shoots, and is a very remarkable variety. Grafted near the ground, the shoots run along the surface, like those of a trailing plant, to a very great distance from the main stem; in good soil, a shoot extending itself 6 or 8 feet in one season. Grafted at the height of 10 or 20 feet or upwards, the shoots hang down, and form one of the most ornamental of pendulous trees, both in summer and winter.

294. Sophòra japónica.

A round-headed tree, readily distinguished in winter by the fine, smooth, dark green bark of its young wood and smaller branches; and, in summer, by the dark blue green of its foliage. In deep free soil, it grows with great rapidity, seedlings attaining the height of 10 or 12 feet in 4 or 5 years; and in 20 or 30 years, in the neighbourhood of London, 30 or 40 feet. There are large specimens in England, which flower freely; but they have never yet ripened seeds: indeed, the tree ripens seeds in France only in the very warmest seasons. The wood is very hard and compact, as much so, it is said, as that of the box. The bark exhales a strong odour, which, it is stated in the *Nouveau Du Hamel*, produces colic and purging on those who prune the tree, or otherwise work with the wood in a green state. Little appears to be known of the uses of the tree in China and Japan: but it is said that the fruit is employed to dye a fine yellow; and the flowers for dyeing a yellow of so superior a hue, that it is exclusively reserved for dyeing stuffs to be worn by the members of the imperial family. None of the arboreous Leguminàceæ are equal to this tree in beauty of foliage and bark. Its flowers, when they are produced, are also in large terminal compound spikes, and very conspicuous, though much smaller than those of the Robínia viscòsa. One remarkable property in the foliage of the sophora is, that the very hottest and driest seasons do not turn it pale, or cause it to drop off, as heat does that of most of the other pinnated-leaved Leguminàceæ. The pendulous variety is well deserving of culture as an object of singularity and beauty; and, where it is desired to cover a surface with intense green foliage during summer, for example, a dry hillock, a plant of this variety, placed on the centre, will accomplish the purpose effectually. The tree will thrive in any free soil; but, in cold climates, it ought to be placed in one rather poor and dry that it may be compelled to make shorter shoots; which, of course, being less succulent, are more easily ripened. It is generally propagated by seeds imported from France; but, where it is desired to have trees that will soon come into flower, seedling plants should be grafted with scions from a flowering tree. It will grow by cuttings, more especially of the roots, and also by layers.

■ 2. *S.* HEPTAPHY'LLA *L.* The 7-leafleted Sophora.

Identification. Lin. Spec., 533.; Dec. Prod., 2. p. 98.; Don's Mill., 2. p. 110.
Engravings. Rumph. Am., 4. p. 50. t. 22.; and our fig. 295.

Spec. Char., &c. Leaflets 7, glabrous. (*Don's Mill.*) A deciduous shrub. China. Height 6 ft. Introduced in 1830. Flowers yellow; October.

There are plants of *S.* heptaphýlla in the Hort. Soc. Garden, which have flowered and appear to be quite hardy, but as they do not exactly agree with Rumphius's figure, especially in the number of leaflets, we wish our engraving to be considered as of doubtful authenticity. The living plants alluded to are sufficiently distinct, and deserve a place in collections.

295. *Sophòra heptaphýlla.*

Genus II.

VIRGI'LIA L. The Virgilia. *Lin. Syst.* Decándria Monogýnia.

Identification. Lam. Ill., t. 346.; Pers. Ench., 1. p. 453.; R. Brown in Hort. Kew., ed. 2. vol. 3. p. 4.; Dec. Prod., 2. p. 98.; Don's Mill., 2. p. 111.
Derivation. Named by Lamarck in honour of the poet *Virgil*, whose *Georgics* entitle him to botanic commemoration.

Gen. Char. Calyx 5-cleft. *Petals* 5, about equal in length. *Vexillum* with the edges not reflexed. *Stigma* beardless. *Legume* compressed, oblong, 2-valved, many-seeded. (*Don's Mill.*)

Leaves compound, impari-pinnate, deciduous; with 9—11 leaflets. *Flowers* yellow, in racemes. — There is only one hardy species, a deciduous low tree.

¶ 1. V. LU'TEA *Michx.* The yellow-*wooded* Virgilia, *or Yellow Wood.*

Identification. Michx. Fil. Arb. Amer., 3. p. 266. t. 3.; Dec. Prod., 2. p. 98.; Don's Mill., 2. p. 112.
Engravings. Delaun. Herb. Amat., t. 197.; Michx. Fil. Arb. Amer., 3. p. 226. t. 3.; the plate of this tree in Arb. Brit., 1st edit., vol. v.; and our *fig.* 296.

Spec. Char., &c. Leaves pinnate; leaflets 9—11; alternate, ovate, pointed, smooth. A deciduous tree. North America. On the mountains of Cumberland, and the Mississippi. Height in America 40 ft.; 10 ft. to 20 ft. in England. Introduced in 1812. Flowers yellowish white, in pendulous racemes; June to August. Pods never produced in England. Decaying leaves rich yellow. Naked young wood yellowish brown.

The leaves, on young trees, are from 1 ft. to 1½ ft. in length, and on old trees not above half that size. The flowers form white pendulous racemes, a little larger than those of the Robínia Pseùd-Acàcia, but not so odoriferous. The seeds are like those of the robinia, and, in America, ripen about the middle

296. Virgília lùtea.

of August. In Britain, the tree has flowered in the Chelsea Botanic Garden, and at Hylands in Essex, but has not yet produced pods. An open airy situation is desirable, in order that the tree may ripen its wood; and, to facilitate the same purpose where the climate is cold, the soil ought to be dry rather than rich. In the London nurseries, it is propagated chiefly by American seeds, but it will doubtless grow by cuttings of the roots.

GENUS III.

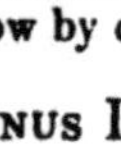

PIPTA'NTHUS *Swt.* THE PIPTANTHUS. *Lin. Syst.* Decándria Monogýnia.

Identification. Swt. Fl.-Gard., 264.; Don's Mill., 2. p. 112.
Derivation. From *piptō*, to fall, and *anthos*, a flower; from the flowers falling off very soon.

Gen. Char. *Calyx* bilabiate; lower lip trifid, upper lip 2-lobed; segments soon falling off. *Petals* deciduous. *Vexillum* large, obcordate, rufescent. *Wings* cuneated. *Keel* cucullate, accumbent. *Stamens* 10; free, deciduous. *Stigma* minute. *Legume* broad-linear, compressed, 6-seeded, stipitate. (*Don's Mill.*)

Leaves compound, trifoliate, stipulate, sub-evergreen; leaflets elliptical-oblong, acute, broad. *Flowers* large, yellow. — One species only in British gardens.

1. P. NEPALE'NSIS *Swt.* The Nepal Piptanthus.

Identification. Swt. Fl.-Gard., 264.; Dec. Prod.; Don's Mill., 2. p. 112.
Synonymes. Thermópsis laburnifòlia *D. Don Prod. Fl. Nep.* p. 239.; *Anagỳris* indica *Wall. MSS.*; Baptìsia nepalénsis *Hook. Exot. Fl.* t. 131.
Engravings. Hook. Exot. Fl., t. 131.; Swt. Fl.-Gard., t. 264.; and our *figs.* 297. and 298.

Spec. Char., &c. Leaves trifoliolate; leaflets elliptical-oblong, acute, broad. Stipules 2, large. A sub-evergreen shrub. Nepal. Height 8 ft. to 10 ft. Introduced in 1821. Flowers rich yellow; May and June. Pod green, turning to brown; ripe in October. Decaying leaves yellow and green. Naked young wood dark green.

The young leaves are silky; and the flowers are of a bright yellow, and are much larger than those of the common laburnum, to which they, and also the leaves and the shoots, bear a general resemblance. In British gardens it may be considered as rather tender, and not of many years' duration; nevertheless, in fine seasons, it ripens abundance of seeds. It may be propagated by cuttings of the roots, and of the shoots, as well as by seeds or layers. In most of the counties north of London, the safest situation for it will be against a wall; and it well deserves a place there, on account of its luxuriant deep green foliage, and large bright yellow flowers. *Anagỳris* indica *Wall.*, Mr. Gordon considers as differing a little from the species.

297. P. nepalénsis.

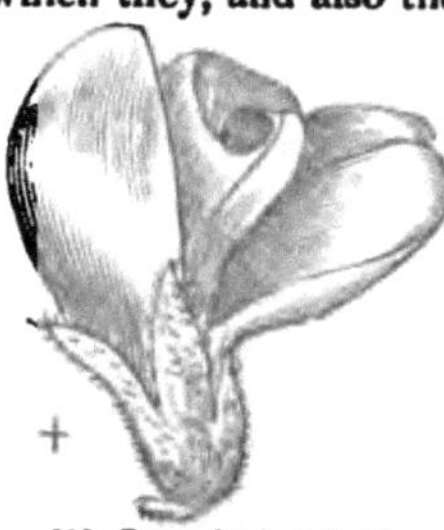

298. P. nepalensis, nat. size.

Sect. II. *Lo'teæ.*

Genus IV.

U'LEX *L.* The Furze. *Lin. Syst.* Monadélphia Decándria.

Identification. Lin. Gen., No. 881.; Lam. Ill., t. 621.; Dec. Prod., 2. p. 144.; Don's Mill., 2. p. 148.
Synonymes. Ajonc, *Fr.*; Heckssame, *Ger.*
Derivation. Said to be derived from *ac*, Celtic, a point; in reference to the prickly branches.

Gen. Char. *Calyx* bibracteate, bipartite, one of the lips 3-toothed, the other bidentate. *Stamens* all connected. *Legume* oval-oblong, turgid, many-ovulate, but few-seeded, hardly longer than the calyx.

Leaves simple, linear, caducous, often changing into spines. *Flowers* solitary, yellow.

Branchy spinous shrubs, evergreen from the colour of the bark, with yellow flowers, natives of Europe, which will grow in any tolerably good soil that is dry; and are readily propagated by seeds, or by cuttings planted in sand.

▲ 1. U'LEX EUROPÆ'A *L.* The European, *or common*, Furze, *or Whin.*

Identification. Lin. Spec., 1045. var. α; Dec. Prod., 2. p. 144.; Don's Mill., 2. p. 14?.
Synonymes. Genísta spinòsa *L'Obel*; *U.* grandiflòra *Pour.*; *U.* vernàlis *Thore*; Whin, Gorse, Prickly Broome; Ajonc common, Jonc marin, Jomarin, or Genêt épineux, *Fr.*
Engravings. Eng. Bot., t. 742.; and our *figs.* 299. and 300.

Spec. Char., &c. Leaves lanceolate, linear. Branchlets villous. Bracteas ovate, loose. Calyx pubescent. An erect compact bush, evergreen, from the colour of the bark. Middle and South of Europe, on gravelly soils; and in Britain on hills. Height 2 ft. to 5 ft.; in sheltered woods, 10 ft. Flowers rich yellow; February to May, and in mild winters September to May. Pod brown; ripe in August.

Varieties.

▲ U. *e.* 2 *flòre plèno* has double flowers, and is a splendid plant when profusely covered with blossoms, well adapted for small gardens, and easily increased by cuttings.

U. *provinciàlis* and U. *strícta* are probably only varieties of *U.* europæ'a, but, as they may possibly belong to *U.* nàna, we have kept them distinct, and treated them as botanical species or races.

299. *U'lex europæ'a.*

300. *U'lex europæ'a.*

The common furze, in Caernarvonshire, grows to the height of 1500 ft. above the sea, in open, airy, warm situations; but in damp shaded valleys, not higher than 600 ft. In the North of England, according to Winch, it forms fine fox covers at 800 or 900 feet; and grows, in warm sheltered situations, at 2000 ft. At Inverness, it is found to the height of 1150 ft. About Tongue, in the north-west of Sutherland, where it was introduced, but is now naturalised, it scarcely attains 350 ft. of elevation. The young branches, bruised, and given to cattle and horses in a green state, are found highly nutritive as fodder; and for this purpose the variety *U.* (e.) strícta is preferable, on account of the absence of prickles. The use of furze for hedges is chiefly desirable in situations where the hawthorn or the holly will not thrive; because the furze is not a plant of long duration. As a shelter to young trees, it is sometimes sown where acorns, beech masts, or chestnuts are to be sown, or young trees are to be planted. The use of furze in a dead state is chiefly as fuel for bakers' ovens, for brick, tile, and lime kilns, and for lighting fires. In Scotland, it is sometimes used in kilns for drying oats. In England, a common use of it is to weave into the sides of hovels for sheltering cattle, to prevent them from rubbing against them. In gardens, the points of the shoots are chopped into pieces of about 1 in. in length, and dropped into the drills in which peas are sown, before the seeds are covered; and, the earth being drawn over them and trod down, they are found effectually to resist the attacks of mice and small birds. In France, the chopped branches are mixed with cow-dung, and the mixture afterwards formed into bricks, which are dried in the sun, and used as fuel. The seeds, if they could be procured in sufficient quantity, would, if ground into flower, form a nutritive food both for cattle and swine: they retain their vital property for several years. In Britany, large heaps are formed of alternate layers of turf and dried furze branches; and, the whole being set fire to, the ashes are preserved as manure. In many parts of both France and England, the ashes of dry furze branches are used as a lye for washing linen. A pound of seeds, which, in London, costs from 8*d.* to 1*s.*, will sow an acre broad-cast, or a drill of a mile in length as a hedge. The double-flowered and the fastigiate varieties are propagated by cuttings: the latter, when wanted for agricultural purposes, may be bedded in, like box, in a sandy soil rather moist, in the beginning of September; and by the following spring they will be fit to transplant.

2. *U.* (E.) NA'NA *Forst.* The dwarf Furze.

Identification. Smith Eng. Fl., 3. p. 266.; Eng. Bot., t. 743.; Dec. Prod., 2. p. 144; Don's Mill. 2. p. 148.; Webb Iter Hispan., 48.
Synonymes. *U.* minor *Roth Cat.* 1. p. 83.; *U.* europæ'us β *Lin. Spec.* 1045.
Engravings. Eng. Bot., t. 743.; and our *figs.* 301. and 302.

Spec. Char., &c. Branches and leaves smooth, the latter linear. Calyx glabrous, with spreading narrow teeth. According to Smith, the essential character consists in the more distinct and spreading calyx teeth, and the more minute, rounded, close-pressed, and often hardly discernible, bracteas. An evergreen, compact, low, spiny shrub. Britain and the western parts of France, on poor gravelly soils. Height 2—3 ft. Flowers rich yellow; August to December. Pods brown; ripe in December.

301. *U.* (e.) nàna.

302. *U.* (e.) nàna.

A very distinct sort, though, from the very different and more luxuriant habit which the plant has when cultivated in gardens on rich soils, we have no doubt of its being only a variety of *U.* europæ'a. In its native habitats, it is easily distinguished from that species by its low growth, seldom exceeding 2 ft. in height; by its being much smaller in all its parts; by its decumbent habit; and by its flowering from the end of August till the beginning of December, and seldom at any other season. Very neat low hedges and edgings may be formed of it.

3. *U.* (E.) PROVINCIA'LIS *Lois.* The Provence Furze.

Identification. Lois. Not., 105.; Dec. Prod., 2. p. 144.; Don's Mill., 2. p. 148.; Webb Iter Hispan., 48.
Synonyme. *U'*lex austràlis *Clement.*
Engravings. Lois. Not., t. 6. f. 2.; and our *fig.* 303.

Spec. Char., &c. Calyx rather pubescent, with lanceolate distant teeth. An erect, evergreen, compact shrub; intermediate, in all its parts and in its habit, between *U.* europæ'a and *U.* nàna. Provence, Andegavany, and Mauritania. Height 2 ft. to 4 ft. Introduced in 1820. Flowers rich yellow; August to December. Pod brown; ripe in December.

303. *U.* provinciàlis.

Whatever doubts there may be as to *U.* nàna being a distinct species, there can be none as to this sort being only a variety. As an evergreen shrub, flowering freely; it well deserves a place in collections.

4. *U.* (E.) STRI'CTA *Mackay.* The upright-*growing, or Irish*, Furze.

Identification. Mackay's List of Irish Plants; Hook. Brit. Fl., p. 317.
Synonymes. *U.* hibérnica *Don's Mill.* 2. p. 148.; *U.* fastigiàta *Hort.*
Engraving. Our *fig.* . in p. .

Spec. Char., &c. Habit erect, narrow, and compact. Spines few or none; and what there are, weak, branched, leafy, and pubescent. An erect, compact, evergreen shrub. Ireland. Height 6 ft. to 10 ft. Introduced in 1815. Flowers yellow, rarely produced; August to December. Pod brown; ripe in December.

Discovered in the Marquess of Londonderry's Park, in the county of

Down, in 1815, or before. It is very upright in its growth, and attains the height, in good soils, of from 6 ft. to 10 ft. in as many years. Its branches are so soft and succulent that sheep and cattle eat them without injuring their mouths, and are very fond of them. It forms excellent garden hedges, and, in rather moist climates, is a most excellent forage plant, as has been already stated under *U.* europæ'a. It only rarely flowers, and has very seldom produced seeds; but it is easily propagated by cuttings.

Other Species of U'lex. — U. *genistöìdes* Brot., *U.* mìtis *Hort.*, Stauracánthus aphýllus *Link*, is a leafless shrub, with the habit of *U'*lex; a native of Portugal in sandy pine woods; and differing from *U'*lex nàna chiefly in the spines branching into two small ones at the sides. It was introduced in 1823; and grows to the height of 1 ft. to 2 ft. It is rather tender in the climate of London, but sometimes stands the winter among rockwork.

Genus V.

SPA'RTIUM *Dec.* The Spartium, *or* *Spanish Broom.* *Lin. Syst.* Monadélphia Decándria.

Identification. Lin. Sp., 995.; Dec. Prod., 2. p. 145.; Don's Mill., 2. p. 148.
Synonymes. Spartiánthus *Link Enum.* 2. p. 223.; Genista sp. *Lam.* and *Mœnch*; Sparzio, *Ital.*
Derivation. From *sparton*, cordage; in allusion to the use of the plant in early ages generally, and in Spain, even to the present day, for making ropes.

Gen. Char., &c. *Calyx* membranous, spathaceous, cleft above, 5-toothed at the apex, somewhat labiate. *Corolla* with a roundish complicated vexillum, and an acuminated keel. *Petals* a little agglutinated, but partable. *Stamens* monadelphous. *Legume* compressed, many-seeded, glandless. (*Don's Mill.*)

Leaves simple, alternate, exstipulate, caducous; lanceolate. *Flowers* in terminal racemes, large, distant, and yellow. — A shrub, a native of Spain and Portugal.

■ 1. *S.* JU'NCEUM *L.* The Rush-*like* Spartium, *or* *Spanish Broom.*

Identification. Lin. Sp., 995.; Dec. Prod., 2. p. 145.; Don's Mill., 2. p. 148.
Synonymes. Genista júncea *Lam.* and *Du Ham.*; *G.* odoràta *Mœnch*; Spartiánthus júnceus *Mœnch*; Genêt d'Espagne, *Fr.*; Binsenartige Pfriemen, *Ger.*; Ginestra di Spagna, *Ital.*
Engravings. N. Du Ham., 2. t. 22.; Bot. Mag., t. 85.; and our *fig.* 305.

Spec. Char., &c. Branches upright, round, of a deep green colour, smooth, and with but few leaves, which are lanceolate, and soon drop off. An upright shrub, evergreen from the colour of its numerous shoots. Spain, Portugal, and the South of France, in gravelly soils. Height 5 ft. to 8 ft.; in British gardens 8 ft. to 12 ft. Introduced in 1548. Flowers dark yellow, large; July to September. Pods brown; ripe in October. Naked young wood smooth and dark green.

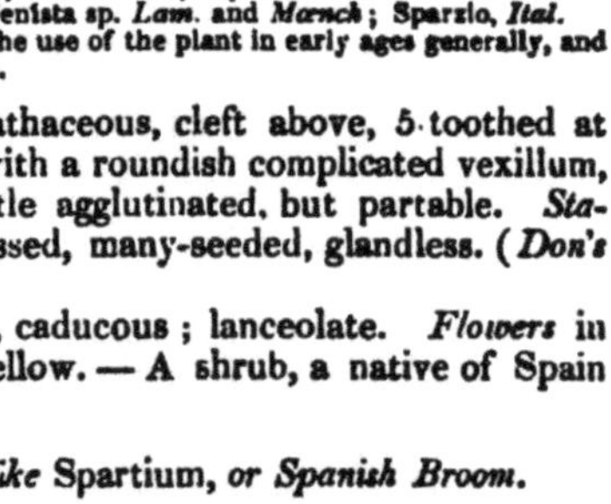

304. Spártium júnceum odoratíssimum.

Varieties.

■ S. j. 2 *odoratíssimum* (*S.* odoratíssimum *D. Don Brit. Fl. Gard.* 2. st. 390.; *S.* acutifòlium *Lindl. Bot. Reg.*; and our *fig.* 304.) has the flowers sweet-scented, and the leaves more acute than those of the species. Raised from Turkish seeds.

■ S. j. 3 *flòre plèno* has double flowers.

In Italy and the South of France a very good cloth is manufactured from the fibres of this plant. Both in Spain and France, the shoots are used for forming baskets, and for tying up vines and other fruit trees. The bees are said to be very fond of the flowers; and the seeds are eaten with great avidity by poultry, partridges, &c. Medicinally, the flowers and leaves, in infusion, act as an emetic, or, in a larger quantity, as an aperient. In Britain, the plant is solely regarded as an ornamental shrub. Seeds are produced in abundance, and they will come up in any soil that is tolerably dry. In the nursery, they ought to be transplanted every year, as they are apt to form long taproots and very few fibres.

305. *Spártium júnceum.*

Genus VI.

GENI'STA *Lam.* THE GENISTA. *Lin. Syst.* Monadélphia Decándria.

Identification. Lam. Dict., 2. p. 616.; Ill t. 619.; Dec. Prod., 2. p. 145.; Don's Mill., 2. p. 148.
Synonymes. Genista, et Spártium, spec. *Lin.*; Genêt, *Fr.*; Ginster, *Ger.*; Ginestra, *Ital.*

Gen. Char. *Calyx* bilabiate, upper lip bipartite, lower one tridentate, or 5-lobed, the three lower lobes nearly joined to the apex. *Vexillum* oblong-oval. *Carina* oblong, straight, not always containing the stamens and pistils. *Stamens* monadelphous. *Legume* compressed, many-seeded. (*Don's Mill.*)

Leaves simple or compound, alternate, rarely opposite, stipulate, deciduous or sub-evergreen; lanceolate, linear, or trifoliolate. *Flowers* terminal or axillary, yellow.

The hardy species are deciduous or sub-evergreen shrubs, generally with trifoliolate leaves and yellow flowers; there is a great sameness of character among them, and, though many are quite distinct, yet it is highly probable that the greater number now recorded as species are only varieties. They are chiefly natives of Europe; but a few are found in the North of Africa. As they grow rapidly, and flower freely, especially on soils not wet at bottom, they are desirable plants for newly formed shrubberies, but in general they are not of long duration. A number of the species were formerly included under the genus *Spártium* and some under *Cýtisus*, from which they have been separated by Lamarck, whose arrangement, as modified by DeCandolle, we have adopted in the following enumeration.

§ 1. *Unarmed. Leaves all, or for the most part, trifoliolate.*

■ 1. *G.* PARVIFLO'RA *Dec.* The small-flowered Genista.

Identification. Dec. Prod., 2. p. 145.; Don's Mill., 2 p. 148.
Synonyme. Spártium parviflórum *Vent. Hort. Cels.* t. 87.
Engravings. Vent. Hort. Cels., t. 87.; and our *fig.* 306.

Spec. Char., &c. Leaf trifoliolate, the petiole very short; and the leaflets usually deciduous, very narrow, glabrous. Flowers in lengthened terminal racemes. Legumes compressed, 1—3-seeded, rather pubescent, being covered with minute closely pressed down, slightly spreading. (*Dec. Prod.*) A deciduous shrub. Levant, near the Gulf of Mundania.

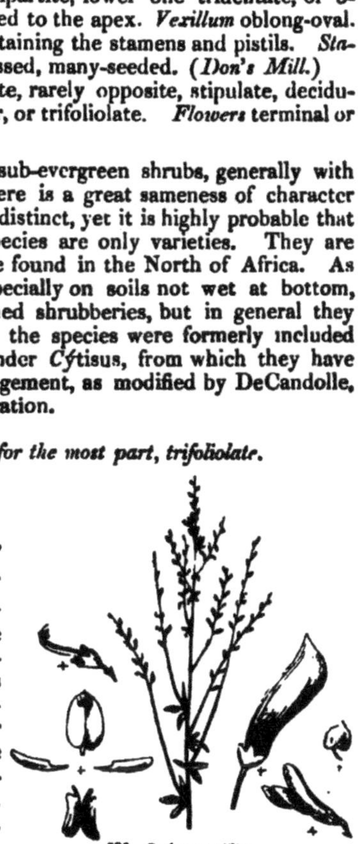

306. Genista parviflòra.

Height 6 ft. to 7 ft. Introduced in 1817. Flowers yellow; May to August. Legume ?. H. S.

307. *G. cándicans.*

2. *G.* CA'NDICANS *L.* The whitish Genista.

Identification. Lin. Amœn.; Dec. Prod., 2. p. 145.; Don's Mill., 2. p. 149.; Webb Iter Hispan., 50.
Synonymes. Cýtisus cándicans *Lin. Sp.*; *C.* pubéscens *Mœnch.*
Engravings. Deud. Brit., t. 80.; and our *fig.* 307.

Spec. Char., &c. Leaf trifoliolate, petiolate; leaflets obovate, pubescent, with appressed down. Branches angled. Flowers in terminal heads, few in a head. Legume hairy. (*Dec. Prod.*) A hoary sub-evergreen shrub, of short duration. Mogador, Italy, and the Levant. Height 4 ft. to 6 ft. Introduced in 1735. Flowers large, yellow, scentless; April to July. Legume ?.

The great advantage of this species is, that it grows rapidly, and flowers freely.

3. *G.* PA'TENS *Dec.* The spreading Genista.

Identification. Dec. Prod., 2. p. 145.; Don's Mill., 2. p. 149.; Webb Iter Hispan., p. 50.
Synonyme. Spártium pàtens *Cav. Icon.* 2. p. 58., exclusive of the synonyme.
Engravings. Cav. Icon., 2. p. 58. t. 176.; and our *fig.* 308.

308. *Genista pàtens.*

Spec. Char., &c. Branches striated, twiggy, glabrous. Leaves stalked, trifoliolate. Leaflets obovate, pubescent beneath. Flowers in fours, pedicellate, nearly terminal. Legume glabrous, 3—6-seeded. (*Don's Mill.*) A spreading shrub. Spain. On mountains near Albayda, and found by P. B. Webb on Monte Santo in Catalonia. Introduced in ?1830. Height 4 ft. to 8 ft. Flowers yellow; April to July.

It differs from Cýtisus pàtens, in the upper lip of the calyx being acutely bipartite; lower lip of three bristles, not with the lips nearly equal and entire.

4. *G.* TRI'QUETRA *Ait.* The triangular-*stemmed* Genista.

Identification. Ait. Hort. Kew., 3. p. 14.; Dec. Prod., 2. p. 146.; Don's Mill., 2. p. 149.
Synonyme. *G.* triquetra *Lam.* ?
Engravings. Bot. Mag., t. 314.; Dend. Brit., t. 79.; and our *fig.* 309.

Spec. Char., &c. Branches 3-sided, decumbent, the younger ones villose. Leaves trifoliolate, simple about the extremities of the branches; leaflets ovate-lanceolate, villose. Flowers in short terminal racemes. (*Dec. Prod.*) A trailing shrub, evergreen from the colour of its shoots. Spain, Italy, and France. Height 6 in. Introduced in 1748. Flowers yellow; April to July. Legume ?.

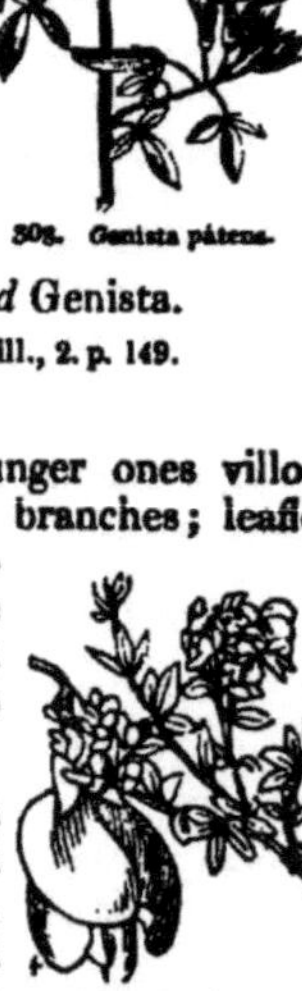
309. *G. triquetra.*

No shrub is more ornamental on rockwork; and, when trained to a stake and allowed to form a head, or grafted standard high on a laburnum, it forms a singular object, and, when in flower, a most magnificent one. It is also an admirable plant for training against a wall, particularly in dry situations, where it is exposed to the sun.

5. *G.* UMBELLA'TA *Poir.* The umbellate-*flowered* Genista.

Identification. Poir. Suppl., 2. p. 715.; Dec. Prod., 2. p. 146.; Don's Mill., 2. p. 149.; Webb Iter Hispan., p. 51.
Synonymes. Spártium umbellàtum *Desf. Atl.* 2. p. 133., *L'Hérit. Stirp.* 183.; Bolina, *in Andalusia.*
Engraving. Our *fig.* . in p. .

Spec. Char., &c. Leaf trifoliolate, its petiole short, its leaflets linear-lanceolate, and rather silky. Flowers in terminal heads. Calyx hairy, in a silky manner. Corolla and legume silky. Branches glabrous. (*Dec. Prod.*) A low shrub. Barbary, on arid hills; and Spain, in Andalusia, on hills. Height 1 ft. to 2 ft. Introduced in 1779. Flowers yellow; April to June.

Variety.

G. *u.* 2 *capitàta* Dec. Spártium capitàtum *Cav. Annal.* 1801, p. 63. — Branches and leaves covered with silky villi. Native of Mogador.

§ 2. *Spinosæ. Leaves all, or some of them, trifoliolate.*

6. *G.* LUSITA'NICA *L.* The Portugal Genista.

Identification. Lin. Sp., 999., exclusive of the synonymes of Clus. and J. Bauh.; Lam. Dict., 2. p. 662., exclusive of the synonymes; Dec. Prod., 2. p. 146.; Don's Mill., 2. p. 450.
Engravings. Andr. Bot. Rep., t. 419.; and our *fig.* 310.

Spec. Char., &c. Branches spiny, round, becoming striate. Leaves trifoliolate, opposite, upon short petioles; the leaflets linear, folded, somewhat silky. Flowers few, terminal. Calyx very hairy. (*Dec. Prod.*) A very spiny shrub, evergreen from the colour of its young shoots. Portugal. Height 4 ft. Introduced in 1771. Flowers yellow; March to May. Legume ?.

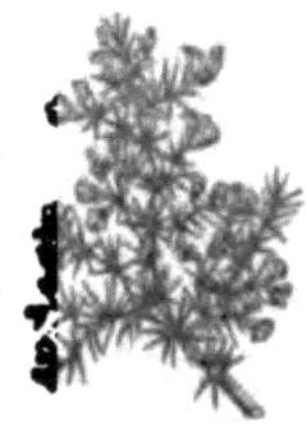

310. G. lusitánica.

Remarkable for having opposite leaves and branches; a character not common among Leguminàceæ.

7. *G.* (L.) RADIA'TA *Scop.* The rayed-*branched* Genista.

Identification. Scop. Carn., No. 871.; Dec. Prod., 2. p. 146.; Don's Mill., 2. p. 149.
Synonymes. Spártium radiàtum *Lin. Sp.* 996., *Mill. Icon.*, *Sims Bot. Mag.*; G. ilvénsis *Dalech.*
Engravings. Mill. Icon., t. 249. f. 1.; Bot. Mag., t. 2260.; and our *fig.* 311.

Spec. Char., &c. Branches angled, grouped, glabrous. Leaf trifoliolate, almost sessile, opposite, the leaflets somewhat silky. Flowers in terminal heads, 2—4 in a head. Corolla and legume silky. The old branches show a tendency to become spiny. The legumes are oval, short, compressed, pointed with the style, and include two seeds. (*Dec. Prod.*) A low shrub, of short duration, evergreen from the colour of its young shoots. Italy, Carniola, and the Vallais. Height 2 ft. to 4 ft. Introduced in 1758. Flowers yellow; June and July. Legume ?.

Variety.

G. (*l.*) *r.* 2 *umbellàta*, G. umbellàta *Poir.*, Spártium umbellàtum *Desf.*, appears, from a plant that was in the Hort. Soc. Garden in 1837, to belong to this species.

311. G. (l.) radiàta.

Differing from *G.* lusitánica principally in being without spines, and having its leaves somewhat longer. Both *G.* radiàta and *G.* lusitánica have a very singular appearance when without their leaves; and, in that point of view, they may be considered as almost as interesting in winter as they are in summer.

8. *G.* EPHEDRÖI'DES *Dec.* The Ephedra-like Genista.

Identification. Dec. Légum. Mém., 6.; Dec. Prod., 2. p. 147.; Don's Mill., 2. p. 150.
Engravings. Dec. Légum Mém., 6. t. 36.; Maund's Botanic Garden, t. 498.; and our *fig.* 312.

312. G. ephedröides.

Spec. Char., &c. Leaves some trifoliolate, some simple, few sessile; leaflets linear, almost glabrous. Branches rigid, round, becoming striated and spiny. Flowers in spikes, alternate, yellow. Calyx somewhat pubescent. (*Dec. Prod.*) A shrub, evergreen from the colour of its young slender shoots. Sardinia. Height 2 ft.; in British gardens 4 ft. Introduced in 1832. Flowers small, yellow; June to September. Legume ?.

The whole plant is glabrous, and resembles in appearance *E'phedra distàchya.* Cuttings strike readily.

313. *G. triacánthos.*

9. *G.* TRIACA'NTHOS *Brot.* The three-spined Genista.

Identification. Brot. Phyt., 130. t. 54.; Dec. Prod., 2. p. 147.; Don's Mill., 2. p. 150.
Synonyme. G. rostràta *Poir. Suppl.* 2. p. 719.
Engravings. Brot. Phyt., t. 54.; and our *fig.* 313.

Spec. Char., &c. Leaves sessile, trifoliolate and simple, glabrous. Leaflets linear-lanceolate. Branchlets spiny, branched. Flowers in terminal racemes, few in a raceme. Calyx, corolla, and legume glabrous; legume 1-seeded. The spines are simple, trifid, or branched. (*Dec. Prod.*) A deciduous undershrub. Portugal, on mountains and in woods. Height 2 ft. to 3 ft. Introduced in 1821. Flowers yellow; May to July. Legume ?.

Variety.

G. *t.* 2 *interrúpta* Dec., Spártium interrúptum *Cav. Annal.*, 1801, vol. iv. p. 58., has linear leaflets, and branches usually simple, and shorter than those of the species. It is found wild about Tangier.

314 Genista hórrida.

10. *G.* HO'RRIDA *Dec.* The horrid Genista.

Identification. Dec. Fl.-Fr., 4. p. 500.; Dec. Prod., 2. p. 146.; Don's Mill., 2. p. 149.; Webb. Iter Hispan., 51.
Synonymes. Spártium hórridum *Vahl Symb.* 1. p. 51., exclusive of the synonyme; *G.* erinàcea *Gilib. Bot. Prat.* 2. p. 239.
Engravings. Gilib. Bot. Prat., 2. p. 239. icon.; and our *fig.* 314.

Spec. Char., &c. Branches grouped, angled, spiny, opposite. Leaves trifoliolate, opposite; the leaflets linear, folded, somewhat silky. Flowers few, almost terminal. Calyx pubescent. (*Dec. Prod.*) A native of the Pyrenees. Height 4 ft. Introduced in 1821. Flowers yellow; May and June. Legume ?.

§ 3. *Spinosæ. Leaves all simple.*

11. *G.* SYLVE'STRIS *Scop.* The wood Genista.

Identification. Scop. Carn., No. 875.; Dec. Prod., 2. p. 148.; Don's Mill., 2. p. 151.
Synonyme. G. hispánica *Jacq. Icon. Rar.* t. 557.
Engravings. Jacq. Icon. Rar., t. 557.; and our *fig.* 315.

315. Genista sylvéstris.

Spec. Char., &c. Leaves simple, linear-awl-shaped, glabrous above, villose in a closely pressed manner beneath. Spines axillary, branched, slender. Flowers glabrous, disposed in a terminal spiked raceme. Teeth of the calyx almost spiny. The keel longer than the standard and wings. (*Dec. Prod.*) A deciduous undershrub. Carniola and Croatia, on hills. Height 1 ft. to 2 ft. Introduced in 1818. Flowers yellow; June and July. Legume ?.

12. *G.* SCO'RPIUS *Dec.* The Scorpion Genista.

Identification. Dec. Fl. Fr., 4. p. 498.; Dec. Prod., 2. p. 148.; Don's Mill., 2. p. 151.; Webb Iter Hispan., 51.
Synonymes. Spártium Scórpius *Lin. Sp.* 995.; *G.* spinifòra *Lam. Dict.* 2. p. 621.; Scorpion Furze *Gerard.*
Engravings. Dend. Brit., t. 78.; and our *fig.* 316.

Spec. Char., &c. Spiny; spines branched, spreading, striated, glabrous. Leaves simple, very few, oblong, somewhat silky. Flowers glabrous, upon short pedicels, in groups disposed somewhat racemosely; the keel as long as the standard. Legume containing 2—4 seeds. (*Dec. Prod.*) An upright, deciduous, spiny shrub. South of Europe and Barbary, in arid places. Height 2 ft. to 3 ft. Introduced in 1570. Flowers yellow; April and May. Legume ?.

316. *G. Scórpius.*

A spiny shrub, almost leafless when the shoots are full grown. This species is commonly thought to be the *Scórpius* of Theophrastus.

13. *G.* HISPA'NICA *L.* The Spanish Genista.

Identification. Lin. Sp., 999.; Dec. Prod., 2. p. 148.; Don's Mill., 2. p. 151.; Webb Iter Hispan., 68.
Synonyme. Spanish Furze, *Hort.*
Engravings. Cav. Icon., 3. t. 211.; Lam. Ill., t. 619. f. 3.; and our *fig.* 317.

Spec. Char., &c. Spiny, except in the flower-bearing branches; spines branched, rigid. Leaves simple, lanceolate, villose. Flowers in a terminal subcapitate raceme. Keel villose, the length of the glabrous standard. Legume oval, including 2—4 seeds; when ripe, rather glabrous. (*Dec. Prod.*) A diminutive undershrub, evergreen from the colour of its shoots. Spain and the South of France. Height ½ ft. to 1 ft. Introduced in 1759. Flowers yellow; June and July. Legume ?.

317. *G. hispánica.*

14. *G.* A'NGLICA *L.* The English Genista, *or Petty Whin.*

Identification. Lin. Sp., 999.; Dec. Prod., 2. p. 149.; Don's Mill., 2. p. 151.
Synonyme. *G.* minor *Lam. Fl. Fr.* 2. p. 615.
Engravings. Eng. Bot., t. 132.; Lobel Icon., 2. p. 93. f. 2.; and our *fig.* 318.

Spec. Char., &c. Spiny, except in the flower-bearing branches; spines simple; the whole plant glabrous. Leaves simple, ovate-lanceolate. Flowers in terminal racemes, few in a raceme; the keel longer than the standard and wings. Legume ovately cylindrical, including many seeds. (*Dec. Prod.*) A prostrate deciduous shrub, with woody stems. Native of the Middle and North of Europe; and frequent in Britain, on moist, boggy, heathy commons. Height 1 ft. Flowers yellow; May and June. Legume brown; ripe in August.

318. *Genísta ánglica.*

Cultivated in collections, where it forms a spiny bush about 2 ft. in height.

15. *G.* GERMA'NICA *L.* The German Genista.

Identification. Lin. Sp., 995.; Dec. Prod., 2. p. 149.; Don's Mill., 2. p. 151.
Synonymes. Scórpius spinòsus *Mœnch Meth.* 134.; Vóglera spinòsa *Fl. Wett.* 2. p. 500.; Bulimacola di Bosco, *Ital.*
Engravings. Fuchs Hist., 220. icon.; Hayne Abbild., t. 122.; and our *fig.* 319.

Spec. Char., &c. Spiny, except in the flower-bearing branches; spines simple or branched. Leaves simple, lanceolate, slightly hairy. Flowers somewhat villose, in terminal racemes. Keel longer than the standard and wings. Legume ovate, slightly hairy, including 2—4 seeds. (*Dec. Prod.*) A spiny shrub. Europe, in woods and on heaths. Height 2 ft. to 3 ft. Introduced in 1773. Flowers yellow; June to August. Legume brown; ripe in September.

319. *Genísta germánica.*

Variety.

G. g. 2. *inérmis* Dec. is almost without spines.

§ 4. *Unarmed. Leaves all simple.*

16. *G.* PU'RGANS *L.* The purging Genista.

Identification. Lin. Sp., 999.; Bull. Herb., 115.; Dec. Prod., 2. p. 149.; Don's Mill., 2. p. 151.
Synonyme. Spártium purgans *Lin. Syst.* 474.
Engravings. Bot. Cab., 1117.; and our *fig.* 320.

320. Genista púrgans.

Spec. Char., &c. Upright, much branched. Branches round, striate. Leaves simple, very few, lanceolate, almost sessile, somewhat silky. Flowers axillary, solitary, scarcely pediceled. Petals equal, glabrous. The young legume adpressedly pubescent. (*Dec. Prod.*) An upright shrub, evergreen from the colour of its shoots. France, on hills. Introd. 1768. Height 3 ft. to 6 ft. Flowers yellow; June and July. Legume brown; ripe in September.

17. *G.* SERI'CEA *Wulf.* The silky Genista.

Identification. Wulf. in Jacq. Coll., 2. p. 167.; Dec. Prod., 2. p. 149.; Don's Mill., 2. p. 151.
Engravings. Jacq. Icon. Rar., 3. t. 556.; and our *fig.* 321.

321. Genista sericea.

Spec. Char., &c. Decumbent, with upright round branches. Leaves simple, linear-lanceolate, silky beneath. Flowers terminal, 3 or 4 together, in a sort of raceme. Petals silky, nearly equal. Lobes of the calyx oblong-acuminate; the floral leaves equalling the calyx in length. (*Dec. Prod.*) A decumbent shrub. Height 6 in. Austria and Croatia, in subalpine places near the shore. Introduced in 1812. Flowers yellow; May and June. Legume brown; ripe August.

18. *G.* APHY'LLA *Dec.* The leafless Genista.

Identification. Dec. Prod., 2. p. 149.; Don's Mill., 2. p. 152.
Synonymes. Spártium aphýllum *Lin. Fil. Suppl.* 320.; *G.* virgàta *Lam. Dict.* 2. p. 616.
Engravings. Pall. Itin. ed. Gall. Append., No. 357. t. 99. f. 2.; and our *fig.* 322.

322. Genista aphýlla

Spec. Char., &c. Branched, upright. Leaves simple, very few, linear, very short. Flowers disposed distantly, in lengthened terminal racemes. Legumes compressed, including 2 seeds; when young, tomentose; when adult, glabrous. (*Dec. Prod.*) Height 3 ft. to 4 ft. in British gardens. Found in Siberia, in deserts, about the Volga. Introd. 1800. Flowers violaceous; June and July. Legume brown; ripe in September.

323. Genista monospérma.

19. *G.* MONOSPE'RMA *Lam.* The one-seeded Genista.

Identification. Lam. Dict., 2. p. 616.; Dec. Prod., 2. p. 150.; Don's Mill., 2. p. 152.; Webb Iter Hispan., 51.
Synonymes. Spártium monospérmum *Lin. Sp.* 995., *Curt. Bot. Mag.* t. 683.; *G. Rætam* Forsk.
Engravings. Bot. Mag., t. 683.; and our *fig.* 323.

Spec. Char., &c. Branched, upright. Leaves simple, very few, linear-oblong, adpressedly pubescent. Flowers in lateral racemes, few in a raceme. Petals silky, almost equal. Legumes ovate, inflated, membranaceous, glabrous, including 1—2 seeds. (*Dec. Prod.*) An erect shrub, with numerous slender, twiggy, flexile branches. On the Mediterranean shores, where, in many places, it serves to retain and consolidate the drifting sand. Height 2 ft. to 4 ft. Introduced in 1670. Flowers white; June and July. Legume brown; ripe in September.

The leaves and young branches are, in these countries, eaten by sheep and goats; and the twigs are used for tying vines to stakes, or tying up faggots; and they are also twisted into ropes.

20. *G.* SPHÆROCA'RPA *Lam.* The round-fruited Genista.

Identification. Lam. Dict., 2. p. 616.; Dec. Prod., 2. p. 150.; Don's Mill., 2. p. 152.; Webb Iter Hispan., 50.
Synonyme. Spártium sphærocárpon *Lin. Mant.* 571.
Engravings. Clus. Hist., 1. p. 102. f. 2.; and our *fig.* 324.

324. Genista sphærocárpa.

Spec. Char., &c. Twiggy, branched. Leaves simple, few, linear, almost glabrous. Flowers in lateral racemes, many in a raceme. Petals glabrous, equal. Legumes ovate, in some measure fleshy, containing 1—2 seeds. Flowers small, and pale yellow. (*Dec. Prod.*) A twiggy shrub. Native of the South of Europe and North of Africa. Height 3 ft. to 4 ft. Introd. 1731. Flowers yellow; June and July. Legume brown; ripe in September.

21. *G.* ÆTHNE'NSIS *Dec.* The Mount Etna Genista.

Identification. Dec. Prod., 2. p. 150.; Don's Mill., 2. p. 152.
Synonymes. Spártium æthnénse *Biv. St. Sic. Mant.* 2., *Rafin. Specch.* 1. p. 17., *Sims Bot. Mag.* 2674.; Spártium trispérmum *Smith* in *Rees's Cycl.* vol. 32. No. 5.
Engravings. Bot. Mag., t. 2674.; and our *fig.* 325.

325. Genista æthnénsis

Spec. Char., &c. Upright, very much branched. Leaves simple, few, linear, silky. Flowers in terminal racemes. Petals almost glabrous, nearly equal in length. Legumes obliquely ovate, compressed, containing 2—3 seeds; when young, pubescent. (*Dec. Prod.*) An erect twiggy shrub. Native of the wooded region of Mount Etna, 3000 ft.—6000 ft. elevation; growing with A'cer monspessulànum. (*Presl*, in *Comp. Bot. Mag.*, vol. i. p. 91.) Height 2 ft. to 4 ft. Introduced in 1816. Flowers yellow; June and July. Legume brown; ripe in September.

Resembles the preceding species, except that the flowers are twice the size.

22. *G.* ANXA'NTICA *Ten.* The Anxantic Genista.

Identification. Ten. Fl. Nap. Prod., p. 41.; Dec. Prod., 2. p. 150.; Don's Mill., 2. p. 152.
Synonyme. *G.* amsántica *Tenore.*
Engravings. Fl. Nap., 2. p. 127. t. 66.; Swt. Fl.-Gar., 2d ser. t. 266.; and our *figs.* 326. and 327.

326. Genista anxántica.

327. Genista anxántica.

328. Genista a. scariósa.

Spec. Char., &c. The whole plant is perfectly glabrous. Stems spreading. Branches angled. Leaves simple, ovate-elliptical, rather coriaceous, veiny. Flowers in racemes. Corolla thrice as long as the calyx; and about 8 lines long. Legume containing 8—10 seeds. (*Dec. Prod.*) A diffuse shrub. Naples. Height 3 ft. to 4 ft. Introduced in 1818. Flowers yellow; June and July. Legume brown; ripe in September.

Variety.

✱ G. *a.* 2 *scariòsa.* G. scariòsa *Vin.* (Frag. Fl. Ital. 1. t. 8.; and our *fig.* 328.)—An upright shrub, closely resembling the species. Introduced in 1821, and flowering in the Hort. Soc. Garden in June and July. It deserves a place in collections.

23. *G.* TINCTO'RIA *L.* The Dyer's Broom, *or Green Weed.*

Identification. Lin. Sp., 998.; Dec. Prod., 2. p. 151.; Don's Mill., 2. p. 152.
Synonymes. G. itálica *Lod. Cat.*; Base Broom, Green Wood, Dyer's Weed, and Wood-waxen; Genêt des Teinturiers, Genêt de Sibérie, *Fr.*; färber.der Ginster, *Ger.*; Bacellina *Ital.*
Engravings. Eng. Bot., t. 44.; and our *fig.* 329.

Spec. Char., &c. Root creeping. Stems almost upright. Branches round, striated, upright. Leaves simple, lanceolate, rather glabrous. Flowers glabrous, in spiked racemes. Legume glabrous. (*Dec. Prod.*) A creeping-rooted low shrub. Common in Europe, in grassy fields, and in woods and copses, particularly in dry gravelly or sandy soils. Height 1 ft. to 3 ft. Flowers yellow; July. Legume brown; ripe in September.

Varieties.

✱ G. *t.* 2 *flòre plèno.*—There are plants in the Epsom Nursery and the Hort. Soc. Garden.

✱ G. *t.* 3 *latifòlia* Dec.—Leaves broad-lanceolate. A native of Auvergne, on the Mont d'Or.

G. *t.* 4 *hirsùta* Dec.—Leaves somewhat villose. Branches upright. A native of sunny meadows.

✱ G. *t.* 5 *praténsis* Poll.—Leaves oblong-lanceolate, rather hairy. Branches ascending. Inhabits the mountainous parts of Upper Italy.

329. Genista tinctòria.

It is very common in pastures, in many places, both in England and Scotland; but, when cows feed on it, it is said by Ray to give a bitter taste to their milk. All parts of this plant, and especially the branches and leaves, have long been used by dyers for producing yellow, especially for dyeing wool that is afterwards to be dyed green with woad (*Isàtis* tinctòria *L.*). The plant is not now in cultivation for this purpose; but, in Norfolk and Suffolk, it is still collected in quantities from sandy wastes and commons, and sold to the dyers.

✱ 24. *G.* (T.) SIBI'RICA *L.* The Siberian Genista.

Identification. Lin. Mant., 571.; Dec. Prod., 2. p. 151.; Don's Mill., 2. p. 152.
Synonymes. Genistöides elàta *Mœnch Meth.* 132.; Genista tinctòria var. *N. Du Ham.*
Engravings. Jac. Hort. Vind., t. 190.; and our *fig.* 330.

Spec. Char., &c. Stems erect; and the whole plant more slender and taller than *G.* tinctòria, of which it is evidently only a variety. An erect shrub. Siberia. Height 6 ft. Introduced in 1785. Flowers yellow; June to August. Legume brown; ripe in September.

330. Genista (t.) sibírica

25. *G.* (T.) OVATA *Waldst.* The ovate-*leaved* Genista.

Identification. Waldst. et Kit. Pl. Hung., 1. t. 84.; Balb.; Bert.; Tav.; Ten.; Dec. Prod. 2. p. 151.; Don's Mill., 2. p. 153.
Synonyme. *G. nervàta Kit.* in *Litt.*
Engravings. Waldst. et Kit. Hung., 1. t. 84.; Dend. Brit., t. 77.; and our *fig.* 331.

Spec. Char., &c. Stems numerous, hairy, erectish, somewhat herbaceous, striated, terete. Leaves ovate, or ovate-oblong, and are, as well as the legumes, hairy. Racemes short. Corolla smooth. (*Don's Mill.*) A shrub. Sclavonia and Hungary; and on the hills of Italy, from Piedmont to Naples. Height 2 ft. to 4 ft. Introduced in 1819. Flowers yellow; June to August. Legume brown; ripe in September.

331. Genista (t.) ovàta.

26. *G.* TRIANGULARIS *Willd.* The triangular-*stemmed* Genista.

Identification. Willd. Sp., 3. p. 939.; Dec. Prod., 2. p. 151.; Don's Mill., 2. p. 153.
Synonyme. *G. triquetra Waldst. et Kit. Hung.* 2. p. 165. t. 153., but not of Aiton.
Engravings. Waldst. et Kit. Hung., 2. t. 153.; and our *fig.* 332.

Spec. Char., &c. Branches smooth, 3-angled, and, as well as the stems, ascending. Leaves lanceolate, and mucronate. Flowers axillary Legume compressed, and mucronate. (*Dec. Prod.*) A shrub. Hungary, on calcareous rocks. Height 1 ft. Introd. in 1815. Flowers yellow; May and June. Legume brown; ripe in August.

332. Genista triangulàris.

Closely resembling *G. triquetra*, of which, notwithstanding its simple leaves, it may possibly be only a variety; the change not being greater than what takes place in *Fráxinus* excélsior simplicifòlia.

27. *G.* SAGITTALIS *L.* The arrow-*jointed* Genista.

Identification. Lin. Sp., 998.; Dec. Prod., 2. p. 151.; Don's Mill., 2. p. 153.
Synonymes. *G.* herbàcea *Lam. Fl. Fr.*; Genistélla racemòsa *Mœnch Meth.*; Saltzwedèl*ia* sagittàlis *Fl. Wett.* 2. p. 498.
Engravings. Jacq. Fl. Aust., t. 209.; Hayne Abbild., t. 117.; and our *fig.* 333.

Spec. Char., &c. Stems prostrate. Branches herbaceous, ascending, 2-edged, membranous, somewhat articulated. Leaves ovate-lanceolate. Flowers disposed in an ovate, terminal, leafless spike. Corolla smooth; but the keel is furnished with a villous line on the back. (*Don's Mill.*) A prostrate shrub. Continental Europe, in mountain pastures. Height 6 in. Introduced in 1750. Flowers yellow; May and June. Legume brown; ripe in August.

Variety.

G. *s.* 2 *minor* Dec. — A small shrub, having the branches clothed with adpressed pubescence at the apex, as well as the leaves.

333. Genista sagittàlis.

For practical purposes, this may be considered as a herbaceous plant. It is a very distinct, ornamental, and hardy sort; growing and flowering freely.

⁂ 28. *G.* DIFFU'SA *Willd.* The diffuse Genista.

Identification. Willd. Sp., 3. p. 942.; Dec. Prod., 2. p. 152.; Don's Mill., 2. p. 153.
Synonymes. G. humifùsa *Wulf.* in *Jacq. Coll.* 2. p. 169.; Spártium procúmbens *Jacq. Icon. Rar.* 3. t. 555., but not of Aiton.
Engravings. Jacq. Icon. Rar., t. 555.; and our *fig.* 334.

Spec. Char., &c. Branches procumbent from the neck, triquetrous. Leaves lanceolate, and smooth, a little ciliated. Peduncles axillary, erect, and disposed in interrupted fascicles. Corollas and legumes glabrous. (*Dec. Prod.*) A procumbent shrub. Italy and Styria, in exposed places. Height 6 in. Introduced in 1815. Flowers yellow; May and June. Legume brown; ripe in August.

334. Genista decúmbens.

⁂ 29. *G.* PROSTRA'TA *Lam.* The prostrate Genista.

Identification. Lam. Dict., 2. p. 618.; Dec. Prod., 2. p. 152.; Don's Mill., 2. p. 153.
Synonymes. G. pedunculàta *L'Hérit. Stirp.* 184.; G. decúmbens *Dur. Bourg.* 1. p. 299.; G. Hálleri *Reyn. Mem.* 1. p. 211. icon.
Engravings. Lodd. Bot. Cab., 718.; and our *figs.* 335, 336.

Spec. Char., &c. Stems diffuse, prostrate. Branches angular, striated, rather hairy. Leaves ovate-oblong, somewhat hairy beneath, Flowers axillary, on long erect pedicels. Corolla glabrous. Legumes hairy, 3—4-seeded. (*Dec. Prod.*) A prostrate shrub. Burgundy, and the Alps of Jura. Height 1 ft. Introduced in 1775. Flowers yellow; May and June. Legume brown; ripe in August.

335. Genista p ostràta

336. Genista prostràta.

⁂ 30. *G.* PROCU'MBENS *Waldst. et Kit.* The procumbent Genista.

Identification. Waldst. et Kit. in Willd. Sp., 3. p. 940., Dec. Prod., 2. p. 152.; Don's Mill., 2. p. 153.
Engravings. Bot. Reg., t. 1150.; and our *fig.* 337

Spec. Char., &c. Branches procumbent, round, striated, rather downy. Leaves lanceolate, acute, and, as well as the calyxes, downy beneath. Flowers pedicellate, axillary, in threes. Corolla glabrous. (*Dec. Prod.*) A procumbent shrub. Hungary and Moravia. Height 1 ft. Introduced in 1816. Flowers yellow; June to August. Legume brown; ripe in September.

337. G. procúmbens

Most likely only a variety of the preceding species.

⁂ 31. *G.* PILO'SA *Lin.* The hairy Genista.

Identification. Linn. Sp. 999.; Smith's Eng. Fl., 3. p. 263.; Hayne Abbild. der deut. Holz., p. 161.; Dec. Prod., 2. p. 152; Don's Mill., 2. p. 153.
Synonymes. G. rèpens *Lam. Fl. Fr.*; Genistöides tuberculàta *Mœnch Meth.*
Engravings. Jacq. Fl. Austr., t. 208.; Hayne Abbild., t. 120.; and our *fig.* 338.

Spec. Char., &c. Stems procumbent, striated, branched, tuberculated. Leaves obovate-lanceolate, obtuse, folded, and having beneath a close-pressed silky down. Flowers axillary, on short pedicels. Calyx and pedicels silky. Legumes pubescent, and 3—4-seeded. (*Dec. Prod.*) A procumbent shrub. South of France, Switzerland, Germany, &c.; and Britain, on dry elevated downs or heaths, in Suffolk, Cornwall, and North Wales. Height 1 ft. Flowers yellow; May and June. Legume brown; ripe in Sept.

The specific name, pilòsa, is certainly not very appropriate, for there are other species, such as *G.* cándicans, much more hairy.

338. *Genista pilòsa.*

Other Species of Genísta.— G. *spinòsa*, in the Hort. Soc. Garden, is a young plant with trifoliolate leaves, and the side shoots terminating in spines. There are various other names in collections, and a great many in books; but the whole genus is in such a state of confusion, that nothing can be determined with certainty respecting the species, till they are all collected together and cultivated in the same garden and examined.

Genus VII.

CY'TISUS *Dec.* The Cytisus. *Lin. Syst.* Monadélphia Decándria.

Identification. Dec. Prod., 2. p. 153.; Don's Mill., 2. p. 154.
Synonymes. Cýtisus and Spártium sp. *Lin.*, *Lam.* &c.; Cytise, *Fr.*; Bohnenbaum, *Ger.*; Citiso, *Ital.*
Derivation. From *Cythnus*, one of the Cyclades, the first of the species known having been found there.

Gen Char. *Calyx* bilabiate. *Upper Lip* usually entire; lower one somewhat tridentate. *Vexillum* ovate, large. *Carina* very obtuse, including the stamens and pistils. *Stamens* monadelphous. *Legume* compressed, many-seeded, glandless. (*Don's Mill.*)

Leaves trifoliolate, alternate, stipulate. *Flowers* of nearly all the species yellow.— Deciduous or sub-evergreen shrubs of short duration, or low trees; natives chiefly of the Middle and South of Europe.

All the species have trifoliolate leaves, and the flowers are for the most part yellow. The shrubs have the habit of Genísta or of Spártium, to both which genera they are nearly allied. They are all ornamental, some of them eminently so; and those which have their flowers in terminal racemes are decidedly more elegant than those which have them in close terminal, or in axillary heads. The wood of the laburnum is valuable in turnery and cabinet-work. All the species produce seeds in abundance, by which they are almost exclusively propagated. The species recorded in books are numerous; but, if they were all brought together, and cultivated in the same garden, we question much if a tithe of them would be found specifically distinct.

§ 1. *Alburnöìdes* Dec.

Derivation. From the word *alburnum*, signifying the white inner sap-wood of trees; and applied to this section from the flowers of the species being white.

Sect. Char. Calyx campanulate. Pod 1—4-seeded, not dilated at the upper suture. Flowers white. Leaves very few. Branches unarmed. (*Dec. Prod.*, ii. p. 153.)

■ 1. *C.* A'LBUS *Link.* The white Cytisus, *or Portugal Broom.*

Identification. Link Enum., 2. p. 241.; Dec. Prod., 2. p. 153.; Don's Mill., 2. p. 154.
Synonymes. Genista álba *Lam. Dict.* 2. p. 623.; Spártium álbum *Desf. Fl. Atl.* 2. p. 132.; Spártium multiflòrum *Ait. Hort. Kew.* 3. p. 11.; Spártium dispérmum *Mœnch Meth.* p. 130.; Genista multiflòra *N. Du Ham.* 2. p. 76.; Spartium à Fleurs blanches, *Fr.*; weisse Pfriemen, *Ger.*
Engravings. N. Du Ham., 2. t. 23.; and our *fig.* 339.

Spec. Char., &c. Branches terete, twiggy. Leaves simple, and trifoliolate, sessile. Leaflets linear-oblong, and silky. Flowers in fascicles, disposed in long racemes. Legume 2-seeded, very villous. (*Don's Mill.*) An upright shrub; evergreen, from the colour of its numerous straight parallel young shoots. Portugal and the Levant. Height 5 ft. to 7 ft. Introduced in 1752. Flowers white; May and June. Legume brown; ripe in August.

339. Cytisus álbus.

A very handsome shrub, more especially when covered with its white flowers in May, and when surrounded by hundreds of bees, busily occupied in extracting the honey. In good soil, it is of very rapid growth, attaining the height of 5 or 6 feet in 3 or 4 years; and, in 6 or 8 years, growing as high as 15 or even 20 feet, if in a sheltered situation. Placed by itself on a lawn, it forms a singularly ornamental plant, even when not in flower, by the varied disposition and tufting of its twiggy thread-like branches. When in flower, it is one of the finest ornaments of the garden. Trained to a single stem, its effect is increased; and, grafted on the laburnum, a common practice about Paris, it forms a very remarkable combination of beauty and singularity. Plants are easily raised from seeds.

Variety.

- C. *a.* 2 *incarnàtus* has flesh-coloured flowers, or flowers very slightly tinged with reddish purple. This variety was introduced in 1818; and reproduces itself from seeds, but it varies much in the quantity of colour in the flowers.

§ ii. *Labúrnum* Dec.

Derivation. A name applied by Pliny to some species of *Cýtisus*.

Sect. Char. Calyx campanulate. Pod many-seeded, not dilated at the upper suture. Flowers yellow. Branches leafy and unarmed. (*Dec. Prod.*, ii. p. 153.)

2. *C. Labu'rnum L.* The *common* Laburnum.

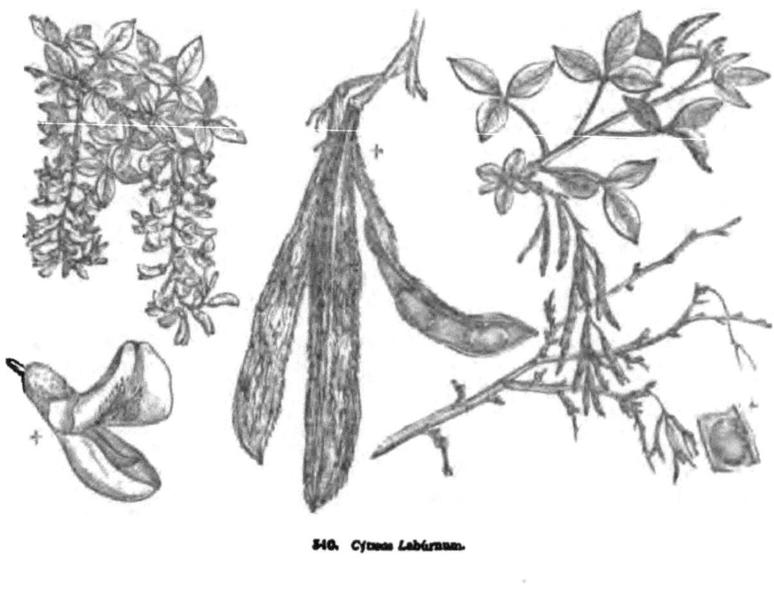

340. Cýtisus Labúrnum.

Identification. Lin. Sp., 1041.; Dec. Prod., 2. p. 153.; Don's Mill., 2. p. 154.
Synonymes. *C. alpinus Lam. Fl. Fr.* 2. p. 621.; Bean-trefoile Tree, and Peascod Tree, *Gerard*; Pea Tree, *Scotch*; Golden Chain; l'Aubours, faux E'bénier, Arbois, or Arc-Bois, *Fr.*; gemeine Bohnenbaum, *Ger.*; Aborniello, *Ital.*
Derivation. The name of L'Aubours, which is given to this tree in Dauphiné and Switzerland, is supposed by Du Hamel to be a corruption of the Latin word *laburnum*. The word Arbois is a corruption of *arc-bois*, the wood of this tree having been used by the ancient Gauls to make their bows; and being still so employed by the country people, in some parts of the Mâconnois, where these bows are found to preserve their strength and elasticity during half a century. The name of Faux E'bénier is applied to the wood, from the blackness of its heart-wood. The German name signifies Bean Tree; and both it and the English and Scotch names of Bean-trefoile and Peascod Tree have reference to the shape of the leaves and the legumes. The name of Golden Chain alludes to the length of the drooping racemes of flowers, which, as Cowper elegantly describes them, are "rich in streaming gold."
Engravings. Jacq. Aust., t. 306.; Bot. Mag., t. 176.; N. Du Ham., 5. t. 44.; the plate of this tree in Arb. Brit., 1st edit., vol. v.; and our *fig.* 340.

Spec. Char., &c. Branches terete, whitish. Leaves petiolate; leaflets ovate-lanceolate, pubescent beneath. Racemes pendulous, simple. Pedicels and calyxes clothed with closely pressed pubescence. Legume linear, many-seeded, clothed with closely pressed pubescence. A low deciduous tree. Native of Europe, on the lower mountains of the South of Germany, and of Switzerland. Height 20 ft. or upwards. Introduced in 1596. Flowers yellow; May and June. Legume dark brown; ripe in October. Decaying leaves yellow. Naked young wood green.

Varieties.

- ♈ C. L. 2 *péndulum* Hort. has slender pendulous branches.
- ♈ C. L. 3 *quercifòlium* Hort., *C. L.* 2 incìsum, has sinuated leaflets, not unlike the leaves of the common oak. (See the plate of this variety in *Arb. Brit.*, 1st edit., vol. v.; and our *fig.* 341.)

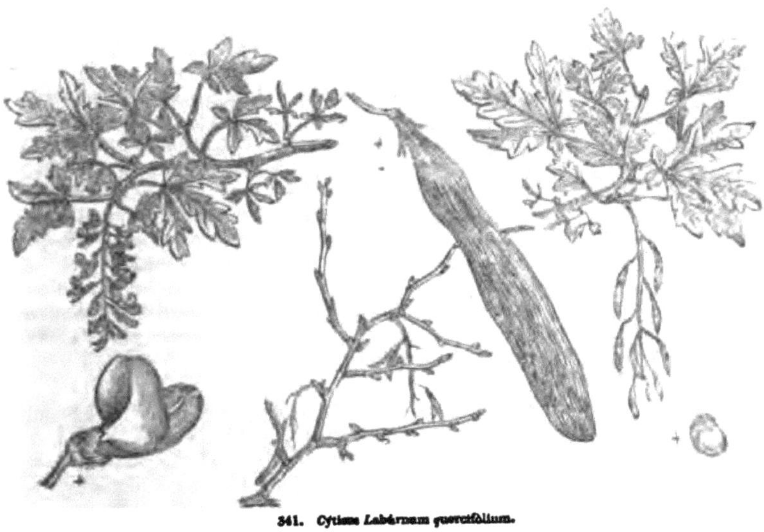

341. *Cytisus Labúrnum quercifòlium.*

- ♈ C. L. 4 *fòliis variegàtis* has variegated leaves; but it is a plant of no beauty, and rarely seen in collections.
- ♈ C. L. 5 *fràgrans* Hort. — Flowers fragrant. Wherever a number of laburnums are found in flower together, whether of this or the other species, the scent of the blossoms will be found to differ very considerably, and occasionally one may be found which may be termed fragrant; hence the origin of this variety.

♈ 3. *C.* (*L.*) ALPI'NUS *Mill.* The Alpine, *or Scotch*, Laburnum.

Identification. Mill. Dict., No. 2.; Dec. Prod. 2. p. 153.; Don's Mill., 2. p. 154.
Synonymes. *C. Labúrnum β Ait., Lam., Dec., Fl. Fr.*; Cýtisus angustifolius *Mœnch Meth.* 145.;

C. Labúrnum var. latifòlium *Pers.* and *Du Mont*; Cytise des Alpes, l'Aubours, *Fr.*; Alpen Bohnenbaum, *Ger.*; Maggio Ciondolo, *Ital.*

Engravings. Waldst. et Kit. Hung., 3. t. 260.; the plate of this tree in Arb. Brit., 1st edit., vol. v. and our *fig.* 342.

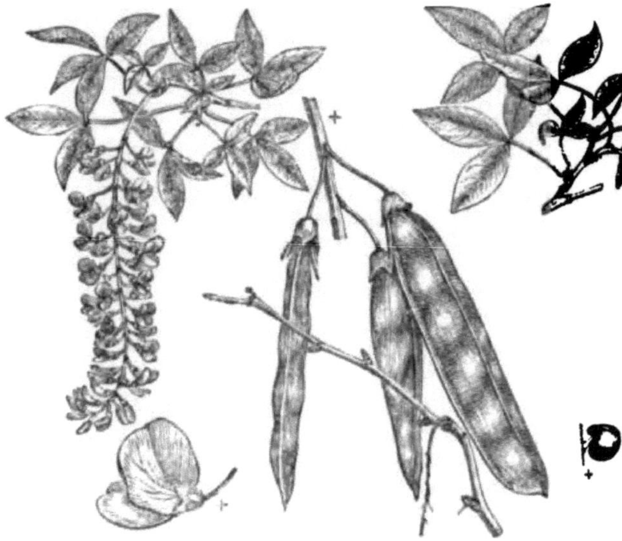

342. *Cýtisus (Labúrnum) alpìnus.*

Spec. Char., &c. Branches glabrous and terete. Leaves petiolate; leaflets ovate-lanceolate, rounded at the base. Racemes pendulous. Pedicels and calyxes puberulous. Legumes glabrous, few-seeded, marginate. (*Don's Mill.*) A deciduous low tree. Found in Carinthia, in the Alps of Jura, on Mount Cenis, and on the Apennines. According to some, it is also found wild in Scotland; but, though it is much cultivated in some parts of Fifeshire and Forfarshire, it is far from being indigenous there. Height 20 ft. to 30 ft., sometimes much higher in a state of cultivation. It was introduced into Britain about the same time as the other species, viz. 1596; and was, probably, for a long time confounded with it; for which reason we shall treat of the two species, or races, together. Flowers yellow; May and June. Legume brown; ripe in October.

Varieties.

¥ C. (L.) *a.* 2 *péndulus* has pendulous branches, and, in the foliage and legumes, seems intermediate between *C. Labúrnum* and *C. (L.)* alpìnus. This is very obvious in a fine specimen of this variety in the arboretum of the Messrs. Loddiges, as shown in the plate in *Arb. Brit.*, 1st edit., vol. v. The pendulous variety of *C. Labúrnum* is a much less robust plant.

¥ C. (L.) *a.* 3 *purpuráscens* Hort., *C. L.* purpùreum *Hort.*, C. Adàmi Poir., *C. L.* coccíneum *Baum. Cat.*, the purple Laburnum, the scarlet Laburnum, is not a hybrid between *C. Labúrnum* and *C.* purpùreus, as was at first supposed, but a sport from a bud of *Cýtisus* purpùreus inserted in *C.* alpìnus, in 1825, by D. Adam, a nurseryman at Vitry, near Paris. The flowers are of a reddish purple, slightly tinged with buff, and are produced in pendent spikes, 8 in. or more long. A few years after this sport was originated, it was found that it had a strong tendency to return to the original kinds; and that from one bud or graft, branches were produced of the true *Cýtisus* purpùreus, of the true *Labúrnum* (either the Alpine or the common, according to which of these may have been chosen as the stock) with yellow flowers, and of the purple laburnum. This was soon observed both in France and England. (See *Gard. Mag.*, vol. xii. p. 225., vol. xv. p. 122.; and *Arb. Brit.*, 1st edit., p. 590.) It is a very vigorous, and somewhat erect and fastigiate, growing variety, having produced shoots from 6 ft. to 9 ft. long in one season; but, though it has been highly spoken of by some cultivators, in point of beauty, it cannot be recommended.

♀ C. (L.) *a.* 4 *fràgrans* Hort. — Blossoms fragrant. There are plants in the Hackney Arboretum.

Miller recognised *C. Labúrnum* and *C.* alpìnus as species; but Linnæus did not. Whether they are species or varieties, they are certainly very distinct; as much so, perhaps, as the *Quércus Ròbur* pedunculàtum, and *Q. R.* sessiliflòrum, and like these two oaks they come true from seed. Both sorts, being highly ornamental, have been extensively propagated and cultivated in British gardens and plantations. There are trees at Syon of *C.* alpìnus above 40 ft. high, and some at Alnwick Castle with trunks 3 ft. in diameter. The heart-wood of the laburnum is of a dark colour; and, though of rather a coarse grain, it is very hard and durable: it will take a polish, and may be made to resemble ebony. A cubic foot weighs 52 lb. 11 oz. in a dried state. The colour and grain of the heart-wood vary much, according to the soil, and the age of the tree. It is darkest in the *C. Labúrnum*, when grown on poor calcareous soil; and lightest in the *C.* (*L.*) alpìnus, when grown in deep rich soil: in which last case its colour is a sort of greenish black. It is in much demand among turners and cabinet-makers. The ordinary use of the wood in the North of Scotland, is to form alternate staves with the wood of the holly, or the spindle tree, in making small noggins, or bickers; but it is also used for the bowls of punch-ladles; for flutes, and other musical instruments. Hares and rabbits being remarkably fond of the bark of the laburnum, it has been suggested to sow laburnum seeds, in order to produce an undergrowth in plantations liable to be infested with these animals; for, though the plants are eaten to the ground every winter, yet they will spring up again the next season, and thus yield a regular supply of winter's food for these kinds of game. As an ornamental tree, the laburnum has few rivals. The shape of the head is irregular and picturesque; its foliage is of a smooth, shining, and beautiful green; and, what is a great recommendation to every ornamental plant, it is not liable to be preyed on by insects. Though the laburnum will grow in a very indifferent soil, it requires a deep fertile sandy loam to attain a large size. In regard to situation, as the tree puts out few horizontal roots, and has rather a spreading head, when it grows rapidly it is apt to be blown aside by high winds; but, for the same reason, it is less injurious to plants growing near it, than some other ornamental trees. For producing timber, it should be placed in masses in a sheltered situation, or in a plantation among other trees, so as to be drawn up with a clear straight stem; and when so circumstanced, in good soil, *C.* (*L.*) alpìnus will grow to the height of from 35 ft. to 45 ft. Both *C. Labúrnum* and *C.* (*L.*) alpìnus are invariably raised from seed, and the pendulous and other varieties are propagated by grafting or budding on either of the common sorts. The seeds are fit to gather in October; and they may be kept in the pod, in a dry airy loft, till the March following, when they should be sown in beds of light soil, at about an inch apart every way, and covered about half an inch or three quarters of an inch thick. Half the plants which come up will be fit for transplanting into nursery lines in the November following.

● 4. *C.* WELDENII Vis. Welden's Cytisus.

Identification. Visiani Pl. Dalm. Ex. Bot. Zeit., Jan. 1830., p. 52.; Don's Mill., 2. p. 155.; Hort. Fl. Austr., 2. p. 339.; Bot. Reg., 1839, Month. Reg., No. 122.
Engraving. Our *fig.* 343. from a drawing kindly sent us by the late Baron Jacquin.

Spec. Char., &c. Erect. Leaves ternate, petiolate; leaflets elliptic, entire, cuneated at the base, and obtuse at the apex, smooth. Racemes terminal, stalked, pyramidal, straight; pedicels hoary and villous. Calyxes campanulate, 3-lobed; lobes tomentosely ciliated. Corolla glabrous, but the carina is clothed with silky villi. Legume glabrous, mucronate by the style. (*Don's Mill.*) An erect woody shrub, resembling a laburnum. Dalmatia, in woods on mountains. Height 2 ft. to 3 ft.; 6 ft. to 8 ft. in gardens. Introduced in 1837. Flowers yellow, fragrant; June and July. Legumes brown; ripe in October.

343. *Cytisus Weldènii.*

The racemes are erect, and do not droop even when in fruit. The seeds are still more poisonous than those of the common laburnum, and the scent of the flowers causes headach. The milk of the goats which feed upon the flowers, Baron Welden observes, produces the same effect, only more severely, upon those who drink it.

♠ 5. *C.* NI'GRICANS *L.* The black Cytisus.

Identification. Lin. Sp., 1041.; Dec. Prod., 2. p. 153.; Don's Mill., 2. p. 155.
Engravings. Jacq. Austr., t. 378.; Bot. Reg., t. 802.; and our *fig.* 344.

Spec. Char., &c. Branches round, twiggy. Leaves stalked, and clothed with closely pressed down beneath, as well as the branches, calyxes, and pods; leaflets elliptic. Racemes elongated, terminal, erect. Calyxes without bracteas. (*Dec. Prod.*) A handsome deciduous shrub. Piedmont, Vallais, and Bohemia. On hills and along waysides. Height 3 ft. to 6 ft. Introduced in 1730. Flowers yellow; June and July. Legume black; ripe in October. The whole plant turns black when drying; whence the specific name.

344. *Cytisus nigricans.*

It ripens seed in abundance; and it may also be propagated by grafting on *C. Labúrnum*, thus forming a handsome standard.

♠ 6. *C.* SESSILIFO'LIUS *L.* The sessile-leaved Cytisus.

Identification. Lin. Sp., 1041.; Dec. Prod., 2 p. 153.; Don's Mill., 2. p. 155.
Engravings. Lam. Ill., t. 618. f. 2.: Bot. Mag., t. 255.; and our *figs.* 345. and 346.

Spec. Char., &c. The whole plant quite smooth. Branches round. Floral leaves almost sessile, and leaflets ovate. Racemes terminal, short, and erect; each calyx having a 3-leaved bractea under it. (*Dec. Prod.*) A shrub, with upright branches, and smooth shining leaves. Native of France and Piedmont. Height 4 ft. to 7 ft. Introduced in 1569. Flowers yellow; May and June. Legume black; ripe in October.

In very general cultivation in British gardens, generally as a bush, but some-

345. Cytisus sessilifolius.

times grafted standard high on the laburnum; when it forms a very formal, symmetrical, round-headed, small tree, which, however, is highly beautiful when in flower. We have given two figures of this species, both drawn to the same scale, to show how much it varies in the magnitude and general appearance of its foliage, according to soil and situation. Plants grafted standard high are common in the London nurseries.

346. Cytisus sessilifolius.

7. *C.* TRIFLO'RUS *L'Hérit.* The three-flowered Cytisus.

Identification. L'Hérit. Stirp., 184.; Desf. Fl. Atl., 2. p. 139.; Dec. Prod., 2. p. 154.; Don's Mill. 2. p. 155.; Webb Iter Hispan., 51.
Synonyme. C. villòsus *Pour. Act. Toul.* 3. p. 317.
Engravings. Clus. Hist., 1. p. 94. f. 3.; Duh., t. 5. f. 452.; our *fig.* 347.

Spec Char., &c. The whole plant hairy. Branches round. Leaves petiolate; leaflets ovate-elliptic. Flowers axillary, pedicellate, terete, and somewhat racemose at the tops of the branches. (*Dec. Prod.*) A straggling hairy shrub, closely resembling *C.* capitàtus and *C.* hirsùtus. South of France, Italy, Sicily, and Mauritania. Height 3 ft. to 4 ft. Introduced in 1640. Flowers yellow; June and July. Legume black; ripe in October.

347. Cytisus triflòrus.

Frequent in gardens, sometimes grafted standard high; but neither as a standard nor as a dwarf is it of great duration. It should be planted in an airy situation.

8. *C.* PA'TENS *L.* The spreading Cytisus.

Identification. Lin. Syst. Vég. 555., according to L'Hérit. Stirp., 184.; Dec. Prod., 2. p. 154.; Don's Mill., 2. p. 155.
Synonymes. C. pendulìnus *Lin. Fil. Supp.* 328.; *C.* grandiflòrus *Dec. Prod.* 2. p. 156.; Genista tomentòsa *Poir. Supp.* 2. p. 719.; Spártium pàtens *Lin. Syst.* 535., *Brot. Fl. Lus.* 2. p. 83., but not of Cav.; Spártium grandiflòrum *Brot. Fl. Lus.*, 2. p. 80.; Sarothámnus pàtens *Webb Iter Hispan.* 51.
Engraving. Our *fig.* 348.

Spec. Char., &c. Branches striated and pubescent. Leaves trifoliolate, petiolate; the upper ones simple, and obovate, as are the leaflets; covered with closely pressed down. Flowers axillary, usually in pairs, pedicellate, nodding. Pods very hairy. (*Dec. Prod.*) A spreading shrub. Native of Portugal. Height 4 ft. to 6 ft. Introduced in 1752. Flowers yellow; June and July. Legume dark brown, or black; ripe in October.

A very handsome shrub, especially when grafted standard high, not so common in collections as it ought to be.

348. Cytisus pàtens.

9. *C.* SCOPA'RIUS *Link.* The *common* Broom.

Identification. Link Enum., 2. p. 241.; Dec. Prod., 2. p. 154.; Don's Mill., 2. p. 155.
Synonymes. Spártium scopàrium *Lin. Sp.* 998., *Smith Eng. Bot.* 1339.; Genista scopària *Lam. Dict.* 2. p. 623., but not of Vill.; *G.* hirsùta *Mœnch Meth.* 144.; Genêt à Balais, Genêt commun, *Fr.*; gemeine Pfriemen, *Ger.*
Engravings. Œd. Fl. Dan., t. 313.; Smith Engl. Bot., t. 1339.; and our *fig.* 349.

Spec. Char., &c. Branches angled, glabrous. Leaves petioled, trifoliolate; the uppermost simple; these and the leaflets oblong. Flowers axillary, pediceled, solitary. Legumes pilose at the margins. (*Dec. Prod.*) A shrub, evergreen from the colour of its numerous young shoots. Native of dry sandy or gravelly soils, throughout Europe. Height 3 ft. to 12 ft. according to the soil and situation. Flowers large, yellow; May and June. Legume black; ripe in September. Naked young wood green.

Varieties.

■ C. s. 2 *álbus* Hort. has the flowers white, or of a very pale yellow.
■ C. s. 3 *flòre plèno* Hort. has flowers slightly double.

The roots are straight, and penetrate perpendicularly to a great depth. The leaves are trifoliolate or simple; the branches numerous, long, straight, angular, dark green, smooth, and tough. The flowers are of a deep golden yellow, sometimes tinged with orange, and occasionally of a uniform pale lemon colour: they are succeeded by pods above an inch long, black when ripe, and each containing 15 or 16 seeds. The flowers are larger than those of any other species of the genus; and, were the plant not so common in a wild state, it would, doubtless, be considered the most ornamental. The whole plant is exceedingly tough, and bitter to the taste, and has a strong disagreeable smell. Though it is at present comparatively neglected, yet in former times it was one of very great importance in rural and domestic economy. The branches are eaten by sheep and cattle; and, on poor gravelly soils, formed, before the general improvement of grass lands which has taken place within the last century, the principal herbage. One of the principal modern uses of the broom, both in Britain and on the Continent, is to form brooms, or besoms; for which purpose, as the specific name would imply, it appears to have been used from time immemorial. The young shoots were formerly used as a substitute for hops in brewing beer; and the flower-buds, just before they become yellow, were pickled in the manner of capers. The tops and leaves are purgative and diuretic. In the North of Scotland, a decoction of the recent shoots is used by shepherds, for dressing the backs of sheep, instead of tobacco water. The broom produces abundance of seeds, which, according to M. Hartig, retain their germinating quality for a very long time: some that he kept 25 years, in a room which was occupied, having come up as readily as new seed.

849. Cytisus scoparius.

§ iii. *Calycótome* Link.

Derivation. From *kalyx*, a calyx, and *tomē*, a cutting; in reference to the calyx, the upper part of which, after some time, falls off, in such a manner as to give the remainder the appearance of being cut round.

Sect. Char. Calyx campanulate, somewhat bilabiate, at length becoming truncate. Pod thickened on the upper suture. Shrubs with spiny branches and yellow flowers. (*Dec. Prod.*)

■ 10. *C.* SPINO'SUS *Lam.* The spiny Cytisus.

Identification. Lam. Dict., 2. p. 247.; Dec. Prod., 2. p. 154.; Don's Mill., 2. p. 185.; Webb Iter Hispan., 51.
Synonyme. Spártium spinòsum *Lin. Sp.* 997.
Engravings. J. Bauh. Hist., 1. p. 2. p. 376., icon.; Lob. Icon., 2. t. 95.; and our *fig.* 850.

Spec. Char., &c. Branches angled, spiny. Leaves trifoliolate; leaflets obovate-oblong. Legumes perfectly smooth. (*Dec. Prod.*) An upright spiny shrub. Upon hills and rough places from Perpignan to Genoa, in Corsica, and in the Algerine country. Height 2 ft. to 10 ft. Introduced in 1596. Flowers yellow; June and July. Legume black; ripe in October.

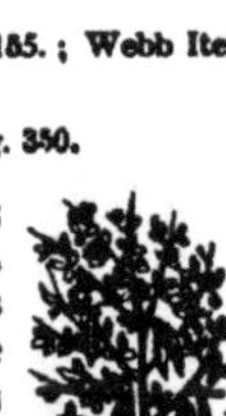
850. C. spinòsus.

There are plants in the Hort. Soc. Garden.

■ 11. *C.* TRIBRACTEOLA'TUS *Webb.* The three-bracted Cytisus.

Identification. Iter Hispan., p. 51.; Otia Hispan., p. 3.
Engravings. Otia Hispan., t. 3. and our *fig.* 351.

351. Cytisus tribracteolàtus.

Spec. Char., &c. Decumbent. Branches tetragonal, divaricate, rigid, obtuse. Leaves trifoliolate, verticillately sub-opposite; leaflets ovate-elliptic, slightly obtuse at the apex, retuse, with ash-coloured silky down, petiolulate. Common petiole none, or cohering with the branch. Flowers axillary, clustered, pedunculate. Calyx bilabiate, hairy; upper lip cut to the middle in narrow acute segments; lower lip longer, narrow, 3-toothed; middle tooth longest, supported at the base by three ovate closely pressed bracts. (*Webb, Otia Hisp.*) A decumbent shrub, evergreen from the colour of its bark. Spain near Medina Sidonia, on the summits of mountains. Height 1 ft. to 3 ft. Introduced in 1838. Flowers yellow; May. Legume ?.

■ 12. *C.* LANI'GERUS *Dec.* The wool-bearing Cytisus.

Identification. Dec. Prod., 2. p. 154.; Don's Mill., 2. p. 155.; Webb Iter Hispan., 51.
Synonymes. Spártium lanígerum *Desf. Fl. Atl.* 2. p. 135.: Calycótome villòsa *Link Enum.*; Spártium villòsum *Brot. Fl. Lus.* 2. p. 85., and *Poir. Voy.* 2. p. 207.
Engraving. Our *fig.* 352. from a specimen in the British Museum.

352. C. lanígerus.

Spec. Char., &c. Branches furrowed, spiny. Leaves trifoliolate; leaflets obovate-elliptical. Legumes very hairy in a woolly manner. (*Dec. Prod.*) A spiny shrub. Found wild on hills and in rough places in Corsica, Crete, the Archipelago, Mauritania, Gibraltar, and Portugal. Height 2 ft. to 10 ft. Introduced in 1821. Flowers yellow; June and July. Legume dark brown, or black; ripe in October.

Variety.

■ C. *l.* 2 *rígidus* Dec. — Spines very strong.

Not common in collections, and in all probability it is nothing more than a variety of the preceding species.

§ iv. *Tubocýtisus* Dec.

Derivation. From *tubus*, a tube, and *cytisus*; in reference to the tubular shape of the calyx.

Sect. Char. Calyx tubular, with the apex toothed-lipped. Thornless shrubs. (*Dec. Prod.*, ii. p. 155.)

A. *Flowers white or whitish.*

■ 13. *C.* LEUCA'NTHUS *Waldst. et Kit.* The white-flowered Cytisus.

Identification. Waldst. et Kit., 2. p. 141.; Dec. Prod., 2. p. 155.; Don's Mill., 2. p. 156.
Engravings. Bot. Mag., t. 1438.; and our *fig.* 353.

353. C. leucánthus.

Spec. Char., &c. Stem erect. Branches round, and, as well as the leaves, clothed with closely pressed pubescence. Leaflets elliptic and acute. Flowers at the points of the branches; heads of flowers bracteated by two leaves. (*Dec. Prod.*) A downy shrub. Croatia, in woods. Height 3 ft. to 4 ft. Introduced in 1806. Flowers yellowish white; June and July. Legume black; ripe in October.

Very ornamental, and well deserving a place among other species of the genus. It forms a handsome object grafted standard high.

B. *Flowers purple.*

☆ 14. *C.* PURPU'REUS *Scop.* The purple-*flowered* Cytisus.

Identification. Scop. Carn., No. 905. t. 43.; Dec. Prod., 2. p. 155.; Don's Mill., 2. p. 156.
Engravings. Jacq. Aust. Append., t. 48.; Lodd. Bot. Cab., t. 892.; Bot. Mag., t. 1176.; and our *figs.* 354, 355.

Spec. Char., &c. Stems procumbent, twiggy. Leaves, calyxes, and legumes glabrous. Leaflets oblong. Flowers axillary, solitary, on short pedicels. (*Don's Mill.*) A procumbent shrub. Native of Carniola in exposed places. Height 1 ft. Introduced in 1792. Flowers purple; May to August. Legume black; ripe in October.

355. *C. purpùreus.*

Varieties.

☆ C. *p.* 2 *flòre álbo* Hort. has the flowers of a pure white.

☆ C. *p.* 3 *flòre ròseo.*—Flowers rose-coloured. Plants in the Horticultural Society's Garden.

354. *C. purpùreus*

Of all the different species of *Cý-tisus*, when grafted on the laburnum standard high, this forms the most graceful tree; and a plant of it covered with its purple flowers, placed on a lawn, or in a border near a standard of *Genísta tríquetra*, covered with its golden yellow flowers, will produce a very striking effect. The singular hybrid or sport formed between this plant and the laburnum has been already described, p. 216.

C. *Flowers yellow.*

▬ 15. *C.* ELONGA'TUS *Waldst. et Kit.* The elongated Cytisus.

Identification. Waldst. et Kit. Hung., 2. p. 200. t. 183.; Dec. Prod., 2. p. 155.; Don's Mill., 2. p. 155.
Engravings. Waldst. et Kit. Hung., t. 183.; and our *fig.* 356.

Spec. Char., &c. Stems erect. Branches elongated and round; young ones hairy. Leaflets obovate, clothed beneath with closely pressed hairs. Flowers lateral, usually in fours, on short pedicels. Calyxes hairy. (*Dec. Prod.*) An erect pubescent shrub. Native of Hungary, in woods. Height 3 ft. to 4 ft. Introduced in 1804. Flowers yellow; May and June. Legume dark brown, or black; ripe in September.

In H. S. Garden, and at Messrs. Loddiges's.

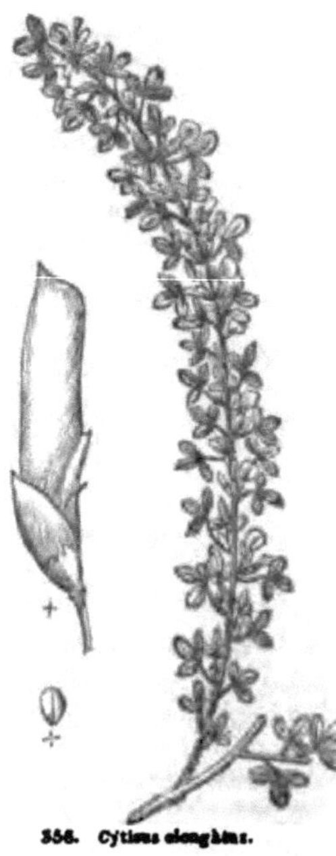

356. *Cytisus elongàtus.*

▬ 16. *C.* MULTIFLO'RUS *Lindl.* The many-flowered Cytisus.

Identification. Lindl. Bot. Reg., t. 1191.; Don's Mill., 2. p. 156.
Synonymes. *C.* elongàtus *Hortul.*, but not of Kit.; *C.* elongàtus β multiflòrus *Dec. Prod.* 2. p. 155.
Engravings. Lindl. Bot. Reg., t. 1191.; and our *fig.* 357.

Spec. Char., &c. Stems erect. Branches elongated, terete, younger ones villous. Leaflets oblong, tapering to the base, villous beneath, and of the same colour on both surfaces. Flowers usually ternary. Pedicels about equal in length to the petioles

Vexillum emarginate, undulated. (*Don's Mill.*) A downy shrub. Native of Europe. Height 2 ft. to 3 ft. Introduced in 1800. Flowers yellow; May and June. Legume black; ripe in September.

It appears to us to be only a variety of the preceding species.

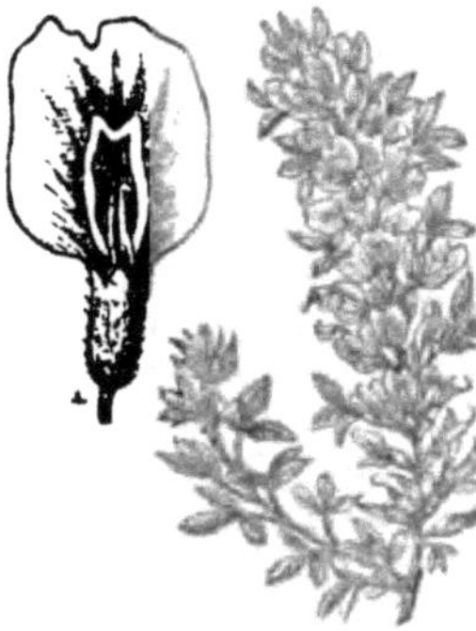
357. Cytisus multiflòrus.

358. Cytisus falcàtus.

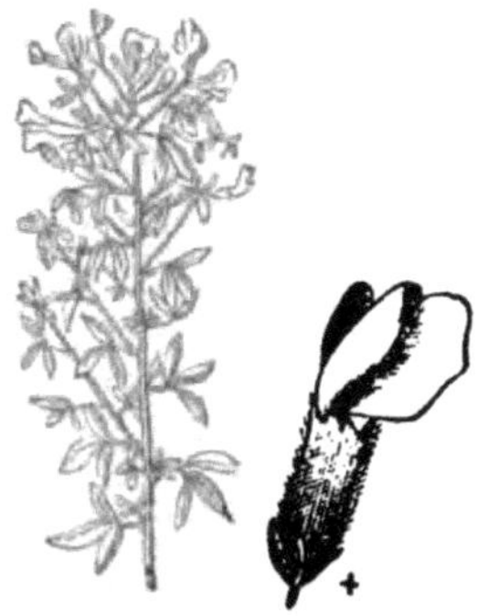
359. Cytisus austrìacus.

17. *C.* FALCA'TUS *Waldst. et Kit.* The sickle-*like-podded* Cytisus.

Identification. Waldst. et Kit. Hung., 3. p. 264.; Dec. Prod., 2. p. 155.; Don's Mill., 2. p. 156.; Lod.
Engravings. Lodd. Bot. Cab., t. 520.; Waldst. et Kit. Hung., 3. t. 238.; and our *fig.* 358.

Spec. Char., &c. Stems declinate Branches round and twiggy; the young ones, as well as the leaves, clothed with closely pressed hairy down. Petioles hairy. Flowers usually in threes, lateral, and on short peduncles. Calyxes clothed with closely pressed hairs. (*Dec. Prod.*) A downy shrub. Native of Croatia, the South of Russia, and Gallicia. Height 2 ft. to 4 ft. Introd. 1816. Flowers yellow; June to August. Legume black; ripe in October.

Varieties. C. triflòrus *Lod.*, C. ruthénicus *Lod.*, C. decúmbens *Lod.*, are apparently all varieties of this species.

18. *C.* AUSTRI'ACUS *L.* The Austrian Cytisus.

Identification. Lin. Sp., 1042.; Dec. Prod., 2. p. 156.; Don's Mill., 2. p. 156.
Engravings. Mill. Icon., 117. f. 2.; Pall. Itin., ed. Gal., t. 100. f. 3.; Jacq. Austr., t. 21.; and our *fig.* 359. above.

Spec. Char., &c. Stems upright. Branches round and twiggy, and, as well as the leaves, clothed with closely pressed strigose pubescence. Leaflets lanceolate, attenuated at both ends. Flowers terminal, somewhat umbellate. Calyxes and legumes rather hairy. (*Dec. Prod.*) An upright downy shrub. Found in woods and rough places in Austria, Upper Italy, the Ukraine, and Siberia. Height 2 ft. to 4 ft. Introduced in 1741. Flowers yellow; July to September. Legume black; ripe in November.

Variety.

C. *a.* 2 *nòva* Lod. has the leaves much smaller than the species, and seems to be an erect, and very distinct variety.

19. *C.* SUPI'NUS *Jacq.* The supine Cytisus.

Identification. Jacq. Fl. Austr., 1. t. 20.; Dec. Prod., 2. p. 156.; Don's Mill., 2. p. 156.
Synonyme. *C. lotöides Pour. Act. Toul.* 3. t. 318.
Engravings. Clus. Hist., p. 96., No. 7., icon.; Jacq. Fl. Austr., 1. t. 20.; and our *fig.* 360.

Spec. Char., &c. Stems branched and decumbent. Branches round, and, when young, rather hairy; adult ones smooth. Leaflets obovate, hairy beneath. Flowers 2—4, usually terminal and pedunculate. Calyxes and pods slightly hairy.

360. Cytisus supìnus.

(*Dec. Prod.*) A decumbent hairy shrub. Native of Belgium, Austria, Pannonia, Siberia, Turkey, and Dauphiné, both on exposed hills, and in sheltered bushy places. Height 1 ft. Introduced in 1755. Flowers pale yellow, with the standard reddish; May to August. Legume dark brown, or black; ripe in November.

20. *C.* HIRSU'TUS *L.* The hairy Cytisus.

Identification. Lin. Sp., 1042.; Dec. Prod., 2. p. 156.; Don's Mill., 2. p. 156.
Synonymes. *C.* supìnus *Bertol. Pl. Gen.*, but not of Lin.; *C.* triflòrus *Lam. Dict.* 2. p. 250., but not of L'Hérit.; *C.* Tournefortiànus Loisel. in N. Du Ham., 5. p. 157.
Engraving. Our *fig.* . in p. .

Spec. Char., &c. Stems decumbent. Branches round and twiggy; when young hairy, but smooth when old. Leaflets ovate, hairy beneath. Flowers lateral on very short pedicels, aggregate. Calyxes and pods hairy. (*Dec. Prod.*) A decumbent hoary shrub. Found in rugged places from Genoa to Hungary. Height 1 ft. Introduced in 1739. Flowers yellow; June to August. Legume black; ripe in October. Closely resembling the *C.* capitàtus, and *C.* triflòrus of Loddiges's arboretum.

21. *C.* CAPITA'TUS *Jacq.* The headed-*flowered* Cytisus.

Identification. Jacq. Fl. Austr., t. 33.; Dec. Prod. 2. p. 156.; Don's Mill., 2. p. 156.
Synonymes. *C.* hirsùtus *Lam. Dict.* 2. p. 250.; *C.* supìnus *Lin. Sp.* 1040.
Engravings. Lodd. Bot. Cab., t. 497.; and our *fig.* 361.

Spec. Char., &c. Stems and branches erect, the latter hispid. Leaflets ovate-elliptic, hairy. Flowers numerous, and forming heads at the points of the branches; but sometimes lateral in the autumn. Calyxes and pods covered with short hairs. (*Dec. Prod.*) An upright hoary shrub. Found wild on the edges of woods in Burgundy, Italy, and Austria. Height 2 ft. to 4 ft. Introduced in 1774. Flowers yellow; June and July. Legume dark brown, or black; ripe in October.

361. *C.* capitàtus.

Varieties or Synonymes. Cýtisus austrìacus *Lod.*, *C.* canéscens *Fisch. of Göt.*, *C.* uralénsis *Lod.*, *C.* calycìnus *Lod.*, *C.* parvifòlius *Lod.*, *C.* hirsùtus *Lod.*, *C.* supìnus *Lod.*, appear to be all varieties of *C.* capitàtus, or in some cases, perhaps, identical with that species.

22. *C.* CILIA'TUS *Wahlenb.* The ciliated-*podded* Cytisus.

Identification. Wahlenb. Fl. Carp., 219.; Dec. Prod., 2. p. 156.; Don's Mill., 2. p. 156.
Engraving. Our *fig.* . in *fig.* .

Spec. Char., &c. Stems upright, Branches smooth when old, but when young hispid. Leaflets obovate, clothed beneath with closely pressed hairs, Flowers approximate in threes, at length lateral. Pods glabrous and ciliated. (*Dec. Prod.*) A hispid shrub. Native of the Carpathian Mountains. Height 2 ft. to 4 ft. Introduced in 1817. Flowers yellow; June and July. Legume black; ripe in October.

23. *C.* POLY'TRICHUS *Bieb.* The many-haired Cytisus.

Identification. Bieb. Fl. Taur. Suppl., 477.; Dec. Prod., 2. p. 156.; Don's Mill., 2. p. 156.
Engraving. Our *fig.* . in p. .

Spec. Char., &c. Stems declinate. Branches hispid. Leaflets obovate-elliptic. Flowers lateral, usually in pairs, pedicellate. Calyxes and pods hairy. (*Dec. Prod.*) A recumbent shrub. Found in pine forests, on high mountains, in Tauria. Height 2 ft. to 4 ft. Introduced in 1818. Flowers yellow; June and July. Legume black; ripe in October. DeCandolle observes of it, that it has the hairiness of *C.* capitàtus, the disposition of the flowers of *C.* hirsùtus, and the habit of *C.* supìnus.

§ v. *Lotöides* Dec.

Derivation. From *lotos*, the lotus, and *eidos*, appearance; from the general resemblance of the species to the genus *Lotus*.

Spec. Char., &c. Tube of the calyx short, obconical; the upper lip 2-parted, the lower 3-toothed. Corolla hardly longer than the calyx. Many-stemmed decumbent shrubs, deciduous, with few flowers, generally capitate and terminal, and all yellow. (*Dec. Prod.*)

⁎ 24. *C.* ARGE'NTEUS *L.* The silvery Cytisus.

Identification. Lin. Sp., 1043.; Dec. Prod., 2. p. 156.; Don's Mill., 2. p. 157.
Synonyme. Lotus argénteus *Brot. Fl. Lus.* 2. p. 119.
Engravings. Lob. Icon., 2. p. 41. f. 2.; and our *fig.* 362.

362. *C. argénteus.*

Spec. Char., &c. Stems decumbent. Leaves, calyxes, corollas, and pods clothed with a closely pressed silky down. Leaves petiolate, trifoliolate; leaflets oblong-lanceolate. Flowers 3—4, produced at the points of the shoots. (*Dec. Prod.*) A decumbent shrub. Native of Carniola, the South of France, and Mauritania. Height 1 ft. Introduced in 1739. Flowers yellow; August. Legume black; ripe in October.

A silky silvery-looking shrub, from the prevalence of closely pressed silky down over all its parts; noticed in the specific character, and whence it derives its specific name.

⁎ 25. *C.* CALYCI'NUS *Bieb.* The *large*-calyxed Cytisus.

Identification. Bieb. Fl. Taur., 2. p. 166.; Dec. Prod., 2. p. 157.; Don's Mill., 2. p. 157.
Synonyme. C. pauciflòrus *Willd. Sp.* 3. p. 1126.
Engravings. Lodd. Bot. Cab., t. 673.; and our *fig.* 363.

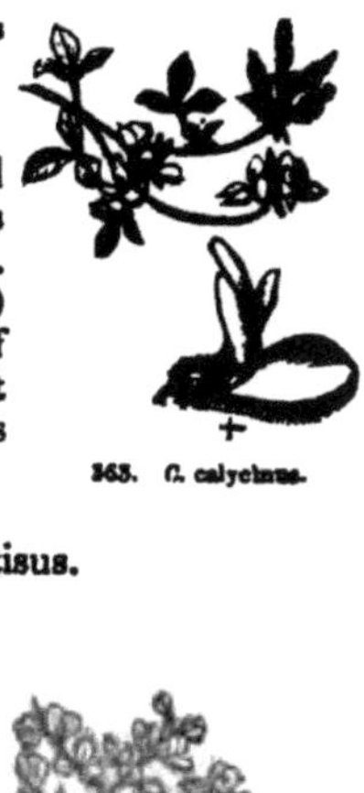
363. *C. calycinus.*

Spec. Char., &c. Stems ascending. Leaves, calyxes, and pods somewhat hairy from spreading down. Leaves trifoliolate and petiolate. Leaflets roundish, obovate. Flowers terminal, from 2 to 8 together. (*Dec. Prod.*) A prostrate shrub, with trailing branches, the ends of which grow upright. Found in stony places on Mount Caucasus. Height 1 ft. Introduced in 1820. Flowers yellow; August. Legume black; ripe in October.

⁎ 26. *C.* NA'NUS *Willd.* The dwarf Cytisus.

Identification. Willd. Enum., 769.; Don's Mill., 2. p. 157.
Engravings. Wats. Dend. Brit., t. 81.; and our *fig.* 364.

364. *Cytisus nànus.*

Spec. Char., &c. Stems cylindrical. Leaves trifoliolate, obovate, clothed with strigose pubescence beneath, and smooth above. Raceme terminal, secund, usually 4-flowered. Calyx deeply 3-parted; hairs on the stems and peduncles adpressed. (*Don's Mill.*) A procumbent shrub. Native of the Levant. Height 1 ft. Introd. in 1816. Flowers yellow; June and July. Legume black; ripe in October.

This is a beautiful little shrub for rockwork; and if planted in dry sandy soil, covered with broad flat stones to retain the moisture during the hot weather of July, it will continue flowering during the whole of that month, and produce abundance of seeds; which may be sent to any distance in the pods.

§ vi. *Chronánthus* Dec.

Derivation. From *chronos*, a year, and *anthos*, a flower; applied to this section because the petals remain attached to the calyxes all the year.

Sect. Char. Calyx with the upper lip bifid, and the lower one trifid; lobes acute, of the same length as the tube. Petals permanent. Legume oval, much compressed, 2-seeded. (*Dec. Prod.*, ii. p. 157.)

⁂ 27. C. ORIENTA'LIS *Lois.* The Oriental Cytisus.

Identification. Lois. in N. Du Ham., 5. p. 156.; Don's Mill., 2. p. 157.
Synonyme. C. orientàlis, &c., *Gerard* and *Vail. Herb.*
Engravings. Pluk. Phyt., t. 31. f. 3.; and our *fig.* 365.

365. C. orientàlis.

Spec. Char., &c. Stems erect, hairy. Leaves almost sessile, trifoliolate, hairy; leaflets linear, acute. Flowers large and yellow, subterminal, on short pedicels, and few. The flowers and pods are both glabrous. Calyx hairy, more 5-cleft than bilabiate. (*Dec. Prod.*, ii. p. 157.) An erect hairy shrub. Native of the Levant. Height 2 ft. to 3 ft. Introduced in 1818. Flowers yellow, large and persistent; June and July. Legume black; ripe in October.

Other Species of Cýtisus.—This genus, in British gardens, is in such a state of confusion, that nothing can be done in it satisfactorily till all the kinds are collected together, and cultivated for two or three years till they show their flowers and fruit. Perhaps two thirds of the alleged species in the London gardens are only varieties. In the mean time, all that a cultivator can do is to procure as many kinds as he can; and in the collection of Messrs. Loddiges he will find the greater number of those above described, though some of them have been killed by the winter of 1837–8. Among the species probably hardy, by far the handsomest in point of foliage is the C. *æólicus* Guss. (Bot. Reg. t. 1902., and our *fig.* 366.) It is a tall, very hoary shrub. A native of Stromboli. Height 5 ft. to 8 ft. Introd. in 1835. Flowers pubescent, yellow, campanulate, and the young legumes glabrous. It has the appearance, Dr. Lindley observes, of being an intermediate species between C. *Labúrnum* and C. triflòrus. (See *Arb. Br.*, 1st ed., p. 2551.) C. *racemòsus* Marnock (Flor. Mag., vol. ii. t. 18.; and our *fig.* 367.) is a handsome shrub, of moderately robust habit; a native of the Peak of Teneriffe. Height 3 ft. to 4 ft. Introduced in 1835. The flowers are terminal, in a spike about 6 in. in length, and of a bright yellow. There were plants in the Epsom Nursery in 1838. Many other species of Cýtisus are described by authors, for which we refer to Don's *Miller*, Webb's *Iter Hispaniense*, Boissier's *Elenchus Plantarum*, Duby and DeCandolle's *Botanicon Gallicum*, and the first edition of this *Arboretum*. Many genera of ligneous plants require to be cultivated together, in the same garden, in order to settle their nomenclature: but while some of these, as Quércus, Pinus, &c., would consume the greater part of a lifetime in procuring them from the different quarters of the world, and waiting till they came into flower, the genera Genísta, Cýtisus, and Adenocárpus are almost exclusively European, and might be collected in the course of one year; while, in three years after the seeds were sown, the plants would in most cases come into flower. It is surprising, therefore, that some amateur of leisure does not undertake their arrangement.

366. Cýtisus æólicus.

367. Cýtisus racemòsus.

Genus VIII.

ADENOCA'RPUS *Dec.* The Adenocarpus. *Lin. Syst.* Monadélphia Decándria.

Identification. Dec. Fl. Fr. Supp., 549.; Lég. Mém., 6.; Prod., 2. p. 158.; Don's Mill., 2. p. 157.
Derivation. From *adēn*, a gland, and *karpos*, fruit; in reference to the legumes being beset with pedicellate glands.

Gen. Char. *Calyx* obconical, usually beset with glands, bilabiate; upper lip bipartite, lower one longer and trifid. *Carina* obtuse, enclosing the stamens and pistils. *Stamens* monadelphous. *Legume* oblong, compressed. (*Don's Mill.*)
Leaves compound, trifoliolate, alternate, stipulate, deciduous. *Flowers* yellow in all the species. — Shrubs, straggling, of short duration, somewhat evergreen from the colour of their young shoots; natives chiefly of Europe.

Branches divergent; leaves trifoliolate, with petiolar stipules, and folded leaflets, and usually grouped; flowers upon bracteolate pedicels, and disposed in terminal racemes. Culture as in *Cýtisus*, from which genus most of the species have been separated.

■ 1. A. hispa'nicus *Dec.* The Spanish Adenocarpus.

Identification. Dec. Fl. Fr. Suppl., 549.; Lég. Mém., 6.; Prod., 2. p. 158.; Don's Mill., 2. p. 158.
Synonymes. Cýtisus hispánicus *Lam. Dict.* 2. p. 248.; *C. anagýrius L'Hérit. Stirp.* 184.. *N. Du Ham.* 5. p. 149.
Engraving. Our *fig.* 368.

Spec. Char., &c. Calyx glandulose and villose; lower lip with three equal segments, that are barely longer than the upper lip. Branchlets hairy. Flowers grouped. Standard rather glabrous. (*Dec. Prod.*) A hairy shrub. Spain and Portugal, in shady and moist places. Height 2 ft. to 4 ft. Introduced in 1816. Flowers yellow; June and July. Legume dark brown, or black; ripe in October.

368. A. hispánicus.

■ 2. A. Boissie'ri Webb. Boissier's Adenocarpus.

Identification. Webb's Iter Hisp., p. 52.; Otia Hispan., p. 4.
Synonymes. A. decórticans *Boiss. Not. sur l'Abies Pinsapo*, p. 9.; Raca vieja, *Span.*
Engravings. Otia Hispanica, t. 4.; and our *fig.* 369.

Spec. Char., &c. Arborescent. Bark scaly. Branches purplish, ash-coloured, clothed with soft hairy pubescence, and with numerous leaves. Petioles elongate, terete. Leaves with adpressed pubescence. Leaflets linear, with revolute margins, somewhat obtuse. Calyx villous, the lower lip somewhat longer than the upper. Vexillum pubescent at the apex and middle. Legumes elongate, obtuse, whitish, with purple glands. Seed greenish black. (*Webb, Otia Hispan.*) A large deciduous shrub. Spain, in warm valleys of the mountains of Granada, 4500 ft. to 5000 ft. above the sea. Height 15 ft. to 20 ft. Introduced ?. Flowers reddish yellow, fragrant; June and July. Legume whitish, covered with numerous purple glands; ripe in August.

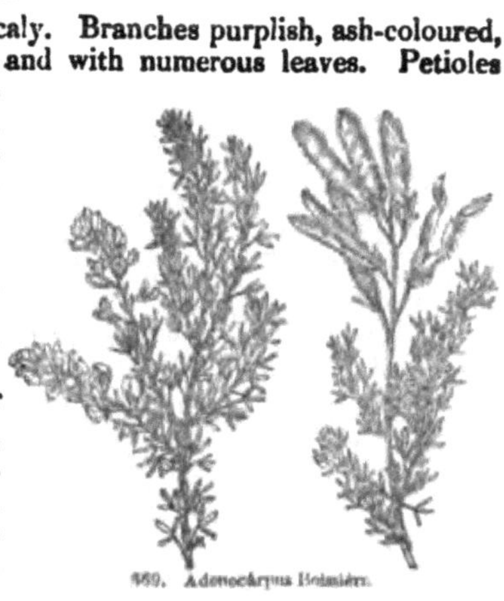
369. Adenocarpus Boissièri.

A most beautiful species when in flower; but, when out of flower, of a gloomy ashy hue. The trunk is disfigured by the old ragged bark, whence the vernacular name. It is truly astonishing, Mr. Webb observes, that this splendid European plant, of almost arboreous stature, with spikes of flowers sometimes a foot in length, should so long have escaped detection. It closely resembles *A.* hispánicus, but, according to Mr. Webb, it is "very entirely distinct." Plants are, or soon will be, in the Milford Nursery.

3. A. INTERME'DIUS *Dec.* The intermediate Adenocarpus.

Identification. Dec. Prod., 2. p. 158.; Don's Mill., 2. p. 158.
Synonyme. Cýtisus complicàtus *Brot. Fl. Lus.* 2. p. 92.
Engravings. Clus. Hist., 1. p. 94. f. 1.; and our *fig.* 370.

Spec. Char., &c. Calyx pubescent; pubescence glandulated; the middle of the three segments of the lower lip of the calyx longer than the side ones, and than the upper lip. Branchlets rather villose. Flowers rather distant. Standard rather glabrous. (*Dec. Prod.*) A pubescent shrub. Native of sunny gravelly places in Portugal and Old Castile, and Mount Scuder, in Sicily, and of Mongiana, in the kingdom of Naples. Height 4 ft. Year of introduction unknown. Flowers yellow; May to July. Legume black; ripe in September.

A very handsome species, and one that is much admired for its fine terminal spikes of flowers, which, in favourable seasons, and in a dry soil, ripen abundance of seeds.

370. A intermedius.

4. A. PARVIFO'LIUS *Dec.* The small-leaved Adenocarpus.

Identification. Dec. Lég. Mém. 6., and Prod. 2. p. 158.; Don's Mill., 2. p. 158.
Synonymes. Cýtisus parvifòlius *N. Du Ham.* 5. p. 147., *Lam. Dict.* 2. p. 248., exclusive of the synonymes; Cýtisus divaricàtus *L'Hérit. Stirp.* 184.; Cýtisus complicàtus *Dec. Fl. Fr.* No. 3821.; Spártium complicàtum *Lois. Fl. Gall.* 441
Engravings. N. Du Ham., 5. t. 47. f. 1.; and our *fig.* 371.

Spec. Char., &c. Calyx somewhat pubescent, with glandulous pubescence; the central segment of the lower lip longer than the side segments, and much exceeding the upper lip in length. Branches glabrous. Flowers distant. Standard pubescent. (*Dec. Prod.*) An erect shrub, whitish from the down on its branches. Native of sunny heaths in the West of France. Height 2 ft. to 10 ft. Introduced in 1800. Flowers yellow; May to July. Legume black; ripe in October.

371. A. parvifolius.

5. A. TELONE'NSIS *Dec.* The Toulon Adenocarpus.

Identification. Dec. Fl. Fr. Suppl. 54., Lég. Mém. 6., Prod. 2. p. 158.; Don's Mill., 2. p. 158.
Synonymes. Cýtisus telonénsis *Lois. Fl. Gall.* 446., and in *N. Du Ham.* 5. p. 155.; Spártium complicàtum *Gouan Hort. Monsp.* 356., exclusive of the synonyme.
Engravings. N. Du Ham., 5. t. 47. f. 2.; and our *fig.* 372.

Spec. Char., &c. Calyx not glandulose, pubescent; the segments on the lower lip nearly equal, exceeding a little the upper lip in length. Branches almost glabrous. Flowers distant. Standard pubescent. (*Dec. Prod.*) An erect shrub. Native of sterile places and heaths in the Pyrenees, in Cevennes, in Provence, and in Rome. Height 2 ft. to 4 ft. Introduced in 1800. Flowers yellow; June and July. Legume dark brown, or black; ripe in October.

It well deserves a place in British gardens; where, when judiciously treated, it will, owing to the moisture of our climate, attain double the height that it does in the South of France.

372. A. telonénsis.

Genus IX.

ONO'NIS *L.* The Restharrow. *Lin. Syst.* Monadélphia Decándria.

Identification. Lin. Gen., No. 863.; Lam. Ill., t. 616.; Dec. Prod., 2. p. 158.; Don's Mill., 2. p. 158.
Synonymes. Anònis and Nàtrix *Mœnch Méth.* 157. and 158.; Arrête-bœuf, or sometimes Bugrane, *Fr.*; Hauhechel, *Ger.*
Derivation. Said to be from *onos*, an ass; because only asses would feed upon so prickly a plant. Restharrow is a corruption of *arrest*, that is, stop, *harrow*; from the long and deeply seated roots opposing a serious impediment to the plough or harrow.

Gen. Char. *Calyx* campanulate, 5-cleft, with linear segments. *Vexillum* large, striated. *Stamens* monadelphous, the tenth one sometimes almost free. *Legume* usually turgid, sessile, few-seeded. (*Don's Mill.*)

Leaves trifoliolate, stipulate, alternate, deciduous. *Flowers* yellow, purplish and red, or rarely white. — Shrubs, very low, suffruticose; natives of Europe. Two species are hardy.

The peduncle is, in many instances, furnished with an awn, which is the petiole of an abortive floral leaf. The two specimens here described are well adapted for rockwork or flower-borders, on account of their lively flowers, which are red, or reddish purple; colours not frequently met with in the ligneous Leguminàceæ, by far the greater part of which have yellow flowers. They are readily propagated by seeds or by division, and will grow in any soil that is tolerably dry.

■ 1. *O.* fruticò'sa *L.* The shrubby Restharrow.

Identification. Lin. Sp., 1010.; Dec. Prod., 2. p. 167.; Don's Mill., 2. p. 160.
Engravings. N. Du Ham., 1. t. 58.; Mill. Icon., t. 36.; Bot. Mag., t. 317; and our *fig.* 373.

Spec. Char., &c. Leaves trifoliolate. Leaflets sessile, lanceolate, serrated. Stipules connate into one, sheathing, and 4-awned; and, in the uppermost parts of the plant, occupying the places of leaves which are absent. Pedicels 3-flowered, disposed in a raceme. (*Dec. Prod.*) A low shrub. Alps of Dauphiné, &c. Height 1 ft. to 4 ft. Introduced in 1680. Flowers purplish red; May and June. Legume brown; ripe in September.

Variety.

■ O. *f.* 2 *microphýlla* Dec., *O.* fruticòsa *Asso.* — Leaflets small, obovate, and serrated. Native of the mountains of Aragon.

373. *O. fruticòsa*

■ 2. *O.* rotundifo'lia *L.* The round-leaved Restharrow.

Identification. Lin. Sp. ed. 1. p. 719., but not ed. 2.; Dec. Prod., 2. p. 161.; Don's Mill., 2. p. 160.
Synonymes. *O.* latifòlia *Asso Syn.* 97., *Lin. Mant.* t. 11. f. 1.; Nàtrix rotundifòlia *Mœnch.*
Engravings. Jacq. Fl. Austr. Append., t. 49.; Bot. Mag., t. 335.; and our *fig.* 374.

Spec. Char., &c. Leaves trifoliolate; leaflets ovate, and toothed. Peduncles 3-flowered, and without bracteas. (*Dec. Prod.*) A low shrub. Native of the Pyrenees, and the Alps. Height 1 ft. to 2 ft. Introduced in 1570. Flowers purplish red; May to Sept. Legume brown; ripe in October.

Other Kinds of Onònis. — O. *tribracteàta* Dec., a suffruticose plant with pink flowers, supposed to be a native of Carinthia, differs little from *O.* rotundifòlia. Several other species are hardy, but not sufficiently ligneous for our purpose.

374. *O. rotundifòlia.*

Genus X.

AMO'RPHA *L.* The Amorpha, *or Bastard Indigo.* *Lin. Syst.* Monadélphia Decándria.

Identification. Lin. Gen., 869.; Lam. Ill., t. 621.; Dec. Prod., 2. p. 256.; Don's Mill., 2. p. 234.
Synonymes. Bonafidia *Neck. Elem.* No. 1364.; Faux Indigo, *Fr.*; Unform, *Ger.*; Amorfa, *Ital.*
Derivation. From *a*, privative, and *morphē*, form; in reference to the deformity of the corolla, from the want of the wings and keel.

Gen. Char. *Calyx* 5-toothed, obconically campanulate. *Vexillum* ovate, concave. *Wings* and *Keel* wanting. *Style* filiform, straight, glabrous. *Stamens* exserted, monadelphous at the very base. *Legume* compressed, 1-celled, 1—2-seeded. (*Don's Mill.*)

Leaves compound, impari-pinnate, alternate, stipulate, deciduous; stipules deciduous. *Flowers* of a blue violet colour, in spicate racemes.

Shrubs, large, deciduous; natives of North America. Leaves having many pairs of leaflets that have transparent dots in their disks, and usually minute stipules at their base. The flowers are disposed in racemes, usually grouped at the tips of the branches. The species are highly ornamental on account of their leaves, and more especially of their long spikes of flowers; which, though, when taken separately, they are small, and imperfect in regard to form, are yet rich from their number, and their colours of purple or violet, spangled with a golden yellow. The plants are not of long duration; and are liable to be broken by wind; for which reason they ought always to be planted in a sheltered situation. They produce abundance of suckers, from which, and from cuttings of the root, they are very readily propagated.

♣ 1. A. frutico'sa *Lin.* The shrubby Amorpha, *or Bastard Indigo.*

Identification. Lin. Sp., 1003.; Dec. Prod., 2. p. 256.; Don's Mill., 2. p. 234.
Synonymes. Wild Indigo, *Amer.*; faux Indigo, *Fr.*, Strauchartiger Unform, *Ger.*; Indaco bastardo, Barba di Giove, *Ital.*
Engravings. Schkuhr Handb., t. 197.; Bot. Reg., 427.; and our *fig.* 375.

Spec. Char., &c. Rather arborescent, somewhat villose or glabrous. Leaflets elliptic-oblong, the lowest distant from the base of the petiole. Calyx somewhat villose; 4 of its teeth obtuse, 1 acuminate. The standard glandless Legume few-seeded. (*Dec. Prod.*) An erect glabrous shrub. Carolina and Florida, on the banks of rivers. Height 9 ft. to 12 ft. Flowers very dark bluish purple; June and July. Legume brown; ripe Oct. Naked young wood greyish brown.

375. Amórpha fruticòsa.

Varieties.

- ♣ *A. f.* 2 *angustifòlia* Pursh has the leaflets linear-elliptic.
- ♣ *A. f.* 3 *emarginàta* Pursh has the leaflets notched, and the calyx hoary.
- ♣ *A. f.* 4 *Lewísii Lodd. Cat.*, 1830, appears to have rather larger flowers and leaves than the species.
- ♣ *A. f.* 5 *cærùlea* Lodd. Cat., 1830, has the flowers of somewhat a paler blue. Perhaps only a variation of A. cròceo-lanàta.

♣ 2. A. (f.) gla'bra *Desf.* The glabrous Amorpha, *or Bastard Indigo.*

Identification. Desf. Cat. Hort. Par., 192.; Dec. Prod., 2. p. 256.; Don's Mill., 2. p. 234.
Engraving. Our *fig.* 376. from a specimen in Dr. Lindley's herbarium.

Spec. Char., &c. Rather arborescent, glabrous. Leaflets elliptic-oblong, the lowest distant from the base of the petiole. Calyx glabrous, four of its teeth obtuse, one acuminate. Standard glanded on the outside. Legume containing few seeds. (*Dec. Prod.*) A glabrous shrub. North America. Height 3 ft. to 6 ft. Introduced in 1810. Flowers bluish purple; July and August. Legume brown; ripe in October.

376. A. (f.) glàbra.

377. A. (f.) nàna.

378. A. (f.) fràgrans.

■ 3. A. (F.) NA'NA *Nutt.* The dwarf Amorpha, *or Bastard Indigo.*

Identification. Nutt. in Fras. Cat., 1813.; Nov. Gen. Amer., 2. p. 91.; Dec. Prod., 2. p. 256.; Don's Mill., 2. p. 234.
Synonyme. A. microphýlla *Pursh Fl. Amer. Sept.* 2. p. 466.
Engravings. Bot. Mag., 2112.; and our *fig.* 377.

Spec. Char., &c. Shrubby, dwarf, rather glabrous. Leaflets elliptical, mucronulate. Calyx glabrous, all its teeth setaceously acuminate. Legume 1-seeded. (*Dec. Prod.*) A low glabrous shrub. Native of herbage-covered hills near the Missouri. Height 1 ft. to 2 ft. Introduced 1811. Flowers purple, fragrant; July and August. Legume brown; ripe in October.

■ 4. A. (F.) FRA'GRANS *Sweet.* The fragrant Amorpha, *or Bastard Indigo.*

Identification. Swt. Fl.-Gard., t. 241.; Don's Mill., 2. p. 234.
Synonyme. A. nàna *Sims* in *Bot. Mag.* t. 2112., but not of others.
Engravings. Swt. Fl.-Gard., t. 241.; Bot. Mag., t. 2112.; and our *fig.* 378.

Spec. Char., &c. Shrubby, pubescent. Leaves with 6—8 pairs of elliptic-oblong mucronate leaflets, obtuse at both ends, young ones pubescent. Calyx pubescent, pedicellate; superior teeth obtuse, lower one acute. Style hairy. (*Don's Mill.*) A pubescent shrub. North America. Height 7 ft. to 8 ft. Introduced in 1800. Flowers dark purple; June and July. Legume brown; ripe in September.

■ 5. A. (F.) CRO'CEO-LANA'TA *Wats.* The Saffron-coloured-woolly Amorpha, *or tawny Bastard Indigo.*

Identification. Wats. Dend. Brit.; Don's Mill., 2. p. 234.
Engravings. Wats. Dend. Brit., t. 139.; and our *fig.* 379.

379. A. (f.) cròceo-lanàta.

Spec. Char., &c. Plant clothed with tawny pubescence. Racemes branched. Leaves with 6—8 pairs of oblong-elliptic, mucronulate, downy leaflets; the 3 upper teeth of calyx ovate, acute, the 2 lower ones very short, and rounded. (*Don's Mill.*) A pubescent shrub. North America. Height 3 ft. to 5 ft. Introd. 1820. Flowers purple or purplish blue; July and August. Legume dark brown, or brown; ripe in October.

6. A. (F.) CANE'SCENS *Nutt.* The canescent Amorpha, *or Bastard Indigo.*

Identification. Nutt. in Fras. Cat., 1813, and Gen. Amer., 2. p. 92.; Pursh Fl. Amer. Sept., 2. p. 467.; Dec. Prod., 2. p. 256.; Don's Mill., 2. p. 234.
Synonyme. ? A. pubéscens *Pursh* 2. p. 467.
Engraving. Our *fig.* 380. from Pursh's specimen in the Lambertian herbarium.

380. A. (f.) canescens.

Spec. Char., &c. Suffruticose, dwarf, all over whitely tomentose. Leaflets ovate-elliptic, mucronate, the lowest near the base of the petiole. Calyx tomentose; its teeth ovate, acute, equal. Ovary 2-ovuled. Legume 1-seeded. (*Dec. Prod.*) A low tomentose shrub; Louisiana, on the banks of the Missouri and the Mississippi. Height 3 ft. Introd. 1812. Flowers dark blue; July and August. Legume brown; ripe in October.

GENUS XI.

EYSENHA'RDT*IA* H. et B. THE EYSENHARDTIA. *Lin. Syst.* Diadélphia Decándria.

Identification. H. B. et Kunth Nov. Gen., vi. p. 489.; Dec. Prod., 2. p. 257.; Don's Mill., 2. p. 234.
Synonyme. Dalbérg*ia* Spreng. Syst. App. 263.
Derivation. Named in honour of *Charles William Eysenhardt*, M.D., a professor in the University of Königsberg, in Prussia.

Gen. Char. *Calyx* obconically campanulate, 5-toothed; upper teeth rather remote, lower one longest. *Petals* 5, disposed in a papilionaceous manner. *Vexillum* oblong, and the two keel petals distinct. *Stamens* diadelphous. (*Don's Mill.*)

Leaves compound, impari-pinnate, alternate, stipulate, deciduous; beset with glands. *Flowers* in terminal racemes, white. — An unarmed shrub or low tree; native of Mexico.

1. E. AMORPHÖI'DES *H. et B.* The Amorpha-like Eysenhardtia.

Identification. H. B. et Kunth; Dec. Prod., and Don's Mill.; Bot. Reg. Chron., 1839, No. 55.
Synonyme. Dalbérg*ia* amorphöides *Spreng.*
Engravings. H. B. et Kunth, 6. t. 592.; and our *fig.* 381.

381. Eysenhárdtia amorphöides.

Spec. Char., &c. An unarmed low tree or shrub, with impari-pinnate leaves, composed of many pairs of stipulate leaflets, and these are, as well as the calyxes, beset with glands. Racemes terminal, cylindrical, flowers white. (*Don's Mill.*, ii. p. 234.) A deciduous shrub or low tree. Mexico, on mountains. Height 6 ft. to 10 ft; in British gardens 4 ft. to 6 ft. Introduced in 1837. Flowers white. Legume ?.

This plant was raised in the Hort. Soc. Garden, and has proved quite hardy. The leaves are distinctly marked with glandular dotting; a very unusual case among leguminous plants. The twigs are short, and so closely set upon the branches as to form a dense mass of foliage. Each is terminated by an erect compact spike, from 2 in. to 3 in. long, of white or pale yellow flowers, which, although not larger than those of a spiræa, nevertheless, from their abundance, must produce a beautiful appearance. (*Bot. Reg.*

Genus XII.

ROBI'NIA Lin. The Robinia, or Locust Tree. *Lin. Syst.* Diadélphia Decándria.

Identification. Dec. Mém. Lég., 6.; Prod., 2. p. 261.; Don's Mill., 2. p. 237.
Synonymes. Pseudacàcia *Tourn. Inst.* t. 417., *Mœnch Meth.* 145.; Robinier, *Fr.*; Robinie, *Ger.*
Derivation. Named in honour of *Jean Robin*, a French botanist, once herbalist to Henry IV. of France, author of *Histoire des Plantes*, 12mo, Paris, 1620; printed with the second edition of Lonicer's *History of Plants*. His son Vespasian was sub-demonstrator at the Jardin des Plantes in Paris, and was the first person who cultivated the Robínia Pseud-Acàcia in Europe.

Gen. Char. *Calyx* 5-toothed, lanceolate, 2 upper ones shorter and approximate. *Corolla* papilionaceous. *Vexillum* large. *Keel* obtuse. *Stamens* diadelphous, deciduous. *Ovaries* 16—20-ovulate. *Style* bearded in front. *Legume* compressed, almost sessile, many-seeded, with the valves thin and flat, margined at the seminiferous suture. (*Don's Mill.*)

Leaves compound, unequally pinnate, alternate, deciduous; leaflets generally ovate or obovate, petiolate. *Flowers* white or rose-coloured, in axillary usually nodding racemes. — Trees, deciduous, natives of North America, where one of the species is highly valued for its timber.

The species are prized, partly for their use, but chiefly for their beauty. They are readily propagated by seeds, large truncheons of the stem and branches, cuttings of the roots, or by grafting; and they will grow in any soil that is not too wet. Their roots are creeping, and their branches very brittle: they grow rapidly, but are generally not of long duration. Their rapid growth is a property that they have in common with all trees and plants the principal roots of which extend themselves close under the surface; because there the soil is always richest: but the same cause that produces this rapidity at first occasions the tree to grow slowly afterwards, unless the roots are allowed ample space on every side; since, as they never penetrate deep, they soon exhaust all the soil within their reach. They are, therefore, highly objectionable among ornamental shrubs, or in flower borders. For this reason, also, such trees are objectionable as hedgerow trees, or as scattered groups in arable lands; their roots proving a serious impediment to the plough, and the suckers thrown up by them choking the corn crops. Roots, on the other hand, which penetrate perpendicularly as well as horizontally, belong to more slowly, but more steadily, growing trees, which always attain a larger size in proportion to the extent of ground they occupy.

‡ 1. R. Pseu'd-Aca'cia *Lin.* The *common* Robinia, *or* False Acacia.

Identification. Lin. Sp., 1043.; Dec. Prod., 2. p. 261.; Don's Mill., 2. p. 237.
Synonymes. Æschynómene Pseudacàcia *Rosb.*; Pseudacàcia odoràta *Mœnch Meth.* 145.; Locust Tree, *Amer.*; the Bastard Acacia; Robinier faux Acacia, Acacia blanc, Carouge des Américains, *Fr.*; gemeine Acacie, Schotendorn, *Ger.*
Derivation. This tree, when first introduced, was supposed to be a species of the Egyptian acacia, (*Acàcia vèra*), from its prickly branches and pinnated leaves, which resembled those of that tree. It was named the locust tree by the missionaries, who fancied that it was the tree that supported St. John in the wilderness. It is not, however, a native of any other part of the world than North America. The name Carouge is the French word for carob bean, the locust tree of Spain; which, being also a native of Syria, is, probably, the true locust of the *New Testament*. The German name of Schotendorn is composed of *schote*, a pod, or legume, and *dorn*, a thorn.
Engravings. Lam. Ill., t. 666.; N. Du Ham., 2. t. 16.; the plate of this species in Arb. Brit., 1st edit., vol. v.; and our *fig.* 382.

Spec. Char., &c. Prickles stipular. Branches twiggy. Racemes of flowers loose and pendulous; and smooth, as are the legumes. Leaflets ovate. The flowers are white and sweet-scented; the roots creeping, and their fibres sometimes bearing tubercles. (*Dec. Prod.*) A deciduous tree, above the middle size. North America. Canada to Carolina. Height 70 ft. to 80 ft. Introduced in 1640. Flowers white; May and June. Legume compressed, dark purplish brown; ripe in September. Decaying leaves yellow. Naked young wood purplish brown.

Varieties. The plant varying much in its different native localities, and also having been long cultivated from seeds in Europe, the varieties are numerous. Some of those included in the following list appear in our *Hortus Britannicus*, and in Don's *Miller*, as species; while some hybrids, such as R. hýbrida and R. intermèdia, might also have been considered as varieties, but we have preferred keeping them apart.

- ♆ *R. P.* 2 *flòre lùteo* Dumont 6. p. 140. has the flowers yellow.
- ♆ *R. P.* 3 *inérmis* Dec. Prod. ii. p. 261., Dec. Cat. Hort. Monsp. 136.— Prickles wanting, or nearly obsolete. Leaflets flat. Plant of free growth, in which respect it differs from R. P. umbraculífera.
- ♆ *R. P.* 4 *críspa* Dec. Prod. ii. p. 261.—Prickles wanting. Leaflets all, or for the most part, undulately curled.
- ♆ *R. P.* 5 *umbraculífera* Dec. Prod. ii. p. 261., Cat. Hort. Monsp. 157. R. inérmis *Dum. Cours.* vi. p. 140.—Prickles wanting. Branches much crowded, and smooth. Head orbicular. Leaflets ovate. This variety is said to have been raised from the seed of R. Pseùd-*Acàcia*; and, according to Dumont de Courset, to have yellow flowers. It has been common in British gardens since 1820, but has not yet flowered in this country.
- ♆ *R. P.* 6 *tortuòsa* Dec. Prod. ii. p. 261.; and the plate in Arb. Brit., 1st edit., vol. v.—Branches much crowded, and twisted. Racemes similar to those of R. Pseùd-*Acàcia*, but smaller and fewer-flowered.
- ♆ *R. P.* 7 sophoræ*fòlia* Lodd. Cat. 1830 has the leaves large, and somewhat like those of *Sophòra* japónica.
- ♆ *R. P.* 8 *amorphæfòlia* Lk. has leaves somewhat like those of Amórpha fruticòsa.
- ♆ *R. P* 9 *strícta* Lk. has the general tendency of the shoots upright; but still the plant is not so fastigiate as the Lombardy poplar.
- ♆ *R. P.* 10 *procèra* Lodd. Cat. 1830.—A tall vigorous-growing variety.
- ♆ *R. P.* 11 *péndula* Ort. Dec. p. 26.—The shoots are somewhat drooping, but not very decidedly so.
- ♆ *R. P.* 12 *monstròsa* Lodd. Cat. 1830.—The leaves are large, and twisted.
- ♆ *R. P.* 13 *macrophýlla* Lodd. Cat. 1830 has the leaves long, and the leaflets broad.
- ♆ *R. P.* 14 *microphýlla* Lodd. Cat. 1830, R. angustifòlia *Hort.*, has the leaves small, and the leaflets narrow.
- ♆ *R. P.* 15 *spectábilis* Dum. has large leaves and is without prickles: it produces straight vigorous shoots, which are angular when young. It was raised from seed by M. Descemet, at St. Denis, and was formerly known in the French nurseries by the name of Acacia agaçante (enticing).
- ♆ *R. P.* 16 *latisíliqua*, the broad-podded locust, is mentioned in Prince's *Catalogue* for 1829.

Most of these varieties are tolerably distinct in the foliage when the plants are young. Those best worth cultivating, for the shape of the tree, are R. P. umbraculífera, the parasol acacia; R. P. péndula; R. P. strícta, the upright-growing sort; and R. P. spectábilis. With regard to the yellow-flowered variety, it may be worth continuing by grafting or suckers; but, to make quite sure of having white flowers, the trees producing them ought to be propagated by grafting also; as plants raised from seed, though for the most part they have white flowers, yet occasionally produce yellowish ones.

The wood, which is commonly of a greenish yellow colour, marked with brown veins, is hard, compact, and susceptible of a bright polish: it has a good deal of strength, and is very durable; but it has not much elasticity, and is somewhat liable to crack. A cubic foot of locust wood, newly cut, weighs 63 lb. 3 oz.; half-dry, 56 lb. 4 oz.; and, when quite dry, only 48 lb. 4 oz. Its value for fuel, when compared with that of the beech, is as 12 to 15. For duration Hartig places it immediately after the oak, before the larch and the

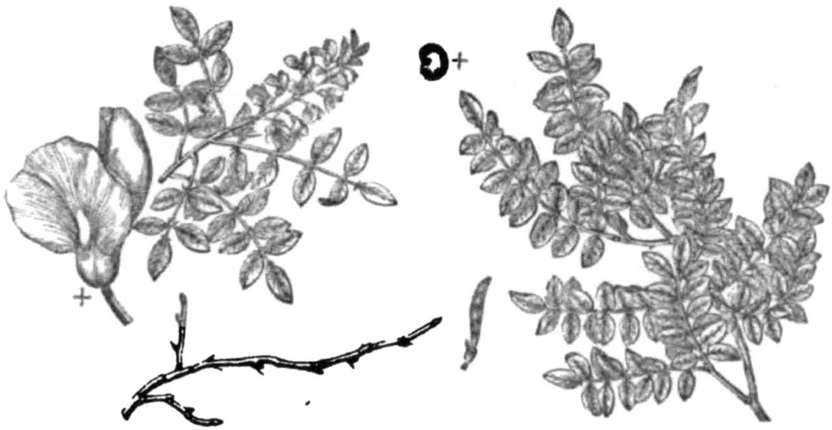

382. Robínia Pseùd-Acàcia.

Scotch pine. According to Barlow, the strength of acacia timber, as compared with fine English oak, is as 1867 to 1672; the strength of ash being as 2026; beech, 1556; elm, 1013; Riga fir, 1108; Norway spar (spruce fir), 1474; and teak, 2462. The tree has one property almost peculiar to it, that of forming heart-wood at a very early age, viz. in its third year; whereas the sap-wood of the oak, the chestnut, the beech, the elm, and most other trees, does not begin to change into heart, or perfect, wood, till after 10 or 15 years' growth. The trees of this species, and of several of its varieties, in the garden of the Horticultural Society, and in the arboretum of Messrs. Loddiges, have attained the height of 30 ft. and upwards, in 10 years from the time they were planted. There is, perhaps, no American tree respecting which so much has been said and done, in Europe, as the locust. It was one of the first trees that we received from America, and it has been more extensively propagated than any other, both in France and England. It has been alternately extolled and neglected in both countries; and even at the present time, though the beauty of its foliage and flowers is generally acknowledged, and though it has, at different periods, been enthusiastically praised by different writers, for the valuable properties of its wood, it cannot be considered as holding a high rank as a timber tree, or as being worth planting with a view to profit. We pass over many curious and historical facts respecting the locust tree, for which we refer to the 1st edit. of this work, and come to the year 1823, when an extraordinary sensation was excited in Britain respecting this tree by Cobbett. This writer, while in America, from 1817 to 1819, "was convinced that nothing in the timber way could be so great a benefit as the general cultivation of this tree." On his return to England he commenced nurseryman, and the name of locust, as applied to this tree, being, before Cobbett's time, almost forgotten in England, many persons, in consequence, thought it was a new tree. Hence, while quantities of plants of Robínia Pseùd-Acàcia stood unasked for in the nurseries, the locust, which every one believed could only be had genuine from Mr. Cobbett, could not be grown by him in sufficient quantities to supply the demand. After creating a prodigious sensation for a few years, the locust mania entirely subsided, and the tree is now, as it was before Cobbett's time, planted only, or chiefly, for ornament.

♀ 2. R. VISCO'SA *Vent.* The clammy-*barked* Robinia.

Identification. Vent. Hort. Cels., t. 4.; Dec. Prod., 2. p. 262.; Don's Mill., 2. p. 238.
Synonymes. R. glutinòsa *Bot. Mag.* 560.; R. montàna *Bartram*; the Rose-flowering Locust.
Engravings. Vent. Hort. Cels., t. 4.; Bot. Mag., t. 560., as R. glutinòsa; the plate of this tree in Arb. Brit., 1st edit., vol. v.; and our fig. 383.

Spec. Char., &c. Branches and legumes glandular and clammy. Racemes crowded, erect. Bracteas concave, deciduous, each ending in a long bristle.

383. Robinia viscòsa.

The 3 lower teeth of the calyx acuminated. Roots creeping. (*Dec. Prod.*) A clammy-barked tree. South Carolina and Georgia, near rivers. Height 30 ft. to 40 ft. Introduced in 1797. Flowers scentless, pale pink mixed with white, varying to pale purple or violet colour according to the soil; June to August. Legume brown; ripe in October. Naked young wood purplish and clammy.

The bark, particularly that of the young shoots, which is of a dull red, is covered with a clammy substance, which, when touched, sticks to the fingers. In every other respect, both in natural characters and artificial culture, this tree strongly resembles the common R. Pseùd-*Acàcia*.

♀ 3. R. DU'BIA *Fouc.* The doubtful Robinia, *or False Acacia.*

Identification. Fouc. in Desv. Journ. Bot., 4. p. 204., but not of Poir.; Dec. Prod., 2. p. 261.; Don's Mill., 2. p. 238.
Synonymes. R. hŷbrida *Audib.*; R. ambigua *Poir. Suppl.* 4. p. 690.; and, perhaps, R. echinàta *Mill. Dict.*, No. 2.; R. intermèdia *Soulange-Bodin* in *Ann. d'Hort. de Paris*, 2. p. 43.
Engraving. Our *fig.* . in p. .

Spec. Char., &c. Spines very short. Branches, petioles, peduncles, and calyxes furnished with a few glands, rarely clammy. Leaflets ovate. Racemes loose and pendulous. Bracteas concave, caducous, ending each in a long bristle. (*Don's Mill.*) A deciduous tree, rather under the middle size. Hybrid? between R. Pseùd-*Acàcia* and R. viscòsa. Originated in ? 1730. Flowers sweet-scented, pale rose-coloured; June to August. Pods brown, thickly beset with short prickles; ripe in October.

♀ ● 4. R. HI'SPIDA *Lin.* The hispid Robinia, *or Rose Acacia.*

Identification. Lin. Mant., 101.; Dec. Prod., 2. p. 262.; Don's Mill., 2. p. 238.
Synonymes. R. ròsea *N. Du Ham.* 1. t. 18.; R. montàna *Bartr. Voy.* 2. p. 128.; Æschynómene híspida *Roxb.*
Engravings. Mill. Ic., t. 244.; Bot. Mag., 311.; and our *fig.* 384.

Spec. Char., &c. Spines wanting. Leaflets obovate. Branches and legumes hispid. Racemes loose; the 3 lower teeth of the calyx acuminated. (*Dec. Prod.*) A shrub, or low tree. Carolina, in pine woods. Height 6 ft. to 20 ft. Introduced in 1758. Flowers large, dark rose-coloured; June to October. Legume brown; ripe in October and November. Naked young wood purplish brown, thickly beset with prickles.

384. Robinia hispida.

Varieties.

- *R. h.* 2 *nàna* Dec. is a plant hardly a foot high, found in pine woods in Carolina.
- *R. h.* 3 *ròsea* Pursh has the leaflets for the most part alternate, and the branches smoothish. In its native habitats, on the high mountains of Virginia and Carolina, it grows, according to Pursh, to a considerable shrub, whereas the species is a low straggling plant.
- *R. h.* 4 *macrophýlla* Dec., R. grandiflòra *Hort.*, figured in Arb. Brit., 1st edit., vol. v., has the leaflets large, and ovate-roundish; and the branches and peduncles glabrous, and without prickles.

The species, and the different varieties, are shrubs or low trees, with tortuous and very brittle branches, and leaves and flowers nearly twice the size of those of Robínia Pseùd-*Acàcia*. They form singularly ornamental shrubs for gardens; but, as standards or bushes, they can be only planted with safety in the most sheltered situations. When grafted standard high, and trained to a wire parasol-like frame, supported on a rod or post 6 or 8 feet high, few plants are equal to R. h. macrophýlla in point of brilliant display.

Genus XIII.

CARAGA'NA Lam. The Caragana, *or Siberian Pea Tree.* *Lin. Syst.* Diadélphia Decándria.

Identification. Lam. Dict., 1. p. 611.; Dec. Prod., 2. p. 268.; Don's Mill., 2. p. 243.
Synonyme. Robínia sp. *L.*
Derivation. *Caragan* is the name of *C.* arboréscens among the Mongol Tartars.

Gen. Char. *Calyx* short, tubulous, 5-toothed. *Corolla* obtuse, straight; the wings and vexillum about equal in length. *Stamens* diadelphous. *Style* glabrous. *Stigma* terminal, truncate. *Legume* sessile, young ones compressed, at length somewhat cylindrical and many-seeded, mucronate by the style. *Seeds* somewhat globose. (*Don's Mill.*)

Leaves compound, abruptly pinnate, alternate, stipulate, deciduous; stipules usually spinescent. *Flowers* mostly yellow.

Trees or shrubs, natives of Siberia and of the East. Leaves with the leaflets mucronate, and the petioles either with a bristly or a spiny point; their flowers axillary, each on a distinct pedicel, usually several together, pale yellow, except in *C.* jubàta, in which they are white tinged with red; their stipules usually become spines. They are all ornamental or curious, and of the easiest culture in any common soil; propagated by cuttings of the roots or by seeds. The dwarf and pendulous-growing species, when grafted standard high on *C.* arboréscens, form very singular trees.

1. *C.* ARBORE'SCENS *Lam.* The arborescent Caragana, *or Siberian Pea Tree.*

Identification. Lam. Dict., 1. p. 615.; Dec. Prod., 2. p. 268.; Don's Mill., 2. p. 243.
Synonymes. Robínia *Caragàna* Lin. Sp. 1044., N. Du Ham. 2. t. 19., Pall. Fl. Ross. 1. t. 42.; *Caragàna* sibírica *Ray*; fausse Acacie de Sibérie, Robinie de Sibérie, Arbre aux Pois des Russes, *Fr.*; Sibirische Erbsenbaum, *Ger.*; Gorochoik, *Russ.*
Engravings. N. Du Ham., 2. t. 19.; Pall. Fl. Ross., 1. t. 42., middle figure; the plate of this species in Arb. Brit., 1st edit., vol. v.; and our *fig.* 385.

Spec. Char., &c. Leaves with 4—6 pairs of oval-oblong villous leaflets. Petiole unarmed. Stipules spinescent. Pedicels in fascicles. (*Don's Mill.*) A low tree. Siberia, in woods, and upon the banks of rivers. Height 15 ft. to 20 ft. Introduced in 1752. Flowers yellow; April and May. Legume brown; ripe in August.

Variety.

¥ C. *a.* 2 *inérmis* Hort. has the branches without spines. Plants in the Horticultural Society's Garden.

Pallas informs us that upon the banks of rivers it grows to the height of 18 ft. or more; but in arid places it is only a small shrub; in the latter state forming, as we think, the varieties *C.* (a.) *Altagàna*, and *C.* (a.) microphýlla. *C.* arboréscens forms an erect stiff tree, with numerous upright-growing branches. The flowers are axillary, one on a pedicel; the pods are oblong-taper, and each contains 3 or 4 seeds. The wood is hard, compact, and very tough; yellow on the outside; and within, waved and striped with red, and with reddish brown.

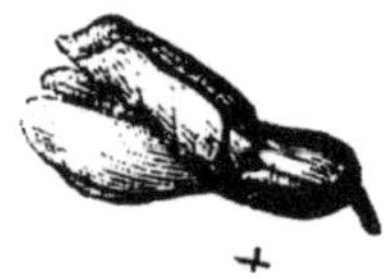

385. C. arboréscens.

■ 2. *C.* (A.) *ALTAGA'NA* Poir. The Altagana Caragana, *or Siberian Pea Tree.*

Identification. Poir. Sup., 2. p. 89.; Dec. Prod., 2. p. 268.; Don's Mill., 2. p. 243.
Synonymes. Robínia *Altagàna* Pall. Fl. Ross. t. 42., L'Hérit. Stirp. t. 76.; *Caragàna* microphýlla *Lam. Dict.* 1. p. 615.
Derivation. *Altagana* is the name of the shrub in Siberia.
Engravings. Pall. Fl. Ross., t. 42., under the name of Robínia *Altagàna*; L'Hérit. Stirp., t. 76.; and our *fig.* 386.

386. C. (a.) Altagàna.

Spec. Char., &c. Leaves having 6 or 8 pairs of glabrous, obovate-roundish, retuse leaflets. Petiole unarmed. Stipules spinescent. Pedicels solitary. Legumes rather compressed. (*Don's Mill.*) A low shrub. Siberia, in arid plains. Height 3 ft. to 4 ft. Introd. 1789. Flowers yellow; April to July. Legume brown; ripe in September.

Usually propagated by grafting on *C.* arboréscens.

3. *C.* (A.) MICROPHY'LLA *Dec.* The small-leaved Caragana, *or Siberian Pea Tree.*

Identification. Dec. Prod., 2. p. 268.; Don's Mill., 2. p. 243.
Synonymes. Robínia microphýlla *Pall. Fl. Ross.* t. 42., f. 1, 2.; *Caragàna Altagàna* var. *Poir. Suppl.* 2. p. 89.
Engravings. Pall. Fl. Ross., t. 32. f. 1, 2., under the name of Robínia microphýlla; and our *fig.* 387.

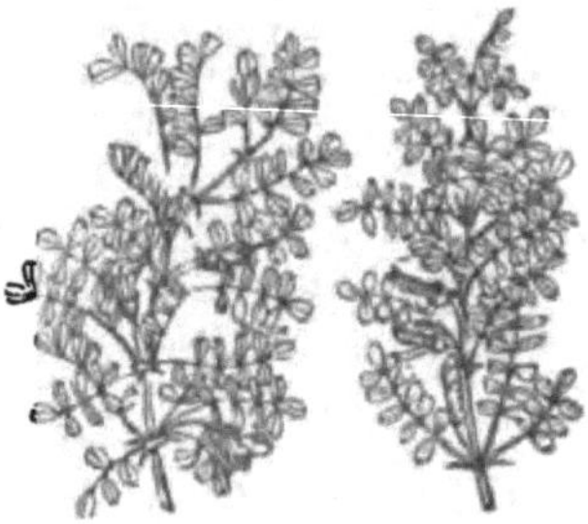

387. C. (a.) microphýlla.

Spec. Char., &c. Leaves with 6—7 pairs of hoary retuse leaflets. Petioles and stipules rather spinescent at the apex. Root creeping. (*Don's Mill.*) A low shrub. Siberia, in the Desert of Baraba, and in other arid places. Height 2 ft. to 3 ft. Introduced in 1819. Flowers yellow; April to June. Legume brown; ripe in August.

■ 4. *C.* (A.) REDO'WSKI Dec. Redowski's Caragana, *or Siberian Pea Tree.*

Identification. Dec. Légum., t. 11.; Don's Mill., 2. p. 243.
Engraving. Dec. Légum., t. 11. f. 45, in the seedling state.

Spec. Char., &c. Leaves with two pairs of ovate, acute, smooth leaflets. Stipules spinose. Flowers yellow. (*Don's Mill.*) A low shrub. Siberia. Height 4 ft. to 5 ft. Introduced in 1820. Flowers yellow; April and May. Legume brown; ripe in August.

Variety.

C. (A.) *R.* 2 *præ'cox* Fisch. only differs from *C.* Redówski in coming into flower earlier. The specimen in the Hort. Soc. Garden was in full leaf, and in flower, on April 30. 1836, when *C.* frutéscens and *C.* arboréscens had not a single leaf expanded.

In general appearance and habit of growth, it resembles *C. Altagàna*, of which it is probably only a variety. H. S.

5. *C.* (A.) ARENA'RIA *Donn* and *Sims.* The Sand Caragana.

Identification. Donn Hort. Cant.; Sims Bot. Mag., t. 1886.
Engravings. Sims Bot. Mag., t. 1886.; and our *fig.* 388.

388 C. (A.) arenària.

Spec. Char., &c. Leaves with, usually, 4 or more pairs of obcordate leaflets. Pedicels usually twin, and shorter than the flowers. Stipules subulate. Flowers yellow. (*Don's Mill.*) A low shrub. Siberia. Height 1 ft. to 2 ft. Introd. 1802. Flowers yellow; April and May. Legume dark brown; ripe in August.

Probably only another variety of *C.* arboréscens. It ripens seeds in England, but is generally propagated by grafting.

6. *C.* FRUTE'SCENS *Dec.* The shrubby Caragana.

Identification. Dec. Prod., 2. p. 268.; Don's Mill., 2. p. 843.
Synonymes. Robínia frutéscens *Lin. Spec.* 1044., *Pall. Fl. Ross.* t. 43.; *C.* digitàta *Lam. Dict.* 1. p. 616.
Engravings. Swt. Fl.-Gard., t. 227.; Pall. Fl. Ross., t. 43., as Robínia frutéscens; and our *fig.* 389.

389. C. frutéscens.

Spec. Char., &c. Leaves having 2 pairs of leaflets, which approximate near the top of the petiole: they are obovate-cuneated. Stipules membranous. Petiole furnished with a short spine at the apex. Pedicels solitary, twice the length of the calyx. Flowers yellow, resupinate. Leaves with a yellow hue. (*Don's Mill.*) A low shrub. Native of Russia, on the banks of the Wolga and other rivers. Height in open situations 5 ft.; in woods and gardens 9 ft. to 10 ft. Introduced in 1752. Flowers yellow; May. Legume brown; ripe in August.

Varieties. DeCandolle mentions two forms: —

C. *f.* 1 *latifòlia*, which has glabrous broadly obovate leaflets, and is frequent in gardens; there being a subvariety, with 2-flowered peduncles; and

C. *f.* 2 *angustifòlia*, which has glabrous oblong cuneated leaflets, and is found near Odessa.

A handsome shrub, sometimes grafted standard high on *C.* arboréscens.

390. C. (F.) móllis.

7. *C.* (F.) MO'LLIS *Bess.* The soft Caragana.

Identification. Bess. Enum. Pl. Volh., p. 29.; Don's Mill., 2. p. 243.
Synonymes. Robínia móllis *Bieb. Fl. Taur Suppl.* 477.; Robínia tomentòsa *Fisch. Hort. Gorenk.* 1818; *Caragàna* frutéscens var. móllis *Dec. Prod.* 2. p. 268.
Engraving. Our *fig.* 390. from a specimen in Dr. Lindley's herbarium.

Spec. Char., &c. Leaves with 2 pairs of oblong, cuneated, approximate leaflets, near the tip of the petiole, clothed with soft hair. Petiole ending in a short spine. Pedicels solitary. Flowers yellow. (*Dec. Prod.*) A low shrub. Native of Tauria and Podolia. Height 2 ft. to 3 ft. Introduced in 1818. Flowers yellow; April and May. Legume brown; ripe in August.

8. *C.* PYGMÆ'A *Dec.* The pygmy Caragana.

Identification. Dec. Prod., 2. p. 268.; Don's Mill., 2. p. 243.
Synonyme. Robinia pygmæ'a *Lin. Sp.* 1044., *Pall. Fl. Ross.* 1. t. 45., *Amm. Ruth.* t. 35.
Engravings. Pall. Fl. Ross., 1. t. 45.; Amm. Ruth., t. 35., as Robínia pygmæ'a; and our *fig.* 391.

Spec. Char., &c. Leaves with 2 pairs of linear, glabrous, approximate leaflets near the tip of the petiole, which is very short. Stipules and petioles spinescent. Pedicels solitary, and nearly the length of the calyx. Calyx nearly equal at the base. Leaflets acute, crowded, usually in the axils of trifid spines. Flowers yellow. (*Dec. Prod.*) A low shrub. Native of the Altaic Mountains. Height 2 ft. to 6 ft. Introduced in 1751. Flowers yellow; April and May. Legume brown; ripe in August.

391. C. pygmæ'a.

Variety.

C. *p.* 2 *arenària* Fisch. has linear-cuneate leaflets, and pedicels of the flowers generally longer than the calyx.

This species has large trifid spines, slender leaves, and small flowers. The leaflets are remarkable for being in fours, disposed in the form of a star, in the axils of the spines. The young shoots are of a fine yellow, very tough, and fit for being used as withs. When grafted as a standard on *C.* arboréscens, it forms a small tree of very singular appearance. Increased by suckers or by grafting.

9. *C.* SPINO'SA *Dec.* The spiny Caragana.

Identification. Dec. Prod., 2. p. 269.; Lindl. Bot. Reg., 1021.; Don's Mill., 2. p. 243.
Synonymes. Robínia spinòsa *Lin. Mant.* 269.; Robínia fèrox *Pall. Fl. Ross.* 1. t. 44., *Itin.* t. E. e. f. 2. and 3.; Robínia spinosissima *Laxm. Nov. Act. Pet.* 15. t. 30. f. 4.; *Caragàna* fèrox *Lam. Dict.* 1. p. 315.
Engravings. Pall. Fl. Ross., 1. t. 44.; Bot. Reg., t. 1021.; and our *fig.* 392.

392. C. spinòsa.

Spec. Char., &c. Leaves with 2—4 pairs of cuneate-linear glabrous leaflets. Stipules small, spinose. Adult petioles permanent, strong, and spinose, twice the length of the leaflets. Flowers solitary, almost sessile, and of a bright yellow. Legume rather compressed. (*Don's Mill.*) A spiny shrub. Siberia, in gravelly arid situations; and also said to be plentiful in China, about Pekin, where branches of it are stuck in clay upon the tops of the walls, in order that its spines may prevent persons from getting over them. (*Pall. Fl. Ross.*) Height 3 ft. to 4 ft. Introduced in 1755. Flowers yellow; April and May. Legume brown; ripe in August.

Seeds, cuttings, or grafting. Standard high it makes a very singular plant.

393. C. tragacanthöides.

10. *C.* TRAGACANTHÖI'DES *Poir.* The Goat's-thorn-like Caragana.

Identification. Poir. Suppl., 2. p. 90.; Dec. Prod., 2. p. 269.; Don's Mill., 2. p. 241.
Synonymes. Robínia tragacanthöides *Pall. Nov. Act. Pet.* 13. t. 7., *Astr.* 115. t. 86.; Robínia macracántha *Lodd. Cat.*
Engravings. Pall. Nov. Act. Pet., 10. t. 7.; Astr., 115. t. 86.; and our *fig.* 393.

Spec. Char., &c. Leaves with 2—4 pairs of oblong-lanceolate silky leaflets, ending in a little spine. Stipules spinescent. Adult petioles permanent, strong, and spinose, twice the length of the leaflets. Pedicels solitary, short. Legume hoary-villous. (*Don's Mill.*) A low spiny shrub. Native of Siberia, among granite rocks. Height 6 in. to 1 ft. Introduced in 1816. Flowers yellow, drooping; April and May. Legume brown; ripe in August. When grafted standard high, it forms a very singular object.

♣ 11. *C.* JUBA`TA *Poir.* The crested Caragana.

Identification. Poir. Suppl., 2. p. 89.; Dec. Prod., 2. p. 269.; Don's Mill., 2. p. 244.
Synonyme. Robínia jubàta *Pall.* in *Act. Pet.* 10. t. 6., *Astr.* p. 113. t. 85., *Lodd. Bot. Cab.* t. 522.
Engravings. Pal. Act. Pet., 10. t. 6.; Lod. Bot. Cab., t. 522., as Robínia jubàta; and our *fig.* 394.

Spec. Char., &c. Leaves with 4 or 5 pairs of oblong-lanceolate lanuginously ciliated leaflets. Stipules setaceous. Petioles somewhat spinose; adult ones deflexed, filiform, permanent. Pedicels solitary, very short. Legume glabrous. Flowers few and white, suffused with red. (*Don's Mill.*) A low shrub; native of Siberia near Lake Baikal. Height 1 ft. to 1 ft. 6 in. Introduced in 1796. Flowers white, tinged with red; April and May. Legume brown; ripe in August.

394. C. jubàta.

This species has a curious shaggy appearance, occasioned by the footstalks of the leaves being bristly or thorny, and remaining on long after the leaflets have dropped off. Increased by grafting on *Caragàna arboréscens*, and as a standard forming a very singular-looking object.

♣ 12. *C.* GRANDIFLO`RA *Dec.* The great-flowered Caragana.

Identification. Dec. Prod., 2. p. 268.; Don's Mill., 2. p. 243.
Synonyme. Robínia grandiflòra *Bieb. Fl. Taur.* 1. p. 168.
Engraving. Our *fig.* 395.

Spec. Char., &c. Leaves with 2 pairs of oblong-cuneated approximate leaflets, near the tip of the petiole, which is very short. Stipules and petioles spinose. Pedicels solitary, almost the length of the calyx, which is gibbous at the base. Legume terete, acute, brown, glabrous. Flowers 1 in. long, yellow. (*Don's Mill.*) A low shrub; native of Georgia, near Teflis. Height 2 ft. to 4 ft. Introduced in 1823. Flowers yellow, 1 in. long; June and July. Legume brown; ripe in September. H. S.

395. C. grandiflòra.

♣ 13. *C. CHAMLA`GU* Lam. The Chamlagu, *or Chinese* Caragana.

Identification. Lam. Dict., 1. p. 616.; Dec. Prod., 2. p. 268.; Don's Mill., 2. p. 243.
Synonyme. Robínia *Chamlàgu* L'Hérit. Stirp., N. Du Ham.
Engravings. L'Hérit. Stirp., t. 77.; N. Du Ham. Arb., 2. t. 21.; and our *fig.* 396.

Spec. Char., &c. Leaves with 2 pairs of distant, oval, or obovate, glabrous leaflets. Stipules spreading, and, as well as the petioles, spinose. Pedicels solitary. Flowers pendulous, large, and yellow, at length becoming reddish. Root smelling like liquorice. (*Don's Mill.*) A diffuse smooth shrub. Native of China. Height 2 ft. to 4 ft. Introduced in 1773. Flowers yellow, at length becoming reddish, large; May and June. Legume brown; ripe August.

396. C. Chamlàgu.

This species has a thick root and branching stem, with grey bark. The branches are alternate; at first upright, and then decumbent. The whole plant has a singular appearance, more especially when just going out of flower. It is generally propagated by separating the offsets, or by seeds, or it may be grafted on *C.* arboréscens. Grafted on this species, especially when the stock is 10 or 12 feet high, it forms a singularly picturesque pendulous tree; beautiful not only when it is in leaf or in flower, but from the graceful lines formed by its branches, even in the midst of winter, when they are completely stripped of their leaves.

Genus XIV.

HALIMODE'NDRON *Fisch.* The Halimodendron, *or Salt Tree.* *Lin. Syst.* Diadélphia Decándria.

Identification. Fisch. in Litt.; Dec. Légum. Mém., 6.; Prod., 2. p. 269.; Don's Mill., 2. p. 244.
Synonyme. Halodéndron *Dec. Mém. ined.* in *Soc. Phys. Gen.* March 1824, but not of Petit Thouars.
Derivation. From *halimos*, maritime, and *dendron*, a tree; in reference to the habitat of the shrubs, which grow in dry naked salt fields, by the river Irtis, in Siberia.

Gen. Char. *Calyx* urceolately campanulate, with 5 short teeth. *Keel* obtuse, straightish. *Wings* very acute, and auricled. *Stamens* diadelphous, about equal in length. *Style* filiform, glabrous. *Stigma* terminal. *Legume* stipitate, inflated, bladdery, hard, ovate, and few-seeded, depressed at the seminiferous suture. *Seeds* oval. (*Don's Mill.*)

Leaves compound, alternate, stipulate, deciduous; abruptly pinnate, with 2 pairs of leaflets: petioles and stipules spinose. *Flowers* bluish pink, or purplish, in 2—3-flowered peduncles.—Shrubs, deciduous; natives of Europe and Asia.

Propagated by seeds, by cuttings of the roots, or by grafting on the common laburnum, or on the *Caragána* arboréscens.

■ 1. H. argénteum *Dec.* The silvery-*leaved* Halimodendron, *or Salt Tree.*

Identification. Dec. Prod., 2. p. 269.; Don's Mill., 2. p. 244.
Synonymes. Robínia Halodéndron *Lin. fil. Suppl.* 330., *Pall. Fl. Ross.* t. 36.; *Caragána argéntea Lam.* in *Pall. Itin. ed. Gall.* App. ii. 360. t. 83. f. 1.
Engravings. Pall. Fl. Ross., t. 36.; and our *fig.* 397.

Spec. Char., &c. Leaves hoary. Peduncles 2-flowered. (*Don's Mill.*) A shrub, known from the colour of its leaves and branches. Native of Siberia in saline steppes near the river Irtis. Height 4 ft. to 6 ft. Introduced in 1779. Flowers bluish pink, fragrant; May to July. Legume inflated, brown; ripe in September.

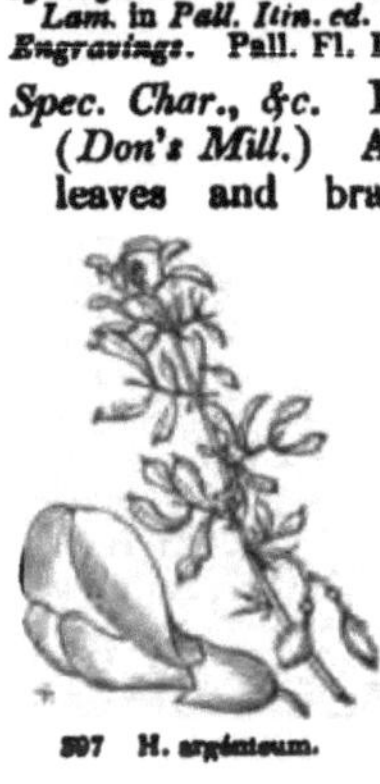
397 H. argénteum.

Varieties. DeCandolle mentions two forms of this species:—

■ *H. a.* 1 *vulgáre* Dec. Prod.—Leaves hoary or silvery. Standard the same length as the keel.

■ *H. a.* 2 *brachysèma* Dec. Prod. (Bot. Mag., 1016.; and our *fig.* 398.)—Leaves hoary or silvery. Standard shorter than the wings and keel. Style short.

398. H. a. brachysèma

H. a. 3 *Sievérsii*, H. Sievérsii Fisch., is a dwarf variety, very hardy, named by some as a species. H. S.

An irregular, much-branched, rigid shrub, with a strigose grey bark, and leaves clothed with a whitish silky down. The flowers are numerous, resembling those of *Láthyrus tuberòsus*, both in colour and size; and they smell sweet. According to Pallas, it is much frequented by insects, especially of the genus Méloe *L.*, many species of which are peculiar to Siberia. It flowers freely from May to July, and, in moist seasons, later; and, when grafted standard high on the common laburnum, it forms one of the most graceful drooping trees that can adorn a lawn.

2. H. (A.) SUBVIRE'SCENS *G. Don.* The greenish Halimodendron, or *Salt Tree.*

Identification. Don's Mill., 2. p. 244.
Synonymes. Robínia triflòra *L'Hérit. Stirp. Nov.* 162.; H. argénteum β subviréscens *Dec. Prod.* 2. p. 169.
Engraving. Our *fig.* . in p. .

Spec. Char., &c. Leaves greenish. The standard of the same length as the keel. Pedicels 3-flowered. (*Don's Mill.*) A shrub, like the preceding one, of which it is, without doubt, only a variety.

GENUS XV.

CALO'PHACA *Fisch.* THE CALOPHACA. *Lin. Syst.* Diadélphia Decándria.

Identification. Fisch. ined.; Dec. Prod., 2. p. 270.
Derivation. From *kalos*, beautiful, and *phakē*, a lentil; in allusion to the beauty of the plant, and to its being one of the leguminaceous kind.

Gen. Char. *Calyx* 5-cleft, the lobes acuminated. *Keel* obtuse. *Stamens* diadelphous. *Style* villous and straight at the base, but glabrous incurved at the apex. *Stigma* terminal. *Legume* sessile, oblong, somewhat cylindrical, mucronate, 1-celled. *Valves* concave, beset with soft hairs, as well as with stiff glandular bristles, mixed. (*Don's Mill.*)

Leaves compound, alternate, stipulate, deciduous; impari-pinnate: stipules lanceolate. *Flowers* yellow.—A shrub, native of Siberia.

1. *C.* WOLGA'RICA *Fisch.* The Wolga Calophaca.

Identification. Fisch. in Litt.; Dec. Prod., 2. p. 270.; Don's Mill., 2. p. 244.
Synonymes. Cýtisus nígricans *Pall. Itin.* 3. p. 764. t. G. g. f. 3., *ed. Gall. Append.* No. 358. t. 101. f. 1.; Cýtisus pinnàtus *Pall. Fl. Ross.* 1. t. 47.; Cýtisus wolgàricus *Lin. fil. Suppl.* 327., *N. Du Ham.* 1. t. 48.; Colùtea wolgárica *Lam.*; Adenocárpus wolgénsis *Spreng. Syst.* 3. p. 226.
Engravings. Pall. Fl. Ross., 4. t. 47.; N. Du Ham., 5. t. 48.; and our *fig.* 399.

399. Calóphaca wolgárica.

Spec. Char., &c. Leaflets 6 or 7 pairs, orbicular, velvety beneath, as well as the calyxes. (*Don's Mill.*) A deciduous shrub. Siberia, in desert places near the rivers Don and Wolga, in a gravelly or sandy soil. Height 2 ft. to 3 ft. Introduced in 1780. Flowers yellow; June. Legume reddish; ripe in August.

Being somewhat difficult to propagate except by seeds, which, however, in fine seasons, it produces in abundance, it is not so common as it ought to be in British gardens. Grafted standard high on the common laburnum, it forms an object at once singular, picturesque, and beautiful, whether when covered with blossoms, or with its fine reddish pods.

GENUS XVI.

COLU'TEA *R. Br.* THE COLUTEA, *or BLADDER SENNA.* *Lin. Syst.* Diadélphia Decándria.

Identification. R. Br. in Hort. Kew., ed. 2, vol. 4. p. 325.; Dec. Prod., 2. p. 270.; Don's Mill., 2. p. 244.
Synonymes. Baguenaudier, *Fr.*; Senna falsa, *Ital.*; Blasenbaum, *Ger.*
Derivation. From *kolouō*, to amputate. The shrubs are said to die if the branches are lopped off. *Koloutea* is also the name of a plant mentioned by Theophrastus.

Gen. Char. *Calyx* 5-toothed. *Vexillum* flat, bicallose, larger than the keel, which is obtuse. *Stamens* diadelphous. *Stigma* lateral, hooked under the top of the style. *Style* bearded longitudinally behind. *Legume* stipitate, ovate, boat-formed, inflated, scarious. (*Don's Mill.*)

Leaves compound, alternate, stipulate, deciduous; impari-pinnate: stipules small. *Flowers* usually yellow, axillary, the racemes shorter than the leaves, and succeeded by bladdery legumes. — Shrubs, deciduous; natives of the Middle and South of Europe, the North of Africa, and Nepal.

All the kinds that have hitherto been introduced into Europe are probably only varieties of one species. They are readily increased by seeds or cuttings of the roots on any common soil.

■ 1. *C.* ARBORE'SCENS *Lin.* The arborescent Colutea, *or Bladder Senna.*

Identification. Lin. Sp. 1045.; Dec. Astr., No. 1.; Dec. Prod., 2. p. 270.; Don's Mill., 2. p. 245.
Synonyme. C. hirsùta *Roth Fl. Germ.* 1. p. 305.
Engravings. N. Du Ham., 1. t. 22.; Schmidt Arb., t. 117.; Bot. Mag., t. 81.; and our *fig.* 400.

Spec. Char., &c. Leaflets elliptical, retuse. Peduncles bearing about 6 yellow flowers. Callosities of the standard short. Legumes closed. (*Dec. Prod.*) A rapid-growing shrub. Middle and South of Europe, Italy, and on Mount Vesuvius is found even on the ascent to the crater, where there are scarcely any other plants. Height 12 ft. to 14 ft. Introduced in 1570. Flowers yellow; June to August. Legume bladder-like, reddish; ripe in September.

400. Colùtea arboréscens.

The bladders, when pressed, explode with a crackling noise. On the Continent, the leaves have been recommended as a substitute for senna, and they are also said to afford a grateful food for cattle. The seeds, in doses of a drachm or two, are said to excite vomiting. In British gardens, the plant is chiefly valuable as a bulky fast-growing shrub, of the easiest culture, and fit for almost any situation.

■ 2. *C.* (A.) CRUE'NTA *Ait.* The bloody-*flowered* Colutea, *or Oriental Bladder Senna.*

Identification. Ait. Hort. Kew., 3. p. 55.; Dec. Astr., No. 3.; Don's Mill., 2. p. 245.
Synonymes. C. orientàlis *Lam. Dict.* 1. p. 353., *Ill.* 624. f. 3., *N. Du Ham.* 1. t. 23.; C. sanguínea *Pall.*; C. àptera *Schmidt Arb.* t. 119.; *C.* hùmilis *Scop.*
Engravings. Lam. Ill., 624. f. 3.; N. Du Ham., 1. t. 23.; Schmidt Arb., t. 119.; Krause, t. 106.; and our *fig.* 401.

Spec. Char., &c. Leaflets obovate, emarginate, glaucous. Peduncles bearing 4—5 flowers. Callosities of the standard obtuse, very small. Legumes opening at the tip. Corolla, in colour, between red and saffron-coloured, with a yellow spot at the base of the standard. (*Dec. Prod.*) A glaucous

401. Colùtea (A.) cruénta.

shrub. Archipelago, Georgia, and the Levant. Height 4 ft. to 8 ft. Introd. 1731. Flowers reddish copper-coloured; June and July. Legume reddish; ripe in August.

Resembles *C.* arboréscens, but of smaller dimensions, and with leaflets more glaucous, and more retuse.

■ 3. *C.* (A.) ME'DIA *Willd.* The intermediate Colutea, *or Bladder Senna.*

Identification. Willd. Enum., 771.; Dec. Prod., 2. p. 270.; Don's Mill., 2. p. 245.
Engravings. Wats. Dend. Brit., t. 140.; and our *fig.* 402.

Spec. Char., &c. Leaflets obcordate, glaucescent. Peduncles usually 6-flowered. Legumes closed at the apex. Flowers orange-coloured. — (*Don's Mill.*)

402. C. (A.) média.

A shrub rather larger than *C.* (A.) cruénta, and differing from it chiefly in having orange-coloured flowers. Perhaps, a hybrid between the two preceding sorts.

■ 4 *C.* (A.) HALE'PPICA *Lam.* The Aleppo Colutea, *or Bladder Senna.*

Identification. Lam. Dict., 1. p. 353.; Ill., t. 624. f. 2.; Dec. Astr., No. 2.; Don's Mill., 2. p. 245.
Synonymes. *C.* Pocóckii Ait. Hort. Kew. 3. p. 55, Schmidt Arb. t. 129.; *C.* ístria *Mill. Dict.* No. 2. t. 100.; *C.* procúmbens *L'Hérit. Stirp. Nov.* 2. t. 42.
Engravings. Schmidt Arb., t. 120.; and our *fig.* 403.

Spec. Char., &c. Leaflets roundishly elliptical, very obtuse, mucronate. Peduncles bearing 3 yellow flowers. Callosities of the standard lengthened, ascending. Legumes closed. Smaller than *C.* arboréscens. (*Dec. Prod.*) A low shrub. Fields about Aleppo. Height 3 ft. to 6 ft. Introduced in 1752. Flowers yellow; May to October. Legume reddish; ripe Oct.

Closely resembling *C.* arboréscens, of which it appears to be a variety, but rather more tender in British gardens.

403. C. (A.) halèppica.

404. C. nepalénsis.

■ 5. *C.* NEPALE'NSIS *Hook.* The Nepal Colutea, *or Bladder Senna.*

Identification. Hook. Bot. Mag., 2622.; Don's Mill., 2. p. 245.
Engravings. Hook. Bot. Mag., t. 2622.; and our *fig.* 404.

Spec. Char., &c. Leaflets roundish-elliptic, retuse. Racemes drooping, few-flowered. Callosities of standard papilliform. Legumes rather coriaceous, pubescent. (*Don's Mill.*) A branchy shrub. Nepal. Height 3 ft. to 10 ft. Flowers yellow; August and September. Legume reddish; ripe in October.

The leaflets are small, and, being more imperfectly developed than in any of the other kinds, give it a singular appearance.

GENUS XVII.

ASTRA'GALUS *Dec.* THE MILK VETCH. *Lin. Syst.* Diadélphia Decándria.

Identification. Dec. Astrag., No. 5.; Prod., 2. p. 291.; Don's Mill., 2. p. 253.
Synonyme. Astrágalus sp. of Lin. and others; Astragale, *Fr.*; Tragant, *Ger.*; Astragalo, *Ital.*
Derivation. From *astragalos*, a vertebra; the seeds in the legumes of some species being squeezed into a squarish form, so as to look something like the joints of the backbone; or, perhaps, from *astēr*, a star, and *gala*, milk. It is also the name given to a shrub by Greek writers.

Gen. Char. *Calyx* 5-toothed. *Keel* obtuse. *Stamens* diadelphous. *Legume* bilocular or half-bilocular, from the upper suture being bent in so much. (*Don's Mill.*)

Leaves compound, alternate, stipulate, deciduous; pinnate: petioles permanent. *Flowers* in the only hardy ligneous species purplish or white. — Shrubs, low; natives of the South of Europe and Asia.

ᵴ 1. *A.* TRAGACA'NTHA *L.* The Goat's-Thorn Milk Vetch, *or Great Goat's Thorn.*

Identification. Lin. Sp., 1073.; Woodv. Med. Bot., 276. t. 98.; Wats. Dend. Brit., 84.
Synonyme. *A.* massiliénsis *Lam. Dict.* 1. p. 320., *Dec. Astr.* No. 96., *Don's Mill.* 2. p. 266.
Engravings. Pall. Astr., t. 4. f. 1, 2.; Duh. Arb., 2. t. 100.; Woodv. Med. Bot., 276. t. 98.; Wats. Dend. Brit., t. 84.; and our *fig.* 405.

Spec. Char., &c. Peduncles usually 4-flowered, about equal in length to the leaves. Calyxes cylindrical, with 5 short blunt teeth. Leaves with 9—11 pairs of elliptic hoary leaflets. (*Don's Mill.*) A low prickly glaucous shrub, with sub-persistent leaves, and persistent petioles. Marseilles and Narbonne, in sandy places, as well as Corsica and Mauritania. Height 6 in. to 1 ft. Introduced in 1640. Flowers purplish or white; May to July. Legumes tomentose, brown; ripe in September. General aspect whitish.

405. *Astrágalus Tragacántha.*

The flowers are disposed on axillary peduncles, so short as to prevent them from being at all conspicuous above the leaves. After the leaflets drop off, the petioles become indurated, so as to give the plant the appearance of being densely covered with spines. Propagated by seeds, which it sometimes ripens in England, or by cuttings. It requires a dry soil, and a sunny situation; and is well adapted for rock-work.

Other ligneous Species of Astrágalus. — A. *altàicus* Lodd. Cab.; A. *aristàtus* L'Hérit. Stirp. 170., with yellow flowers, which is figured in Bot. Cab. t. 1278., and our *fig.* 406.; A. *brevifòlius*, with a purplish flower, figured in Bot. Cab. t. 1388., and our *fig.* 407.; and A. *massiliénsis* Lam., which is probably a variety of *A.* Tragacántha with white flowers instead of purplish

406. *Astrágalus aristàtus.*

407. *Astrágalus brevifòlius.*

ones, are in British collections; but, though technically ligneous, are usually treated as rock herbaceous plants.

Sect. III. *Hedysa'reæ.*

Genus XVIII.

CORONI'LLA *Neck.* The Coronilla. *Lin. Syst.* Diadélphia Decándria.

Identification. Neck. Elem., No. 1319.; Lam. Ill., t. 630.; Dec. Prod., 2. p. 309.
Synonyme. Coronilla sp. of Lin. and others.
Derivation. From *corona*, a crown; in reference to the disposition of the flowers in crowns, or umbels, at the tops of the peduncles.

Gen. Char. *Calyx* campanulate, short, 5-toothed, the two superior teeth approximate, and joined together higher up than the rest. *Claws* of petals usually longer than the calyx. *Carina* acute. *Stamens* diadelphous. *Legume* nearly terete, slender, at length separating into oblong 1-seeded joints. *Seeds* ovate or cylindrical. (*Don's Mill.*)

Leaves compound, alternate, stipulate, deciduous; impari-pinnate. *Flowers* in axillary peduncles, bearing at their tops umbels of pedicellate flowers, usually yellow.—Shrubs, deciduous or sub-evergreen; natives of the South of Europe or Asia.

They are all highly ornamental, and most of them produce seeds in England, by which, or by cuttings, they are easily propagated in common soil.

■ 1. C. E'merus *L.* The Scorpion-Senna Coronilla.

Identification. Lin. Sp., 1046.; Dec Prod., 2. p. 309.; Don's Mill., 2. p. 274.
Synonymes. E'merus màjor *Mill. Icon.* t. 132. f. 1., and E. minor, f. 2.; C. pauciflòra *Lam. Fl. Fr.*
Engravings. Bot. Mag., t. 445.; N. Du Ham., 4. t. 131.; and our *fig.* 408.

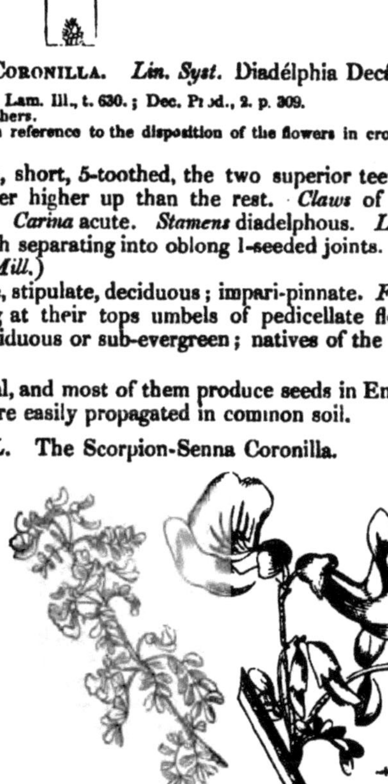

408. Coronilla E'merus.

Spec. Char., &c. Shrubby, glabrous. Its leaves are attended by minute stipules, and have 5—7 obovate leaflets. Its flowers are yellow, disposed 3 upon a peduncle. The claws of the petals are thrice as long as the calyx. The legume is rather cylindrical than compressed, and its joints separate slowly and unobviously, but they do separate. (*Dec. Prod.*) A bushy glabrous shrub. Middle and Southern Europe, in thickets and hedges. Height 4 ft. to 10 ft. Introduced in 1596. Flowers yellow; April to June. Legume brown; ripe in September.

Before the flowers are expanded, the corolla is partly red externally, mostly so towards the tips of the petals; and the mingling of the yellow flowers, with flower-buds more or less red, and the elegant foliage, produce a fine effect. It flourishes most in a sunny sheltered situation, and a dry soil. It bears clipping, and would form a beautiful garden hedge.

2. C. JU'NCEA *L.* The rushy-*branched* Coronilla.

Identification. Lin. Sp., 1047.; Dec. Prod., 2. p. 309.; Don's Mill., 2. p. 274.
Engravings. Bot. Reg., t. 820.; Bot. Cab., t. 235.; and our *fig.* 409.

Spec. Char., &c. Shrubby, glabrous. Branches rush-like, round, bearing but few leaves; the latter are attended by minute stipules, and have 3—7 leaflets, that are linear oblong, obtuse, and rather fleshy; the lowest leaflets being rather distant from the base of the petiole. The flowers are yellow, 5—7 in an umbel. The claws of the petals are scarcely longer than the calyx. The legume is rather compressed, and its joints separate obviously. (*Dec. Prod.*) An erect glaucous shrub. South of France. Height 2 ft. to 3 ft. Introduced in 1756. Flowers bright yellow; June and July. Legume brown; ripe in September.

409. C. júncea.

It deserves a place in collections, on account of the singularity of its rush-like slender branches, which, like those of *Spártium júnceum*, are partly destitute of leaves.

Sect. IV. *Phaseo'leæ.*

Genus XIX.

WISTA'R*IA* Nutt. The Wistaria. *Lin. Syst.* Diadélphia Decándria.

Identification. Nutt. Gen. Amer., 2. p. 115.; Dec. Prod., 2. p. 389.; Don's Mill., 2. p. 348.
Synonymes. Glýcine sp. *L.*, Thyrsánthus *Elliot*, Kraúnhia Rafin.
Derivation. Named in honour of *Caspar Wistar*, late Professor of Anatomy in the University of Pennsylvania. (*Don's Mill.*, ii. p. 348.) Nuttall first characterised and named this genus, from the American species, which he denominated W. speciósa; but which DeCandolle has changed to W. frutéscens. In DeCandolle's *Prodromus*, and some other works, Wistària is erroneously spelled Wistèria.

Gen. Char. *Calyx* campanulate, somewhat bilabiate, upper lip with two short teeth, lower one with 3 subulate teeth. *Corolla* papilionaceous. *Vexillum* bicallose. *Wings* conforming to the keel, which is 2-edged. *Stamens* diadelphous. Nectariferous tube girding the stipe of the ovary. *Legume* standing on a short stipe, coriaceous, 2-valved, 1-celled, rather torulose at the seeds. (*Don's Mill.*)

Leaves compound, opposite, exstipulate, deciduous; impari-pinnate. *Flowers* bluish purple, in terminal and axillary racemes.—Shrubs, deciduous, twining; natives of North America and China; of vigorous growth, forming, when in flower, some of the most splendid ornaments of British gardens.

The species are quite hardy, will grow in any soil, and are generally propagated by layers of the young shoots, which will root at every joint if laid down during summer as they grow. They may also be propagated by cuttings of the roots; or by seeds.

1. W. FRUTE'SCENS *Dec.* The shrubby Wistaria.

Identification. Dec. Prod., 2. p. 390.; Don's Mill., 2. p. 348.
Synonymes. Glýcine frutéscens *Lin. Sp.* 1067.; A'pios frutéscens *Ph. Fl. Am. Sept.* 2. p. 474.; Anónymos frutéscens *Walt. Fl. Car.* 186.; Wistària speciòsa *Nutt. Gen. Amer.* 2. p. 115.; Thyrsánthus frutéscens *Elliot Journ. Acad. Sci. Philad.*; Phaseolöîdes *Hort. Angl.* 55.; the Kidneybean Tree.
Engravings. Bot. Mag., t. 2103.; and our *fig.* 410.

410. Wistària frutéscens.

Spec. Char., &c. Wings of the corolla each with two auricles. Ovary glabrous. Flowers odorous. (*Dec. Prod.*) An elegant deciduous climber. Virginia, Carolina, and the Illinois, in boggy places. Stem 20 ft. to 30 ft. Introd. 1724. Flowers bluish purple, sweet-scented, the standard having a greenish yellow spot at the base; July to September. Legume brown; ripe in October.

It is readily propagated by cuttings of the root and by layers, and forms a very ornamental climber, especially when trained against a wall.

2. W. CHINE'NSIS *Dec.* The Chinese Wistaria.

Identification. Dec. Prod., 2. p. 390.; Don's Mill., 2. p. 348.
Synonymes. Glýcine chinénsis *Bot. Mag.* t. 2083.; G. sinénsis *Bot. Reg.* t. 650.; Wistària Consequàna Loudon Gard. Mag. vol. ii. p. 422., and Hort. Brit.
Engravings. Swt. Brit. Fl.-Gard., t. 211.; Bot. Mag., t. 2083.; Bot. Reg., t. 650.; Bot. Cab., t. 773.; and our *fig.* 411.

411. Wistària chinénsis

Spec. Char., &c. Wings of the corolla each with one auricle. Ovary villose. Flowers large. (*Dec. Prod.*) A vigorous-growing deciduous twiner. China. Stem 50 ft. to 120 ft. Introduced in 1816. Flowers pale bluish purple; May and June, and sometimes producing a second crop of flowers in August. Legume ?.

The flowers are larger than those of W. frutéscens: they are disposed in longer and looser racemes, and are somewhat paler in colour. On established plants they are produced in great abundance; but they have not yet been succeeded by legumes in England. This plant may truly be considered the most magnificent of all our hardy deciduous climbers. It will grow wherever the common laburnum will flourish; but, as its flowers are somewhat more tender than those of that tree, they are more liable to be injured by frosts in very late springs. A plant in the Hort. Soc. Garden, against a wall, extends its branches above 100 ft. on each side of the main stem; one at Coughton Hall covers 905 superficial feet of walling.

Sect. V. *Cassie'æ.*

Genus XX.

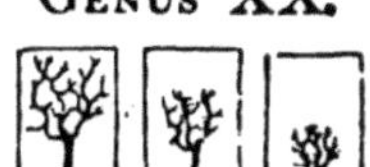

GLEDI'TSCHIA L. THE GLEDITSCHIA. *Lin. Syst.* Polygàmia Diœ'cia.

Identification. Lin. Gen., 1159.; Lam. Ill., p. 857.; Dec. Prod., 2. p. 479.; Don's Mill., 2. p. 428.
Synonymes. Acàcia sp. *Pluk.*; Févier, *Fr.*; Gleditschie, *Ger.*; Gleditsia, *Ital.*

Derivation. In honour of *Gottlieb Gleditsch*, of Leipsic, once a professor at Berlin, and defender of Linnæus against Siegesbeck; author of *Methodus Fungorum* (1753), *Systema Plantarum a Staminum situ* (1764), and many other smaller works.

Gen. Char. *Flowers* unisexual from abortion, or hermaphrodite. *Calyx* of 3—4—5 equal sepals, which are connected together at the base into a cupule. *Petals* equal in number to the sepals, rising from the tube of the calyx. *Stamens* equal in number to the sepals. *Style* short. *Stigma* pubescent above. *Legume* continuous, furnished with more or less pulp. *Seeds* compressed. (*Don's Mill.*)

Leaves compound, abruptly pinnate, and bipinnate, on the same tree; or, rarely, by the coalition of the leaflets, almost simple, alternate, stipulate, deciduous. *Flowers* greenish, in spikes.—Trees, deciduous, of the 1st, 2d, and 3d ranks; natives of North America or China. Branchlets supra-axillary, and often converted into branched spines. Decaying leaves yellow. Naked young wood purplish or brownish green.

The species are of easy culture in good free soil; and, in Britain, generally propagated by imported seeds, or grafting. The species appear to be in a state of great confusion in British gardens; and, judging from the trees in the Hort. Soc. Garden, and in the arboretum of Messrs. Loddiges, we should conjecture that there are, probably, not more than two species, the American and the Chinese. The Chinese species is distinguished by its trunk being more spiny than its branches.

♆ 1. G. TRIACA'NTHOS *Lin.* The three-thorned Gleditschia, *or Honey Locust.*

Identification. Lin. Sp., 1509.; Dec. Prod., 2. p. 479.; Don's Mill., 2. p. 428.
Synonymes. G. triacánthos var. α polyspérma *Mart. Mill.*; G. melilòba *Walt.*; G. spinòsa *Du Ham.*; *Acàcia* triacánthos *Hort.*; *Acàcia* americàna *Pluk.*; Févier d'Amérique, *Fr.*; Fava americana, *Ital.*; Thorny Acacia, Sweet Locust, *United States*; Carouge à Miel, *Canada.*
Engravings. Michx. fil. Arb., 2. p. 164. t. 10.; Wats. Dend. Brit., t. 138.; the plates of this species in Arb. Brit., 1st edit., vol. v.; and our *fig.* 412.

Spec. Char., &c. Spines simple or trifid; stout, at the very base compressed, in the upper part cylindrical, but tapered. Leaflets linear-oblong. Legumes flattish, rather crooked, many-seeded, and more than ten times as long as broad. (*Dec. Prod.*) A large tree. Carolina and Virginia. Height 50 ft. to 80 ft. Introduced in 1700. Flowers greenish; June and July. Legume brown; ripe in October. Decaying leaves yellow. Naked young wood purplish brown.

♆ *G. t.* 2 *inérmis* Dec., G. læ'vis *Hort.* (Dec. Leg. Mém., 2. t. 22. fig. 109.; Catesb. Carol., 1. t. 43.; Pluk. Alm., t. 123. fig. 3.; and the plates of this variety in Arb. Brit., 1st edit., vol. v.), has the stem and branches not spiny, or but very sparingly so.

♆ *G. t.* 3 *brachycárpa.* G. brachycárpa *Pursh*, G. triacánthos var. β *Michx.* — Leaflets oblong obtuse. Spines thick, short, not rarely 3 together. Legumes oblong, short.

The trunk and branches, when the tree is young, are covered with large prickles, which, though they are not ligneous, become hard, and remain on for several years, and offer a formidable defence. These prickles are not only produced by the young wood, but occasionally protrude themselves from the trunk, even when the tree is of considerable bulk and age. In general, the trunk presents a twisted appearance, and the branches proceed from it rather horizontally than in an upright direction. The pinnated foliage is particularly elegant, and of an agreeable light shining green: it appears late in spring, the trees in the neighbourhood of London sometimes not being fully clothed till the middle or end of June; and it begins to turn yellow, and drop off, early in autumn. The flowers are inconspicuous; the male flowers being in the form of catkin-like racemes of nearly the same colour as the leaves. Some trees in the grounds at Syon have ripened seeds, the pods containing which, being 1 ft. to 2 ft. in length, and remaining on the trees after the leaves are off, have a singular appearance. These crooked pendulous pods are of a reddish

412. Gledítschia triacánthos.

brown colour; they contain hard, smooth, brown seeds, enveloped in a pulpy substance, which, for about a month after the maturity of the seeds, is very sweet, but after a few weeks becomes extremely sour. The rate of growth of this tree, for the first 15 or 20 years, is generally about the average of a foot a year; but in favourable situations it will grow at double that rate. In the garden of the Hort. Soc., and in the arboretum of the Messrs. Loddiges, plants 10 years planted were, in 1835, from 20 ft. to 25 ft. in height. The wood of this tree, when dry, weighs at the rate of 52 lb. the cubic foot: it is very hard, and splits with great facility, resembling in this and other respects the wood of the robinia; but its grain is coarser, and its pores more open. In Britain, this, and all the kinds of the genus, can only be considered as ornamental trees; but in that character they hold the first rank; their delicate acacia-like foliage, and the singularly varied, graceful, and picturesque forms assumed by the tree, more especially when young or middle-aged, together with the singular feature afforded by its spines, will always recommend it in ornamental plantations. It requires a deep, rich, free soil, and a situation not exposed to high winds; and it requires the South of England or France to ripen the seeds. The species is always propagated by seeds imported from America, or from France or Italy. The plants are best transplanted to where they are finally to remain when quite young; as they make but few fibrous roots, and these take, for the most part, a descending direction. The variety G. t. inérmis can only be insured by grafting on the species. In general, however, abundance of plants without spines may be selected from beds of seedlings of G. triacánthos.

¶ 2. G. (T.) MONOSPE'RMA *Walt.* The one-seeded Gleditschia, *or Water Locust.*

Identification. Walt. Car., 254.; Dec. Prod., 2. p. 479.; Don's Mill., 2. p. 428.
Synonymes. G. carolinénsis *Lam. Dict.* 2. p. 464.; G. aquática *Marsh.*; G. triacántha *Gært. Fruct.* 2. p. 149.
Engravings. Mill. Icon., 5.; and our *fig.* 413.; in which the male flower, the pod, and the seed, are of the natural size.

Spec. Char., &c. Spines slender, not rarely trifid, few. Leaflets ovate-oblong, acute. Legumes flattish, roundish, 1-seeded. (*Dec. Prod.*) A tree of the largest size. Native of Carolina, Florida, and the Illinois, in moist woods. Height 60 ft. to 80 ft. Introduced in 1723. Flowers greenish; June and July. Legume not seen in England.

Closely resembles the honey locust, from which, in England, where neither of them ripens seeds, it is almost impossible to distinguish it. The bark, though smooth when the tree is young, yet cracks and scales off when the tree grows old, as in G. triacánthos. The ieaves, Michaux says, differ from those of G. triacánthos, in being a little smaller in all their proportions. The branches are armed with thorns, which are also less numerous, and somewhat smaller than those of G. triacánthos. The tree is treated in all respects like G. triacánthos; of which it has, till lately, been considered only a variety. It is raised in the nurseries from imported seed; but whether the plants really turn out perfectly distinct, with respect to the form of their fruit, is uncertain; from their not having yet, as far as we know, fruited in England.

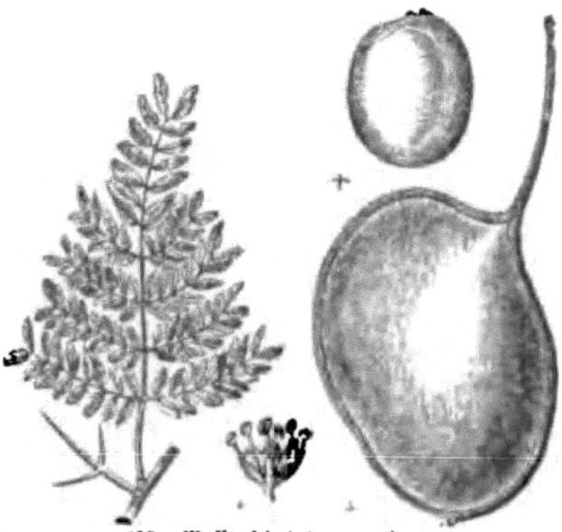

413. Gleditschia (t.) monospérma.

♀ 3. G. SINE'NSIS *Lam.* The Chinese Gleditschia.

Identification. Lam. Dict., 2. p. 465.; Dec. Prod., 2. p. 479.; Don's Mill., 2. p. 428.
Synonymes. G. hórrida *Willd. Sp.* 4. p. 1098.; Févier de la Chine, *Fr.*
Engravings. Dec. Légum. Mém., l. t. 1.; the plate of this species in Arb. Brit., 1st edit., vol. v.

Spec. Char., &c. Spines stout, conical; those on the branches simple or branched; those on the stem grouped and branched. The leaflets ovate-elliptical, obtuse. Legumes compressed, long. The spines in this species are axillary, not distant from the axil. (*Dec. Prod.*) A middle-sized tree. China. Height 30 ft. to 50 ft. Introduced in 1774. Flowers greenish; June and July. Legume not seen in England.

414. Gleditschia s. inérmis.

Varieties.

♀ *G. s.* 2 *inérmis* N. Du Ham., G. japónica *Lodd. Cat.*, G. javánica *Lam.* (see the plate of this tree in Arb. Brit., 1st edit., vol. v.; and our *fig.* 414.), differs from G. sinénsis in being without spines, of much less vigorous growth, and in having the foliage of a much deeper green. It seems a very desirable variety for small gardens.

♀ *G. s.* 3 *màjor* Hort., G. hórrida màjor *Lodd. Cat.*, seems scarcely to differ from the species.

♀ *G. s.* 4 *nàna* Hort., G. h. nàna in Hort. Soc. Gard. (see the plate of this tree in Arb. Brit., 1st edit., vol. v.; and our *fig* 415.), is a tree of somewhat lower growth than the species, but scarcely, as it appears to us, worth keeping distinct.

♀ *G. s.* 5 *purpùrea* Hort., G. h. purpùrea *Lodd. Cat.* (see the plate in Arb. Brit., 1st edit., vol. v.; and our *fig.* 416.), is a small tree of compact upright growth, very suitable for gardens of limited extent.

Other Varieties of G. sinénsis.—In Loddiges's arboretum there is a plant

415. Gledítschia s. nàna

416. Gledítschia s. purpùrea.

marked *G. chinénsis* (Potts), which was imported from China by the Hort. Soc. It is, at present, a low bush, and may, perhaps, prove something distinct. There were also, in 1835, in the Hort. Soc. Garden, some plants without names, apparently belonging to this species; but, as we have already observed, the genus is in great confusion, and nothing perfectly satisfactory can be stated respecting it.

The spines, which are very strong and branchy, are more abundant on the trunk than on the branches, and are frequently found in bundles.

The leaves are bipinnate, and the leaflets are elliptic obtuse, notched on the edges, smooth, shining, and much larger than those of any other species. (*Desf. Arb.*, ii. p. 248.) The pods are rarely above 6 in. long. The tree stands the cold better than the honey locust, and has ripened its fruit in Paris, in the Jardin des Plantes, and in the nursery of M. Cels. (*Dict. des Eaux et des Forêts*, vol. ii. p. 150.)

♆ 4. G. (s.) MACRACA'NTHA *Desf.* The long-spined Gleditschia.

Identification. Desf. Arb., 2. p. 246.; Don's Mill., 2. p. 428.
Synonymes. G. fèrox *Baudr.*; Févier à grosses E'pines, *Fr.*
Engraving. The plate of this species in Arb. Brit., 1st edit., vol. v.

Spec. Char., &c. Prickles strong, branchy, numerous. Leaflets lanceolate, somewhat rigid, notched, dentated. Pods elongated. (*Desf. Arb.*) A middle-sized tree, with a prickly trunk. China ?. Height 40 ft. to 50 ft. Flowers greenish; June and July

The prickles are axillary and large. The leaves are twice winged; the leaflets large, coriaceous, dark green, and shining on the upper surface. The young shoots are covered with extremely short hairs, and are of a purplish brown colour. On the whole, it bears a close resemblance to G. sinénsis, of which it is, probably, only a variety. It is very hardy; and Desfontaines says that it fruits freely in France. The fruit ripens in the autumn; and the pods are long, pendulous, swelled, and rather cylindrical. They are filled with a sharp acrid pulp, somewhat resembling that of tamarinds, but the emanations from which, when inhaled, occasion sneezing.

♆ 5. G. (s.) FE'ROX *Desf.* The ferocious-*prickled* Gleditschia.

Identification. Desf. Arb., 2. p. 247.; Don's Mill., 2. p. 428.
Synonymes. G. orientàlis *Bosc*; Févier hérissé, *Fr.*
Engraving. Our *fig.* . in p. .

Spec. Char., &c. Prickles large, robust, much compressed, trifid. Leaflets lanceolate, notched. (*Desf. Arb.*)

A middle-sized branching deciduous tree, the trunk of which is thickly beset with strong branchy prickles, and which is supposed to grow from 30 ft. to 50 ft. in height; but of which the native country, and year of introduction into Britain, are unknown. Judging from the plants in the Hort. Soc. Garden, and those in the arboretum of Messrs. Loddiges, we should say it was only a variety of G. sinénsis; though Desfontaines states the foliage and habit of growth to be somewhat different. It has not yet flowered in Europe.

♆ 6. G. CA'SPICA *Desf.* The Caspian Gleditschia.

Identification. Desf. Arb., 2. p. 247.; Don's Mill., 2. p. 428.
Synonyme. G. caspiàna *Bosc.*
Engraving. Our *fig.* 417.

417. G. cáspica.

Spec. Char., &c. Prickles slender, trifid, compressed. Leaflets elliptic-lanceolate, obtuse. (*Desf. Arb.*) A middle-sized tree Persia, and found also near the Caspian Sea. Height 30 ft. to 40 ft. Introduced in 1822.

Nothing is known of its flowers and fruit; but it strongly resembles G. sinénsis (of which it is, probably, only a variety) in its leaves, general appearance, and habit.

Variety.

♆ *G. c.* 2 *subviréscens* Hort., Févier verdâtre, *Fr.*, is mentioned in the *Bon Jardinier* for 1836, as a variety of this species.

Other Sorts of Gleditschia.—Every modification of the species of this genus is so interesting, both in point of the elegance of its foliage, and the singularity of its prickles, that new varieties have been eagerly sought after by cultivators; and the genus seems particularly favourable to this desire, from the tendency of seedling plants to sport. Hence there are several names in collections, of which it is difficult to say any thing satisfactory in the present young and immature state of the plants. In the Hort. Soc. Garden, there were in 1837 *G. micracántha*, *G.* Bòqui, and *G. præ'cox*; and in Messrs. Loddiges's arboretum were plants marked *G. aquática*, which are evidently the same as G. monospérma, *G. orientàlis*, evidently G. fèrox, *G. chinénsis* (already mentioned); and some young plants without names.

Genus XXI.

GYMNO'CLADUS *Lam.* The Gymnocladus. *Lin. Syst.* Diœ'cia Decándria.

Identification. Lam. Dict., 1. p. 733.; Ill., t. 828.; Dec. Prod., 2. p. 479.
Derivation. From *gumnos*, naked, and *klados*, a branch; from the naked appearance of the branches during winter, when they seem, unless perhaps at the points of the shoots, totally devoid of buds.

Gen. Char. *Flowers* diœcious from abortion. *Calyx* tubular, 5-cleft. *Petals* 5, equal, oblong, exserted from the tube. *Stamens* 10, enclosed. *Legume* oblong, thick, filled with pulp inside. (*Don's Mill.*)

Leaves compound, alternate, stipulate, deciduous; bipinnate. *Flowers* in terminal racemes, white. — A tree, deciduous, with upright branches and inconspicuous buds; native of North America.

⚲ 1. G. canade'nsis *Lam.* The Canada Gymnocladus, *or Kentucky Coffee Tree.*

Identification. Lam. Dict., 1. p. 733., and Ill., t. 823.; Michx. Fl. Bor. Amer., 2. p. 241.; Dec. Prod., 2. p. 480.; Don's Mill., 2. p. 429.
Synonymes. Guilandina dioica *Lin. Sp.* 546.; Hyperanthèra dioica *Vahl Symb.* 1. p 31., *Duh. Arb.* 1. t. 103.; Nicker Tree, Stump Tree, *United States*; Bonduc, Chiquier, *Fr.*; Chicot, *Canadians*; Canadischer Schusserbaum, *Ger.*
Engravings. Reich. Mag., t. 40.; Duh. Arb., t. 103.; our plates of this tree in Arb. Brit., 1st edit., vol. v.; and our *fig.* 418.

Spec. Char., &c. Branches blunt at the tip, bipinnate leaves, flowers in racemes, and whitish petals. The leaf has 4—7 pinnæ; the lower of which consist each of but a single leaflet, the rest each of 6—8 pairs of leaflets. (*Dec Prod.*) A singular tree. Canada. Height 30 ft. to 60 ft. Introduced in 1748. Flowers white; May to July. Decaying leaves yellow. Naked young wood of a mealy white, without any appearance of buds.

The branches have almost always an upright direction; and the appearance of the head, in the winter season, is remarkable, from being fastigiate, and from the points of the branches being few, and thick and blunt, as compared with those of almost every other tree. They are also wholly without the appearance of buds; and this latter circumstance, connected with the former, gives the tree, during winter, the appearance of being dead; and hence the Canadian name of chicot, or stump tree. The leaves, on young thriving trees, are 3 ft. long, and 20 in. wide; but, on trees nearly full grown, they are not half that size. The leaflets are of a dull bluish green, and the branches of the petioles are somewhat of a violet colour. It is very hardy, and flowers freely in the neighbourhood of London, but does not produce pods. The wood is hard, compact, strong, tough, and of a fine rose colour. In America, it is used both in cabinet-making and carpentry, and, like the wood of the robinia, it has the remarkable property of rapidly converting its sap-wood into heart-wood; so that a trunk 6 in. in diameter has not more than six lines of sap-wood, and may, consequently, be almost entirely employed for useful purposes. The seeds were, at one time, roasted and ground as a substitute for coffee in Kentucky and Tennessee; but their use in this way has been long since discontinued. The pods, preserved like those of the tamarind (to which

418. Gymnócladus canadénsis.

this genus is nearly allied), are said to be wholesome, and slightly aperient. In Britain, the only use of the tree is for ornamental purposes; and, considered as an object of curiosity and beauty, no collection ought to be without it. A rich, deep, free soil is essential to the thriving of this tree; and such a soil is never met with naturally in exposed situations. The tree is generally propagated by imported seeds; but it will grow freely from cuttings of the roots, care being taken in planting to keep that end upwards which is naturally so.

Genus XXII.

CE'RCIS *L.* The Judas Tree. *Lin. Syst.* Decándria Monogýnia.

Identification. Lin. Gen., No. 510.; Dec. Prod., 2. p. 518.; Don's Mill., 2. p. 463.
Synonymes. Siliquástrum *Tourn. Inst.* t. 414., *Mœnch Meth.*; Gainier, *Fr.*; Judasbaum, *Ger.*; Albero de Giuda, *Ital.*
Derivation. From *kerkis*, a shuttlecock, the name given by Theophrastus to this tree.

Gen. Char. *Calyx* urceolate at the base and gibbous, bluntly 5-toothed at the apex. *Petals* 5, unguiculate, all distinct, disposed in a papilionaceous manner; the wings or side petals the largest. *Stamens* 10, free, unequal. *Ovary* on a short stipe. *Legume* oblong, slender, compressed, 1-celled, many-seeded, somewhat winged on the seminiferous suture. *Seeds* obovate. (*Don's Mill.*)

Leaves simple, alternate, stipulate, deciduous; heart-shaped at the base, many-nerved, rising after the flowers have decayed. *Flowers* in 1-flowered pedicels, rising from the trunk and branches in fascicles.—Trees, deciduous, of the third rank; natives of Europe, or North America. Decaying leaves yellowish purple. Propagated by seeds or grafting.

♆ 1. *C.* SILIQUA'STRUM *L.* The *common* Judas Tree

Identification. Lin. Sp., 534.; Dec. Prod., 2. p. 518.; Don's Mill., 2. p. 463.
Synonymes. Siliquástrum orbiculàtum *Mœnch Meth.*; Love Tree; Gainier commun, Arbre de Judée, *Fr.*; Arbol d'Amor, *Span.*; Judasbaum, *Ger.*
Engravings. N. Du Ham., t. 7.; Bot. Mag., t. 1138.; the plates of this species in Arb. Brit., 1st edit., vol. v.; and our *fig.* 419.

Spec. Char., &c. Leaves very obtuse, and wholly glabrous. (*Dec. Prod.*) A low tree. South of Europe, in Greece, in Asiatic Turkey, and more especially in Judea. Height 20 ft. to 30 ft. Introduced in 1596. Flowers purplish pink; May. Legume brown; ripe in September. Naked young wood purplish, with small white spots.

Varieties.

- ♣ C. S. 2 *parviflòrum* Dec.—A shrub; its branches spotted with white, its flowers smaller by half than those of the species. A native of Bokhara.
- ♆ C. S. 3 *flòre álbido.*—Flowers whitish. H. S.
- ♆ C. S. 4 *ròsea.*—A seedling, raised from foreign seeds, which has flowered in the Botanic Garden at Kew; has numerous flowers, which are brighter, and a shade darker, than those of the species; and they also appear about a fortnight later; but it is, perhaps, hardly worth noticing as a variety.

419. Cércis Siliquástrum

The common Judas tree, in the South of Europe, forms a handsome low tree. with a flat spreading head, in the form of a parasol; and it is a singularly beautiful object in spring, especially when it is covered with its numerous bright purplish pink flowers, which appear before the leaves, and are produced not only from the young wood, but from wood of 6 or 8 years growth, and even from the trunk. The leaves are not liable to be attacked by insects. The flowers are succeeded by flat, thin, brown pods, nearly 6 in. in length, which remain on the tree all the year, and give it a very singular appearance in the winter season. The rate of growth is about 18 in. a year, for the first 10 years. The wood is very hard, and agreeably veined, or rather blotched or waved, with black, green, and yellow spots, on a grey ground. It takes a beautiful polish, and weighs nearly 48 lb. to the cubic foot. The flowers, which have an agreeable acid taste, are mixed with salads, or fried with batter, as fritters; and the flower-buds are pickled in vinegar. In British gardens, the tree grows about the same height, and flowers about the same time, as the laburnum, the Guelder rose, and the hawthorn, and enters into beautiful combination with these and other trees. The foliage is hardly less beautiful and remarkable than the flowers; the leaves being of a pale bluish green on the upper surface; and of a sea-green underneath, and of a cordate reniform shape, apparently consisting of two leaflets joined together; which circumstance, combined with others, brings the genus in close alliance with that of Bauhínia. Like most of the Leguminàceæ, this tree prefers a deep, free, sandy soil, rich rather than poor; and it will only thrive, and become a handsome tree, in sheltered situations. In the northern parts of the island, it requires to be planted against a wall; and few ornamental trees better deserve such a situation. The species is propagated by seeds, and the varieties by grafting. The seeds are sown on heat early in spring, and come up the same season; and the plants will produce flowers in three or four years.

♀ 2. *C.* CANADE'NSIS *L.* The Canada Judas Tree.

Identification. Lin. Sp., 534.; Dec. Prod., 2. p. 518.; Don's Mill., 2. p. 463.
Synonymes. Silliquástrum cordàtum *Mœnch Meth.*; red Bird Tree, *Amer.*; Gainier de Canada, Bouton rouge, *Fr.*
Engravings. Mill. Icon., t. 2.; the plate of this species in Arb. Brit., 1st edit., vol. v.; and our *fig.* 420.

420. *Cércis canadénsis.*

Spec. Char., &c. Leaves acuminate, villose beneath at the axils of the veins. As compared with *C.* Siliquástrum, its flowers are of a paler rose colour, the legume is on a longer pedicel, and tipped with a longer style. A low tree. Canada to Virginia. (*Dec. Prod.*) Height 10 ft. to 20 ft. Introd. 1730. Flowers red; May and June.; Legume brown; ripe in August.

Variety.

¶ C. c. *pubéscens* Ph. — Leaves pubescent on the under surface. (*Dec.*)

This tree bears a general resemblance to the preceding species; but it is more slender and smaller in all its parts; and it seldom rises higher than 20 feet. It is at once distinguished from *C.* Siliquástrum by its leaves being heart-shaped and pointed; they are also much thinner, more veined, and of a lighter green; and the flowers are generally produced in smaller numbers than in the other species. The flowers are used by the French Canadians in salads and pickles, and the young branches to dye wool of a nankeen colour. The wood resembles that of the other species. Propagated by imported seeds, and considered more tender than *C.* Siliquástrum.

Order XXVI. ROSA'CEÆ.

The term Rosàceæ has been applied to this order, because all the species belonging to it agree more or less with the genus Ròsa, in essential characters. It includes many genera belonging to the Linnæan class Icosándria.

Ord. Char. *Flowers* regular. *Calyx*, in most cases, with 5 lobes, the odd one posterior to the axis of inflorescence. *Petals* and *stamens* arising from the calyx. *Stamens*, for the most part, numerous. *Ovaries* many, several, or solitary; each of one cell that includes, in most cases, 1 ovule; in some, 1 to many ovules. *Style* lateral or terminal.

Leaves simple, or pinnately divided, alternate, in nearly all stipulate. *Flowers* showy, with numerous stamens. *Fruit*, in many of the genera, edible. —Trees and shrubs, mostly deciduous; natives of Europe, Asia, and America.

The ligneous species which constitute this order include the finest flowering shrub in the world, the rose; and the trees which produce the most useful and agreeable fruits of temperate climates, viz. the apple, the pear, the plum, the cherry, the apricot, the peach, and the nectarine. The plants are, for the most part, deciduous low trees or shrubs, all producing flowers more or less showy; and the greater number fruits which are edible. They are chiefly natives of Europe and Asia; but several of them are also found in North America, and some in South America, and the North of Africa. The fruit-bearing species, and the rose, have followed man from the earliest period of civilisation, and are, perhaps, better known to mankind in general than any other ligneous plants. The medical properties of several of the species are remarkable, from the circumstance of their yielding the prussic acid; while others produce a gum nearly allied to the gum Arabic, which indicates a degree of affinity between this order and Leguminàceæ. The bark of some species, as of Cérasus virginiàna, is used, in North America, as a febrifuge; and that of others, as the capollin cherry (Cérasus *Capóllin*), for tanning, in Mexico. The leaves of Cratæ'gus Oxyacántha, Prùnus spinòsa, Cérasus sylvéstris, and Ròsa rubiginòsa, have been used as a substitute for tea, or for adulterating tea. The leaves and bark of the spiræas are said to be at once astringent and emetic. There are two characteristics of this order, with reference to its cultivation, which are of great importance to the gardener: the first is, the liability of almost all the species to sport, and produce varieties differing, in many cases, more from one another, than they differ from other species; and the second is, that they are remarkably subject to the attacks of insects and diseases. They almost all require a free loamy soil, not overcharged with moisture, and rich rather than poor; and, while all the species are increased by seeds, which, for the most part, are produced freely in Britain, or by cuttings of the roots, almost all the varieties are best increased by grafting or budding; and not, as in some other orders, with equal ease by cuttings of the shoots, or by layers.

With reference to landscape-gardening, all the rosaceous trees have three properties which deserve to be kept constantly in view by the improver of grounds: 1st, they never attain a large size; 2d, they attain their natural size and shape in a very few years, in good soil not requiring more than from 10 to 20 years; and 3d, they sooner take the character of old trees than the trees of any other natural order of ligneous plants. A few exceptions may be taken from different orders, such as the common field maple, the common laburnum, &c.; but we know of no natural order, in which, like the *Rosàceæ*, all the trees are low or middle-sized, and all take the character of age while comparatively young. Hence their value in laying out small places, where the object is to make a new place appear old, or a small place appear large, and at the same time to combine character of form with beautiful blossoms in spring, and showy (*Cratæ`gus*, Cotoneáster, and *Amelánchier*) or useful (*Pýrus* and *Prùnus*) fruit in autumn.

The genera are included by DeCandolle and G. Don in five tribes; and the following are their names and distinctions: —

Sect. I. *Amygda`leæ Juss.*

Sect. Char. Fruit a drupe; the nut 2-ovuled, 1—2-seeded. Style terminal. Calyx deciduous. Leaves feather-nerved, undivided, serrate, with the lower serratures or the petioles glanded. Stipules not attached to the petiole. Kernel containing more or less of hydrocyanic acid: chiefly fruit trees.

Amy'gdalus Tourn. Covering of nut not fleshy; nut even, or perforated Young leaves folded flatwise. Flowers almost sessile, solitary or twin, protruded before the leaves.

Pe'rsica Tourn. Covering of nut very fleshy; nut wrinkled. The characters of the other parts described under *Amýgdalus* are the same in *Pérsica*.

Armeni'aca Tourn. Covering of nut fleshy; nut furrowed at both edges, in the other parts even. Young leaves with their edges rolled inwards. Flowers almost sessile, solitary or a few together, protruded before the leaves.

Pru`nus Tourn. Covering of nut fleshy; nut indistinctly furrowed at the edges, in the other parts even. Young leaves with the edges rolled inwards. Flowers upon pedicels, in groups resembling umbels, and produced before or after the leaves.

Ce'rasus Juss. Nut subglobose, even, its covering fleshy. Young leaves folded flatwise. Flowers upon pedicels, either in groups resembling umbels, and produced before the leaves, or in racemes terminal to the shoots, protruded along with them.

Sect. II. *Spiræe`æ Dec.*

Sect. Char. Fruit of 5, or fewer, capsular carpels, which are distinct from the calyx (which is persistent in *Spiræ`a*, and, perhaps, in the other genera), and, in most cases, from each other: each contains 1—6 seeds. Style terminal. Low deciduous shrubs.

Pu'r*shia* Dec. Stamens about 20. Carpels 1—2, ovate-oblong.

Ke'r*ria* Dec. Stamens about 20. Carpels 5—8, distinct.

Spiræ`a L. Stamens 10—50. Carpels 1 to several, distinct; stipitate; each includes 2—6 seeds, affixed to the inner suture.

Sect. III. Potenti'lleæ *Juss.* (Synon. Dryàdeæ *Vent.*)

Sect. Char. Fruit an aggregation of carpels; their integuments dry or succulent; the carpels distinct from one another, and from the calyx, which is persistent, and surrounds them, and, in many, is subtended by as many bracteas as it has lobes; the bracteas alternate with the lobes. Style proceeding from a little below the tip of the carpel. Leaves, in most cases, pinnately divided. Stipules attached to the petiole. Shrubs bearing fruit, or ornamental.

RU'BUS L. Integuments of carpels juicy.

POTENTI'LLA *Nestl.* Integuments of carpels dry.

Sect. IV. *RO'SEÆ Dec.*

Sect. Char. Fruit a hip; that is, with the tube of the calyx fleshy, of a pitcher shape, contracted at the mouth; and including an aggregation of carpels attached to its inner face. Style proceeding from the inner side of the carpel. Shrubs eminently ornamental.

RO'SA Tourn. Leaf impari-pinnate. Stipules attached to the petiole. Prickles simple.

LO'WEA Lindl. Leaf simple. Stipules none. Prickles usually compound.

Sect. V. PO'MEÆ *Lindl.*

Sect. Char. Fruit a pome; that is, with the tube of the calyx become very fleshy, and including, and connate with, the carpels. Carpels normally 5, with gristly or bony walls, including 1—2 seeds; in Cydònia, several. Habit spiny or not; leaves, in most cases, undivided, in some pinnate. Stipules not connate with the petiole. Ornamental low trees, or large shrubs, with showy flowers, in some genera spiny, and in others bearing some of our best hardy kitchen and dessert fruits.

CRATÆ'GUS Lindl. Carpels 1—5 prismatic nuts with bony shells, each including 1 seed. Leaves angled or toothed; in most cases deciduous. Flowers in terminal corymbs. Spiny shrubs or low trees.

PHOTI'NIA *Lindl.* Carpels 2. Petals reflexed. Evergreen. Flowers in terminal panicles. Leaves simple, leathery, serrated or entire. In P. integrifòlia the ovaries are 3, and each includes 2 ovules.

COTONEA'STER *Medik.* Carpels 2—3; ovules 2 in each cell. Leaves simple, entire, woolly beneath. Flowers in lateral spreading corymbs.

AMELA'NCHIER Medik. Ovaries 5, each divided by a partition; ovules 1 in each cell. Ripe pome including 3—5 carpels. Petals lanceolate. Small trees. Leaves simple, serrate, deciduous. Flowers in racemes.

ME'SPILUS Lindl. Carpels 2—5 compressed nuts with bony shells, each including 1 seed. Leaves lanceolate, serrulate, deciduous. Flowers large, subsessile, subsolitary.

PY'RUS Lindl. Carpels 5, or 2—5. Seeds 2 in each carpel. Leaves simple or pinnate, deciduous. Flowers in spreading terminal cymes or corymbs.

CYDO'NIA *Tourn.* Carpels 5, each including many seeds.

Sect. I. *AMYGDA'LEÆ Juss.*

GENUS I.

AMY'GDALUS Tourn. THE ALMOND TREE *Lin. Syst.* Icosándria Monogýnia.

Identification. Tourn. Inst., t. 402.; Dec. Prod., 2. p. 330.; Don's Mill., 2. p. 482.

Synonymes. Amygdalóphora *Neck.*; Amandier, *Fr.*; Mandelbaum, *Ger.*; Mandorio, *Ital.*
Derivation. From *amussō*, to lacerate, in reference to the fissured shell of the nut. Martinius suspects that it comes from a Hebrew word which signifies vigilant; because its early flowers announce the return of spring.

Gen. Char. *Drupe* clothed with velvety pubescence, having a dry rind, which separates irregularly, containing a pitted or smooth putamen or nut. (*Don's Mill.*)

Leaves simple, conduplicate when young, alternate, stipulate, deciduous. *Flowers* nearly sessile, usually pink or rose coloured, rising either singly or by pairs from the scaly buds, earlier than the leaves. — Shrubs or trees of the middle size, deciduous. Natives of the North of Africa, and the mountains of Asia; also of Russia, and the Levant.

The fruit-bearing species are cultivated in the Middle and South of Europe and the Levant, and are propagated chiefly by grafting; and the others by grafting, layers, suckers, or cuttings of the root. The almond was included by Linnæus in the same genus with the peach and nectarine, of both of which it is, doubtless, the parent, as trees have been found with almonds in a state of transition to peaches, and with both peaches and nectarines on the same branch.

☙ 1. *A.* NA'NA *L.* The dwarf, *or shrubby*, Almond.

Identification. Lin. Mant., 396.; Dec. Prod., 2. p. 530.; Don's Mill., 2. p. 482.
Synonymes. Prùnus inérmis *Gmel.*; *A.* nàna var. *a* vulgàris *Dec.*; Amandier nain, *Fr.*; Zwerchemandel, *Ger.*; Peschino della China, *Ital.*
Engravings. Bot. Mag., t. 161.; N. Du Ham., 4. t. 30.; and our *figs.* 422, 423.

Spec. Char., &c. Leaves oblong-linear, tapered at the base, serrated, glabrous; Flowers solitary, rose-coloured. Calyx cylindrically bell-shaped. Fruit of the same shape as that of *A.* commùnis, but much smaller. (*Dec. Prod.*) A deciduous low shrub. Calmuc Tartary, very frequent on the banks of the Volga, and about Odessa. Height 2 ft. to 3 ft. Introduced in 1683. Flowers rose coloured; March and April.

Varieties.

☙ A. *n.* 2 *geórgica* Dec. *A.* geórgica *Desf. Arb.* 2. p. 221., and *Lodd. Cat.* — It differs from the species in having the lobes of the calyx lanceolate, and as long as its tube; and the styles only tomentose at the base, being scarcely so there, and not protruded. A native of Georgia, which has been cultivated in the Geneva Botanic Garden.

☙ A. *n.* 3 *campéstris* Ser. *A.* campéstris *Besser Enum.* p. 46. No. 1425., *Hort. Fl. Aust.* 2. p. 2., and *Lodd. Cat.*; *A.* Besseriàna Schott in *Cat. Hort. Vindob.* 1818, and *Lodd. Cat.* — Leaves broader. Lobes of the calyx as long as the tube. Petals narrower, longer, and white. Styles tomentose at the base. The form of the nut, according to Besser, is various. Supposed to be a native of the South of Podolia. (*Dec. Prod.*) This variety is in the Hort. Soc. Garden, where it was raised from seeds received from Dr. Fischer of Petersburg.

☙ A. *n.* 4 *sibírica* Lodd. Cat., and Lodd. Bot. Cab. 1599., and our *fig.* 421., is extant in some British botanical collections, where it is an upright shrub, about 6 ft. high, with wand-like shoots, clothed with fine, long, willow-like, glossy, serrate leaves; on account of which, and its upright habit of growth, the latter being different from that of all the other species and varieties of almond, it is valuable in every collection where variety of character is desired. H. S.

421. *A. n. sibirica.*

All the different forms of the dwarf almond are low shrubs, seldom exceeding 2 or 3 feet in height. The leaves bear a general resemblance to those of some

422. A. nàna.

of the species of willow, but are of a darker and more shining green, at least in the original species. The stems are not of long duration; but the plant throws up abundance of travelling suckers, by which it is continued naturally, and also propagated. It is common through all the plains of Russia, from 55° N. lat. to the south of the empire. In British gardens it is valuable on account of its early flowering, the gracefulness of the slender twigs, on which its flowers are produced before the leaves appear, and of its easy culture in any dry soil. Its fruit resembles that of *A.* commùnis, but is much smaller, and rarely seen in England. The plant, which is usually called the dwarf double-blossomed almond in British gardens, is *Cérasus* japónica flòre plèno, or, as it is frequently named in the nurseries, *Amýgdalus* pùmila.

423. A. nàna.

2. *A.* INCA'NA *Pall.* The hoary *dwarf* Almond.

Identification. Pall. Ross., 1. p. 13.; Smith Fl. Græc., t. 497.; Don's Miller, 2. p. 482.; Lindl. in Bot. Reg., 1839, t. 58.
Synonymes. *A.* nàna var. incàna *Guldenstäd*, and *Arb. Brit.* 1st edit. vol. ii. p. 674.; *A.* tomentósula *Lodd. Cat.*
Engravings. Pall. Ross., 1. t. 7.; Flor. Græc., t. 477.; Bot. Reg., 1839, t. 58.; and our *fig.* 424.

Spec. Char., &c. Leaves obovate serrated, clothed with white tomentum beneath. Flowers solitary. Drupe compressed, pubescent. (*Don's Mill.*) A low deciduous shrub. Caucasus and the Levant, between Smyrna and Bursa. Height 2 ft. to 3 ft. Introduced in 1815. Flowers red; March and April.

Readily known from *A.* nàna by its leaves being covered with hoariness beneath. Increased by budding on the common plum.

424. A. incàna.

3. *A.* COMMU'NIS *L.* The common Almond Tree.

Identification. Lin. Sp., 677.; Dec. Prod., 2. p. 530.; Don's Mill., 2. p. 482.
Engravings. N. Du Ham., 4. t. 29.; and the plate of this tree in Arb. Brit., 1st edit., vol. v.; and our *fig.* 425.

Spec. Char., &c. Leaves oblong-lanceolate, serrulate. Flowers solitary. Calyx of a bell-shape. Fruit compressed, and rather egg shaped. (*Dec. Prod.*) A middle-sized deciduous tree. Mauritania, and in the mountainous parts of Asia. Height 20 ft. to 30 ft. Introduced in 1538. Flowers white or rose-coloured; March and April. Drupe brownish; ripe in October. In fine seasons the fruit ripens on standards as far north as Derby, and on walls at Edinburgh.

Varieties.

A. c. 1 *amàra* Dec. *The bitter Almond.* Amandier amer, *Fr.*; gemeine Mandelbaum, *Ger.* — Flowers large. Petals pale pink, with a tinge of rose colour at the base. Styles nearly as long as the stamens, and tomentose in the lower part. Seeds bitter. There are two forms of the bitter almond; one with a hard shell, and the other with a brittle one. The tree is cultivated in the South of France, in Austria, in Italy, in Greece, &c., for its fruit, which is preferred for some purposes in medicine and in domestic economy to that of the sweet almond, particularly for giving a flavour; and for stocks for grafting the other varieties on, and the peach, apricot, and even the plum. Bitter almonds are generally mixed with sweet ones, in very small proportions, for making blancmange, &c.

425. *Amýgdalus commùnis.*

♈ A. *c.* 2 *dúlcis* Dec., Lam. Ill. t. 430. *The sweet Almond.* Amandier à petits Fruits, Amande douce, *Fr.*; susse Mandel, *Ger.* — Leaves grey-green. Flowers protruded earlier than the leaves. Styles much longer than the stamens. Fruit ovate-compressed, acuminate. Shell hard. Kernel sweet-flavoured. Cultivated in the same places as the preceding sort, and generally propagated by grafting standard high on the bitter almond, or any strong-growing seedling almonds, in order to make sure of the fruit being sweet.

♈ A. *c.* 3 *flòre plèno* Baum. Cat. has double flowers.

♈ A. *c.* 4 *fòliis variegàtis* Baum. Cat. has variegated leaves.

♈ A. *c.* 5 *frágilis* Ser., Dec. Prod. 2. p. 531. *A.* frágilis *Hell.* 1. p. 500.; Amandier des Dames, *N. Du Ham.* 4. p. 113., *Noisette Jard. Fruit.* p. 7.; Coque molle, Amandier à Coque tendre, *Fr.*; Abellan, *Provence.*—Flowers protruded at the same time as the leaves, and of a pale rose colour. Petals broader, and deeply emarginate. Leaves shorter; the petioles thick. Fruit acuminate; shell soft; kernel sweet-flavoured. Cultivated for its fruit.

♈ A. *c.* 6 *macrocárpa* Ser., Dec. Prod. 2. p. 531. Amandier à gros Fruits, *N. Du Ham.* 4. p. 112., *Noisette Jard. Fruit.* p. 7.; Amandier Sultane, Amandier des Dames, Amandier Pistache, *Fr.*— Leaves broader, acuminate, scarcely grey. Peduncles short, turgid. Flowers of a very pale rose colour, large, protruded before the leaves. Petals broadly obcordate, waved. Fruit large, umbilicate at the base, acuminate at the tip; shell hard. There are two subvarieties, one with the fruit rather smaller, called, commonly, in France, Amandier Sultane; and another, with the fruit still smaller, called there Amandier Pistache; the kernels of both of which are considered remarkably delicate, and are preferred for the table. The flowers of this variety are always produced earlier than those of any other; and the kernels of the fruit are always sweet. In British gardens, the *A.* c. macrocárpa has much the largest flowers of any of the varieties. It is a vigorous large tree, of rapid growth, somewhat more fastigiate than the species.

⚘ A. c. 7 *persicöides* Ser., Dec. Prod. 2. p. 531 Amandier-Pêcher, *N. Du Ham.* 4. p. 114., *Noisette Jard. Fruit.* p 7. — Leaves similar to those of the peach tree. Fruit ovate, obtuse; its husk slightly succulent; the shell of a yellowish dark colour, and the kernel sweet-flavoured. Du Hamel has stated that its fruits vary upon the same branch, from ovate, obtuse, with the husk rather fleshy, to ovate, compressed, acuminate, and the husk dry. Cultivated in France and Italy for its fruit, but rarely found in British gardens.

Other Varieties. The almond, considered as a fruit tree, has given rise to some other varieties, which will be found treated of at length in French works on gardening, in the *Nouveau Du Hamel*, and the *Nouveau Cours d'Agriculture*

There are several varieties of the almond in cultivation on the Continent for their fruit; and two or three in this country, partly for the same purpose, but chiefly for their flowers. The common almond, in a wild state, is found sometimes with the kernels bitter, and at other times with them sweet; in the same manner as the *Quércus* hispánica, which, in Spain, though it generally bears sweet and edible acorns, yet sometimes produces only such as are bitter. For this reason, in the case of the almond, instead of giving one form as the species, we have followed DeCandolle, and described both the bitter and the sweet almond separately, either of which may be considered as the species, and classed them with the varieties.

⚘ 4. *A.* ORIENTA`LIS *Ait.* The Eastern Almond Tree.

Identification. Ait. Hort. Kew., ed. 1. 1. p. 162., ed. 2. 3. p. 195.; Dec. Prod., 2. p. 530.; Don's Mill., 2. p. 482.
Synonyme. *A.* argéntea *Lam. Dict.* 1. p. 103., *N. Du Ham.* 3. p. 115.
Engravings. Lodd. Bot. Cat., t. 1137.; and our *fig.* 426.

Spec. Char., &c. Imperfectly evergreen. Branches and leaves clothed with a silvery tomentum; petiole of the leaf short, the disk lanceolate and entire. Flowers rose-coloured, and rather longer than those of *A.* nàna. Calyx cylindrically bell-shaped. Fruit tipped with a point. (*Dec. Prod.*) A tall shrub or low tree. Levant. Height 8 ft. to 10 ft.; and, according to Bosc, 15 ft. to 20 ft. Introduced in 1756 Flowers rose-coloured; March and April.

426. *A.* orientàlis.

Very striking, from the hoary, or rather silvery, appearance of its leaves; and it makes a handsome plant when budded standard high on the common almond or the plum. It flowers much less freely than the common almond; notwithstanding which, it well deserves a place in collections, on account of its fine silvery foliage.

GENUS II.

PE'RSICA *Tourn.* THE PEACH TREE. *Lin. Syst.* Icosándria Monogýnia.

Identification. Tourn. Inst., t. 400.; Mill. Dict.; Dec. Fl. Fr., 487.; Don's Mill., 2. p. 483.
Synonymes. *Amýgdalus* sp. of Lin. and Juss.; Trichocárpus *Neck. Elem.* No. 718.; Pêcher, *Fr.*; Pfirschenbaum, *Ger.*; Pesco, *Ital.*
Derivation. So named from the peach coming originally from *Persia*.

Gen. Char. *Drupe* fleshy, with a glabrous or velvety apicarp, and having the putamen wrinkled from irregular furrows. (*Don's Mill.*)

Leaves simple, alternate, stipulate, deciduous; conduplicate when young.

Flowers almost sessile, solitary, or twin, rising from the scaly buds earlier than the leaves.—Tree, deciduous, beneath the middle size, and not of long duration. Persia.

The peach and the nectarine are by some botanists made distinct species; but there can be no doubt of their being only varieties of one kind, which kind is itself nothing more than an improved or fleshy almond; the almond being to the peach and nectarine what the crab is to the apple, and the sloe to the plum.

♀ 1. *P.* VULGA'RIS *Mill.* The common Peach Tree.

Identification. Mill. Dict., No. 1.; Dec. Prod., 2. p. 531.; Don's Mill., 2. p. 483.
Synonymes. *Amygdalus Pérsica Lin. Sp.* 677.; Pêche duveteuse, *Fr.*; Pfirsche, *Ger.*
Engravings. N. Du Ham., 1. 2—8.; Nois. Jard. Fruit. Icon.; the plate of this tree in Arb. Brit., 1st edit., vol. vi.; and our *fig.* 427.

427. *Pérsica vulgàris.*

Spec. Char., &c. Fruit clothed with velvety tomentum. A deciduous tree Persia. Height 20 ft. to 30 ft. Cultivated in 1562, or probably long before. Flowers rose-coloured; March and April. Fruit red and yellow; ripe in September.

Varieties.

♀ P. *v.* 1, *the freestone common Peach,* Pêche, *Fr.*, has the flesh of the fruit parting from the shell of the nut (the stone).

♀ P. *v.* 2, *the clingstone common Peach,* Pavie, *Fr.*, has the flesh of the fruit adhering to the shell of the nut.

♀ P. *v.* 3 *flòre plèno* Hort.—Flowers double.

♣ P. *v.* 4 *álba* Lindl.—Flowers white. A hardy ornamental shrub, with the habit of an almond. Its fruit has little merit.

♀ P. *v.* 5 *fòliis variegàtis* Hort. —Leaves variegated.

♣ P. *v.* 6 *compréssa* Hort., the flat Peach of China (Hort. Trans. iv. t. 19.; and our *fig.* 428.), is chiefly remarkable for the form of its fruit, and for being nearly evergreen in its leaves. In the Hort. Soc. Garden, against a wall, it keeps growing throughout the winter, when the weather is not too severe.

428. *P. v. compréssa.*

♀ 2. *P.* (v.) LÆ'VIS *Dec.* The smooth-*skinned* Peach, *or Nectarine Tree.*

Identification. Dec. Fl. Fr., 4. p. 487.; Don's Miller, 2. p. 493.
Synonymes. Amýgdalus Pérsica *Lam. Dict.*; *A. Pérsica* Nectarina *Ait. Hort. Kew.*; Pêche lisse, Brugnon, *Fr.*; Pesco noce, *Ital.*
Engravings. Nois. Jard. Fruit., t. 20. f. 2, 3. t. 21. f. 3, 4.

Spec. Char., &c. Fruit smooth. A deciduous tree. Persia. Height 15 ft. to 20 ft. Cultivated in 1562, or probably long before. Flowers rose-coloured; March and April. Fruit red and yellow; ripe in September.

Varieties.

♀ P. (*v*) *l.* 1, *the freestone Nectarine* (with the fruit parting from the nut). Pêche lisse, *Fr.* — The Elruge is the best variety.

♀ P. (*v.*) *l.* 2, *the clingstone Nectarine* (with the flesh adhering to the nut). Brugnon, *Fr.* — The Orange is the best variety.

The different varieties of peach and nectarine, when treated as standard trees in the open garden, assume the general form and character of the almond; but, as they are more delicate, in consequence of being farther removed from their aboriginal state, they are of slower growth, form trees of less size, and are of shorter duration. The nectarine, as a standard in the open garden, forms a smaller and more delicate tree than the peach; and the double-flowered peach is of less vigorous growth than most of the single-flowered varieties, but very ornamental.

GENUS III.

ARMENI'ACA *Tourn.* THE APRICOT. *Lin. Syst.* Icosándria Monogýnia.

Identification. Tourn. Inst., t. 399.; Juss. Gen., 341.; Dec. Fl. Fr., 4. p. 485., Prod. 2. p. 531.; Don's Mill., 2. p. 495.
Synonymes. Prúnus sp. of Lin. and others; Abricotier, *Fr.*; Aprikosenbaum, *Ger.*; Albicocco, *Ital.*
Derivation. The genus is named *Armeniaca*, from the apricot being originally from *Armenia*. The popular English name was originally præcocia, from the Arabic, *berkoche*; whence the Tuscan bacoche, or albicocco; and the English, abricot, or apricock, eventually corrupted into apricot. Some persons derive the name from *præcox*, from this fruit ripening sooner than most others.

Gen. Char. *Drupe* ovate globose, fleshy, covered with a velvety skin, containing a nut or stone, which is acute at one end, and blunt at the other, with a furrow on both sides; the rest smooth, not wrinkled. (*Don's Mill.*) *Leaves* simple, alternate, stipulate, deciduous; when young, convolute. *Flowers* almost sessile, solitary, or few together, rising before the leaves from scaly buds.—Trees, low, deciduous, or shrubs; natives of Europe and Asia.

The common apricot is a fruit tree in general cultivation throughout the temperate regions of the globe, distinguished at first sight from the almond, peach, and nectarine, by its heart-shaped, smooth, shining leaves, and white flowers. There are several wild varieties, bearing flowers of different shades of pink, chiefly cultivated as ornamental. The great beauty of both the wild and the cultivated sorts of apricot is, that they come into bloom in Britain before almost every other tree; the Siberian apricot flowering a fortnight, or more, before the common sloe or almond.

♀ 1. *A.* VULGA'RIS *Lam.* The common Apricot Tree.

Identification. Lam. Dict., 1. p. 2.; Dec. Prod., 2. p. 532.; Don's Mill., 2. p. 495.
Synonyme. Prúnus *Armeniaca Lin. Sp.* 679.; Albicocco americano, *Ital.*
Engravings. N. Du Ham., 1. t. 49.; the plate in Arb. Brit., 1st edit., vol. v.; and our *fig.* 431.

Spec. Char., &c. Flowers sessile. Leaves heart-shaped or ovate. (*Dec. Prod.*) A middle-sized tree. Armenia, Caucasus, the Himalayas, China, &c. Height 20 ft. to 30 ft. Cultivated in 1548. Flowers pinkish white; February and March. Drupe orange and red; ripe in August and Sept. Decaying leaves reddish yellow. Naked young wood smooth, purplish.

Varieties.

♀ A. *v.* 1 *ovalifòlia* Ser. Abricot Angoumois, A. précoce, A. blanc, *Fr.* (N. Du Ham., 5. t. 50. f. 6.; and our *fig.* 429.)—Leaves oval; fruit small.

429. *A. v.* ovalifòlia.

♀ A. *v.* 2 *cordifòlia* Ser. (N. Du Ham., 5. p. 167. t. 49.; and our *fig.* 430.)—Leaves heart-shaped, broad. Fruit larger.

430. *A.* v cordifòlia

♀ A. *v.* 3 *fòliis variegàtis* Hort. —Leaves variegated. Flowers double. The Breda variety is generally that which has variegated leaves in British gardens.

♀ A. *v.* 4 *flòre plèno* Hort.—Grossier says that the Chinese have a great many varieties of double-blossomed apricots, which they plant on little mounts.

Very few trees attain the appearance of maturity so soon as the apricot; a standard 10 or 12 years planted, in good loamy rich soil, will grow to the height of 20 ft., with a head 25 ft. in diameter, presenting all the appearance

431. *Armeniaca vulgaris.*

of a tree of 20 or 30 years' growth, or of a tree arrived at maturity. The best variety for producing fruit, as a standard, is the Breda apricot. It is also a very handsome-growing plant, and its blossom buds, before they are expanded, are of a most beautiful and brilliant scarlet.

♀ 2. *A.* DASYCA'RPA *Pers.* The rough-fruited Apricot Tree.

Identification. Pers. Syn., 2. p. 36.; Dec. Prod., 2. p. 532.; Don's Mill., 2. p. 497.
Synonymes. *A.* atropurpùrea *Lois.* in *N. Du Ham.* 5. p. 172.; Prùnus dasycárpa *Ehrh. Beitr.* 6. p. 90.; *P. Armeniaca* nigra *Desf. Cat.* ed. 2. p. 206.; the black Apricot.
Engravings. N. Du Ham., 5. t. 51. f. 1.; Lodd. Bot. Cab., t. 1250.; and our *figs.* 432, 433.

Spec. Char., &c. Leaves ovate, acuminate, doubly serrate. Petioles glanded. Flowers upon thread-shaped pedicels. In the flowers of a plant in the Geneva Botanic Garden, the calyx was purple, and 6-lobed; the petals were 6; and the stamens 24. (*Dec. Prod.*) A tree with a twisted trunk, resembling the common apricot, but smaller. Levant ?. Height 10 ft. to 15 ft. Introduced in 1800. Flowers white; April. Drupe purple or black; ripe in August and September.

432. A. dasycárpa.

433. A dasycárpa.

434 A. d. persicifòlia

Variety.

♀ A. *d.* 2 *persicifòlia* Lois. *A.* persicifòlia *Don's Mill.*, ii. p. 498. Abricot noir à Feuilles de Pêcher, *Fr.* (N. Du Ham., 5. p. 172. t. 52. f. 1.; and our *fig.* 434.)—Leaves ovate and short, or lanceolate, with small lobes. Flesh of the fruit red, variegated with pale yellow. In the *Nouveau Du Hamel*, it is stated to be a very slight variety, which can only be continued by budding.

The rough-fruited apricot merits cultivation for its flowers, which are generally white, but which, in this country, from the earliness of their appearance, are not often succeeded by fruit, unless the tree is planted against a wall, where it can be protected by netting from the spring frosts.

435. Armeniaca (v.) sibírica.

♀ 3. *A.* (v.) SIBI'RICA *Pers.* The Siberian Apricot Tree.

Identification. Pers. Syn., 2. p. 36.; Dec. Prod., 2. p. 532.; Don's Mill., 2. p. 498.
Synonyme. Prùnus sibírica *Lin. Sp.* 179.
Engravings. Ammann Stirp. Ruth., 272. t. 29.; Pall. Fl. Ross., 1. t. 8.; and our plate in Arb. Brit., 1st edit., vol. v.; and our *fig.* 435.

Spec. Char., &c. Leaves ovate acuminate, of the form of those of the beech. The petioles long and glandless. Fruit small. A native of mountainous districts in the most remote parts of Siberia. Persoon has stated that it varies with leaves linear-lanceolate. (*Dec. Prod.*) A low tree, having the general appearance of the common apricot, but smaller in all its parts. Dahuria, on mountains, growing upon the face of perpendicular rocks exposed to the sun. Height 6 ft. to 8 ft.; in England 8 ft. to 20 ft. Introduced in 1788. Flowers rose-coloured; May. Drupe ?.

This tree, on the mountains of Dahuria, does not attain a greater height than that of a man; but it has a trunk the thickness of the wrist, a rough and black bark, and hard wood. It flowers about the same time as the *Rhododéndron dáuricum*; growing on the south sides of the mountains, while the latter grows on the north sides. When both these plants are in flower, Pallas observes, the north sides of the mountains appear of a purple colour, and the south of a rose colour. (*Fl. Ross.*, i. p. 13.) In British gardens, the Siberian apricot forms a tree of nearly the same height as the common apricot, of which it appears to us to be the wild form.

⚴ 4. *A.* (v.) BRIGANTI'ACA *Pers.* The Briançon Apricot Tree.

Identification. Pers. Syn., 2. p. 36.; Dec. Prod., 2. p. 532.; Don's Mill., 2. p. 498.
Synonyme. Prùnus brigantìaca *Vill. Dauph.* 3. p. 535., *Dec. Fl. Fr.* No. 3789, *Lois.* in *N. Du Ham.* 5. p. 185.
Engravings. N. Du Ham., 5. t. 59.; and our *fig.* 436.

Spec. Char., &c. Leaves nearly heart-shaped, toothed with numerous sharp subimbricate teeth. Flowers in groups, almost sessile, scarcely protruded before the leaves. (*Dec. Prod.*) A low tree. Dauphiné, in only one locality, and in another in Piedmont, where an oil, called huile de marmotte, has for a long time been expressed from the seeds. Height in British gardens 14 ft. to 15 ft. in 10 or 12 years; in its native habitats, 6 ft. to 8 ft. Introduced in 1819. Flowers white or pink; March and April. Drupe ?.

Seringe suggested that this kind may be the same as *A.* sibírica, and we think it not unlikely to be only another variety of the common apricot in its wild state, with toothed leaves.

436. *A.* (v.) brigantìaca.

Other Species of Armeniaca.—*A.* pedunculàta *Led.* has been raised in the Hort. Soc. Garden, from seeds received under this name from Dr. Ledebour.

GENUS IV.

PRU'NUS Tourn. THE PLUM. *Lin. Syst.* Icosándria Monogýnia.

Identification. Tourn. Inst., t. 358.; Dec. Prod., 2. p. 532.; Don's Mill., 2. 498.
Synonymes. Prunóphora *Neck. Elem.* No. 719.; *Prùnus* sp. of Lin. and others; Pruno, *Ital.*
Derivation. Said to be a word of Asiatic origin; the wild plant, according to Galen, being called *proumnos* in Asia. The Greek name for the plum is *proumē*: it occurs in Theophrastus.

Gen. Char. *Drupe* ovate or oblong, fleshy, quite glabrous, covered with a glaucous bloom; containing a compressed nut or putamen, which is acute at both ends, and a little furrowed on the margin, the rest smooth. (*Don's Mill.*)
Leaves simple, alternate, stipulate, deciduous; convolute when young. *Flowers* usually disposed in umbellate fascicles, solitary on the pedicels, rising generally before the leaves.—Trees or shrubs; natives of Europe, Asia, and North America.

Many of the species are spiny in a wild state; most of them bear edible fruits; and all of them have showy blossoms. In British gardens, they are chiefly propagated by grafting, but some of them by layers; and they will grow in any soil that is tolerably free, and not overcharged with moisture, but a calcareous soil is found best. The epidermis of the bark of the plum, as well as that of the cherry, and perhaps that of some of the other genera of Amygdàleæ, is readily divisible transversely, and may frequently be seen divided in this manner into rings on the tree.

♆ 1. *P.* SPINO'SA *L.* The spiny Plum Tree, *or common Sloe Thorn.*

Identification. Lin. Sp., 3. 681.; Dec. Prod., 2. p. 532.; Don's Mill., 2. p. 496.,
Synonymes. *P.* sylvéstris *Fuch. Hist.* p. 404., *Ray Syn.* p. 462.; Blackthorn; Prunier épineux, Prunellier, E'pine noire, or Mère-du-Bois, *Fr.*; Schleadorn, or Schlen Pflaum, *Ger.*; Prugno, or Prunello, *Ital.*
Engravings. Vahl Fl. Dan., t. 926.; the plate in Arb. Brit., 1st edit., vol. v.; and our *fig.* 437.
Derivation. The name of Mère-du-Bois is applied to the sloe thorn in France, in the neighbourhood of Montargis, because it has been remarked there, that, when it was established on the margins of woods, its underground shoots, and the suckers which sprang up from them, had a constant tendency to extend the wood over the adjoining fields.

Spec. Char., &c. Branches spiny. Leaves obovate, elliptical, or ovate; downy beneath, doubly and sharply toothed. Flowers produced before the leaves or with them, white, and solitary. Calyx campanulate; with lobes blunt, and longer than the tube. Fruit globose; the flesh austere. (*Dec. Prod.*) A low tree or shrub. Europe from Upsal to Naples, and the West of Asia and North of Africa. Height 10 ft. to 15 ft. Flowers white; March and April. Drupe black; ripe in October.

Varieties.

♆ P. *s.* 1 *vulgàris* Ser. *P.* spinòsa *Lois.* (N. Du Ham., 5. p. 185. t. 54. f. 1.) — Leaves obovate-elliptical. Fruit dark purple. This may be considered as the normal form of the species.

♆ P. *s.* 2 *fòliis variegàtis* Ser. — Found wild; but a plant of no beauty.

♆ P. *s.* 3 *microcárpa* Wallr. (Exs. Cent. 1. No. 45.) — Leaves elliptic, narrow, bluntish. Fruit smaller than that of the species.

♆ P. *s.* 4 *macrocárpa* Wallr. (Exs. Cent. 1. No. 45.) — Leaves obovate, bluntish. Fruit large, dark purple. This has been found wild in Germany; but Seringe doubts whether it be not identical with *P.* doméstica Juliàna, or with *P.* insitítia.

♆ P. *s.* 5 *ovàta* Ser. (Blackw. Herb., t. 494.) — Leaves ovate, roundish.

♆ P. *s.* 6 *flòre plèno.* — This is a very beautiful variety, said to be in cultivation, and highly prized, in China and Japan; and also found wild some years ago at Tarascon. The flowers are white, and are produced in such abundance as to entirely cover the branches.

The sloe, or blackthorn, is much more frequently seen as a large spiny shrub, than as a tree; but, when the suckers are removed from it, and all the strength of the plant is allowed to go into one stem, it forms a small scrubby tree of the most characteristic kind. The stems of the sloe differ from those of the hawthorn, in growing to the height of 3 or 4 feet before they branch off. The bark is black, whence the name of blackthorn; and the leaves are dark green. The roots are creeping, and, in every soil and situation, throw up numerous suckers; so much so, that a single plant, in a favourable soil, would cover an acre of ground in a very few years. In hedges, in Britain, it is seldom seen above 20 ft. in height; but in woods and in parks, as single trees, we have seen it above 30 ft. high: for example, in Eastwell Park, in Kent. The wood is hard, and in colour resembles that of the peach, though without its beauty: it takes a fine polish; but it is so apt to crack, that little use can be made of it, except for handles for tools, teeth for hay-rakes, swingles for flails, and walking-sticks. The wood weighs, when dry, nearly 52 lb. per cubic foot. The branches, from being less spreading than those of the common hawthorn, make better dead hedges than those of that species; and, for the same reason, they are particularly well adapted for forming guards to the stems of trees

437. *Prùnus spinòsa.*

planted in grass fields or in parks, to protect them from cattle. They are in general use for this purpose in France. They are also used as a substitute for stones and tiles in draining; and, formed into faggots, they are sold for heating bakers' ovens, and for burning lime or chalk in kilns, &c. The living plant cannot be recommended for hedges, on account of the rambling habit of its roots, and the numerous suckers they throw up; and because it is apt to get naked below, from the tendency of the shoots to grow upright and without branches. These upright shoots make excellent walking-sticks, which, accordingly, throughout Europe, are more frequently taken from this tree than from any other. Leaves of the sloe, dried, are considered to form the best substitute for Chinese tea which has yet been tried in Europe; and they have been extensively used for the adulteration of that article. The juice of the ripe fruit is said to enter largely into the manufacture of the cheaper kinds of port wine; and, when properly fermented, it makes a wine strongly resembling new port. In planting groups and masses in parks, by the addition of a few plants of the sloe, a degree of intricacy may be given sooner and more effectively, than by the use of the common thorn; but, at the same time, the sloe produces a degree of wildness from its numerous suckers, and the want of control which they indicate, which is not displayed by any of the species of *Cratæ'gus*, which do not throw up suckers. For producing wildness and intricacy, therefore, in park scenery, the sloe is of great value, and its effect is much heightened by the addition of the common furze or the broom. The sloe prefers a strong calcareous loam. It may be propagated freely by suckers, or by seeds: the latter should be gathered in October, when the fruit is dead ripe, mixed with sand, and turned over two or three times in the course of the winter; and, being sown in February, they will come up in the month of May.

♆ 2. *P.* INSITI'TIA *L.* The engrafted Plum Tree, *or Bullace Plum.*

Identification. Lin. Sp., 680.; Dec. Prod., 2. p. 532.; Don's Mill., 2. p. 498.
Synonymes. *P. sylvéstris præ'cox áltior Tourn.*; *P. sylvéstris májor Ray*; Prunier sauvage, *Fr.*; Alfatous, *in Dauphiny*; Kirschen Pflaume, *Ger.*
Engravings. Eng. Bot., t. 841.; Hayne Abbild., t. 65.; the plate in Arb. Brit., 1st edit., vol. v.; and our *fig.* 438.

Spec. Char., &c. Branches becoming spiny. Flowers in pairs. Leaves ovate or lanceolate; villose beneath, not flat. Fruit roundish. (*Dec. Prod.*) A low tree. England, Germany, and the South of France, and also Barbary. Height 10 ft. to 20 ft. Flowers white; March and April. Drupe black; ripe in October.

Varieties.

♆ P. *i.* 1 *frúctu nìgro* Hort. The black-fruited, or common, Bullace.
♆ P. *i.* 2 *frúctu lùteo-álbo* Hort.—Fruit yellowish-white.

♆ P. *i.* 3 *frúctu rùbro* Hort. — Fruit red.
♆ P. *i.* 4 *flòre plèno* Descemet in *Mém. de la Russie Méridionale*, 1. p. 63. — Flowers double.

The fruit, which is globular, and usually black, is sometimes yellowish or waxy, with a red tint, and sometimes red; it is also so much less austere than the sloe, as to make excellent pies and puddings, and a very good preserve.

438. Prûnus insititia.

The fruit of this plum in Provence is called prune sibarelle, because it is impossible to whistle after having eaten it, from its sourness. The wood, the branches, the fruit, and the entire plant are used, throughout France, for the same purposes as that of the sloe.

♆ 3. *P.* DOME'STICA *L.* The domestic *cultivated* Plum Tree.

Identification. Lin. Sp., 680.; Dec. Prod., 2. p. 533.; Don's Mill., 2. p. 499.
Synonymes. *P.* sativa *Fuchs* and *Ray*; Prunier domestique, *Fr.*; gemeine Pflaume, *Ger.*; Susino domestico, *Ital.*
Engravings. Wood. Med. Bot., t. 85.; E. Bot., t. 1783.; and our *fig.* 439.

Spec. Char., *&c.* Branches spineless. Flowers mostly solitary. Leaves lanceolate-ovate, concave on the surface, not flat. (*Dec. Prod.*) A low tree. South of Europe, on hills; in England, found sometimes in hedges, but never truly wild. Height 15 ft. to 20 ft. Flowers white; April and May. Drupe various; ripe August to October.

439. Prûnus doméstica.

Varieties.
♆ P. *d.* 2 *flòre plèno* Hort. — Flowers large, double.
♆ P. *d.* 3 *fòliis variegàtis* Hort. — Leaves variegated.
♆ P. *d.* 4 *armenioìdes* Ser. — Leaves and fruit like those of *Armeniaca* brigantiaca.

The cultivated plum resembles the common sloe, but is larger in all its parts, and without thorns. There are numerous varieties and subvarieties; but, as they belong more to pomology than to arboriculture, we have here only noticed those that have some pretensions to distinctness in an ornamental point of view.

The apricot-like plum seems intermediate between the wild plum and the wild apricot. The varieties cultivated for their fruit have, in general, much larger leaves, and stronger young shoots, than the other sorts; they flower later, their blossoms are larger, and their fruit, particularly such sorts as the magnum bonum and the diamond plum, several times as large; the latter being upwards of 2½ in. long. These fruit-bearing varieties are in universal cultivation in temperate climates; and for every thing of interest relating to them, as such, we refer to our *Encyclopædia of Gardening*, edit. 1835, p. 920. The use of the fruit in domestic economy, in Britain, for the dessert, and for making tarts and puddings, is well known. In France, plums are used principally dried, as an article of commerce, and they are known under the name of brignoles, prunes, and French plums. The different modes of preserving plums in France will be found detailed at length in the 1st edit. of this work, and in our *Suburban Horticulturist.*

♆ 4. P. (D.) MYROBA'LANA *L.* The Myrobalan, *or Cherry*, Plum.

Identification. Lin. Sp., 680.; Dec. Prod., 2. p. 533.; Don's Mill., 2. p. 500.
Synonymes. *P.* Myrobàlan *Du Ham.*; *P.* myrobálana *Lois.*; *P.* cerasifera *Ehrh. Beitr.* 4. p. 17.; Virginian Cherry; Early Scarlet Plum; Prunier myrobalan, or Cerisette, *Fr.*; Kirschpflaume, *Ger.*
Engravings. Du Ham. Arb. Fr. 2. p. 111. t. 2. fig. 15.; the plate of this tree in Arb. Brit., 1st edit., vol. v.; and our *fig.* 440.

440. *Prùnus* (d.) *myrobálana.*

Spec. Char., &c. Sepals narrow. Fruit globose, depressed at the base; umbilicus depressed; nut with a small point. (*Dec. Prod.*) A low tree. Europe, or, according to some, North America. Height 15 ft. to 30 ft. Cultivated in gardens for an unknown period. Flowers white; March and April. Fruit cordate, red, rarely produced in England.

Variety.

♆ P. (*d.*) *m.* 2 *fòliis variegàtis* N. Du Ham. has variegated leaves.

Though we consider this nothing more than a variety of the common plum, yet it is so distinct, both in the habit of the tree and the colour of the fruit, that we think it more convenient to keep it apart. Its flowers are produced as early as those of the sloe; and, the plant being more tender than that species, it seldom produces fruit in England, except when the blossoms are protected. It forms a good stock for varieties intended to be kept dwarf. In India the fruit is sold to dye black.

5. P. CA'NDICANS *Balb.* The white Plum Tree.

Identification. Balb. Cat. Taur., 1813. p. 62.; ? Willd. Enum. Suppl., p. 32.; Dec. Prod., 2. p. 533.; Don's Mill., 2. p. 498.
Engravings. Bot. Reg. t. 1135.; and our *figs.* 441, 442.

441. *P. cándicans.*

442. *P. cándicans.*

Spec. Char., &c. Branches pubescent. Flowers 2 or 3 together, upon short pubescent peduncles. Calyx bell-shaped. Leaves broadly ovate, whitish beneath. Stipules of the length of the petiole, very narrow, and cut in a toothed manner. (*Dec. Prod.*) A low shrub. ? Tauria. Height 6 ft. to 8 ft. Introd. in 1825. Flowers white; April.

It is not known of what country it is a native. It is quite hardy, easily cultivated, and so laden with white blossoms in spring as to appear a mass of snow, whence the name.

6. P. COCOMI'LLA Tenore. The Cocomilla Plum Tree.

Identification. Tenore Prod. Suppl., 2. p. 67.; Cat., 1819, p. 46.; Dec. Prod., 2. p. 533.; Don's Mill., 2. p. 498.
Engraving. Our *fig.* 443. from a living specimen.

443. *P. Cocomilla.*

Spec. Char., &c. Flowers upon short peduncles, in pairs. Leaves obovate, crenulate, glabrous on both surfaces; the crenatures glanded. Fruit ovate-oblong, with a small point, yellow, ? bitter or ? acid. (*Dec. Prod.*) A low shrub. Calabria, in hedges. Height 2 ft. to 3 ft. Introduced in 1824. Flowers white; April. Drupe yellow; ripe in August.

The bark is febrifugal, and, in Calabria, is considered to be a specific for the cure of the pestilential fevers common in that country.

7. P. MARI'TIMA *Wangenheim.* The sea-side-*inhabiting* Plum Tree.

Identification. Wangenh. Amer., 103., according to Willd. Enum., p. 519.; Pursh Fl. Amer. Sept., 1. p. 332.; Dec. Prod., 2. p. 533.; Don's Mill., 2. p. 499.
Synonyme. ?P. acumináta *Michx. Fl. Bor. Amer.* p. 284.
Engravings. Our *fig.* 444., from a living specimen; and *fig.* 445. from Pursh's specimen in the Lambertian herbarium.

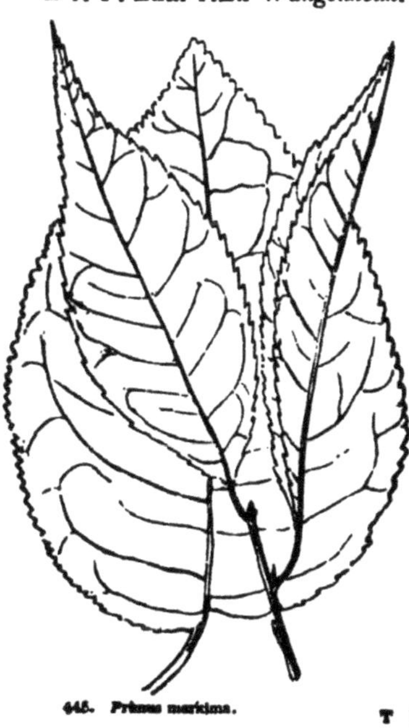

445. *Prunus maritima.*

444. *Prunus maritima.*

Spec. Char., &c. Leaves lanceolate-ovate, serrate. Flowers in pairs. Fruit small, round, sweet, dark blue.

(*Dec. Prod.*) A middle-sized shrub. North America, in sandy soils, on the sea coast, from New Jersey to Carolina. Height 6 ft. to 8 ft. Introduced in 1818. Flowers white; April and May. Fruit, of the size of a pigeon's egg, dark purple, and, according to Pursh, very good to eat; ripe ?.

There are plants in British gardens, but they have never yet set fruit.

▲ 8. *P.* PUBE'SCENS *Poir.* The pubescent-*leaved* Plum Tree.

Identification. Poir. Suppl., 4. p. 584., not of Pursh; Dec. Prod., 2. p. 533.; Don's Mill., 2. p. 499.
Engraving. Our *fig.* 2084 in p. 1106.

Spec. Char., &c. Leaves with short pubescent petioles, and disks that are slightly pubescent, ovate, thickish, rounded, or shortly acuminate and unequally toothed. Flowers mostly solitary and nearly sessile. Fruit oval. (*Dec. Prod.*) A shrub. Native country unknown. Cultivated in 1818. Height 2 ft. to 3 ft. Flowers white; May.

▲ 9. *P.* DIVARICA'TA *Led.* The divaricated-*branched* Plum Tree.

Identification. Ledeb. Ind. Hort. Dorp. Suppl. 1824, p 6.; Fl. Alt., t. 13.; Dec. Prod., 2. p. 534.; Don's Mill., 2. p. 504.
Engravings. Led. Flor. Alt., t. 13.; and our *fig.* 446.

Spec. Char., &c. Branches spineless. Leaves with glandless petioles, and disks oblong-elliptical, tapered to both ends, concave above, serrate, glabrous, with the midrib bearded beneath. Flowers solitary, very numerous. Calyx reflexed. Fruit elliptical, yellow. (*Dec. Prod.*) A middle-sized shrub. Caucasus. Height 8 ft. to 10 ft. Introduced in 1820. Flowers white; April.

446. Prùnus divaricàta.

Other Species of Prùnus Juss. — In consequence of many species of the genus Prùnus being removed to Cérasus; and also because of the close resemblance of one species to another in both genera, there is a good deal of confusion, which cannot be cleared up till the plants are studied in a living state. Prùnus effùsa was raised in 1838, in the Hort. Soc. Garden, from seeds presented by Baron Jacquin.

GENUS V.

CE'RASUS *Juss.* THE CHERRY. *Lin. Syst.* Icosándria Monogýnia.

Identification. Juss. Gen., 340.; Dec. Fl. Fr. 4. p. 479.; Prod., 2. p. 535; Don's Mill., 2. p. 504.
Synonymes. Laurocérasus *Tourn.*; Prùnus sp. *Lin.*; Cerisier, *Fr.*; Kirsche, *Ger.*; Ciliegio, *Ital.*
Derivation. From *Cerasus*, the ancient name of a town of Pontus in Asia, whence the cultivated cherry was first brought to Rome, by Lucullus, a Roman General, 68 B.C.

Gen. Char. *Drupe* globose, or umbilicate at the base, fleshy, quite glabrous, destitute of bloom, containing a smooth, rather globose compressed stone. (*Don's Mill.*)

Leaves simple, alternate, stipulate, deciduous, or evergreen; when young, conduplicate. *Flowers* white. *Pedicels* 1-flowered, rising before the leaves, in fascicled umbels, from scaly buds; but sometimes rising after the evolution of the leaves, in racemes, from the tops of the branches.—Trees and shrubs, almost all deciduous, with smooth serrated leaves, and white flowers; and, generally, with light-coloured bark. Natives of Europe, Asia, and North America.

Some of them are cultivated for their fruit, and the others as ornamental. In British nurseries, the deciduous species are generally propagated by grafting or budding on the Cérasus sylvéstris, and the evergreens are propagated by cuttings or seeds; they will grow iu any common soil that is tolerably dry. There is much confusion in all the species, more particularly as regards those which are natives of North America; and which, as Sir W. J. Hooker judiciously observes, can only be "removed by carefully studying the plants in a living state, both during the season of the blossom and that of the fruit." (*Flor. Bor. Amer.*, i. p. 167.)

§ i. *Cerasóphora* Dec. *The Cherry-bearing Kinds.*

Sect. Char. Flowers produced from buds upon shoots not of the same year; and, in many instances, disposed umbellately. Leaves deciduous.

A. *Species cultivated for their Fruit.*

The Cherries cultivated in Gardens, according to Linnæus, and almost all botanists to the time of DeCandolle, have been referred to *Prùnus* àvium *L.* and *Prùnus Cérasus L.*; the former being the mérisier of the French, and corresponding with the small wild black bitter cherry of the English (the *C.* sylvéstris of Ray); and the latter the cerisier of the French, and corresponding with the common red sour cherry of the English (the *C.* vulgàris of Miller). To these two species DeCandolle has added two others: *Cérasus* Juliàna, which he considers as including the guigniers; and *Cérasus* durácina, under which he includes the bigarreaus, or hard cherries. Under each of these four species, Seringe, in DeCandolle's *Prodromus*, has arranged a number of varieties, with definitions to each group: but, as neither the species nor the groups appear to us distinct, we have adopted the arrangement of the author of the article on *Cérasus* in the *Nouveau DuHamel*, as much more simple and satisfactory; and have referred all the cultivated varieties to the same species as Linnæus; substituting for *Prùnus* àvium *L.*, *Cérasus* sylvéstris, the synon. of Ray; and for *Prùnus Cérasus L.*, *Cérasus* vulgàris, already used to designate the same species in *Mill. Dict.*, and by Loiseleur in the *Nouveau Du Hamel*. The arrangement of the varieties, and general culture of the cherry in the kitchen-garden and orchard, will be found at length in our *Encyclopædia of Gardening;* and, in a more condensed form, in our *Suburban Horticulturist.*

⚴ 1. *C.* SYLVE'STRIS *Bauh.* and *Ray.* The wild *black-fruited* Cherry Tree.

Identification. Bauh. Hist., 1. 1. 2. p. 220.; Ray Hist. 1539.; Pers. Syn., 2. p. 35.

Synonymes and Garden Names. *C.* àvium *Mœnch*, *N. Du Ham.* 5. p. 10., *Don's Mill.* 2. p. 505.; *C.* nigra *Mill. Dict.* No. 2., not of Ait.; *Prùnus* àvium *Lin. Sp.* 680.; *P.* àvium var. α and β *Willd. Baum.* ed. 2. p. 308.; *Prùnus* àvium var. β and γ *Eng. Flor.* 2. p. 355.; *P.* nigricans and *P.* vària *Ehr. Beitr.* 7. p. 126, 127.; Gean, Bigarreau, Corone, Coroon, Small Black, Black Hertfordshire, Black Heart, Black Mazzard, the Merry Tree *of the Cheshire peasants*, the Merries *in Suffolk;* Mérisier, Merise grosse noire, Guignier, Bigarreautier, Heaumier, *Fr.*; Süsse Kirsche, *Ger.*; Ciregiolo, *Ital.*

Derivation. This cherry is called Corone, or Coroon, in some parts of England, from *corone*, a crow, in reference to its blackness. Merry Tree and Merries are evidently corruptions of the word Mérisier; and Mérisier is said to be derived from the words *amère*, bitter, and *cerise*, a cherry. Bigarreau is derived from *bigarrée* party-coloured, because the cherries known by this name are generally of two colours, yellow and red; and Heaumier is from the French word *heaume*, a helmet, from the shape of the fruit.

Engravings. Du Ham. Tr. Arb., 1. p. 156.; Arb. Brit., 1st edit., vol. vi.; and our *fig.* 447.

Spec. Char., &c. Branches vigorous and divaricate; the buds from which the fruits are produced oblong-acute. Flowers in umbel-like groups, sessile, not numerous. Leaves oval-lanceolate, pointed, serrated, somewhat pendent, slightly pubescent on the under side, and furnished with two glands at the base. (*Dec. Prod.*, *N. Du Hamel.*) A middle-sized tree. Europe, in woods and hedges. Height, in dry fertile soils, 40 ft. to 50 ft. or upwards. Flowers white; April and May. Drupe red or black; ripe in July. Decaying leaves of a fine red, or rich yellow and red.

Varieties.

1. *Mérisiers* or *Merries*, with black or yellow fruit.
2. *Guigniers* or *Geans* (*C.* Juliàna Dec.), with red or black fruit, early or late, and including the tobacco-leaved guignier, or gean, of four to the pound (the *C.* decumàna of Delauny).

3. *Heaumiers, the Helmet-shaped Cherries* (*C.* Juliàna var. *heaumiàna* Dec.), somewhat resembling the bigarreau, but with less firm flesh.

Variety of this race used for ornamental purposes.

♅ *C. s. durácina* 2 *flòre plèno* Hort., the double flowered wild black Cherry; Mérisier à Fleurs doubles, or Mérisier Renunculier, *Fr.*; is a very beautiful variety, known, in the garden of the Hort. Soc., as the double French white.

4. *Bigarreautiers, the Bigarreaus, or hard-fleshed Cherries* (*C.* durácina *Dec.*) with white, flesh-coloured, and black fruit, generally heart-shaped.

447. Cérasus sylvéstris.

The colour of the fruit of the wild species is a very deep dark red, or black; the flesh is of the same colour, small in quantity, austere and bitter before it comes to maturity, and insipid when the fruit is perfectly ripe. The nut is oval or ovate, like the fruit, firmly adhering to the flesh, and very large in proportion to the size of the fruit. The juice is mostly coloured; and the skin does not separate from the flesh.

♅ 2. *C.* VULGA'RIS *Mill.* The common Cherry Tree.

Identification. Mill. Dict., No. 1.; N. Du Ham., 5. p. 18.

Synonymes and Garden Names. Cérasus àvium *Mœnch*; Prùnus Cérasus *Lin. Sp.* 679.; *C.* horténsis *Pers. Syn.* 2. p. 34.; *C. capronidna* Dec. Prod. 2. p. 536., Don's Mill. 2. p. 507.; *P.* austèra and *P.* ácida *Ehr. Beitr.* 7. p. 129. and 130.; *P. Cérasus* var. *a Eng. Flor.* 2. p. 354.; Cherry, Kentish or Flemish Cherry, Morello, May Duke; Cerise de Montmorency, Cerise de Paris, Cerise à Fruits ronds, Cerisier du Nord, Cerisier, and Griottier in some provinces, *Fr.*; Saure Kirsche, *Ger.*; Marasca, or Ciliegio, *Ital.*

Derivation. *Capronidna* is said to be derived from *capron*, the hautbois strawberry, probably from this cherry possessing so much more flavour than *C.* sylvéstris. Morello is either from *morel* (*Morchélla* esculénta), the flesh being of the same consistency as the flesh of that fungus; or, perhaps, from the French word *morelle*, a female negro. May Duke is a corruption of *Médoc*, the province of France where the variety is supposed to have been originated. Griottier is said to be derived from *aigreur*, sourness, or sharpness, and applied to this cherry from the acidity of its fruit.

Engravings. Eng. Bot., t. 706.; Arb. Brit., 1st edit., vol. vi., as *Cérasus* àvium; and our *fig.* 448.

Spec. Char., &c. Tree small, branches spreading. Flowers in subsessile umbels, not numerous. Leaves oval-lanceolate, toothed, glabrous. A deciduous tree. Europe and Britain, in gardens and plantations. Height 30 ft. to 40 ft. Flowers white; May. Drupe red; ripe in July. Decaying leaves red and yellow.

Varieties. — There are numerous cultivated varieties, which are classed by Loiseleur in the *Nouveau Du Hamel* in three groups, including in the first

448. Cérasus vulgàris.

of these the four following varieties, which we particularise on account of their being purely ornamental : —

♀ C. *v.* 2 *flòre semiplèno* Hort.—Flowers semidouble.

♀ C. *v.* 3 *flòre plèno* Hort.—All the stamens of this variety are changed into petals; and the pistillum into small green leaves, which occupy the centre of the flower. The flower is smaller and less beautiful than that of the double mérisier; but, as the tree does not grow so high, and as it can be grown as a shrub, it is suitable for planting in situations where the other cannot be introduced. It is commonly grafted on the Prùnus *Mahàleb.*

♀ C. *v.* 4 *persiciflòra* Hort.—The flowers are double, and rose-coloured. This variety was known to Bauhin and to Tournefort, but is at present rare in gardens.

♀ C. *v.* 5 *fòliis variegàtis* Hort. has variegated leaves.

The fruit-bearing varieties are arranged in the *Nouveau Du Hamel,* under the following heads : —

1. Flesh whitish, and more or less acid; including the Montmorency cherry.
2. Flesh whitish, and only very slightly acid; including the English duke cherries.
3. Flesh red, including the griottiers, or morellos.

The following selection has been made by Mr. Thompson, with a view of exemplifying the different forms which the varieties of the cultivated cherries assume, as standard trees : —

The Bigarreau is a tree of vigorous growth, with large pale green leaves, and stout divergent branches.

Buttner's Yellow is a vigorous-growing tree, like the preceding, but with golden-coloured fruit.

The Kentish Cherry is a round-headed tree, with slender shoots, somewhat pendulous.

The May Duke is a middle-sized or low tree with an erect fastigiate head.

The Morello is a low tree, with a spreading head, somewhat pendulous; most prolific in flowers and fruit; the latter ripening very late, and, from not being so greedily eaten by the birds as most other sorts, hanging on the trees a long time.

D'Ostheim is a dwarf weeping tree, a great bearer.

♇ C. *v.* 6 Maráscha, Prùnus *Maráscha* Jacq., is the variety from the fruit of which the liqueur called Maraschino is made. Plants of it have been raised in the Hort. Soc. Garden in 1837.

The flowers are smaller than those of *C.* sylvéstris. The fruit is round, melting, full of a watery sap, more or less flavoured, and almost always sensibly acid. The skin of the fruit is commonly red, but, in the numerous varieties in cultivation, passing into all the shades between that colour and dark purple or black. The skin of all the varieties of *C.* vulgàris separates easily from the flesh, and the flesh parts readily from the stone; while, in all the varieties of *C.* sylvéstris, the skin is more or less adhering to the flesh, and the flesh to the stone. (*N. DuHam.*, v. p. 18.) This species forms a tree of less magnitude than that of *C.* sylvéstris: it is never found in a truly wild state in Europe, and the aboriginal form is unknown.

Remarks referring to both Species. — The cherry trees in cultivation, whether in woods or gardens, may, in point of general appearance, be included in three forms: large trees with stout branches, and shoots proceeding from the main stem horizontally, or slightly inclining upwards, and when young and without their leaves, bearing a distant resemblance to gigantic candelabra, such as the geans, and many of the heart cherries; fastigiate trees of a smaller size, such as the dukes; and small trees with weak wood, and branches divergent and drooping, such as the Kentish or Flemish cherries, and the morellos. The leaves vary so much in the cultivated varieties, that it is impossible to characterise the sorts by them; but, in general, those of the large trees are largest, and the lightest in colour, and those of the slender-branched trees the smallest, and the darkest in colour; the flowers are also largest on the large trees. The distinction of two species, or races, is of very little use, with reference to cherries as fruit-bearing plants; but, as the wild sort, *C.* sylvéstris, is very distinct, when found in its native habitats, from the cherry cultivated in gardens, it seems worth while to keep them apart, with a view to arboriculture and ornamental planting. For this reason, also, we have kept *Cé*rasus semperflòrens, *C.* Pseùdo-*Cé*rasus, *C.* serrulàta, and *C. Ch*amæcérasus apart, though we are convinced that they are nothing more than varieties. In consequence of its rapid growth, the red-fruited variety ought to be preferred where the object is timber, or where stocks are to be grown for fruit trees of large size. As a coppice-wood tree, the stools push freely and rapidly; and, as a timber tree, it will attain its full size, in ordinary situations, in 50 years. Its rate of growth, in the first 10 years, will average, in ordinary circumstances, 18 in. a year. There are various trees in the neighbourhood of London upwards of 60 ft. high; one on the Cotswold Hills, on the estate of the Earl of Harrowby, is 85 ft. high. The wood of the wild cherry (*C.* sylvéstris) is firm, strong, close-grained, and of a reddish colour. It weighs, when green, 61 lb. 13 oz. per cubic foot; and when dry, 54 lb. 15 oz.; and it loses in the process of drying about a 16th part of its bulk. The wood is soft and easily worked, and it takes a fine polish. It is much sought after by cabinetmakers, turners, and musical instrument makers, more particularly in France, where mahogany is much less common than in Britain. The fruit of the cherry is a favourite with almost every body, and especially with children. The distillers of liqueurs make great use of ripe cherries: the spirit known as kirschewasser is distilled from them after fermentation; and both a wine and a vinegar are made by bruising the fruit and the kernels, and allowing the mass to undergo the vinous fermentation. The ratafia of Grenoble is a celebrated liqueur, which is made from a large black gean; from which, also, the best kirschewasser is made; and the maraschino from a variety of the tree found in Dalmatia. The preparation of these will be found in our first edition.

B. *Species or Varieties cultivated as ornamental or curious.*

3. *C.* (v.) SEMPERFLO'RENS *Dec.* The ever-flowering Cherry Tree.

Identification. Dec. Fl. Fr., 4. p. 481., and Prod., 2. p. 537.; Don's Mill., 2 p. 13.
Synonymes. Prùnus semperflòrens *Ehrh. Beitr.* 7. p. 132.; *P.* serótina *Roth Catal.* 1. p. 58.; the weeping Cherry, the Allsaints Cherry; Cerise de la Toussaint, Cerise de St. Martin, Cerise tardive, *Fr.*
Engravings. N. Du Ham., p. 30. No. 18. t. 5. f. A; the plate in Arb. Brit., 1st edit., vol. v.; and our *fig.* 449.

449. Cérasus (v.) semperflòrens.

Spec. Char., &c. Branches drooping. Leaves ovate, serrated. Flowers protruded late in the season, axillary, solitary. Calyx serrated. Fruits globose, and red. Its native country not known. (*Dec. Prod.*) A low pendulous tree. Height 10 ft. to 20 ft. A garden production. Cultivated in ? 1700. Flowers white; May. Drupe red; July and August.

An ornamental tree, usually grafted standard high on the common wild cherry, or gean; growing rapidly for 8 or 10 years, and forming a round head, 8 or 10 feet high, and 10 or 12 feet in diameter, with the extremities of the branches drooping to the ground; and flowering and fruiting almost the whole summer. It forms a truly desirable small single tree for a lawn.

450. Cérasus serrulàta.

4. *C.* SERRULA'TA *G. Don.* The serrulated-*leaved* Cherry Tree.

Identification. Hort. Brit., p. 480.; Don's Mill., 2. p. 514.
Synonymes. Prùnus serrulàta *Lindl. Hort. Trans.* 7. p. 238.; double Chinese Cherry; Yung-To, *Chinese.*
Engraving. Our *fig.* 450. from a living specimen.

Spec. Char., &c. Leaves obovate, acuminated, setaceously serrulated, quite glabrous. Petioles glandular. Flowers in fascicles. (*Don's Mill.*) A low erect tree, or rather tree-like shrub. China. Height in China 4 ft. to 6 ft.; in British gardens 6 ft. to 10 ft. Introduced in 1822. Flowers double white, tinged with red, though not so much so as the double French cherry.

This tree resembles the common cherry tree, but is not of such vigorous growth; and only the double-flowered variety of it has been yet introduced. A very ornamental plant.

♀ 5. *C.* Pseu'do-Ce'rasus *Lindl.* The False-Cherry Tree.

Identification. Hort. Brit., No. 12663.; Don's Mill., 2. p. 514.
Synonymes. Prùnus Pseùdo-Cèrasus *Lindl. Hort. Trans.* 6. p. 90.; *P.* paniculàta *Ker Bot. Reg.* 800., but not of Thunb.; Yung-To, *Chinese.*
Engravings. Bot. Reg., t. 100.; and our *fig.* 451.

Spec. Char., &c. Leaves obovate, acuminate, flat, serrated. Flowers racemose. Branches and peduncles pubescent. Fruit small, pale red, of a pleasant subacid flavour, with a small smooth stone. (*Don's Mill.*) A low tree. China. Height 8 ft. to 10 ft. Introduced in 1819. Flowers white; March and April. Fruit pale red; ripe in June.

451. *C.* Pseùdo-Cérasus.

This tree is readily known from the other cherry trees, even when without its leaves, by its rough gibbous joints, at which it readily strikes root; and is, consequently, very easily propagated. It has been tried by Mr. Knight, as a fruit tree; and he finds that it forces in pots better than any other variety. Desirable for small gardens, on account of its very early flowering.

⁂ 6. *C.* Chamæce'rasus *Lois.* The Ground-Cherry Tree, *or Siberian Cherry.*

Identification. Lois. in N. Du Ham., 5. p. 29.; Dec. Prod., 2. p. 537.; Don's Mill., 2. p. 513.
Synonymes. *C.* intermèdia *Lois.* in *N. Du Ham.* 5. p. 30.; *Prùnus* intermèdia *Poir. Dict.* 5. p. 674.; *P.* fruticòsa *Pall.*, according to *Besser*; Cérasus pùmila *C. Bauh.*, according to *Pall. Fl. Ross.*; Chamæcérasus fruticòsa *Pers. Syn.* 2. p. 34.
Engravings. N. Du Ham., 5. p. 29. t. 5. f. A; Hayne Abbild., t. 61.; and our *fig.* 452.

Spec. Char., &c. Leaves ovate-oblong, glabrous, glossy, crenate, bluntish, rather coriaceous, scarcely glanded. Flowers in umbels, which are usually on peduncles, but short ones. Pedicels of the fruit longer than the leaves. Fruit round, reddish purple, very acid. (*Dec. Prod.*) A neat little shrub. Siberia and Germany. Height 3 ft. to 4 ft. Introd. in 1587. Flowers white; May. Drupe reddish purple; ripe in August.

It forms a neat little narrow-leaved bush, which, when grafted standard high, becomes a small round-headed tree with drooping branches, at once curious and ornamental. It does not grow above a fourth part of the size of *C.* semperflòrens; and, like it, it flowers and fruits during great part of the summer.

452. *C.* Chamæcérasus.

⁂ 7. *C.* prostra'ta *Ser.* The prostrate Cherry Tree.

Identification. Seringe in Dec. Prod., 2. p. 538.; Don's Mill., 2. p. 514.
Synonymes. Prùnus prostràta *Hort. Kew.* ed. 2 3. 199., and *Lab. Pl. Syr Dec.* 1. p. 15.; Prùnus incàna *Stephen* in *Mém. Soc. Mosq.* 3. p. 263.
Engravings. Pall. Fl. Ross., 1. t. 7.; Bot. Reg., t. 106.; and our *fig.* 453.

Spec. Char., &c. Decumbent. Leaves ovate, serrately cut, glandless, tomentose, and hoary beneath. Flowers mostly solitary, nearly sessile. Calyx tubular. Petals ovate, retuse, rose-coloured. Fruit ovate, red; flesh thin.

(*Dec. Prod.*) A prostrate shrub. Native of the mountains of Candia, of Mount Lebanon, and of Siberia. Height ½ ft. to 1 ft. Introduced in 1802. Flowers rose-coloured; April and May. Drupe red; ripe July.

A very desirable species for grafting standard high on the common cherry. The red colour of the flowers is very uncommon in this genus.

453. *Cerasus prostrata.*

Ɨ 8. *C.* PERSICIFO'LIA *Lois.* The Peach-tree-leaved Cherry Tree.

Identification. Lois. in N. Du Ham., 5. p. 9.; Dec. Prod., 2. p. 537.; Don's Mill., 2. p. 513.
Synonyme. Prùnus persicifòlia *Desf. Arb.* 2. p. 205.
Engraving. Our *fig.* 000. in *fig.* 000.

Spec. Char., &c. Leaves ovate-lanceolate, acuminate, unequally serrate, glabrous, with two glands upon the petiole. Flowers numerous, upon slender peduncles, and disposed umbellately. (*Dec. Prod.*) A deciduous tree. ? America. Height 30 ft. to 50 ft. Introduced in 1818. Flowers white; May. Drupe small, black; ripe in July.

A rapidly growing tree, attaining the height of the common wild cherry, and bearing so close a resemblance to it in almost every respect, that it is probably only a variety of it. There are trees of this kind of cherry in the Jardin des Plantes at Paris, of a pyramidal form, with a reddish brown smooth bark, flowers about the size of those of *C. Mahàleb*, and fruit about the size of peas. The wood is said to be harder and redder than that of the common wild cherry. It was raised from seeds sent from America by Michaux.

Ɨ 9. *C.* BOREA'LIS *Michx.* The North-*American* Cherry Tree.

Identification. Michx. Fl. Bor. Amer., 1. p. 286.; Lois. in N. Du Ham., 5. p. 32. No. 22.; Dec. Prod., 2. p. 538.; Don's Mill., 2. p. 513.
Synonymes. Prùnus boreàlis *Poir. Dict.* 5. p. 674.; the Northern Choke Cherry, *Amer.*
Engravings. Michx. Arb. Amer., 3. t. 8.; and our *fig.* 454.

Spec. Char., &c. Leaves oval-oblong, acuminate, membranaceous, glabrous, denticulate and almost in an eroded manner: they resemble those of the common almond tree, but have the serratures inflexed, protuberant, and tipped with minute glandulous mucros. Flowers on longish pedicels, and disposed nearly in a corymbose manner. Fruit nearly ovate, small; its flesh red. (*Dec. Prod.*) A small tree. Northern parts of North America. Height 20 ft. to 30 ft., with a trunk 6 in. to 8 in. in diameter. Introduced in 1822. Flowers white; May. Drupe red; ripe in July.

Of all the cherries of North America, Michaux observes, the *C.* boreàlis is the one that has the greatest analogy with the cultivated cherry of Europe. Pursh describes it as a very handsome small tree, the wood exquisitely hard and fine-grained; but the cherries, though agreeable to the taste, astringent in the mouth, and hence called choke cherries.

454. *Cérasus boreàlis.*

⁂ 10. *C.* PU'MILA *Michx.* The dwarf Cherry Tree.

Identification. Michx. Fl. Bor. Amer., 2. p. 286.; Dec. Prod., 2. p. 537.; Don's Mill., 2. p. 513.
Synonymes. Prùnus pùmila *Lin. Mant.* 75., *Pursh Fl. Amer. Sept.* 1. p. 331.; Cérasus glauca *Mœnch Meth.* 672.; Ragouminier, Nega, Menel du Canada, *Fr.*

Engravings. Mill. Icon., t. 80. f. 2.; and our *fig.* 455.

455. Cérasus pùmila.

Spec. Char., &c. Branches twiggy. Leaves obovate-oblong, upright, glabrous, indistinctly serrulated, glaucous beneath. Flowers upon peduncles, disposed rather umbellately. Calyx bell-shaped, short. Fruit ovate, black. (*Dec. Prod.*) A low somewhat procumbent shrub. North America, in Pennsylvania and Virginia, in low grounds and swamps. Height 3 ft. to 4 ft. Introduced in 1756. Flowers white; May. Drupe black; ripe in July.

A curious and rather handsome tree, when grafted standard high; and a fit companion for the other dwarf sorts, when so grafted. Sir W. J. Hooker suspects this to be the same as *C.* depréssa. It has been compared, Sir W. J. Hooker observes, in its general habit, to *Amýgdalus* nàna; and such a comparison is equally applicable to *C.* depréssa. (*Fl. Bor. Amer.*, i. p. 167.)

✻ 11. *C.* (P.) DEPRE'SSA *Ph.* The depressed, *or prostrate*, Cherry Tree.

Identification. Pursh Fl. Amer. Sept., 1. p. 332.; Dec. Prod., 2. p. 538.; Hook. Fl. Bor. Amer., 1. p. 168.; Don's Mill., 2. p. 514.
Synonymes. *C.* pùmila *Michx. Fl. Bor. Amer.* 1. p. 286., not *Prùnus* pùmila *L.*; *P.* Susquehànæ *Willd. Enum.* 519., *Baumz.* ed. 2. p. 286.; Sand Cherry, *Amer.*
Engraving. Our *fig.* 456. from living plant in Loddiges's arboretum.

456. C. (p.) depréssa.

Spec. Char., &c. Branches angled, depressed, prostrate. Leaves cuneate-lanceolate, sparingly serrate, glabrous, glaucous beneath. Flowers in grouped sessile umbels, few in an umbel. Fruit ovate. (*Dec. Prod.*) A prostrate shrub. North America, from Canada to Virginia, on the sandy shores of rivers and lakes. Height 1 ft. Introduced in 1805. Flowers white; May. Drupe black, small, and agreeably tasted; ripe in July.

In America it is called the sand cherry, and said to be distinguished at sight from all the other species, not less by its prostrate habit, than by its glaucous leaves, which bear some resemblance in shape to those of *Amýgdalus* nàna; and, according to Sir W. J. Hooker, to those of *C.* pùmila.

✻ 12. *C.* PYGMÆ'A *Lois.* The pygmy Cherry Tree.

Identification. Lois. in N. Du Ham., 5. p. 32. and 21.; Dec. Prod., 2. p. 538.; Don's Mill., 2. p. 513.
Synonyme. *Prùnus* pygmæ'a *Willd. Sp.* 2. p. 993., *Pursh Fl. Amer. Sept.* 1. p. 331.
Engraving. Our *fig.* 457. from a specimen in the Lambertian herbarium.

457. C. pygmæ'a.

Spec. Char., &c. Leaves ovate-elliptical, but tapered to the base, and rather acute at the tip, sharply serrated, glabrous on both surfaces, and with 2 glands at the base. Flowers of the size of those of *P.* spinòsa, disposed in sessile umbels, a few in an umbel. (*Dec. Prod.*) A low shrub. Western parts of Pensylvania and Virginia. Height 4 ft. to 5 ft. Introduced in 1823. Flowers white; May. Drupe black, of the size of a large pea, a little succulent, and very indifferent to the taste; ripe in July.

✻ ⸸ 13. *C.* NI'GRA *Lois.* The black Cherry Tree.

Identification. N. Du Ham., 5. p. 32.; Dec. Prod., 2. p. 538.; Don's Mill., 2. p. 513.
Synonymes. *Prùnus* nìgra *Ait. Hort. Kew.* 2d ed. 3. p. 193., *Pursh Fl. Amer. Sept.* 1. p. 331.; *P.* americàna *Darlington* in *Amer. Lyc. N. H. of New York.*
Engravings. Bot. Mag., t. 1117.; and our *figs.* 458, 459.

Spec. Char., &c. Leaf with 2 glands upon the petiole, and the disk ovate-acuminate. Flowers in sessile umbels, few in an umbel. Calyx purple; its lobes obtuse, and their margins glanded. (*Dec. Prod.*) A tall shrub or

458. C. nigra.

459. C. nigra.

low tree. Canada and the Alleghany Mountains. Height 6 ft. to 10 ft. Introduced in 1773. Flowers white, with purple anthers. Drupe red; April, May.

The fruit, which, as far as we know, has not been produced in England, is described by Sir W. J. Hooker as being as large as a moderate-sized cherry, and, apparently, red. In British gardens this forms a very handsome small tree, distinguished even in winter by the smoothness and dark colour of its young wood, and in this respect resembling more a plum than a cherry.

■ 14. *C.* HYEMA'LIS *Michx.* The winter Cherry Tree.

Identification. N. Du Ham., 5. p. 194.; Dec. Prod., 2. p. 538.; Don's Mill., 2. p. 514.
Synonymes. *P.* hyemàlis *Michx. Fl. Bor. Amer.* 1. p. 284., *Pursh Fl. Amer. Sept.* 1. p. 331., *Elliot Carol.* 1. p. 542.; the black Choke Cherry, *Amer.*
Engraving. Our *fig.* 460. from a specimen in the museum of the Jardin des Plantes.

460. C. hyemàlis.

Spec. Char., &c. Leaves oblong-oval, or oval, abruptly acuminate. Flowers glabrous, disposed umbellately. Lobes of the calyx lanceolate. Fruit nearly ovate, and blackish. (*Dec. Prod.*) A shrub. Western mountains of Virginia and Carolina. Height 3 ft. to 4 ft. Introduced in 1805. Flowers white; May. Drupe small, black, and extremely astringent, but eatable in winter; ripe in October.

■ 15. *C.* CHI'CASA *Michx.* The Chicasaw Cherry Tree.

Identification. Michx. Fl. Bor. Amer., 1. p. 284.; Lois. in N. Du Ham., 5. p. 183.; Dec. Prod., 2. p. 538.; Don's Mill., 2. p. 514.
Synonymes. Prùnus chicasa *Pursh Fl. Amer. Sept.* 1. p. 332.; *P.* insititia *Walt. Carol.*; Chickasaw Plum, *in Carolina.*
Engraving. Our *fig.* 461. from a living specimen in Loddiges's arboretum.

461. C. chicasa.

Spec. Char., &c. Branches glabrous, becoming rather spiny. Leaves oblong-oval, acute, or acuminate. Flowers upon very short peduncles, and mostly in pairs Calyx glabrous, its lobes very short. Fruit nearly globose, small, yellow. (*Dec. Prod.*) A shrub. Carolina and Virginia. Height 6 ft. Introd. 1806. Flowers white; April and May. Drupe small, yellow, and agreeably tasted; ripe in July.

Sir W. J. Hooker observes that a plant which he received under this name appeared to him identical with *C.* boreàlis; the plants in the London gardens are very different, resembling much more closely the common sloe, as will appear by our figure.

■ 16. *C.* PUBE'SCENS *Ser.* The pubescent Cherry Tree.

Identification. Seringe in Dec. Prod., 2. p. 538.; Don's Mill., 2. p. 514.
Synonymes. Prùnus pubéscens *Ph. Fl. Amer. Sept.* 1. p. 331., and *Lodd. Cat.*; *P.* sphærocárpa *Michx. Fl. Bor. Amer.* 1. p. 284., not of Swartz.
Engravings. Our *fig.* 462. from a living specimen in Loddiges's arboretum; and *fig.* 463. from a specimen in the Lambertian herbarium.

462. C. pubéscens. 463.

Spec. Char., &c. Young branches pubescent. Leaves with

the disk shortly oval, serrulated, and usually with 2 glands at its base. Flowers in sessile umbels, few in an umbel; pedicels and calyxes pubescent. Fruit upon a short pedicel, globose, brownish purple, austere. (*Dec. Prod.*) A low shrub. Western parts of Pennsylvania, on the borders of lakes. Height 1 ft. to 3 ft. Introduced in 1820. Flowers white; April and May. Drupe brownish purple, very astringent; ripe in July.

■ 17. *C.* PENNSYLVA'NICA *Lois.* The Pennsylvanian Cherry Tree.

Identification. Lois. in N. Du Ham., 5. p. 9., Dec. Prod., 2. p. 539.; Don's Mill., 2. p. 514.
Synonymes. Prùnus pennsylvánica *L. fil. Suppl.* p. 252., *Pursh Fl. Amer. Sept.* 1. p. 331.; *P.* pennsylvánica *Willd. Baum.* edit. 1811, p. 310.; *P.* lanceolàta *Willd. Abb.* p. 240.
Engravings. Abb. Georg. Ins., vol. 1. p. 89. t. 45.; and our *fig.* 464.

Spec. Char., &c. Leaves with two glands at the base of the disk, which is oblong-lanceolate, acuminate, and glabrous. Flowers disposed in grouped sessile umbels, which have something of the character of panicles. (*Dec. Prod.*) A shrub. North America, from New England to Virginia, in woods and plantations. Height 6 ft. to 8 ft. Introduced in 1773. Flowers white; May. Drupe black, small, but agreeable to eat; ripe in July.

464. C. pennsylvánica.

Sir W. J. Hooker considers this sort as synonymous with *C.* boreàlis *Michx.*, in which he may probably be correct. We have, however, kept them distinct; not only because the whole genus appears in a state of confusion, but because, though *C.* pennsylvánica is said to have been introduced in 1773, we have never seen the plant in a healthy state, and, consequently, feel unable to give any decided opinion respecting it.

■ 18. *C.* JAPO'NICA *Lois.* The Japan Cherry Tree.

Identification. Lois. in N. Du Ham., 5. p. 83.; Don's Mill., 2. p. 514.
Synonymes. Prùnus japónica *Thunb. Fl. Jap.* p. 201., and *Lindl.* in *Bot. Reg.* t. 1801.; *P.* sinénsis *Pers. Ench.* 2. p. 36.
Engravings. Bot. Reg., t. 1801., from a plant grown in a green-house; and our *figs.* 465, 466. from a plant grown in the open air.

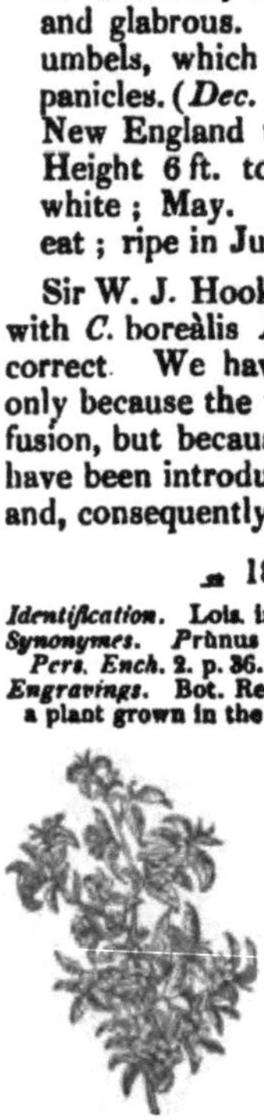
465. C. japónica.

Spec. Char., &c. Leaves ovate, acuminated, glabrous, shining. Peduncles solitary. Lobes of calyx shorter than the tube. (*Don's Mill.*) A slender shrub, somewhat tender. China. Height 2 ft. to 4 ft. Introd. 1810. Flowers pale blush-coloured, produced in profusion on numerous slender purplish or brownish red twigs; March to May. Drupe ?. Naked young wood brownish red.

466. C. japónica.

Variety.

■ C. *j.* 2 *múltiplex* Ser. Amýgdalus pùmila *Lin. Mant.* 74., *Bot. Mag.* t. 2176., and of the Hammersmith and other nurseries. (Our *figs.* 467, and 468.) — Flowers semidouble, pink like those of the species.

467. C. j. multiplex.

There are two shrubs in British nurseries often confounded under the name of *A.* pùmila. The one is that now described, which may be known at any season by the purplish or brownish red colour of the bark of its young shoots; and, in summer, by its glabrous finely serrated leaves, which have a

468. C. j. múltiplex.

reddish tinge on their margins, and on the midribs. The other, *C. sinénsis* described below, the *Prùnus japónica* of Ker, and of the Hammersmith and other nurseries, may be known in the winter season by the light green or greyish colour of the bark of its young shoots; by its larger, paler-coloured, and comparatively rugose leaves, doubly or coarsely serrated; and by its more compact habit of growth. The flowers of this sort are also on longer peduncles, resembling those of a cherry; while the flowers of *C. jap. múltiplex*, the *Amýgdalus pùmila* or double dwarf almond of the nurseries, have much shorter peduncles, and are sometimes nearly sessile, giving the plant more the appearance of a *Prùnus* than that of a *Cérasus*. The *C. japónica múltiplex* has been in cultivation in British gardens, under the name of *Amýgdalus pùmila*, since the days of Bishop Compton; and, though it is stated in books to have been introduced from Africa, there can be little doubt of its being of Asiatic origin. The great confusion which exists respecting these two plants, in botanical works, has induced us to examine, with particular attention, the plants of them that are in the Horticultural Society's Garden, and in the Hammersmith Nursery. In the former garden, there was (June 10. 1837) a *Cérasus japónica* in its single state, but not in its double state; the plant bearing the name of *C. japónica flòre plèno* being unquestionably the *C. sinénsis* described below, the *Prùnus japónica* of the nurseries. In the Hammersmith Nursery, there were then some dozens of plants of *C. japónica múltiplex*, there called *Amýgdalus pùmila*, or the double dwarf almond, growing in parallel nursery lines, with some dozens of plants of *C. sinénsis*, there called *Prùnus japónica*, or the double Chinese almond. We have considered it necessary to be thus particular, to justify us for having deviated from the *Bot. Mag.* and *Bot. Reg.*

19. *C.* SINE'NSIS *G. Don.* The Chinese Cherry.

Identification. Don's Mill., 2. p. 514.
Synonyme. Prùnus japónica *Ker* in *Bot. Reg.* t. 27.
Engravings. Bot. Reg., t. 27.; and our *fig.* 469.

Spec. Char., &c. Leaves ovate-lanceolate, doubly serrated, wrinkled from veins beneath. Peduncles sub-aggregate. (*Don's Mill.*) A highly ornamental low shrub. China. Height 2 ft. to 3 ft. Introduced in 1810. Flowers semi-double, red above, and white underneath, produced in great profusion on the preceding year's branches; April and May.

There is no single state of this species in Britain, but there was in 1836 a plant of the double variety against a wall in the Hort. Soc. Garden, named *C. japónica flòre plèno*; and, as noticed under the preceding species, there were many plants in the Hammersmith Nursery, under the name of *P. japónica*, or the double Chinese almond. The plant is somewhat more tender than *C. j. múltiplex*, which is well known in gardens as a hardy border shrub; and, except in favourable situations, it requires to be planted against a wall. Though *C. sinénsis* and *C. japónica* are quite distinct, there is nothing in that distinctness, as it appears to us, to determine that they are not varieties of the same species.

469. *Cérasus sinénsis.*

20. *C.* SALI'CINA *G. Don.* The Willow-*leaved* Cherry Tree.

Identification. Don's Mill., 2. p. 514.
Synonymes. Prùnus salicina *Lindl.* in *Hort. Trans.*; Ching-Cho-Lee, or Tung-Choh-Lee, *Chinese.*
Engraving. *Fig.* 470. from a specimen in the herbarium of Dr. Lindley.

Spec. Char., &c. Flowers usually solitary, shorter than the leaves. Leaves

470. *Cérasus salicina.*

obovate, acuminated, glandularly serrated, glabrous Stipules subulate, glandular, length of the petiole. Petiole glandless. (*Don's Mill.*) A shrub. China. Height 4 ft. to 6 ft. Introd. in 1822. The flowers small, and white. Drupe about the size of that of the myrobalan plum.

471. *Cérasus Púddum.*

Species belonging to the preceding Subdivision (B.), *not yet introduced.* — C. Phóshia *Hamilt.*, Prùnus cerasöìdes *D. Don.*, Cérasus *Púddum* Roxb. (Wall. Pl. Rar., ii. t. 143.; and our *fig.* 471.), is a native of Nepal, producing fruit like that of the common cherry, and wood which is considered valuable as timber. The flowers are of a pale rose colour, and the tree grows to the height of 20 or 30 feet. C. *glandulòsa*, C. *áspera*, and C. *incìsa* Lois., are Japan shrubs, with rose-coloured flowers, described by Thunberg; and C. *hùmilis* Moris., a native of Sardinia.

§ ii. Pàdi vèri Ser. *The true Bird-Cherry Kinds of* Cérasus.

Sect. Char. Flowers produced upon the shoots of the same year's growth as the flowers; the latter disposed racemosely. Leaves deciduous.

A. *Species of Bird-Cherry Trees already in Cultivation in Britain.*

♈ 21. *C. Maha'leb* Mill. The Mahaleb, *or perfumed*, Cherry Tree.

Identification. Mill. Dict., No. 4.; Dec. Prod., 2. p. 539.; Don's Mill., 2. p. 514.
Synonymes. Prùnus *Mahàleb* L. Sp. 678.; Bois de Sainte Lucie, or Prunier odorant, *Fr.*; Mahalebs-kirsche, *Ger.*; Ciliegio canino, *Ital.*
Engravings. N. Du Ham., 5. t. 2.; Jacq. Fl. Austr., t. 227.; the plate of this species in Arb. Brit. 1st edit., vol. v.; and our *fig.* 472.

Spec. Char., &c. Leaves cordately ovate, denticulate, glanded, curved. Flowers in leafy subcorymbose racemes. Fruit black, between ovate and round. (*Dec. Prod.*) A small tree. Middle and South of Europe; common in France, especially in the mountainous districts; very common near St. Lucie, whence the French name. Height 10 ft. to 20 ft.; in British gardens 20 ft. to 30 ft. Introduced in 1714. Flowers white; April and May. Drupe black; ripe in July.

Varieties. Besides one with variegated leaves, there are: —

♈ C. M. 2 *frúctu flàvo* Hort. — Fruit yellow. There is a plant of this variety in the garden of the Horticultural Society.
♈ C. M. 3 *latifòlium* Hort. — Leaves broader than in the species.

A handsome small tree, with a white bark, and numerous branches. The leaves somewhat resembling those of the common apricot, but of a paler green. The wood, the leaves, the flowers, and the fruit, are powerfully scented; the flowers so much so as not to be supportable in a room. The wood is hard, brown, veined, and susceptible of a high polish. Its smell is less powerful, and more agreeable, when it is dry, than when the sap is in it. In a dry state it weighs 59 lb. 4 oz. per cubic foot. In France, it is much sought after by cabinetmakers, on account of its fragrance, hardness, and the fine polish which it receives. In Austria it is used for forming the twisted tubes of tobacco pipes. In France the mahaleb is used as a stock on which to graft the different kinds of fruit-bearing cherries: for which it has the advantages of growing on a very poor soil; of coming into sap 15 days later than

472. Cérasus Mahaleb.

the common wild cherry, by which means the grafting season is prolonged; and, lastly, of dwarfing the plants grafted on it. In British gardens, it is partly used for this purpose, but principally as an ornamental shrub or low tree. As in the case of other dwarf species of a genus which will unite to a tall robust-growing species, the mahaleb, when grafted on the common wild cherry (*C.* sylvéstris), grows to a larger tree than when on its own roots. The mahaleb will grow in any poor soil that is dry, even in the most arid sands and naked chalks; and, as it forms a low bushy tree which is capable of resisting the wind, it may be planted in an exposed situation. When young plants are to be raised from seed, the fruit is sown as soon as ripe, or preserved among sand till the following spring, in the same manner as that of the cherry. Seedling plants generally grow 1 ft. in length the first year, and 1 ft. to 18 in. the second. The tree may also be propagated by layers; by slips from the stool, taken off with a few roots attached; and by suckers, or by cuttings from the roots.

♀ 22. *C. Pa'dus Dec.* The Bird-Cherry Tree.

Identification. Dec. Fl. Fr., 4. p. 580.; Prod., 2. p. 539.; Don's Mill., 2. p. 515.
Synonymes. Prûnus Pàdus *Lin. Sp.* 677., *Hook. Brit. Flora*, p. 220., *Smith Eng. Flora*, 2. p. 354.; Bird Cherry, Fowl Cherry; Hag-berry, *Scot.*; Cerasier à Grappes, Merisier à Grappes, Laurier-Putier, or Putiet, faux Bois de Ste. Lucie, *Fr.*; Hag-bier, *Swedish*; Traubeden Kirsche, *Ger.*; Ciliegio ramoso, *Ital.*
Engravings. Eng. Bot., t. 1383.; the plate of this species in Arb. Brit., 1st edit., vol, v.; and our *fig.* 474.

Spec. Char., &c. Leaves ovate-lanceolate, somewhat acuminate, thin, serrulate, with the teeth rather spreading. Racemes long, leafy. Fruit round, bitter. (*Dec. Prod.*) A low tree. Indigenous in most parts of Central Europe, and as far north as Lapland. Height 12 ft. to 40 ft. Flowers white; April and May. Drupe black; ripe in July. Decaying leaves greenish yellow, or reddish. Naked young wood purplish, with white spots.

473. Cérasus Pàdus rùbra.

Varieties.

♀ C. P. 1 *vulgàris* Ser. *C. Pàdus Dec., N. Du Ham.* v. t. 1. — This kind has large flowers loosely disposed upon long pedicels, and black fruit.

♀ C. P. 2 *parviflòra* Ser. (Œd. Fl. Dan., t. 205.)—This has smaller flowers, upon shorter pedicels, which are disposed more densely; and black fruit.

C. P. 3 *rùbra* Ser. (Our *fig.* 473.) — This has red fruit. It is the *C. Pàdus* frúctu rùbro of Dec. and of Loiseleur; and, according to *Ait. Hort. Kew.*, 2d ed. p. 299., it is the *Prùnus* rùbra of *Willd. Arb.*, 237. t. 4. f. 2.

‡ C. P. 4 *bracteòsa* Ser. *Pàdus* racemòsus *Hort.* — A very beautiful variety, distinguished by its long racemes of flowers, with their pedicels furnished with long bracteas at the points of the shoots, by which the latter are bent down, both when in blossom and when the fruit is ripe, so as to give the whole tree a pendulous appearance.

A very handsome small tree or large bush. The leaves are finely serrated, smooth, and somewhat glaucous; and their scent, when bruised, resembles that of rue. The flowers are of a pure white, in copious, long, terminal racemes, making an elegant appearance in spring, but scarcely lasting a fort-

474. *Cérasus Pàdus.*

night. The fruit is small, black, austere, and bitter, with a large corrugated nut. "Birds of several kinds soon devour this fruit, which is nauseous, and probably dangerous to mankind; though, perhaps, like that of the cherry laurel, not of so deadly a quality as the essential oil or distilled water of the leaves." (*Eng. Flora*, ii. p. 354.) The tree grows rapidly when young, attaining the height of 10 or 12 feet in 5 or 6 years; and, as it has a loose head, and bears pruning, it allows the grass to grow under it. The wood is hard and yellowish, and, in a green state, it has a disagreeable bitter odour and taste; whence the French name *putiet*, from *puer*. It is much sought after in France by the cabinetmakers and turners, who increase the beauty of its veining by sawing out the boards diagonally, that is, obliquely across the trunk, instead of parallel with its length. The fruit, though nauseous to the taste when eaten fresh from the tree, gives an agreeable flavour to brandy; and is sometimes added to home-made wines. In Sweden and Lapland, and also in some parts of Russia, the bruised fruit is fermented, and a powerful spirit distilled from it. In Britain, the principal use of the *Cérasus Pàdus* is as an ornamental tree; and few make a finer appearance than it does, either when in flower, in April and May; or in August, when covered with its pendent racemes of black fruit. It comes into flower a little before the ornamental crab trees, and about the same time as the *Sórbus* aucupària and the *A'cer* platanöídes. The bird cherry prefers a dry soil; but it will not thrive on such poor ground as the perfumed cherry. It will grow in almost any situation; but, to attain a timber-like size, it requires the shelter either of a favourable locality, or of adjoining trees. The species is propagated by seeds, which should be treated in all respects like those of *C. Mahàleb.* The red-fruited variety will generally come true from seed; as, doubtless, will the early-flowering and late-flowering varieties, which may be observed in copse woods where this tree abounds. *C. P.* bracteòsa *Ser.*, which is a very remarkable variety, and one which deserves a place in every collection, both on account of its large racemes of flowers and its fruit, will be continued with most certainty by grafting or budding. The leaves are more infested and injured by the larvæ of moths and butterflies, than those of any other European tree or shrub.

♀ 23. *C.* VIRGINIA'NA *Michx.* The Virginian Bird-Cherry Tree.

Identification. Michx. Fl. Bor. Amer., 1. p. 285.; Dec. Prod., 2. p. 539.; Don's Mill., 2. p. 515.
Synonymes. Prùnus rùbra *Ait. Hort. Kew.* 1st ed. 2. p. 162., *Willd. Abb.* 238. t. 5. f. 1.; *P.* argùta *Bigelow* in Litt.; Cerisier de Virginie, *Fr.*; Virginische Kirsche, *Ger.*; Wild Cherry Tree, *Amer.*
Engravings. Willd. Abb., 238. t. 5. f. 1.; Michx. Fl. Arb. Amer., 2. t. 88.; and our *fig.* 475.

Spec. Char., &c. Leaves oblong, acuminate, doubly toothed, smooth; the petiole bearing about 4 glands. Racemes straight, petals round. Fruit red. Different from the *Prùnus* virginiàna of Miller, which is *C.* (v.) serótina. (*Dec. Prod.*) A tree attaining a large size. Virginia, Carolina, and Canada. Height, in England, 30 ft. to 40 ft.; in some parts of North America, 80 ft. to 100 ft. Introd. 1724. Flowers white; May. Drupe red; ripe in July. Leaves remaining on late in the season, and dropping, green. Naked young wood slender, purplish, but not spotted with white like *C. Pàdus.*

Readily distinguished from *Cérasus Pàdus* by the slender drooping character of its branches. The fruit is frequently ripened in the neighbourhood of London, and plants in copse woods, which have risen from selfsown seeds, are to be met with in different parts of Surrey. The wood of the Virginia bird cherry is of a light red tint, which deepens with age. It is compact, fine-grained, and takes a brilliant polish; it is also not liable to warp when perfectly seasoned. In America, it is extensively used by cabinet-makers for every species of furniture. In Europe, *C.* virginiàna is planted solely as an ornamental tree; and, as such, it well deserves a place in every collection. It should be planted in every shrubbery or wood where it is desirable to attract frugivorous singing birds. For soil, situation, propagation, culture, &c., see *C. Pàdus.*

475. Cérasus virginiàna.

♀ 24. *C.* (v.) SERO'TINA *Lois.* The late-*flowering*, or *American*, Bird-Cherry Tree.

Identification. Lois. in Du Ham., 5. p. 3.; Dec. Prod., 2. p. 540.; Don's Mill., 2. p. 515.
Synonymes. Prùnus serótina *Willd. Abb.* 239.; *P.* virginiàna *Mill. Dict.* No. 2.
Engravings. Willd. Abb., 239. t. 5. f. 2; Wats. Dend. Brit., t. 48.; and our *figs.* 476, 477.

Spec. Char., &c. Leaves ovate-lanceolate, rather coriaceous, glossy, serrated; the teeth imbricate, very numerous, and the lowest ones indistinctly glanded. Midrib downy at its base. Floral leaves narrowed at the base. Racemes loose. Fruit black. (*Dec. Prod.*) A middle-sized tree. A native of North America, in Canada and Newfoundland. Height 10 ft. to 20 ft. Introd. 1629. Flowers white; May and June. Drupe black; ripe in August. Leaves retained late, and dropping green. Naked young wood slender, purplish.

476. Cérasus (v.) serótina.

Variety.

♆ C. *s.* 2 *retùsa* Ser. — Leaves obovate, round, very obtuse, almost retuse, slightly villose beneath; midrib hairy above and below. A native of South America.

477. Cérasus (v.) serótina.

C. (v.) serótina so closely resembles *C.* virginiàna, that we have no doubt whatever of their being one and the same species.

♆ 25. *C.* MO'LLIS. *Doug.* The soft Bird-Cherry Tree.

Identification. Dougl. MSS.; Hook. Fl. Bor. Amer., 1. 169.; Don's Mill., 2. p. 515.
Engraving. Our *fig.* 478., from a specimen in the British Museum.

478. C. móllis.

Spec. Char., &c. Racemes short, pubescently tomentose, as well as the calyxes; calycine segments reflexed; leaves obovate oblong, crenated, pubescent beneath; fruit ovate. (*Don's Mill.*) A tree. America, near the mouth of the Columbia, and on subalpine hills, near the source of the river. Height 12 ft. to 24 ft. Introd. 1838. Flowers white. Naked young wood dark brown and downy, and the general habit said to be that of *C.* pubéscens.

Young plants have been raised in the Hort. Soc. Garden, from seeds sent home by Douglas.

479. Cérasus Capóllin.

♆ ♆ 26. *C.* *CAPO'LLIN* Dec. The Capollin Bird-Cherry Tree.

Identification. Dec. Prod., 2. p. 539.; Don's Mill., 2. p. 515.
Synonymes. Prùnus virginiàna *Flor. Mexic. Ic.* and *MSS.*; *P.* canadénsis *Moc. et Sesse Pl. Mex. Ic. ined.*, *Hern. Mex.* 95.

Engravings. Fl. Mex. Icon. ined.; Pl. Mex. Icon. ined.; Hern. Mex. Icon., 95.; and our *fig.* 479.

Spec. Char., &c. Leaves lanceolate, serrated, and glabrous, resembling in form, and nearly in size, those of Sàlix frágilis. Racemes lateral and terminal. Fruit globose, resembling, in form and colour, that of *C.* sylvéstis. (*Dec. Prod.*) A handsome sub-evergreen low tree or shrub. Mexico, in temperate and cold places. Height 6 ft. to 12 ft. Introduced in 1820. Flowers white; May. Drupe red; ripe in August.

The plant bearing this name in the Hort. Soc. Garden has leaves broader than those of Sàlix frágilis; and, taken altogether, it is of more luxuriant growth than *C.* virginiàna. It is, however, less hardy as a plant in the open garden, and was killed to the ground in the winter of 1837–8. In favourable situations, however, it is a very desirable species, being a remarkably free flowerer, and ripening abundance of fruit, which have stones as large as those of the wild cherry.

■ 27. *C.* NEPALE'NSIS *Ser.* The Nepal Bird-Cherry Tree.

Identification. Seringe in Dec. Prod., 2. p. 540.; Don's Mill., 2. p. 515.
Synonyme. Prùnus glaucifòlius *Wall. MSS.*
Engraving. Our *fig.* 480., from a specimen in Dr. Lindley's herbarium.

Spec. Char., &c. Leaves resembling in form those of Sàlix frágilis; long, lanceolate, acuminate, serrate, with blunt teeth, glabrous, whitish beneath; the veins much reticulated; and the axils of the larger of them hairy. Peduncle short, and, as well as the rachis, slightly villose. Calyx glabrous. (*Dec. Prod.*) A deciduous shrub or low tree. Nepal. Height 6 ft. to 12 ft. Introd. 1820. Flowers white; May. Drupe ?.

480. Cérasus nepalénsis.

B. *Species of Bird-Cherry Trees which have not yet been introduced, or of which we have not seen Plants.*

† C. *acuminàta* Wall. (Pl. Rar. Asiat., ii. p. 78. t. 181.; and our *fig.* 481.) is a Nepal tree, growing to the height of 20 or 30 feet, with the flowers in axillary racemes, and nodding, a little shorter than the leaves.

■ C. *emarginàta* Dougl. (*Hook. Fl. Bor. Amer.*, p. 169.) is a shrub, growing to the height of 6 or 8 feet, with its flowers in corymbose racemes; having oval, serrulated, glabrous leaves; and globose fruit, astringent to the taste. The leaves are 2 in. long; the flowers are white; and the wood red, with white spots. It is found wild about the upper part of the Columbia River, especially about the Kettle Falls.

481. Cérasus acuminàta.

† C. *capricìda* G. Don. *The Goat-killing Bird Cherry.* Prùnus capricìda *Wall.*; *P.* undulàta *Hamilt.* in *D. Don's Prod. Nepal.* p. 239.; *C.* undulàta *Dec. Prod.* ii. p. 540.—Leaves elliptic, acuminated, coriaceous, glabrous, quite entire, with undulately curled margins. Petioles glandulous. Racemes either solitary or aggregate by threes, many-flowered, glabrous, shorter than the leaves. (*Don's Mill.*, ii. p. 515.) A handsome showy tree, probably evergreen, a native of Nepal, at Narainhetty; where the leaves are found to contain so large a quantity of prussic acid as to kill the goats which browse upon them. Royle seems to consider *C.* undulàta and *C.* capricìda as distinct species; and

he observes that these, and "*C.* cornùta, remarkable for its pod-like monstrosity, are handsome showy trees, growing on lofty mountains, and worthy of introduction into England." (*Royle's Illust.*, p. 205.)

C. canadénsis *Lois.*, *C.* ellíptica *Lois.*, *C.* paniculàta *Lois.*, and some other hardy species, are mentioned in our first edition.

§ iii. *Laurocérasi.* *The Laurel-Cherry Trees.*

Sect. Char. Evergreen. Flowers in racemes.

■ ♀ 28. *C.* LUSITA'NICA *Lois.* The Portugal *Laurel*-Cherry, *or common Portugal Laurel.*

Identification. Lois. in N. Du Ham., 5. p. 5.; Dec. Prod., 2. p. 540.; Don's Mill., 2. p. 516.
Synonymes. Prúnus lusitánica *Lin. Sp.* 678.; the Cherry Bay; Cerisier Laurier du Portugal, *Fr.*; Azareiro, *Portuguese.*
Engravings. Mill. Ic., 131. t. 196. f. 1.; Dill. Elth., 193. t. 159. f. 193.; the plate of the species in Arb. Brit., 1st edit., vol. v.; and our *figs.* 483, 484.

Spec. Char., &c. Evergreen. Leaves coriaceous, ovate-lanceolate, serrate, glandless. Racemes upright, axillary, longer than the leaves. (*Dec. Prod.*) An evergreen low tree. Portugal, and the Azores. Height 10 ft. to 20 ft.; in British gardens sometimes 30 ft. Introduced in 1648. Flowers white; June. Drupe dark purple; ripe in September. Young wood purplish black.

482. *Cérasus l. Híxa.*

Variety.

■ ♀ C. *l.* 2 Híxa *Ser.* Prùnus *Híxa* Broussonet; *P.* multiglandulòsa *Cav.*; *C. Híxa* Webb et Bert. Hist. Can. t. 38. (Our *fig.* 482.) — Leaves larger, with their lowest teeth glanded. Racemes elongate. Flowers more loosely disposed. Spontaneous in the islands of Teneriffe, Grand Canary, and Palma. Mr. P. B. Webb informs us that this tree, in its native localities, attains the height of 60 or 70 feet.

The Portugal laurel is generally seen as an immense bush, but when trained up to a single stem it forms a very handsome tree with a conical head. It is not of rapid growth, seldom making shoots more than 9 or 10 inches in length; but, when planted in good free soil, and trained to a single stem, plants, in the neighbourhood of London, will reach the height of from 12 ft. to 15 ft. in 10 years. It is generally planted solely as an ornamental evergreen; but sometimes hedges are formed of it in nursery-grounds and flower-gardens. The berries are greedily eaten by birds, and form a favourite food for pheasants. What renders the tree particularly valuable, Miller

483. *Cérasus lusitanica.*

484. Cérasus lusitánica.

observes, is its being "so very hardy as to defy the severest cold of this country; for, in the hard frost of 1740, when almost every other evergreen tree and shrub was severely pinched, the Portugal laurels retained their verdure, and seemed to have felt no injury." In the winter of 1837–8, it was severely hurt in all low moist situations in the climate of London; but in dry gravelly soil, there, and in most parts of England, it escaped uninjured. In British nurseries, it is propagated by seeds, which, before and after sowing, are treated like those of the common wild cherry (*C.* sylvéstris), or those of the bird cherry (*C. P*àdus).

■ ● 29. *C.* LAUROCE'RASUS *Lois.* The Laurel-Cherry, *or common Laurel.*

Identification. Lois. in Du Ham., 5. p. 6.; Dec. Prod., 2. p. 540.; Don's Mill., 2. p. 516.
Synonymes. Prùnus Laurocérasus *Lin. Sp.* 678.; Cherry Bay, Cherry Laurel; Laurier au Lait, Laurier Cerisier, Laurier Amandier, *Fr.*; Kirsche Lorbeer, *Ger.*; Lauro di Trebisonda, *Ital.*
Engravings. Blackw. Herb., t. 512.; Du Ham. Arb., 1. p. 346. t. 133.; and our *fig.* 485.

Spec. Char., &c. Evergreen. Leaves coriaceous, ovate-lanceolate, remotely serrate, bearing upon the under surface of the disk 2—4 glands. Racemes shorter than the leaves Fruit ovate-acute. (*Dec. Prod.*) A large, rambling, evergreen, sub-prostrate shrub. Trebisond in Asia Minor; and found in Caucasus, Persia, and the Crimea. Height 6 ft. to 20 ft. Introduced in 1629. Flowers white; April and May. Drupe dark purple; ripe in October. Young shoots of a light green.

Varieties.

● C. *L.* 2 *variegàta* Hort.—Leaves variegated with either white or yellow.
■ C. *L.* 3 *angustifòlia* Hort., with leaves about a third part of the width of

those of the species, and a more dwarf-growing plant. A very distinct variety, which seldom, if ever, flowers. In some nurseries, it is called Hartògia capénsis, though this latter is a totally different plant.

The common laurel, though it will grow as high as the Portugal laurel, is, in its habit, decidedly a shrub, though it is occasionally seen trained to a single stem as a low tree, and in France it is grafted standard high on the common cherry for this purpose, though such plants, from the stock being deciduous, only last a year or two. The growth of the common laurel is rapid for an evergreen, being at the rate of from 1 ft. to 3 ft. a year; but, as the shoots extend in length, they do not increase proportionately in thickness, and hence they recline; so that plants with branches 30 or 40 feet in length, though gigantic in size, still retain the character of prostrate shrubs. Notwithstanding the rapid and vigorous growth of this plant in ordinary seasons, it suffers a great deal more from very severe frosts than the Portugal laurel, and is sometimes killed down to the ground, which the latter rarely is in England. In Britain, the common laurel is considered one of the most ornamental of our evergreen shrubs; and it is also used for covering walls, and for hedges, to afford shelter; for which last purpose it is extensively used in the market-gardens about Isleworth. It is also extensively used as undergrowth in sandy soil. Laurel leaves have a bitter taste, and the peculiar flavour of prussic acid, which is common to bitter almonds, and to the kernels generally of the Amygdàleæ. The flowers have a similar flavour; and the powdered leaves excite sneezing. The leaves, in consequence of their flavour, are used in a green state in custards, puddings, blancmange, and other culinary and confectionary articles, but always in very small quantities. Any soil tolerably dry will suit the common laurel; but, to thrive, it requires a sheltered situation, and a deep free soil. It thrives better as an undergrowth than, perhaps, any other ligneous plant, with the exception of the box and the holly.

485. C. Laurocérasus.

■ ‡ 30. C. CAROLINIA'NA *Michx.* The Carolina Bird-Cherry Tree.

Identification. Michx. Fl. Bor. Amer., 1. p. 285.; Lois. in N. du Ham., 5. p. 5.; Don's Mill., 2. p. 516.
Synonymes. Prùnus caroliniàna *Ait. Hort. Kew.*, 2. p. 163.; *P.* sempervirens *Willd. Enum.*; Pàdus caroliniàna *Mill. Dict.* No. 6.; Wild Orange, *Amer.*
Engravings. Michx. Arb. d'Amer., 3. t. 7.; and our *fig.* 486.

Spec. Char., &c. Evergreen. Leaves, with the petiole short; and the disk lanceolate-oblong, mucronate, even, rather coriaceous, mostly entire. Flowers densely disposed in axillary racemes, that are shorter than the leaves. Fruit nearly globose, mucronate. (*Dec. Prod.*) An evergreen shrub or low tree; in England a tender shrub. North America, from Carolina to Florida, and the Bahama Islands. Height

486. C. caroliniàna.

20 ft. to 30 ft. Introduced in 1759. Flowers white; May. Drupe dark purple; ripe ?.

This tree Michaux considers as one of the most beautiful vegetable productions of the southern parts of the United States; and it is generally selected by the inhabitants to plant near their houses, not only on this account, but because it grows with rapidity, and affords an impenetrable shade. Pursh describes it as a handsome evergreen shrub, resembling *C.* lusitánica; but he says nothing of the flowers, which, from the figure in Michaux, from which ours was copied, appear to be almost without petals. Seeds are frequently imported from America, and abundance of young plants reared; but, as they are rather tender, and, north of London, would require the protection of a wall, they are very seldom seen in British gardens. The largest plant which we know of is in Hampshire, at Swallowfield, where, in 1833, it formed a bush 10 ft. high, with a head about 12 ft. in diameter, flowering and fruiting occasionally. Culture as in *C.* virginiàna, but north of London it requires the protection of a wall.

Sect. II. Spiræ'æ.

Genus VI.

PU'RSHIA Dec. The Purshia. *Lin. Syst.* Icosándria Monogýnia.

Identification. Dec. in Trans. of Linn. Soc., 12. p. 157.; Prod., 2. p. 541.; Lindl. in Bot. Reg. t. 1446.; Don's Mill., 2. p. 517.
Synonyme. *Tigàrea* Ph. Fl. Amer. Sept., 1. p. 333., not of Aublet.
Derivation. *Frederick Pursh* first characterised the only known species in his *Flora Americæ, Septentrionalis*, and named it *Tigàrea* tridentàta. The generic name, however, having been preoccupied by Aublet, De Candolle has named the present genus after Pursh himself.

Gen. Char. *Calyx* 5-cleft; lobes ovate, obtuse. *Petals* 5, obovate. *Stamens* about 25, rising with the petals from the calyx. *Carpels* 1—2, ovate oblong, pubescent, tapering into the style at the apex, at length opening by a longitudinal chink. *Seed* 1, inserted in the base of the carpel. (*Don's Mill.*)

Leaves simple, grouped together, cuneate, 2—3-toothed at the apex, stipulate or exstipulate, deciduous. *Flowers* yellow. — Shrub, of which there is only one species known.

⚘ 1. P. tridenta'ta *Dec.* The 3-toothed-*leaved* Purshia.

Identification. Dec. in Lin. Trans., 12. p. 157.; Prod., 2. p. 541.
Synonyme. *Tigàrea* tridentàta *Ph. Fl. Amer. Sept.*, 1. p. 333. t. 15., not of Aublet.
Engravings. Ph. Fl. Amer. Sept., t. 15.; Bot. Reg., t. 1446.; and our *figs.* 487, 488.

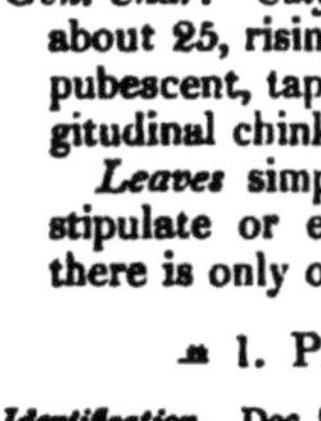

487. P. tridentàta.

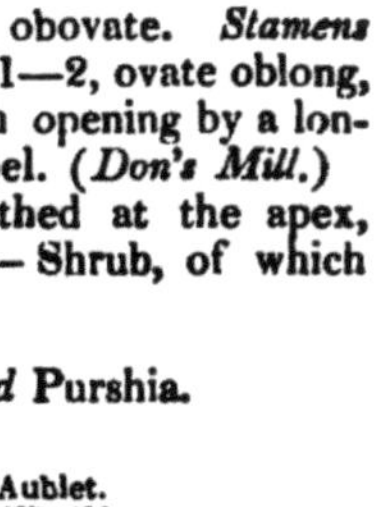

488. P. tridentàta.

Spec. Char., &c. Leaves obcuneate, 3-dentate, crowded on the points of the shoots, hairy above, and tomentose beneath. Flowers terminal on short peduncles. (*Ph. Fl. Amer.*) A spreading shrub. North America, on pastures by the river Columbia. Height 2 ft. to 3 ft. Introduced in 1826. Flowers yellow; July. Carpels ?.

Almost the only shrub to be seen through an immense tract of barren sandy soil, from the head source of the Missouri, to the Falls of the Columbia. The plants in the London gardens were all killed in the winter of 1837-8.

Genus VII.

KE'RRIA Dec. The Kerria. *Lin. Syst.* Icosándria Polygýnia.

Identification. Dec. in Trans. of Lin. Soc., 12. p. 156.; Prod., 2. p. 541.; Don's Mill., 2. p. 417.
Synonymes. Rùbus *L.*, Córchorus *Thunb.*, Spiræ'a *Camb.*
Derivation. Named in honour of *W. Ker*, a collector of plants for the Kew Gardens.

Gen. Char. *Calyx* 5-cleft; lobes ovate, 3 of which are obtuse, and the other two callously mucronate at the apex; imbricate in æstivation. *Petals* 5, orbicular. *Stamens* about 20, arising from the calyx with the petals, exserted. *Carpels* 5—8, globose, free, glabrous, each ending in a filiform style. *Seed* solitary. (*Don's Mill.*)

Leaves simple, ovate, lanceolate, alternate, stipulate, deciduous; coarsely and unequally serrated, feather-nerved, conduplicate. *Flowers* yellow.

1. K. japo'nica *Dec.* The Japan Kerria.

Identification. Dec. in Trans. of Lin. Soc., 12. p. 156.; Prod., 2. p. 541.
Synonymes. Rùbus japónicus *Lin. Mant.* 245.; Córchorus japónicus *Thunb. Fl. Jap.* 227.; Spiræ'a japónica *Camb. Ann. Sci. Nat.* 1. p. 389.; Spirée du Japon, *Fr.*
Engravings. Swt. Brit. Fl. Gard., 2d ser. t. 337.; and our *fig.* 489.

Spec. Char., &c. Leaves ovate, lanceolate, coarsely and unequally serrated, feather-nerved; stipules linear, subulate A deciduous shrub. Japan. Height 3 ft. to 5 ft. Introduced in 1835. Flowers yellow; March to June, and often all the summer. Carpels ?.

Variety.

K. j. 2 *flòre plèno* (Bot. Reg., t. 587.; Bot. Mag., t. 1296.; and our *fig.* 490.)—Flowers double. Introduced in 1700, and in very general culture in British gardens.

489. Kérria japónica.

490. Kérria j. flòre plèno.

It has soft, and not very persistent, wood, clothed with a smooth greenish bark; twig-like branches; leaves that are ovate-lanceolate, and serrated with large and unequal teeth, feather-veined, and concave on the upper surface; stipules that are linear-subulate. The single-flowered variety was, until 1835, only known through a solitary specimen received from Thunberg by Linnæus, and preserved in the herbarium of that great botanist, now in the possession of the Linnæan Society. It was after examining this specimen that De Candolle removed it from the genus Córchorus, and formed that of Kérria. The double variety is generally planted against a wall, more especially north of London. It is easily and rapidly propagated by its suckers, and grows freely in any common soil.

GENUS VIII.

SPIRÆ'A L. THE SPIRÆA. *Lin. Syst.* Icosándria Di-Pentagýnia.

Identification. Lin. Gen., No. 630.; Gærtn. Fruct., 1. p. 337. t. 69.; Dec. Prod., 2. p. 541.; Don's Mill., 2. p. 517.
Synonymes. Spiræ'a sp. *Cambessedes Mon. Spir.* in *Ann. Sci. Nat.* 1. p. 227.; Spirée, *Fr.*; Spierstaude, *Ger.*
Derivation. From *speira*, a cord, in reference to the supposed flexibility of the branches of some of the species; or, according to some, from *speiraō*, to wreath; in allusion to the fitness of the flowers to be twisted into garlands. *Spiræon* is Pliny's name for a plant the blossoms of which were used, in his time, for making garlands; but that plant is thought by some to have been the Vibúrnum Lantàna.

Gen. Char. *Calyx* 5-cleft, permanent. *Stamens* 10—50, inserted in the torus, lining the calyx along with the petals. *Carpels* solitary, or several together, rarely connected at the base, ending in short points, sessile, rarely stipitate. *Seeds* 2—6. (*Don's Mill.*)

Leaves usually simple, but sometimes pinnately cut, having pinnate, or palmately ternate, nerves; alternate, stipulate, deciduous. *Flowers* white or reddish, never yellow. — Shrubs, low, deciduous. Europe, Asia, America.

Generally of erect growth, with conspicuous flowers of considerable elegance and beauty. The naked young wood, in almost all the species, is of a cinnamon brown; and, in those kinds in which the shoots are numerous so as to produce a mass, the effect is conspicuous in the winter season. They are all readily propagated by suckers, which, in general, they produce in abundance, and they will grow in any common soil.

§ i. *Physocárpos* Camb.

Derivation. From *phusa*, a bladder, and *karpos*, a fruit; in reference to the bladdery carpels.

Sect. Char. Ovaries connected at the base. Torus lining the calycine tube. Carpels bladdery, rather membranous. Ovula 2—3, fixed to the seminiferous margin of the carpel, ovoid, at first horizontal, but at length suspended. Flowers hermaphrodite, disposed in umbels. Pedicels 1-flowered. Leaves toothed, or somewhat lobed, usually stipulate. (*Don's Mill.*, p. 517.)

1. *S.* OPULIFO'LIA *L.* The Guelder-Rose-leaved Spiræa, *or Virginian Guelder Rose.*

Identification. Lin. Sp., 702.; Dec. Prod., 2. p. 542.; Don's Mill., 2. p. 517.
Synonymes. Nine Bark, *Amer.*; Evonimo del Canadà, *Ital.*
Engravings. N. Du Ham., 6. t. 14.; and our *figs.* 491, 492.

Spec. Char., &c. Leaves lobed, or 3-lobed, and partaking of an ovate figure, doubly serrated, petioled, and many of them stipuled. Flowers white, numerous, disposed in stalked hemispherical corymbs; the pedicel of each flower slender and glabrous. Sepals spreading. Torus wholly connate with the tube of the calyx. Ovaries connate with each other at the base. Ovules in each 2—3, affixed to the margin, egg-shaped, at first horizontal, at length the one pendulous, the rest ascending. Carpels bladdery, rather membranaceous, large and diverging. Seeds obovate, glossy, and yellow. (*Dec. Prod.*) A large shrub. North America, from Canada to Carolina. Height 8 ft. to 10 ft. Introd. in 1690. Flowers

491. *S. opulifolia.*

492. *S. opulifolia.*

white; June and July. Capsule inflated red; ripe in September. Decaying leaves purplish red, mixed with yellow. Naked young wood light brown.

Hardy, and very ornamental, from its abundance of white flowers, which are produced in corymbs, and resemble those of the Guelder rose; and from the numerous inflated reddish capsules which succeed the flowers. Propagated by division of the root; but sometimes by layers, or by cuttings of the young wood put, in autumn, in a shady border, in a sandy soil.

Varieties.

■ S. o. 2 *tomentélla* Ser. has the peduncles and calyx tomentose. (*Dec. Prod.*) It is found at the Grand Rapids of the Columbia River.

■ S. o. 3 *monógyna*. *S.* monógyna *Torrey*, Don's Mill. 2. p. 518.—A native of the Rocky Mountains, where it grows to the height of 3 or 4 feet. It is considered by Sir W. J. Hooker as a variety of *S.* opulifòlia.

■ 2. *S.* CAPITA'TA *Ph.* The capitate-*corymbed* Spiræa.

Identification. Ph. Fl. Amer. Sept., 1. p. 342.; Camb. Monog.; Dec. Prod., 2. p. 542.; Don's Mill., 2. p. 518.
Synonyme. S. opulifòlia var. *Hook.*
Engraving. Our *fig.* 349. from a specimen in the Lambertian herbarium.

Spec. Char., &c. Leaves ovate, doubly toothed, almost lobed; beneath reticulate and tomentose. Flowers disposed in terminal subcapitate corymbs placed on very long peduncles. Calyx tomentose. (*Dec. Prod.*) A deciduous shrub. N. America, on its eastern coast by the River Columbia. Height 4 ft. to 6 ft. Introduced in 1827. Flowers white; June and July.

93. *S.* capitàta.

§ ii. *Chamæ'dryon* Ser.

Derivation. From *Chamæ'drys*, the name of the germander; from a similarity in the form of the leaves.

Sect. Char. Ovaries distinct. Torus with its base connate with the tube of the calyx, but with its tip separate. Carpels not inflated. Flowers each upon a distinct pedicel, and disposed in umbels or corymbs. Leaves entire, or toothed, without stipules. (*Dec. Prod.*, ii. p. 542.)

■ 3. *S.* CHAMÆDRIFO'LIA *L.* The Germander-leaved Spiræa.

Identification. Lin. Sp., 701.; Camb. Monog.; Dec Prod., 2. p. 542.; Don's Mill., 2. p. 518.
Synonyme. S. cantoniénsis *Lour.*
Engravings. Pall. Fl. Ross., t. 15.; and our *fig.* 495.

Spec. Char., &c. Leaves ovate, cut at the tip in a serrated manner, glabrous. Flowers upon long slender pedicels, in hemispherical corymbs. Sepals veiny, reflexed. (*Dec. Prod.*) An erect shrub. Siberia, Kamtschatka, Dahuria, the N. W. coast of N. America, China, and Japan. Height 2 ft. to 8 ft. Introduced in 1789. Flowers white; June and July. Capsule reddish; ripe Sept. Naked young wood light brown.

Varieties. Seringe enumerates the first four of the following forms of this species; to which, we think, might be added *S. u*lmifòlia, *S.* flexuòsa, *S. c*ratægifòlia, *S. be*tulæfòlia, and, perhaps, some others.

■ S. c. 1 *vulgàris* Camb. Monog. — Leaves with the disks broad and glabrous; the petioles ciliated.

■ S. c. 2 *mèdia* Ph. Fl. Amer. Sept. i. p. 342., Camb. Monog., and our *fig.* 494.—Leaves smaller, slightly villose upon both surfaces. Flowers smaller. Wild in Canada, and upon the rocks of Dahuria.

494. *S. c.* mèdia.

■ S. c. 3 *oblongifòlia* Camb. Monog. *S.* oblongifòlia *Waldst. et Kit. Pl. Hung.* iii. p. 261. t. 235. — Leaves narrower, and less serrated.

■ S. c. 4 *subracemòsa* Ser. — Flowers distantly disposed along a lengthened rachis.

■ S. c. 5 *incìsa* Hort. (*S. c*hamædrifòlia latifòlia *Hort.*) has been raised from seeds received from Germany through Mr. Hunneman; and it appears to be only a variety of this species.

495. S. chamædrifòlia.

In Kamtschatka the leaves are used as a substitute for tea; and the shoots, when straight, are bored for tobacco-pipes. In its wild state, it varies exceedingly in the magnitude of the entire plant, in the largeness or smallness of its leaves, and in their being more or less cut or serrated, and more or less smooth or pubescent. A very ornamental hardy shrub, producing its corymbs of white flowers, which are tolerably large, in June and July. It is said to make beautiful garden hedges. Though the seeds ripen in England, plants can seldom be raised from them; and, as this species does not produce suckers freely, it is generally raised by layers or cuttings.

■ 4. *S.* (*c.*) ULMIFO'LIA *Scop.* The Elm-leaved Spiræa.

Identification. Scop. Fl. Carn., ed. 2. vol. 1. p. 349.; Dec. Prod., 2. p. 542.; Don's Mill., 2. p. 518.
Synonyme. S. chamædrifòlia *Jacq. Hort. Vindob.* t. 140.
Engravings. Bot. Reg., t. 1222.; Bot. Cab., 1042.; and our *fig.* 496.

Spec. Char., &c. Leaves ovate-lanceolate, acute, flat, sharply serrated, ciliated. Flowers terminal, in rather hemispherical corymbs. Sepals reflexed. (*Dec. Prod.*) An erect shrub. Carinthia and Siberia. Height 3 ft. to 5 ft. Introduced in 1790. Flowers white; June and July. One of the handsomest species of this section.

496. S. (c.) ulmifòlia.

497. S. (c.) u. phyllántha.

Variety.

■ S. (c.) u. 2 *phyllántha* Ser. (Our *fig.* 497.) — In this variety a whorl of distinct leaves, that are petioled, lanceolate, and sharply serrated, occupies the place of the sepals, and is described as being these transformed. Petals and stamens are either not present, or deformed. (*Dec. Prod.*)

■ 5. *S.* (*c.*) FLEXUO'SA *Fisch.* The flexible-*branched* Spiræa.

Identification. Fisch. in Litt.; Camb. Monog.; Dec. Prod., 2. p. 542.; Don's Mill., 2. p. 518.
Synonymes. S. alpina *Hort. Par.*, according to Camb. and Fisch. in Litt. 3.; *S.* sibírica *Hort.*
Engravings. Camb. Monog. Spir. in Ann. Sci. Nat., 1. t. 36.; and our *fig.* 498.

Spec. Char., &c. Leaves lanceolate, glabrous; from the tip to the middle

498. S. (c.) flexuòsa.

dentately serrate. Flowers in corymbs. (*Dec. Prod.*) A slender-branched shrub. Native country unknown. Height 4 ft. to 6 ft. Cultivated in 1820. Flowers white; June and July. Capsule reddish; ripe in September.

Varieties. *S.* flexuòsa latifòlia *Hort.*; *S.* däùrica *Hort.*; *S.* *u*lmifòlia, *S.* *c*arpinifòlia, *S.* *b*etulæfòlia, in Messrs. Loddiges's collection, are identical with, or very slight variations of, this species.

■ 6. *S.* (*c.*) CRATÆGIFO'LIA *Lk.* The Cratægus-leaved Spiræa.

Identification. Lk. Enum., 2. p. 40.; Dec. Prod., 2. p. 546.; Don's Mill., 2 p. 518.
Engraving. Our *fig.* 499.

Spec. Char., &c. Leaves obovate, obtuse; in the terminal part doubly serrate; glabrous, reticulate on the under surface. Flowers white, disposed in terminal compound corymbs, whose composite parts are rather capitate. (*Dec. Prod.*) An erect shrub. Native country unknown. Cultivated in 1823. Flowers white; June and July.

499. S. (c.) cratægifòlia.

500. S. (c.) betulæfòlia.

■ 7. *S.* (*c*) *B*ETULÆFO'LIA *Pall.* The Birch-leaved Spiræa.

Identification. Pall. Fl. Ross., 1. p. 75.; Dec. Prod., 2. p. 544.; Don's Mill., 2. p. 519.
Synonymes. ? *S.* corymbòsa *Raf.* in *Desv. Journ.*; ? *S.* cratægifòlia *Lk. Enum.* 2. p. 42.
Engravings. Pall. Fl. Ross., 1. t. 16.; and our *fig.* 500.

Spec. Char., &c. Leaves broadly ovate, serrated, glabrous; the petiole very short. Flowers in fastigiate panicles. Carpels 5, upright, glabrous. (*Dec. Prod.*) An erect shrub. Siberia; and North America, on the Blue Mountains, and in various other places on the western coast. Height 3 ft. to 6 ft. Introd. in 1812. Flowers white; June and July. Capsule red; ripe Sept.

8. *S.* CA'NA *Waldst. et Kit.* The hoary-*leaved* Spiræa.

Identification. Waldst. et Kit. Pl. Rar. Hung., 3. p. 252. t. 227.; Camb. Monog.; Dec. Prod., 2. p. 542.; Don's Mill., 2. p. 518.
Engravings. Waldst. et Kit. Pl. Rar. Hung., 3. t. 227.; and our *fig.* 501.

Spec. Char. &c. Leaf ovate, of about the size of that of *S*àlix rèpens or *S.* argéntea, acute, perfectly entire, or slightly toothed, hoarily villose. Corymbs somewhat racemose; the lateral ones peduncled, of few flowers, and lax. Sepals spreading. Styles thick. Carpels divergent, rather villose. (*Dec. Prod.*) A low shrub. Croatia,

501. Spiræ'a càna.

on high rocks. Height 1 ft. to 2 ft. Introduced in 1825. Flowers white; June and July. Capsule reddish; ripe in September.

A very distinct little species approaching *S. vacciniifòlia*, *D. Don*. H. S.

■ 9. *S.* TRILOBA'TA *L.* The 3-lobed-*leaved* Spiræa.

Identification. Lin. Mant., 244.; Camb. Monog.; Dec. Prod., 2. p. 543.
Synonyme. *S. triloba* *Don's Mill.* 2. p. 518.
Engravings. Pall. Ross., 1. t. 17.; Wats. Dend. Brit., t. 68., as *S. triloba*; and our *fig.* 502.

Spec. Char., &c. Leaves roundish, lobed, crenated, glabrous, reticulately veined. Flowers in umbel-like corymbs. Sepals ascending. Carpels glabrous. (*Dec. Prod.*) A low erect shrub. Alps of Altai. Height 1 ft. to 2 ft. Introduced in 1801. Flowers white; May. Capsule reddish; ripe in September.

502. *S. trilobàta.*

This species is very handsome, with branches spreading horizontally, and bearing, in the flowering season, numerous compact corymbs of pure white flowers; which, combined with the neat appearance of the plant, and its glaucous leaves, rounded in their outline, and yet lobed, render the species a very interesting and ornamental one.

■ 10. *S.* ALPI'NA *Pall.* The *Siberian* alpine Spiræa.

Identification. Pall. Fl. Ross., 1. p. 82. t. 20., according to Camb. Monog.; Dec. Prod., 2. p. 543.; Don's Mill., 2. p. 518.
Engravings. Pall. Fl. Ross., 1. t. 20.; and our *fig.* 503.

Spec. Char., &c. Leaves lanceolate-oblong, sessile, serrulated, glabrous; the midrib pinnately branched. Flowers in terminal, stalked, and, in many instances, leafless, corymbs. Sepals ascending. (*Dec. Prod.*) An erect shrub. Siberia, in wooded alps. Height 4 ft. to 6 ft. Introduced in 1806. Flowers white; June and July. Capsule reddish; ripe in September.

The corymbs of flowers being large in proportion to the leaves renders this species very ornamental when in bloom; and its beauty is farther heightened by the plant being of erect growth.

503. *S. alpina.*

■ 11. *S.* HYPERICIFO'LIA *Dec.* The Hypericum-leaved Spiræa.

Identification. Dec. Fl. Fr., 5. p. 645.; Prod., 2. p. 543.; Don's Mill., 2. p. 519.
Synonymes. *Hypéricum frùtex Hort.*; Italian May.
Engravings. See *Varieties*.

Spec. Char., &c. Leaves obovate-oblong, 3—4-nerved, entire or toothed, glabrous, slightly downy; primary veins pinnately branched. Flowers in either peduncled corymbs, or sessile umbels. Pedicels glabrous, or slightly downy. Sepals ascending. A species that presents diversified appearances. (*Dec. Prod.*) An erect shrub. Supposed by some to be a native of Canada; but Smith and Hooker think that, like most of the species of the section to which it belongs, it is only to be found wild in the Old World. Height 4 ft. to 6 ft. Introduced in 1640. Flowers white; June and July. Capsule reddish; ripe in September.

Varieties. Seringe has characterised six forms of this species, which he describes as follows: —

■ S. h. 1 *uralénsis* Ser. *S. crenàta Lin., Fisch.* in Litt., and *Don's Mill.* ii. p. 519.; *S. hypericifòlia Camb. Monog.* (Our *fig.* 504.) — Branches rigid,

504. *S. h. uralénsis.*

thickish. Leaves ovate-rounded; the whole margin crenated. A native of the Ural Mountains.

S. h. 2 *Plukenetiàna Ser.* *S. hypericifòlia Lin. Sp. Pl.* 701., and *Don's Mill.* ii. p. 519.; *S. h.* var. β *Dec. Fl. Fr.* v. p. 645. (Pluk. Phyt., t. 218. f. 5.) — Leaves perfectly entire, glabrous. Flowers in sessile corymbs. A native of Canada.

S. h. 3 *acùta* Ser. *S.* acutifòlia *Willd. Enum.* 540., *Camb. Monog.*, and *Don's Mill.* ii. p. 519.; *S.* sibírica *Hort. Par.*, according to *Camb. Monog.*; *S.* ambígua *Pall.* (Our *fig.* 505.) — Leaves spathulate, elongate, acute, perfectly entire, or rarely 3—5-toothed, rather glabrous. Flowers in sessile corymbs.

505. *S. h.* acùta.

S. h. 4 *crenàta* Ser. S. obovàta *Waldst. et Kit.*? in *Willd. Enum.* 541., *Camb. Monog.*, *Barr. Icon. Rar.* n. 1376. t. 564.; *S. hypericifòlia* γ *Dec. Fl. Fr.* v. p. 645.; *S.* crenàta *Lin. Sp.* 701., *Camb. Monog.*, *Don's Mill.* ii. p. 519., *Lodd. Cat.* (Our *fig.* 506.) — Leaves obovate.

506. *S. h.* crenàta.

507. *S. h.* savránica.

508. *S. h.* Besseriàna.

S. h. 5 *savránica* Ser. *S.* savránica *Besser* in Litt., *Don's Mill.* ii. p. 519.; *S.* crenàta *Pall. Fl. Ross.* i. p. 35. t. 19.; *S. hypericifòlia* var. β longifòlia *Led. Fl. Ross. Alt. Ill.* t. 429. (Our *fig.* 507.) — All parts pubescent. Leaves entire, or, at the tip, toothed. Flowers minute, disposed in dense terminal corymbs. It is wild about Barnaoul, and in Podolia.

S. h. 6 *Besseriàna Ser.* *S.* crenàta *Besser* in Litt.; *S.* savránica β Besseriàna Don's Mill. ii. p. 519. (Lodd. Bot. Cab., t. 1252; and our *fig.* 508.)—All parts rather glabrous. Leaves mostly entire. Flowers disposed in rather lax terminal corymbs. A native of Podolia and Caucasus.

Other Varieties or Synonymes. The following kinds, in Messrs. Loddiges's collection, and in the Hort. Soc. Garden, appear to be either identical with, or varieties of, *S. hypericifòlia*: — *S.* infléxa *Wendland* (Hort. Soc. Gard.), *S.* obovàta *Wendland* (Hort. Soc. Gard.), *S.* argéntea, *S.* cuneàta, *S.* nàna, *S.* alpìna, *S.* acutifòlia, *S.* decúmbens.

This species has small hard stems, with numerous side branches, clothed with a dark green bark, and with numerous wedge-shaped leaves, like those of St. John's wort, with glands in their substance, which give them the appearance of being punctured on the surface; whence the name. The flowers are produced in great abundance; and, when the shrub is allowed space to expand on every side, it forms a very beautiful bush in the flowering season. It makes handsome garden hedges, and will bear the shears, which were formerly applied to it, to shape it into artificial forms, when topiary work was fashionable in garden scenery. It is readily propagated by layers, or by detaching its suckers.

509. S. (h.) thalictröides.

⸗ 12. *S.* (*H.*) THALICTRÖI'DES *Pall.* The Meadow-Rue-*leaved* Spiræa.

Identification. Pall. Fl. Ross., 1. p. 34.; Dec. Prod., 2. p. 543.; Don's Mill., 2. p. 518.
Synonymes. *S. aquilegifòlia Pall. Itin.* 3. App. 734. No. 94.; *S. hypericifòlia* var. *flàva*; and *S. alpìna latifòlia.*
Engravings. Pall. Fl. Ross., 1. t. 18.; and our *fig.* 509.

Spec. Char., &c. Leaves obovate, obtuse, indistinctly 3-lobed, inconspicuously 3-nerved, glaucous beneath. Flowers in lateral sessile umbels. (*Dec. Prod.*) A low erect shrub. Alps of Dahuria. Height 1 ft. to 3 ft. Introduced in 1816. Flowers white; May and June. Capsule reddish; ripe in September.

⸗ 13. *S.* CUNEIFO'LAI *Wall.* The wedge-leaved Spiræa.

Identification. Wall. Cat., 699.; Bot. Reg. M. Chron. 839, No. 87.
Synonymes. *S. canéscens*; Don Prod.; Dec. Prod., 2. p. 544.; Don's Mill., 2. p. 519.; *S. argéntea Hort.*
Engraving. Our *fig.* 510.

Spec. Char., &c. Leaves oval or obovate, obtuse, stalked, quite entire, villous; corymbs crowded, and as well as the branches tomentose. (*Don's Mill.*) An erect-branched canescent shrub, with the habit of *S. hypericifòlia.* Nepal, at Sirinagur. Height ?. Introduced in 1837. Flowers white, downy, in close corymbose panicles.

The leaves are small, thick, downy, wedge-shaped, and either crenated near the point, or undivided; they are bright green on the upper side, and glaucous beneath, with nothing of a canescent appearance, which is only visible when they are dried.

510. S. cuneifòlia.

14. *S.* PIKOWIE'NSIS *Besser.* The Pikow Spiræa.

Identification. Besser Enum. Pl. Pod., p. 46. No. 1428.; Dec. Prod., 2. p. 543.; Don's Mill., 2. p. 519.
Engraving. Our *fig.* 000. in p. 000.

Spec. Char., &c. Leaves with three primary veins, and serrate at the tip, cuneate-lanceolate, obtuse, rarely pointed. Flowers in peduncled corymbs. (*Dec. Prod.*) A deciduous shrub. Podolia at Pikow. Height 3 ft. to 4 ft. Introd. in 1807. Flowers white; June and July.

⸗ 15. *S.* CEANOTHIFO'LIA *Horn.* The Ceanothus-leaved Spiræa.

Identification. Horn. Hort. Hafn., p. 2. 466.; Dec. Prod., 2. p. 544.
Engraving. Our *fig.* 511.

Spec. Char., &c. Leaves ovate, rounded at the base, serrated from the middle to the tip with sharp unequal teeth. Flowers in indistinctly peduncled terminal corymbs. (*Dec. Prod.*) An erect shrub. Native country unknown. Height 3 ft. to 4 ft. Introduced in 1823. Flowers white; June and July. Capsule reddish; ripe in September.

511. S. ceanothifòlia.

■ 16. *S.* CORYMBO'SA *Raf.* The corymbose-*flowering* Spiræa.

Identification. Rafin. Précis des Découvertes Somiologiques, ou Zoologiques et Botaniques (Palerme, 1814), p. 36.; and in Desv. Journ. Bot., 1814, p.168.; Dec. Prod. 2. p. 544.
Engravings. Lod. Bot. Cab., t. 671.; and our *fig.* 512.

512. *S. corymbosa.*

Spec. Char., &c. Leaves oval-oblong, unequally serrated, glabrous, green above, hoary below. Flowers trigynous, disposed in terminal corymbs. (*Dec. Prod.*) A handsome shrub. Virginia. Height 4 ft. to 5 ft. Introd. in 1819. Flowers white, produced in great abundance; June and July. Capsule reddish; ripe in Sept.

Variety.

■ S. *c.* 2 *sorória*, *S.* sorória *Penny* in *Hort. Brit.*, is a smaller plant, seldom growing higher than 2 ft., and flowers rather later than the species.

A very desirable species, on account of its large corymbs of white flowers, and its distinctness in external character.

■ 17. *S. VACCINIIFO'LIA D. Don.* The Vaccinium-leaved Spiræa.

Identification. Don Prod. Fl. Nep., 1. p. 227.; Dec. Prod., 2. p. 546.; Don's Mill., 2. p. 518.
Synonyme. *S.* adiantifólia *Hort.*
Engravings. Bot. Cab., t. 1430.; and our *fig.* 513.

513. *S. vacciniifolia.*

Spec. Char., &c. Upright. Branchlets hairy. Leaves elliptical, acute, serrated at the tip, glabrous, glaucous on the under surface. Flowers disposed in terminal tomentose cymes, a few in a cyme. (*Dec. Prod.*) An upright shrub. Nepal. Height 2 ft. to 3 ft. Introduced in 1820. Flowers white; July and August. Capsule reddish; ripe in October.

■ 18. *S.* LAXIFLO'RA *Lindl.* The loose-flowered Spiræa.

Identification. Lindl. Bot. Reg. M. Chron. 1830, No. 89.
Engraving. Our *fig.* 000. in p. 000.

Spec. Char., &c. Branches weak, round, downy. Leaves smooth, ovate-crenate, long-stalked, glaucous beneath. Panicles loose, villose. Petals reflexed. (*Lindl.*) A shrub resembling *S. vacciniifólia* in the form of the leaves, and the colour of their under surface. Nepal. Height 1 ft. to 2 ft. Introduced in 1838. Flowers white, in large, loose, shaggy panicles; July and August.

■ 19. *S.* BE'LLA *Sims.* The beautiful Spiræa.

Identification. Sims; Dec. Prod., 2. p. 542.; Don's Mill., 2. p. 518.
Engravings. Bot. Mag., t. 2426.; and our *fig.* 514.

514. *S. bélla.*

Spec. Char., &c. Stems erect, branched, glabrous, and reddish. Leaves ovate, acute, sharply serrated, whitishly tomentose on the under surface. Flowers pretty, rose-coloured, in corymbs laxly disposed. Lobes of the calyx deflexed. (*Dec. Prod.*) An erect, loose-growing shrub. Nepal. Height 3 ft. to 4 ft. Introd. 1820. Flowers beautiful rose-coloured; May and June. Capsule reddish; ripe in September.

This species is as hardy, and as easily propagated, as that very common shrub, *S. salicifólia*, from which species it differs in its loose branchy manner of growth, and in the flowers being in corymbs. One of the most beautiful species of the genus.

§ iii. *Spiraria* Ser.

Sect. Char. Ovaries distinct. Torus with its base connate with the tube of the calyx; its tip separate. Carpels not inflated. Inflorescence a panicle. Leaves serrate, without stipules. (*Dec. Prod.*, ii. p. 544.)

20. *S.* SALICIFO'LIA *L.* The Willow-leaved Spiræa.

Identification. Lin. Sp., 700.; Camb. Monog.; Dec. Prod., 2. p. 544.
Synonymes. Spiræ'a frùtex *Hort.*; Bridewort, Queen's Needlework.
Engraving. Gmel. Fl. Sibir., 3. t. 49.

Spec. Char., &c. Stem and peduncles glabrous. Leaves lanceolate, serrated, and, more or less, doubly so; glabrous. Lobes of the calyx triangular, spreading. Carpels glabrous. (*Dec. Prod.*) An erect shrub. Siberia, Tartary, and Bohemia; also of Canada, and, perhaps, of Britain. Height 4 ft. to 6 ft. Cultivated in 1665. Flowers red; July and August. Capsule red; ripe in September.

515. S. salicifòlia cárnea.

Varieties. Seringe has characterised four forms of this species as follows:—

S. s. 1 *cárnea* Ait. Hort. Kew. ed. iii. p. 254., Camb. Mon. (Eng. Bot., t. 1468.; and our *fig.* 515.) — Leaves lanceolate. Panicles consisting of racemes more or less spicated. Petals of a flesh colour. Bark of the branches yellowish. This is the form found wild in Britain: whether it be indigenous or not, botanists are not agreed. Professor Henslow considers it "possibly introduced by the agency of man."

S. s. 2 *alpéstris* Pall. Fl. Ross. i. p. 36. t. 22., Camb. Monog. *S.* alpéstris *Don's Mill.* ii. p. 519. (Our *fig.* 516.)—A small shrub. Leaves shorter than those of *S. s.* cárnea. Branches very short.

516. S. s. alpéstris.

S. s. 3 *paniculàta* Willd. Sp. ii. p. 1055., Ait. Hort. Kew. ed. 2. iii. p. 254. *S.* álba *Ehrh. Beitr.* vii. p. 137. (Our *fig.* 517.) — Leaves ovate-oblong. Petals white. Bark of the branches red. It is likely that this is also the *S.* álba of *Miller*, of *Wats. Dend. Brit.* t. 133., and of *Muhlenberg's Catalogue of North-American Plants.*

517. S. s. paniculàta.

518. S. s. latifòlia.

519. S. s. grandiflòra.

S. s. 4 *latifòlia* Willd. Sp. ii. p. 1055. *S.* obovàta *Raf.* in Litt., not of Waldst. et Kit., according to Willd. Enum. 541.; *S.* carpinifòlia

Willd. Enum. p. 540., *Don's Mill.* ii. p. 520. (Wats. Dend. Brit., t. 66.; and our *fig.* 518.)—Leaves ovate-oblong. Petals white. Bark of branches rather reddish.

■ S. s. 5 *grandiflòra* ; S. grandiflòra *Lodd. Bot. Cab.* t. 1988., and our *fig.* 519. ; has its pink flowers nearly twice as large as those of the species ; and is a very ornamental free-growing shrub. Raised from seeds sent from Kamtschatka, in 1826.

■ S. s 6 *taúrica.* S. taúrica *Hort.*—An upright shrub, 8 ft. to 10 ft. high, tolerably distinct, and coming into flower before any other variety. Hort. Soc. Garden.

Other Varieties or Synonymes. The following kinds, in Messrs. Loddiges's collection and in that of the Hort. Soc., are either varieties of, or identical with, *S. salicifòlia* : — S. canadénsis, S. urticæfòlia, S. laciniàta, S. chamædrifòlia, S. lanceolàta, S. carpinifòlia, S. refléxa, S. incarnàta.

This species sends up numerous straight rod-like stems, and these and the lateral branches terminate in large, conical, spiked panicles, of pale red, or flesh-coloured, flowers. In deep moist soils, a sucker will attain the height of 4 ft. in one season, and flower. These suckers are produced in such abundance, that, in order to keep the shrub in a vigorous state, they ought to be cut down when they have flowered two years, in the same manner as is practised with raspberries; and the entire plant ought also to be taken up every three or four years, and separated; otherwise the old shoots are apt to die, and render the bush unsightly. It is one of the hardiest of garden shrubs, and is, also, very beautiful, from its long spicate panicles full of light feathery-looking flowers.

■ 21. S. Menzie'sii Hook. Menzies's Spiræa.

Identification. Hook. Fl. Bor. Amer., p. 173. ; Don's Mill., 2. p. 520.
Engraving. Our fig. 520. from a specimen in Dr. Hooker's herbarium.

520. S. Menziesii.

Spec. Char., &c. Branches pubescent at the apex, as well as the peduncles and calyces. Sepals reflexed. Leaves elliptic, coarsely and unequally serrated towards the apex, glabrous, the same colour on both surfaces; panicle crowded with flowers, oblong, obtuse. Flowers small. Stamens twice the length of the corolla; ovaries 5, glabrous. (*Don's Mill.*) An erect shrub. North America, on the west coast. Height 2 ft. to 3 ft. Introduced in 1838. Flowers rose-coloured; June and July.

■ 22. S. tomento'sa *L.* The downy Spiræa.

Identification. Lin. Sp., 701. ; Camb. Monog. ; Dec. Prod., 2. p. 544. ; Don's Mill., 2. p. 230.
Synonyme. S. Douglàsii Hook. Fl. Bor. Amer. p. 172.
Engravings. Pluk. Phyt., t. 321. f. 5. ; Schmidt Arb., 1. t. 51. ; and our *fig.* 521.

521. S. tomentòsa.

Spec. Char., &c. Nearly all the parts of this plant are more or less clothed with tomentum, the under surface of the leaves most so. The tomentum upon the stem and peduncles, and perhaps elsewhere, is of a reddish colour. The leaves are ovate and serrated, the latter partly doubly so. Lobes of the calyx triangular and deflexed. Carpels divaricate. (*Dec. Prod.*) An erect shrub. Canada, on mountains. Height 2 ft. to 3 ft. Introduced in 1814. Flowers white; June and July. Capsule reddish ; ripe in September.

This species, or subspecies, in its mode of growth, resembles S. salicifòlia ; but differs from it in having rather smaller and more deeply serrated leaves, which are very tomentose beneath. The flowers are much smaller, and of a deeper red.

23. *S.* LÆVIGATA *L.* The smooth-*leaved* Spiræa.

Identification. Lin. Mant., 244.; Camb. Monog.; Dec. Prod., 2. p. 544.; Don's Mill., 2. p. 519.
Synonymes. *S.* altaicènsis *Laxm. Nov. Act. Petrop.* 15. p. 555. t. 29. f. 2.; *S.* altàica *Pall. Fl. Ross.* 1. p. 272.
Engravings. Nov. Act. Petrop., t. 29. f. 2.; Pall. Fl. Ross., 1. t. 23.; and our *fig.* 522.

522. *S.* lævigàta.

Spec. Char., &c. Leaves obovate-oblong, perfectly smooth, entire, sessile, tipped with a small mucro. Branchlets of the panicle cylindrical. Bracteas linear, rather shorter than the calyx. Lobes of the calyx triangular, ascending. (*Dec. Prod.*) A spreading shrub. Siberia, in valleys at the foot of the more lofty of the Altaian Mountains. Height 2 ft. to 4 ft. Introd. in 1774. Flowers white; May and June. Capsule reddish; ripe in September.

A very interesting and handsome species, with a habit exceedingly dissimilar to that of spiræas in general.

24. *S.* ARIÆFOLIA *Smith.* The White-Beam-tree-leaved Spiræa.

Identification. Smith, in Rees's Cyclop., vol. 33.; Bot. Reg., t. 1365.; Dec. Prod., 2. p. 544.; Don's Mill., 2. p. 520.
Engravings. Bot. Reg., t. 1365.; and our *figs.* 523, 524.

523. *S.* ariæfòlia.

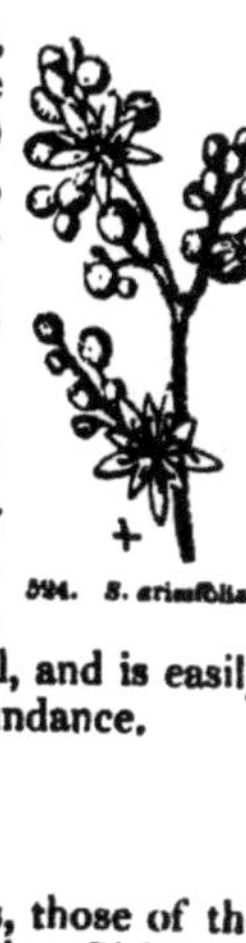

524. *S.* ariæfòlia.

Spec. Char., &c. Leaves elliptical, oblong, more or less lobed, toothed, pale, villose beneath. Panicle villose. (*Dec. Prod.*) An erect bushy shrub. North America, principally on the north-west coast. Height 6 ft. to 8 ft. Introduced in 1827. Flowers white; June and July. Capsule reddish; ripe in September.

A free-growing dense bush, prolific both in leaves and flowers; and, as the latter appear at a season when the flowering of shrubs is comparatively rare, it is justly considered as a most valuable addition to British gardens. It is perfectly hardy, will grow in any free soil, and is easily propagated either by division or by seeds, which it ripens in abundance.

§ iv. *Sorbària* Ser.

Sect. Char. Leaves pinnate, resembling, as the name implies, those of the mountain ash, or other species of *Pỳrus* belonging to the section *Sórbus*.

25. *S.* SORBIFOLIA *L.* The Sorbus-leaved Spiræa.

Identification. Lin. Sp., 702.; Camb. Monog.; Dec. Prod., 2. p. 545.
Synonyme. *S.* pinnàta *Mœnch Meth.* 633.
Engravings. Gmel. Fl. Sib., 3. p. 190. t. 40.; Schmidt Baum., 1. t. 56.; and our *fig.* 525.

Spec. Char., &c. Leaves stipuled, pinnate; the leaflets sessile, opposite, lanceolate, doubly and sharply serrated. Inflorescence a thyrse-like panicle. Torus wholly connate with the tube of the calyx. Ovaries connate, 5. (*Dec. Prod.*) A thick stiff-branched shrub. Siberia, in moist places. Height 3 ft. to 6 ft. Introd. in 1759. Flowers white; July and August. Capsule reddish; ripe in September.

Variety.

S. s. 2 *alpìna* Pall. Fl. Ross. i. p. 34. and 89. t. 25. *S.* grandiflòra *Sweet's Hort. Brit.* p. 194.; *S.* Pallàsii Don's Mill. ii. p. 520. — Suffruticose. Flowers twice the size of those of the species, and disposed in corymbs. Leaves smaller, and serrately incised. A native of Eastern Siberia and of Kamtschatka. (*Dec. Prod.*) This is very different from *S. s.* grandiflòra, the *S.* grandiflòra of *Lodd.*, described above, among the varieties of *S. s*alicifòlia.

525. S. sorbifòlia.

*S. s*orbifòlia is a branchy shrub, growing to the height of 6 or 8 feet, with a round, brown-coloured, warty stem; the wood of which is brittle, and hollow within, with a soft ferruginous pith. The leaves are thin in texture, and bright green on both sides. The flowers are in terminating panicles and small: they are odorous, but not agreeably so. In dry rocky situations, it does not rise above 1 ft. in height, and is subherbaceous. It deserves a place in every collection, from its marked character, and from the beauty both of its foliage and its flowers. It throws up abundance of suckers, by which it is easily propagated.

♠ 26. S. Lindleya'na *Wall.* Lindley's Spiræa.

Identification. Wall. Cat., and Gard. Mag. 1840, p. 3.
Engraving. Our *fig.* 526. from the Linnean herbarium.

Spec. Char., &c. Leaflets ovate lanceolate, acuminate, finely and sharply serrated. A large shrub. Nepal. Height 6 ft. to 8 ft. Flowers white; July and August.

The leaves are larger than those of any other species of the genus; and, altogether, it forms a remarkably handsome plant, well deserving a place in collections.

526. S. Lindleyàna.

A Selection of Species.—The following kinds, in the London gardens, appear distinct:—

1. S. *tomentòsa*, comprehending *S.* Tobólski.
2. S. *chamædrifòlia*, comprehending *S. b*etulæfòlia, *S.* dáùrica, *S.* sibírica, *S.* laciniàta.
3. S. *salicifòlia*, comprehending *S.* canadénsis, *S.* grandiflòra, *S.* paniculàta, *S. u*rticæfòlia, *S.* lanceolàta, *S. c*arpinifòlia, *S.* refléxa, *S.* incarnàta, *S.* taúrica.
4. S. *flexuòsa*, comprehending *S. u*lmifòlia, *S. c*arpinifòlia, *S. b*etulifòlia, and, perhaps, some others.
5. S. *bélla*.
6. S. *corymbòsa*.
7. S. *cuneifòlia*.
8. S. *vacciniifòlia*.
9. S. *laxiflòra*.
10. S. h*ypericifòlia*, comprehending *S.* crenàta, *S.* infléxa *Wendland* (H. S. Gard.), *S.* obovàta *Wendland* (H. S. Gard.), *S.* argéntea, *S.* cuneàta, *S.* nàna, *S.* alpìna, *S.* acutifòlia, *S.* decúmbens.
11. S. *oblongifòlia* Wendland, apparently an upright fastigiate variety of *S. h*ypericifòlia.
12. S. *càna*.
13. S. *trilobàta*.
14. S. *sorbifòlia*, comprehending *S.* picowiénsis of Loddiges, which is a totally different plant from the *S.* pikowiénsis of Besser, our No. 14. in p. 305.
15. S. *Lindleyàna*.

Sect. III. Potenti'lleæ.

Genus IX.

RU'BUS *L.* The Bramble. *Lin. Syst.* Icosándria Polygýnia.

Identification. Lin. Gen., No. 364.; Dec. Prod., 2. p. 556.; Don's Mill., 2. p. 59.
Synonymes. Ronce, Framboisier, *Fr*; Himbeere, Brombeerstrauch, *Ger.*
Derivation. From *rub*, red in Celtic; in reference to the colour of the fruit in some of the species.

Gen. Char. Calyx flattish at the bottom, 5-cleft. *Petals* 5. *Stamens* numerous, inserted in the calyx along with the petals. *Carpels* or *Achenia* numerous, fleshy, disposed in a head upon an elevated torus. *Styles* lateral, near the apex of the carpel. (*Don's Mill.*)

Leaves compound, digitate, pinnate or lobed, stipulate, deciduous or sub-evergreen; with the leaflets usually stalked. *Flowers* white or pink, in terminal racemes. *Fruit* edible.

Shrubs, deciduous, subligneous, with prickly stems; for the most part prostrate, but a few of them growing upright. Some of them, such as *R.* fruticòsus, may be considered as sub-evergeen, as they retain the greater part of their leaves in a green state through the winter. All the kinds popularly called brambles may be considered as gigantic strawberry plants; and all their shoots are used by thatchers, and makers of beehives, straw mats, &c. No less than 48 supposed species of the genus are described and figured in the *Rubi Germanici* of Weihe and Nees von Esenbeck. The number of species in *English Botany* is, in Dr. Lindley's *Synopsis of the British Flora*, 21; which, he says, may be reduced to 5, or possibly to 2, exclusive of the herbaceous species. In *Don's Miller*, 147 are given as the total number described by botanists. We shall only notice such as are tolerably distinct, and which are in cultivation in British gardens. The propagation of the shrubby, or raspberry-like, species of Rùbus is effected by suckers or seeds; that of the bramble division of the genus by pegging down the points of the shoots to the soil, when they will root, and throw out other shoots, which may again be pegged down; so that plants are procured from brambles much in the same way as from strawberries.

§ i. *Leaves pinnate, of* 3—7 *Leaflets.*

⁂ 1. *R.* suberé'ctus *Anders.* The sub-erect Bramble.

Identification. Anders. in Linn. Soc.; Dec. Prod., 2. p. 556.; Don's Mill., 2. p. 534.; Smith Eng. Flora, 2. p. 406.
Synonymes. Lindley, in his *Syn. of the Brit. Flora*, has given the following: — *R.* nessénsis *Hall*; *R.* plicàtus *W. & N.*, not of *Suppl. to Eng. Bot.* t. 2714., which is a smaller form of *R.* affìnis *W. & N.*; *R.* corylifòlius *Wahlenb.*
Engravings. Eng. Bot., t. 2572.; and our *fig.* 527.

Spec. Char., &c. Stem erect. Leaf of never more than 5 leaflets, digitate, occasionally pinnate, thin, shining, and plaited. Flowers in simple corymbose racemes. Prickles weak. (*Lindl.*) A sub-erect shrub. Britain, in moist woods and by the sides of rivulets, chiefly in the northern counties. Stems 3 ft. to 4 ft. Flowers white; June to September. Fruit pale purple; ripe in August.

527. Rùbus suberéctus.

x 4

The stems are biennial, and flower the second year, like those of the common raspberry, afterwards dying off. The fruit consists of a small number of dark red, or blood-coloured, aggregate grains, agreeably acid, with some flavour of the raspberry; whence it has been recommended by some as perhaps not unworthy of cultivation.

2. *R.* AFFI`NIS *Weihe & Nees.* The related Bramble.

Identification. Weihe and Nees's Rubi Germanici, p. 22. t. 3. and 36.; Dec. Prod., 2. p. 560.; Lindl. Synops. Br. Fl., 2d ed. p. 92, 93.
Synonymes. Lindley mentions the following: — *R.* collinus *Dec.*; *R.* nitidus *Smith* 'n *Eng. Flor.*, *Lindley* in *Syn. Br. Fl.* ed. 1.; *R.* plicàtus *Borrer* in *Eng. Bot. Suppl.* t. 2714.
Engravings. Weihe and Nees's R. G., t. 3. and 36.; Eng. Bot. Suppl., t. 2714.; and our *fig.* 528.

Spec. Char., &c. Stem arched, angled, prickly with strong recurved prickles, glabrous. Leaflets 3—5 in a leaf, ovate with a heart-shaped base, cuspidate, sharply serrated, flat at the base, a little waved towards the tip, having downy tomentum beneath. Flowers in a compound panicle, the component ones cymose. Sepals ovate-acuminate, externally naked, reflexed. Carpels large, blue-black. (*Dec. Prod.*) A low bramble. Germany, also of barren hills of Montpelier, and of Britain, in boggy places. Flowers white; July and August.

528. Rùbus affìnis.

Variety.

✱ R. *a.* 2 *bracteòsus* Ser. *R. a.* γ and δ, *Weihe and Nees's Rubi Germ.* t. 3. b. — Bracteas very broad, undivided.

3. *R.* MICRA'NTHUS *D. Don.* The small-flowered Bramble.

Identification. Don Prod. Fl. Nepal., p. 235.; Dec. Prod., 2. p. 557.; Don's Mill., 2. p. 530.
Synonyme. *R.* pauciflòrus *Lindley* in *Bot. Reg.*, *Hort. Brit.* n. 13500.
Engravings. Bot. Reg., t. 854., as *R.* pauciflòrus *Lindl.*; and our *fig.* 529. representing a sprig to the usual scale, and *figs.* 530. and 531. representing the flowers and fruit of the natural size.

Spec. Char., &c. Upright. Stem round, branched, and bearing awl-shaped inflexed prickles, or straight prickles, and the branches recurved ones. Young branchlets rather glaucous at the extremity. Leaf pinnate, of 5—7 leaflets, that are ovate or oblong, mucronate, doubly serrated, plaited; green and glossy above; whitishly tomentose, or else glaucous, beneath. Petiole and rachis bearing prickles here and there. Petiole pilose. Stipules lanceolate, acuminate, membranaceous. Flowers small, reddish purple, disposed in a corymbose panicle. Petals clawed, shorter than the sepals. (*Dec. Prod.*) A gigantic bramble. Nepal. Stems 8 ft. to 20 ft. Introd. 1822. Flowers bright reddish purple; May to August. Fruit black; ripe in August. Naked young wood of a dark mahogany colour.

529. Rùbus micránthus.

530. Rùbus micránthus.

531. R. mic.

It is easily distinguished from all the other brambles in British gardens, by its nearly erect, strong, smooth, dark mahogany-coloured shoots, and by its very long pinnate leaves. The flowers are small, and the petals are of a bright reddish purple, and shorter than the sepals. The fruit is of a blackish purple, of the middle size; depressedly spherical, and covered with a fine bloom. The grains are fleshy, with a sweet subacid taste. This species throws up suckers sparingly; but its magnificent shoots arch over after they get to 6 or 8 feet in height, and grow branching and flowering on every side, till they reach the ground, when their extreme points strike root, and form new plants. Horticultural Society's Garden.

■ 4. *R.* OCCIDENTA`LIS *L.* The Western, *or American*, Bramble.

Identification. Lin. Sp., 706.; Dec. Prod., 2. p. 558.; Don's Mill., 2. p. 531.
Synonymes. *R.* virginiànus *Hort.*; *R.* idæ`us frúctu nigro *Dill.*
Engravings. Sloane Jam., t. 213. f. 1.; Dill. Hort. Elt., t. 247. f. 319.; and our *fig.* 532.

Spec. Char., &c. The whole plant is pretty glabrous. Stems round and whitish. Prickles recurved. Leaves of the barren branches pinnate; of the fertile branches trifoliolate. Leaflets ovate, incisely serrated, whitely tomentose beneath. Stipules very narrow, and bristle-like. Flowers in umbels. Peduncle prickled. Sepals lanceolate-linear, tomentose, longer than the petals, which are obovately wedge-shaped, two-lobed, and spreading. Fruit black, acid, of the form of that of *R.* idæ`us. Carpels numerous, rather glabrous; becoming, by drying, rugged with little hollows. (*Dec. Prod.*) A sub-erect shrub. Canada and the West Indies. Height 4 ft. to 6 ft. Introduced in 1696. Flowers white; May and June. Fruit black; ripe in August. Horticultural Society's Garden.

532. *R.* occidentàlis.

■ 5. *R.* IDÆ`US *L.* The Mount Ida Bramble, *or common Raspberry.*

Identification. Lin. Sp., 706.; Dec. Prod., 2. p. 558.
Synonymes. *R. frambæsiànus* Lam. Fl. Fr., 3. p. 135.; Framboisier, *Fr.*; gemeine Brombeere, *Ger.* "The Raspis is called in Greeke BATOS IDAIA; in Latin, *Rubus Idæa*, of the mountaine Ida, on which it groweth; in English, Raspis, Framboise, and Hinde-berry." (*Johns. Ger.*, p. 1274.)
Engravings. Eng. Bot., t. 2442.; and our *fig.* 533.

Spec. Char., &c. Villose. Stem round, bearing slender recurved prickles. Leaves pinnate; those of the fertile stems of 3 leaflets, those of the sterile stems of 5, rather palmately disposed. Leaflets ovate, incisely serrated, whitely tomentose beneath. Stipules very narrow and bristle-like. Flowers in a corymbose panicle. Sepals ovate-lanceolate, whitely tomentose, ending in a point. Petals obovate-wedge-shaped, entire, conniving, shorter than the calyx. Carpels numerous, tomentose. (*Dec. Prod.*) An upright shrub, with a creeping root and biennial stems. Europe, and, probably, Asia, Africa, and America. Found in every part of Great Britain, and in Ireland, in the agricultural and subalpine regions, in woods, and in moist wastes. Height 4 ft. to 8 ft. Flowers white; May, June, and July. Fruit red; ripe in July, August, and September.

Varieties.

■ R. *i.* 2 *microphýllus* Wallr. Sched. p. 256. — Leaves all of 3 leaflets. Stem suffruticose; dwarfer and more bushy than the species. (*Dec. Prod.*)

Garden Varieties. There are varieties with red fruit, yellow fruit, and white fruit; and one which bears twice in the year.

The fruit of the species, in a wild state, is crimson, amd consists of numerous juicy grains, beset with the permanent styles, and highly fragrant; with a very deliciously sweet, and yet slightly acid flavour, when eaten. Improved varieties of it have long been in cultivation in gardens, for the fruit, which is delightfully fragrant, and grateful to the palate in itself, and is used in numerous culinary and confectionery articles, as well as in liqueurs. The raspberry requires a vegetable soil, rather moist, soft, and not very deep; because most of the roots, like those of all other plants that throw up numerous suckers, keep near the surface; and the situation should be shaded, rather than fully exposed to the meridian sun. In a wild state, it is almost always found more or less shaded by trees, but not under their drip; and in woods, the situation of which is rather low and moist, than hilly and rocky or dry. The root belongs to that description which is called travelling; that is, the suckers extend themselves all round the central plant, so as every year to come up in fresh soil. Hence, as Miller observes, a raspberry plantation requires to be renewed every five or six years.

533. *R. idæ'us.*

§ ii. *Leaves digitate, of 3—5 Leaflets.*

⁎ 6. *R.* LACINIA'TUS *W.* The cut-*leaved* Bramble.

Identification. Willd. Hort. Berol., p. 82. and t. 82.; Dec. Prod., 2. p. 558.; Don's Mill., 2. p. 532.
Engravings. Willd. Hort. Berol., t. 82.; Wats. Dend. Brit., t. 69.; and our *fig.* 534.

Spec. Char., &c. Stem rather round, bearing stout recurved prickles, compressed at the base. Leaflets 3—5, pinnately cut, sharply serrate, a little downy beneath. Flowers in loose panicles, white or rose-coloured. Sepals lanceolate, leafy at the tip, tomentose, prickled, reflexed. Petals wedge-shaped, 3-lobed at the tip. Carpels roundish, dark-coloured. (*Dec. Prod.*) A large and handsome bramble. Stems 4 ft. to 10 ft. Flowers white, or rose-coloured; June to September. Fruit black; ripe in August. Naked young wood of a fine purple colour.

534. *R. laciniàtus.*

The appearance of this plant is that of the common bramble, except in the leaflets, which, from their being deeply cut, are strikingly different. Where it was first found is unknown; but it is, in all probability, only a variety of the common bramble, analogous to the cut-leaved variety of the elder (*Sambùcus nìgra laciniàta.*) H. S.

⁎ 7. *R.* CÆ'SIUS *L.* The grey Bramble, *or Dewberry.*

Identification. Lin. Sp., 706.; Dec. Prod., 2. p. 558.; Don's Mill., 2. p. 533.
Engravings. N. Du Ham., 5. t. 22.; Hayne Abbild., t. 100.; Eng. Bot., t. 826.; and our *fig.* 536.

Spec. Char., &c. Stem trailing, round, in many instances suffused with a grey bloom, bearing slender and a little recurved prickles. Leaflets 3 in a leaf, ovate, doubly serrated or crenated, glabrous, or obscurely ciliated. Panicle almost simple. Sepals ovate-acuminate. Petals white. Fruit sweet. Carpels large, few, greyish. (*Dec. Prod.*) A low straggling bramble. Europe and the North-East of Asia, in

535. *R. c. parviflòrus.*

woods and hedges. Stem 4 ft. to 8 ft. Flowers white; June and July. Fruit black; ripe in August.

Varieties.

- ✱ R. c. 2 *arvénsis* Wallr. Sched. p. 288. *R.* pseùdo-cæ'sius *Weihe*, according to Ern. Meyer in Litt. — Leaflets tomentose beneath. Branches puberulous.
- ✱ R. *c.* 3 *grandiflòrus* Ser. — Pubescent. Petals and sepals long.
- ✱ R. *c.* 4 *parvifòlius* Wallr. Sched., p. 228. (Our *fig.* 535.) — Stem ascending, purplish, ultimately naked. Leaves small, incisely lobed. Peduncle 1—3-flowered. A native of herbage-covered hills.
- ✱ R. *c.* 5 *fòliis variegàtis* Hort. has variegated leaves.

536. *R. cæ'sius.*

A low, weakly, straggling, prostrate plant, having the flowers with blush-coloured petals, and the fruit small, with few grains; but these large, juicy, black, with a fine glaucous bloom, and very agreeably acid. This species varies exceedingly in the size of its flowers and leaves in different situations, whence have arisen many varieties.

✱ 8. *R.* CORYLIFO'LIUS *Smith.* The Hazel-leaved Bramble.

Identification. Smith Fl. Brit., p. 542.; Smith in Eng. Bot., t. 827.; Don's Mill., 2. p. 533.
Synonymes. *R.* vulgàris *Weihe & Nees*, according to Lindley, *Synopsis of Brit. Flora*, ed. 2. p. 94.; *R.* nemoròsus *Heync*, according to Sprengel and Goldbach.
Engravings. Eng. Bot., t. 827.; and our *fig.* 538.

Spec. Char., &c. Stem angled, bearing straightish prickles. Leaflets 3—5, cordate-ovate, firm, doubly serrated, pilose beneath. Panicle nearly simple. Flowers white. Sepals ovate-acuminate, ultimately reflexed. Carpels purplish-blue, and large. (*Dec. Prod.*) A large bramble. Native of Europe, especially in the southern part; and frequent in Britain, in hedges and thickets. Stems 6 ft. to 12 ft. Flowers white; June and July. Fruit purplish blue or black, large; ripe in August.

Varieties.

- ✱ R. c. 2 *cànus* Wallr. Sched. p. 231. — Leaflets all similar in form, roundish heart-shaped, whitishly tomentose upon both surfaces.

537. *R. c. glandulòsus.*

538. *R. corylifòlius.*

- ✱ R. c. 3 *glandulòsus* Wallr. Sched. p. 231. *R.* glandulòsus *Spreng.*, according to Wallr. (Our *fig.* 537.) — Stems, petioles, and peduncles glandulous.

The stems are long and trailing, sometimes arching, glaucous and purplish in the sun, and green in the shade: they are brittle and full of pith. The flowers are large, and appear earlier than those of most o fthe British species. The berry is large, agreeably acid, of larger and fewer grains than in *R.* fruticòsus, and of a browner black: they are ripened before those of *R.* fruticòsus and its allies.

According to Dr. Lindley, the following British kinds of *R*ùbus may be associated with *R.* corylifòlius *Smith*, either as related species, or as varieties: — R. *macrophýllus* Weihe & Nees (Eng. Bot. Suppl., t. 2625.); R. *carpinifòlius* Weihe & Nees; R. *fúsco àter* Weihe & Nees; R. *Kö'hleri Weihe & Nees* (Eng. Bot. Suppl., t. 2605.), *R.* pállidus *Weihe & Nees*; R. *glandulòsus* Smith; R. *rùdis* Weihe & Nees, *R.* echinàtus of ed. 1. of *Lindl. Synops.*, and our *H. B.* No. 28335.; R. *diversifòlius* Lindl. Synops. ed. 1., *R.* diversifòlius *Weihe, Hort. Brit.* No. 28330.

♠ 9. *R.* SPECTA'BILIS *Ph.* The showy-*flowered* Bramble.

Identification. Ph. Fl. Amer. Sept., 1. p. 348. t. 16.; Dec. Prod., 2. p. 559.; Don's Mill., 2. p. 533.
Synonyme. *R. ribifòlius Willd. Herb.*, according to Steven.
Engravings. Ph. Fl. Amer. Sept., t. 16.; Bot. Reg., 1424.; Bot. Cab., t. 1602.; and our *fig.* 539.

Spec. Char., &c. Stem not bearing prickles, glabrous. Leaf of three leaflets, that are ovate, acute, doubly and unequally serrated, downy beneath. Flowers of an agreeable purplish colour, produced singly on terminal peduncles. Sepals oblong, rather abruptly acuminate, shorter than the petals. (*Dec. Prod.*) An elegant shrubby bramble. Native of North America, on the banks of the Columbia River, and the north-west coast. Height 4 ft. to 5 ft. Introduced in 1827. Flowers rosy purple, odoriferous; April and May. Fruit large, dark yellow; ripe in July.

539. *R. spectábilis.*

Branches subflexuose, round, smooth; with large odoriferous flowers, succeeded by large dark-yellow fruit, of an acid and somewhat astringent taste, which make excellent tarts. It merits a place in every collection, both as a flowering shrub, and for its fruit.

♠ 10. *R.* FRUTICO'SUS *L.* The shrubby Bramble, *or common Blackberry.*

Identification. Lin. Sp., 707.; Weihe and Nees Rubi Germanici, p. 25.; Dec. Prod., 2. p. 561.; Don's Mill., 2. p. 534.
Synonymes. *R.* discolor and *R.* abrúptus, in *Lindl. Synops. of Brit. Flora*, ed. 1.
Engravings. Eng. Bot., t 715.; and our *fig.* 541.

Spec. Char., &c. Stem erect, and afterwards decurved, 5-angled, rather tomentose, bearing recurved prickles. Leaflets 3—5, ovate-oblong, acute, glabrous, beneath greyly tomentose, each on a secondary petiole. Panicle decompound, narrow, straight. Sepals reflexed, almost without prickles. (*Dec. Prod.*) A large bramble. Native of Europe, in hedges, thickets, and woods; in Britain abounding in the agricultural zone, and tolerably frequent in the upland zone; with, according to Mr. Winch, a limit similar to that of *U'*lex europæ'a. Stem 6 ft. to 12 ft. Flowers white, or rose-coloured; June to August. Fruit purplish black; ripe August to September or October.

540. *R. f. pompònius.*

Varieties.

R. *f.* 2 pompònius *Ser.* *R.* fruticòsus δ *Weihe & Nees.* (*fig.* 540.) — Flowers semidouble or double. Leaves pale green; leaflets obovate. Cultivated in gardens. This variety may be considered as highly ornamental, from the large size and numerous petals of its flowers, and from its very vigorous growth. Though it will thrive at the roots of trees, and in places where other ornamental plants will hardly grow, yet it produces most effect when it is trained against a wall.

⁂ R. *f.* 3 *taúricus* Hort. is a vigorous-growing plant, which produces by far the best fruit of any variety of bramble. H. S.

⁂ R. *f.* 4 *flòre ròseo plèno* Baum. Cat.—Flowers double pink. H. S.

⁂ R. *f.* 5 *fòliis variegàtis*.—Leaves variegated, and not liable to the objections made to most variegated plants.

⁂ R. *f.* 6 *leucocárpus* Ser. — Fruit white. (*Dec. Prod.*)

This species is considered as being more common than any of the other brambles, and also as attaining a greater size. It is always found to prosper best on a soil somewhat dry and gravelly; and, accordingly, Switzer, when speaking of choosing a soil and situation for a vineyard, recommends looking out for one where the bramble is abundant and vigorous. The fruits have been eaten by children, in every country where they grow wild, since the time of Pliny. They have also been used, both in France and England, to produce a subacid drink; an inferior description of wine; by fermentation and distillation, a strong spirit; and, boiled with sugar, a very good jam.

541. R. fruticòsus.

⁂ 11. *R.* HI'SPIDUS *L.* The hispid-*stemmed* Bramble.

Identification. Lin. Sp., 706.; Dec. Prod., 2. p. 564.; Don's Mill., 2. p. 537.
Synonymes. R. triviàlis *Michx. Fl. Bor. Amer.* p. 296.; *R.* procúmbens *Mühl.*; *R.* flagellàris *Willd.*, according to *Spreng. Syst.* 2. p. 530.
Engravings. Hayne Abbild., t. 71.; and our *fig.* 542.

Spec. Char., &c. Stem procumbent, round, very hispid, bearing scattered bristles and recurved prickles. Leaflets 3—5, wedge-shaped at the base, unequally toothed, pretty glabrous above. Flowers solitary, upon longish peduncles. Petals obovate. Carpels black. (*Dec. Prod.*) A bramble with procumbent stems. Canada. Stems 6 ft. to 8 ft. Introduced in 1759. Flowers white; August. Fruit black, edible; ripe in September and October.

542. R. híspidus.

§ iii. *Leaves lobed, not pinnate or digitate.*

⁂ 12. *R.* ODORA'TUS *L.* The sweet-scented-*leaved* Bramble.

Identification. Lin. Sp., 707.; Dec. Prod., 2. p. 566.; Don's Mill., 2. p. 539.
Synonymes. R. occidentàlis *Hort.*, but not of Lin.; the Virginian Raspberry, the flowering Raspberry.
Engravings. Mill. Ic., t. 223.; Bot. Mag., t. 323.; and our *fig.* 543.

Spec. Char., &c. Stem upright. Petioles, peduncles, and calyxes bearing glanded hairs. Disks of leaves 5-lobed, unequally toothed. Inflorescence subcorymbose. Flowers large, showy, red. Sepals ovate, longly acuminate, shorter than the petals. Carpels numerous, ovate, velvety. Style funnel-shaped. (*Dec. Prod.*) An upright shrub. North America, in the woods of Canada, and on the Alleghany Mountains. Height 4 ft. to 6 ft. Introduced in 1700. Flowers purplish red; June to Sept. Fruit yellow, rarely seen in England.

543. R. odoràtus.

The flowers are not succeeded by fruit in this country; but Pursh informs us that, in a wild state, the fruit is yellow, and of a very fine flavour, and a large size. "Cornutus, who first figured and described this plant, gave it the name of odoràtus, on account of the very grateful fragrance of its foliage." (*Bot. Mag.*, t. 323.)

13. *R.* NUTKA'NUS *Moc.* The Nootka Sound Bramble.

Identification. Mocino Pl. Nutk.; Dec. Prod., 2. p. 566.; Bot. Reg., t. 1368.; Don's Mill., 2. p. 540.
Synonyme. *R.* odoràtus *Hort.*, but not of Lin.
Engravings. Mocino Pl. Nutk.; Bot. Reg., t. 1368.; and our *fig.* 544.

544. Rùbus nutkànus.

Spec. Char., &c. Stem glutinous. Branches round, glabrous, rufous. Leaves 5-lobed, unequally toothed. Inflorescence subcorymbose. Flowers about 4 in a corymb, white. Sepals ovate, longly acuminate, glabrous, as long as the petals. Allied to *R.* odoràtus, but the peduncle and calyx are glabrous. (*Dec. Prod.*) An upright shrub. North-west coast of North America, from New California to Nootka Sound, and at various places between north latitude 43° and 52°, in mountains and woods. Height 5 ft. to 6 ft. Introduced in 1826. Flowers white; May to October. Fruit yellow or reddish; ripe in August.

The general aspect and appearance of *R.* odoràtus, except being of a paler green. The flowers are succeeded by large yellow or reddish berries, which are found to make excellent tarts; and the plant will probably soon be ranked as a fruit shrub. Horticultural Society's Garden.

Species and Varieties of Rùbus *best deserving of Cultivation in British Gardens, as ornamental Shrubs.*

A. *Erect Raspberry-like Sorts.*

R. *occidentàlis, the Western,* or black, *Raspberry,* No. 4.; and fig. 532. in p. 313.
R. *nutkànus, the Nootka Sound Raspberry,* No. 13.; and fig. 544. in p. 318.
R. *odoràtus, the sweet-scented,* or Virginian *Raspberry,* No. 12.; and fig. 543. in p. 317.
R. *spectábilis, the showy*-flowered *Raspberry,* No. 9.; and fig. 539. in p. 316.
R. *idæ'us, the Mount Ida,* or common, *Raspberry,* No. 5.; and fig. 533. in p. 314. — The varieties of this species which are recommended as being most suitable for planting in an arboretum are, the red Antwerp, the white Antwerp, and the smooth cane.

B. *Shrubby Brambles.*

R. *suberéctus, the sub-erect Bramble,* No. 1.; and fig. 527. in p. 311.
R. *micránthus, the small-flowered,* or Nepal, *Bramble,* No. 3.; and figs. 530, 531. in p. 312.
R. *fruticòsus, the shrubby Bramble,* or common Blackberry, No. 10.; and fig. 540. in p. 316.—The varieties recommended are, the double-flowered, the double pink-flowered (if it can be got), the variegated-leaved, and *R.* f. taúricus, on account of its large and excellent fruit.
R. *laciniàtus, the cut*-leaved *Bramble,* No. 6.; and fig. 534. in p. 314.
R. *corylifòlius, the Hazel-leaved Bramble,* No. 8.; and fig. 537. in p. 315.

C. *Prostrate Brambles*

R. *cæ'sius, the grey Bramble,* or Dewberry, No. 7.; and fig. 535. in p. 314.—The variety recommended, in addition to the species, is that with variegated leaves.

Remark. The plants in the last two groups are propagated by division of the roots, or by encouraging the points of the shoots to root, like the runners of a strawberry; and the plants in the first group by division of the root, or by suckers.

Other Sorts of shrubby Rubuses.—R. *macropétalus* Doug. MS. in *Hook. Fl. Bor. Amer.* p. 178. t. 59., and our *fig.* 545., is a native of low woods in the valley of the Columbia, with white flowers, and the general habit of *R.* spectábilis.

R. *deliciòsus* Torrey in *Ann. Lyc.* ii. p. 196. is a native of North America, among the Rocky Mountains; with purple flowers, succeeded by a very delicious fruit. It is a shrubby bramble, 5 or 6 feet high.

R. *tiliàceus* Smith in *Rees's Cycl.* vol. xxx. is a native of Upper Nepal, with white flowers, and leaves like those of *T*ília álba. *R.* cordifòlius *D. Don* appears to be the same species, or perhaps a variety. Hort. Soc. in 1834.

545. *R. macropétalus.*

Genus X.

POTENTI'LLA *L.* The Potentilla, *or Shrubby Cinquefoil.* *Lin. Syst.* Icosándria Polygýnia.

Identification. Lin. Gen., No. 865.; Nestl. Pot. Diss., 4to; Lehm. Pot. Diss., 4to; Dec. Prod., 2. p. 571.; Don's Mill., 2. p. 549.
Derivation. From *potens*, powerful; in allusion to the supposed medicinal qualities of some species.

Gen. Char. *Calyx* 10-parted, the 5 outer segments accessory. *Petals* 5. *Stamens* numerous. *Carpels* numerous with lateral styles, seated on a dry permanent, elevated receptacle. (*Don's Mill.*)

Leaves compound, alternate or opposite, stipulate, sub-evergreen; pinnately cut. *Flowers* white or yellow.

Shrubs low, natives of Europe and America, and of easy culture in a dry soil. They are propagated by seeds or cuttings; and, except the common species, P. fruticòsa, are not much in cultivation.

1. P. fruticò'sa *L.* The shrubby Potentilla, *or Cinquefoil.*

Identification. Lin. Sp., 709.; Dec. Prod., 2. p. 579.; Don's Mill., 2. p. 560.; Nestl. Pot.; Lehm. Pot.; Smith's Eng. Flora, 2. p. 416.
Engravings. Eng. Bot., t. 88.; Nestl. Pot., 30. t. 1.; and our *fig.* 546.

Spec. Char., &c. Stem shrubby. Leaves pinnately cut, hairy; the lobes oblong, lanceolate, entire, approximate, of nearly the same colour on both surfaces. Stipules lanceolate, membraneous, acute. Inflorescence rather corymbose. Flowers yellow. Sepals pilose, lanceolate, acute, broad at the base. Bracteas linear-lanceolate, indistinctly petioled. Corolla longer than the calyx. Receptacle very hairy. (*Dec. Prod.*) A roundish bush. England, Germany, the Pyrenees, and other places; in England, in Middleton, Teesdale; and in Rock Forest, Clare, in Ireland. Height 2 ft. to 4 ft. Flowers yellow; July and August. Fruit brown; ripe in October.

546. Potentilla fruticòsa.

Varieties, according to Seringe, in *Dec. Prod.*

P. *f.* 2 *dahùrica* Ser. P. dahùrica *Nestl. Pot.* 31. t. 1.; P. fruticòsa β *Lehm. Pot.* 32. — Glabrous. Lobes of the leaves 3—5, pinnately cut. Sepals shorter and broader than the bracteoles. Dahuria. Introduced in 1824; and producing its yellow flowers in July. Plants in the Hort. Soc. Garden, and in the Epsom Nursery.

P. *f.* 3 *tenuìloba* Ser.; P. fruticòsa β *Nestl. Pot.* 30., *Lehm. Pot.* 32. var. γ; P. floribúnda *Ph. Fl. Amer. Sept.* 1. p. 355, *Watson's Dend. Brit.* t. 70.; P. tenuifòlia *Schlectend. Berl. Mag.*, according to *Lehm. Pot.* 32. (Our *fig.* 547.) — Sepals and lobes of the leaves narrow, and with a slight hoary silkiness. North America; where it is a low-growing shrub, not above 18 in. high. It was also found by Pallas in Siberia. The handsomest variety of the species.

547. P. f. tenuìloba.

2. P. GLA'BRA *Lodd.* The glabrous Potentilla.

Identification. Lodd. Bot. Cab., t. 914.; Dec. Prod., 2. p. 584.; Don's Mill., 2. p. 561.
Synonymes. P. fruticòsa álba *Busch*, according to *Lodd. Bot. Cab.*, t. 914.
Engravings. Lodd. Bot. Cab., t. 914.; and our *fig.* 548.

Spec. Char., &c. Frutescent. Branches pendulous, purple. Leaves pinnately cut into 5 entire lobes. Flowers terminal, white, of the size of those of the wood strawberry (Fragària vésca). (*Dec. Prod.*) A beautiful little shrub. Siberia. Height 2 ft. to 4 ft. Introduced in 1818. Flowers white; June to August. Fruit brown; ripe in September.

548. P. glábra.

It differs from P. fruticòsa in being perfectly smooth in all its parts, and in having pendulous branches and undulated leaves. It thrives best in a mixture of loam and peat, is of slow growth, and difficult to increase, except by seed.

3. P. SALESO'VII Steph. Salesovius's Potentilla.

Identification. Steph., according to Willd. Enum., p. 552.; Nestl. Pot., 31.; Lehm. Pot., 35. t. 1.; Dec. Prod., 2. p. 583.
Engraving. Lehm. Pot., 35. t. 1.; and our *fig.* 549.

Spec. Char., &c. Habit resembling that of Cómarum palústre. Stem suffruticose. Leaves pinnately cut, coriaceous. Lobes oblong, acutely serrate, pubescent above upon the veins, whitely tomentose beneath. Stipules lanceolate, very acute, entire, rather filmy at the edge. Flowers large, white, upon short peduncles, and grouped. Sepals lanceolate, very acute, broad, almost as long as the petals, which are obovate. Bracteoles very narrow, smaller than the sepals. Receptacle lanuginose. (*Dec. Prod.*) A low shrub. Siberia. Height 1 ft. to 2 ft. Introduced in 1823. Flowers white; June and July. Fruit brown; ripe in September.

549. P. Salesòvii.

Cómarum palústre L. (Eng. Bot., t. 172.), Potentílla Cómarum *Scop.*, a well known British plant, found in boggy soils, with somewhat ligneous shoots, and showy flowers of a deep dingy purple, may be added to this order, and may prove useful in particular situations, on the margins of ponds.

Genus XI.

COWA'NIA D. Don. The Cowania. *Lin. Syst.* Icosándria Polygýnia.

Identification. D. Don, in Sw. Fl. Gard., 2d series.
Derivation. In honour of *James Cowan*, a London merchant, who several times visited Mexico and Peru, whence he introduced a great many plants.

Gen. Char. *Calyx* 5-cleft. *Petals* 5. *Ovaries* 5—14. *Ovule* erect. *Styles* terminal, continuous. *Achenia* adorned with the plumose persistent style. (*D. Don.*)

Leaves simple, alternate, stipulate, evergreen; wedge-shaped, oblong, pinnatifid, plaited. *Flowers* terminal, solitary, almost sessile, red. — Shrub; native of Mexico; very ornamental, but somewhat tender in British gardens.

1. C. plica'ta *D. Don.* The plaited-*leaved* Cowania.

Identification. Swt. Brit. Flor. Gard., t. 400.; Gard. Mag., 13. p. 452.
Engravings. Swt. Brit. Fl. Gard., t. 400.; and our *fig.* 550.

Spec. Char., &c. Leaves wedge-shaped, oblong, pinnatifid, plaited. (*D. Don.*) A rigid, evergreen, decumbent, much-branched shrub. Mexico. Height 1 ft. to 2 ft. Introduced in 1835. Flowers dark red; June and July. Naked young wood dark brown.

550. Cowanse plicata.

Branches copiously clothed with stalked glands. Petioles of the leaves very short, slightly channeled above, sheathing at the base. Stipules adherent. Flowers, when protruding from the bud exactly like those of *Ròsa*. This promising evergreen shrub, being extremely difficult to propagate, has been lost, for the present, to British gardens.

Sect. IV. *Ro'seæ Dec.*

Genus XII.

RO'SA *Tourn.* The Rose Tree. *Lin. Syst.* Icosándria Polygýnia.

Identification. Tourn. Inst., 1. p. 636. t. 408.; Lin. Gen., 631., Lam. Ill., t. 440.; Lindl. Ros. Mon., 8vo, 1820; Pronville's Nomen. Rais., 1818, Monog. Ros., 1824; Don's Mill., 2. p. 564.
Synonymes. Rhodóphora *Neck. Elem.*; Rosier, *Fr.*; Rosenstock, *Ger.*; Rooseboom, *Dutch*; Rosajo, *Ital.*; Rosal, *Span.*; and Roseira, *Portuguese*.
Derivation. From *rhos*, red, Celtic; in reference to the colour of the flowers of most of the species.

Gen. Char. *Calyx* with the tube contracted at the mouth, and with the segments usually pinnately divided. *Petals* 5. *Stamens* numerous. *Carpels* numerous, bony, inserted on the inside of the tube of the calyx, which at length becomes baccate and encloses them. They are dry and indehiscent, bearing each a style on the inner side. (*Don's Mill.*)

Leaves compound, alternate, stipulate, deciduous or evergreen; imparipinnate. *Stipules* attached to the petiole. *Prickles* simple. *Flowers* large, showy; red, white, or yellow; fragrant.

Shrubs, for the most part deciduous; natives of Europe, and of the temperate regions of Africa, Asia, and America, but not of Australia; and they have been in cultivation in the Old World, for the beauty and fragrance of their flowers, from time immemorial. As the culture of roses belongs more to floriculture than to arboriculture, it will be found given at length in our *Encyclopædia of Gardening*, and in the first edition of this work. All the species may be propagated by cuttings of the roots, cuttings of the young wood in a growing state, by layers, or by budding or grafting; and they will all thrive in loamy soil, dry and rich, rather than poor. The genus *Ròsa* is in a state of confusion still greater than that which subsists among the different kinds of *Rùbus*; nor can it well be otherwise, when we consider that the greater number of kinds in cultivation are garden productions, and that the wild kinds differ exceedingly according to soil and situation, and have been chiefly described by botanists from dried specimens. We have adopted the arrangement in Don's *Miller*, with the exception of omitting the first section, Simplicifòliæ, now made a separate genus by Dr. Lindley. Where the species arranged under a section are natives of different countries, it may fairly be presumed, that there is at least one in each country entitled to be considered a species; or, at least, it may be convenient to do so, in the present state of our knowledge. Nature, it is observed in the *Nouveau Du Hamel*, "appears scarcely to have placed any limit between the different species of the rose; and, if it is already very difficult to define the wild species, which have not yet been modified by culture, it is almost impossible to refer to their original type the numerous varieties which culture has made in the flowers of species already so nearly resembling each other."

§ i. *Feròces* Lindl. Mon. p. 3.

Derivation. From *ferox*, fierce; in reference to the branches being thickly beset with prickles.

Sect. Char. Branches clothed with permanent tomentum. Fruit naked. The plants contained in this section are a truly natural group; they are low shrubs, losing their leaves early in autumn, and are then remarkable for their hoary branches, bristles, and numerous prickles. Their fruit is perfectly smooth, which separates them from the next section, in which the fruit is downy. Sepals usually toothed. (*Don's Mill.*) — Deciduous low bushes, natives of Caucasus and Kamtschatka.

1. *R.* FE'ROX *Lawr.* The fiercely-*prickled* Rose.

Identification. Lawr. Ros., t. 42.; Don's Mill., 2. p. 565.
Synonymes. *R.* kamtschática *Red. Ros.* 1. p. 47.; *R.* kamtschática β fèrox *Ser.* in *Dec. Prod.* 2. p. 607.; *R.* echinàta *Dupont.*
Engravings. Lawr. Ros., t. 42.; Red. Ros., 1. p. 47. t. 12.; and our *fig.* 551.

Spec. Char., &c. Prickles all alike in shape, and much crowded. Flowers large, red. Fruit globose, scarlet. (*Don's Mill.*) A very prickly shrub. Caucasus. Height 3 ft. to 4 ft. Introduced in 1796. Flowers large, red; July and August. Fruit scarlet; ripe in September.

Variety.

R. *f.* 2 *nìtens* Lindl. in *Bot. Reg.* t. 824, *Ser.* in *Dec. Prod.* ii. p. 607., has shining pale green glabrous leaves, and pale crimson flowers. Deserving a place in collections on account of its singularity.

551. Rosa fèrox.

2. *R.* (F.) KAMTSCHA'TICA *Vent.* The Kamtschatka Rose.

Identification. Vent. Cels., t. 67.; Don's Mill., 2. p. 565.
Engravings. Vent. Cels., t. 67.; N. Du Ham., vol. 7. t. 10. f. 2.; and our *fig.* 552.

Spec. Char., &c. Prickles infra-stipular, falcate, large. Leaves opaque. Flowers solitary, deep red. Fruit spherical, scarlet, less than that of *R.* fèrox. (*Don's

Mill.) Kamtschatka, in dry rocky places. Height 3 ft. to 4 ft. Introduced in 1791. Flowers deep red; June and July. Fruit scarlet; ripe in September.

552. R. (f.) kamtschatica.

From the appearance of the plants bearing this name in the extensive collection in Messrs. Loddiges's arboretum, we should consider it to be only a variety of *R.* fèrox. It is, however, very distinct, and, from the rich colour of its flowers and fruit, well deserving a place in collections.

§ ii. *Bracteàtæ.*

Sect. Char. Branches and fruit clothed with permanent tomentum. This section is readily distinguished from the last by the woolliness of the fruit. Leaves dense, usually shining, and prickles placed under the stipules in pairs. Sepals simple, or nearly so. (*Don's Mill.*) — Evergreen, or sub-evergreen bushes. Natives of China and Nepal.

■ 3. *R.* BRACTEA'TA *Wendl.* The *large*-bracted Rose.

Identification. Wendl. Obs., 50.; Red. Ros., 1. p. 35., ic.; Lindl. Ros. Mon., p. 10.; Dec. Prod., 2. p. 6 2.; Don's Mill., 2. p. 565.
Synonyme. Lord Macartney's Rose.
Engravings. Wendl. Hort. Herrenhaus., fasc. 4. t. 22.; N. Du Ham., 7. t. 13.; and our *fig.* 553.

Spec. Char., &c. Evergreen. Branches upright, tomentose. Prickles stout, recurved, in many instances in pairs. Leaflets 5—9, obovate, subserrate, coriaceous, glossy, glabrous. Stipules scarcely attached to the petiole, bristle-shaped, but fringed. Flowers solitary, terminal, white, large. Peduncles and calyxes tomentose. Fruit globose, large, orange red. (*Dec. Prod.*) Evergreen. China. Height 5 ft. to 6 ft. Introduced in 1795. Flowers large, showy, white, solitary, nearly sessile; June to October. Fruit orange red; ripe in October.

553. R. bracteata.

It flowers abundantly, but is rather tender, on which account it succeeds best when trained against a wall.

Varieties.

- ■ R. *b.* 2 *scabricaúlis* Lindl. Rosar. Monog. p. 10. (Sims Bot. Mag., t. 1377.) — Branches bristly. Prickles smaller than in the species, and rather straight. China, province of Tchetchiang. (*Dec. Prod.*)
- ■ R. *b.* 3 *flòre plèno* Hort.—Flowers double, but never expanding freely.
- ■ R. *b.* 4 *Maria Leonida* Hort.—Flowers double, white, yellowish pink in the centre, expanding freely. One of the finest of autumnal roses.

The species and the varieties, being somewhat tender, succeed best when trained against a wall. They are very ornamental from their shining evergreen foliage, as well as from their flowers.

■ 4. *R.* MICROPHY'LLA *Roxb.* The small-leafleted Rose.

Identification. Roxb. Fl. Ind. ined., according to Lindl. Rosar. Monog., p. 9. 145, 146.; Dec. Prod., 2. p. 602.; Don's Mill., 2. p. 565.
Synonyme. Hoi-tong-hong, *Chinese*.
Engravings. Bot. Reg., t. 919.; and our *fig.* 554.

Spec. Char., &c. Stem almost without prickles. Leaflets glossy, sharply serrated, veiny beneath, with densely netted anastomosing veins. Stipules very narrow, unequal. Flowers double, of a delicate rose colour. Calyx densely invested with prickles. Sepals short, broadly ovate, echinate, ending in a point. Prickles having at the base two longitudinal furrows.

554. R. microphylla.

(*Dec. Prod.*) Sub-evergreen. China Height 2 ft. to 3 ft. Introduced in 1828. Flowers very large, double, and of a delicate blush colour; August to October. Fruit orange red; ripe in October.

An interesting little shrub, but somewhat tender, like *R.* bracteàta. There is a variety in the Hort. Soc. Garden called *R.* m. álba.

n. 5. *R.* INVOLUCRA'TA *Roxb.* The involucred-*corymbed* Rose.

Identification. Roxb. Fl. Ind. ined., according to Lindl. Rosar. Monog., p. 8.; Dec. Prod., 2. p. 602; Don's Mill., 2. p. 565.
Synonymes. *R.* Lindleyàna Tratt. Ros. 2. p. 109; *R.* palústris Buchan. (*Ham.*) *MSS.*
Engravings. Bot. Reg., t. 739.; and our *fig.* 555.

555. *R.* involucràta.

Spec. Char., &c. Shoots long, tender. Branches pale brown, tomentose, scarcely prickly. Leaflets 3—9, elliptical-lanceolate, tomentose beneath. Stipules hardly attached to the petiole, bristle-like-fringed. Flowers terminal, mostly solitary, white. Peduncles and calyxes tomentose. (*Dec. Prod.*) Deciduous, branchy. Nepal and China. Height 3 ft. to 4 ft. Introd. 1818. Flowers white; June and July. Fruit orange red; ripe in October.

Seringe seems to consider this as a variety of *R.* bracteàta. The flowers are in corymbs, surrounded by three or four approximate leaves. The plants are rather tender, and succeed best against a wall, where they flower magnificently. Not common in collections. Lodd.

§ iii. *Cinnamòmeæ* Lindl. Ros. p. 13.

Sect. Char. Plants setigerous or unarmed, bracteate. Leaflets lanceolate glandless. Disk thin, never thickened. This section is distinguished by its long lanceolate leaflets without glands, its upright shoots, and compact habit. Flowers red, never solitary, except by abortion, and always supported by bracteas. Fruit round, small, red (soon losing its long narrow sepals), and with small, smooth, shining carpels. The shoots are usually setigerous next the ground; but rarely so towards the apex, except in one or two instances. *R.* alpìna and *R.* aciculàris, of the following division, sometimes have bracteas; but their sepals never fall off till the fruit is decayed. Sepals simple, entire, or nearly so, unless when mentioned otherwise. (*Don's Mill.*) — Plants of most of the species are in cultivation in British gardens. Deciduous rambling bushes; natives of Continental Europe and North America, and some of them of Britain.

A. *Species Natives of North America.*

6. *R.* LU'CIDA *Ehrh.* The shining-*leaved* Rose.

Identification. Ehrh. Beitr., 4. p. 22.; Lindl. Rosar. Monog., p. 17.; Dec. Prod., 2. p. 602.; Don's Mill., 2. p. 565.
Synonymes. *R.* rùbra lùcida *Rossig. Ros.*, t. 7. and t. 25. f. 1.; *R.* lùcida *Jacq. Fragm.* 71.; Rose Turneps; Rosier à Feuilles de Frêne, *Fr.*
Engravings. Red. and Thor. Ros., 1. p. 45. ic.; N. Du Ham., vol. 7. t. 7.; and our *fig.* 556.

556. *R.* lùcida.

Spec. Char., &c. Prickles recurved, or none. Leaflets 5—9, lanceolate-elliptical, coriaceous, bluntly serrated, glossy. Stipules dilated, large, finely serrated, and extended as far as to the leaflets. Peduncles somewhat hispid. Flowers red, and opening late in the season. Sepals almost entire, appendicled, spreading. Fruit oblately globose, a little hispid or glabrous, scarlet. (*Dec. Prod.*) An erect shining leaved shrub. North America, from New York to Carolina; near Boston, in bogs, and on the edges of marshes; and in Newfoundland. Height 4 ft. to 6 ft. Introduced in 1724. Flowers red, overtopped by the leaves and young branches; June to August. Fruit bright red; ripe in October.

A handsome species, on account of its shining foliage, and one which is very hardy; but the flowers have a very disagreeable smell.

7. *R.* NI'TIDA *W.* The glossy-*leaved* Rose.

Identification. Willd. Enum., 544.; Lindl. Rosar. Monog., p. 13.; Dec. Prod., 2. p. 603.; Don's Mill., 2. p. 565.
Synonymes. *R.* Redutèa rufescens *Thory* in *Red. Ros.* 1. p. 103. ic.; the dwarf Labrador Rose
Engravings. Lindl. Rosar. Monog., p. 13. t. 2.; Redouté Ros. 1. p. 103. ic.; and our *fig.* 557.

557. *R.* nitida.

Spec. Char., &c. Dwarf and reddish in aspect. Stem and branches almost covered with slender, rather equal, prickles. Leaflets 5—9, rather rigid, lanceolate, glossy. Stipules large, finely serrated, extending as far as to the leaflets. Flowers red. Peduncle bristly. Sepals spreading. Fruit bristly, shining, and scarlet. (*Dec. Prod.*) A shrub beset with straight red spines. Newfoundland. Height 2 ft. to 3 ft. Introduced in 1807. Flowers deep red; June to August. Fruit depressed, spherical, bright scarlet; ripe in September.

This is an interesting plant, from its dwarf stature, its abundant reddish prickles, its glossy leaves, its flowers, and its fruit.

8. *R.* RA'PA *Bosc.* The Turnip-*fruited* Rose.

Identification. Bosc Dict. d'Agric.; Lindl. Rosar. Monog., p. 15.; Dec. Prod., 2. p. 602.; Don's Mill., 2. p. 565.
Synonymes. *R.* túrgida *Pers. Ench.* 2. p. 49.; *R. fraxinifòlia Dumont* in *Cours. Bot. Cult.*
Engravings. Red. and Thor. Ros., 2. p. 7. ic.; and our *fig.* 558.

558. *R.* Ràpa.

Spec. Char., &c. Taller than *R.* lùcida, and spreading. Branches without prickles. Leaflets oblong, undulate, shining. Fruit hemispherical. Closely allied to *R.* lùcida, of which it is very likely a variety. (*Dec. Prod.*) Petals always multiplied, smaller than those of *R.* lùcida; bright red. Fruit deep red. Sepals compound. A tall straggling bush. North America, in the warmer states. Height 3 ft. to 4 ft. Introduced ?. Flowers double, bright red; June to August. Fruit deep red; ripe in October.

Only known in its double-flowered state in British gardens, where it is a freely growing hardy plant, with large double flowers.

9. *R.* PARVIFLO'RA *Ehrh.* The small-flowered, *or Pennsylvanian*, Rose.

Identification. Ehrh. Beitr., 4. p. 21.; Don's Mill., 2. p. 565.
Synonymes. *R.* hùmilis *Marsh Arb.* 136.; *R.* caroliniàna *Mich. Fl. Bor. Amer.* 1. p. 295.; *R.* carolina γ et δ *Ait. Hort. Kew.* ed. 2. vol. 3. p. 260.; Pennsylvanian Rose, *Lawr. Ros.* t. 3. and t. 66., and of the nurseries.
Engraving. Smith Insect. Georg., 1. p. 49. t. 25.

Spec. Char., &c. Dwarf. Stipules linear. Prickles acicular. Leaflets lanceolate, smoothish, sharply serrated. Calyxes clammy. Flowers usually in pairs. (*Don's Mill.*) A low weak shrub. North America, on the declivities of hills, in the states of New York and Carolina. Height 2 ft. Introduced in 1724. Flowers pale blush; June to August. Fruit red; ripe in October.

559. *R. parviflora flòre plèno.*

Variety.

- R. *p.* 2 *flòre plèno* Red. Ros. 2. p. 73., and our *fig.* 559. — Flowers double, pale blush, unexpanded. A neat little rose, but not in very general cultivation.

10. *R.* FRAXINIFO'LIA *Börk.* The Ash-leaved Rose.

Identification. Börk. Holz., 301.; Don's Mill., 2. p. 566.
Synonymes. *R.* virginiàna *Mill. Dict.* No. 10.; *R.* blánda *a Sol. MSS. Jacq. Fragm.* 70. t. 105.; *R.* corymbòsa *Bosc Dic. d'Agr.* ex *Desf. Cat. Hort. Par.* p. 272.?; *R.* alpina β *Ait. Hort. Kew.* ed. 2. vol. 3. p. 265.; *R.* alpina læ'vis *Red. Ros.* 1. p. 57., *Lawr. Ros.* t. 75.
Engravings. Bot. Reg., t. 458.; Lawr. Ros., t. 75.; and our *fig.* 560

560. *R. fraxinifòlia.*

Spec. Char., &c. Tall, unarmed. Branches straight, glaucescent. Leaflets opaque, undulated, and glabrous. (*Don's Mill.*) Branches dark purple, with a pale blue bloom. Flowers small, red, in few-flowered cymes. Fruit naked, small, round or ovate, of a dull pale red. A glabrous shrub. Native of Newfoundland, and on the north-west coast of America. Height 4 ft. to 6 ft. Introduced ?. Flowers red; May and June. Fruit dull pale red; ripe in September.

Other North-American Species.—*R.* Woódsii Lindl., *R.* carolìna *Lin.*, *R.* Lindlèyi Spreng., are described in our first edition; and the first two are in Messrs. Loddiges's collection.

B. *Species Natives of Nepal.*

11. *R.* MACROPHY'LLA *Lindl.* The long-leaved Rose.

Identification. Lindl. Ros. Monog., p. 35.; Don's Mill., 2. p. 566.
Engravings. Lindl. Ros. Monog., t. 6.; and our *fig.* 561.

561. *R. macrophylla.*

Spec. Char., &c. Unarmed. Leaves very long; leaflets 5—11, lanceolate. Petioles with a few glands, which, as well as the leaflets, are woolly beneath. Sepals narrow, longer than the petals, which are apiculated. (*Don's Mill.*) A smooth shrub. Gossainthan. Height 5 ft. to 6 ft. Introduced in 1830. Flowers red, on villous peduncles, and furnished with a few unequal setæ, as well as the fruit.

Dr. Lindley observes of this rose, that its leaves are the largest he has ever seen; that it cannot be confounded with any thing else; and that it may be considered the link between *Cinnamòmeæ* and Pimpinellifòliæ. Horticultural Society's Garden.

C. *Species Natives of Continental Europe.*

12. *R.* CINNAMO'MEA *Besl.* The Cinnamon-*scented* Rose.

Identification. Besl. Hort. Eyst. Vern. Ord., 6. p. 5.; Lin. Sp., 703.; Don's Mill., 2. p. 566.
Synonymes. *R.* fœcundíssima *Munch. Hausv.* 5. p. 279., *Fl. Dan.* t. 1214.; *R.* majàlis *Herm. Diss.* 8.
Engravings. Lindl. Ros., t. 5.; Fl. Dan., t. 1214.; and our *fig.* 562.

Spec. Char., &c. Tall, cinereous. Branches straight. Prickles stipular, straightish. Stipules dilated, undulated. Leaflets oblong, obtuse, wrinkled, tomentose beneath. (*Don's Mill.*) Flowers solitary, or 2—3 together,

562. R. cinnamòmea.

pale or bright red. Fruit round, naked, and crimson. The double-flowered variety is most common in gardens. An upright shrub. Native of most parts of Europe, and a doubtful native of England. Height 5 ft. to 6 ft. Flowers pale or bright red; May and June. Fruit crimson; ripe in September.

A very desirable sort, on account of its fragrance, which resembles that of cinnamon. There is a semi-double variety; and the single state is supposed to be identical with *R.* majàlis below.

Other European Species not Natives of Britain. — *R.* frutetòrum *Bess.*, *R.* taúrica *Bieb.*, and *R.* dahùrica *Pall.*, are described in our first edition, and the first two are in Messrs. Loddiges's collection.

D. *Species Natives of Britain.*

13. *R.* (*c.*) MAJA'LIS *Retz.* The May Rose.

563. R. majàlis.

Identification. Retz. Obs. Bot., 3. p. 33.; Don's Mill., 2. p. 566.
Synonymes. *R.* mùtica *Fl. Dan.* 688.; *R.* spinosíssima *Gorter. Ingr.* 78.; *R.* collíncola *Ehr. Beitr.* 2. p. 70.; *R.* cinnamòmea *Eng. Bot.* 2388.
Engravings. Fl. Dan., t. 688.; Eng. Bot., t. 2388.; and our *fig.* 563.

Spec. Char., &c. Dwarf grey. Branches straight, coloured. Prickles scattered, nearly equal. Stipules linear. Leaflets oblong, flat, glaucous, and tomentose beneath. (*Don's Mill.*) A nearly smooth shrub. Flowers usually solitary, pale red. Fruit orange red, spherical, and naked. Native of Sweden and Lapland; and of Britain, near Pontefract, in Yorkshire. Height 3 ft. to 4 ft. Flowers pale red; May and June.

This is supposed by some to be the single state of *R.* cinnamòmea.

14. *R.* DICKSONIA'NA Lindl. Dickson's Rose.

564. R. Dicksoniàna.

Identification. Lindl. Hort. Trans., 7. p. 224.; Eng. Bot., vol. iv. p. 51.; Don's Mill., 2. p. 566.
Engravings. Eng. Bot., t. 2707.; and our *fig.* 564.

Spec. Char., &c. Branches flexuous, setigerous, armed with a few slender scattered prickles. Leaflets folded together, unequal, with coarse double serratures. Stipules, petioles, and sepals compound. Styles stretched out, glabrous. (*Don's Mill.*) A large prickly shrub. Ireland. Height 5 ft. to 6 ft. Flowers white or pale pink; June and July.

§ iv. *Pimpinellifòliæ* Lindl.

Sect. Char., &c. Plants bearing crowded, nearly equal, prickles, or unarmed. Bractless, rarely bracteate. Leaflets ovate or oblong. Sepals connivent, permanent. Disk almost wanting. This section is essentially different from the last in habit, but in artificial characters they approach very nearly. It, however, may be distinguished by the greater number of leaflets; which vary from 7 to 13, and even to 15, instead of from 5 to 7. The flowers are also universally without bracteas; except in the *R.* alpìna, *R.* Sabìni, *R.* Doniàna, and, perhaps, *R.* marginàta. These, having connivent permanent sepals, cannot be confounded with the preceding division; nor, on account of their disk, with the following. There is no instance of stipular prickles in the present tribe. The sepals are entire, or nearly so, unless when mentioned otherwise. (*Don's Mill.*)—Deciduous: forming bushes partly low and dense, and partly large and rambling. Natives of Britain, Continental Europe, Asia, and North America.

A. *Species Natives of Europe.*

■ 15. *R.* ALPI'NA *Lin.* The Alpine Rose.

Identification. Lin. Sp., 703.; Don's Mill., 2. p. 567.
Synonymes. *R.* rupéstris *Crantz Austr.* 85.; *R.* monspeliaca *Govan Monsp.* 255.; *R.* inérmis *Mill. Dict.* No. 6.; *R.* hýbrida *Vill. Dauph.* 3. p. 554.; *R.* lagenària *Vill.* l. c. p. 563.; *R.* biflòra *Krok. Fl. Sil.* 2. p. 157.
Engravings. Jacq. Fl. Austr., t. 279.; Lindl. Bot. Reg., t. 474.; and our *fig.* 565.

Spec. Char., &c. Unarmed. Fruit elongated, pendulous. Peduncles hispid. (*Don's Mill.*) Flowers erect, blush-coloured, solitary. Fruit orange red, oblong or obovate, with long sepals, generally pendulous. An unarmed shrub. Alps of Austria, hills in the South of France, Silesia, Bohemia Dauphiné, Switzerland, &c. Height 5 ft. to 8 ft. Introduced in 1683 Flowers blush-coloured; June and July. Fruit orange red; ripe in September.

565. *R.* alpina.

566. *R.* a. læ'vis

Varieties.

- ■ R. *a.* 2 *læ'vis* Ser., but not of Desv. or Red.; *R.* Sanguisórba majòris, &c., *Dill. Elth.*; *R.* alpìna glàbra *Desv.*; *R.* a. vulgàris *Red. Ros.* 2. p. 111., and our *fig.* 566.; has the stem, peduncles, and calyx quite glabrous, and the fruit oblong.
- ■ R. *a.* 3 *speciòsa* Hort. Drummond's Thornless Rose. — A very beautiful climbing variety, raised by Mr. Drummond in the Cork Botanic Garden, about 1820.

Other Varieties. Fourteen are described in the first edition of this work, but they are chiefly of botanical interest.

■ 16. *R.* SUA'VIS *Willd.* The sweet Rose.

Identification. Willd. Enum. Suppl., p. 37.; Link Enum., 2. p. 57.; Don's Mill., 2. p. 567.
Engravings. Hayne Abbild., t. 40.; and our *fig.* 567.

Spec. Char., &c. Stem hispid. Leaves glabrous, glaucescent beneath. Peduncles and petioles clothed with glandular bristles. (*Don's Mill.*) Petals deep purple, deeply 2-lobed. Fruit oblong, glabrous. A hispid shrub. Native country unknown, most probably Europe. Height 3 ft. to 4 ft. Introduced in 1818 Flowers deep purple; June and July. Fruit scarlet, ripe in September.

567. *R.* suavis.

This very distinct variety, or perhaps species, of rose is probably at present wanting in British collections; for it must not be confounded with *Ròsa* suavèolens or with *Ròsa* suavifòlia, both described in *Le Botaniste Cultivateur* as varieties of *R.* rubiginòsa, or synonymes to that species.

17. *R.* SULPHU'REA *Ait.* The sulphur-*coloured-flowered* Rose.

Identification. Ait. Hort. Kew, 2. p. 201., Lindl. Ros., t. 77.; Don's Mill., 2. p. 568.
Synonymes. *R.* hemisphǽrica *Herm. Diss.* 18.; *R.* glaucophýlla *Ehrh. Beitr.* 2. p. 69.; *Ròsa* lùtea flòre plèno *Rai. Hist.* 1475. No. 31.; *R.* lùtea *Brot. Fl. Lus.* 1. p. 337.; the double yellow Rose.
Engravings. Lindl. Ros. t. 77.; Bot. Reg., t. 46.; and our *fig.* 568.

Spec. Char., &c. Stipules linear, divaricate, dilated at the apex. Leaflets glaucous, flattish. Tube hemispherical. (*Don's Mill.*) Stem prickles unequal, scattered. A deciduous shrub. Levant. Height 4 ft. to 10 ft. Introduced before 1629. Flowers fine transparent yellow, double; July.

568. *R.* sulphùrea.

This sort does not flower freely, except in open airy situations; and, if trained against a wall, exposed to the north or east rather than to the south. Its flower buds are apt to burst on one side before they expand, and consequently to become deformed; to prevent this, the blossom buds should be thinned, and care taken that they have abundance of light and air. Watering it freely in the flowering season is found advantageous; and the shoots, in general, ought not to be shortened. This beautiful species is said to flower freely, if grafted on the musk cluster at 8 or 10 feet from the ground; or it will do well on the China rose.

18. *R.* SANGUISORBIFO'LIA *Donn.* The Burnet-leaved Rose.

Identification. Donn Hort. Cant., ed. 8. p. 169.; Don's Mill., 2. p. 569.
Synonymes. *R.* spinosíssima var. *s* sanguisorbifòlia *Lindl. Ros.* p. 51.; *R.* spinos. var. macrophýlla *Ser.* in *Dec. Prod.* 2. p. 609.
Engraving. Our *fig.* ., in p.

Spec. Char., &c. Tall. Prickles nearly equal. Leaflets 9—11, oblong, glabrous, simply serrated. Fruit globose, depressed, dark. (*Don's Mill.*) An erect shrub. Habitat unknown, most probably Europe. Height 3 ft. to 5 ft. Flowers white; May and June. Fruit black; ripe in September.

Easily distinguished by the number of its leaflets, the shortness of its peduncles, and by its globose depressed fruit.

B. *Species Natives of Siberia.*

19. *R.* GRANDIFLO'RA *Lindl.* The large-flowered Rose.

Identification. Lindl. Ros., p. 53.; Don's Mill., 2. p. 569.
Synonyme. *R.* pimpinellifòlia *Bieb. Fl. Taur.* 2. p. 394.
Engravings. Bot. Reg., t. 888.; and our *fig.* 569.

Spec. Char., &c. Branches without bristles. Prickles nearly equal, distant. Leaflets flat, glabrous, simply serrated. (*Don's Mill.*) A prickly shrub. Siberia. Height 4 ft. to 6 ft. Introduced in 1818. Flowers white; May and June. Fruit dark; ripe in Sept.

569 *R.* grandiflòra.

Differs from *R.* spinosíssima, though scarcely so much as to render it a distinct species.

R. aciculàris *Lindl.*, and *R.* oxyacántha *Bieb.*, are described in our first edition.

C. *Species Natives of North America and Siberia.*

20. *R.* LUTE'SCENS *Pursh.* The yellow *American* Rose.

Identification. Pursh Fl. Amer. Sept., vol. 2. p. 735.; Lindl. Ros., p. 47.; Don's Mill., 2. p. 568.
Synonyme. *R.* híspida *Curt. Bot. Mag.* t. 1570.
Engravings. Lindl. Ros., t. 9.; Bot. Mag., t. 1570.; and our *fig.* 570.

570. *R. lutéscens.*

Spec. Char., &c. Prickles of the branches crowded, unequal, slender, reflexed; of the branchlets, small and nearly equal. Leaflets flat, glabrous, simply serrated. (*Don's Mill.*) An erect shrub. North America and Siberia. Height 4 ft. to 6 ft. Introduced in 1780. Flowers pale yellow; May and June. Fruit large, ovate, black.

A very distinct variety, or probably species, well deserving a place in botanical collections.

21. *R.* MYRIACA'NTHA *Dec.* The myriad-prickled Rose.

Identification. Dec. Fl. Fr., 4. p. 439.; Don's Mill., 2. p. 562.
Synonymes. *R. parvifòlia Pall. Ros.* 62. ?; *R. provinciàlis Bieb. Fl. Taur.* 1. p. 396. ?; *R. spinosíssima* var. η myriacántha *Ser.* in *Dec. Prod.* 2. p. 608.
Engravings. Lindl. Ros., t. 10.; and our *fig.* 571.

Spec. Char., &c. Prickles unequal, larger ones dagger-formed. Leaflets glandular, glabrous, orbicular. (*Don's Mill.*) A diminutive spiny shrub. Siberia, Tauria, and also Dauphiné, and near Montpelier. Height 1 ft. to 2 ft. Flowers white; May and June. Fruit dark; ripe in September.

Shoots simple and erect, resembling, in many respects, *R. spinosíssima* in a stunted state.

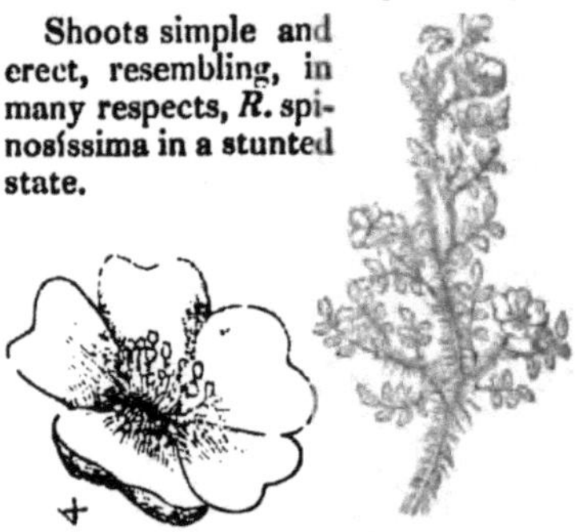

571. *R. myriacántha.*

572. *R. revérsa.*

22. *R.* REVE'RSA *Waldst. et Kit.* The reversed-*prickled* Rose.

Identification. Waldst. et Kit. Hung., 3. p. 293.; Don's Mill., 2. p. 569.
Engravings. Waldst. et Kit. Hung., 3. t. 264.; our *fig.* 572.

Spec. Char., &c. Prickles setaceous, nearly equal, reflexed. Leaves doubly serrated, pubescent. Fruit hispid. (*Don's Mill.*) A large rambling shrub. Hungary, on the mountains of Matra, in stony places. Height 2 ft. to 5 ft. Introduced in 1816. Flowers solitary, white, tinged with pink; June and July. Fruit ovate, dark purple; ripe in September.

D. *Species Natives of Britain.*

23. *R.* SPINOSI'SSIMA *L.* The most spiny, *or Scotch*, Rose.

Identification. Lin. Fl. Suec., 442.; Sp., 491.; Don's Mill., 2. p. 568.
Engravings. Eng. Bot., t. 187.; Hayne Abbild., t. 37.; and our *fig.* 573.

Spec. Char., &c. Prickles unequal. Leaflets flat, glabrous, simply serrated. (*Don's Mill.*) A dwarf compact bush, with creeping suckers. Flowers small, solitary, white or blush-coloured. Fruit ovate, or nearly round, black or dark purple. A very spiny shrub. Europe; plentiful in Britain. Height 1 ft. to 2 ft. Flowers white or blush; May and June. Fruit purple or black; ripe in Sept.

73. *R. spinosíssima.*

Varieties. A great many varieties, cross-breds, and hybrids have been raised of this rose, with flowers double, semidouble, white, purple, red, and even yellow. The first double variety was found in a wild state, in the neighbourhood of Perth, by Mr. Brown of the Perth Nursery, who raised a number of others from seed. Mr. Austin of the Glasgow Nursery also raised upwards of 50 select varieties; and, subsequently, the number of these varieties for sale in the nurseries has become so great, and they are changing their names so often, that it would be useless to attempt to give a list of them in this work. In Mr. Rivers's Abridged List of Roses, in the *Rose Amateur's Guide*, the following sorts are recommended: Erebus, Guy Mannering, La Neige, Lady Baillie, Queen of May, True yellow, William the Fourth, and Venus.

■ 24. *R.* RUBE'LLA *Smith.* The reddish Rose.

Identification. Smith Eng. Bot., 2521.; Don's Mill., 2. p. 567.
Engravings. Eng. Bot., 2521.; and our *fig.* 574.

574. *R. rubélla.*

Spec. Char., &c. Prickles slender, straight, crowded. Fruit globose. Leaflets glabrous. Peduncles bristly. (*Don's Mill.*) A low shrub, with divaricating branches. England, in Northumberland, on the sandy sea coast. Height 2 ft. to 3 ft. Flowers either blush-coloured, or white blotched with pink, delicately fragrant; July. Fruit bright scarlet; ripe in September.

A rare species, nearly allied to *R.* spinosíssima

■ 25. *R.* HIBE'RNICA *Smith.* The Irish Rose.

Identification. Smith in Eng. Bot., 2196.; Don's Mill., 2. p. 569.
Engravings. Eng. Bot., t. 2196.; and our *fig.* 575.

Spec. Char., &c. Prickles unequal, slightly hooked, smaller ones bristle-formed. Leaflets ovate, acute, simply serrated, with the ribs hairy beneath. Sepals pinnate. Fruit nearly globular, smooth, as well as the peduncles. (*Don's Mill.*) A prickly shrub. Ireland, in the counties of Derry and Down, in thickets. Height 4 ft. to 6 ft. Flowers small, light bluish; June to November. Fruit orange-coloured; ripe Sept.

575. *R. hibérnica*

■ 26. *R.* WI'LSONI Borr. Wilson's Rose.

Identification. Hook. Brit. Flor., p. 228.; Eng. Bot. Suppl., 2723.; Don's Mill., 2. p. 570.
Engravings. Eng. Bot., t. 2723.; and our *fig.* 576.

Spec. Char., &c. Prickles crowded, unequal, straight, intermixed with setæ. Leaflets simply serrated, hairy, their disks glandless. Sepals simple. Fruit nearly globular. (*Don's Mill.*) England, near Bangor Ferry. Height 2 ft. to 4 ft. Flowers beautiful dark pink; June and July.

576. *R. Wilsoni.*

■ 27. *R.* INVOLU'TA *Smith.* The involute-*petaled* Rose.

Identification. Smith in Eng. Bot., 2068.; Don's Mill., 2. p. 569.
Synonyme. *R. nivàlis Donn Hort. Cant.* ed. 1. p. 170.
Engravings. Eng. Bot., t. 2068.; and our *fig.* 577.

Spec. Char., &c. Prickles very unequal, and very much crowded. Leaflets doubly serrated, pubescent. Petals convolute. Fruit prickly. (*Don's Mill.*) Petals pale red, concave. Hebrides, in the Isle of Arran, and in Glen Lyon. A low shrub. Height 2 ft. to 3 ft. Flowers pale red; June. Fruit black; ripe in September.

577. *R. involùta.*

28. *R.* Sabi'ni Woods. Sabine's Rose.

Identification. Woods in Lin. Trans., 12. p. 188.; Don's Mill., 2. p. 569.
Engravings. Borr. in Eng. Bot. Suppl., t. 2595.; and our *fig.* 578.

578. *R.* Sabíni.

Spec. Char., &c. Peduncles, calyx, fruit, and branches bristly. Prickles scattered, straightish. Leaflets doubly serrated, nearly smooth, with hairy ribs. Sepals pinnate. (*Don's Mill.*) Flower stalks rather aggregate. Petals fine red. Fruit ovate, bright scarlet. An upright branchy shrub. Native of Scotland, near Dunkeld; of England, in Cumberland, Northumberland, and Yorkshire. Height 5 ft. to 8 ft. Flowers red; July. Fruit bright scarlet; ripe in September.

Variety. A plant in the Horticultural Society's Garden is named *R.* S. grácilis.

29. *R.* Donia'na Woods. Don's Rose.

Identification. Woods in Lin. Trans., 12. p. 185.; Don's Mill., 2. p. 579.
Synonyme. *R.* Sabíni β *Lindl. Ros.* p. 59.
Engravings. Eng. Bot. Sup., t. 2601.; and our *fig.* 579.

579. *R.* Doniána.

Spec. Char., &c. Peduncles bractless, bristly, as well as the globular fruit and calyx. Stem bristly and prickly, like the downy petioles. Leaflets elliptical, doubly and sharply serrated, hairy on both sides. Petals spreading. (*Don's Mill.*) Segments of the calyx simple. A large shrub. Highlands of Scotland, particularly on the mountains of Clova, Angusshire. Height 4 ft. to 5 ft. Flowers pink; June and July. Fruit red; ripe in September.

Variety. There is a variety in the Hort. Soc. Garden, called *R.* D. hórrida; and Dr. Lindley remarks that *R.* Wílsoni Borr. (Eng. Bot. Supp. t. 2723.), *R.* Sabiniàna, and *R.* involùta are all some of the endless varieties of *R.* móllis, our *R.* villòsa, No. 35. (*Comp. Bot. Mag.*, i. p. 189.)

This rose was named in honour of Mr. Don of Forfar.

§ v. *Centifòliæ* Lindl.

Derivation. From *centum*, a hundred, and *folium*, a leaf; because the species contained in this section agree in character with the hundred-leaved rose, which is so double as to seem to have a hundred petals.

Sect. Char., &c. Shrubs all bearing bristles and prickles. Peduncles bracteate. Leaflets oblong or ovate, wrinkled. Disk thickened, closing the throat. Sepals compound. — This division comprises the portion of the genus *Ròsa* which has most particularly interested the lover of flowers. It is probable that the earliest roses of which there are any records of being cultivated belonged to this section; but, to which particular species those of Cyrene or Mount Pangæus are to be referred, it is now too late to enquire. The attar of roses, which is an important article of commerce, is either obtained from roses belonging to this division indiscriminately, as in the manufactory at Florence, conducted by a convent of friars; or from some particular kind, as in India. (*Don's Mill.*, ii. p. 571. adapted.) Deciduous bushes, generally erectish; natives of Syria, Caucasus, and Middle of Europe.

30. *R.* damasce'na *Mill.* The Damascus, *or Damask*, Rose.

Identification. Mill. Dict., No. 15.; Don's Mill., 2. p. 571.
Synonymes. *R.* bélgica *Mill. Dict.* No. 17.; *R.* calendàrum *Munch. Hausv.* ex *Bork. Holz.* 330., *Rossig Ros.* t. 8. and t. 33.; *R.* bífera *Poir. Suppl.* 6. p. 276., *Red. Ros.* 1. p. 107. and p. 121.; Rose à quatre Saisons.
Engravings. Redout. Ros., 1. t. 58.; our *fig.* 580. of *R.* d. coccínea; and our *fig.* 581. of *R.* d. subálba.

Spec. Char., &c. Prickles unequal, larger ones falcate. Sepals reflexed

Fruit elongated. (*Don's Mill.*) A bushy shrub. Syria. Height 2 ft. to 8 ft. Introduced in 1573. Flowers large, white or red, single or double; June and July. Fruit red; ripe in September.

Varieties. There are nearly 100 varieties which are classed under this species; but it is very doubtful, whether many of them are not hybrids between this and other sorts. Among the names of the varieties classed under this head are, the monthly blush; the blush damask; the red and white damask; the red and white monthly; the incomparable; the crimson perpetual; and, perhaps, the handsomest variety of the species, the quatre saisons, of which there are six or eight subvarieties; the royal; and the York and Lancaster.

580. *R.* d. ceocinea.

581. *R.* damascèna subálba.

The present species may be distinguished from *R.* centifòlia by the greater size of the prickles, the greenness of the bark, the elongated fruit, and the long reflexed sepals. The petals of this species, and all the varieties of *R.* centifòlia, as well as those of other species, are employed indiscriminately for the purpose of making rose-water. *R.* damascèna is extremely beautiful, from the size and brilliant colour of its flowers.

ꝏ 31. *R.* CENTIFO'LIA *Lin.* The hundred-petaled, *Provence, or Cabbage,* Rose.

Identification. Lin. Sp., 704.; Don's Mill., 2. p. 571.

Synonymes. *R.* provinciàlis *Mill. Dict.* No. 18.; *R.* polyánthos *Rossig. Ros.* t. 35.; *R.* caryophýllea *Poir. Suppl.* 6. p. 276.; *R.* unguiculàta *Desf. Cat.* 175.; *R.* vàrians *Pohl Bohem.* 2. p. 171.

Engravings. Rossig. Ros., t. 1.; Red. Ros., 1. p. 25. t. 1.; and our *fig.* 582. of the double-flowered variety.

582. *R.* centifòlia.

Spec. Char., &c. Prickles unequal, larger ones falcate. Leaflets ciliated with glands. Flowers drooping. Calyxes clammy. Fruit oblong. (*Don's Mill.*) A bushy shrub. Eastern Caucasus, in groves. Height 3 ft. to 6 ft. Introduced in 1596. Flowers white or red; single, but most commonly double; June and July. Fruit red; ripe in September.

Varieties. Above 100 varieties are assigned to this species, which are classed in three divisions: —

ꝏ R. *c.* 1 *provinciàlis* Mill.; *the Provence,* or Cabbage, *Roses;* among which are the royal and cabbage blush; the carmine; the cluster; the Duchesse d'Angoulême, a very handsome white rose; the Provence, of which there are upwards of twenty subvarieties; the prolific; the striped nosegay; and the Versailles.

583. *R. c.* cristàta.

- R. *c.* 2 *muscòsa* Mill., *the Moss Roses;* among which are the common single (*fig.* 583.), the common double, the blush, the dark, the striped, the white, and the crested moss (*R.* c. *m.* cristàta), and many others.
- R. *c.* 4 pompònia *Dec.*, *the Pompone Roses* N. Du Ham.; *R. pompònia* Redouté Ros. p. 65.; among which are the well-known rose de Meaux, an old inhabitant of the gardens; the mossy de Meaux, the dwarf, and small Provence; the rose de Rheims; and the common and proliferous pompone. These roses should be cut down every year, when they have done flowering, that they may send up new shoots every spring to produce flowers. If this be not done, the principal branches will dry up, and become bare like those of the bramble.

This species is distinguished from *R.* damascèna by the sepals not being reflexed, and the flowers having their petals curved inwards, so as, in the double state, to give the flower the appearance of the heart of a cabbage, whence the name of the cabbage rose. Its fruit is either oblong or roundish, but never elongated. From *R.* gállica it is distinguished by the flowers being drooping, and by the larger size of the prickles, with a more robust habit.

32. *R.* GA'LLICA *Lin.* The French Rose.

Identification. Lin. Sp., 704.; Dec. Prod., 2. p. 603.
Synonymes. *R.* centifòlia *Mill. Dict.* No. 41.; *R.* sylvática *Gater. Mont.* p. 94.; *R.* rùbra *Lam. Fl. Fr.* 3. p. 130.; *R.* holosericea *Rossig. Ros.* t. 18.; *R.* bélgica *Brot. Fl. Lus.* 1. p. 338.; *R.* blánda *Brot.* l. c.; Rose de Provins, *Fr.*; Essig Rose, *Ger.*
Engravings. Mill. Ic., t. 221. f. 2.; Red. Ros., 1. t. 25.; our *fig.* 584. of the species; and *fig.* 855., which is of the variety called the Bishop Rose.

584. *R.* gállica.

Spec. Char., &c. Prickles unequal. Stipules narrow, divaricate at the tip. Leaflets 5—7, coriaceous, rigid, ovate or lanceolate, deflexed. Flower bud ovate-globose. Sepals spreading during the time of the flowering. Fruit subglobose, very coriaceous. Calyx and peduncle more or less hispid with glanded hairs, somewhat viscose. A species allied to *R.* centifòlia *L.*, but with round fruit, and very coriaceous leaflets, with more numerous nerves, that are a little prominent, and are anastomosing. (*Dec. Prod.*) A bushy shrub. Middle of Europe and Caucasus, in hedges. Height 2 ft. to 3 ft. Introduced ?. Flowers red, crimson, or white, single or double; June and July. Fruit red; ripe in August.

Varieties. The varieties of this species are very numerous; some of the principal are, the cramoisie, royal crimson, black damask, Fanny Bias, Flanders, giant, gloria mundi, grand monarque, the Dutch, the blush, the bishop (*fig.* 585.), and Singleton's, all old favourites in our gardens; Malta, marbled, several subvarieties; mignonne, six or eight sorts; Morocco, negro, mottled black, Ninon de l'Enclos, Normandy; officinal, or the rose of the shops, several varieties; purple, 14 sorts; poppy; velvet, several

585. *R. gállica.*

kinds; ranunculus, rosa mundi, sultana; Tuscany; the *Ròsa g.* parvifòlia, our *fig.* 586. The village maid, a striped rose, introduced by Mr. Rogers of Southampton, probably belongs to this species. Besides these, there are numerous distinct varieties, which will be found described in our first edition.

586. *R. g.* parviflòra.

The petals of some of the varieties of this rose are used in medicine, particularly of that called officinal; which, though not so fragrant as those of the Dutch hundred-leaved rose, another variety, are preferred for their beautiful colour and their pleasant astringency.

§ vi. *Villòsæ.*

Derivation. From *villosus*, villous; in allusion to the hairiness of the species.

Sect. Char. Surculi erect. Prickles straightish. Leaflets ovate or oblong, with diverging serratures. Sepals connivent, permanent. Disk thickened, closing the throat. — This division borders equally close upon those of Canìnæ and Rubiginòsæ. From both it is distinguished by its root-suckers being erect and stout. The most absolute marks of difference, however, between this and Canìnæ, exist in the prickles of the present section being straight, and the serratures of the leaves diverging. If, as is sometimes the case, the prickles of this tribe are falcate, the serratures become more diverging. The permanent sepals are another character by which this tribe may be known from Canìnæ. Rubiginòsæ cannot be confounded with the present section, on account of the unequal hooked prickles, and glandular leaves, of the species. Roughness of fruit and permanence of sepals are common to both. (*Don's Mill.*, ii. p. 576.) Deciduous shrubs, mostly with erectish branches. Natives of Middle Europe, or Britain.

A. *Natives of Middle Europe, not of Britain.*

■ 33. *R.* TURBINA'TA *Ait.* The turbinate-*calyxed, or Frankfort*, Rose.

Identification. Ait. Hort. Kew., ed. 1. vol. 2. p. 206.; Dec. Prod., 2. p. 603.; Don's Mill., 2. p. 576.

Synonymes. *R.* campanulàta *Ehrh. Beitr.* 6. p. 97.; *R.* francofortiàna *Munch. Hausv.* 5. p. 24.: *R.* francfurténsis *Rossig. Ros.* t. 11.

Engravings. Jacq. Schönbr., 4. t. 415.; Lawr. Ros., t. 69.; and our *fig.* 587.

587. *R. turbinàta.*

Spec. Char., &c. Stem nearly without prickles. Branches smooth. Leaflets 5—7, ovate-cordate, large, wrinkled in a bullate manner, serrate, approximate, a little villous beneath. Stipules large, clasping the stem or branch. Flowers disposed subcorymbosely, large, violaceous red. Peduncles wrinkled and hispid. Calyx turbinate, smoothish. Sepals undivided, subspathulate. (*Dec. Prod.*) An erectish shrub. Germany. Height 4 ft. to 6 ft. Introduced in 1629. Flowers large, red, and loose; June and July.

Varieties. R. *t.* 1 *francofurtàna* Ser., and R. *t.* 2 *orbessàna* Ser., are the commonest forms of this species.

34. *R.* A'LBA *Lin.* The *common* white Rose.

Identification. Lin. Sp., 705.; Lawr. Ros., t. 23. 25. 32. 37.; Œd. Fl. Dan., t. 1215.; Red. Ros., 1. p. 97. and p. 17.; Don's Mill., 2. p. 577.
Synonyme. *R.* usitatissima *Gat. Montaub.* t. 94.
Engravings. Lawr. Ros., t. 23. 25. 32. 37.; Œd. Fl. Dan., t. 1215.; Red. Ros., 1. p. 17. and p. 97.; Gat. Montaub., t. 94.; our *fig.* 588. of the species; and *fig.* 589. of the double variety common in gardens.

588. *R.* álba.

589. *R.* álba.

Spec. Char., &c. Leaflets oblong, glaucous, rather naked above, simply serrated. Prickles straightish or falcate, slender or strong, without bristles. Sepals pinnate, reflexed. Fruit unarmed. (*Don's Mill.*) A large shrub. Piedmont, Denmark, France, and Saxony. Height 4 ft. to 10 ft. Introduced in 1597. Flowers large, either white, or of the most delicate blush colour, with a grateful fragrance; June and July. Fruit oblong, scarlet, or blood-coloured; ripe in September.

Varieties. The garden varieties are very numerous; and some of the most beautiful are the double, semidouble, and single blush; the celestial, a well-known favourite; the great, small, and cluster maiden's blush; the double thornless; and the double, semidouble, and single white. The rose blanche à cœur vert, the bouquet blanc, and the blanche de la Belgique. are well-known and beautiful varieties of this species.

B. *Natives of Europe and Britain.*

35. *R.* VILLO'SA *Lin.* The villous-*leaved* Rose.

Identification. Lin. Sp., 704.; Don's Mill., 2. p. 576.
Synonymes. *R.* móllis *Smith* in *Eng. Bot.* t. 2459.; *R.* tomentòsa β *Lindl. Ros.* p. 77; *R.* heterophýlla *Woods* in *Lin. Trans.* 12. p. 195.; *R.* pulchélla *Woods* l. c. p. 196.; *R.* pomífera *Herm. Diss.* 16.
Engravings. Eng. Bot., t. 2459.; and our *fig.* 590.

Spec. Char., &c. Leaflets rounded, bluntish, downy all over. Fruit globose, rather depressed, partly bristly. Sepals slightly compound. Branches without bristles. (*Don's Mill.*) A rambling shrub. Europe, in hedges; in Britain, in bushy rather mountainous situations in Wales, Scotland, and the North of England. Height 5 ft. to 6 ft. Flowers red or pink; July. Fruit purple; ripe Sept.

Varieties. *R.* grácilis *Woods*, *R.* Sherárdi Davies, *R.* sylvéstris *Lindl.*, are described in our first edition.

A very variable plant. (See No. 29. p. 332.)

590. *R.* villòsa.

36. *R.* TOMENTO'SA *Smith.* The tomentose, *or woolly-leaved*, Rose.

Identification. Smith Fl. Brit., 539.; Eng. Bot., 990.; Don's Mill., 2. p. 576.
Synonymes. *R.* villòsa *Ehrh. Arb.* p. 45., *Du Roi Harbk.* 2. p. 341., *Fl. Dan.*, t. 1458., *R.* mollíssima *Börk. Holz.* p. 307.; *R.* dùbia *Wibel Wirth.* p. 268.; *R.* villòsa β *Huds.* 219.
Engravings. Red. Ros., 2. p. 39. and p. 88.; Eng. Bot., t. 990.; and our *fig.* 591.

591. *R.* tomentòsa.

Spec. Char., &c. Leaflets ovate, acute, more or less downy. Fruit elliptical, hispid. Sepals pinnate.

Prickles slightly curved. Petals white at the base. A rambling shrub. (*Don's Mill.*) Europe, in hedges and thickets; plentiful in Britain. Height 6 ft. Flowers pink; June and July. Fruit scarlet; ripe in September.

§ vii. *Rubiginòsæ* Lindl.

Derivation. From *rubiginosus*, rusty; the leaves of the species being usually furnished with rust-coloured glands beneath.

Sect. Char., &c. Prickles unequal, sometimes bristle-formed, rarely wanting. Leaflets ovate or oblong, glandular, with diverging serratures. Sepals permanent. Disk thickened. Root-shoots arched. The numerous glands on the lower surface of the leaves will be sufficient to prevent anything else being referred to this section; and although *R.* tomentòsa has sometimes glandular leaves, the inequality of the prickles of the species of Rubiginòsæ, and their red fruit, will clearly distinguish them. (*Don's Mill.*, ii. p. 577.) — This division includes all the eglantine, or sweet-briar roses, which are for the most part erect or erectish bushes with deciduous leaves. Natives of Britain, Middle Europe, and Caucasus.

A. *Species Natives of Britain.*

■ 37. *R.* RUBIGINO'SA *Lin.* The rusty-*leaved* Rose, *Sweet Briar, or Eglantine.*

Identification. Lin. Mant., 2. p. 594.; Dec. Prod., 2. p. 604.; Don's Mill., 2. p. 577.
Synonymes. *R.* suavifòlia *Lightf. Scot.* 1. p. 261., *Fl. Dan.* t. 870.; *R. Eglantèria* Mill. Dict. No. 4., Lin. Sp. edit. 1. p. 491.; *R.* agréstis *Savi Fl. Pis.* p. 475.; *R.* rubiginòsa parviflòra *Rau. Enum.* 135.
Engravings. Eng. Bot., t. 991.; Lawr. Ros., t. 41. 61. 65. 72. and 74.; and our *fig.* 592.

Spec. Char., &c. Prickles hooked, compressed, with smaller straighter ones interspersed. Leaflets elliptical, doubly serrated, hairy, clothed beneath with rust-coloured glands. Sepals pinnate, and bristly, as well as the peduncles. Fruit obovate, bristly towards the base. (*Don's Mill.*) A rambling shrub. Europe, and Caucasus; in Britain, in bushy places, on a dry gravelly or chalky soil. Height 4 ft. to 6 ft. Flowers pink; June and July. Fruit scarlet, obovate or elliptic; ripe in September. Leaves sweet-scented when bruised.

592. *R.* rubiginòsa.

Varieties. Eleven are described in our first edition. Some of the best for a rosarium are, the blush, cluster, double, dwarf, semidouble, mossy, scarlet, tree double, and white semidouble.

■ 38. *R.* MICRA'NTHA *Sm.* The small-flowered Rose, *or Sweet Briar.*

Identification. Smith in Eng. Bot., t. 2490.; Don's Mill., 2. p. 578.
Synonyme. *R.* rubiginòsa β micrántha *Lindl. Ros.* p. 87., with erroneous synonymes.
Engravings. Eng. Bot., t. 2490.; and our *figs.* 593, 594.

Spec. Char., &c. Prickles hooked, scattered, nearly uniform. Leaflets ovate, doubly serrated, hairy, glandular beneath. Sepals pinnate. Fruit elliptic, rather bristly, contracted at the summit. Stems straggling. (*Don's Mill.*) A bush with arched shoots, and straggling branches. Britain, in hedges and thickets, chiefly in the South of England. Height 5 ft. to 6 ft. Flowers small, pale red; June and July. Fruit coral red, ovate; ripe in September.

593. 594 *R.* micrántha.

■ 39. *R.* SE'PIUM *Thuil.* The Hedge Rose, *or Briar.*

Identification. Thuil. Fl. Par., 252.; Borr. in Eng. Bot. Suppl., t. 2653.; Don's Mill., 2. p. 578.
Synonymes. *R.* helvética and *R.* myrtifòlia *Hall*; *R.* canina β *Dec. Fl. Fr.* ed. 3. No. 3617.; *R.* agréstis *Savi Fl. Pis.* 1. p. 474.; *R.* biserràta, *R.* macrocárpa, and *R.* stipulàris *Mer. Fl. Par.* 190. ex *Desv.* f. 75.
Engravings. Eng. Bot. Suppl., t. 2653.; and our *fig.* 595.

595. *R. sèpium.*

Spec. Char., &c. Prickles slender. Branches flexuous. Leaflets shining, acute at both ends. Flowers usually solitary. Fruit polished. Sepals pinnate, with very narrow segments. (*Don's Mill.*) A densely branched bush. Europe, in hedges; in England, near Bridport, Warwickshire. Height 4 ft. to 6 ft. Flowers small, pale yellowish pink; June and July. Fruit small, oblong-ovate, scarlet; ripe in September.

40. *R.* INODO'RA. The scentless Rose.

Identification. Eng. Bot. Suppl., 2610.; Hood ed. 2. 232.
Synonymes. *R.* dumetòrum *Eng. Bot.* 2579.; *R.* Bórreri Smith Eng. Fl. 2398., Don's Mill. 2. p. 580.; *R.* rubiginòsa var. inodòra *Lindl. Ros. Monog.* 101.
Engravings. Eng. Bot. 2579.; and our *fig.* 596.

Spec. Char., &c. Prickles hooked. Leaflets ovate; doubly serrated, without glands. Sepals pinnate, often doubly pinnate, deciduous. Flower stalks aggregate, hairy. Fruit elliptical, smooth. (*Don's Mill.*) A stout bush. Britain. Not very unfrequent in hedges and thickets. Height 6 ft. to 7 ft. Flowers pink; June and July. Fruit elliptic, or nearly globose, scarlet.

596. *R. inodòra.*

The foliage has, notwithstanding the specific name, a scent more or less faint, according to the number of glands developed in different individuals; but it resembles rather the turpentine odour of the plants of the preceding section than the fragrance of the sweet briar.

B. *Species Natives of Middle Europe.*

41. *R.* LU'TEA *Dodon.* The yellow *Eglantine* Rose.

Identification. Dodon. Pempt., 187.; Mill. Dict., No. 11.; Don's Mill., 2. p. 577.
Synonymes. *R. Eglantèria* Lin. Sp. 703., Red. Ros. 1. p. 69.; *R.* fœ'tida *Herm. Diss.* 18.; *R.* chlorophylla *Ehrh. Beitr.* 2. p. 69.; *R.* cèrea *Rossig. Ros.* t. 2.
Engravings. Lawr. Ros., t. 12.; Bot. Mag., t. 363.; Red. Ros., 1. p. 69.; Rossig. Ros., t. 2.; and our *fig.* 597.

Spec. Char., &c. Prickles straight. Leaflets deep green. Sepals nearly entire, setigerous. Petals flat, concave. Flowers deep yellow, large, cup-shaped, solitary. Fruit unknown. A branchy shrub. Germany and the South of France. Height 3 ft. to 4 ft. Introduced in 1596. Flowers deep yellow; June.

597. *R. lùtea.*

Varieties.

- R. *l.* 2 *subrùbra* Red. Ros. iii. p. 73., with a fig. — Peduncles rather hispid and glandular. Leaves and petioles glabrous. Stem prickly at the base. Prickles unequal, scattered. Petals of a lurid red above, and yellowish beneath. Stigmas yellow. (*Don's Mill.*)
- R. *l.* 3 *punìcea* Lindl. Ros. p. 84.; *R.* punìcea *Mill. Dict.* No. 12., *Rossig. Ros.* t. 5.; *R.* cinnamòmea *Roth Fl. Germ.* i. p. 217.; *R.* lùtea bícolor *Jacq. Vind.* i. t. 1., *Lawr. Ros.* t. 6., *Bot. Mag.* t. 1077.; *R. Eglantèria* punícea *Red. Ros.* i. p. 71. t. 24.; *R. Eglantèria* bícolor *Dec. Fl. Fr.* iv. p. 437.; and our *fig.* 598.; has the petals scarlet above, and yellow beneath.

598. *R. l. punìcea.*

- R. *l.* 4 *flòre plèno.* Williams's *double yellow Sweet Briar.* — A very beautiful variety, and a free flowerer, raised from seeds by Mr. Williams of Pitmaston. Horticultural Society's Garden.
- R. *l.* 5. *Hóggii D. Don* in *Swt. Brit. Fl. Gard.* t. 410. *Hogg's yellow*

American *Rose*.—Pretty, and a free flowerer. Raised by Mr. Hogg, nurseryman, in New York, from seeds of the single yellow rose. Horticultural Society's Garden.

Other Species belonging to this Section.—*R.* ibérica *Smith*, native of Eastern Iberia; *R.* glutinòsa *Smith*, native of Greece; *R.* Klùkii Bess., native of Tauria; *R.* suavèolens *Pursh*, native of North America; and *R.* Montezùmæ Humb., native of Mexico, are described in our first edition.

§ viii. *Caninæ* Lindl.

Derivation. From *caninus*, belonging to a dog; because *R.* canina is commonly called the dog rose. The name is applied to this section, because all the species contained in it agree in character with *R.* canina.

Sect. Char., &c. Prickles equal, hooked. Leaflets ovate, glandless or glandular, with the serratures conniving. Sepals deciduous. Disk thickened, closing the throat. Larger suckers arched. (*Don's Mill.*) Deciduous, but some sub-evergreen.—Chiefly bushes, but partly sarmentose and procumbent. Natives of Britain, Middle Europe, and Asia.

A. *Species Natives of Britain.*

42. *R.* CANI'NA *Lin.* The *common* Dog Rose.

Identification. Lin. Sp., 704.; Don's Mill., 2. p. 579.
Synonymes. *R.* dumàlis *Bechst. Forstb.* 241. and 939. ex *Rau.*; *R.* andegavénsis *Bat. Fl. Main. et Loir.* 189., *Red. Ros.* 2. p. 9. t. 3.; *R.* glaùca *Lois.* in *Desv. Journ.*; *R.* arvénsis *Schrank Fl. Mon.*; *R.* glaucéscens *Mer. Par.*; *R.* nitens *Mer.* l. c.; *R.* teneriffénsis *Donn Hort. Cant.* ed. 8. p. 169.; *R.* senticòsa *Achar. Acad. Handl.* 34. p. 91. t. 3.
Engravings. Eng. Bot., t. 992.; Lawr. Ros., t. 21. 29.; and our *fig.* 599.

599. *R.* canina.

600. *R. c.* aciphylla.

Spec. Char., &c. Prickles strong, hooked. Leaflets simply serrated, pointed, quite smooth. Sepals pinnate. Fruit ovate, smooth, or rather bristly, like the aggregate flower stalks. (*Don's Mill.*) A rambling shrub. Europe generally, and the North of Africa; plentiful in Britain, in hedges, woods, and thickets. Height 6 ft. to 10 ft. Flowers rather large, pale red, seldom white; June and July. Fruit scarlet; ripe in September.

Variety.

R. c. 2 *aciphýlla* Lindl. Ros. p. 99.; *R.* aciphýlla *Rau.* 69. with a fig., *Red. Ros.* ii. p. 31. t. 13.; and our *figs.* 600, 601.; is a very remarkable variety, from the straightness of its shoots, and its singular habit of growth. The leaves are smooth on both surfaces, and the flowers are smaller than those of the species.

601. *R. c.* aciphylla.

Other Varieties. Seventeen are described in our first edition.

Fruit ovate, bright scarlet, of a peculiar and very grateful flavour, especially if made into a conserve with sugar. The pulp of the fruit besides saccharine matter, contains citric acid, which gives it an acid taste. The pulp, before it is used, should be carefully cleared from the nuts or seeds. Numerous varieties.

602. *R.* Försteri.

43. *R.* FO'RSTERI Sm. Forster's *Dog* Rose.

Identification Smith Eng. Fl., 2. p. 392.; Bor. in Eng. Bot. Suppl., 2611.; Don's Mill., 2. p. 580.
Synonyme. *R.* collina β and γ *Woods* in *Lin. Trans.* 12. p. 392.
Engravings. Eng. Bot Suppl., t. 2611.; and our *fig.* 602.

Spec. Char., &c. Prickles scattered, conical, hooked.

Leaflets simply serrated, smooth above, but hairy on the ribs beneath. Sepals doubly pinnate. Fruit elliptical, smooth, like the aggregate flower stalks. (*Don's Mill.*) A large shrub. Native of Europe, in hedges; plentiful in England. Height 6 ft. to 8 ft. Flowers pale red; June and July. Fruit scarlet; ripe in September.

❧ 44. *R.* DUMETO'RUM *Thuill.* The Thicket *Dog* Rose.

Identification. Thuil. Fl. Par., 250.; Bor. in Eng. Bot. Suppl., t. 2610.; Don's Mill., 2. p. 580.
Synonymes. *R.* leucántha β acutifòlia *Bast.* in *Dec. Fl. Fr.* 5. p. 535.; *R.* sèpium *Borkh.* ex *Rau. Enum.* 79.; *R.* solstitiàlis *Bess. Prim. Fl. Gall.* 324.; *R.* corymbífera *Gmel. Fl. Bad. Als.* 2. p. 427.
Engravings. Eng. Bot. Suppl., t. 2610.; and our *fig.* 603.

603. R. dumetòrum.

Spec. Char., &c. Prickles numerous, scattered, hooked. Leaflets simply serrated, hairy on both surfaces. Sepals pinnate, deciduous. Peduncles aggregate, slightly hairy Fruit elliptical, smooth, as long as the bracteas. (*Don's Mill.*) A large shrub. Europe, in hedges; and found in England, in the southern counties, but seldom in any abundance. Height 4 ft. to 6 ft. Flowers reddish; June and July. Fruit scarlet; ripe in September.

❧ 45. *R.* SARMENTA'CEA *Swartz.* The sarmentaceous *Dog* Rose.

Identification. Swartz MSS.; Woods in Lin. Trans., 12. p. 213.; Don's Mill., 2. p. 580.
Synonymes. *R.* glaucophýlla *Winch Geogr. Distrib.* 45.; *R.* canìna *Roth Fl. Germ.* 2. p. 560.
Engravings. Curt. Lond., fasc. 5. t. 34.; and our *fig.* 604.

604. R. sarmentàcea.

Spec. Char., &c. Prickles hooked. Leaflets ovate, doubly serrated, smooth, glandular. Peduncles aggregate, smooth or minutely bristly. Sepals pinnate, deciduous. Fruit broadly elliptic, naked. (*Don's Mill.*) A rambling shrub. Europe, common in hedges and bushy places; plentiful in Britain. Height 8 ft. to 10 ft. Flowers pink, and fragrant; June and July. Fruit scarlet; ripe in September.

The fruit is as grateful to the palate, probably, as that of *R.* canìna, with which this equally common plant is generally confounded. This is the species most commonly made choice of as a stock for garden roses.

❧ 46. *R.* CÆ'SIA *Sm.* The grey *Dog* Rose.

Identification. Smith Eng. Bot., t. 2367.; Don's Mill., 2. p. 580.
Synonymes. *R.* canìna pubéscens *Afz. Ros. Suec. Tent.* 1. p. 2.; *R.* canìna ζ cæ'sia *Lindl. Ros* p. 99.
Engravings. Eng. Bot., t. 2367.; and our *fig.* 605.

605. R. cæ'sia.

Spec. Char., &c. Prickles hooked, uniform. Leaflets elliptical, somewhat doubly serrated, glaucous, hairy beneath, without glands. Sepals distantly pinnate, deciduous. Flower stalks smooth, solitary. Fruit elliptical, smooth. (*Don's Mill.*) A rambling shrub. Scotland, in the Highland valleys, but rare; at Taymilt, in Mid-Lorn, Argyleshire; and in Strath Tay, between Dunkeld and Aberfeldie, and by the side of Loch Tay. Height 4 ft. to 5 ft. Flowers generally of a uniform carnation hue, but occasionally white; July. Fruit scarlet; ripe in September.

B. *Species Natives of Middle Europe.*

❧ 47. *R.* RUBRIFO'LIA *Vill.* The red-leaved *Dog* Rose.

Identification. Vill. Dauph., 3. p. 549.; Don's Mill., 2. p. 581.
Synonymes. *R.* multiflòra *Reyn. Act. Laus.* 1. p. 70. t. 6.; *R.* rubicúnda *Hall. fil.* in *Roem. Arch.* 3. p. 376.; *R.* lùrida *Andr. Ros.*; *R.* cinnamòmea γ rubrifòlia *Red. Ros.* 1. p. 134.
Engravings. Bell. in Act. Taur., 1790, p. 229. t. 9.; Jacq. Fragm., 70. t. 106.; Red. Ros., 1 p. 32. t. 4.; Lindl. in Bot. Reg., t. 430.; and our *fig.* 606.

Spec. Char., &c. Prickles small, distant. Leaflets ovate, and, as well as the branches, glabrous, opaque, discoloured. Sepals narrow, entire. Fruit ovate, globose, smooth. Flowers corymbose. Peduncles smooth. (*Don's Mill.*) A large shrub. Dauphiné, Austria, Savoy, Pyrenees, and Auvergne, in woods. Height 5 ft to 6 ft. Flowers red; June and July. Fruit scarlet; ripe in September.

Stems red. Leaves red at the edges. Flowers small. Sepals narrow, longer than the petals. A shrub, producing a pleasing effect in a shrubbery, from the pinkness of its foliage. At the funeral of Villars, who first named and described this rose, branches and flowers of it were cut and strewed over his grave. There are several varieties described in our first edition.

606. *R. rubrifòlia.*

C. *Species Natives of Asia.*

● 48. *R.* CAUCA'SEA *Pall.* The Caucasian *Dog* Rose.

Identification. Pall. Ross., t. 11.; Lindl. Ros., p. 97.; Don's Mill., 2. p. 579.
Synonyme. *R.* leucántha *Bieb. Fl. Taur. Suppl.* 351.?
Engravings. Lind. Ros., t. 11.; and our *fig.* 607.

Spec. Char., &c. Prickles strong, recurved. Leaflets soft, ovate, glaucous. Calyx and peduncles hispid. Sepals simple. Fruit smooth. Flowers large, growing in bunches. (*Don's Mill.*) A robust shrub. Siberia. Height 10 ft. to 12 ft. Introduced in 1798. Flowers white or pale red; June and July. Fruit red; ripe in September.

607. *R. caucàsea.*

608. *R. indica*

609. *R. indica Noisettiàna.*

● 49. *R.* I'NDICA *L.* The Indian, *or common China*, Rose.

Identification. Lin. Sp., 705.; Don's Mill., 2. p. 581.
Synonymes. *R.* sinica *Lin. Syst. Veg.* ed. 13. p. 398.; *R.* semperflòrens cárnea *Rossig. Ros.* t. 19.; *R.* índica chinénsis semiplèna *Ser. Mel.* 1. p. 31.; *R.* reclinàta flòre submultiplici *Red. Ros.* p. 79.; the monthly Rose, the blush China Rose, the Tea-scented Rose; Rosier Indien, Rose Thé, *Fr.*; Indische Rose, *Ger.*
Engravings. Lawr. Ros., t. 26.; Red. Ros., 1. p. 51. t. 142., p. 35. t. 15.; and our *fig.* 608. above.

Spec. Char., &c. Stem upright, whitish, or green, or purple. Prickles stout, falcate, distant. Leaflets 3—5; ovate-acuminate, coriaceous, shining, glabrous, serrulate; the surfaces of different colours. Stipules very narrow, connate with the petiole, almost entire or serrate. Flowers solitary, or in panicles. Stamens bent inwards. Peduncle sub-articulate, mostly thickened upwards, and with the calyx smooth, or wrinkled and bristly. (*Dec. Prod.*) Sub-evergreen. China, near Canton. Height 4 ft. to 20 ft. Introd. in 1789

Flowers red, usually semidouble; April to November. Fruit red; ripe in September.

One of the most valuable of garden roses.

Varieties. There are numerous varieties and hybrids of this beautiful rose. The following are quite distinct; and may each be considered the type of a long list of subvarieties: —

- R. *i.* 2 *Noisettiàna Ser.* in *Dec. Prod.* ii. p. 600., *Don's Mill.* ii. p. 581.; *fig.* 609. above. — Stem firm, and, as well as the branches, prickly. Stipules nearly entire. Flowers panicled, very numerous, semidouble, pale red. Styles exserted. Raised in North America, from an accidental cross between *R.* índica and *R.* moschàta, and the plants being sold to Philip Noisette, his name was given to this variety. It was first brought to England by Mr. Fraser. This well-known and very beautiful rose is almost invaluable in a shrubbery, from its free and vigorous growth, and the profusion of its flowers, which are continually being produced during the whole summer. Numerous subvarieties have been raised of the Noisette rose, some of the most distinct of which are, the R. *i. N. purpùrea* of Redouté, which has red flowers; R. *i. N. nívea*, the Aimé Vil ert of the French nurseries, which has double white flowers; R. *i. N.* Smíthii, Smith's yellow Noisette rose, the flowers of which are very double, of a deeper yellow than the double yellow China rose (*R.* i. ochroleùca), and disposed in clustered corymbs of from 10 to 22, and are highly fragrant.

610. R. i. odoratíssima.

- R. *i.* 3 *odoratíssima* Lindl. Ros. p. 106., Bot. Reg. t. 864., Don's Mill. ii. p. 582.; *R.* odoratíssima *Swt. Hort. Sub. Lond.*; *R.* índica fràgrans *Red. Ros.* i. p. 6. t. 19.; and our *fig.* 610.; the sweetest, or tea-scented, China Rose; Rose à Odeur de Thé, *Fr.*; has semidouble flowers, of a most delicious fragrance, strongly resembling the scent of the finest green tea. There are numerous subvarieties.

611. R. i. longifòlia.

- R. *i.* 4 *longifòlia* Lindl. Ros. p. 106.; *R.* longifòlia *Willd. Enum.* ii. 1079., *Red Ros.* ii. t. 27.; *R.* semperflòrens var. 7. *N. Du Ham.* vii. p. 22.; *R.* salicifòlia *Hort.*; and our *fig.* 611.; has the stems nearly unarmed, and long lanceolate leaflets.
- R. *i.* 5 *pùmila* Lindl. Ros. p. 106. is a dwarf variety, with purplish flowers, having ovate petals.
- R. *i.* 6 *caryophýllea* Red. Ros. iii. p. 59. has the flowers in a kind of panicle, and the leaflets large and thin.
- R. *i.* 7 *pannòsa* Red. has the stems and branches prickly; the leaflets ovate, and red beneath, with the stipules so finely denticulated as to give them somewhat of a fringed or pannose appearance. Flowers drooping a little, purple on the outside, and with the inner petals rose-coloured.
- R. *i.* 8 *cruénta* Red., and Don's Mill. ii. p. 582.; differing from the above principally in having the stems and branches almost unarmed, and the stipules almost entire.
- R. *i.* 9 *Fraseriàna Hort. Brit.* p. 211., and *Don's Mill.* ii. p. 582. — A hybrid, with double pink flowers.

612. R. i. rùga.

■ R. *i.* 10 *rùga* Lindl. Bot. Reg. t. 1389., and our *fig.* 612., has double, blush changging to white, sweet-scented flowers. It is a hybrid between *R.* i. odoratíssima and *R.* arvénsis, brought from Italy, where it was raised by Mr. Clare. It grows freely, making shoots 10 or 12 feet long in a season.

■ R. *i.* 11 *ochroleùca* Bot. Reg. has large cream-coloured flowers, deepening almost into yellow in the centre. It was introduced by Mr. Parks in 1824, and appears to have been since lost.

■ R. *i.* 12 *flavéscens.* — This, Mr. Gordon assures us, is the true tea-scented yellow China Rose, and not the preceding variety, which is generally considered as such, and confounded with it.

■ R. *i.* 13 *Blaírii D. Don* in *Swt. Brit. Fl. Gard.* t. 405., and our *fig.* 613. —Raised in 1830, by Mr. Blair, from seeds of the yellow China rose, which had been fecundated by the pollen of the Tuscan rose. A robust plant, remarkable for the size of its leaves and flowers. Petals purple, but yellow at the base, especially towards the centre of the flower. Fragrant, and a free flowerer.

613. R. i. Blaírii.

■ 50. *R.* SEMPERFLO'RENS *Curt.* The everflowering *China* Rose.

Identification. Curt. Bot. Mag., t. 284.; Smith Exot. Bot., 2. p. 91.; Don's Mill., 2. p. 582.

Synonymes. *R.* diversifòlia *Vent. Cels.* t. 35.; *R.* bengalénsis *Pers. Ench.* 2. p. 50.; *R.* indica *Red. Ros.* 1. p. 49. t. 13.

Engravings. Law. Ros., t. 23.; N. Du Ham., vii. t. 18.; and our *fig.* 614. of a double French variety.

614. R. semperflòrens.

Spec. Char., &c. Branches dark green, armed with scattered, compressed, hooked prickles, and a very few glands. An erectish sub-evergreen shrub. Leaflets 3—5, ovate-lanceolate, crenate-serrated, shining above, but glaucous and slightly setigerous beneath. Sepals compound, narrow. Fruit spherical. (*Don's Mill.*) China. Height 8 ft. to 10 ft. Introduced in 1789. Flowers solitary, single, or semidouble, deep crimson; April to November. Fruit red; ripe September.

There are some very splendid varieties of this species, with semidouble crimson flowers. They are all free growers, and abundant flowerers; and few plants are more ornamental against the walls of a cottage.

ɷ 51. *R.* LAWRENCEA'NA Swt. *Miss* Lawrence's *China* Rose.

Identification. Sweet Hort. Suburb.; Lindl. Ros., p. 110.; Don's Mill., 2. p. 582.

Synonymes. *R.* semperflòrens minima *Sims Bot. Mag.* t. 1762.; *R.* indica var. α acuminàta *Red. Ros.* 1. p. 53.; *R.* indica Lawrenceana Red. Ros. 2. p. 38.

Engravings. Red. Ros., 1. p. 53.; and our *fig.* 615.

615. R. Lawrenceàna.

Spec. Char., &c. Dwarf. Prickles large, stout, nearly straight. Leaflets ovate acute, finely serrated. Petals acuminated. (*Don's Mill.*) A low shrub. China. Height

1 ft. Introduced in 1810. Flowers small, single or semidouble, pale blush; April to November.

The beautiful little plants called Fairy Roses, or Miniature Roses, in Rivers's *Abridged List*, are nearly all varieties of *R.* Lawrenceàna: and they are well worthy of culture, from their extreme dwarfness (often flowering when not more than 6 in. high), and the beautiful colour of their miniature rosebuds, the petals of which appear of a much darker hue than those of the expanded flower. Rivers enumerates five select varieties, of which he says the Gloire des Lawrencianas is one of the prettiest.

616. *R. sericea.*

⚘ 52. *R.* SERI'CEA *Lindl.* The silky Rose.

Identification. Lindl. Ros. Monog., p. 105.; Don's Mill., 2. p. 581.
Engravings. Lindl. Ros. Monog., t. 12.; and our *fig.* 616.

Spec. Char., &c. Prickles stipular, compressed. Leaflets 7—11, oblong, obtuse, serrated at the apex, silky beneath. Flowers solitary, bractless. Sepals entire, ending in long points. (*Don's Mill.*) A shrub. Gossainthan. Height 4 ft. to 6 ft. Flowers ? pale red. Fruit in peduncles, naked.

§ ix. *Sýstylæ* Lindl.

Derivation. From *sun*, together, and *stulos*, a stile; in reference to the styles being connected.

Sect. Char. Styles cohering together into an elongated column. Stipules adnate. The habit of this section is nearly the same as that of the last division. The leaves are frequently permanent. (*Don's Mill.*) Deciduous, evergreen, or sub-evergreen, and mostly climbing. Natives of Britain, Middle Europe, Asia, Africa, and North America.

A. *Species Natives of Britain, and other Parts of Europe.*

⚘ ✶ 53. *R.* SY'STYLA *Bat.* The connate-styled Rose.

Identification. Bat. Fl. Main. et Loir. Suppl., 31.; Don's Mill., 2. p. 582.
Synonymes. *R.* collina *Smith* in *Eng. Bot.* t. 1895.; *R.* stylòsa *Desv. Journ. Bot.* 2. p. 317.; *R.* brevistyla *Dec. Fl. Fr. Suppl.* p. 537.; *R.* bibracteàta *Dec.* l. c.; *R.* sýstyla α ovàta *Lindl. Ros.* p. 111.
Engravings. Eng. Bot., t. 1895.; and our *fig.* 617.

617. *R. sýstyla.*

Spec. Char., &c. Shoots assurgent. Prickles strong, hooked. Peduncles glandular. Sepals pinnate, deciduous, Styles smooth. Floral receptacle conical. (*Don's Mill.*) A rambling shrub. France and England, in hedges and thickets; common in Sussex; in the South of Scotland, on hills. Height 6 ft. to 12 ft. Flowers fragrant, pink or almost white; May to July. Fruit scarlet, ovate oblong.

There are several varieties, but they do not differ materially in appearance from the species.

⚘ ✶ 54. *R.* ARVE'NSIS *Huds.* The Field Rose.

Identification. Huds. Fl. Angl., ed. 1. p. 192.; Dec. Prod., 2. p. 596.
Synonymes. *R.* sylvéstris *Hem. Diss.* p. 10.; *R.* scándens *Mœnch Weiss. Pl.* p. 118.; *R.* herpéthodon *Ehrh. Beitr.* 2. p. 69.; *R.* Hálleri Krok. Siles. 2. p. 150.; *R.* fúsca *Mœnch Meth.* p. 688.; *R.* sérpens *Ehrh. Arbor.* p. 35.; *R.* sempervirens *Rossig. Ros.*; *R.* rèpens *Gmel. Fl. Bad. Als.* 2. p. 418., *Jacq. Fragm.* p. 69. t. 104.; *R.* rámpans *Reyn. Mém. Laus.* 1. p. 69. t. 5.
Engravings. Eng. Bot., t. 188.; Bot. Mag., t. 2054.; and our *fig.* 618.

Spec. Char., &c. Shoots cord-like. Prickles unequal and falcate. Leaves deciduous, and composed of 5—7 glabrous, or indistinctly ciliated, leaflets, glaucescent beneath. Stipules diverging at the tip. Flowers solitary or globose.

Sepals almost entire, short. Styles cohering into an elongated glabrous column. Fruit ovate, or ovate-globose, coriaceous, crimson, glabrous, or a little hispid, as well as the peduncles. (*Dec. Prod.*) Trailing, or climbing, deciduous; in some situations sub-evergreen. Europe, in many places; in England, in hedges and thickets and the borders of fields, chiefly in the midland counties. Stems 20 ft. to 40 ft. Flowers white; July. Fruit dark blood-coloured; ripe in September.

618. *R. arvénsis.*

Varieties. Several varieties are enumerated in catalogues; the following appear distinct, and of general interest: —

⅃ ⁎ ⅃ ⁎ R. *a.* 2 *ayreshírea* Ser. *R.* capreolàta *Neill* in *Edin. Phil. Journ.* No. 3. p. 102. — Sub-evergreen. Prickles slender, very acute. Leaflets ovate, sharply serrated, thin, nearly of the same colour on both surfaces. Peduncles hispid with glanded hairs, or wrinkled. A vigorous-growing climber, producing shoots sometimes 20 ft. in length in one season, and flowering profusely from the middle of May to the middle of September. One of the hardiest of climbing roses, and particularly useful for covering naked walls, or unsightly roofs. Cultivated in British gardens under the name of the Ayrshire Rose.

⅃ ⁎ R. *a.* 3 *hýbrida* Lindl. Ros. 113. has semidouble flowers, of a most delicate flesh colour, and is called, in the nurseries, the double hip rose; the term hip rose being applied by gardeners to the commonest wild roses.

In open situations, a trailing plant, sometimes rooting at the joints; but, in hedges and among bushes, a climber by elongation; reaching to their tops, and covering them with tufts of foliage and flowers; the leaves remaining on late in the season; and the fruit often remaining on all the winter. The shoots are, in general, feeble, much divided, and entangled; and they generally produce, here and there, rugged excrescences, which readily take root.

B. *Species Natives of Middle Europe.*

⅃ ⁎ 55. *R.* (A.) SEMPERVI'RENS *Lin.* The evergreen (Field) Rose.

Identification. Lin. Sp., 704.; Dec. Prod., 2. p. 597.; Don's Mill., 2. p. 583.
Synonymes. *R.* scándens *Mill. Dict.* No. 8.; *R.* baleárica *Desf. Cat. Pers. Ench.* 2. p. 49.; *R.* atrovirens *Viv. Fl. Ital.* 4. t. 6.; *R.* sempervirens globòsa *Red. Ros.* 2., with a fig.; *R.* sempervirens var. *a* scándens *Dec. Fl. Fr.* 5. p. 533.
Engravings. Lawr. Ros., t. 45.; Bot. Reg., t. 459.; and our *fig.* 619.

Spec. Char., &c. Evergreen. Shoots climbing. Prickles pretty equal, falcate. Leaves of 5—7 leaflets, that are green on both sides, coriaceous. Flowers almost solitary, or in corymbs. Sepals nearly entire, longish. Styles cohering into an elongate pilose column. Fruit ovate or ovate globose, orange-coloured. Peduncles mostly hispid with glanded hairs. Closely allied to *R.* arvénsis, but differing in its being evergreen, in its leaves being coriaceous; and in its stipules being subfalcate, and more acute at the tip. (*Dec. Prod.*) A vigorous evergreen climber. France, Portugal, Italy, at Pæstum and other places, Greece, and the Balearic Islands. Stem 20 ft. to 40 ft. Introduced in 1629. Flowers white or pale rose-coloured; June to August. Fruit orange-coloured; ripe in September.

619. *R. sempervirens.*

Varieties. Several varieties are enumerated in catalogues; those which we consider best worth mentioning are, —

⅃ ✱ R. (*a.*) *s.* 2 *Russelliàna*, raised from seed by Mr. Sinclair of the New Cross Nursery. A very strong-growing variety, quite deciduous, with blush flowers.

⅃ ✱ R. (*a.*) *s.* 3 *Clàrei*. *The Rose Clare*. (Bot. Reg., t. 1438.) —An elegant variety, with deep red flowers. Both these varieties are as much entitled to be considered species, as many so designated in this enumeration.

Used for the same purposes as the Ayrshire rose; from which it differs in retaining its leaves the greater part of the winter, and in its less vigorous shoots.

C. *Species Natives of Asia, and One of them of Africa.*

⅃ 56. *R.* MULTIFLO'RA *Thunb.* The many-flowered Rose.

Identification. Thunb. Fl. Jap., 214.; Dec. Prod., 2. p. 598.; Don's Mill., 2. p. 583.
Synonymes. *R.* flàva *Donn Hort. Cant.* ed. 4. p. 121.; *R.* flórida *Poir. Suppl.*; *R.* diffùsa *Roxb.*
Engravings. Bot. Mag., t. 1059.; Bot. Reg., t. 425.; and our *fig.* 620.

Spec. Char., &c. Branches, peduncles, and calyxes tomentose. Shoots very long. Prickles slender, scattered. Leaflets 5—7, ovate-lanceolate, soft, finely wrinkled. Stipules pectinate. Flowers in corymbs, and, in many instances, very numerous. Buds ovate globose. Sepals short. Styles protruded, incompletely grown together into a long hairy column. (*Dec. Prod.*) A deciduous climbing shrub. Japan and China. Stems 10 ft. to 30 ft. Introduced in 1822. Produces a profusion of clustered heads of single, semi-double, or double, white, pale red, or red, flowers in June and July. Fruit bright red; ripe in September.

620. *R. multiflòra.*

Varieties.

⅃ R. *m.* 2 *Grevíllei Hort.* *R.* Roxbúrghii Hort.; *R.* platyphýlla *Red. Ros.* p. 69.; The Seven Sisters Rose. (Our *fig.* 621.) —A beautiful variety, with much larger and more double flowers,

621. *R. multiflòra Grevíllei.*

of a purplish colour. No climbing rose better deserves cultivation against a wall. It is easily known from *R.* multiflòra by the fringed edge of the stipules; while those of the common *R.* multiflòra (*fig.* 621. *a*) have much less fringe, and the leaves are smaller, with the leaflets much less rugose. The form of the blossoms and corymbs is pretty nearly the same in both. A rapid-growing variety, producing shoots 18 ft. to 20 ft. long in a season, flowering profusely for two or three months, but only of three or four years duration.

 R. *m.* 3 *Boursaúlti Hort.*, *Boursault's Rose*, is placed, in Don's *Miller*, under this species; though it differs more from the preceding variety than many species do from each other. It is comparatively a hard-wooded durable rose, and valuable for flowering early and freely This is a very remarkable rose, from its petals having a reticulated appearance.

The species is very distinct, and produces numerous blossoms, which continue expanding for two months. The first variety, when well grown against a wall, forms one of the most beautiful of wall roses. This variety and the species may be considered as rather tender, but they will not thrive under glass.

57. *R.* BRUNO'NII Lindl. Brown's Rose.

Identification. Lindl. Ros. Monog. p. 120. t. 14.; Dec. Prod., 2 p. 598.
Synonyme. *R.* Brównii Spreng. Syst., 2. p. 556.
Engravings. Lindl. Rosar. Monog., t. 14.; and our *fig.* 622.

622. *R.* Brunònii.

Spec. Char., &c. Shoots trailing. Prickles of the stem stout and arched. Leaflets 5—7, lanceolate, pilose on both surfaces; the under one glandulous, and of a different colour from the upper one. Stipules narrow, acute. Inflorescence corymbose. Peduncles and calyxes pilose, and a little hispid. Sepals entire, narrow, and longish. Styles cohering into a very long pilose column. Fruit ovate. Leaves simply serrated. Flowers in terminal bunches. (*Dec. Prod.*) A rambling shrub. Nepal. Stems 10 ft. to 12 ft. Introduced in 1822. Flowers white or pale red; June and July.

58. *R.* MOSCHA'TA *Mill.* The Musk Rose.

Identification. Mill. Dict., No. 13.; Red. et Thor. Ros., 1. p. 33. ic., and p. 99. ic.; Lindl. Rosar. Monog. p. 121.; Dec. Prod., 2. p. 598.; Don's Mill., 2. p. 583.
Synonymes. *R.* opsostémma *Ehrh. Beitr.* 2. p. 72.; *R.* glandulifera *Roxb.*
Engravings. Red. et Thor. Ros., 1. p. 33. ic., and p. 99. ic.; and our *fig.* 623.

623. *R.* moschàta.

Spec. Char., &c. Shoots ascending. Prickles upon the stem slender, recurved. Leaflets 5—7, lanceolate, acuminate, nearly glabrous, the two surfaces of different colours. Stipules very narrow, acute. Flowers, in many instances, very numerous; white, with the claws of the petals yellow; very fragrant. Lateral peduncles jointed, and, as well as the calyx, pilose, and almost hispid. Sepals almost pinnately cut, long. Fruit red, ? ovate. (*Dec.*) A rambling shrub. North of Africa, extending across the continent from Egypt to Mogador; and in Madeira. Stems 10 ft. to 12 ft. Introduced in 1596. Flowers white; July to October.

Varieties

 R. *m.* 2 *flòre plèno* G. Don.—Flowers double.

624. *R. m. nivea.*

♣ R. *m.* 3 *nivea* Lindl. (Bot. Reg., t. 861.; and our *fig.* 624.) *R.* nívea *Dupont*, not of Dec.; *R.* m. ? var. ròsea *Ser.* in *Dec. Prod.* — Leaflets 3—5, ovate-cordate, subacuminate, large. Flowers disposed in an imperfectly corymbose manner. Peduncle and calyx a little hispid. Petals white, or pale rose-coloured, large, obcordate. This is a very beautiful variety: the petals are white, with a most delicate, yet rich, tinge of blush.

♣ ✶ R. *m.* 4 *nepalénsis* Lindl. (Bot. Reg., t. 829.; and our *fig.* 625.) differs from the species in having longer and acuminated sepals. Raised at Claremont, from Nepal seed, in 1824

625. *R. m. nepalénsis.*

Other Varieties. In Rivers's *Abridged List of Roses*, published in 1840, the kinds recommended are, the Fringed, Princesse de Nassau, and Tea-scented; the latter a hybrid, with large flowers of a pure white, with a peculiar habit and perfume.

The branches of the musk rose are generally too weak to support, without props, its large bunches of flowers, which are produced in an umbel-like manner at their extremities; and hence the plants require very little pruning. Being rather tender, it does best against a wall. The musky odour is very perceptible, even at some distance from the plant, particularly in the evening; and this musk gives the peculiar odour to the Persian attar of roses.

D. *Species Natives of North America.*

♣ 59. *R.* rubifo'lia *R. Br.* The Bramble-leaved Rose.

Identification. R. Brown in Ait. Hort. Kew., ed. 2. vol. 3. p. 260.; Lindl. Rosar. Monog., p. 123. ic.; Dec. Prod., 2. p. 598.
Engravings. Lindl. Rosar. Monog., t. 15.; and our *fig.* 626.

Spec. Char., &c. Stems ascending. Branches glabrous. Prickles scattered, falcate. Leaves pubescent beneath. Leaflets 3, ovate-lanceolate, serrate. Stipules narrow, entire. Flowers very small, of a rosy colour, mostly solitary. Buds ovate. Sepals ovate, short, simple. Peduncles and calyxes a little hispid. Styles cohering into a tomentose club-shaped column, as long as the stamens. Fruit pea-shaped. (*Dec.*) A shrub. North America. Height 3 ft. to 4 ft. Introduced in 1830. Flowers pale red; August and September. Fruit scarlet; ripe in October.

626. *R. rubifòlia.*

§ x. *Banks*iànæ *Lindl.*

Identification. Lindl. Ros. p. 125.; Don's Mill., 2. p. 584
Derivation. So called in consequence of all the species contained in this section agreeing in character with *R. Bánksiæ*, a rose named in honour of Lady Banks.

Sect. Char., &c. Stipules nearly free, subulate, or very narrow, usually deciduous. Leaflets usually ternate, shining. Stems climbing. The species of this section are remarkable for their long, graceful, and often climbing shoots, drooping flowers, and trifoliolate shining leaves. They are particularly distinguished by their deciduous, subulate, or very narrow stipules. Their fruit is very variable. (*Don's Mill.*) — Rambling shrubs, deciduous, or sub evergreen; somewhat tender in British gardens, where they only succeed when planted against a wall. Natives of China.

ⱶ 60. *R.* SI'NICA *Ait.* The *trifoliate-leaved* China Rose.

627. *R.* sinica.

Identification. Ait. Hort. Kew., ed. 2. vol. 3. p. 261.; Lindl. Ros., p. 126. t. 16.; Don's Mill., 2. p. 584.
Synonymes. *R.* trifoliàta *Bosc Dict.* ex *Poir.*; *R.* ternàta *Poir. Suppl.* 6. p. 284.; *R.* cherokeénsis *Donn Hort. Cant.* ed. 8. p. 170.; *R.* nívea *Dec. Hort. Monsp.* 137., *Red. Ros.* 2. p. 81. with a fig.; *R.* hýstrix *Lindl. Monog.*; *R.* lævigàta *Michx.*
Engravings. Lindl. Ros., t. 16.; Hook. Bot. Mag., 2847.; Bot. Reg., 1922.; our *fig.* 627. after Redouté; and *fig.* 628. after Bot. Reg.

628. *R.* sinica

Spec. Char., &c. Stipules setaceous, deciduous. Cauline prickles equal, falcate. Petioles and ribs of leaves prickly. Peduncles and fruit beset with straight bristles. Sepals entire, permanent. Flowers white, solitary. Fruit elliptic, orange-red. Disk conical. (*Don's Mill.*) A rambling sub-evergreen shrub. China. Height 6 ft. to 8 ft. Introduced in 1759. Flowers white; May and June. Fruit orange red; ripe in September.

ⱶ 61. *R.* BA'NKSIÆ R. Br. *Lady* Banks's Rose.

Identification. R. Br. in Ait. Hort. Kew., ed. 2. vol. 3. p. 256.; Lindl. Rosar. Monog., p. 131.; Dec. Prod., 2. p. 601.
Synonymes. *R.* Banksiàna Abel Chin. 160.,; *R.* inérmis *Roxb.*?
Engravings. Bot. Mag., t. 1954.; Red. et Thor. Ros., 2. p. 43. ic.; and our *fig.* 629.

Spec. Char., &c. Without prickles, glabrous, smooth. Leaflets 3—5, lanceolate, sparingly serrated, approximate. Stipules bristle-like, scarcely attached to the petiole, rather glossy, deciduous. Flowers in umbel-like corymbs, numerous, very double, sweet-scented, nodding. Tube of the calyx a little dilated at the tip. Fruit globose, black. (*Dec. Prod.*) A climbing deciduous shrub. China. Stems 10 ft. to 30 ft. Introduced in 1807. Flowers white; June and July. Fruit black.

629. *R.* Bánksiæ.

630. *R.* Bánksiæ lùtea.

Variety.

ⱶ R. *B.* 2 *lùtea* Lindl. (Bot. Reg., t. 1105., and our *fig.* 630.) has the flowers of a pale buff colour, and is a very beautiful variety.

This is an exceedingly beautiful and very remarkable kind of rose; the flowers being small, round, and very double, on long peduncles, and resembling in form the flowers of the double French cherry, or that of a small ranunculus, more than those of the generality of roses. The flowers of *R.* Bánksiæ álba are remarkably fragrant; the scent strongly resembling that of violets. Plants of neither variety thrive in the atmosphere of the metropolis.

Some which had attained a large size at Kew, and other places in the neighbourhood of London, were killed by the winter of 1837–8.

62. *R.* MICROCA'RPA *Lindl.* The small-fruited Rose.

Identification. Lindl. Ros. Mon., 130. t. 18.; Doc. Prod., 2. p. 601.
Synonyme. *R.* cymòsa *Tratt. Ros.* 1. p. 87.
Engravings. Lindl. Rosar. Monog., t. 18.; and our *fig.* 631.

631. *R. microcárpa.*

Spec. Char., &c. Prickles scattered, recurved. Leaflets 3—5, lanceolate, shining, the two surfaces different in colour. Petioles pilose. Stipules bristle-shaped or awl-shaped, scarcely attached to the petiole, deciduous. Flowers disposed in dichotomous corymbs. Peduncles and calyxes glabrous. Styles scarcely protruded higher than the plane of the spreading of the flower. Fruit globose, pea-shaped, scarlet, shining. Allied to *R. Bánksiæ.* (*Dec. Prod.*) A rambling sub-evergreen shrub. China, in the province of Canton. Height 8 ft. to 10 ft. Introd. 1822. Flowers white, numerous, small; May to September.

63. *R.* HY'STRIX *Lindl.* The Porcupine Rose.

Identification. Lindl. Ros Monog., p. 129.; Don's Mill., 2. p. 594.
Engravings. Lindl. Ros. Monog., t. 17.; and our *fig.* 632.

632. *R. hýstrix.*

Spec. Char., &c. Prickles on branches unequal, crowded, larger ones falcate, small ones straight. Stipules very narrow, united half way, the free part deciduous. Leaflets three, smooth, ovate, shining, simply serrated, with a few prickles on the middle nerve. Sepals nearly entire; permanent. Fruit bristly. (*Don's Mill.*) A rambling shrub, with flagelliform branches. China and Japan. Flowers large, solitary. Fruit oblong purple.

Other Species and Varieties of Ròsa. — In the catalogue of Messrs. Loddiges, 147 species are registered, of all of which, with one or two exceptions, there are living plants. The garden varieties in the same collection amount to about 1500. There is indeed no end to the garden varieties, new ones being every year raised from seed, and old varieties every year disappearing. New species are also occasionally introduced, and several have been lately raised in the Hort. Soc. Garden from Himalayan seeds. For species we recommend the cultivator to have recourse to the collection of Messrs. Loddiges, and for garden varieties to the most fashionable nurserymen of the time. In Rivers's *Abridged List of Roses*, 1840, he recommends, as a selection for small gardens: — Provence roses, 7; moss roses, 8; hybrid Provence roses, 7; hybrid China roses, 20; French roses, 16; *Ròsa* álba, 9; damask roses, 6; Scotch roses, 8; sweet briars, 5; Austrian briars, 2; Ayrshire roses, 6; *Ròsa* multiflòra, 3; evergreen roses, 7; Boursault roses, 4; Banksian roses, 2; hybrid climbing roses, 4; perpetual roses, 12; Bourbon roses, 7; China roses, 15; tea-scented roses, 11; miniature roses, 5; Noisette roses, 12; musk roses, 3; Macartney roses, 3; *Ròsa* microphýlla, 3. In all, 185 sorts; which would form a very efficient rosarium.

Soil and Situation adapted for Roses.

The common wild roses will grow in very poor soil, provided it be dry; but all the cultivated sorts require a soil naturally light and free, and more or less enriched. The situation should be open and airy, exposed to the east, or, in warm situations, to the north, rather than to the south; because the intensity of the sun's rays accelerates too rapidly the expansion of the flowers, and also diminishes the colour and fragrance of the petals. A rose-garden, fully exposed to the sun during the whole day, may have a useful degree of shade given to it by the distribution of a few standard roses of not less than 8 or 10 feet in height; or by the introduction of frames of wood or wire, in the forms of obelisks, gnomons, crosses, columns surmounted by globes, or cones, on which climbing roses may be trained. These would produce no bad effect by their drip, and yet by their shadow, which would vary with the position of the sun, they would afford a salutary protection to the dwarf roses by which they were surrounded; and thus produce, in some degree, the same object as a cool situation and exposure. The rose is one of those plants that will not thrive in the neighbourhood of towns where the prevailing fuel is pit-coal; hence the roses grown within a circle of ten miles of the metropolis are much inferior in beauty to those grown at double that distance.

In country residences, roses are generally distributed in the margins of shrubberies along with other flowering shrubs: but, considering the culture they require, it is impossible they can thrive in such a situation; and, even if they did thrive, the kind of beauty which they would produce would be of a character so different from that of a general shrubbery, as to require their exclusion from it. The only roses fit to be planted in a shrubbery are the single kinds, in their wild state. Roses, and all other kinds of shrubs or trees, that are far removed from a state of nature, and valued for something produced by art, either in their flowers, fruit, habit, or leaves, should be grown in situations where the art which produced the artificial effect can be employed. Hence all fruit-bearing trees and shrubs should be grown in orchards, in kitchen-gardens, or in some place by themselves, so as to admit of properly cultivating the soil, and managing the plants. Roses, and all double-flowering shrubs, ought, in like manner, to be grown by themselves; and the same principle will apply to shrubs having any peculiarity in their foliage, and even in their mode of growth. The continuation of the peculiarity may not always require a rich soil; on the contrary, it will generally be found to have been produced by a soil and situation of a peculiar nature: but that peculiarity of soil it is as much the object of art to imitate, as it is to form the rich soil, and favourable situation, which produce large or double flowers, or large and succulent fruit, or variegated leaves. Hence, to cultivate roses properly, they must be grown either in groups by themselves on a lawn, or in a flower-garden; or be connected into a system of groups or beds, in a rosarium, or rose-garden. On this subject, and on the pruning, and general treatment of roses, we must refer to the first edition of this work, where it will be found given at great length, illustrated with numerous diagrams, having reference to propagation, training, pruning, the formation of rosariums, for which several plans are given, and the destruction of insects.

Rosarium, or Rosetum.—Where it is intended to plant a collection of roses, the best effect will be produced by devoting a group to each section; such as one to moss roses, another to Noisettes, a third to Scotch roses, &c. These groups ought generally to be planted with dwarfs rather than standards; because the former are more conveniently looked upon by the spectator: but a handsome standard may, frequently, occupy the centre of each group, if it is a circle or a square; and two or three in a line, or radiating from a point, if it is of a long or an irregular form. Sometimes a group may be surrounded by a row of standards, which, in that case, should have clear stems, not less than 7 ft. high, through which the dwarf roses may be seen by persons walking round the group. Standard roses, in general, have the best effect when formed into an avenue along the margin of a walk; and for this purpose they are very suitable for common flower-gardens, where the groups, instead of being planted with dwarf roses, are filled with herbaceous plants. The sizes of the different groups in a rosarium ought to be proportioned to the number of varieties belonging to the section to be planted in each, the bulk which they attain, and their habit of growth. For these purposes, the *Abridged List* of Messrs. Rivers may be taken as a basis; and, as it contains 27 groups, these may be represented by 27 beds of different dimensions.

Genus XIII.

LO'WEA Lindl. The Lowea. *Lin. Syst.* Icosándria Polygýnia.

Identification. Lindl. Bot. Reg., t. 1261.
Synonyme. Ròsa sp. *Pall.* and *Lindl.* in *Ros. Monog.*
Derivation. In compliment to the *Rev. Mr. Lowe*, Travelling Bachelor of the University of Cambridge. (*Lindley* in *Bot. Reg.* t. 1261.)

Gen. Char., &c. *Calyx* with the tube contracted at the mouth. *Petals* 5. *Stamens* and *Carpels* numerous as in *Ròsa*. *Leaf* simple, exstipulate. *Prickles* often compound. (*Lindl.*)

Leaves simple, alternate, exstipulate, deciduous; serrated. *Flowers* yellow, marked with purple. — An undershrub. Native of Persia.

■ 1. L. BERBERIFO'LIA *Lindl.* The Berberry-leaved Lowea.

Identification. Lindley in Bot. Reg., t. 1261.
Synonymes. Ròsa simplicifòlia *Sal. Hort. Allert.* 359., *R. berberifòlia Pall.*, *Lindl. Rosarum Monog.* p. 1. French edition, p. 23., *Dec. Prod.* 2. p. 602., *Spreng. Syst.* 2. p. 546., *Wallroth Monog.* p. 25.
Engravings. Bot. Reg., t. 1261.; Redouté Ros., 1. t. 2.; and our *fig.* 633.

Spec. Char., &c. Leaves undivided, without stipules, obovate-cuneated, serrated at the tip. Prickles decurrent, and of the colour of ivory. Sepals entire, subspathulate. Petals yellow, marked with purple at the base. (*Dec. Prod.*) An undershrub. Persia, near Amadan, where it abounds in saltish soil; and also in fields at the bottom of Mount Elwend, and in the Desert of Soongaria. Height 2 ft. Introd. in 1790. Flowers yellow and purple; June and July. Somewhat difficult of culture, and not a free flowerer; but it is readily propagated by budding on the dog rose, or by seeds, which it produces on the Continent in abundance in common soil.

633. L. berberifòlia.

Varieties. Several are described in *Dec. Prod.*, and some hybrids have recently been raised between this species and some kinds of *Ròsa*.

Sect. V. Po'meæ *Lindl.*

Genus XIV.

CRATÆ'GUS *Lindl.* The Thorn. *Lin. Syst.* Icosándria Di-Pentagýnia.

Identification. Lindl. in Lin. Trans., 13. p. 105.; Dec. Prod., 2. p. 626.; Don's Mill., 2. p. 596.
Synonymes. Cratæ'gus and Méspilus sp. of *Lin.* and others; Néflier, Alisier, and Aubépine, *Fr.*; Doorn, Uzbeer, and Mispel, *Ger.*; Doorn, *Dutch*; Cratægo, *Ital.*; and Espino, *Span.*
Derivation. From *kratos*, strength; in reference to the hardness and strength of the wood.

Gen. Char. *Calyx* with an urceolate tube, and a 5-cleft limb. *Petals* orbicular, spreading. *Ovarium* 2—5-celled. *Styles* 2—5, glabrous. *Pome* fleshy, ovate, closed; the calycine teeth, or the thickened disk, containing a bony putamen. (*Don's Mill.*)

Leaves simple, alternate, stipulate, chiefly deciduous, but in part evergreen; angular or toothed. *Flowers* in corymbs, usually white. *Bracteas* subulate. deciduous. *Fruit* red, yellow, or black. Decaying leaves yellow, or reddish yellow.

Trees or shrubs, small, deciduous, sometimes evergreen; mostly natives of Europe and North America, and some of them of Asia and the North of Africa. One of them, the common hawthorn, is well known throughout the Middle and North of Europe, as a hedge plant. The species all flower and fruit freely; and the wood of all of them is hard and durable, and the plants of considerable longevity. Almost all the flowers are white, and the fruit is generally red; though in some sorts it is yellow, purple, black, or green. All the species ripen fruit in the neighbourhood of London, most of them abundantly; by which, or by grafting or budding on the common hawthorn, they are generally propagated. When the species which have naturally a dwarf habit of growth are intended to assume the character of low trees, they are grafted standard high upon *C. Oxyacántha*, *C.* coccínea, or on some other of the strong-growing kinds; in consequence of which practice, this genus furnishes a greater number of handsome small trees for ornamental grounds than any other ligneous family whatever. All the species will grow on any soil that is tolerably dry; but they will not grow vigorously in a soil that is not deep and free, and rich rather than poor. Whether as small trees or as shrubs, they are all admirably adapted for planting grounds of limited extent; and especially for small gardens in the neighbourhood of large towns.

§ i. *Coccíneæ.*

Sect. Char., &c. Leaves cordate, lobed, acutely serrated. Flowers and fruit large. The plants also large, and of free and vigorous growth.

♀ 1. *C.* COCCI'NEA *L.* The scarlet-*fruited* Thorn.

Identification. Lin. Sp., 682.; Pursh Amer. Sept., 1. p. 337.; Dec. Prod., 2. p. 627.; Don's Mill., 2. p. 599.
Synonymes. *C.* æstivàlis *Booth*; *Méspilus* æstivàlis *Walt. Fl. Car.*; *M.* coccinea *Mill. Nouv. Du Ham.*; thornless American Azarole; Néflier écarlate, *Fr.*; scharlachrothe Mispel, *Ger.*; Lazzeruolo rosso, *Ital.*
Engravings. Pluk., t. 46. f. 4.; Dend. Brit., t. 62; Bot. Mag., t. 3432.; our *fig.* 677. in p. 386.; the plate in Arb. Brit., 1st edit., vol. vi.; and our *fig.* 634.

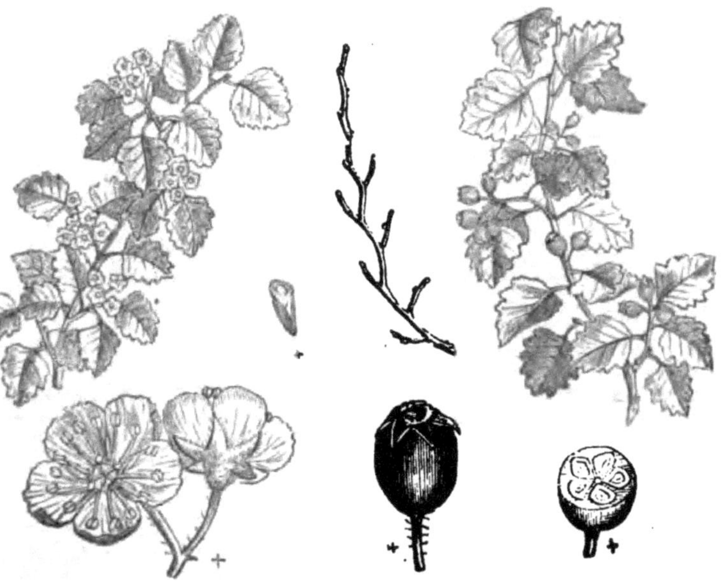

631. Cratæ'gus coccinea.

Spec. Char., &c. Disks of leaves cordate-ovate, angled with lobes, acutely serrated, glabrous. Petioles and calyxes pubescent, glanded. metals or.

biculate. Styles 5. Fruit scarlet, eatable. (*Dec. Prod.*) A low tree. North America, from Canada to Carolina, in hedges and woods. Height 15 ft. to 20 ft. Introduced in 1683. Flowers white; May and June. Fruit large, round, or somewhat pear-shaped, scarlet; ripe in September. Decaying leaves yellow, inclining to scarlet. Naked young wood dark-coloured; old wood with a whitish bark.

Varieties. It would be easy to procure as many varieties of this species as there are of the common hawthorn, by raising some thousands of plants every year from seed, and selecting from the seed-beds plants indicating any peculiarity of leaf or of habit; but, as in the nurseries the most rapid way of producing saleable plants of this, and all the other species and varieties of *Cratæ`gus*, is found to be by grafting on the common hawthorn, very few seedlings are raised, and the varieties in cultivation are only the three or four following: —

- ❦ C. *c.* 2 *corállina.* *C.* corállina *Lodd. Cat.*; the *C.* pyrifórmis and *C.* pectinàta of some collections. (*fig.* 678. in p. 387.) — The leaves and the entire plant are, perhaps, rather smaller than in the species; the habit of the tree is decidedly more upright and fastigiate; and the fruit is smaller, long, and of a fine coral red; whence the name is probably derived, though, in the first edition of the Hort. Soc. *Catalogue*, it is called the red-branched hawthorn. The plants at Messrs. Loddiges's, however, exhibit only a slight degree of redness in the branches of the young wood.
- ❦ C. *c.* 3 *indentàta.* *C.* indentàta *Lodd. Cat.*; *C.* geórgica *Doug.* (*fig.* 678. in p. 387.) — The leaves are smaller, and less lobed, than those of the species; the plant is also weaker, of upright habit, and with a smooth clear bark. It is very prolific in flowers and fruit.
- ❦ C. *c.* 4 *máxima* Lodd. Cat. *C.* c. spinòsa *Godefroy*; *C.* acerifòlia *Hort.*; *C.* ? flabellàta *Hort.* — The leaves are larger than those of any other variety; and the fruit is also large. As we have not seen living plants of *C.* flabellàta, but only dried specimens sent from Terenure and the Humbeque Nursery, we are not absolutely certain that *C.* flabellàta and *C.* c. máxima are the same; but we feel quite certain that they both belong to *C.* coccínea. We are informed that the *C.* flabellàta of some nurseries is *C.* tanacetifòlia; which certainly has its leaves more flabellate, or fan-like, than any variety of *C.* coccínea.
- ❦ ❦ C. *c.* 5 *neapolitàna* Hort. *Méspilus* constantinopolitàna *Godefroy.* — Plants were in Messrs. Loddiges's collection in 1837.

❦ 2. *C.* GLANDULO`SA *W.* The glandular Thorn.

Identification. Willd. Sp., 2. p. 1002., not of Michx.; Pursh Amer. Sept., 1. p. 337.; Dec. Prod., 2. p. 627.; Don's Mill., 2. p. 599.
Synonymes. ? *C.* sanguinea *Pall. Fl. Ros.* 1. t. 11.; ? *Méspilus* rotundifòlia *Ehrh. Beitr.* 3. p. 20.: *Pýrus* glandulòsa *Mœnch*; *C.* rotundifòlia *Booth.*
Engravings. ? Pall. Fl. Ross., 1. t. 11.; Lod. Bot. Cab., t. 1012; Dend. Brit., t. 58.; our *fig.* 680. in p. 388.; the plate of this species in Arb. Brit., 1st edit., vol. vi.; and our *fig.* 636.

Spec. Char., &c. Leaves with the disk obovate-wedge-shaped, angled, glabrous, glossy. Petioles, stipules, and sepals glanded. Fruit oval, scarlet; nuts 4—5; flesh hard and dry. (*Dec. Prod.*) A low tree. North America, in Canada and on the Alleghany Mountains, and also found on the Rocky Mountains. Height 12 ft. to 15 ft. Introduced in 1750. Flowers white; May and June. Fruit scarlet; ripe in September.

Varieties.

- ❦ C. *g.* 2 *succulénta* Fisch., *Méspilus* succulénta *Booth*, has the fruit larger than that of the species, and succulent, juicy, and eatable. We have seen only one plant of this variety; but we were assured by the late M. Fischer of Göttingen, that there are several in the botanic garden there, and in various other collections in Germany.

635. C. g. subvillòsa.

♆ *C. g.* 3 *subvillòsa*, *C.* subvillòsa *Fisch.*, (our *fig.* 636., and *fig.* 681. in p. 388.) is apparently another variety of the preceding sort, or, perhaps, of *C.* coccínea. It is very distinct in appearance, from its villous twisted leaves, and stunted tortuous shoots; but, from its having been only three or four years in the country, very little is known of its habit of growth, which seems to be rather more loose than that of *C.* glandulòsa. There are plants in the Horticultural Society's Garden, and in Messrs. Loddiges's arboretum.

Differs from the preceding sort in the stipules and calyxes being glandular, and in the head of the tree forming a dense mass of small twigs. Being a small compact tree, of somewhat conical or fastigiate habit, and of comparatively low growth, and yet very prolific in flowers and fruit, it is well adapted for small gardens; and, being at the same time full of branches and very spiny, it is better calculated than many other kinds of American *Cratæ'gus* for forming field hedges.

636. C. glandulòsa.

§ ii. *Punctàtæ*.

Sect. Char. Leaves not lobed, large, with many nerves. Bark white, or ash-coloured. Fruit large, or small.

♆ 3. *C.* PUNCTA'TA *Ait.* The dotted-*fruited* Thorn.

Identification. Ait. Hort. Kew., 2. p. 169.; Jacq. Hort. Vind., 1. and 28.; Pursh Fl. Amer. Sept., 1. p. 338.; Dec. Prod., 2. p. 627.; Don's Mill., 2. p. 598.

Synonymes. *C.* Crús-galli *Du Roi*; Méspilus cuneifòlia *Ehrh. Beitr.* 3. p. 21.; *M.* punctàta *Link Enum.*; *M.* cornifòlia *Lam. Encyc.* 4. p. 444.

Engravings. Jac. Hort., 1. t. 28.; our *figs.* 682. and 683. in p. 389.; the plate in Arb. Brit., 1st edit., vol. vi.; and our *fig.* 637.

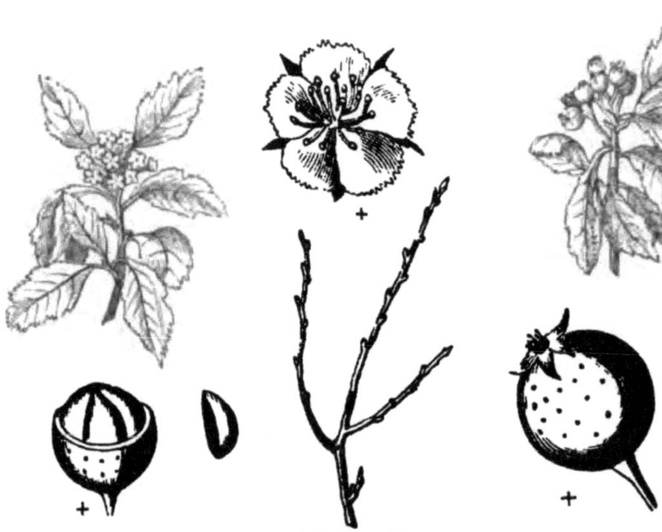

637. C. punctàta.

Spec. Char., &c. Leaves obovate-wedge-shaped, glabrous, serrated. Calyx a little villose; its sepals awl-shaped, entire. Fruit usually dotted. (*Dec. Prod.*) A small tree. North America, in the woods and swamps of Virginia and Carolina; where, according to Pursh, it grows to a handsome size, particularly the variety having yellow fruit. Height 15 ft. to 30 ft. Introduced in 1746. Flowers white; May and June. Fruit scarlet; ripe in September. Leaves dropping yellow. Naked young wood grey.

Varieties. There are four forms of this species in British gardens.

- ♈ C. *p.* 2 *rùbra* Pursh, *C.* edùlis *Ronalds* (*fig.* 682. in p. 389.) is the most common, and is a spreading tree, growing to the height of from 15 ft. to 30 ft., with red fruit, and, when old, with few thorns.
- ♈ C. *p.* 3 *rùbra strícta* Hort., *C.* p. strícta *Ronalds*, has the fruit red, like the preceding sort; but the general habit of the plant is fastigiate, like that of the following sort.
- ♈ C. *p.* 4 *aúrea* Pursh, *C.* p. flàva *Hort.*, *C.* dúlcis *Ronalds*, *C.* edùlis *Lodd. Cat.*, *C.* pentágyna flàva *Godefroy* (*fig.* 682. in p. 389.), is a tree like *C.* p. rùbra, with yellow fruit, and also, when old, with few thorns.
- ♈ C. *p.* 5 *brevispìna* Doug., and our *fig.* 638. — A very handsome fastigiate tree, with large, very dark purplish red fruit. Hort. Soc. Garden.

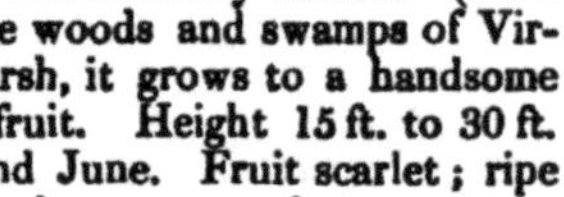

638. *C. p.* brevispìna.

The wood is so hard that the Indians of the west coast of America make wedges of it for splitting trees.

♈ 4. *C.* PYRIFO'LIA *Ait.* The Pear-tree-leaved Thorn.

Identification. Ait. Hort. Kew., 2. p. 168.; Pursh Fl. Amer. Sept., 1. p. 337.; Dec. Prod., 2. p. 627. Don's Mill., 2. p. 599.

Synonymes. C. leucophlœ'os (white-barked) *Mœnch Weiss.* p. 31. t. 2.; *C.* radiàta *Lodd. Cat.* edit. 1836; *C.* tomentòsa *Du Roi Harbk.* 1. p. 183.; *C.* latifòlia *Pers.*; *Méspilus* latifòlia *Lam. Encyc.* 4. p. 444.; *M.* Calpodéndron *Ehrh. Beitr.*; *M.* pyrifòlia *Link Enum.*; *M.* cornifòlia *Poir.*; *C.* latifòlia *Ronalds*; *C.* cornifòlia *Booth*; Lazzarollo perino, *Ital.*

Engravings. Mœnch Weiss., p. 31. t. 2.; Wats. Dend. Brit., t. 61.; Bot. Reg., t. 1877.; our *fig.* 684. in p. 389.; the plate of the species in Arb. Brit., 1st edit., vol. vi.; and our *fig.* 639.

Spec. Char., &c. In some instances spiny, in some without spines. Leaves ovate-elliptical, incisely serrated, obscurely plaited, a little hairy. Flowers 3-styled. Calyx slightly villose; its sepals linear-lanceolate, serrated. (*Dec. Prod.*) A low tree, generally spineless. North America, from Pennsylvania to Carolina, in woods and rocky places. Height 20 ft. to 25 ft. Introduced in 1765. Flowers white; June and July, rather later than C. punctàta. Fruit small, yellowish red; ripe early in September, and more eagerly sought after by birds than those of any other species.

639. C. pyrifòlia.

The leaves of young trees are larger, and the fruit smaller, than those of most other species; the leaves are also more strongly plaited, having the appearance of being furrowed from the midrib to the margin. When the fruit is not eaten by birds, it shrivels, turns black, and remains on the tree throughout the winter. The leaves drop early, of a rich yellow.

§ iii. *Macracánthæ.*

Sect. Char. Leaves large, ovate-oblong, slightly lobed and serrated, with numerous nerves, and subplicate. Fruit small. Spines very long. Tree vigorous and spreading.

‡ 5. C. MACRACA'NTHA *Lodd. Cat.* The long-spined Thorn.

Synonymes. C. glandulòsa β macrántha *Lindl.*, Bot. Reg., t. 1912.; C. spina longíssima in the Hammersmith Nursery; C. pyriflòra *Torrey*; see *Bot. Reg.* t. 1957.
Engravings. Bot. Reg., t. 1912.; our *fig.* 685. in p. 390.; the plate of this species in our Arb. Brit., 1st edit., vol. vi.; and our *fig.* 640.

Spec. Char., &c. Spines longer than the leaves, and numerous. Leaves ovate-oblong, somewhat acuminate, slightly lobed and bluntly serrated. nerved, and subplicate. Fruit small, or middle-sized, of a shining red, and very succulent when ripe. Tree open, spreading, and of very vigorous

640. *C. macracántha*

growth. The shoots straight, and tending upwards at an angle of 45°. North America, and the most common species in the northern states. Height 10 ft. to 30 ft. Introduced in 1819. Flowers white; May and June. Fruit scarlet, rather smaller than that of *C.* coccínea; ripe in September.

Variety.

¥ C. *m.* 2 *minor* (*fig.* 686. in p. 390.) only differs from the species in having smaller fruit. There are plants at Somerford Hall, Staffordshire.

Raised from American seed, in 1819, in the nursery of Messrs. Falla, at Gateshead, near Newcastle; whence it was sent to the Edinburgh Botanic Garden, under the name of the large American azarole.

§ iv. *Crús-gálli.*

Sect. Char. Leaves without lobes, obovate-oblong or obovate-lanceolate, more or less serrated, and of a dark shining green, with petioles margined by the decurrence of the leaf. Fruit small, or middle-sized, round, dark green till nearly ripe, and, when ripe, scarlet. Spines very long, and bent like the spur of a cock.

¥ 6. *C.* Cru's-ga'lli *L.* The Cock's-spur Thorn.

Identification. Lin. Sp., p. 682.; Dec. Prod., 2. p. 626.; Pursh Fl. Amer. Sept., 1. p. 338.; Don's Mill., 2. p. 598.

Synonymes. Cratæ'gus lùcida *Wang. Am.* t. 17. f. 42., and *Mill. Dict.*; *C.* cuneifòlia *Lodd. Cat.*; *Méspilus* lùcida *Ehrh. Beitr.*; *M.* Crús-gálli *Poir.*; *M.* hyemàlis *Walt*; *M.* cuneifòlia *Mœnch*; Neflier Pied de Coc, *Fr.*; glänzende Mispel, *Ger.*; Lazzarollo spinoso, *Ital.*

Engravings. Wang. Am., t. 17. f. 42.; Dend. Brit., t. 56.; our *fig.* 687. in p. 391.; the plate of the species in Arb. Brit., 1st edit., vol. vi.; and our *fig.* 641.

Spec. Char., &c. Spines long. Leaves obovate-wedge-shaped, nearly sessile, glossy, glabrous, falling off late. Stipules linear. Lobes of the calyx lanceolate, and somewhat serrated. Styles 2. Fruit scarlet. (*Dec. Prod.*) A low tree. North América; common in woods and hedges, and on the banks of rivers, from Canada to Carolina. Height 15 ft. to 20 ft. Intro-

641. C. Crús-gálli.

duced in 1691. Flowers white; May and June. Fruit small, green, and at length scarlet; ripe in September and October. Leaves retained longer than in most of the species; so that in the South of England it appears a sub-evergreen, retaining also its showy fruit through the winter.

arieties.

¥ C. *C.* 2 *spléndens* Dec. Prod., Ait. Hort. Kew. ii. p. 170., Pluk. t. 46. f. 1. *C. arbutifòlia* and *C.* spléndens *Lodd. Cat.* (*fig.* 688. in p. 391.) — Leaves obovate-wedge-shaped, and shining; and, being produced in abundance, the plant has a splendid appearance.

¥ C. *C.* 3 *pyracanthifòlia* Dec. Prod., Ait. Hort. Kew. ii. p. 170 *C. pyracanthifòlia* *Lodd. Cat.*; *Méspilus* lùcida *Dum. Cours. Bot. Cult.* ed. 2. v. p. 448. (*fig.* 693. in p. 391.; the plate in Arb. Brit., 1st edit., vol. vi.; and our *fig.* 642.) — Leaves oblong, with the upper part lanceolate; the lower part tending to wedge-shaped. This, even when only 3 or 4 years grafted, forms a singular little old-looking tree, spreading like a miniature cedar of Lebanon.

642. C. C.-g. pyracanthifòlia.

♈ *C.* C. 4 *salicifòlia* Dec. Prod., Ait. Hort. Kew. ii. p. 170. *C.* salicifòlia. (*fig.* 691. in p. 391.; and the plate in Arb. Brit., 1st edit., vol. vi.; and our *fig.* 643.) — Leaves oblong, with the upper part lanceolate;

643. *C. C.-g.* salicifolia.

the lower part tending to wedge-shaped. This forms a low flat-headed tree, like the preceding variety. A plant in Messrs. Loddiges's arboretum, in 1835, after being five years grafted at a foot from the ground, was not quite 5 ft. high. The miniature trees of this variety are admirably adapted for children's gardens.

♈ C. *C.* 5 *lineàris* Dec. Prod. *Méspilus* lineàris *Desf. Arb.* ii. p. 156., *Poir. Suppl.* iv. p. 70.; *C.* lineàris *Lodd. Cat.* (*fig.* 690. in p. 391.) — Leaves linear-lanceolate. Spines, or thorns, few and shortish. Styles 1—2. Fruit of a yellowish red.

■ C. *C.* 6 *nàna* Dec. Prod. *Méspilus* nàna *Dum. Cours. Suppl.* p. 386. — Branchlets tomentose in some degree. Leaves oval-lanceolate; the under surface paler than the upper. A shrub, or, when trained to a single stem, a miniature tree.

This species, being one of the first introduced into England, has been more cultivated than any other American thorn; and on the whole it is one of the most splendid in appearance, from its smooth, shining, dark green foliage, and the great abundance of its fine white flowers, and dark red fruit which remains long on the tree. In the South of England, and in the climate of London, in warm sheltered situations, where the soil is rich and moist, it retains its leaves and fruit through great part of the winter, so as to appear quite evergreen.

♈ 7. *C.* (C.) OVALIFO'LIA *Horn.* The oval-leaved Thorn.

Identification. Hornem. Hort. Hafn. Suppl., 52.; Dec. Prod., 2. p. 627.; Don's Mill., 2. p. 598.
Synonyme. *C.* elliptica *Lodd. Cat.*; *C.* Crús-gálli ovalifòlia *Bot. Reg.* t. 1860.
Engravings. Bot. Reg., t. 1860.; our *fig.* 692. in p. 391.; the plate of this species in Arb. Brit. 1st edit., vol. vi.; and our *fig.* 644.

Spec. Char., &c. Leaves oval, serrated, a little pilose on both surfaces, and shining on the upper one. Stipules half-heart-shaped, incisely serrated, with glanded serratures. (*Dec. Prod.*) A low tree. North America. Height 15 ft. to 20 ft. Introd. in 1810. Flowers white. A very distinct variety of *C.* Crús-gálli, with a loose spreading habit of growth, and broad leaves.

644. C. C. ovalifòlia.

♀ 8. *C.* (C.) PRUNIFO'LIA *Bosc.* The Plum-leaved Thorn.

Identification. Bosc ined.; Dec. Prod., 2. p. 627.; Don's Mill., 2. p. 598.
Synonymes. Méspilus prunifòlia *Poir. Dict.* 4. p. 443.; *C.* caroliniàna *Lodd. Cat.*; Lazzeruolino, *Ital.*

645. C. C. prunifòlia.

Engravings. Bot. Reg., t. 1868.; our *fig.* 689. in p. 391.; the plate of this tree in Arb. Brit., 1st edit., vol. vi.; and our *fig.* 645.

Spec. Char., &c. Leaves with the disk broadly ovate, unequally serrated, and glabrous; the petioles bearing a few glands. Sepals with glanded serratures. Peduncle and calyx a little villose. Seeds 2 in a pome. (*Dec. Prod.*) A low tree. North America. Height 15 ft. to 20 ft. Introduced in 1818, or before. Flowers white; May and June. Fruit scarlet.

Differs from the preceding variety in having broader and shorter leaves, a more compact and fastigiate habit of growth, and rather more thorns on the branches. The leaves of this and the preceding kinds die off of a much deeper red than the narrow-leaved varieties, which often drop quite green, yellow, or of a yellowish red.

§ v. *Nìgræ.*

Sect. Char. Leaves middle-sized, deeply lobed. Lobes pointed. Fruit round, black or purple. Tree rather fastigiate, with few or no spines. Bark smooth.

♀ 9. *C.* NI'GRA *Waldst. et Kit.* The black-*fruited* Thorn.

Identification. Waldst. et Kit. Pl. Rar. Hung., t. 61.; Dec. Prod., 2. p. 628.; Don's Mill., 2. p. 599.
Synonymes. *Méspilus nígra Willd. Enum.* 524.; *C. carpática Lodd. Cat.*
Engravings. Waldst. et Kit. Pl. Rar. Hung., t. 61.; *fig.* 694. in p. 392.; the plate of this species in Arb. Brit., 1st edit., vol. vi.; and our *fig.* 646.

646. *C. nigra.*

Spec. Char., &c. Leaves sinuately lobed, and serrated, somewhat wedge-shaped, though truncately so, at the base; whitely villose beneath. Stipules oblong, serrately cut. Calyxes villose; the lobes slightly toothed. Styles 5. Fruit black. (*Dec. Prod.*) A low tree. Hungary. Height 15 ft. to 20 ft., throwing up numerous suckers from its widely spreading roots, which soon cover the ground with a forest of bushes. In England, where it is generally propagated by grafting on the common thorn, it forms a very handsome, upright, somewhat fastigiate tree, from 20 ft. to 30 ft. high, putting forth its leaves, in mild seasons, in February or March. Introduced in 1819. Flowers white; April and May. Fruit black; ripe in July and August.

Variety. *C. fúsca Jacq.*, judging from a seedling plant in the Hort. Soc. Garden, appears to belong to this species.

Nightingales are said to be attracted by this tree, probably because it is particularly liable to be attacked by insects, and because numerous caterpillars are to be found on it about the time when the nightingale is in full song. The same property of attracting nightingales is ascribed to the common hawthorn, in *La Théorie du Jardinage, &c.*, published in 1709.

¥ 10. *C.* PURPU`REA *Bosc.* The purple-*branched* Thorn.

Identification. Bosc ined.; Dec. Prod., 2. p. 628.; Don's Mill., 2. p. 509.
Synonyme. *C. sanguinea Hort.*
Engravings. Wats. Dend. Brit., t. 60.; our *fig.* 695. in p. 392.; the plate of this species in Arb. Brit., 1st edit., vol. vi.; and our *fig.* 647.

647. C. purpùrea.

Spec. Char., &c. Branches dark purple. Leaves ovate, cuneate at the base, lobed with broad lobes, serrated, glabrous, or pubescent beneath. Stipules somewhat circular, serrated with glanded serratures. (*Dec. Prod.*) A shrub or low tree. Altaic Mountains. Height 6 ft. to 10 ft. Introduced in 1822. Flowers white; early in April; being the very first species of Cratæ`gus that comes into flower in the neighbourhood of London, excepting always the Glastonbury thorn. Fruit dark red or purple, sometimes yellow, very succulent; ripe in July.

Variety.

¥ C. *p.* 2 *altàica*; *C.* altàica *Led., Lod. Cat.* (*fig.* 696. in p. 393.); has the leaves somewhat larger than the species, and they appear a little earlier.

It forms an upright, rigid, rather slow-growing tree, without thorns. It has a few small branches, and is not densely clothed with leaves. It has a starved and somewhat stunted appearance, and is readily known by the purple colour of its young shoots. The bark of the old wood is of a dark purple or brown colour, and rough and scaly. The fruit is small, round, and most commonly of a dark purple; but it varies to pale yellow, or a milk white, and red, on the same plant. It ripens about the end of July, and is very soft and juicy, but soon drops off. The tree is interesting from its early flowering, and the dark colour of the anthers of its flowers, which contrasts strongly with the whiteness of the petals. The leaves are also large, and of a peculiar shape.

§ vi. *Douglàsii.*

Sect. Char. Leaves small, and not lobed as in the preceding section; furnished with numerous parallel nerves, somewhat like those of *C.* punctàta. Spines rather numerous and rigid. Fruit small, and dark purple; pulp soft and watery.

☨ 11. *C.* Dougla'sii Lindl Douglas's Thorn.

Identification. Bot. Reg., t. 1810.; Lod. Cat., edit. 1832.
Engravings. Bot. Reg., t. 1810.; our *fig.* 697. in p. 393.; the plate of the species in Arb. Brit., 1st edit., vol. vi.; and our *fig.* 648.

648. *C. Douglàsii.*

Spec. Char., &c. Branches ascending. Spines rigid, straightish, now short, now very long. Leaves some obovate, some oval, gashedly serrated, acute; at the base wedge-shaped, glabrous; in the autumn, remarkably leathery, and they then acquire a purplish cast, and are shining. (*Lindl.*) A shrub or low tree. North-West America. Height 10 ft. to 15 ft. Introduced in 1827. Flowers white; May. Fruit small, dark purple; ripe in August. Decaying leaves purplish, leathery, shining, falling off early, like those of *C.* punctàta and *C. pyrifòlia.* Naked young wood purplish.

This is a very distinct sort, more particularly as it respects the colour of the fruit, and the colour and texture of the leaves. The general habit of the plant is fastigiate; and it is one of the latest kinds in leafing in the spring. The flowers and fruit are produced in great abundance, and both are very ornamental.

§ vii. *Flàvæ.*

Sect. Char. Leaves small, obovate, slightly lobed, and serrated. Flowers frequently solitary. Spines numerous, straight, and more slender than in any other division. Fruit top, or pear, shaped; yellow, or greenish yellow.

☨ 12. *C.* fla'va *Ait.* The yellow-*fruited* Thorn.

Identification. Ait. Hort. Kew., 2. p. 169.; Pursh Fl. Amer. Sept., 1. p. 338.; Dec. Prod., 2. p. 628.; Don's Mill., 2. p. 600.
Synonymes. *C.* glandulòsa *Michx. Fl. Bor. Amer.* 1. p. 288., not of Walt.; Méspilus Michaùxii Pers. Syn. 2. p. 38.; *C.* caroliniàna *Poir. Dict.* 4. p. 442.; *C.* flavíssima *Hort.*; *C.* ? turbinàta *Pursh.*
Engravings. Bot. Reg., t. 1939.; *fig.* 698. in p. 394.; the plate of the species in Arb. Brit., 1st edit., vol. vi.; and our *fig.* 649.

Spec. Char., &c. Disks of leaves obovate-wedge-shaped, slightly lobed, crenately serrate, upon short petioles. Stipules glanded. Flowers mostly solitary. Sepals glanded. Fruit top-shaped, yellow, or yellowish green, Nuts 4 in a fruit. (*Dec. Prod.*) A low spreading tree. North America.

649. *C. flàva.*

from Virginia to Carolina. Height 15 ft. to 20 ft. Introduced in 1724. Flowers white; May. Haws yellow; ripe in October. Decaying leaves rich yellow.

The flowers and the fruit are neither produced in abundance, nor make any great show; but the tree has a marked character from its general form, and the horizontal tendency of its branches.

☤ 13. *C.* (F.) LOBA'TA *Bosc.* The lobed-*leaved* Thorn.

Identification. Bosc ined.; Dec. Prod., 2. p. 628.; Don's Mill., 2. p. 599.
Synonymes. Méspilus lobàta *Poir. Suppl.* 4. p. 71.; *C.* lùtea *Poir.*
Engravings. *Fig.* 699. in p. 394.; and our *fig.* 650.

Spec. Char., &c. Branches a little villose. Disks of leaves ovate, unequally serrated, or lobed, slightly downy beneath, upon very short petioles. Stipules cut. Flowers in loose corymbs. (*Dec. Prod.*) A tree closely resembling *C.* flàva in general appearance. Native country supposed to be America. Height 10 ft. to 15 ft. Introduced in 1819. Flowers white; May. Fruit green; ripe in October.

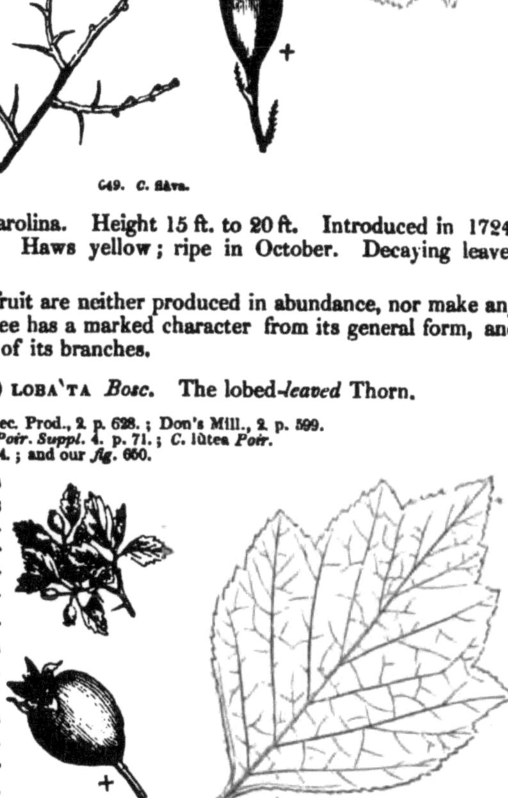

650. *C. f. lobata.*

Differing from *C.* flàva in having some of the leaves with larger lobes, and some of the spines larger. The flowers are sparingly produced, amongst dense tufts of leaves; and the fruit, which is green when ripe, is still less abundant. It is pear-shaped, and very different from that of every other kind of Cratæ'gus, except *C.* flàva and *C.* f. trilobàta.

♈ 14. *C.* (F.) TRILOBA'TA *Lodd. Cat.* The three-lobed-*leaved* Thorn.

Identification. Lodd. Cat., edit. 1832.
Synonyme. *C.* spinosissima *Lee.*
Engraving. *Fig.* 700. in p. 395.

Spec. Char., &c. Leaves ovate-cuneate, notched and serrated. Petioles slender. Surface flat, shining, somewhat veined. Branches small, thickly beset with slender thorns. Habit spreading. A hybrid, raised from seed in the Hammersmith Nursery, about 1820, or before. It forms a tree in general appearance resembling *C.* flàva, but with the branches much less vigorous, and more thorny. The fruit is yellow, slightly tinged with red; and what distinguishes it from the two allied sorts is, that its leaves die off, in autumn, of an intensely deep scarlet.

§ viii. *Apiifòliæ.*

Sect. Char. Leaves deltoid, or somewhat resembling those of the common thorn. The fruit is also of the same colour; but the tree has a totally different habit, having the shoots loose and spreading, weak, and almost without thorns.

♈ 15. *C.* APIIFO'LIA *Michx.* The Parsley-leaved Thorn.

Identification. Michx. Fl. Bor. Amer., 1. p. 287., not of Med.; Dec. Prod., 2. p. 627.; Dou's Mill. 2. p. 599.
Synonymes. *C.* Oxyacántha *Walt. Carol.* 147.; *C.* apiifòlia màjor *Lodd. Cat.*
Engravings. *Fig.* 702. in p. 395.; the plate in Arb. Brit., 1st edit., vol. vi.; and our *fig.* 651.

651. *C.* apiifòlia.

Spec. Char., &c. Leaves deltoid, cut into lobes that are acute and incisely toothed. Pedicels in the corymb villose, mostly simple. Tube of calyx villose. Sepals obscurely serrated. Fruit scarlet. (*Dec. Prod.*) A low spreading tree with flexible branches. Virginia and Carolina, in moist woods. Height 10 ft. to 20 ft. Introduced in 1812. Flowers white; May and June. Haws scarlet; ripe in October. Decaying leaves rich yellow.

Variety.

♈ C. a. 2 *minor*, *C.* apiifòlia *Lodd. Cat.* (*fig.* 701. in p. 395.), has the leaves smaller than those of the species, and more fringed at the edges, like those of the common parsley; but this fringed appearance is by no means constant, either in the variety or in the species. This variety forms a most ornamental low bush; or, when grafted standard high, a beautiful pendent tree.

§ ix. *Microcárpæ.*

Sect. Char. Fruit small, round, red. Flowers small, produced in corymbs, later in the season than in any of the other species. Spines few, but sometimes very large.

♀ 16. *C.* CORDA`TA *Mill.* The heart-shaped-*leaved* Thorn.

Identification. Mill. Ic., t. 179.; Dec. Prod., 2. p. 628.; Don's Mill., 2. p. 599.
Synonymes. *M.* Phænópyrum *Linn.*; *M.* cordàta *Mill.*; *C.* populifòlia *Walt. Car.* 147., and *Pursh Sept.* 1. 337.; Méspilus acerifòlia *Poir. Dict.* 4. p. 442.
Engravings. Mill. Ic., t. 179.; Wats. Dend Brit. t. 63.; Bot. Reg., t. 1151.; *fig.* 703. in p. 396. the plate of this species in Arb. Brit., 1st edit., vol. vi.; and our *fig.* 652.

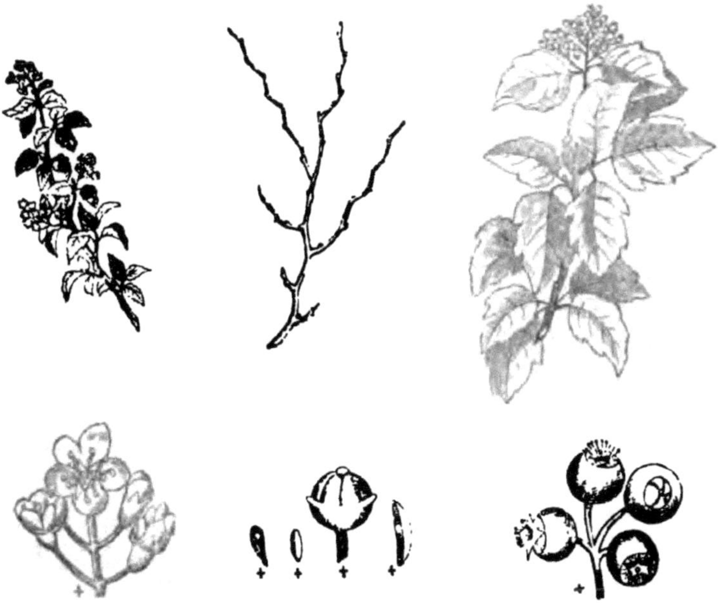

652 C. cordàta.

Spec. Char., &c. Disks of leaves cordate-ovate, angled by lobes, glabrous. Petioles and calyxes without glands. Styles 5 in a flower. (*Dec. Prod.*) A compact, close-headed, small tree, with leaves of a deep shining green. Canada to Virginia, in hedges and rocky places. Height 20 ft. to 30 ft. Introduced in 1738. Flowers white, in numerous terminal corymbs; May and June. Fruit small, scarlet; ripe in October.

A very distinct and handsome species.

♀ 17. *C.* SPATHULA`TA *Elliott.* The spathula-*shaped-leaved* Thorn.

Identification. Elliott Fl. S. Car., 1. p. 552.; Lodd. Bot. Cat., t. 1261.
Synonymes. *C.* microcárpa *Lindl. Bot. Reg.* t. 1846.; *C.* flórida *Godefroy.*
Engravings. Bot. Cab., t. 1261.; Bot. Reg., t. 1846.; *fig.* 704. in p. 396.; the plate of the species in Arb. Brit., 1st edit., vol. vi.; and our *fig.* 653.

Spec. Char., &c. Subspinose. Leaves in fascicles, oblong cuneated, 3-cleft, lobed and crenated, smooth, shining. Corymbs many-flowered. Calyx smooth; segments ovate, quite entire. Fruit ovate, subrotund, smooth,

653. *C. spathulata.*

5-celled; shell thin. (*Lindl.*) A low tree; in England a shrub, unless when grafted standard high. Georgia and Carolina. Height 12 ft. to 15 ft.; in England 6 ft. to 8 ft. Introduced in 180f. Flowers white; May and June. Fruit bright red; ripe in October.

Variety.

⸸ *C. s.* 2 *geórgica*. *C.* geórgica *Lod.*—Leaves 5-lobed, on longer footstalks, and rather smaller than the species. This variety and the species were killed to the ground in the winter of 1837-8. (See *Gard. Mag*, vol. xvi. p. 3.)

A slow-growing, very neat, little bush or tree, with slender, smooth, drooping branches, and something of the habit of *C. Oxyacántha*. Its leaves have a very handsome appearance, and are remarkably shining, and deep green: they usually grow in clusters; have a long stalk, tapering upwards into a blade, which is sometimes nearly entire, with only a tooth or two at the end; sometimes they are 3-lobed, with crenated segments; and occasionally they are deeply 3-parted; their form is always more or less spathulate. The stipules of the more vigorous branches are large and leafy. The flowers are white, and appear at the same time as those of *C. cordàta*. The fruit is rather abundant, but small.

§ x. Azaròli.

Sect. Char. Fruit large, round or pear-shaped; good to eat; yellow or red; the yellow fruit generally produced on fastigiate species or varieties; and the red on trees with a spreading and rather a drooping head. Leaves wedge-shaped, 3-cleft or more, shining, pubescent or hairy. Spines few or none.

⸸ 18. *C. AZARO'LUS* L. The Azarole Thorn.

Identification. Lin. Sp., 683.; Dec. Prod., 2. p. 629.; Don's Mill., 2. p. 600.
Synonymes. Pỳrus *Azaròlus* Scop. Carn. No. 597., J. Bauh. Hist. 1. p. 6'.; *Méspilus Azaròlus* All. Ped., N. Du Ham. 4. p. 158.; Néflier Azarole, Neflier de Naples, E'pine d'Espagne, Pommettes à deux Closes, *Fr.*; Azarol Mispel, *Ger.*; Azzeruolo, *Ital.*
Engravings. N. Du Ham., 4. t. 42.; Bot. Rep., t. 579.; *fig.* 705. in p. 397.; the plate in Arb. Brit., 1st edit., vol. vi.; and our *fig.* 654.

Spec. Char., &c. Leaves pubescent, wedge-shaped at the base, trifid; lobes blunt, and with a few large teeth. Branchlets, corymbs, and calyxes pubescent.

654. *C. Azaròlus.*

Sepals obtuse. Styles 1—3 Fruit globose, scarlet. Seeds usually two; and hence the name, common at Montpelier, *pommettes à deux closes*. (*Dec. Prod.*) A low tree, never found wild as a bush. South of France and Italy, in small woods and in rough places. Height 20 ft. to 30 ft; in England 10 ft. to 20 ft. Introduced in 1656. Flowers white; May and June. Fruit red; ripe in September. Leaves drop with the first frost, without much change of colour.

Varieties. In the *Nouveau Du Hamel*, six varieties are enumerated, viz.: — 1. *Mespilus Arònia*, with the leaves hairy beneath; 2. Azarole, with large deep red fruit; 3. Azarole, with yellowish white fruit; 4. Azarole, with long fruit of a whitish yellow; 5. Azarole, with double flowers; and, 6. the White Azarole of Italy. With the exception of the first-mentioned, none of these varieties, as far as we know, are in British gardens.

The fruit, when ripe, is mealy, and somewhat acid; and, in Italy and the Levant, it is occasionally sent to table.

♀ 19. *C.* (*A.*) MAROCCAˋNA *Pers.* The Morocco Thorn.

Identification. Pers. Syn., 2. p. 37.; Dec. Prod., 2. p. 628.; Don's Mill., 2. p. 600.
Synonymes. DeCandolle expresses a doubt whether *C.* maúra *Lin. fil. Sup.* 253. be not a synonyme of this species; Sarrour, *Arabian*.
Engravings. Bot. Reg., 1855.; *fig.* 707. in p. 397.; the plate in Arb. Brit., 1st edit., vol. vi.; and our *fig.* 655.

Spec. Char., &c. Leaves wedge-shaped, 3-lobed, and pinnatifid, glabrous, glandless. Stipules cut, rather palmately. Flowers upon long peduncles, in terminal glabrous corymbs. Sepals obtuse. Styles 2. (*Dec. Prod.*) A handsome pyramidal low tree, with dark-coloured branches. Palestine, on Mount Sinai and St. Catherine; and ? Morocco. Height 15 ft. to 35 ft. Flowers pure white, very fragrant; May and June. Fruit scarlet; ripe in October.

Closely resembling *C. Azaròlus*, but smaller in all its parts. It produces its leaves very early in the season, in mild winters even in January; and it retains them very late. It is a small, but decided tree, and may be considered one of the handsomest species of the genus. Horticultural Society's Garden

655. C. (A.) maroccàna.

¥ 20. C. Aro'nia *Bosc.* The Aronia Thorn.

Identification. Bosc ined.; Dec. Prod., 2. p. 629.; Don's Mill., 2. p. 601.
Synonymes. Méspilus Arònia *Willd. Enum. Suppl.* and *N. Du Ham.* 4. p. 158.; *C. Azaròlus* β *Willd. Sp.*; *C.* fissa *Lodd. Cat.*
Engravings. Pococke's Travels, t. 85.; Bot. Reg., 1897.; *fig.* 706. in p. 397.; the plate in Arb. Brit., 1st edit., vol. vi.; and our *fig.* 656.

656. C. Arònia.

Spec. Char., &c. Branchlets pubescent. Leaves pubescent on the under surface, wedge-shaped at the base, 3-cleft; lobes obtuse, entire, each ending in 3 obtuse mucronate teeth. Fruit yellow. (*Dec. Prod.*) An erect-branched low tree. Greece and the Levant. Height 15 ft. to 20 ft. Introduced in 1810. Flowers white; May and June. Haws yellow, smooth, large, succulent, agreeable to eat; ripe in August and September, and hanging on the tree till the leaves drop in November or December. Naked young wood dark-coloured.

Remarkable for the abundance of its large yellow fruit, which are good to eat, and have been made into excellent tarts with Siberian crabs.

♀ 21. *C.* ORIENTA'LIS *Bosc.* The Eastern Thorn.

Identification. Bosc ined.; Bot. Rep., t. 890.; Don's Mill., 2. p. 600.
Synonymes. *Méspilus* orientàlis *Tourn.* and *Poir. Suppl.* 4. p. 72.; *C.* odoratissima *Bot. Rep.* and *Lod. Cat.*; *C.* tanacetifòlia var. β taúrica *Dec. Prod.* 2. p. 629.
Engravings. Bot. Reg., 1885.; *fig.* 708. in p. 398.; the plate in Arb. Brit., 1st edit., vol. vi.; and our *fig.* 657.

657. *C.* orientàlis.

Spec. Char., &c. Branches whitely tomentose. Leaves 3-lobed, downy beneath; the two side lobes ovate, and having tooth-like incisions at the tip; the middle lobe trifid. Stipules broad and cut. (*Dec. Prod.*) A low spreading tree. Levant. Height 15 ft. to 20 ft. Introduced in 1810. Flowers white; May and June. Haws numerous, large, yellowish red or coral colour, very agreeable to the taste; ripe in August and September, and remaining on sometimes after the leaves.

Variety.

♀ C. o. 2 *sanguínea*, *C.* sanguínea *Schrader Index Sem. Hort. Acad. Gott.* 1834, *C.* orientàlis *Lindl. Bot. Reg.* t. 1852., and *fig.* 709. in p. 398., has the fruit of a very dark purplish red, or port wine, colour. Dr. Lindley considers this "the genuine *Méspilus* orientàlis of Tournefort, with villous celery-like leaves, and a large, purple, 5-cornered, smooth fruit," which description, we think, indicates rather the following species. It is a native of the Crimea, and the parts bordering on the Black Sea; and was introduced in 1810. On account of the colour of its fruit, and the abundance in which it is produced, it deserves a place in every collection.

Readily distinguished from most other species by its very hoary branches, which are loose, rambling, crossing each other, and somewhat pendulous. It is late in producing its leaves, and also its flowers: the latter generally appear with those of *C.* tanacetifòlia, about the end of May (in 1836, on the 17th of June).

♆ 22. *C.* TANACETIFO'LIA *Pers.* The Tansy-leaved Thorn.

Identification. Pers. Syn., 2. p. 38.; Dec. Prod., 2. p. 629.; Don's Mill., 2. p. 601.
Synonymes. Méspilus tannacetifòlia *Poir. Dict.* 4. p. 440., and *N. Du Ham.*; *M.* pinnàta *Dum. Cours.*, *Sm. Exot.* t. 85.; DeCandolle doubts whether *Méspilus* Cels*iàna* Dum. Cours. Suppl. p. 286. be different from this species; Lazzeruolo turco, *Ital.*
Engravings. Bot. Rep., t. 591.; Sm. Exot. Bot., t. 85.; Bot. Reg., 1884.; *fig.* 710. in p. 398.; the plate in Arb. Brit., 1st edit., vol. vi.; and our *fig.* 658.

658. C. tanacetifolia.

Spec. Char., &c. Leaves pinnatifidly cut, hairy; lobes oblong, acute, having a few teeth. Sepals acutish, reflexed, hairy. Styles 5. Fruit globose, yellowish green. (*Dec. Prod.*) A robust-growing fastigiate tree, with upright rigid branches, commonly terminating in thorny points. Levant. Height 20 ft. to 30 ft. Introduced in 1789. Flowers white; May and June. Haws large, yellow; ripe in October.

Varieties.

♆ C. *t.* 2 *glàbra* Lodd. (*fig.* 711. in p. 398.; the plate in Arb. Brit., 1st edit., vol. vi.; and our *fig.* 659.) has shining leaves, and fruit about half the size of that of the species, of a reddish yellow. A hybrid between *C.* tanacetifòlia and *C.* Oxyacántha. Introduced from Germany about 1810.

♆ *C. t.* 3 *Cels*iàna. *Méspilus* Cels*iàna* Dumont de Cours., vol. vii. p. 286.; *C.* t. 3 Lee*àna* Arb. Brit. 1st ed.; *C.* incìsa *Lee.* (*fig.* 712. in p. 399.; plate in Arb. Brit., 1st edit.; and our *fig.* 660.)—Somewhat resembling *C.* orientàlis; but the leaves much larger, and more deeply cut, and the trees of a more robust, erect, and fastigiate habit. Fruit large, yellow. Cultivated by Cels, and supposed by Dumont de Courset to be a native of Persia or the Levant. Erroneously said to have been raised in the Hammersmith Nursery. A splendid tree,

659. *C. t.* glàbra.

most striking in appearance, from its large and deeply cut foliage, and its strong, upright, vigorous shoots.

660. *C. t. Celsiàna.*

The leaves and calyxes are covered on both sides with long hairs. The globular slightly compressed fruit has somewhat the appearance of being ribbed like a melon ; is larger than that of any other species of the genus, except *C. Arònia* and *C. mexicàna*; greenish yellow when ripe ; and easily distinguishable by the bracteas generally adhering to it. The foliage is the latest in appearing of any of the species, except *C. orientàlis*, frequently equally late.

§ xi. *Heterophýlla.*

Sect. Char. Leaves cuneate, and sub-persistent. Frûit long, middle-sized, and crimson.

♆ 23. *C.* HETEROPHY'LLA *Flugge.* The various-leaved Thorn.

Identification. Flugge Ann. Mus., 12. p. 423. t. 38. ; Dec. Prod., 2. p. 629. ; Don's Mill, 2. p. 600. ; Lindl. in Bot. Reg., t. 1847.
Synonymes. C. neapolitàna *Hort.*
Engravings. Ann. Mus., 12. t. 38. ; Bot. Reg., t. 1161. and t. 1847. ; *fig.* 713. in p. 399. ; the plate in Arb. Brit., 1st edit., vol. vi. ; and our *fig.* 661.

661. C. heterophylla.

Spec. Char., &c. Leaves bright, falling off late, lanceolate-cuneate, toothed at the apex, 3-cleft ; segments serrate. Tube of the calyx fusiform. Cymes many-flowered. Flowers 1-styled. Fruit ovate, including one nut, with a hard bony shell, and one seed. Stipules large, pinnatifid. (*Lindl.*) A low tree, with ascending branches. Native country uncertain, most probably the South of Europe ; possibly a hybrid between the common hawthorn, and the azarole, or some other species. Height 10 ft. to 20 ft. Cultivated in 1816. Flowers white, produced in great abundance ; May. Fruit rich crimson, resembling in shape that of the common hawthorn, but narrower ; ripe in September and October.

A very handsome and most desirable species ; producing its leaves and flowers early in the spring, and retaining its leaves and fruit till the first autumnal frosts.

§ xii. *Oxyacánthæ.*

Sect. Char. Leaves obovate, trifid, or variously cut. Flowers numerous, in corymbs. Fruit generally red.

⁊ 24. *C.* Oxyaca'ntha *L.* The sharp-thorned Cratægus, *or common Hawthorn.*

Identification. Lin. Sp., 683.; Dec. Prod., 2. p. 628.; Don's Mill., 2. p. 600.

Synonymes. The Pyracantha of the Greeks; *Méspilus Oxyacántha Gærtn.* and *N. Du Ham.*; E'pine blanche, noble E'pine, Bois de Mai, Scuelleir Aubépine, Néflier Aubépine, *Fr.*; Hagedorn, gemeiner Weissdorn, *Ger.*; Hagetoan, *Dan.*; Hagotorn, *Swed.*; Acanta da siepe, Azzarolo salvatico, and Bianco Spino, *Ital.*; Espino blanco, *Span.*; White Thorn, Maybush, Quick, Quickset, May.

Derivation. Booth derives the word Haw from *hage*, or *hæg*, a hedge; consequently he makes hawthorn signify hedgethorn. Quick signifies live; and was, probably, applied, from live hedges made of hawthorn being used instead of fences of dead branches of trees. Whitethorn, from the profusion of its white flowers and its being thorny, or possibly from its white bark, as compared with that of the blackthorn, *Prùnus spinòsa.* May and Maybush have reference both to the time of flowering of the plant, and to its use in the May or floral games. The French name Aubépine, refers to its flowering in spring, or in what may poetically be called the morning of the year; *aube* signifying the dawning of the day.

Engravings. Gærtn. Fruct., 2. t. 87.; Eng. Bot., t. 2054.; *fig.* 715. in p. 400.; the plate in Arb. Brit., 1st edit., vol. vi.; and our *fig.* 662.

662. C. Oxyacántha.

Spec. Char., &c. Leaves obovate-wedge-shaped, almost entire, or trifid or cut, glabrous, rather glossy. Corymbs of several flowers. Sepals glandless, acute. Styles 1—3. (*Dec. Prod.*) A shrub or low tree. Europe, common in hedges, and varying much in different situations. Flowers white; May. Haws red or scarlet; ripe in September.

Varieties. These are very numerous, and some of them very distinct. In the following enumeration we have confined ourselves to such as we have actually seen in the Hort. Soc. Garden, or in the arboretum of Messrs. Loddiges.

A. *Varieties differing from the Species in the general Form and Mode of Growth.*

⁊ C. O. 2 *stricta* Lodd. Cat., *C. O.* rígida *Ronalds*, (the plate in Arb. Brit.,

663. C. O. stricta.

1st edit., vol. vi.; and our *fig.* 663.) has the shoots upright, and the general habit as fastigiate as that of a Lombardy poplar. It was discovered in a bed of seedlings in Messrs. Ronald's Nursery, about 1825, and forms a very distinct and desirable variety.

♀ C. O. 3 *péndula* Lodd. Cat. has drooping branches. A very marked variety of this kind, which was selected from a bed of seedlings by General Monckton, is said to be in the collection of thorns at Somerford Hall.

664. C. O. reginæ.

♀ C. O. 4 *reginæ* Hort. *Queen* Mary's *Thorn.* (The plate of the tree in Arb. Brit., 1st edit., vol. vi.; and our *fig.* 664.) — The parent tree is in a garden near Edinburgh, which once belonged to the Regent Murray. It is very old, and its branches have somewhat of a drooping character; but whether sufficiently so to constitute a variety worth propagating as a distinct kind appears to us very doubtful. (*Lodd.*)

♈ C. O. 5 *Celsiàna Hort.* is also somewhat fastigiate in its habit; but it is a much more slender-growing plant; and we have never seen a specimen in a situation where it could display its natural form and mode of growth.

♈ C. O. 6 *capitàta* Smith of Ayr differs from the species chiefly in being of a somewhat more fastigiate habit, and in producing its flowers in close heads, mostly at the extremities of its branches.

♈ C. O. 7 *flexuòsa* Smith of Ayr has the small branches twisted in a zigzag manner. Horticultural Society's Garden.

B. *Varieties differing in the Colour of the Flowers.*

♈ C. O. 8 *ròsea* Hort.; E'pinier Marron, *Fr.* (*fig.* 725. in p. 401.); has the petals pink, with white claws, and is a well-known and very beautiful variety.

♈ C. O. 9 *punícea* Lodd. Cat., *C. O.* ròsea supérba *Hort.*, has larger petals, which are of a dark red, and without white on the claws.

C. *Varieties differing in the Developement or Structure of the Flowers.*

♈ C. O. 10 *múltiplex* Hort., *C. O.* flòre plèno *Hort.* (*fig.* 722. in p. 401.), has double white flowers, which die off of a beautiful pink; and which, being produced in great profusion, and lasting a long time, render this a most desirable variety: accordingly, it is to be found in almost every shrubbery and garden.

♈ C. O. 11 *punícea flòre plèno* Hort.—Flowers double, nearly as dark and brilliant as *C. O.* punícea. Imported in ? 1832, by Mr. Masters of Canterbury.

♈ C. O. 12 *monógyna*, *C.* monógyna *Jacq.*, has flowers with only one style, like *C. O.* sibírica, but does not flower early like that variety.

♈ C. O. 13 *apétala* Lodd. Cat.—This remarkable variety has the flowers without petals, or very nearly so.

D. *Varieties differing in the Time of Flowering.*

♈ C. O. 14 *præ'cox* Hort., the Glastonbury Thorn, comes into leaf in January or February, and sometimes even in autumn; so that occasionally, in mild seasons, it may be in flower on Christmas-day.

♈ C. O. 15 *sibírica*, *C.* sibírica *Lodd. Cat.*, *C.* monógyna *L.* (*fig.* 665.), is an early leafing variety, a native of Siberia. In mild seasons, it begins to put forth its leaves in January; and in dry summers it loses them proportionately soon in the autumn. On account of its early leafing and flowering, it well merits a place in collections. The flowers have only one style; but, as there are other varieties having only one style which do not flower early, we have not adopted Linnæus's name of *C.* monógyna to this variety, but to another, a native of Britain, which does not flower earlier than the common hawthorn.

665. *C. O. sibírica.*

♈ C. O. 16 *transylvánica* Booth, from the plant in the Hort. Soc. Garden, appears to be nearly, if not quite, the same as *C. O.* sibírica.

E. *Varieties differing in the Colour of the Fruit.*

♀ C. O. 17 *melanocárpa, C.* físsa *Lee, C.* Oxyacántha platyphýlla *Lodd. Cat., C.* platyphýlla *Lindl.* in *Bot. Reg.* t. 1874., (*fig.* 718. in p. 400.; the plate in Arb. Brit., 1st edit., vol. vi., and our *fig.* 666.)

666. C. O. melanocárpa.

also has the fruit black, as the name implies. It differs from the preceding variety chiefly in being of more vigorous growth; in having its leaves of a deep rich green, and in flowering a week later. A splendid low tree, deserving a place in every collection.

667. C. O. Oliveriàna.

♀ C. O. 18 *Oliveriàna* · *C.* Oliveriàna Bosc, Dec. Prod. ii. p. 630., and

Don's Mill. ii. p. 601.; *C.* Olivèria Lodd. Cat.; *C.* orientàlis *Lodd. Cat., Bot. Reg.* t. 1953. (*fig.* 719. in p. 400.; the plate in Arb. Brit., 1st edit., vol. vi.; and our *fig.* 667.); has the leaves small and hoary, and the fruit also small and black. A very distinct variety. Hort. Soc.

♆ C. O. 19 *aúrea* Hort., *C.* flàva *Hort.* (*fig.* 723. in p. 401.), has the leaves like *C. O.* obtusàta, and the fruit roundish, and of a golden yellow. This is a very distinct variety, and ought never to be omitted in collections.

♆ C. O. 20 *aurantìaca* Booth is said to have orange-coloured fruit; but there are only small plants of it in the London gardens. Mr. Wilson found, in Ayrshire, a variety with greenish orange fruit. (*Hook.*)

♆ C. O. 21 *leucocárpa*, a variety with white haws, is said to have been discovered in a hedge near Bampton, in Oxfordshire; but we have never seen it.

F. *Varieties differing in having the Fruit woolly.*

♆ C. O. 22 *eriocárpa* Lindl., *C.* eriocárpa *Lodd. Cat.* (*fig.* 720. in p. 400.; the plate in Arb. Brit., 1st edit., vol. vi.; and our *fig.* 668.), is a robust

668. C. O. eriocárpa.

rapidly growing variety, with large leaves, and strong thick shoots; a clear white bark, and few thorns. It is very prolific in flowers, and the fruit which succeeds them is woolly in its young state, but not large. If ever the hawthorn should be cultivated for its timber, to supply the wood-engravers as a substitute for box, this variety and *C. O.* melanocárpa will deserve the preference. Hort. Soc.

G. *Varieties differing in the Form of the Leaves.*

♆ C. O. 23 *obtusàta* Dec. Prod.; *Méspilus Oxyacántha* integrifòlia *Wallr. Sched.* 219.; *C.* oxyacanthöìdes *Thuill. Fl. Par.* 245., *Bot. Reg.* t. 1128., *Dec. Fl. Fr.* iv. p. 433.; *C. Oxyacántha Fl. Dan.* t. 335.; the French hawthorn. (*fig.* 714. in p. 399.; the plate in Arb. Brit., 1st edit., vol. vi.; and our *fig.* 669.) — Leaves rather rhomb-shaped at the base, obovate, undivided, or with three obtuse lobes,

669. C. O. obtusàta.

crenate, of the same colour on both surfaces. Styles 1—3. Fruit containing more than 1 seed. Distinguished from the species by its smaller, obovate, less cut, flat, and shining leaves. *C.* lùcida *Smith of Ayr*, *C.* oxyacanthöìdes lùcida *Sweet*, is scarcely or not at all different from this variety.

♀ C. O. 24 *quercifòlia* Booth (*fig.* 721. in p. 401.) appears very distinct in regard to foliage.

670. C. O. laciniàta.

♀ C. O. 25 *laciniàta*, *C.* laciniàta *Lodd. Cat.* (*fig.* 716. in p. 400.; the plate in Arb. Brit., 1st edit., vol. vi.; and our *fig.* 670.), has finely cut leaves; the shoots are comparatively slender, the plant less robust, and the fruit smaller, than in the species. It is a very distinct and elegant variety.

♈ C. O. 26 *pteridifòlia*, *C.* pterifòlia *Lodd. Cat.* (*fig.* 717. in p. 400.), resembles the preceding, but the leaves are longer in proportion to their breadth, and more elegantly cut.

♈ C. O. 27 *oxyphýlla* Monckton. — Leaves much larger than those of the species. Raised by General Monckton, at Somerford, in ? 1837. Horticultural Society's Garden.

H. *Varieties differing in the Colour of the Leaves.*

♈ C. O. 28 *fòliis aùreis* Lodd. Cat., *C.* lutéscens *Booth*, has leaves variegated with yellow; but they have generally a ragged and diseased appearance, when fully expanded; though, like those of most other variegated deciduous plants, when first opening in spring, they are strikingly showy and distinct.

♈ C. O. 29 *fòliis argénteis* Hort. has leaves variegated with white; but, like the preceding variety, it cannot be recommended as handsome at any other period than when the leaves are first expanding.

♈ C. O. 30 *lùcida*.—We apply this name to a very distinct and very beautiful-leaved variety, which forms a standard in the southern boundary hedge of the Hort. Soc. Garden, and which, we trust, will soon be propagated in the nurseries. The leaves are large, regularly cut, somewhat coriaceous in texture, and of a fine shining green. The plant is of vigorous growth.

The common hawthorn, in its wild state, is a shrub or small tree, with a smooth bark and very hard wood. The rate of growth, when the plant is young, and in a good soil and climate, is from 1 foot to 2 or 3 feet a year, for the first three or four years; afterwards its growth is slower, till the shrub or tree has attained the height of 12 or 15 feet, when its shoots are produced chiefly in a lateral direction, tending to increase the width of the head of the tree rather than its height. In a wild state, it is commonly found as a large dense bush; but, pruned by accident or design to a single stem, it forms one of the most beautiful and durable trees of the third rank that can be planted: interesting and valuable for its sweet-scented flowers in May, and for its fruit in autumn, which supplies food for some of the smaller birds during part of the winter. In hedges, the hawthorn does not flower and fruit very abundantly when closely and frequently clipped; but, when the hedges are only cut in at the sides, so as to be kept within bounds, and the summits of the plants are left free and untouched, they flower and fruit as freely as when trained as separate trees. The plant lives for a century or two, and there are examples of it between 40 ft. and 50 ft. in height, with trunks upwards of 3 ft. in diameter at 1 ft. from the ground.

The wood of the hawthorn is very hard, and difficult to work: its colour is white, but with a yellowish tinge; its grain is fine, and it takes a beautiful polish; but it is not much used in the arts, because it is seldom found of sufficient size, and is, besides, apt to warp. It weighs, when green, 68 lb. 12 oz. per cubic foot; and, when dry, 57 lb. 5 oz. It contracts, by drying, one eighth of its bulk. It is employed for the handles of hammers, the teeth of mill-wheels, for flails and mallets, and, when heated at the fire, for canes and walking-sticks. The branches are used, in the country, for heating ovens; a purpose for which they are very proper, as they give out much heat, and possess the property of burning as readily when green, as in their dry state. They are not less useful in the formation of dead hedges, for the protection of seeds, or of newly planted live hedges or single trees; and they will last a considerable time without decaying; especially when they have been cut in autumn. The leaves are eaten by cattle, which, nevertheless, pay some regard to the spines by which they are defended. The fruit is astringent. By far the most important use of the hawthorn is as a hedge plant. For this purpose, it is planted in single or in double lines, most commonly along the margin of a ditch; though, however convenient this may be with respect to fencing the plants when young, and draining the soil, it is a great drawback to their progress afterwards, by preventing their roots from extending themselves, except on one side; and, by the drainage of the ditch, it also deprives them of their natural share of moisture. Wherever thorn hedges are planted, and intended to thrive, the ground ought to be trenched at least 2 ft. in depth, manured if poor, and the plants inserted on a flat surface, so as to receive and retain the whole of the moisture that falls from the heavens.

The hawthorn will do no good unless planted in a soil naturally dry and fertile, or that has been rendered so by art. The plant is never found naturally on a wet soil; and, if planted on such a soil, it soon becomes stunted, and covered with lichens and moss. The situation should be airy; but it will grow either in exposed places, or in such as are sheltered, and even somewhat shaded, by other trees. In cases of this kind, however, it neither forms a handsome tree, nor a close thick hedge.

The species is almost always propagated by seeds, but sometimes by cuttings of the roots; which, when about half an inch in thickness, and 1 ft. or 18 in. in length, and planted with the root end undermost, speedily make large plants. Where old thorn plants are taken up, the roots may always be used for forming new hedges; but it must be acknowledged that, as they do not all send up shoots equally, some remaining a year in the ground before they do so, the preferable mode is to plant them in a nursery for the first year; or, if this is not done, they ought to be planted thick, so as to make allowance for some not pushing till the second year, and some not pushing at all.

When the hawthorn is to be raised from seed, the haws should not be gathered till they are dead ripe; which will be in October or November. As many haws contain more than one seed, they ought not to be put in the ground entire, but, if they are to be sown immediately, they must be macerated in water till the pulp is separated from the nuts; and the latter should then be mixed with dry sand, to keep them separate, and to enable the sower to scatter them equally over the surface. But, as the seeds do not come up till the second year, a saving of ground is made by keeping them the first year in a heap, technically called a rot-heap, mixed with a sufficient quantity of soil, to prevent them from heating, and to facilitate the decomposition of the pulp. These heaps are kept in the open air, and exposed to the full influence of the weather; care being taken to turn them over frequently, at least once a month, so as to equalise this influence. When the seeds are not to be prepared in a heap, they should be sown in November or December, as soon as separated from the pulp; but, when they are to be separated by decomposition in a heap, they need not be sown till the February, or even the March, of the second year; by which means fifteen or sixteen months' use of the soil is saved. They may be sown thinly in beds, the seeds being scattered so as to lie about 1 in. apart every way, and covered about a quarter of an inch. The nursery culture required is mere routine. Hawthorns ought always to be two years transplanted before they are employed for hedges; younger and untransplanted plants, though cheaper to purchase, are always the most expensive to the planter, as they require temporary protection for a longer period. As stocks, hawthorn plants may be treated like stocks for fruit trees; and the different species and varieties may be budded and grafted on them, either for dwarfs or standard high, in a similar manner. Not only the different species of *Cratæ`gus*, but those of *Méspilus*, *Sórbus*, *Pýrus*, and even *Màlus*, *Cydònia*, *Amelánchier*, Eriobótrya, and others, may be grafted on the common hawthorn; and in this way field hedges might be rendered ornamental, and even productive of useful fruits.

§ xiii. *Parvifòliæ*.

Sect. Char. Leaves small, ovate, serrated or notched, but scarcely lobed. Fruit green, or greenish yellow; rather large, hard.

■ 25. *C.* PARVIFO'LIA *Ait.* The small-leaved Thorn.

Identification. Ait. Hort. Kew., 2. p. 169.; Dec. Prod., 2. p. 627.; Don's Mill., 2. p. 598.
Synonymes. Méspilus axillàris *Pers. Syn.* 2. p. 37.; *M.* tomentòsa *Poir. Dict.* 4. p. 443.; *M.* xanthocárpos *Lin. fil. Suppl.* 254.; *M.* parvifòlia *Wats. Dend. Brit.*; Cratæ'gus tomentòsa *Lin. Sp.* 682., *Trew Ehr.* t. 17.; *C.* uniflòra *Du Roi*; *C.* turbinàta *Pursh*; *C.* viridis, axillàris, betulifòlia, flórida, lineàris, *Lodd. Cat.*; Gooseberry-leaved Thorn; Lord Iley's Thorn.
Engravings. Trew Ehr., t. 17.; Dend. Brit., t. 65.; our *fig.* 671.; and *fig.* 727. in p. 402.

Spec. Char., &c. Leaves oval-lanceolate, incisely serrated, and pubescent. Flowers mostly solitary. Branchlets and calyxes villose. Stipules bristle-like. Sepals serrated. Fruit almost top-shaped, yellow, or yellowish green. Nuts 5. (*Dec. Prod.*) A low shrub. North America, New Jersey to Carolina, in sandy shady woods. Height 4 ft. to 6 ft. Introduced previously to 1713. Flowers white; May and June, rather later than in most other species. Haws large, greenish yellow; ripe in November, often hanging on the tree all the winter.

671. C. parvifòlia.

Varieties.

■ C. p. 2 *flórida*, *C.* flórida *Lodd. Cat.* (*fig.* 726. in p. 402., and our *fig.* 672.), has the leaves and fruit somewhat smaller and rounder than those of the species.

672. C. p. flórida.

■ C. p. 3 *grossulariæfòlia*, *C.* lineàris *Lodd. Cat.* (*fig.* 728. in p. 402., and our *fig.* 673.), has the leaves lobed, and somewhat like those of the gooseberry.

These varieties run so much into one another, that, unless they are seen together in a living state, as in Messrs. Loddiges's arboretum, it is difficult to distinguish them from the species, or from each other; for, however different the leaves may appear in our figures (see p. 402.), all the forms of these may occasionally be found on the same plant: and some plants of each variety are wholly without spines, while in others the spines are very numerous. As all of them are small plants, with flowers large in proportion to the size of the leaves, they are well adapted for exemplifying the genus Cratæ'gus in a miniature arboretum.

673. p. grossulariæfòlia.

♣ 26. *C.* VIRGI'NICA *Lodd.* The Virginian Thorn.

Identification. Lodd. Cat., ed. 1830, and ed. 1836.
Synonymes. *C.* virginiàna *Hort.*, *C.* spathulàta *Michx.* and *Lindl. Bot. Reg.* t. 1890.; *C.* viridis *Hort.*
Engravings. Bot. Reg., t. 1890.; *fig.* 729. in p. 402.; and our *fig.* 674.

Spec. Char., &c. Leaves obovate, cuneate, glabrous, shining, notched, but not lobed; small. Fruit round, rather larger than a common haw, green. A low shrub. Virginia. Height 4 ft. to 5 ft. Introduced in 1812. Flowers white; June. Fruit green; October, often remaining on the branches during the winter.

674. *C.* virginica.

The plant bears a general resemblance to *C.* spathulàta (No. 17.) in its foliage and habit of growth; but the foliage of the latter is lobed, while that of the former is entire. The fruit of *C.* virgínica is, also, six times larger than that of *C.* spathulàta; and is of a dark green, while the other is of a bright red. The blossoms and fruit of *C.* virgínica are, also, produced in corymbs of twos and threes; while those of *C.* spathulàta consist of a considerable number of flowers. The species differ, also, in the foliage; which in *C.* spathulàta has long winged footstalks, while in *C.* virgínica the footstalks are short and slender. (See the leaves of *C.* virgínica *fig.* 729. in p. 402., and of *C.* spathulàta *fig.* 704. in p. 396.)

§ xiv. *Mexicàna.*

Sect. Char. Leaves large, oval-lanceolate, notched and serrated. Fruit large, green or greenish yellow.

♀ ♀ 27. *C.* MEXICA'NA *Moc. et Sessc.* The Mexican Thorn.

Identification. Moc. et Sesse Fl. Mex. icon. inedit.; Dec. Prod., 2. p. 629.; Swt. Brit. Fl.-Gard., 2d ser. t. 300.; Don's Mill., 2. p. 601.
Synonymes. *C.* stipulàcea *Lodd. Cat.*, see *Gard. Mag.* ix. p. 630.; *C.* Lambertiàna Hort.
Engravings. Swt. Brit. Fl.-Gard., 2d ser. t. 300.; Bot. Reg., t. 1910.; our *fig.* 730. in p. 402. the plate of this species in Arb. Brit., 1st edit., vol. vi.; and our *fig.* 675.

Spec. Char., &c. Leaves oval-lanceolate, notched and serrated; acuminate, somewhat ciliated at the base. Petioles short, channeled, and with a winged margin. Stipules stalked. Corymbs terminal. Petals scarcely longer than the calycine teeth. Stamens varying from 10 to 15. Styles 2, or rarely 4. Fruit large, pale green, or yellowish, when ripe; and, with the leaves, remaining on the tree all the winter in sheltered situations. Handsome, and resembling a small apple, but not good to eat. A low tree; evergreen against a wall, and sub-evergreen as a standard in the climate of London, and southwards. Mexico, on table lands. Height 20 ft. to 30 ft.; in British gardens 10 ft. to 15 ft. Introduced in 1824. Flowers large, white; June. Fruit large, pale green, ripening against a wall in October.

675. C. mexicàna.

A remarkable and very handsome species, resembling, in general appearance, *Méspilus grandiflòra.*

§ xv. *Pyracántha.*

Sect. Char. Leaves oval-lanceolate, glabrous, entire, small, evergreen. Fruit numerous, of a bright coral colour.

■ 28. *C. Pyraca'ntha Pers.* The fiery Thorn, *or Pyracantha.*

Identification. Pers. Syn., 2. p. 37.; Pall. Fl. Ross., 1. p. 29.; Dec. Prod., 2. p. 626.; Don's Mill., 2. p. 598.

Synonymes. *Méspilus Pyracántha L.*; evergreen Thorn; Buisson ardent, *Fr.*; immergrüne Mispel, *Ger.*; Agazzino, *Ital.*

Engravings. Lob. Icon., 2. p. 182. f. 1.; Pall. Fl. Ross., t. 18. f. 2.; N. Du Ham., 2. t. 20. No. 2.; and our *fig.* 676.

Spec. Char., &c. Evergreen. Leaves glabrous, ovate-lanceolate, crenate. Lobes of the calyx obtuse. Styles 5. Fruit globose, scarlet, ornamental; continuing a good while upon the plant; which, on account of the colour of its fruit, and of its being a shrub, is called in France Buisson ardent. (*Dec. Prod.*) An evergreen shrub. South of Europe, in rugged places and hedges. Height 4 ft. to 6 ft. Introduced in 1629. Flowers white; May. Fruit red; ripe in September.

676. C. Pyracántha.

Variety.

■ C. P. 2 *crenulàta. C.* crenulàta *Hort.* — A plant bearing this name in the Hort. Soc. Gard. differs very little from the species.

The flowers and fruit are produced in large corymbs, which are very ornamental; and the fruit remains on all the winter, especially when the shrub is trained against a wall. The berries are bitter, and are not so greedily eaten by birds as those of some other kinds, unless in very severe winters. The plant is very hardy, and, in the open garden, forms a handsome evergreen bush; and, if grafted standard high on the common hawthorn, it would form a most desirable evergreen low tree.

Cratæ'gus coccínea. The scarlet-*fruited* Thorn.
Leaves and fruit of the natural size.

677

Cratæ'gus coccínea corállina, and C. *c. indentàta*. The coral-*fruited* Thorn, and the indented-*leaved* Thorn.

Leaves and fruit of the natural size.

678

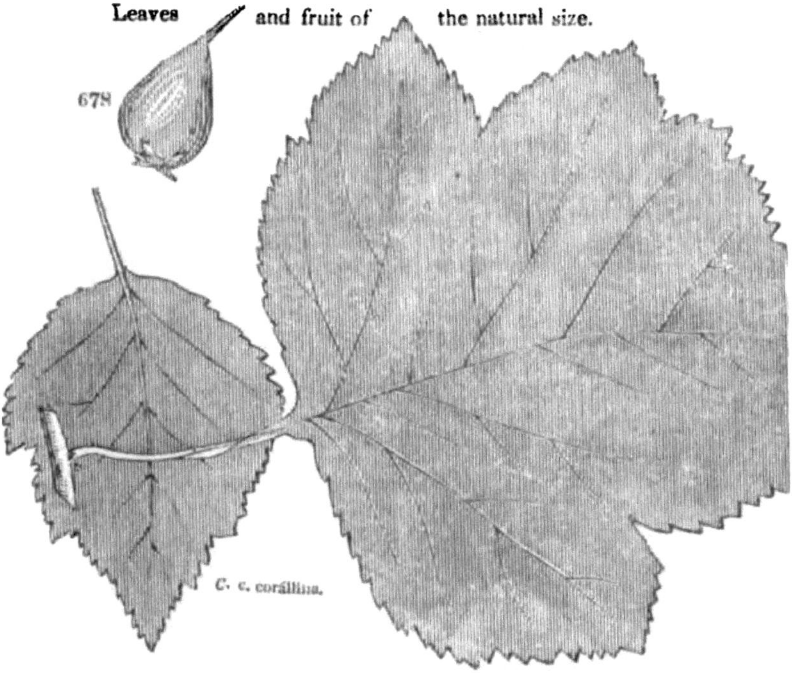

C. c. corállina.

679

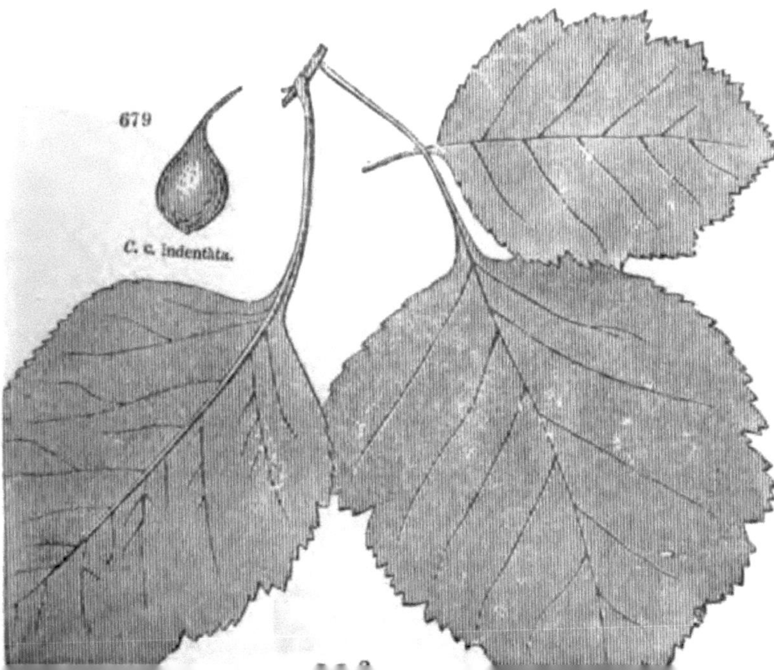

C. c. indentàta.

Cratæ`gus glandulòsa, and C. _g. subvillòsa._ The glandular Thorn, and the subvillose-_leaved_ glandular Thorn.

Leaves and fruit of the natural size.

680

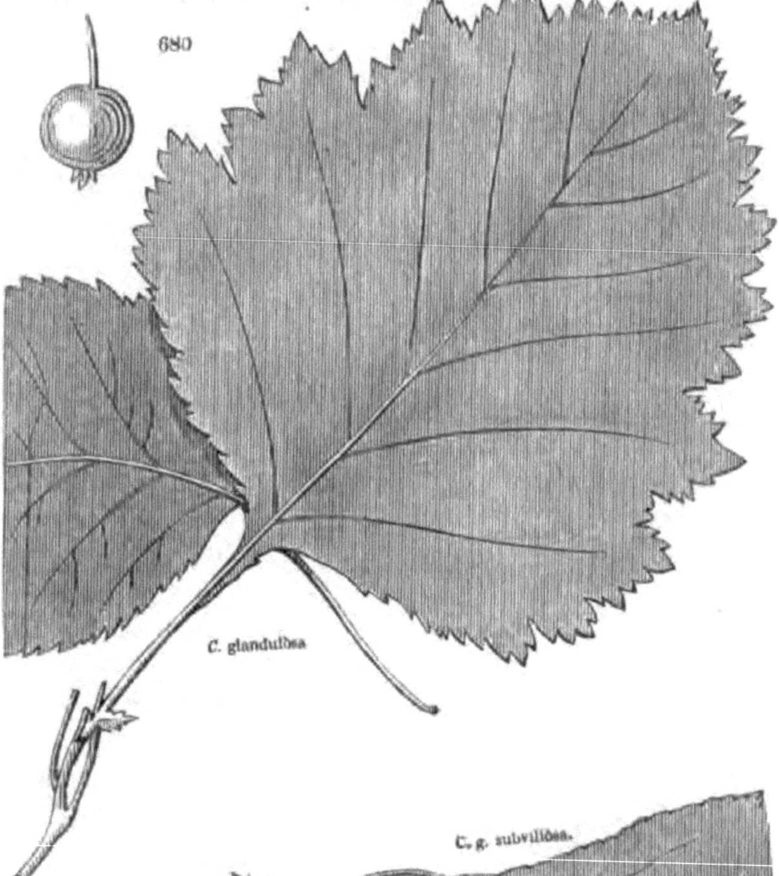

681

Cratæ`gus punctàta, and C. pyrifòlia. The dotted-*fruited* Thorn, and the Pear-leaved Thorn.

Leaves and fruit of the natural size.

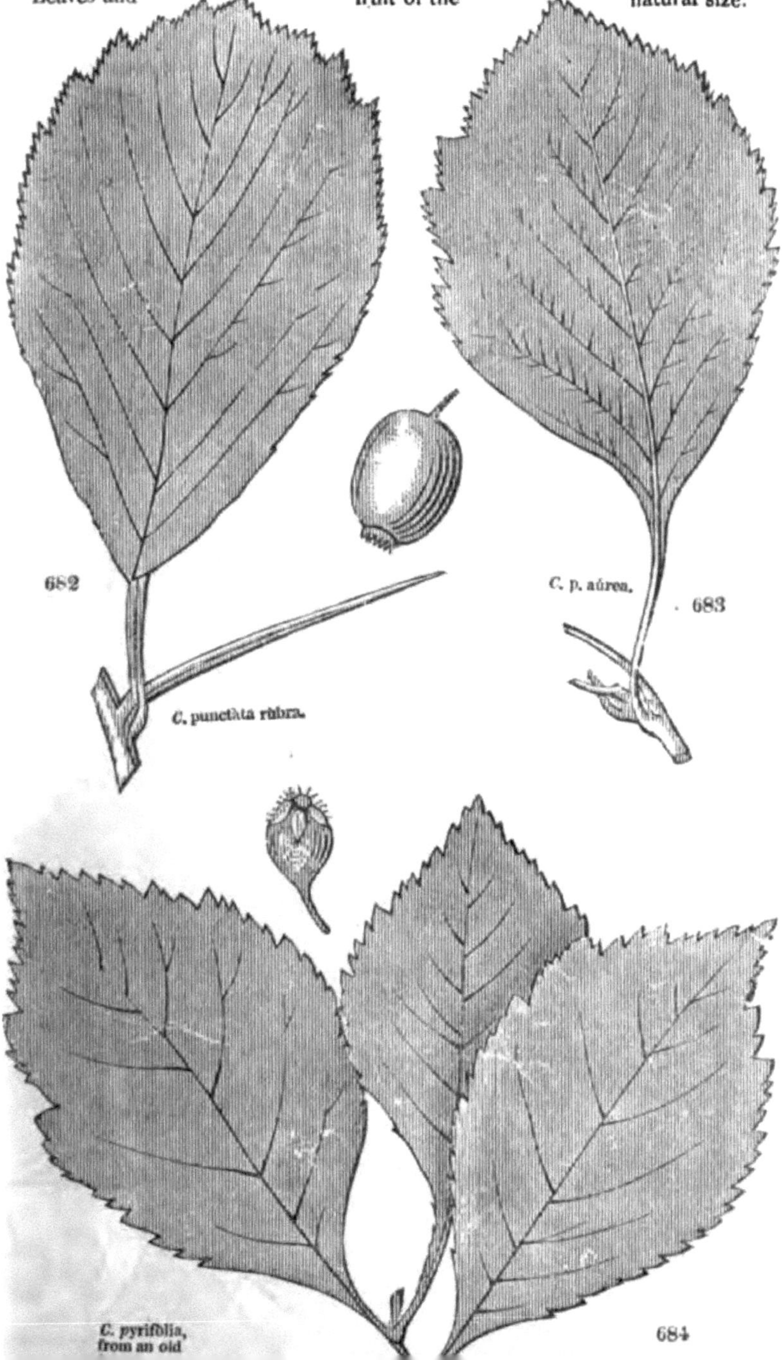

Cratæ`gus macracántha. The long-spined Thorn.
Leaves and fruit of the natural size.

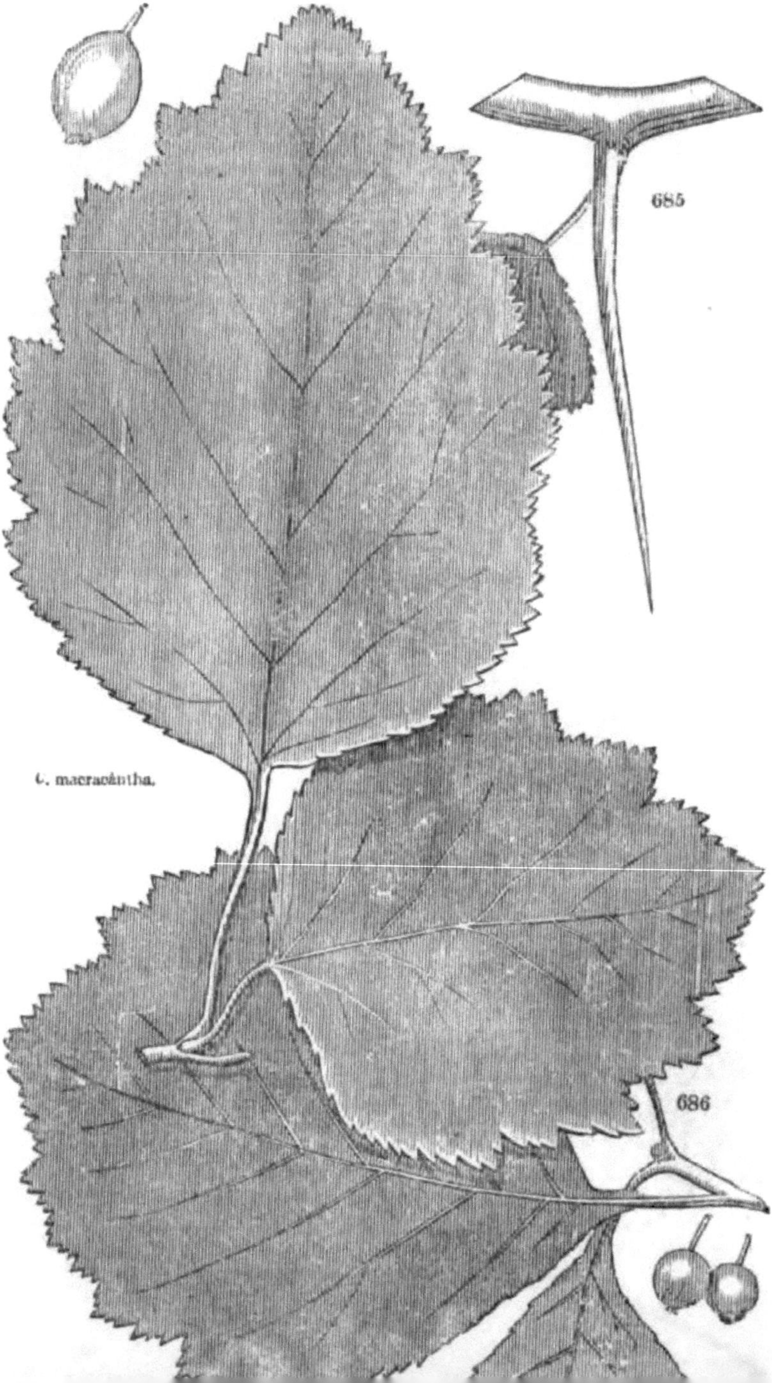

Cratæ'gus Crús-gálli, et var. The Cock's-spur Thorn, and its varieties.
Leaves and fruit of the natural size.

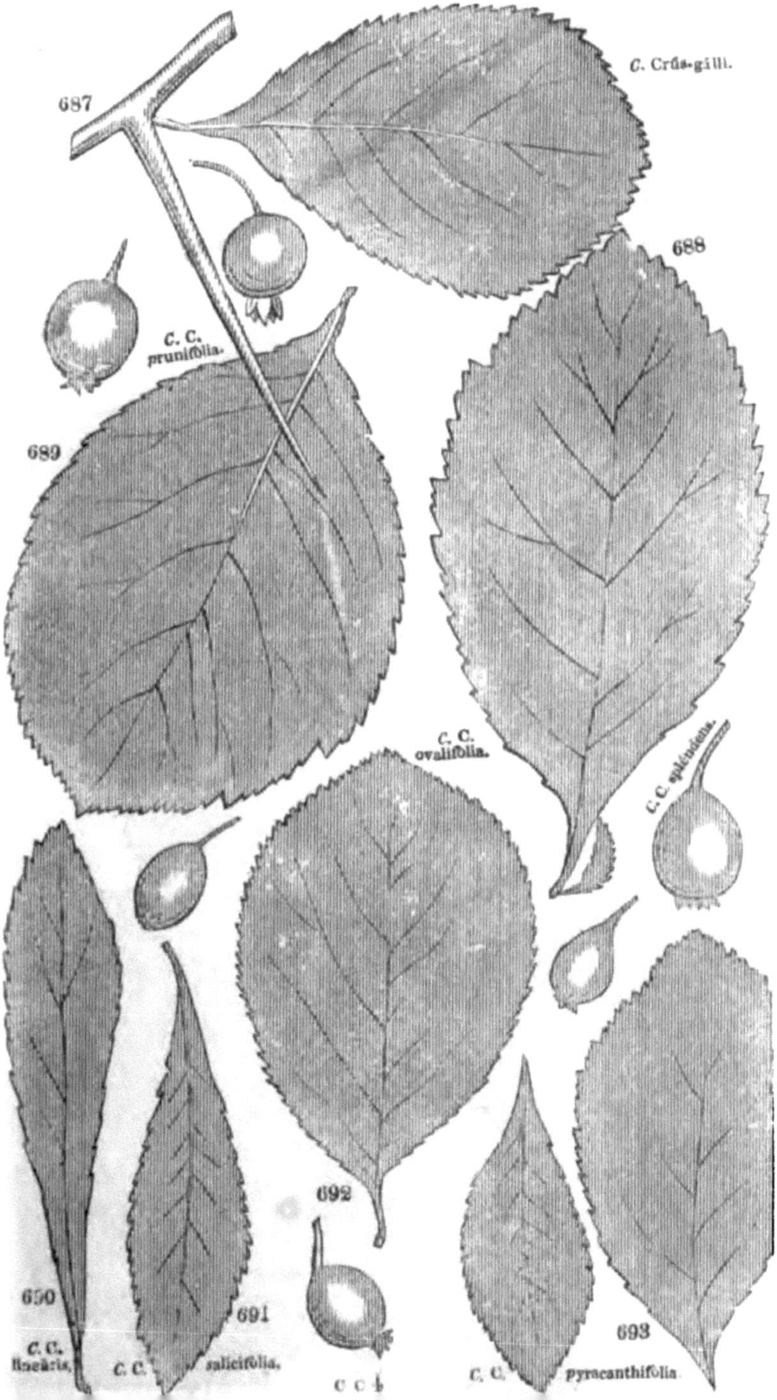

Cratæ̀gus nìgra, and C. *purpùrea*. The black-*fruited* Thorn, and the purple-*branched* Thorn.

Leaves and fruit of the natural size.

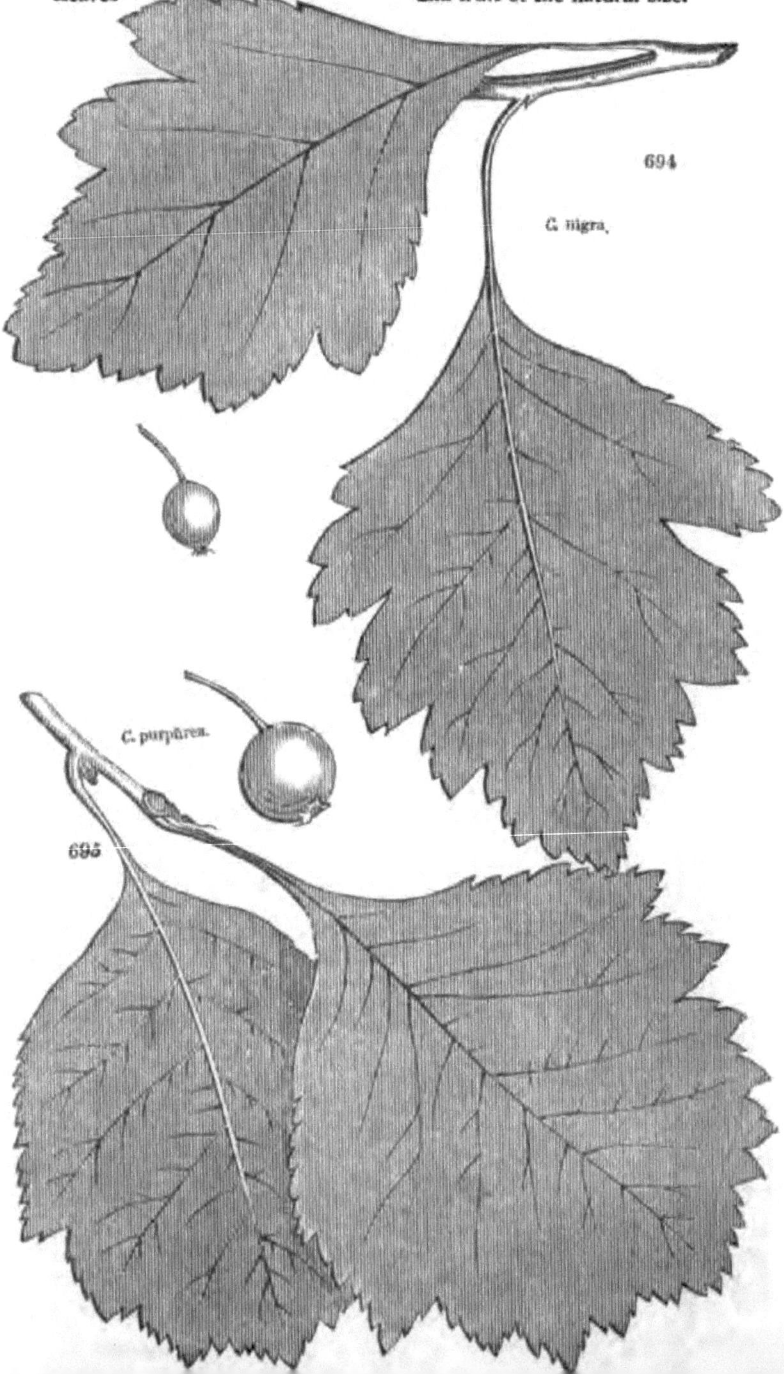

Cratæ`gus *purpùrea altàica*, and C. *Douglàs*ii. The Altaic purple-*branched* Thorn, and Douglas's Thorn.

Leaves and fruit of the natural size.

696

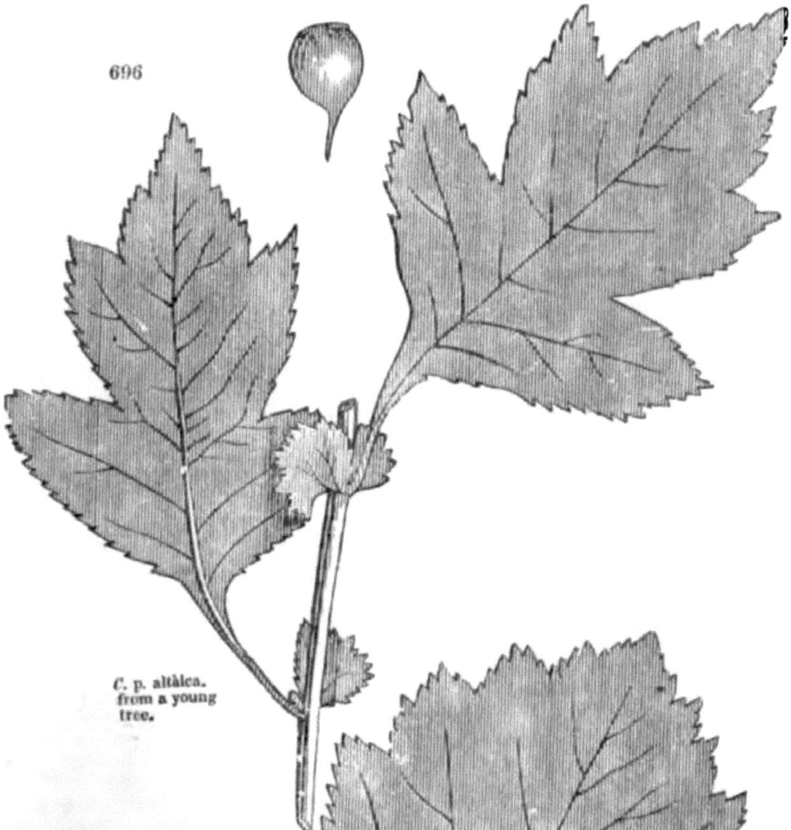

C. p. altàica. from a young tree.

697

Cratæ̀gus flàva, and C. *lobàta*. The yellow-*fruited* Thorn, and the lobed-*leaved* Thorn.

Leaves and fruit of the natural size.

698

C. flàva.

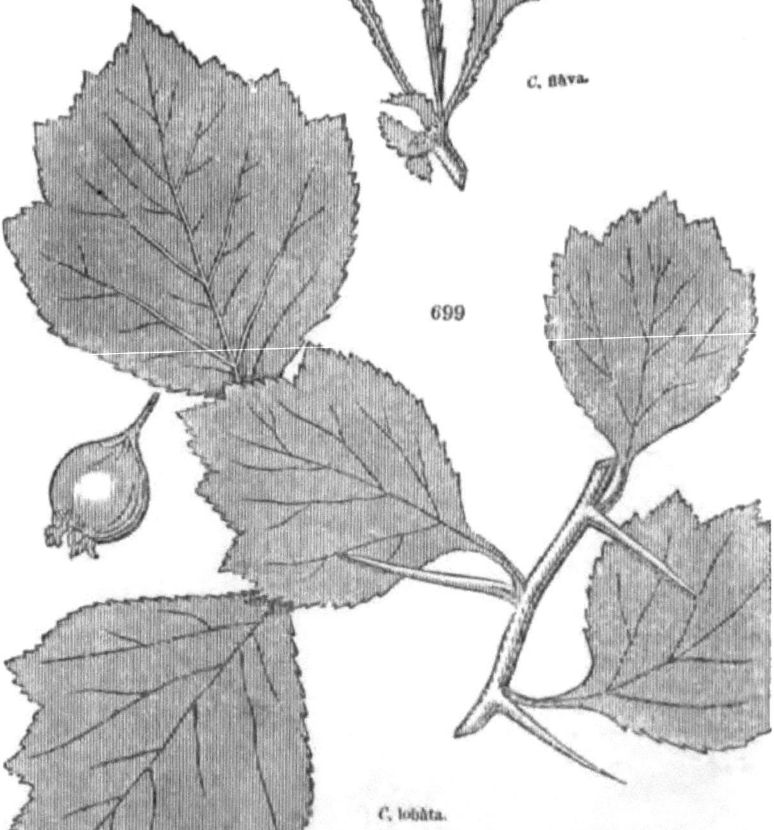

699

C. lobàta.

Cratæ̀gus trilobàta, and C. apiifòlia. The three-lobed-leaved Thorn, and the Parsley-leaved Thorn.

Leaves and fruit of the natural size.

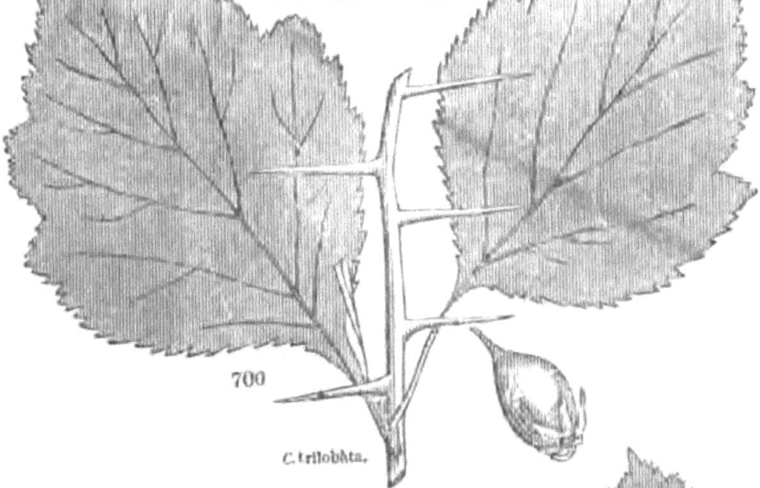

700

C. trilobàta.

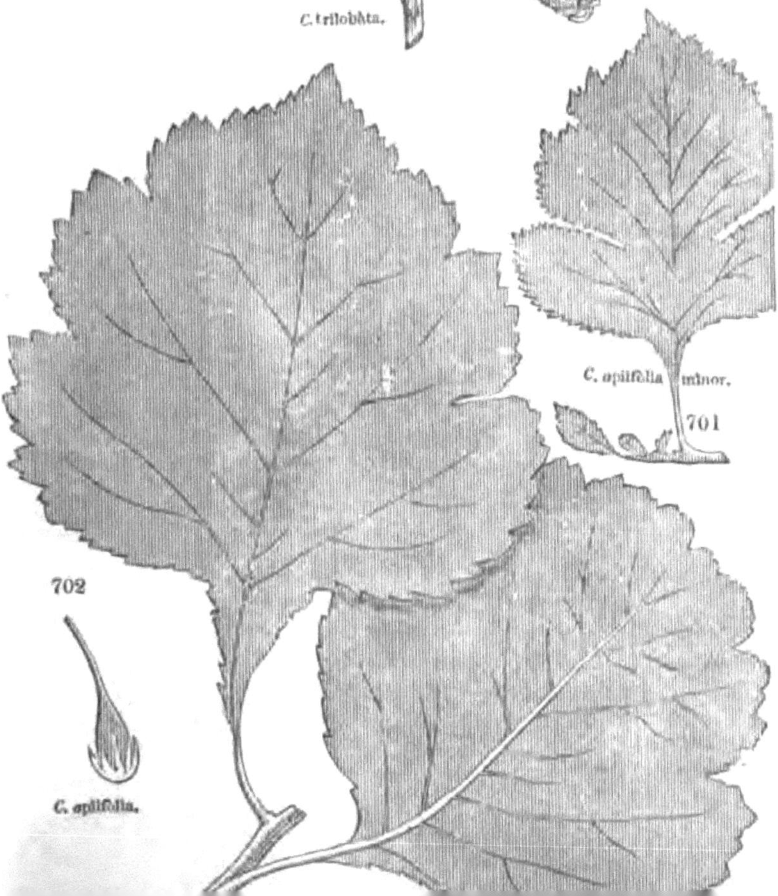

C. apiifòlia minor.

701

702

C. apiifòlia.

Cratæ̀gus *cordàta*, and C. *spathulàta*. The heart-*shaped-leaved* Thorn, and the spathula-*shaped-leaved* Thorn.

Leaves and fruit of the natural size.

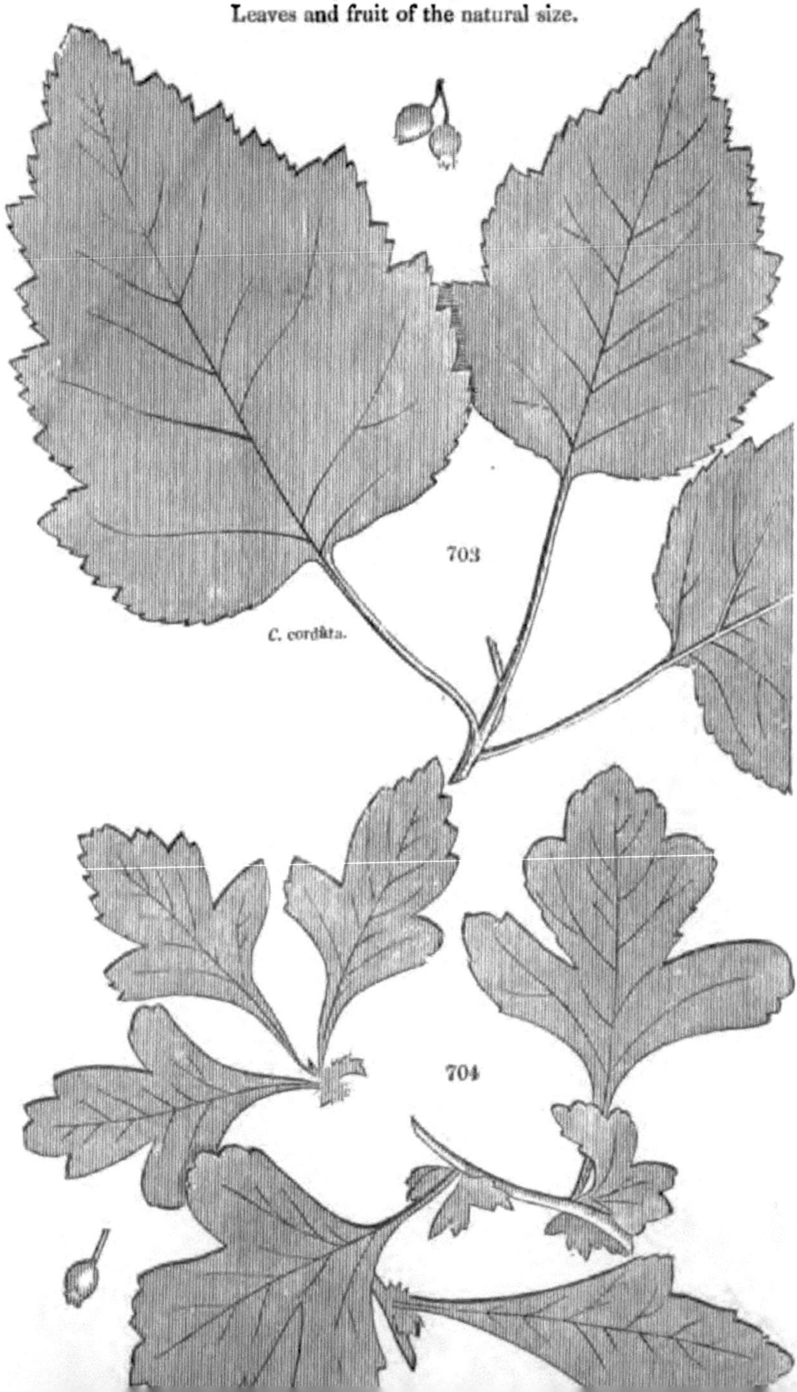

Cratæ'gus Azaròlus, C. Arònia, and C. *maroccàna.* The Azarole Thorn, the Aronia Thorn, and the Morocco Thorn.

Leaves and fruit of the natural size.

705

706

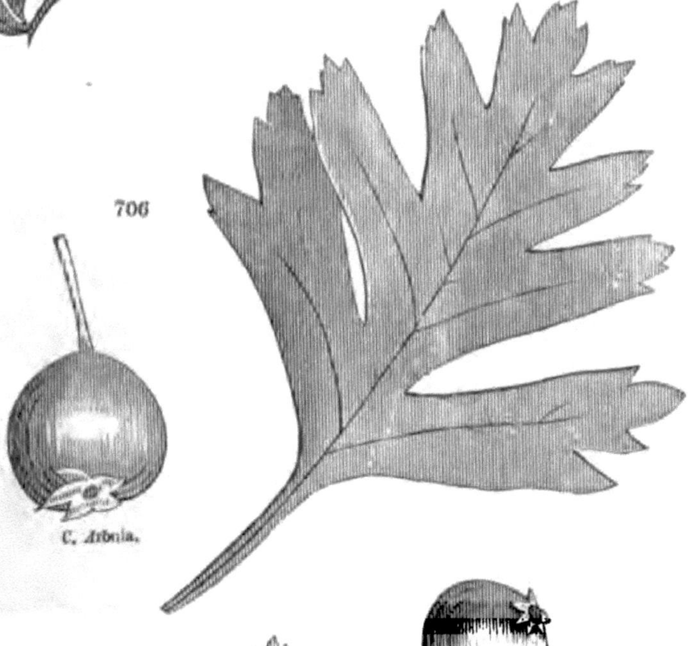

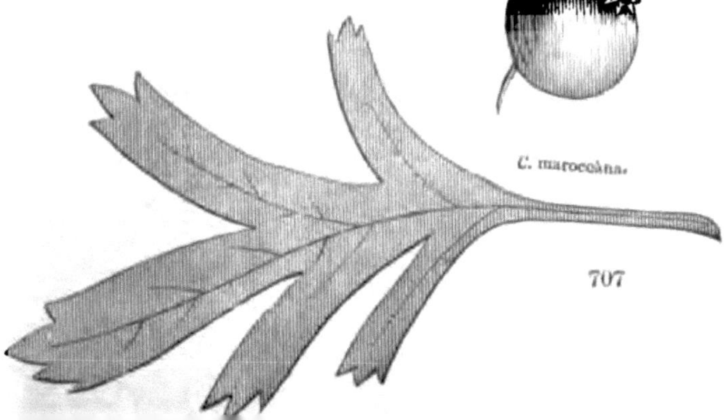

707

Cratæ̀gus orientàlis, et var., and C. *tanacetifòlia, et var.* The Eastern Thorn, and the Tansy-leaved Thorn, with Varieties.

Leaves and fruit of the natural size.

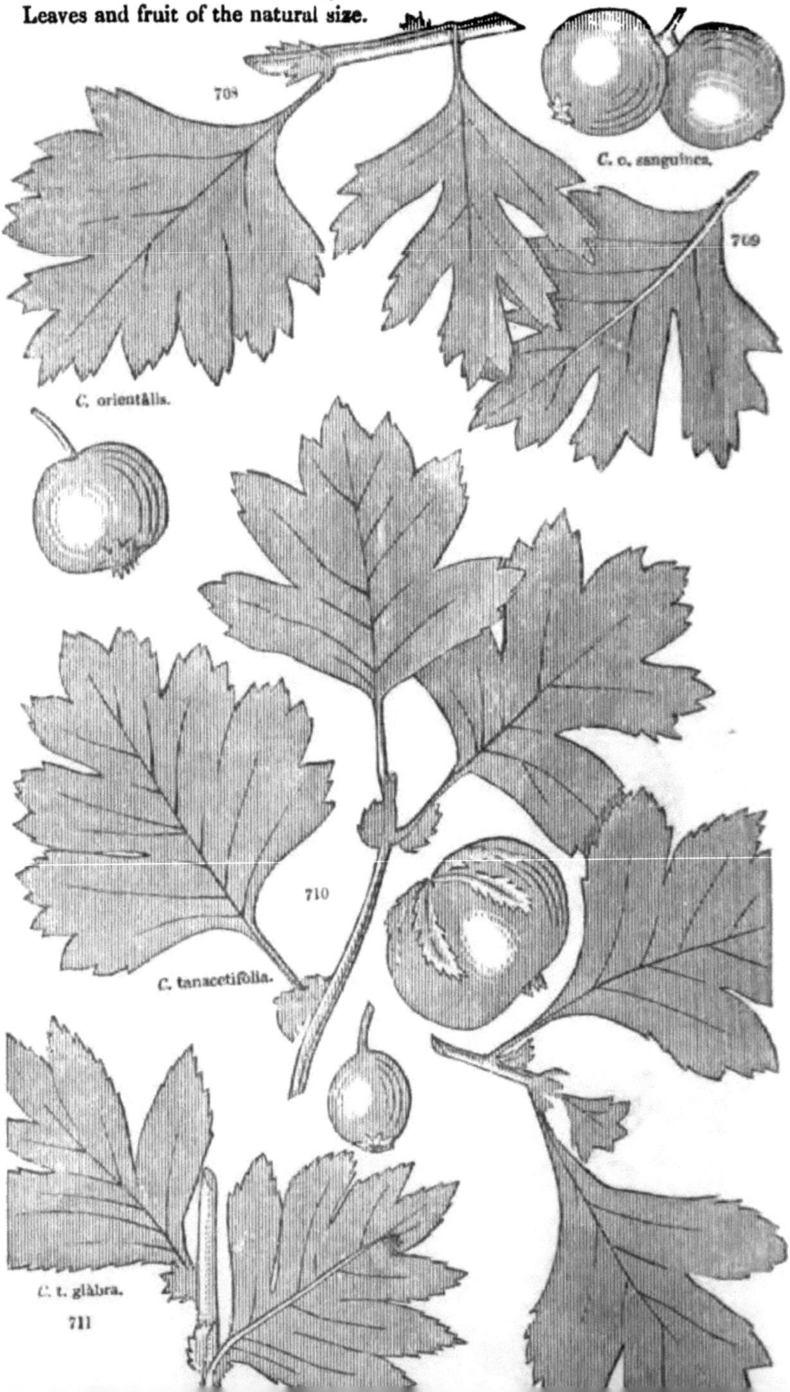

Cratæ'gus *tanacetifòlia Leeàna*, C. *heterophýlla*, and C. Oxyacántha *obtusàta*. Lee's Tansy-leaved Thorn, the various-leaved Thorn, and the obtuse-*leaved* Hawthorn.

Leaves and fruit of the natural size.

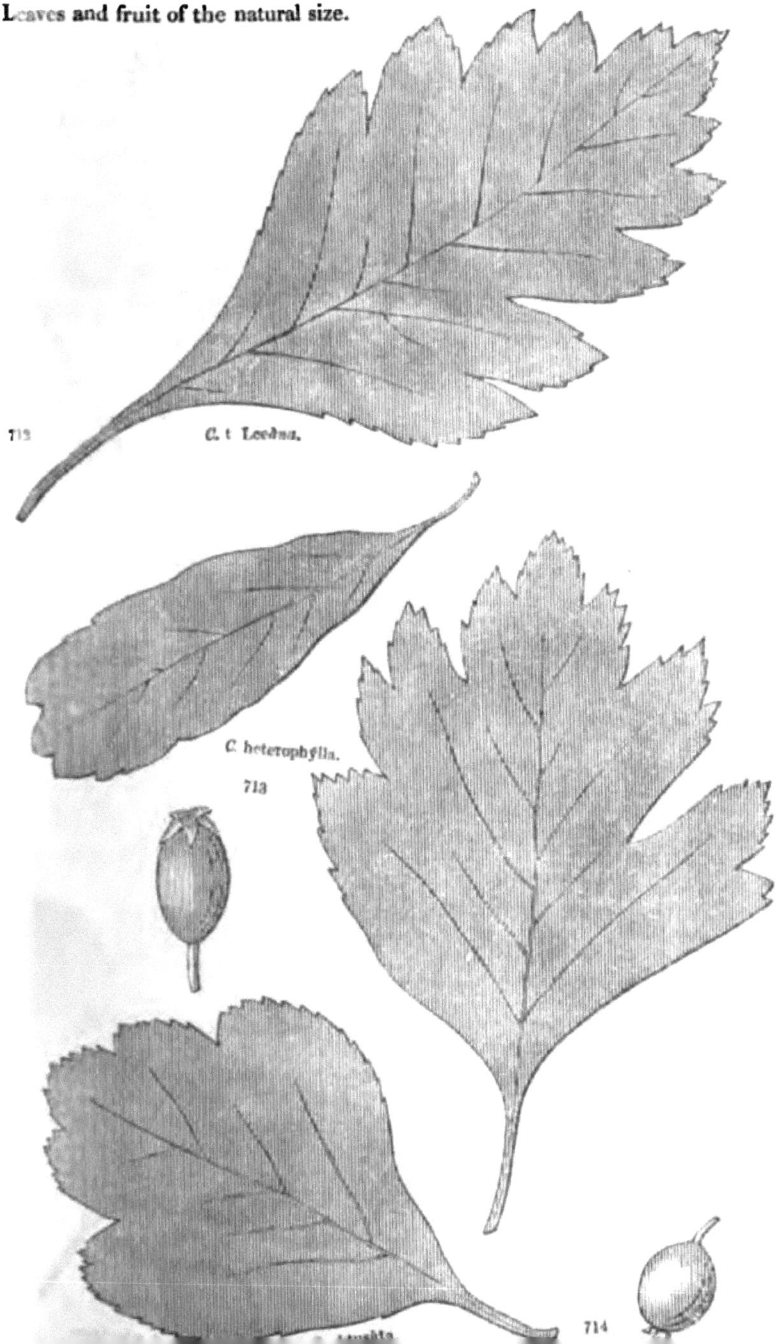

712 C. t. Leeàna.

C. heterophýlla.

713

714

Cratæ̀gus Oxyacántha, et var. The *common* Hawthorn, and Five of its Varieties.

Leaves and fruit of the natural size.

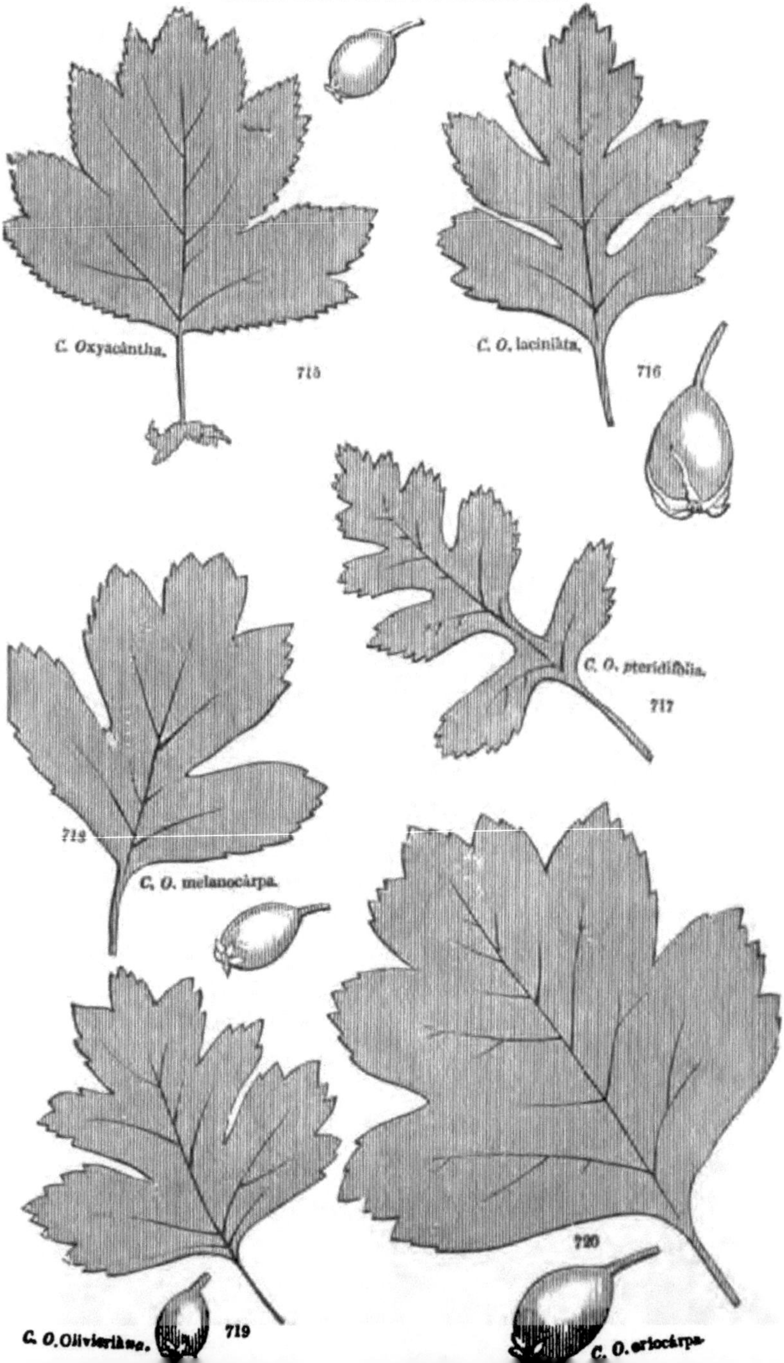

Cratæ`gus Oxyacántha var. Five Varieties of the Hawthorn.
Leaves and fruit of the natural size.

721

C. O. quercifolia.

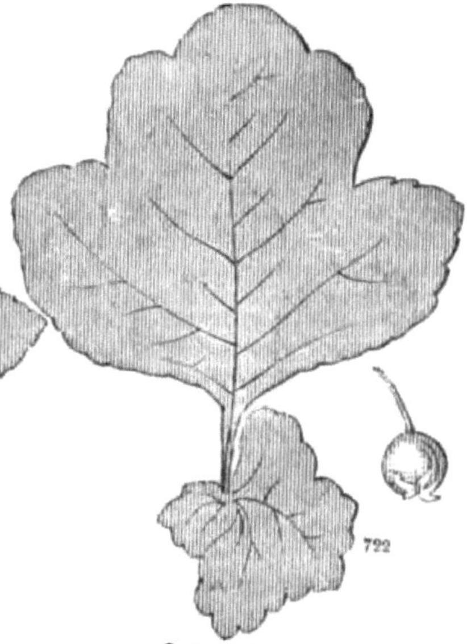

722

C. O. múltiplex.

723

C. O. aúrea.

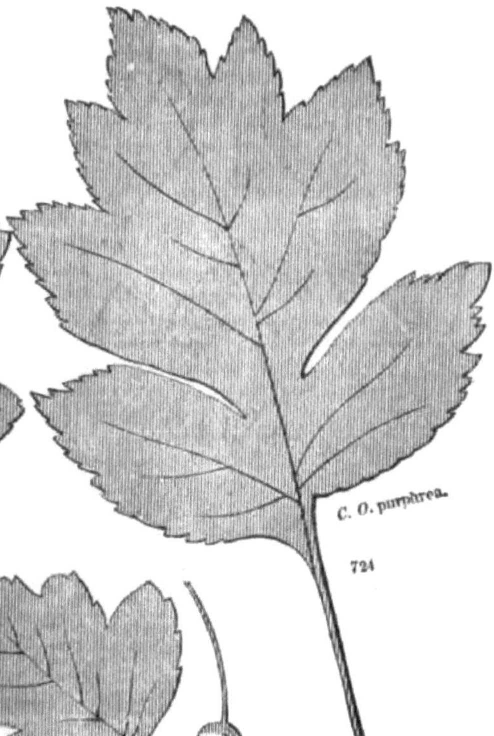

724

C. O. purpùrea.

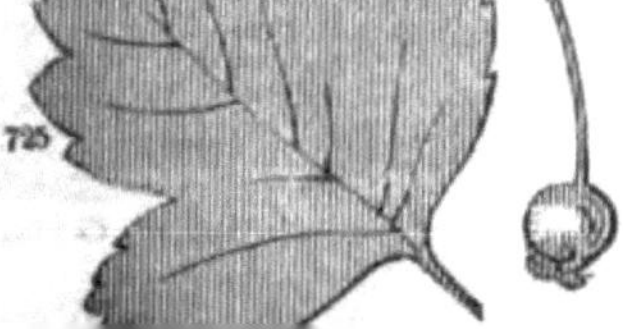

725

Cratæ̀gus parvifòlia, C. *p. flórida*, C. *p. grossulariæfòlia*, C. *virgínica*, C. *mexicàna*. The small-leaved Thorn, the Florida Thorn, the Gooseberry-leaved Thorn, the Virginian Thorn, and the Mexican Thorn. Leaves and fruit of the natural size.

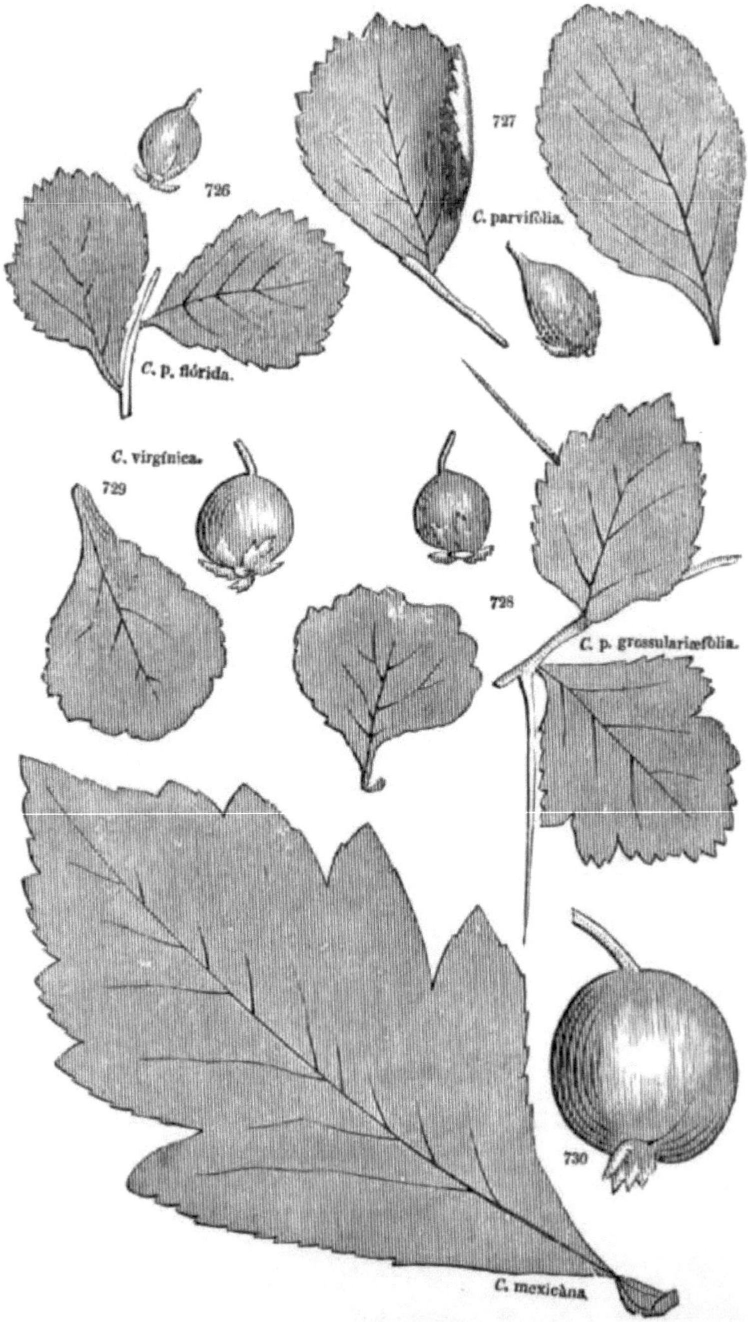

Genus XV.

STRANVÆ'SIA Lindl. The Stranvæ'sia. *Lin. Syst.* Icosándria Di-Pentagýnia.

Identification. Lindl. in Bot. Reg.
Synonyme. Cratæ'gus in part.
Derivation. In honour of the Hon. *W. T. H. Fox Strangways*, F.H.S., &c., a botanical amateur, who possesses a rich collection.

Gen. Char. *Calyx* 5-toothed. *Petals* 5, concave, sessile, spreading, villous at the base. *Stamens* 20, spreading. *Ovary* villous, superior, 5-celled; cells containing 2 ovules. *Fruit* spherical, enclosed by the calyx, containing a superior, 5-valved, hard, brittle, dehiscent capsule. *Seeds* oblong. *Testa* cartilaginous. (*Lindl.*)

Leaves simple, alternate, stipulate, evergreen; lanceolate, serrated. *Flowers* corymbose. — An evergreen tree; native of the temperate parts of Asia; in Britain a rather tender shrub.

■ 1. S. glauce'scens *Lindl.* The glaucous-*leaved* Stranvæsia.

Identification. Lindl. in Bot. Reg., t. 1956.
Synonyme. Cratæ'gus glaúca *Wall. Cat.* 672.
Engravings. Bot. Reg., t. 1956.; and our *figs.* 731. and 732.

731. Stranvæ'sia glaucéscens.

732. S. glaucéscens.

Spec. Char., &c. Leaves lanceolate, coriaceous, serrated, pointed at the base; midrib and nerves on the under side, as well as the young twigs, hairy. Corymbs somewhat woolly. Pedicels 3 or 4 times as long as the bud. (*Lindl.*) An evergreen shrub; in Nepal, a tree 20 ft. high. Introduced in 1828. Flowers white; July. Fruit small, yellowish red; ripe in October.

Somewhat tender when treated as a standard in the open garden; but, when trained against a wall, forming a very handsome evergreen. Propagated by grafting on *Cratæ'gus*.

Genus XVI.

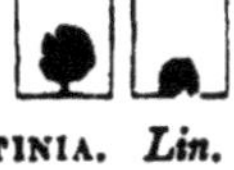

PHOTI'NIA *Lindl.* The Photinia. *Lin. Syst.* Icosándria Di-Pentagýnia.

Identification. Lindl. in Lin. Soc. Trans., 13. p. 103.; Dec. Prod., 2. p. 631.; Don's Mill., 2. p. 602.
Synonyme. Cratæ'gus sp. *L.*
Derivation. From *phōteinos*, shining; in allusion to the lucid surface of the leaves.

Gen. Char. *Calyx* 5-toothed. *Petals* reflexed. *Ovarium* semiadherent, villous, 2-celled. *Styles* 2, glabrous. *Pericarp* 2-celled, enclosed in the fleshy calyx. *Testa* cartilaginous. (*Don's Mill.*)

Leaves simple, alternate, stipulate, evergreen; quite entire, or serrated.

Flowers white, disposed in terminal corymbose panicles. *Fruit* small, smooth. — Evergreen trees; natives of Asia and America, requiring the same garden treatment as *Cratæ'gus* (on which all the species may be grafted), except that the species are somewhat more tender, and are best grown against a wall even in the climate of London.

■ 1. P. SERRULA'TA *Lindl.* The serrulated-*leaved* Photinia.

Identification. Lindl. in Lin. Soc. Trans., 13. p. 103.; Dec. Prod., 2. p. 631.; and Don's Mill., 2. p. 602.
Synonyme. Cratæ'gus glàbra *Thunb. Fl. Jap.* 205., *Bot. Mag.*, *Lodd. Bot. Cab.*, *Colla Hort. Ripul.*
Engravings. Bot. Mag., t. 2105.; Bot. Cab., t. 248.; Colla Hort. Ripul., t. 36.; the plate of this species in Arb. Brit., 1st edit., vol. vi.; and our *fig.* 733.

733. Photinia serrulàta.

Spec. Char., &c. Leaves oblong, acute, serrulated. Pedicels longer than the calyx. Buds large, red. (*Dec. Prod.*) An evergreen shrub or low tree. Japan and China. Height 12 ft. to 15 ft. Introduced in 1804. Flowers white; April and May. Fruit ?; rarely produced in England.

The young shoots and leaves are remarkable for their deep red, or mahogany colour; and the decaying leaves exhibit this colour much more intensely, mixed with a fine yellow or scarlet. Altogether it forms a very splendid plant, when trained against a wall; or, in warm sheltered situations, as a standard. It is commonly grafted or budded on thorn stocks; and it also does well upon quince stocks. In the neighbourhood of London it flowers between the middle of April and the middle of May; but it has not yet produced fruit in England. The largest and oldest plants are at White Knights, where it was planted in 1804; and, in 1835, formed a large bush or tree, nearly 15 ft. high.

■ 2. P. ARBUTIFO'LIA *Lindl.* The Arbutus-*leaved* Photinia.

Identification. Lindl. in Lin. Soc. Trans., 13. p. 103.; Dec. Prod., 2. p. 631.; Don's Mill., 2. p. 602.
Synonyme. Cratæ'gus arbutifòlia *Ait. Hort. Kew.* ed. 2. vol. 3. p. 232.
Engravings. Bot. Reg., t. 491.; and our *fig.* 734.

734. P. arbutifòlia.

Spec. Char., &c. Leaves with the disk oblong-lanceolate, acute, distantly serrated, six times longer than the petiole, which is red. The panicle, in this species, is not corymbose. (*Dec. Prod.*) An evergreen shrub or low tree. California. Height 10 ft. to 20 ft. Introduced in 1796. Flowers white; July and August. Fruit ?.

A very desirable evergreen for walls, hitherto comparatively neglected in British gardens. Horticultural Society's Garden.

735. P. integrifòlia.

† 3. P. INTEGRIFO'LIA *Lindl.* The entire-leaved Photinia.

Identification. Lindl. in Lin. Trans., 13. p. 103.; Don's Mill., 2. p. 602.
Synonyme. Pỳrus integérrima *Wall.* ex *D. Don Prod. Fl. Nep.* p. 237.
Engraving. Our *fig.* 735. from a specimen in the Linnæan herbarium.

Spec. Char., &c. Leaves elliptic, acuminated, quite entire. Panicle diffuse. Pedicels bractless. Ovary 3-celled; cells biovulate. (*Don's Mill.*) An evergreen tree. Nepal. Height 20 ft. Introduced in 1820. Flowers white ?.

Requires the protection of a wall, which it well deserves, on account of the beauty of its foliage.

† 4. P. DU'BIA *Lindl.* The doubtful Photinia.

Identification. Lindl. in Lin. Trans., 13. p. 104.; Don's Mill., 2. p. 602.
Synonymes. Méspilus bengalénsis *Roxb.*; *M.* tinctòria *D. Don Prod. Fl. Nep.* 238.; Cratæ'gus *Shicòla* Ham. MSS.
Engravings. Lin. Trans., 13. t. 10.; and our *fig.* 736.

736. P. dùbia.

Spec. Char., &c. Leaves lanceolate, distantly serrated. Panicle corymbose, pilose. Fruit 2-celled. Seed 1, large, and clothed with a loose testa. Fruit sometimes 1-celled from abortion, and 2-seeded. (*Don's Mill.*) An evergreen tree. Nepal. Height 20 ft. Introduced in 1821. Flowers white ?.

It appears nearly allied to Raphiólepis. Hamilton states that the bark is used, in Nepal, to dye cotton red. Culture and management as in the preceding species.

GENUS XVII.

COTONEA'STER *Med.* THE COTONEASTER. *Lin. Syst.* Icosándria Di-Pentagýnia.

Identification. Med. Gesch. Bot., 1793.; Lindl. in Lin. Soc. Trans., 13. p. 101.; Dec. Prod., 2. p. 632.; Don's Mill., 2. p. 603.
Synonymes. Méspilus sp. *Lin.*; Néflier cotonneux, *Fr.*; Quitten-Mispel, *Ger.*; Cotognastro, *Ital.*
Derivation. *Cotoneaster*, a sort of barbarous word signifying quince-like. The quince was called *Cotonea* by Pliny: and *aster*, a corruption of *ad instar*, is used occasionally to express similitude.

The genus, and C. frígida in particular, is not unlike the quince in its leaves. (*Lindley* in *Bot. Reg.* t. 1187. and 1229.)

Gen. Char. *Flowers* polygamous from abortion. *Calyx* turbinate, bluntly 5-toothed. *Petals* short, erect. *Stamens* length of the teeth of the calyx. *Styles* glabrous, shorter than the stamens. *Carpels* 2—3, parietal, biovulate, enclosed in the calyx. (*Don's Mill.*)

Leaves simple, alternate, stipulate or exstipulate, deciduous or evergreen; generally woolly beneath. *Flowers* in corymbs, lateral, spreading, furnished with deciduous subulate bracteas. *Petals* small, permanent.—Shrubs or low trees. Natives of Europe, America, and Asia.

The species are very desirable garden shrubs or low trees, from the beauty of their foliage, their flowers, and their fruit; the fruit of C. frígida and C. affinis, in particular, being produced in great abundance, and being of an intense scarlet colour, have a very splendid appearance, and remain on the trees the greater part of the winter. The cotoneasters are all readily propagated by seeds, cuttings, layers, or grafting on C. vulgàris, on the common quince, or on the hawthorn. Though the greater part of the species are natives of Asia, yet in Britain they are found to be as hardy as if they were indigenous to the North of Europe, more especially those of them that are true evergreens. Not one of them was killed by the winter of 1837–8 in the Hort. Soc. Garden.

§ i. *Leaves deciduous. Shrubs.*

♠ 1. C. VULGA'RIS *Lindl.* The common Cotoneaster.

Identification. Lindl. in Lin. Soc. Trans., 13. p. 101.; Dec. Prod., 2. p. 632.; Don's Mill., 2. p. 603.
Synonymes. Méspilus Cotoneáster *Lin. Sp.* 686., *Œd. Fl. Dan.* t. 112.; Néflier cotonneux, *Fr.*; Quitten-Mispel, *Ger.*; Salciagnolo, *Ital.*
Engravings. Œd. Fl. Dan., t. 112.; Eng. Bot. Suppl., t. 2713.; and our *fig.* 737.

Spec. Char., &c. Leaves ovate, rounded at the base. Peduncles and calyxes glabrous. (*Dec. Prod.*) A deciduous shrub. Europe and Siberia, on the sunny parts of subalpine hills. Height 2 ft. to 3 ft. in a wild state; in cultivation 4 ft. to 5 ft. Cultivated in 1656. Flowers white, slightly tinged with pink; April and May. Fruit red or black; ripe in July and August.

737. C. vulgàris.

Varieties. The following three forms of this species are to be met with, both in a wild state, and in gardens: —

- ♠ *C. v.* 1 *erythrocárpa* Led. Fl. Alt. ii. p. 219. has the fruit red when ripe.
- ♠ *C. v.* 2 *melanocárpa* Led., *Méspilus* Cotoneáster *Pall. Fl. Ross.* p. 30. t. 14., *M.* melanocárpa *Fisch.*, C. melanocárpa *Lod. Cat.*, has the fruit black when ripe.
- ♠ *C. v.* 3 *depréssa* Fries Nov. Suec. p. 9., Dec. Prod. ii. p. 632., is rather spiny, with lanceolate acutish leaves, and fruit including 4 carpels. It is a native of the rocks of Sweden near Warberg.

♠ 2. C. (v.) TOMENTO'SA *Lindl.* The tomentose, *or woolly*, Cotoneaster.

Identification. Lindl. in Lin. Soc. Trans., 13. p. 101.; Dec. Prod., 2. p. 632.; Don's Mill., 2. p. 603.
Synonymes. Méspilus tomentòsa *Willd. Sp.* 2. p. 1012., not Lam.; *M.* eriocárpa *Dec. Fl. Fr. Synops.* and *Suppl.* No. 3691.
Engraving. Our *fig.* 738. from a specimen in the British Museum.

Spec. Char., &c. Leaves elliptical, obtuse at both ends. Peduncles and calyxes woolly. (*Dec. Prod.*) A deciduous shrub, like the preceding species, of which it appears to us to be only a variety, found wild on the rocks

738. C. (v.) tomentòsa.

of Jura, and in other parts of the Alps of Switzerland; and in cultivation in British gardens since 1759.

■ 3. C. (v.) LAXIFLO'RA *Jacq*. The loose-flowered Cotoneaster.

Identification. Jacq. ex Lindl. Bot. Reg., t. 1305.; Don's Mill., 2. p. 604.
Engravings. Bot. Reg., t. 1305.; and our *figs*. 739. and 740.

739. C. (v.) laxiflòra.

Spec. Char., &c. Leaves oblong, obtuse at both ends, smooth above, and woolly beneath. Cymes panicled, pilose. Calyxes quite smooth. Flowers pink. (*Don's Mill.*) Branches brownish purple, with an ash-coloured cuticle, which peels off. A deciduous shrub, flowering in April, and having the same general appearance and habit as C. vulgàris, but differing from it in having large loose racemes, and in the colour of its flowers, and their greater number. It was raised in the Garden of the Horticultural Society, from seeds sent by Professor Jacquin of Vienna, in 1826. Its native country is unknown.

740. C. (v.) laxiflòra.

Variety.

■ *C.* (*v.*) *l.* 2 *uniflòra* Fischer.—Flowers solitary. Horticultural Society's Garden.

■ ■ 4. C. DENTICULA'TA. The toothed-leaved Cotoneaster.

Identification. H. et B., vol. vi. p. 214.
Engravings. H. et B., vol. vi. t. 556.; and our *fig*. 741.

Spec. Char., &c. Leaves elliptic, or obovate-elliptic, rounded on both sides, mucronate-cuspidate, denticulate at the apex, coriaceous, smooth above, tomentose, pubescent and hoary beneath. Corymbs simple. Flowers sub-dodecandrous, with 1—2 stigmas. Calyx woolly tomentose. (*H. et B.*) A shrub, apparently sub-evergreen, and very like *C.* vulgàris. Mexico, on elevated plains near Actopa, at the height of 6000 ft. Height ? 5 ft. to 6 ft. Introduced in 1839. Flowers white; September. Fruit ?. Raised in the Horticultural Society's Garden from seeds sent home by M. Hartweg.

741. C. denticulàta.

§ ii. *Sub-evergreen or deciduous. Tall Shrubs or low Trees.*

■ 5. C. FRI'GIDA *Wall*. The frigid Cotoneaster

Identification. Wall. ex Lindl. Bot. Reg., t. 1229.; and Don's Mill., 2. p. 604.
Synonyme. *Pyrus Nùssia* Ham. in Prod. Fl. Nep. p. 237., Dec. Prod. 2. p. 634.
Engravings. Bot. Reg., t. 1229.; the plate of this species in Arb. Brit., vol. vi.; and our *fig*. 742.

Spec. Char., &c. Branchlets woolly. Leaves elliptical, mucronate, coriaceous, crenulated, glabrous, woolly beneath when young. Corymbs paniculate, terminal, white, and woolly. Pomes spherical. (*Dec. Prod.*) A sub-evergreen shrub or low tree. Nepal, on the higher mountains of the northern region. Height 10 ft. to 20 ft. Introduced in 1824. Flowers of a snowy white, produced in great abundance; April and May. Fruit crimson, or bright red; ripe in September, and generally remaining on the trees great part of the winter.

A remarkably robust-growing, sub-evergreen, low tree, producing shoots 3 or 4 feet long every season, when young; and, in 3 or 4 years from the seed, becoming very prolific in flowers and fruit. As the fruit, with the greater part

742. C. frígida.

of the leaves, remains on all the winter, the tree makes a splendid appearance at that season; and, in sheltered situations, in the neighbourhood of London, it may be considered as an evergreen. It is very hardy; the specific name of frígida being given to it on account of the coldness of the locality in which it was found. It is propagated by grafting on the common hawthorn.

6. C. (F.) AFFI'NIS *Lindl.* The related (to C. frígida) Cotoneaster.

743. C. (f.) affi'nis.

Identification. Lindl. in Lin. Soc. Trans., 13. p. 101.; Dec. Prod., 2. p. 632.; and Don's Mill., 2. p. 603.
Synonymes. Méspilus integérrima *Hamilt. MSS.*; *M.* affínis *D. Don Prod. Fl. Nep.* 238.; ? C ku-mâna in Lodd. collection.
Engravings. The plate in Arb. Brit., 1st edit., vol. vi.; and our *fig.* 743.

Spec. Char., &c. Leaves ovate, with a small mucro at the tip, and tapered at the base. Peduncles and calyxes woolly. (*Dec. Prod.*) A sub-evergreen shrub or low tree. Nepal, at Chittong, in the lower country. Height 10 ft. to 20 ft. Introduced in 1828. Flowers white; April and May. Fruit bright red, or deep crimson; ripe in September, and remaining on the trees great part of the winter.

A robust shrub or low tree, in general habit and appearance so like the preceding sort, as to induce us to think that they are only different forms of the same species. They are, however, different in foliage, and on that account worth keeping distinct.

♀ ♀ 7. C. ACUMINA'TA *Lindl.* The acuminated-*leaved* Cotoneaster.

Identification. Lindl. in Lin. Soc. Trans., 13. p. 101.; Dec. Prod., 2. p. 632.; Don's Mill., 2. p. 603
Synonyme. Méspilus acumináta *Lodd. Bot. Cab.* t. 919.
Engravings. Lodd. Bot. Cab., t. 919.; Lin. Soc. Trans., 13. t. 9.; the plate of this species in Arb. Brit., 1st edit., vol. vi.; and our *fig.* 744.

744. C. acumináta.

Spec. Char., &c. Leaves ovate, acuminated, rather pilose on both surfaces. Peduncles glabrous, 1—2, rather reflexed, shorter than those of C. vulgàris, C. tomentòsa, or C. affi'nis. Calyxes glabrous. (*Dec. Prod.*) A sub-evergreen shrub or low tree. Nepal. Height 10 ft. to 15 ft. Introduced in 1820. Flowers white; April and May. Fruit scarlet; ripe in September, and remaining on the plants all the winter.

A vigorous-growing, fastigiate, leathery-leaved shrub, or very handsome sub-evergreen low tree; very distinct, and a most desirable species.

♀ 8. C. NUMMULA'RIA *Lindl.* The money-like-*leaved* Cotoneaster.

Identification. Lindl. in Hort. Trans., 6. p. 396.
Synonymes. C. elliptica *Hort.*; Eriobótrya elliptica *Lindl.*; Méspilus *Cuile* Hort.; ? C. læ'vis in Lodd. collection.
Derivation. Probably from the roundness of the leaf, resembling the general form of coins.
Engravings. The plate in Arb. Brit., 1st edit., vol. vi.; and our *fig.* 745.

745. C. nummularia.

Spec. Char., &c. Disk of leaf flat, orbicular, or elliptical, ending in a mucro, in some instances emarginate. Petiole of about the length of the stipules, which are linear-lanceolate, membranous, and soon fall off. Bark, buds, flower buds, stipules, petiole, the under surface of the disk of the leaf, and part of the upper surface of the midrib, tomentosely hairy, while in a young state; the bark, petioles, midrib on its upper surface, and calyx, become glabrous when old. Flowers in axillary cymes, few in a cyme. Style and carpel, which has a bony shell, mostly solitary. Erect, branched in a spreading manner; branchlets straight, slender. An elegant low sub-evergreen tree. Nepal, in the mountainous region. Height 10 ft. to 15 ft. Introduced in 1824. Flowers white; April and May. Berries numerous, black; ripe in September.

A very handsome species, distinguished at first sight from the others by its spreading habit of growth and smaller leaves.

§ iii. *Leaves evergreen, leathery. Low Shrubs, with prostrate Branches; Trailers, but not properly Creepers.*

≏ 9. C. ROTUNDIFO'LIA *Wall.* The round-leaved Cotoneaster.

Identification. Wall. Cat.; Lindl. Bot. Reg., 1229.
Synonymes. C. microphy̆lla β U'va-úrsi *Lindl. Bot. Reg.* t. 1187.; C. U'va-úrsi *Hort.*; the Bearberry-leaved Nepal Cotoneaster.
Engravings. Bot. Reg., t. 1187.; and our *figs.* 746. and 747.

746. C. rotundifolia.

Spec. Char., &c. Leaves roundish, pilose beneath, evergreen. Peduncles 1-flowered. Producing its white flowers in April and May. (*Don's Mill.*) An evergreen shrub. Nepal, on mountains. Height 3 ft. to 4 ft. Introduced in 1825. Flowers white, with the backs of the petals often

747. C. rotundifolia.

pinkish; April and May. Fruit bright scarlet; ripe in August, and remaining on all the winter.

A most desirable shrub for a small garden, for clothing a naked wall, covering rockwork, or grafting standard high, so as to form a pendent evergreen tree.

ꞷ 10. C. (R.) MICROPHY'LLA *Wall.* The small-leaved Cotoneaster.

Identification. Wall. ex Lindl. Bot. Reg., t. 1114.; Don's Mill., 2. p. 604.
Engravings. Bot. Reg., t. 1114.; and our *fig.* 748.

Spec. Char., &c. Leaves oblong, obtuse, pubescent beneath, evergreen. Peduncles usually 1-flowered. (*Don's Mill.*) A prostrate evergreen shrub, closely resembling the preceding species, and in our opinion only a variety of it. Nepal. Height 2 ft. to 3 ft. Introduced in 1824. Flowers white; May and June. Berries bright scarlet; ripe in August, and remaining on all the winter.

748. C. (r.) microphýlla.

It is exceedingly hardy, and forms a fine plant on rockwork, or on a lawn, where it has room to extend itself. A plant of C. microphýlla, at High Clere, of about 10 years' growth, was, in 1835, 6 ft. high, and formed a dense bush, covering a space 21 ft. in diameter. Another, at Redleaf, was, in 1837, nearly as large. Grafted standard high on the thorn, or any of its congeners, this shrub forms a singular and beautiful evergreen drooping tree: or it will cover a naked wall nearly as rapidly as ivy; and it possesses a decided advantage over that plant, and particularly over the variety called the giant ivy, in its shoots, which may be prevented from extending many inches from the face of the wall, and, consequently, being not likely to injure the plants growing near it. Were the practice of training trees and shrubs in architectural or sculptural shapes again to come into fashion, there are few plants better adapted for the purpose than this and the preceding sort of Cotoneáster.

ꞷ 11. C. (R.) BUXIFO'LIA *Wall.* The Box-leaved Cotoneaster.

Identification. Wall. ex Lindl. Bot. Reg., t. 1229.; Don's Mill., 2. p. 604.
Engraving. Our *fig.* 749. from a living specimen.

Spec. Char., &c. Leaves ovate, woolly beneath, evergreen. Peduncles 2—3-flowered, woolly. Flowers white. (*Don's Mill.*) A native of Neelgherry; introduced in 1824; and apparently a variety of C. rotundifòlia, from which it differs in having the peduncles 2- and 3-flowered, but scarcely in any thing else.

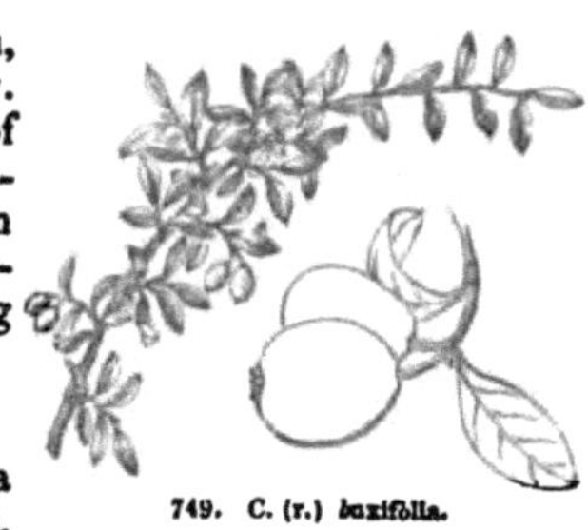

749. C. (r.) buxifòlia.

Variety.

ꞷ *C.* (*b.*) 2 *marginàta*, C. marginàta *Lindl.*, has rather larger leaves, which are covered thickly on the under side and margin with a dense white tomentum. Raised in the Horticultural Society's Garden in 1838, from seeds received from Dr. Falconer of Saharunpore.

Genus XVIII.

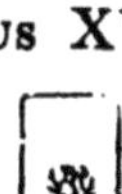

AMELA'NCHIER Med. The Amelanchier. *Lin. Syst.* Icosándria Di-Pentagýnia.

Identification. Med. Gesch., 1793.; Lindl. in Lin. Soc. Trans., 13. p. 100.; Dec. Prod., 2. p. 632.; Don's Mill., 2. p. 604.

Synonymes. Méspilus *L.*; Pŷrus *W.*; Arònia *Pers.*
Derivation. According to Clusius, *Amelancier* is the old Savoy name for *A.* vulgàris. (*E. of Pl.*) *Amelancier* is the Savoy name for the medlar.

Gen. Char. *Calyx* 5-cleft. *Petals* lanceolate. *Stamens* rather shorter than the calyx. *Ovarium* of 10-cells, or of 5 bipartite ones. *Ovula* 10, solitary in the partitions of the cells. *Styles* 5, joined together a little at the base. *Pome*, when mature, 3—5 celled. *Seeds* 3—5; endocarp cartilaginous. (*Don's Mill.*)

Leaves simple, alternate, stipulate, deciduous; serrated. *Flowers* white, in racemes. *Bracteas* linear lanceolate, deciduous. — Small trees, natives of Europe, America, or Asia.

In British gardens, they are cultivated for their flowers, which are white, abundant, showy, and produced early in the season; for their fruit, which ripens in June; and for the deep red, or rich yellow hue, which their foliage assumes in autumn. They are propagated by grafting on the hawthorn or the quince; or the weaker on the stronger-growing species of the genus.

⚘ 1. *A.* VULGA'RIS *Mœnch.* The c⌣mmon Amelanchier.

Identification. Mœnch Meth., 682.; Dec. Prod., 2. p. 632.; Don's Mill., 2. p. 604.
Synonymes. Méspilus *Amelánchier* Lin. Sp. 685., Jacq. Fl. Austr. t. 300.; Pŷrus *Amelánchier* Willd. Sp. 2. p. 1015.; *Arònia* rotundifòlia *Pers. Syn.* 2. p. 39.; Cratæ'gus rotundifòlia *Lam.*; Sórbus *Amelánchier* Crantz; Alisier Amelanchier, Amelanchier des Bois, Néflier à Feuilles rondes, *Fr.*; Felsenbirne, *Ger.*; Pero cervino, *Ital.*
Engravings. Jacq. Fl. Austr., t. 300.; Bot. Mag., t. 2430.; and our *fig.* 750.

Spec. Char., &c. Leaves roundish-oval, bluntish, downy beneath, afterwards glabrous. Fruit dark blue. (*Dec. Prod.*) A deciduous low tree. Continent of Europe, the Alps, Pyrenees, and at Fontainbleau in France. Height 15 ft. to 20 ft. Introduced in 1596. Flowers white; March and April. Fruit black, soft and eatable; ripe in July. Decaying leaves bright yellow.

A most desirable low tree, on account of its early and numerous flowers, which cover the tree like a white sheet, about the middle of April, and, in very mild seasons, even in March.

750. *Amelánchier vulgàris.*

⚘ ⚘ 2. *A.* (v.) BOTRYA'PIUM *Dec.* The Grape-Pear, *or Snowy-blossomed* Amelanchier.

Identification. Dec. Prod., 2. p. 632.; Hook. Fl. Bor. Amer., 1. p. 202.; Don's Mill., 2. p. 604.
Synonymes. Méspilus canadénsis *Lin. Sp.* 185.; *M.* arbòrea *Michx. Arb.* 2. t. 66.; Cratæ'gus racemòsa *Lam. Dict.* 1. p. 84.; Pŷrus Botryàpium *Lin. fil. Suppl.* p. 255.; Arònia Botryàpium *Pers. Syn.* 2. p. 39.; the Canadian Medlar, Snowy Mespilus, June Berry, wild Pear Tree; Alisier de Choisy, Amelanchier de Choisy, Alisier à Grappes, *Fr.*; Traubenbirne, *Ger.*
Engranings. Schm. Arb., t. 84.; Willd. Abbild., t. 79.; Krause, t. 56.; the plates of this species, in a youug and an old state, in Arb. Brit., 1st edit., vol. vi.; our *fig.* 751., from a specimen taken from the tree in the Horticultural Society's Garden, with the leaves and flowers fully expanded; and *figs.* 752. and 753., copied from Michaux's *North American Sylva*; *fig.* 751. showing the plant in spring before the flowers are fully opened; and *fig.* 752. showing the plant in fruit. Both differ in some respects from *fig.* 753. See Sir W. J. Hooker's remarks under *A.* ovàlis, No. 4.

751. *A.* (v.) Botryàpium.

Spec. Char., &c. Leaves oblong-elliptical, cuspidate, somewhat villous when young, afterwards glabrous. (*Dec. Prod.*) A deciduous shrub or low tree, closely

752. *A.* (v.) Botryàpium.

753. *A.* (v.) Botryàpium.

resembling the preceding species; and by some botanists considered as only a variety of it. America. Height in America 30 ft. to 40 ft., with a trunk 10 in. to 12 in. in diameter; in England 20 ft. to 30 ft. Introduced in 1746. Flowers white; April. Fruit purple, agreeable to eat; ripe in the beginning of June. Decaying leaves rich yellow.

A very ornamental tree, from its profusion of blossoms early in April, and from its rich autumnal foliage; and even the fruit is not altogether to be despised, either eaten by itself, or in tarts, pies, and puddings. The wood is white, and it exhibits no difference between the heart and the sap: it is longitudinally traversed by small bright red vessels, which intersect each other, and run together; a physiological peculiarity which, Michaux observes, occurs also in the red birch.

■ ♀ 3. *A.* (v.) SANGUI'NEA *Dec.* The blood-coloured Amelanchier.

Identification. Dec. Prod., 2. p. 633.; Lindl. in Bot. Reg., t. 1171.; Hook. Fl. Bor. Amer., 1. p. 203.; Don's Mill., 2. p. 604.
Synonymes. Pỳrus sanguinea *Pursh Fl. Amer. Sept.* 1. p. 340.; *A*rònia sanguinea *Nutt.*; *Méspilus* canadénsis γ rotundifòlia *Michaux Fl. Bor. Amer.* 1. p. 391.
Engravings. Bot. Reg., t. 1171.; and our *figs.* 754. and 755.

754. *A.* (v.) sanguinea.

755. *A.* (v.) sanguinea.

Spec. Char., &c. Leaves oval, obtuse at both ends, mucronate, finely serrated, somewhat heart-shaped at the base. Flowers few in a raceme. Calyx glabrous. Petals linear, obtuse. Fruit eatable. (*Dec. Prod.*) A deciduous tree or large shrub. Hudson's Bay. Height 10 ft. to 20 ft. Introduced in 1824. Flowers white; April. Fruit dark purple, full of blood-coloured juice; ripe in July. Decaying leaves rich yellow.

This plant differs from *A.* (v.) Botryàpium in the fewer flowers, much shorter raceme, and shorter, broader, and more ovate petals; in the young leaves being perfectly destitute of pubescence, and the head somewhat fastigiate.

■ ♀ 4. *A.* (v.) OVA'LIS *Dec.* The oval-*lvd* Amelanchier.

Identification. Dec. Prod., 2. p. 632.; Don's Mill., 2. p. 604.
Synonymes. Cratæ'gus spicàta *Lam. Dict.* 1. p. 84.?; *Méspilus Amelánchier* Walt. Car. p. 184.; *A.* parviflòra *Doug. MS.*; *M.* canadénsis var. α ovàlis *Michx. Am.* 1. p. 291.; Pỳrus ovàlis *Willd. Sp.* 2. p. 1014.; *A*rònia ovàlis *Pers. Syn.* 2. p. 240.; Amelanchier du Canada, Alisier à E'pi, *Fr.*; rundblättrige Birne, *Ger.*
Engraving. *Fig.* 756. from a living specimen.

Spec. Char., &c. Leaves roundish-elliptical, acute; when young, rather velvety beneath; when adult, glabrous. Raceme coarctate. Petals obovate. Calyx pubescent. (*Dec. Prod.*) A large deciduous shrub or low tree. North America, from Lake Huron to

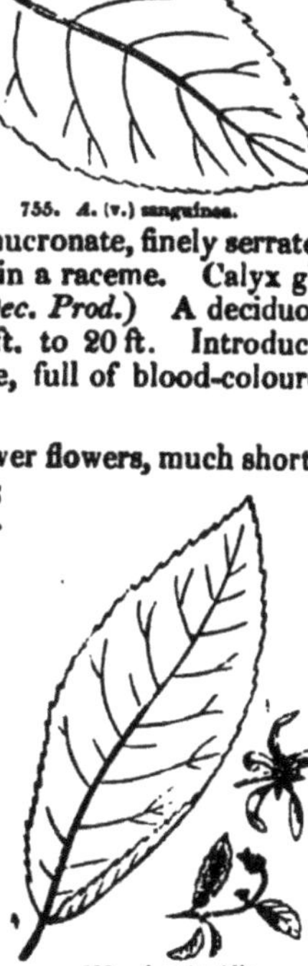

756. *A.* (v.) ovàlis.

the Rocky Mountains. Height 10 ft. to 20 ft. Introduced in 1800. Flowers white; April. Fruit purple; ripe in July. Decaying leaves rich yellow.

Varieties.

- ▲ ♈ A. (*v.*) *o.* 2 *subcordàta* Dec.; *Arònia* subcordàta *Raf.*; *Màlus* microcárpa *Raf.*—A native of mountains near New York. (*Dec. Prod.*)
- ▲ ♈ A. (*v.*) *o.* 3 *semi-integrifòlia* Hook. Fl. Bor. Amer. p. 201.—Leaves for the most part separated at the apex. A native about the Grand Rapids, and at Fort Vancouver, on the Columbia.

Sir Wm. J. Hooker is disposed to agree with Dr. Torrey, who suspects this to be only a variety of *A.* Botryàpium; and he adds that Michaux seems to have included *A.* Botryàpium and *A.* vulgàris under his *A.* canadénsis. The wood of *A.* ovàlis, according to Dr. Richardson, is prized by the Cree Indians for making arrows and pipe stems; and it is thence termed by the Canadian voyagers Bois de flèche. Its berries, which are about the size of a pea, are the finest fruit in the country; and are used by the Cree Indians both in a fresh and in a dried state. They " make excellent puddings, very little inferior to plum-pudding." (*Hook. Fl. Bor. Amer.*, i. p. 203.)

▲ ♈ 5. *A.* (v.) FLO'RIDA *Lindl.* The flowery Amelanchier.

Identification. Lindl. Bot. Reg., t. 1589.; Gard. Mag., vol. ix. p. 484.
Engravings. Bot. Reg., t. 1589.; and our *figs.* 758. to a scale of 2 in. to 1 ft., and *fig.* 757. of the natural size.

Spec. Char., &c. Leaves oblong, obtuse at both ends, coarsely serrate in the terminal portion, glabrous in every state. Bracteas and stipules feathery at the tip, soon falling off. Flowers in upright racemes, many in a raceme. Calyx glabrous externally; its segments longer than, or at least as long as, the stamens. (*Lindl.*) A handsome hardy deciduous shrub or low tree, in habit and general appearance like *A.* (v.) Botryàpium, but at once recognised as distinct by its fastigiate habit of growth, and by the shortness of its stamens. North America, on the north-west coast. Height 10 ft. to 20 ft. Introduced in 1826. Flowers white; May. Fruit purple; ripe in August. Decaying leaves rich yellow.

757. *A.* (v.) flórida.

758. *A.* (v.) flórida.

Variety.

- ▲ A. (*v.*) *f.* 2 *parvifòlia*, the *A.* parvifòlia of the Horticultural Society's Garden, is of a dwarf habit, not growing above 3 or 4 feet high, and has smaller leaves.

The leaves somewhat resemble those of the hornbeam; the petals vary in length, some having measured more than ¾ of an inch. In general habit, it is somewhat more fastigiate than the other sorts, unless we except *A.* sanguínea, to which, Dr. Lindley observes, it is very near akin. Possibly a distinct species, but we doubt it.

Genus XIX.

ME'SPILUS Lindl. The Medlar. *Lin. Syst.* Icosándria Di-Pentagýnia.

Identification. Lindl. in Lin. Trans., 13. p. 99.; Dec. Prod., 2. p. 633.; Don's Mill., 2. p. 604.
Synonymes. Méspilus sp. of Lin. and others; Mespilóphora sp. of Neck.; Neflier, *Fr.*; Mispel *Ger.*; Nespolo, *Ital.*
Derivation. From *mesos*, a half, and *pilos*, a bullet; fruit resembling half a bullet.

Gen. Char. *Calyx* 5-cleft, the segments foliaceous. *Petals* nearly orbicular. *Disk* large, full of honey. *Styles* 2—5, glabrous. *Pome* turbinate, open; 5-celled. *Endocarp* bony. (*Don's Mill.*)

Leaves simple, alternate, stipulate, deciduous; lanceolate, serrulated. *Flowers* large, nearly sessile, usually solitary, white. *Bracteas* permanent.— Trees; in a wild state furnished with spines. Natives of Europe.

The first species is cultivated for its fruit, which is eatable, and the seeds of which are accounted anti-lithic. The second species is an ornamental shrub or low tree, of the general character of a *Cratæ'gus.* Both are propagated by grafting on the quince, the wild pear, or the common hawthorn; and both grow freely in any common soil, rather moist than dry.

1. *M.* GERMA'NICA *L.* The German, *or common,* Medlar.

Identification. Lin. Sp., 684.; Pall. Fl. Ross., t. 13. f. 1.; Dec. Prod., 2. p. 633.; Don's Mill., 2. p. 605.
Engravings. Pall. Fl. Ross., t. 13. f. 1.; the plate of this species in Arb. Brit., 1st edit., vol. vi.; and our *fig.* 759.

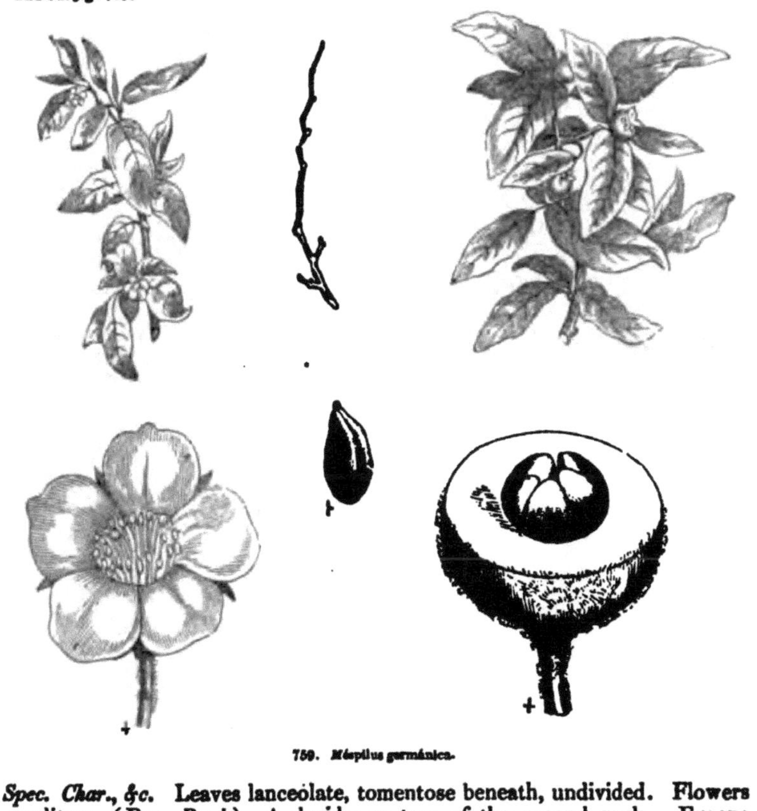

759. *Méspilus germánica.*

Spec. Char., &c. Leaves lanceolate, tomentose beneath, undivided. Flowers solitary. (*Dec. Prod.*) A deciduous tree of the second rank. Europe and the West of Asia, in bushy places and woods; and said to be found, also, in Kent, Sussex, Surrey, and about Chester, in England; apparently in a truly wild state in Sussex. Cultivated in 1596. Flowers white; May and June. Fruit brown; ripe in October and November. Decaying leaves dark brown, or yellow.

Varieties. DeCandolle gives the following forms of this species, which may be considered as natural varieties:—

♈ M. g. 1 *sylvéstris* Mill. Dict. No. 1. — Spiny. Fruit small. It loses its spines in a state of cultivation.

♈ M. g. 2 *stricta* Dec., Ait. Hort. Kew. ii. p. 172., Dod. Pempt. 801. — Spineless. Leaves doubly serrated.

♈ M. g. 3 *diffùsa* Dec., Ait. Hort. Kew. ii. p. 172., Du Ham. Arb. Fr. i. t. 3. — Thornless. Leaves nearly entire. Fruit, in many instances, abortive of seeds.

In the *Horticultural Society's Fruit Catalogue*, the following four cultivated sorts are given, which may be considered as artificial varieties : —

1. *Blake's large-fruited Medlar.*
2. *Dutch Medlar.* — Fruit the largest of any.
3. *Nottingham, or common, Medlar.* — Fruit obovate, middle size, and of the best quality : the only sort worth cultivating for its fruit in England.
4. *The stoneless Medlar.* — Fruit small, and of little merit.

The fruit of the medlar is not eaten till in a state of incipient decay, when it is very agreeable to some palates ; though it is, as Du Hamel observes, more *un fruit de fantaisie*, than one of utility. A number of trees of the different varieties may be seen in the orchard of the Horticultural Society's Garden, where they have taken very picturesque shapes.

♈ 2. *M.* SMI'THII Dec. Smith's Medlar.

Identification. Dec. Prod., 2. p. 633. ; Don's Mill., 2. p. 605.
Synonymes. *M.* grandiflòra *Smith Exot. Bot.* 1. p. 33. ; *M.* lobàta *Poir*, *Hook.* in *Bot. Mag.* t. 3342.
Engravings. Smith Exot. Bot., 1. t. 18. ; Bot. Mag., t. 3442. ; the plate of this species in Arb. Brit., 1st edit., vol. vi. ; and our *fig.* 760.

760. *Méspilus Smithii.*

Spec. Char., &c. Leaves oblong, elliptic, serrated, pubescent on the nerves beneath. Flowers usually solitary. (*Don's Mill.*) A deciduous tree with rambling irregular branches. Native country unknown. Cultivated in 1800. Height 15 ft. to 20 ft. Flowers white; May and June. Fruit reddish brown ; ripe in October.

As hardy as the common medlar, and well deserving a place in ornamental plantations for the beauty of its flowers, which are produced in great profusion. The general aspect and habit of the tree are those of a Cratæ'gus ; and, indeed, it is by many persons considered as more properly belonging to that genus than to *Méspilus*.

GENUS XX.

PY'RUS *Lindl.* THE PEAR TREE. *Lin. Syst.* Icosándria Di-Pentagýnia.

Identification. Lindl. Lin. Soc. Tr., 13. p. 97.; Dec. Prod., 2. p. 633.; Don's Mill., 2. p. 605.
Synonymes. Pýrus Màlus, and Sórbus, *Tourn.*; Pýrus and Sórbus *Lin.*; Pyróphorum and Apyróphorum *Neck.*

Gen. Char. *Calyx* with an urceolate tube, and a 5-lobed limb. *Petals* roundish. *Styles* usually 5, rarely 2 or 3. *Pome* closed, 5-celled. *Putamen* cartilaginous. *Seeds* 2 in each cell. *Testa* cartilaginous. (*Don's Mill.*)
Leaves simple, alternate, stipulate, deciduous, or sometimes sub-evergreen; entire, serrated, or pinnately divided. *Flowers* numerous, in cymes. *Bracteas* subulate, deciduous.

Low trees, and some shrubs; almost all deciduous; natives of Europe, Asia, and North America. Some of them are in great estimation throughout the world for their fruit; and others are cultivated chiefly for their flowers. Under the genus Pýrus, botanists have lately united the Linnæan genera Pýrus and Sórbus, together with several species formerly included under *Méspilus*, Cratæ'gus, and other genera.

Some of the species of Pýrus are, and have been for ages, the most universally cultivated of all ligneous plants; the apple and the pear being highly esteemed fruits, both in the temperate and transition zones of both hemispheres. These, and all the species of the genus, are propagated by grafting on the wild varieties of each division.

§ 1. *Pyróphorum* Dec.

Sect. Char. Petals spreading, flat. Styles 5, distinct. Pome more or less top-shaped, or subglobose, without a concavity at the base. Pedicels simple, umbeled. Leaves simple, not glanded. (*Dec. Prod.*, ii. p. 633.) This section comprehends all the pears, properly so called.

‡ 1. *P.* COMMU'NIS *L.* The common Pear Tree.

Identification. Linn. Sp., 686.; Dec. Prod., 2. p. 633.; Don's Mill., 2. p. 605.
Synonymes. *P. A'chras Gærtn. Fruct.* 2. p. 44. t. 87.; *P. sylvéstris Dod Pempt.* 800.; Pyráster *Ray Syn.* 452.; Poirier, *Fr.*; gemeine Birne, or Birnebaum, *Ger.*; Pero domestico, *Ital.*; Pera, *Span.*; and Gruschka, *Russian.*
Engravings. Blackw. Herb., t. 453.; Eng. Bot., t. 1784.; the plate of this species in Arb. Brit., 1st edit., vol. vi.; and our *fig.* 761.

Spec. Char., &c. Branches and buds glabrous. Leaves ovate, serrated, glabrous upon both surfaces. Flowers corymbose. (*Dec. Prod.*) A deciduous tree of the middle size. Europe, in woods and waste places, from the east of Russia to the west of England. Height 30 ft. to 50 ft. rarely 70 ft. In cultivation from time immemorial. Flowers white, never tinged with pink like those of the apple; April and May. Fruit in a wild state green, turning yellowish in November. Decaying leaves rich yellow or reddish yellow.

Varieties. DeCandolle mentions two forms of the wild species, comparatively permanent; to which we have added several others, the result of cultivation, and which are more or less accidental or temporary. To these we might have subjoined a class of wild pears with hoary leaves, such as *P.* nivàlis, *P.* salicifòlia, &c., which we consider as varieties, or races, though commonly treated as species; but we have preferred giving them afterwards as distinct sorts.

‡ P. *c.* 1 A'*chras* Wallr. Sched. p. 213. — Spiny. Leaves woolly when young, but afterwards glabrous; the disk ovate, acuminate, entire; the petiole long. Tube of the calyx woolly when young, afterwards becoming glabrous. Pome with its basal part long.

761. *Pyrus commùnis.*

♆ P. *c.* 2 *Pyráster* Wallr. Sched. p. 214., Gærtn. Fr. t. 87. f. 2.—Spiny. Leaves roundish, acute, sharply serrated, glabrous even when young. Tube of the calyx, while young, glabrous. Pome rounded at the base.

♆ P. *c.* 3 *fòliis variegàtis* has variegated leaves.

♆ P. *c.* 4 *frúctu variegàto* has the skin of the fruit variegated with yellow and white.

♆ P. *c.* 5. *sanguinolénta, the sanguinole Pear*, has the flesh of the fruit red or reddish ; and, though small and gritty, is not bad to eat when ripe.

♆ P. *c.* 6 *flòre plèno*, Poire de l'Arménie *Bon. Jard.* p. 43., has double flowers.

♆ P. *c.* 7. *jáspida* ; Bon Chrétien à Bois jaspé *Bon. Jard.* edit. 1836, p. 424. ; has the bark of the wood striped with yellow.

♆ P. *c.* 8 *sativa* Dec.—Without spines. This is the cultivated variety, of which there are very numerous subvarieties in gardens. For these DeCandolle refers us to Miller's *Dictionary*, and to Du Hamel's *Des Arbres Fruitiers ;* but, at the present time, by far the most complete collection in the world, of cultivated pears, is in the garden of the London Horticultural Society ; and they are described in the *Fruit Catalogue* (edit. 1831) of that body. From this catalogue Mr. Thompson has made for us the following selection of sorts which are at once deserving of culture as ornamental trees, and as producing fruit of first-rate excellence.

Subvarieties.

Beurré Diel.—Leaves large, and flowers very large. A hardy tree, somewhat fastigiate in its shape ; a great bearer, and deserving of extensive cultivation on account of its fruit, independently altogether of its handsome shape and large flowers.

Beurré de Rans (not Beurrée rance, as commonly written, which means rank, or rancid).—Branches spreading, or pendulous. The best very late pear yet known. It bears very well as a standard.

Bezi de la Motte.—Leaves remarkably narrow.

Glout Morceau. (*fig.* 762.)—Branches spreading. Head pyramidal. A hardy tree, and a great bearer. The fruit of most excellent flavour,

762. P. c. var. Glout Morceau

and hanging late on the tree. There is a plate of this variety in the *Arb. Brit.*, 1st edit., vol. vi.

Napoléon. — Leaves broad and shining. Blossoms large. The tree vigorous, and a good bearer. The fruit excellent.

Swan's Egg. — A handsome pyramidal tree, and an excellent bearer. The fruit roundish, or obovate. This is one of the commonest pear trees in the market-gardens about London; and we have introduced the name here, from having ourselves observed the handsome shapes taken by the trees. The fruit, however, as compared with that of the above sorts, is not worth cultivating.

The following Scotch pears are recommended by Mr. Gorrie, as forms adapted for landscape scenery; but little can be said in favour of their fruit, as compared with that of the new Flemish varieties.

The Benvie, the Golden Knap, and the Elcho take fastigiate forms; the latter more especially, Mr. Gorrie says, may be called the Lombardy poplar of the pear tribe. These trees generally attain the height of from 45 ft. to 50 ft. in as many years, in the Carse of Gowrie, in Perthshire.

The busked Lady and the Pow Meg take spreading orbiculate forms, such as will assort with the *A'cer* Pseùdo-*Plátanus*, and may be called the oaks and elms of the pear family. (See *Gard. Mag.*, vol. iv. p. 11.)

The pear tree, in a wild state, has a pyramidal-shaped head, with thorny branches, at first erect, and afterwards curved downwards and pendulous. The roots are few, and descend perpendicularly, with few lateral ramifications, except in shallow and rich soil. The leaves vary exceedingly in different soils, and in different parts of Europe and Asia: in Britain they are generally green, and slightly tomentose, and do not differ greatly in magnitude; but in the woods of Poland, and in the vast steppes of Russia, the leaves of the wild pear trees are commonly white with down, and vary so exceedingly in their dimensions, as to include what are called the willow-leaved, the sage-leaved, the elæagnus-leaved, and other narrow-leaved varieties, which by many are considered to be species. The fruit of the pear, in a wild state, is seldom more than a fourth part of the size of even the most ordinary cultivated varieties; and it is also austere, and unfit to eat. The plant is always found on a dry soil, and more frequently on plains than on hills or mountains; and solitary, or in small groups, rather than in woods and forests. The rate of growth is 2 or 3 feet a year for the first 6 or 7 years; in 10 years it will attain the height of 20 ft. in gardens; and in 30 years the height of 50 ft., with a trunk from 1 ft. to 18 in. in diameter; which may be considered its average dimensions in Britain. The tree is of great longevity. M. Bosc says that he has seen trees that were considered to be more than 400 years old: and Mr. Knight believes that there are trees of the Teynton squash (a famous perry pear) which existed as early as the beginning of the fifteenth century. All writers on trees, from Theophrastus to the present day, agree that, as the tree grows old, it increases in fruitfulness; which is, indeed, the case with most other trees.

The wood of the wild pear is heavy, strong, compact, of a fine grain, and slightly tinged with red. It weighs, green, 79 lb. 5 oz. per cubic foot; and, when dry, from 49 lb. to 53 lb. This wood, in common with that of all the *Rosàceæ*, is liable to have its natural colour changed by steeping it in water; which ought, therefore, to be avoided when it is intended for particular purposes. It is readily stained black, and then so closely resembles ebony as to be scarcely distinguishable from it. When it can be obtained, it is much used by turners and pattern-makers; also for joiners' tools, and to make various articles which are dyed black in imitation of ebony. As fuel, the wood of the pear is excellent, producing a vivid and durable flame, accompanied by intense heat. It also makes excellent charcoal. The leaves, according to Withering, afford a yellow dye, and may be used to give a green to blue cloths. The great use of the pear tree, however, is as a fruit tree. The fruit is used in the dessert, and for stewing and preserving. It is also occasionally used in tarts, though very inferior for this purpose to apples. In France and Belgium, the fruit is very generally dried in ovens, in which state it forms an article of commerce both domestic and foreign, and will keep a year. It is also dried in this manner in Russia; and, when stewed, is excellent, either as a substitute for pies and puddings, or as forming part of the dessert. It is essential that the soil should be dry; and, where the tree is intended to grow large and be productive, it ought to be deep and good. There are few trees better adapted for being grown in hedgerows than the fastigiate-growing varieties of pear, because their roots descend perpendicularly, and can, therefore, never interfere with the plough; and the heads, whether fastigiate or spreading, it is known from experience, do very little injury to pasture. If, therefore, fastigiate-growing trees, producing excellent sorts of fruit, were planted in all hedges, a very great benefit would result to the proprietors and to the public. The wild pear is continued by seed; and the varieties cultivated for their fruit are budded or grafted on stocks of different kinds. For the poorer soils, and exposed situations, stocks of the wild pear of the given locality must, doubtless, be the best, because they must be the hardiest: but it is found from experience, and it is consistent with physiological principles, that, on good soils, or where the pear is to be cultivated entirely as a fruit tree, both the tree and the fruit will grow larger when the stock is a seedling pear of some vigorous-growing variety. When dwarf trees are required, the pear is grafted on the quince, the medlar, or the thorn; or on the mountain ash, or some other species of *Sórbus*. It grows remarkably well on the common hawthorn; though, unless the graft be made under ground, it does not form a very safe and durable tree; because, as the scion increases faster in diameter than the stock, it is liable to be blown off. When the graft, however, is made close to the surface of the ground, or immediately under the surface, the root swells in nearly the same proportion as the scion, and there is no danger of the tree being blown down, or of its not being sufficiently long-lived.

♀ 2. *P.* (c.) SALVIFOLIA *Dec.* The Sage-leaved, *Aurelian, or Orleans*, Pear Tree.

Identification. Dec. Fl. Fr., 531., in a note; Prod., 2. p. 634.; Don's Mill., 2. p. 622.
Synonyme. Poirier Sauger *D'Ourch* in *Bibl. Phys. Econ. Mai* 1817 p. 299.
Engravings. Bot. Reg., 1482.; and our *fig.* 763.

Spec. Char., &c. Branches thick. Buds tomentose. Leaves lanceolate, entire, tomentose all over when young; when adult, glabrous on the upper surface. Fruit thick, long, fit for making perry. Wild and cultivated about Orleans, in France. (*Dec. Prod.*) Introduced by the London Horticultural Society, in 1826; and, in our opinion, only a variety of the common wild pear.

763. *P.* (c.) salvifolia.

♀ 3. *P.* (c.) NIVALIS *Lin. fil.* The snowy-*leaved* Pear Tree.

Identification. Lin. fil. Suppl., 253.; Jac. Fl. Austr., t. 107.; Dec. Prod., 2. p. 634.; Don's Mill., 2. p. 623.
Engravings. Jacq. Fl. Austr., t. 107.; and our *fig.* 764.

Spec. Char., &c. Leaves oval, entire, obtuse, white and silky beneath. Corymbs terminal. Fruit globose, very acid, except when ripe and beginning to decay, when it becomes very sweet. (*Dec. Prod.*) A native of the Alps of Austria, where it grows to the height of 10 or 12 feet. It was introduced into the Horticultural Society's Garden in 1826, or before; and is already 15 ft. high, forming a very handsome white-foliaged tree; though, as we think, decidedly only a variety, or race, of the common wild pear.

764. *P.* (c.) nivalis.

♀ 4. *P.* (c.) SINAICA *Thouin.* The *Mount* Sinai Pear Tree.

765. *P.* (c.) sinaica.

Identification. Thouin Mém. Mus., 1. 170. t. 9.; Dec. Prod., 2. p. 634.; Don's Mill., 2. p. 622.

Synonymes. P. Sinài *Desf. Arb.* 2. p. 144., *N. Du Ham.* 6. t. 57.; *P. pérsica Pers. Syn.* 2. p. 40.; the Mount Sinai Medlar.
Engravings. Mém. Mus., 1. t. 9.; N. Du Ham., 6. t. 57.; Dend. Brit., t. 49.; the plate in Arb. Brit., 1st edit., vol. vi.; and our *fig.* 765.

Spec. Char., &c. Very much branched, and spreading. Buds whitishly pubescent. Leaves ovate-oblong, subacute, very minutely crenated; whitishly pubescent beneath, above glabrous and almost shining; falling off late. (*Dec. Prod.*) A native of Mount Sinai, whence it was brought to the Paris Garden early in the present century, and introduced into England in 1820. It so closely resembles the preceding sort, as hardly to be distinguishable from it; and we have no doubt that seeds of either, if sown to a considerable extent, would produce plants of these and several other kinds.

❦ 5. *P.* (C.) SALICIFO'LIA *L.* The Willow-leaved Pear Tree.

Identification. Lin. Suppl., 255.; Dec. Prod., 2. p. 635.; Don's Mill., 2. p. 622.
Synonymes. *P. elæagnifòlia Pall.*; *P. orientàlis Horn. Suppl.* 52.; *P.* (C.) *elæagnifòlia Arb. Brit.* 1st edit. p. 889.
Engravings. Pall. Itin., 3. p. 374. t. N. f. 3.; Fl. Ros., 1. t. 9.; and our *fig.* 766.

Spec. Char., &c. Buds whitely tomentose. Leaves linear-lanceolate, acute, entire, hoary, particularly upon the under surface. The disk three times as long as the petiole. Flowers upon short pedicels, disposed in corymbs, a few in a corymb. (*Dec. Prod.*) A native of Siberia, common in the deserts between the rivers Cuma and Terec; and found, also, on Caucasus, and in Persia, generally accompanied by *C. Oxyacántha* and *Prùnus spinòsa.* It was introduced into England in 1780; and forms a very distinct variety; attaining the height of 20 or 25 feet. There are fine trees of this sort, 20 ft. high, at White Knights.

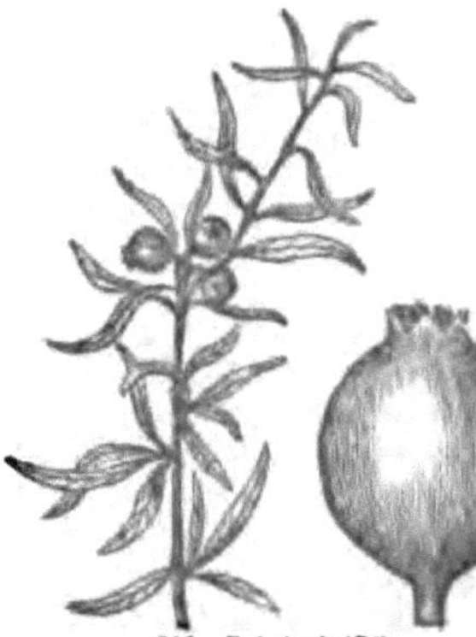
766. *P.* (c.) *salicifòlia.*

❦ 6. *P.* (C.) AMYGDALIFO'RMIS *Vil.* The Almond-shaped Pear Tree.

Identification. Vill. Cat. Strasb., 322.; Dec. Suppl., 531.; Dec. Prod., 2. p. 634.; Don's Mill., 2. p. 622.
Synonymes. *P. sylvéstris Magnol. Bot.* 215.; *P. salicifòlia Lois. Not.* 79.
Engraving. Our *fig.* 767.

Spec. Char., &c. Spiny. Buds tomentose. Leaves oblong, acute, entire; tomentose all over when young; when adult, glabrous on the upper surface. The disk six times longer than the petiole. Flowers in corymbs. (*Dec. Prod.*) Wild in rough places in France, in Provence, Dauphiny, and Languedoc; and very closely resembling the preceding sorts. It was introduced in 1810; and the finest plant that we know of it, in the neighbourhood of London, is at Kenwood; where it is 22 ft. high, with a very irregular picturesque head, and many of the side branches sweeping the ground. In May, it is completely covered with white blossoms, and in autumn with small green fruit, which drop off with the first severe frost.

767. *P.* (c.) *amygdalifórmis*

❦ 7. *P.* SINE'NSIS *Lindl.* The Chinese Pear Tree.

Identification. Lindl. Hort. Trans., 6. p. 396.; Don's Mill., 2. p. 622.
Synonymes. *Pỳrus commùnis Lois. Cochin.* p. 321.; *P. sinica Royle Illustr.* p. 207.; Ri vulgo Nas, *Japanese*, *Kæmpf. Amœn.* fasc. 804.; the Sandy Pear, Snow Pear, Sand Pear; Sha Lee, *Chinese.*
Engravings. Bot. Reg., t. 1148.; the plate in Arb. Brit., 1st edit., vol. vi.; and our *fig.* 768.

Spec. Char., &c. Leaves cordate, apiculated, shining, serrated, and when young, pubescent beneath. Peduncles corymbose. Calyx glabrous inside. Fruit warted and bony. (*Don's Mill.*) A deciduous tree of the middle size.

768. *P. sinénsis.*

China and Cochin China. Height 15 ft. to 20 ft. Introduced in 1820. Flowers white, slightly tinted with pink. Fruit large, edible, yellowish green when ripe; rarely seen in England.

Differs from the common pear in having longer and greener branches, and larger, more lucid, and almost evergreen leaves; insipid, roundish, warted, very gritty fruit; and a calyx, the inside of which is destitute of the down that is found on all the varieties of the European pear. The fruit is perfectly hardy, and it is ornamental; but it is worthless as a fruit tree, It vegetates very early in spring; when it is easily recognised by the deep rich brown of its young leaves and shoots.

Ɏ 8. *P.* BOLLWYLLERIA`NA *Dec.* The Bollwyller Pear Tree.

769. *P. bollwylleriàna.*

Identification. Dec. Fl. Fr. Suppl., p. 530. ; Prod., 2. p. 634. ; N. Du Ham., 6. p. 191.; Don's Mill., 2. p. 522.
Synonymes. P. bollwylleriàna *J. Bauh. Hist.* 1. p. 59. ic. ; *P.* Pollvèria Lin. Mant. 234. ; *P.* auriculàris *Knoop Pomol.* 2. p. 38. t. 4., according to Reichenbach.
Engravings. J. Bauh. Hist., ic. ; Knoop Pomol., 2. p. 38. t. 4., according to Reichenbach ; N. Du Ham., 6. t. 58. ; the plate of this species in Arb. Brit., 1st edit., vol. vi. ; and our *fig.* 769.

Spec. Char., &c. Buds downy. Leaves ovate, coarsely serrated, tomentose beneath. Flowers many in a corymb. Fruit top-shaped, small, yellowish within. (*Dec. Prod.*) A deciduous tree of the middle size, with but few ascending thick coarse branches. France, at Bollwyller on the Rhine, in hedges, but rare; possibly a hybrid between the pear and apple. Height 10 ft. to 20 ft. Introduced in 1786. Flowers white; April and May. Fruit greenish brown; ripe in September.

A very distinct variety, with large rough leaves, having somewhat the appearance of those of the apple. The fruit is turbinate, small, orange yellow, and unfit to eat. The tree produces fewer branches than any other species or variety of pear; and these branches are upright, thick, and rigid.

♆ 9. *P.* VARIOLO'SA *Wall.* The variable-*leaved* Pear Tree.

Identification. Wall. Cat. 680. ; Don's Mill., 2. p. 622.
Synonyme. *P. Páshia* Ham. ex Herb. Lin. Soc.
Engravings. The plate in Arb. Brit., 1st edit., vol. vi. : and our *fig.* 770.

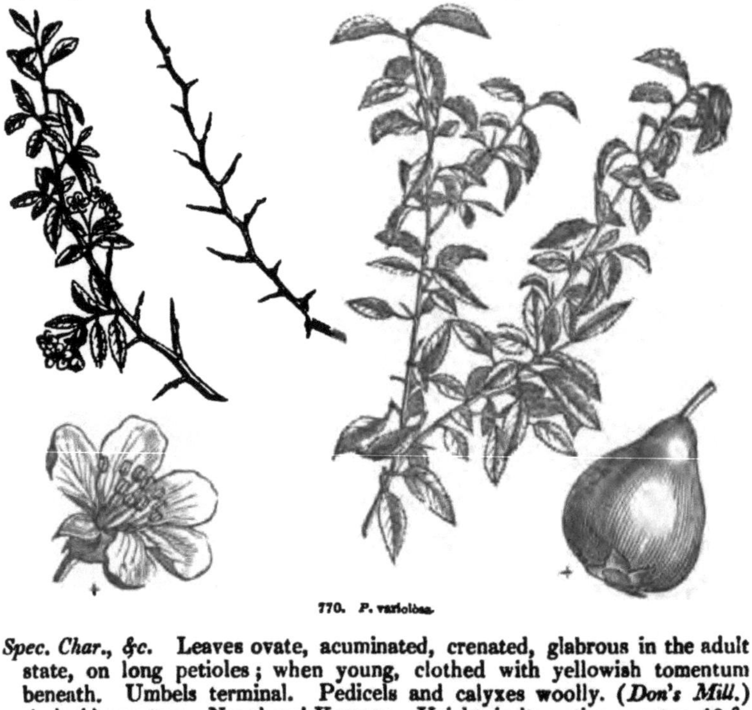

770. *P. variolòsa.*

Spec. Char., &c. Leaves ovate, acuminated, crenated, glabrous in the adult state, on long petioles; when young, clothed with yellowish tomentum beneath. Umbels terminal. Pedicels and calyxes woolly. (*Don's Mill.*) A deciduous tree. Nepal and Kamaon. Height, in its native country, 40 ft. to 50 ft.; in England 10 ft. to 15 ft. Introduced in 1825. Flowers white, slightly tinged with pink; April and May. Fruit pear-shaped; ripe in October, and remaining on the tree in the climate of London all the winter; eatable, like that of the medlar, in a state of incipient decay.

In the open air, in mild winters, this species is sub-evergreen; and, against a wall, in the Horticultural Society's Garden, it is completely so. It forms a very handsome tree, but is rather tender, having been killed to within a few feet of the ground, in several places in the neighbourhood of London, by the winter of 1837–8. It has ripened fruit in the Fulham Nursery, which is brown when ripe, and which, according to Dr. Royle, is not eatable until it is somewhat decayed. The veins of the leaves, and the entire plant, are tinged with reddish brown. The young seedlings of this species, and also the root shoots from plants cut down, have the leaves cut like a *Cratæ\`gus* or *Sórbus*.

♀ 10. *P.* MICHAU′XII Bosc. Michaux's Pear Tree.

771. P. Michaùxii.

Identification. Bosc in Poir Suppl., 4. p. 432.; Don's Mill., 2. p. 623.
Engraving. Our *fig.* 771. from a specimen in Dr. Hooker's herbarium.

Spec. Char., &c. Leaves oval, quite entire, acutish, glabrous on both surfaces, and shining above. Peduncles usually twin, when bearing the fruit thick and woody. Fruit globose. (*Don's Mill.*) A deciduous tree. North America. Height 15 ft. to 20 ft. Introduced in 1837. Flowers ?.

There are plants of this species in the Hort. Soc. Garden, and in some of the London nurseries; but they are too small to enable us to form any judgement as to the kind of tree which they will ultimately form.

♀ 11. *P.* I′NDICA *Colebr.* The Indian Pear Tree.

Identification. Colebr. Wall. Pl. Rar. Asiat., 2. t. 172.; Don's Mill., 2. p. 622.
Engravings. Wall. Pl. Rar. Asiat., 2. t. 172.; and our *fig.* 772.

Spec. Char., &c. Leaves ovate, or ovate-cordate, acute, serrulated, glabrous, white beneath, as well as the petioles and calyxes; in the young plants lobed. Claws of the petals shorter than the calyx. Umbels sessile, few-flowered. Styles villous at the base. (*Don's Mill.*) A deciduous tree. Bengal, on the mountains of Sylhet. Height 15 ft. to 20 ft. Introd. ?. Flowers white; May. Fruit about the size of a wild pear; ripe in October.

772. P. índica.

§ ii. *Màlus.*

Sect. Char. Petals spreading, flat. Styles 5, more or less strictly connate at the base. Pome mostly globose, depressed, and invariably having a concavity at its base. Flowers in corymbs. Leaves simple, not glanded. (*Dec. Prod.*)—This section includes all the apples and crabs.

♀ 12. *P. MA\`LUS L.* The *common, or wild,* Apple Tree.

Identification. Lin. Sp., 686.; Smith's Eng. Flora, 2. p. 362.; Lindl. Synop., 2d edit., p. 105.; Dec. Prod., 2. p. 635.; Don's Mill., 2. p. 623.
Synonymes. P. Màlus mitis *Wallr. Sched.* p. 215.; Màlus communis *Dec. Fl. Fr.*; Pommier commun, *Fr.*; gemeine Apfelbaum, *Ger.*; Pero Melo, and Melo Pomo, *Ital.*
Engravings. Eng. Bot., t. 179.; the plate in Arb. Brit., 1st edit., vol. vi.; and our *fig.* 773.

Spec. Char., &c. Leaves ovate, acute, crenated, woolly on the under surface. Flowers in corymbs. Tube of calyx woolly. Styles glabrous. (*Dec. Prod.*) A deciduous tree. Europe in woods and wastes; frequent in hedges. Height 20 ft. to 30 ft., sometimes 50 ft. In cultivation as a fruit tree from time immemorial. Flowers white, tinged with pink; May. Fruit red and yellow; ripe in October. Decaying leaves brownish.

Cultivated in gardens, it is wholly, or conjointly with other species or races, the parent of innumerable varieties, termed,

773. P. Màlus.

generally, in England, cultivated apple trees; and in France, pommiers doux, or pommiers à couteau. We adopt the specific name *Màlus*, to indicate what may be called the normal form, for the sake of convenience, though many of the cultivated varieties are derived not only from the wild apple, or crab, of Europe, but from the crabs of Siberia. We shall designate these crabs as varieties of *P. Màlus*, and afterwards make a selection from the cultivated sorts, of such as we think suitable for being planted for their timber, or as ornamental trees.

♀ 13. *P.* (*M.*) ACE'RBA *Dec.* The sour-*fruited* Apple Tree, *or common Crab.*

Identification. Dec. Prod., 2. p. 635.; Don's Mill., 2. p. 623.

774. *P.* (*M.*) acérba.

Synonymes. Pýrus Màlus austèra *Wallr. Sched.* 215.; Màlus acérba *Merat Fl. Par.* 187., *Dec. Suppl.* 530.; *M.* commùnis sylvéstris *Desf.*; *P.* Màlus sylvéstris *Fl. Dan.* t. 1101.; *P.* Màlus *Smith Eng. Bot.* t. 179.; Pommier sauvageon, *Fr.*; Holzapfelbaum, *Ger.*; Melo salvatico, *Ital.*
Engravings. Fl. Dan., t. 1101.; the plate in Arb. Brit., 1st edit., vol. vi.; and our *fig.* 774.

Spec. Char., *&c.* Leaves ovate, acute, crenated, glabrous even when young. Flowers in corymbs. Tube of the calyx glabrous. (*Dec. Prod.*) A native of woods and way sides in Europe. This form, according to DeCandolle, yields many subvarieties with sour fruit, called, in Britain, cider apples; and in France, generally, pommiers à cidre.

775. *P.* (*M.*) prunifolia.

♀ 14. *P.* (*M.*) PRUNIFO'LIA *W.* The Plum-tree-leaved Apple Tree, *or Siberian Crab.*

Identification. Willd. Sp., 2. p. 1018.; Dec. Prod., 2. p. 635.; Don's Mill., 2. p. 646.
Synonymes. *P.* Màlus β *Ait. Hort. Kew.* 2. p. 175.; ? Màlus hýbrida *Desf. Arb.* 2. p. 141.
Engravings. Mill. Ic., t. 269.; and our *fig.* 775.

Spec. Char., *&c.* Leaves ovate, acuminated, serrated, glabrous. Peduncles pubescent. Tube of calyx glabrous. Styles woolly at the base; and, as appears from *Mill. Ic.*, t. 269., with the styles twice as long as the stamens, and the fruit subglobose, yellowish, and austere. (*Dec. Prod.*) A native of Siberia; introduced in 1758.

According to Mr. Knight, some of the finest varieties raised by him are from cultivated apples fecundated with the blossoms of this tree. The progeny, he found, formed more hardy trees than any other kinds, and produced earlier and more highly flavoured fruit.

♀ 15. *P.* (*M.*) BACCA'TA *L.* The berry-*like-fruited* Apple Tree, *or Siberian Crab.*

Identification. Lin. Mant., 75.; Amm. Ruth., t. 31.; Pall. Fl. Ross., t. 10.; Dec. Prod., 2. p. 635. Don's Mill., 2. p. 646.
Synonyme. Màlus baccàta *Desf. Arb.* 2. p. 141.
Engravings. Amm. Ruth., t. 31.; Pall. Fl. Ross., t. 10.; the plate in Arb. Brit., 1st edit., vol. vi.; and our *fig.* 776.

776. (*P. M.*) baccàta.

Spec. Char., &c. Disks of leaves ovate, acute, equally serrated, glabrous, the length of the petiole. Flowers grouped. Sepals deciduous. (*Dec. Prod.*) A native of Siberia and Dahuria, and only differing from the preceding sort, of which it is, doubtless, a subvariety, in not having a persistent calyx.

♀ 16. *P.* (*M.*) DIOI'CA *W.* The diœcious-*sexed* Apple Tree.

Identification. Willd. Arb., 263.; Spec. 5., p. 1018.; Dec. Prod., 2. p. 635.; Don's Mill., 2. p. 646.
Synonymes. *P.* apétala *Münch. Hausv.* 5. p. 247., on the authority of Willdenow; Màlus dioica *Audib. Cat.*
Engraving. Our *fig.* 2088. in p. 1106.

Spec. Char., &c. Leaves oval, serrated, tomentose beneath. Flowers, in many instances, solitary. Sexes diœcious, by defect. Calyx tomentose. Petals linear, the length of the sepals. Styles glabrous. (*Dec. Prod.*) Not differing in external appearance from a crab or apple tree. Horticultural Society's Garden.

♀ 17. *P.* (*M.*) ASTRACA'NICA *Dec.* The Astrachan Apple Tree.

Identification. Dec. Prod., 2. p. 635.; Don's Mill. 2. p. 646.
Synonymes. Màlus astracánica *Dum. Cours.* ed. 2. 5. p. 426.; Transparent de Moscovie; Glace de Zélande; the transparent Crab of English nurseries.
Engraving. Our *fig.* 2089 in p. 1106.

Spec. Char., &c. Leaves oval-oblong, acute, partially doubly serrated; pale beneath, and the nerves there villose; above glabrous, except being slightly downy on the midrib. A native of the country around Astrachan, on the testimony of gardeners. (*Dec. Prod.*) A very ornamental tree, from the beauty of its fruit, which it produces freely, and which is also good to eat.

Varieties of P. *Màlus cultivated for their Fruit.* From the above forms, we think it may safely be presumed, that all the apples cultivated for the dessert or the kitchen have been obtained, either by selection from seedlings, or by cross-fecundation; and that no other wild sort has been used, unless, perhaps, we except *Pýrus* coronària; which, however, we have never heard of as being employed in cross-fecundation. These garden, or cultivated, varieties, as will hereafter appear, are very numerous; but the following selection of sorts, which are handsome-growing trees, or have fruit of a particular character, has been made for us by Mr. Thompson, of the Horticultural Society's Garden, from the collection under his care: —

The Red Astrachan. The tree is middle-sized, with a branchy head; the fruit is of a bright red, with a fine bloom like that of a plum. This is one form of the sort which DeCandolle has designated as *P.* astracánica: our No. 19.

The White Astrachan, or transparent Crab of Moscow. The tree resembles the preceding sort, but has the branches tending upwards when young, and afterwards becoming pendulous. The fruit is of a wax colour, with a fine bloom on it, and is almost transparent. This is another form of *P.* (*M.*) astracánica. It is known in English nurseries under the name of the transparent crab.

The Black Crab is a tree of the middle size, with very dark small fruit of no value as such.

The Court pendu plat is a remarkably dwarf-growing tree, and so late in flowering, that the leaves are expanded before, or at the same time as, the flowers; and, consequently, the latter are seldom, if ever, injured by frost: for which reason, it is commonly called by gardeners the wise apple. Grafted on the French paradise stock, the tree may be kept of a size not larger than that of a gooseberry bush; in which state it will bear fruit in abundance and of good flavour.

The Lincolnshire Holland Pippin is remarkable for the large size of its blossoms. Its fruit keeps till February.

The Tulip Apple is a great bearer of fruit which is of a very bright red.

The Violet Apple has fruit of a violet colour, covered with a bloom like that of the plum.

The Cherry Crab, or Cherry Apple, is a subvariety of *P.* (*M.*) baccàta. The tree is spreading, with drooping branches; and the fruit is numerous, and about the size and colour of a large cherry.

The Supreme Crab has fruit rather larger than the preceding sort. The tree is of robust growth, and the branches are somewhat erect.

Bigg's Everlasting Crab was raised in the Cambridge Botanic Garden, by Mr. Biggs, the curator, from seeds received from Siberia in 1814. It is a vigorous-growing tree, with pendulous branches and abundance of fruit, which, in form and character, are intermediate between *P.* (*M.*) prunifòlia and *P.* (*M.*) baccàta, and which remain on the trees long after Christmas. In sheltered situations, and mild winters, this tree appears almost a subevergreen.

The apple tree, whether in a wild state or cultivated, is by no means so handsome in form as the pear tree, though its blossoms are much more ornamental, and are, besides, fragrant. It seldom grows above half the height of the pear tree; the oldest apple trees known in Europe not being above 30 or 35 feet in height. The trunk is generally crooked, and the branches rambling horizontally when young, and when old becoming pendulous. The diameter of the head is also often greater than the height of the tree. The apple tree is much more liable to the canker, and other diseases, than the pear tree. The wood of the apple tree, in a wild state, is fine-grained, hard, and of a brownish colour; and that of the cultivated apple tree is said to be of a still finer and closer grain, which is a result of cultivation contrary to what is usual. The weight of the wood of the apple tree varies much according to the locality in which it is grown. In a green state, it weighs from 48 lb. to 66 lb. per cubic foot; and it loses from an eighth to a twelfth of its bulk in drying, and about a tenth of its weight. The wood of the cultivated tree weighs heavier than that of the wild tree, in the proportion of about 66 to 45. The tree, as an object in landscape scenery, cannot be recommended as harmonising well with other forms; but, as it has a character of its own, and as the fruit is of the greatest use to the poor, as well as to the rich, it deserves introduction into every hedgerow and every orchard. For hedgerows, it is more especially desirable, as, though not so fastigiate as the pear, it does very little injury to the crops by its shade; and it may be added, that, in nurseries and market-gardens, the former more especially, young trees of almost every kind thrive under the shade both of the apple and the pear. The crab is used as a stock for the cultivated apple, and for all the other species and varieties of this division of the genus; but, as we have before observed, it will not serve as a stock for the pear, or any of the plants included in that or the other divisions of *Pỳrus*. In France, and also in some parts of Germany, the thorny wild apple, or crab, is formed into live hedges, the branches of which, according to Agricola, are inarched into each other, in order to give them more strength to resist cattle. The fruit of the crab, in the forests of France, is a great resource for the wild boar; and it is also given in that country to swine and cows. A drink of it, called boisson, is made in some parts of France, as well as in England; and verjuice is a well-known description of vinegar produced from the most austere of the fruit. The bark affords a yellow dye; and the leaves are eaten by horses, cows, sheep, and goats. Pomatum, according to Gerard, was so called from its being anciently made of the pulp of apples beaten up with "swine's grease" (lard) and rose-water. The uses of the apple as an eatable fruit are very numerous, and well known. The apple, as a fruit tree, will do no good, except in a fertile soil and a sheltered situation. All the best apple orchards of England, and more especially those of the cider districts, it has been observed by geologists, follow the tract of red sandstone, which stretches across the island from Dorsetshire to Yorkshire It has been observed in Ireland (see *Dublin Soc. Trans.*), that the best orchards there are on limestone gravel; and, in Scotland, that the few orchards which are to be found in that country are on soils more or less calcareous. On the Continent, the two districts most famous for apple trees are Normandy and the Vale of Stuttgard; and the subsoil, in both countries, is well known to be limestone. In short, every kind of fruit, to be brought to perfection, requires a soil more or less calcareous.

The propagation and culture of the apple are the same as those of the pear tree. Wild crabs, like wild pears, are gathered when they are fully ripe, and either laid in a heap to rot, or passed between fluted rollers, and the crushed fruit pressed for the juice, which is made into an inferior kind of cider or perry, and the seeds are afterwards separated from the pomace by maceration in water and sifting. The apple, like the pear, may be grafted on the common thorn; but it does not form nearly so desirable a tree on that stock as the pear does, and therefore crab stocks are always to be preferred. As a fruit tree, where it is intended to be grown as a dwarf, the paradise stock effects for it what the quince does for the pear, and the *Cérasus Mahàleb* for the cherry. (See *Encyc. of Gard.*, edit. 1835.)

♀ ♀ 18. *P.* CORONA'RIA *L.* The garland-*flowering* Apple Tree.

Identification. Lin. Sp., 687.; Dec. Prod., 2. p. 635.; Don's Mill., 2. p. 647.
Synonymes. *Màlus coronària Mill.*; Crab Apple, the sweet-scented Crab, *Amer.*
Engravings. N. Du Ham., 6. pl. 44. f. 1.; Bot. Mag., t. 2009.; Michx. Arb., 2. t. 65.; the plate in Arb. Brit., 1st edit., vol. vi.; and our *fig.* 777.

Spec. Char., &c. Leaves broadly ovate, rounded at the base, subangulate, serrated, smooth. Peduncles in corymbs, glabrous. Flowers odorous,

777. *P. coronària.*

white, becoming purple before they drop off. The fruit is flatly orbiculate, of a deep green when it falls from the tree, and becoming yellow after lying some time on the ground. (*Dec. Prod.*) A deciduous or sub-evergreen tree. North America, from Pennsylvania to Carolina, and more especially abundant in the back parts of Pennsylvania and Virginia. Height 15 ft. to 18 ft., with a trunk 5 or 6 inches in diameter; sometimes 25 ft. high. Introduced in 1724. Flowers white, tinged with pink, violet-scented; May. Fruit green, in no state fit to eat; ripe in October. Decaying leaves green, violet-scented, dropping off by the first severe frost; but in mild winters remaining on till spring.

In British gardens, the leaves and the fruit are retained much longer on the tree than is the case with the European crab; so much so, that in very mild seasons, and sheltered situations, it might be almost considered sub-evergreen. The deep green and flat round form of the fruit, and the lobed and veined character of the leaves, render this sort of *M*àlus easily distinguished from every other; and this distinctiveness of character, and the fragrance of the blossoms, together with the lateness of their appearance (which is in the end of May), render it a most desirable tree in every shrubbery, however small.

♈ ♈ 19. *P.* (c) ANGUSTIFO'LIA *Ait.* The narrow-leaved Apple Tree.

Identification. Ait. Hort. Kew., 2. p. 276.; Pursh Fl. Amer. Sept., 1. 24.; Dec. Prod., 2. p. 635.; Don's Mill., 2. p. 647.
Synonymes. *P.* coronària *Wang. Amer.* 61. t. 21. f. 47., upon the authority of Willdenow, and Wats. in Dend. Brit.; *M*àlus sempervìrens *Desf. Arb.* 2. p. 141.; *P.* pùmila *Hort.*
Engravings. Wang. Amer., 61. t. 21. f. 47.; N. Du Ham., 6. t. 43. f. 1.; Wats. Dend., t. 132.; Bot. Reg., t. 1207.; the plate in Arb. Brit., 1st edit., vol. vi.; and our *fig.* 778.

Spec. Char., &c. Leaves glossy, lanceolate-oblong, dentately serrated, tapered and entire at the base. Flowers in corymbs. (*Dec. Prod.*) A deciduous tree, sometimes sub-evergreen. Carolina, in woods. Height 15 ft. to 20 ft. Introduced in 1750. Flowers pale blush; May or June. Fruit green; ripe in October.

Differs from the preceding sort in having the leaves narrower and the fruit much smaller, in being more decidedly sub-evergreen, and in having

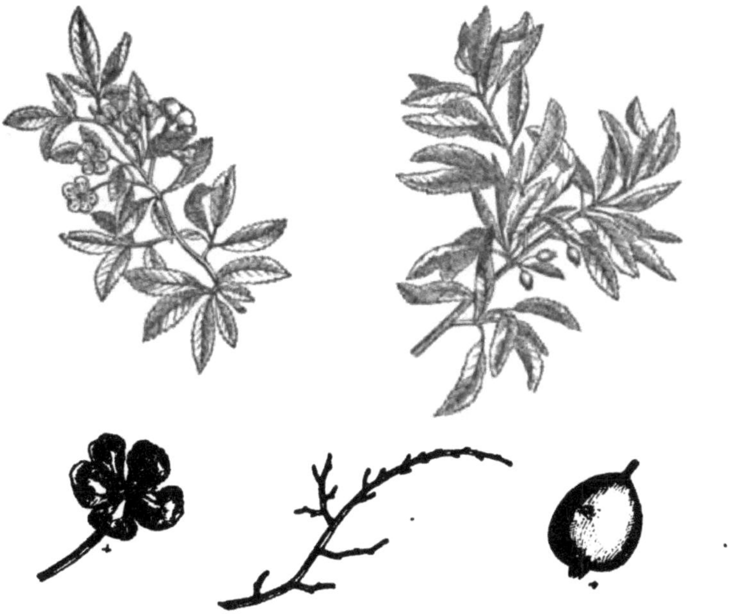

778. P. (c.) angustifòlia.

lead-coloured speckled branches. Notwithstanding all these points of difference, however, it bears such a general resemblance to *P.* coronària, that we cannot doubt its being only a variety of it. The fruit is intensely acid, like that of *P.* coronària ; but it is much narrower and smaller.

☨ 20. *P.* SPECTA'BILIS *Ait.* The showy-*flowering wild* Apple Tree, *or Chinese Crab Tree.*

Identification. Ait. Hort. Kew., 2. p. 175. ; Curt. Bot. Mag., t. 267. ; Dec. Prod., 2. p. 635. ; Don's Mill., 2. p. 646.
Synonymes. Màlus spectábilis *Desf. Arb.* 2. p. 141., *N. Du Ham.* 6. p. 141. ; Màlus sinénsis *Dum. Cours.* ed. 2. 5. p. 429.
Engravings. Bot. Mag., t. 267. ; N. Du Ham., 6. t. 42. f. 2. ; the plate of the species in Arb. Brit., 1st edit., vol. vi. ; and our *fig.* 779.

Spec. Char., &c. Leaves oval-oblong, serrated, smooth. Flowers in sessile umbels, many in an umbel ; large, and very elegant ; at first of an intense rose-colour, but afterwards of a pale one. Tube of calyx smooth. Petals ovate, clawed. Styles woolly at the base. (*Dec. Prod.*) A deciduous tree, thickly crowded with upright branches, which at length become spreading. China. Height 20 ft. to 30 ft. Introduced in 1780. Flowers pink, large, showy ; April and May. Fruit greenish yellow, and unfit to eat till it is in a state of incipient decay.

This is by far the most showy of all the different species of *Pỳrus*, both of this and of the other sections. The flowers are semidouble, and of a pale rose-colour ; but before they are expanded, the flower buds, which are large, appear of a deep red. In this state the tree is extremely beautiful. The stamens and pistils are much more numerous than in the other species ; the former sometimes exceeding 40, and the latter 20. The fruit is small, irregularly round, angular, and about the size of a cherry : it is of a yellow colour when ripe, but is without flavour, and is only fit to eat when in a state of incipient decay ; at which period it takes the colour and taste of the medlar. No garden, whether large or small, ought to be without this tree.

779. *P. spectábilis.*

Species of which there are only very young Plants in British Gardens.

P. *Sievérsii Led. Fl. Alt.* 2. p. 222., *Don's Mill.* 2. p. 647.; *P.* nov. sp. *Sievers* in *Pall. Nord. Beitr.* 7. p. 292.; is a bush, with many stems rising from the same root; with ovate leaves, rather tomentose; and umbellate flowers, succeeded by very acid fruit. A native of Siberia.

P. ? *Schóttii Ledeb.*—Plants in the Hort. Soc. Garden.

P. *stipulàcea* Hort.—Plants in the Hort. Soc. Garden, raised from seeds received from the Himalayas.

§ iii. A'ria Dec.

Sect. Char., &c. Petals spreading, flat. Styles mostly 2—3. Pome globose. Flowers in racemose corymbs; the peduncles branched. Leaves simple, not glanded, whitely tomentose beneath. (*Dec. Prod.*, ii. p. 635.)—Deciduous trees, natives of different parts of Europe, and of Asia; chiefly found on dry, calcareous, or clayey soils, and varying much under cultivation. The species and varieties are in a state of great confusion.

♈ 21. *P. A'ria Ehrh.* The White Beam Tree.

Identification. Ehrh. Beitr., 4. p. 20.; Dec. Prod., 2. p. 636.; Don's Mill., 2. p. 647.
Synonymes. Cratæ'gus *A'ria* var. *α Lin. Sp.* 681.; *Méspilus A'ria Scop.*; Sórbus *A'ria Crantz Austr.* 1. t. 2. f. 2., *Bauh. Hist.* 1. p. 65.; *A'ria* Theophrásti L'Obel; white wild Pear, white Leaf Tree, red Chess-Apple, Sea Ouler, Cumberland Hawthorn, *Gerard*; Alisier Allouchier, Alisier blanc, *Fr.*; Mehlbeerbaum, or Mehlbaum, *Ger.*; Aria, or Sorba montana, *Ital.*; Mostaco, *Span.*; Axelbeer, *Dan.*; Oxilbeer, *Swed.*
Derivation. *A'ria*, the name given to this tree by Theophrastus, is probably from the name of that country in Asia. The White Beam Tree is a pleonasm, *beam* being the Saxon word for tree. The word Allouchier is from *allouchion*, the cog of a wheel, the wood of the tree being much used for that purpose in France. Mehlbaum is literally the meal tree, from the mealy appearance of the under side of the leaves.
Engravings. Crantz Austr., 1. t. 2. f. 2.; Eng. Bot., t. 1858.; Fl. Dan., t. 302.

Spec. Char., &c. Leaves ovate, doubly serrated, tomentose beneath, with appressed white tomentum. Corymbs flat. (*Dec. Prod.*) A deciduous tree.

Varieties.

♀ P. A. 1 *obtusifòlia* Dec. Prod. ii. p. 636., and Fl. Dan. t. 302.; *P. A* ovàlis *Hort.*—Leaves broadly ovate, and obtuse.

♀ P. A. 2 *acutifòlia* Dec. Prod. l. c. Cratæ`gus longifòlia *N. Du Ham.* 4. t. 34.; ? Pỳrus alpìna *Willd. Enum.* 527.—The leaves are ovate-oblong and acute.

♀ P. A. 3 *undulàta* Lindl. Hort. Trans. vii. p. 234., and the plate in Arb. Brit. 1st edit. vol. vi., and our *fig.* 780., has the leaves flat, oval-

780. P. A`ria undulàta.

lanceolate, broad, undulated, unequally and deeply serrated, acuminated, and cobwebbed above.

♀ P. A. 4 *angustifòlia* Lindl. l. c., *P. A.* longifòlia *Hort.*, has the leaves oval, obtuse, concave, somewhat simply serrated, woolly above.

♀ P. A. 5 *rugòsa* Lindl. l. c.—Leaves large, ovate-elliptic, doubly serrated, shining above and wrinkled, white beneath.

♀ P. A. 6 *crética* Lindl. l. c. *P. A.* rotundifòlia *Hort.*; *P.* græ`ca *Hort.*; *P. A.* edùlis *Hort.*; Cratæ`gus græ`ca *Hort.*—Leaves flat, orbicularly elliptic, crenately serrated, retuse, cuneated at the base; smooth above, and hoary beneath. Branches cobwebbed.

♀ P. A. 7 *bullàta* Lindl. Hort. Trans. vii. p. 234., *P. A.* acuminàta *Hort.*, has the leaves concave, elliptic, acuminated, blistered; closely serrated at the apex, but entire at the base.

The rate of growth, when the tree is young and in a good soil, is from 18 in. to 2 ft. a year: after it has attained the height of 15 or 20 feet it grows much slower; and, at the age of twenty or thirty years, it grows very slowly; but is a tree of great duration. The roots descend very deep, and spread very wide; and the head of the tree is less affected by prevailing winds than almost any other. In the most exposed situations, on the Highland mountains, this tree is seldom seen above 10 or 15 feet high; but it is always stiff and erect.

In Britain it is later in coming into leaf than any other indigenous tree, except the ash. It bears lopping, and permits the grass to grow under it.

The wood is very hard, of a fine close grain, yellowish white, and susceptible of a high polish. In a green state, it has a strong smell, which it retains, in a slight degree, even after it is dried. It weighs, in that state, 55 lb. 6 oz. per cubic foot. It may be stained of any colour, and is much used in the smaller manufactures, such as making handles to knives and forks, wooden spoons, &c.; and for musical instruments, and various turnery articles. It was universally employed as cogs for wheels till cast iron became generally substituted for it. The leaves are eaten both by goats and sheep. The fruit is acid and astringent; but it is not disagreeable to eat, when it is in a state of incipient decay. Dried, and reduced to powder, it has been formed into a sort of bread, which has been eaten, both in France and Sweden, in years of great scarcity. Fermented, the fruit affords a beer; or, by distillation, a powerful spirit. It is greedily eaten by small birds; on which account the trees are ordered to be preserved in the French forests, that the number of birds may be increased, in order to keep down the insects. The fruit is also the food of squirrels; and, when it drops, of the wild boar, the deer, the hedgehog, &c.

As an ornamental tree, the white beam has some valuable properties. It is of a moderate size, and of a definite shape; and in summer, when clothed with leaves, it forms a compact green mass, till it is ruffled by the wind, when it suddenly assumes a mealy whiteness. In the winter season, the tree is attractive from its smooth branches, and its large green buds. When the tree is covered with its fruit, it is exceedingly ornamental. Among the different varieties enumerated, *P. A.* crética is by far the most distinct: but all of them are well deserving of cultivation.

A calcareous and dry soil is essential; and the tree will not attain a timber size unless it is placed in an airy situation. The situation may be exposed to the highest and coldest winds that prevail in this country, and yet the tree will never fail to grow erect, and produce a regular head; and, for this reason, no tree is better adapted for sheltering houses and gardens in very exposed situations.

The species may be raised from seed, and the varieties be grafted on stocks of the species of the pear, of the Cratæ'gus, and even of the quince and medlar; which trees, it is almost unnecessary to add, may be reciprocally grafted on the white beam tree. When plants are to be raised from seed, the seeds should be sown as soon as the fruit is ripe; otherwise, if kept till spring, and then sown, they will not come up till the spring following. When it is inconvenient to sow them immediately after they are gathered, they may be mixed with soil, and treated like haws (see *Pýrus* aucupària); and, if sown in the March following, they will come up the same season. The varieties may be propagated by cuttings, or by layering; but they root by both modes, with great difficulty. Layers require to be made of the young wood, and to remain attached to the stool for two years.

♆ 22. *P.* (*A.*) INTERME'DIA *Ehrh.* The intermediate White Beam Tree.

Identification. Ehrh. Beitr., 4. p. 20.; Dec. Prod., 2. p. 636.; Don's Mill., 2. p. 647.
Synonymes. Cratæ'gus A'ria β *Lin. Sp.* 681.; *C.* scándica *Wahlenb. Fl. Ups.* 165.; *C.* suécica *Ait. Hort. Kew.* 167.; Alisier de Fontainebleau, *Fr.*; Schwedischer Mehlbaum, *Ger.*

Spec. Char., &c. Leaves ovate, incisely lobed, tomentose beneath, with white appressed tomentum. Corymbs flat. Fruit eatable. (*Dec. Prod.*) A deciduous tree. Europe, in France, Germany, and Sweden, and also in Wales and Scotland. Height 15 ft. to 20 ft. Flowers and fruit as in the preceding kind.

Varieties. DeCandolle has described the two following forms of this species:—

♆ P. (A.) *i.* 1 *latifòlia.* Cratæ'gus latifòlia *Poir. Dict.* 4. p. 444., *Du Ham. Arb.* 1. t. 80., *N. Du Ham.* 4. t. 35.; *Sórbus* latifòlia *Pers.*; Cratæ'gus dentàta *Thuil. Fl. Par.* — Leaves broadly ovate. A native of the woods of Fontainebleau.

† P. (A.) *i.* 2 *angustifòlia*. *P.* edùlis *Willd. En.* p. 527. (Wats. Dend. Brit., t. 52.; and our *fig.* 781.) — Leaves oblong, wedge-shaped at the base.

781. *P.* (*A.*) intermèdia angustifòlia.

These trees bear so close a resemblance to *P. A'ria*, as to leave no doubt in our mind that they are only varieties and subvarieties of that species. They are found in a wild state in France, Germany, and Sweden; and perhaps also in the Highlands of Scotland, where, according to Sir W. J. Hooker, *P. A'ria* varies in having the leaves more or less cut at the margin. They are all well deserving of culture.

† 23. *P.* VESTI'TA *Wall.* The clothed White Beam Tree.

Identification. Wall. Cat., 679.; Don's Mill., 2. p. 647.
Synonymes. Pýrus nepalénsis *Hort.*; Sórbus vestita *Lodd. Cat.* edit. 1836; *P.* crenàta *D. Don Prod. Fl. Nep.* p. 237., *Bot. Reg.* t. 1655.
Engravings Bot. Reg. t. 1655.; Arb. Brit., 1st edit., vol. vi.; and our *figs.* 782, 783, and 784.

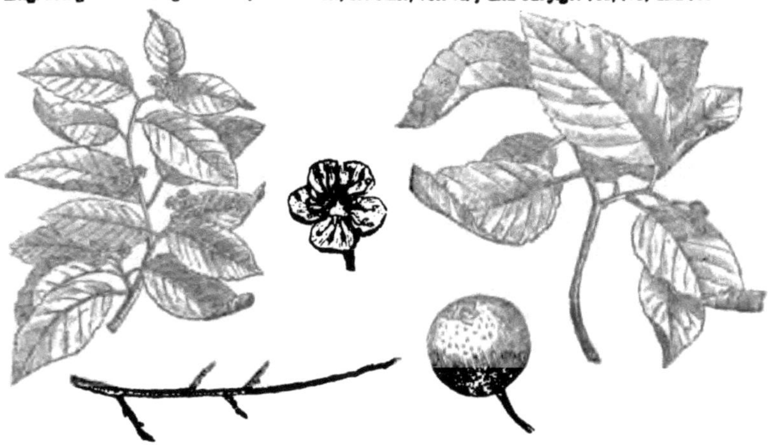

782. *P.* vestita.

Spec. Char., &c. Leaves, cymes, and young branches, clothed with white tomentum. Leaves elliptic, or obovate-elliptic, acuminated, serrated towards the apex. Corymbs branched and terminal. Flowers white. Fruit greenish brown. (*Don's Mill.*) A deciduous tree of the middle size. Nepal and Kamaon, at elevations of from 9,000 to 12,000 ft. Height 20 ft. to 30 ft. Introduced in 1820. Flowers white; May and June. Fruit brown; ripe in October. Decaying leaves of a beautiful straw-coloured yellow or brown.

783 *P.* vestita.

784. *P.* vestita.

This tree is remarkable for the rapidity of its growth, its long broad leaves, and their woolly whiteness; and also for being one of the very latest trees, whether foreign or indigenous, in coming into leaf; being later than either the mulberry or ash. The leaves are conspicuous, on their first expansion, for their whiteness, particularly underneath; and in autumn, before they drop off, for their fine yellow colour.

§ iv. *Torminària* Dec.

Sect. Char., &c. Petals spreading, flat, having short claws. Styles 2—5, connected, glabrous. Pome scarcely at all juicy, top-shaped at the base, truncate at the tip; the sepals deciduous. Leaves angled with lobes; in the adult state glabrous. Flowers in corymbs. The peduncles branched. (*Dec. Prod.*, ii. p. 636.) —Trees of the same general character, in regard to habit and constitution, as *P. A'ria*.

❦ 24. *P.* TORMINA'LIS *Ehrh.* The griping-*fruited* Service Tree.

Identification. Ehrh. Beitr., 6. p. 92.; Dec. Prod., 2. p. 636.; Don's Mill., 2. p. 647.
Synonymes. Cratæ'gus torminàlis *Lin. Sp.* 681., *Smith Eng. Bot.* t. 298., *Fl. Dan.* t. 798., *Jacq. Fl. Austr.* t. 443.; Sórbus torminàlis *Crantz Austr.* p. 85.; the Maple-leaved Service Tree; Allsior de Bois, *Fr.*; Elzbeerbaum, *Ger.*; Clavardello, or Mangiarello, *Ital.*
Engravings. Jacq. Fl. Austr., t. 443.; the plate in Arb. Brit., 1st edit., vol. vi.; and our *fig.* 785.

785. *P. torminàlis.*

Spec. Char., &c. Leaves cordate-ovate, feather-nerved, pinnatifidly lobed; when young, slightly downy beneath; when adult, glabrous; the lobes acuminate and serrated, the lowest divaricate. Seeds cartilaginous. (*Dec. Prod.*) A deciduous tree. Britain in woods, and throughout the Middle and North of Europe, and Western Asia. Height 40 ft. to 50 ft. Flowers white; May and June. Fruit red; ripe in November. Decaying leaves yellowish brown. Naked young wood purplish, marked with white spots.

The leaves, which are on long footstalks, are cut into many acute angles, like those of some species of maple. They are nearly 4 in. long, and 3 in. broad in the middle, bright green above, and slightly woolly underneath. The flowers are produced in large bunches at the end of the branches; and they are succeeded by roundish compressed fruit, not unlike common haws, but larger, and of a brown colour when ripe. The tree is of slow growth, and in this respect, and most others, it resembles *P. A'ria*; but it is less hardy. The wood resembles that of *P. A'ria*, but is without its peculiarly strong smell. It weighs, when newly cut, 65 lb. to the cubic foot, and when dried, 48 lb. 8 oz. It is employed for all the different purposes to which that of *P. A'ria* is applicable, and is considered rather preferable as fuel, and for charcoal. The fruit is brought to market both in England and France; and, when in a state of incipient decay, it eats somewhat like that of a medlar. As an ornamental tree, its large green buds strongly recommend it in the winter time, as its fine large-lobed leaves do in the summer, and its large and numerous clusters of rich brown fruit do in autumn. It will grow in a soil not poorer, but more tenacious and moist, than what is suitable for *P. A'ria*; and it requires a sheltered situation. It seems more liable to the attacks of insects than that species, and does not thrive so well in the neighbourhood of London. It is propagated exactly in the same manner as *P. A'ria*. There being no varieties, it does not require to be continued by grafting.

♀ 25. *P.* RIVULA'RIS *Dougl.* The River-side Wild Service Tree.

Identification. Dougl. MS. Hook. Fl. Bor. Amer., 1. p. 203.; Don's Mill., 2. p. 647.
Synonyme. Pow-itch, the name of the fruit in the language of the Chenook tribe of Indians.
Engravings. Hook. Fl. Bor. Amer., t. 68.; and our *figs.* 786. and 787.

786. *P. rivulàris.*

787. *P. rivulàris.*

Spec. Char., &c. Leaves ovate, entire, and angular, somewhat 3-lobed, rather acuminated, acutely serrated, pubescent beneath. Corymbs terminal, simple. Calyces hairy, and densely tomentose inside. Styles 3—4, connected at the base. (*Don's Mill.*) A low deciduous tree. North-west coast of North America, at Nootka Sound, and other places. Height 15 ft. to 20 ft. Introduced in 1836. Flowers white; April and May. Fruit small, subglobose, red or yellow; ripe ?.

The fruit is used as an article of food, and the wood is employed for making wedges, and is so hard as to be susceptible of a fine polish. Horticultural Society's Garden.

§ v. *Eriólobus* Dec.

Sect. Char. Petals spreading, flat, with short claws, and with about 3 teeth at the tip. Styles 5, long; at the base very hairy, and somewhat connected. Pome globose, glabrous, crowned with the lobes of the calyx, which are tomentose upon both surfaces. Leaves palmately lobed, glabrous. Flowers upon unbranched pedicels, disposed in corymbs. (*Dec. Prod.*, ii. p. 636.)

788. *P. trilobàta.*

♀ 26. *P.* TRILOBA'TA *Dec.* The three-lobed-*leaved* Pear Tree.

Identification. Dec. Prod., 2. p. 636.; Don's Mill., 2 p. 648.
Synonyme. *Cratæ'gus trilobàta* *Labill. Dec.* 4. p. 15. t.10., *Poir. Suppl.* 1. p. 291.
Engravings. Labill. Dec., 4. t. 10.; and our *fig.* 789.

Spec. Char., &c. Leaves glabrous, palmately

lobed; the middle lobe 3-lobed; the side lobes, in many instances, 2-lobed; the secondary lobes serrated. (*Dec. Prod.*) A deciduous tree. A native of Mount Lebanon; growing to the height of 20 ft.; said to have been introduced in 1810; but of which we have not seen a plant.

§ vi. *Sórbus* Dec.

Sect. Char., &c. Pétals spreading, flat. Styles 2—5. Pome globose, or top-shaped. Leaves impari-pinnate, or pinnately cut. Flowers in branched corymbs. (*Dec. Prod.*)—Trees growing to the height of from 20 ft. to 40 ft. or upwards. Natives of Europe, North America, and the Himalayas. For the most part very hardy, and of easy culture in common soil.

¥ 27. *P.* AURICULA'TA *Dec.* The auricled Service Tree.

Identification. Dec. Prod., 2. p. 636.; Don's Mill., 2. p. 648.
Synonyme. Sórbus auriculàta *Pers. Syn.* 2. p. 39.
Engraving. Our *fig.* . in p. .

Spec. Char., &c. Leaves of 3 pairs of leaflets, and an odd one, hirsute beneath; 2—4 of the lowest leaflets distinct, the rest connate with the odd one into an ovate one, which is crenate. Corymb compact. (*Dec. Prod.*) A deciduous tree. Egypt. Height 20 ft. to 30 ft. Introduced in 1800. Flowers white; May. Fruit ?.

We have not seen this plant, which is, perhaps, only a variety of *P.* pinnatífida.

¥ 28. *P.* PINNATI'FIDA *Ehrh.* The pinnatifid-*leaved* Service Tree.

Identification. Ehrh. Beitr., 6. p. 93.; Smith in Eng. Bot., t. 2331.; Dec. Prod., 2. p. 636.
Synonymes. Sórbus hýbrida *Lin. Dec.* 6.; Pýrus hýbrida *Smith Fl. Brit.*, not of Willd.; the Bastard Service Tree.
Engravings. Eng. Bot., t. 2331.; the plate in Arb. Brit., 1st edit., vol. vi.; and our *fig.* 789.

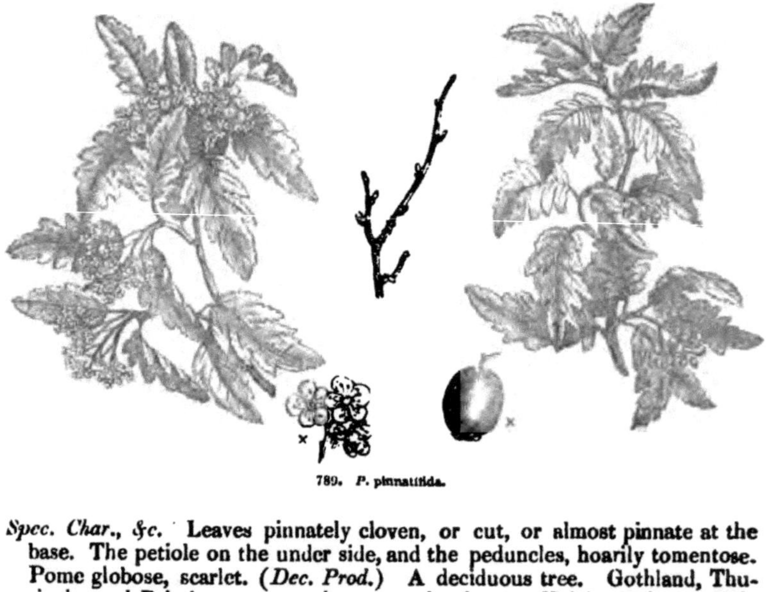

789. *P.* pinnatífida.

Spec. Char., &c. Leaves pinnately cloven, or cut, or almost pinnate at the base. The petiole on the under side, and the peduncles, hoarily tomentose. Pome globose, scarlet. (*Dec. Prod.*) A deciduous tree. Gothland, Thuringia, and Britain, on mountainous woody places. Height 20 ft. to 30 ft. Flowers white; May and June. Fruit red; ripe in September.

Varieties.

¥ P. *p.* 2 *lanuginòsa* has the leaves more woolly than those of the species.

♀ P. *p.* 3 *péndula*, Sórbus hýbrida péndula *Lodd. Cat.*, has the head loose, and the branches somewhat pendulous; the fruit red. H. S.

♀ P. *p.* 4 *arbúscula* Dec., Poir. Suppl. v. p. 144. — Dwarf. Leaflets glabrous in a measure, obtuse; the outermost usually connate. A native of Germany. Perhaps a variety of *P.* aucupària. (*Dec. Prod.*)

P. pinnatífida, according to DeCandolle, is a hybrid between *P.* intermèdia and *P.* aucupària. Culture as in *P. A'*ria.

♀ 29. *P.* AUCUPA'RIA *Gærtn.* The Fowler's Service Tree, *or Mountain Ash.*

Identification. Gærtn. Fruct., 2. p. 45. t. 87.; Dec. Prod., 2. p. 637.; Don's Mill., 2. p. 648.
Synonymes. Sórbus aucupària *Lin. Sp.* 683.; *Méspilus* aucupària *All.*; Quicken Tree, Quick Beam, wild Ash, wild Service, Wichen Tree, Rowan Tree, Rowne Tree, Roan Tree, Roddan, Routry Mountain Service, Witchen, wild Sorb, Whichen, Whitten, Wiggen tree; Sorbier des Oiseleurs, or Sorbier des Oiseaux, *Fr.*; Vogel Beerbaum, *Ger.*; Sorbo salvatico, *Ital.*
Derivation. The Latin name, *P.* aucupària (the Fowler's Pyrus); the French names, Sorbier des Oiseleurs (the Bird-catcher's Service), and Sorbier des Oiseaux (the Bird Service); and the German name, Vogel Beerbaum (the Bird's Berry Tree); are all derived from the use made of the berries by bird-catchers, in all countries where the tree grows wild, and from time immemorial, to bait springes with. It is called the Mountain Ash, from its growing on mountains, and the pinnæ of its leaves bearing some resemblance to those of the common ash. Witchen, and all its derivatives, bear relation to supposed powers of the tree, as a protection against witches and evil spirits.
Engravings. Gærtn. Fruct., 2. t. 87.; Eng. Bot., t. 337.; the plate of this species in Arb. Brit., 1st edit., vol. vi.; and our *fig.* 790.

790. *P.* aucupària.

Spec. Char., &c. Buds softly tomentose. Leaflets serrated, slightly glabrous. Pomes globose. (*Dec. Prod.*) A low tree. A native of almost every part of Europe. Height 20 ft. to 30 ft. Flowers white; May. Fruit red; ripe in September.

Varieties.

♀ P. *a.* 2 *frúctu lùteo* has yellow berries, and is continued by grafting.

♀ P. *a.* 3 *fòliis variegàtis* has variegated leaves.

♀ P. *a.* 4 *fastigiàta* has the branches upright and rigid. Horticultural Society's Garden.

The mountain ash forms an erect-stemmed tree, with an orbicular head. When fully grown, like every other description of *Pỳrus*, it assumes a somewhat formal character; but in a young state, its branches are disposed in a more loose and graceful manner. The tree grows rapidly for the first three or four years; attaining, in five years, the height of 8 or 9 feet; after which it begins to form a head, and in ten years will attain the height of 20 ft. This head will continue increasing slowly, though the tree seldom grows much higher, for the greater part of a century; after which, as it appears by the oldest trees that we have observed or heard of in Scotland, the extremities of the branches begin to decay. The tree will not bear lopping, but grass and other plants grow well under its shade. The wood, when dry, weighs 51 lb. 12 oz. per cubic foot. It is homogeneous, fine-grained, hard, capable of being stained any colour, and of taking a high polish; and it is applied to all the various uses of *P. A'ria* and *P.* torminàlis, when it can be obtained of adequate dimensions. In Britain, the tree forms excellent coppice wood, the shoots being well adapted for poles, and for making excellent hoops; and the bark being in demand by tanners. As it will grow in the most exposed situations, and rapidly, when young, it forms an admirable nurse tree to the oak, and other slow-growing species; and, being a tree of absolute habits; that is, incapable of being drawn up above a certain height by culture, it has this great advantage, that, after having done its duty as a nurse, instead of growing up with the other trees, and choking them, it quietly submits to be over-topped, and destroyed by the shade and drip of those which it was planted to shelter and protect. It may be mentioned, as somewhat singular, that the alpine laburnum, though naturally a much lower tree than the mountain ash, will, when drawn up in woods, attain twice the height of the latter tree. The fruit of the mountain ash is greedily devoured by birds: and, in various parts of the North of Europe, these berries are dried and ground into flour, and used as a substitute for the flour made of wheat, in times of great scarcity. In Livonia, Sweden, and Kamtschatka, the berries of the mountain ash are eaten, when ripe, as fruit; and a very good spirit is distilled from them. As an ornamental tree, the mountain ash is well adapted for small gardens; and it is also deserving of a place in every plantation, where the harbouring of singing-birds is an object. In the grounds of suburban gardens in the neighbourhood of the metropolis, the mountain ash forms almost the only tree that makes a great display by means of its fruit; for, though many species of *Cratæ'gus* would be equally effective in this respect, they have not yet become sufficiently well known to the planters of such gardens. One great advantage of the mountain ash, in all gardens, is, that it never requires pruning, and never grows out of shape. The mountain ash will grow in any soil, and in the most exposed situations, as it is found on the sea shore, and on the tops of mountains, in Forfarshire, as high as 2500 ft. Plants are almost always raised from seed, which should be gathered as soon as it is ripe, to prevent its being eaten by birds, which are so fond of it as to attack it even before it is ripe. When gathered, the fruit should be macerated in water till the seeds are separated from the pulp, and they may be then sown immediately; but, as they will, in that case, remain 18 months in the ground before coming up, the common mode adopted by nurserymen is, to mix the berries with light sandy soil, and spread them out in a layer of 10 in. or 1 ft. in thickness, in the rotting ground; covering the layer with 2 or 3 inches of sand or ashes, and allowing them to remain in that state for a year. They are then separated from the soil by sifting, and sown in beds of light rich soil, being covered a quarter of an inch. The plants having large leaves, the seeds should not be dropped nearer together than 2 in., which will allow the plants to come up with sufficient strength. They may be sown any time from November to February, but not later: they will come up in the June following, and, by the end of the year, the strongest plants will be 18 in. high, and fit to separate from the others, and to plant out in nursery lines.

¥ 30. *P.* AMERICA'NA *Dec.* The American Service.

Identification. Dec. Prod., 2. p. 637.; Hook. Fl. Bor. Amer., 1. p. 204.; Don's Mill., 2. p. 648.
Synonymes. Sórbus americàna *Ph. Fl. Bor. Amer.* 1. p. 341., *Willd. Enum* 520.; *S.* americàna var. β *Michx. Fl. Amer.* p. 290.; *P.* canadénsis *Hort.*
Engravings. Wats. Dend. Brit., t. 54.; the plate of this species in Arb. Brit., 1st edit., vol. vi.; and our *fig.* 791.

791. *P. americàna.*

Spec. Char., &c. Leaflets acute, almost equally serrated, glabrous, as is the petiole. Pomes globose, of a purplish tawny colour. (*Dec. Prod.*) A tree, closely resembling the common mountain ash. Canada and Newfoundland, in woods. Height 15 ft. to 20 ft. Introduced in 1782. Flowers white; May. Fruit bright scarlet; ripe in September. Decaying leaves yellowish green. Naked young wood dark purplish.

It is, apparently, a more robust-growing tree than the European mountain ash, with larger leaves, shining above, and smooth beneath; but, in reality, it is more tender. Though it has been many years in the country, we do not know of a large, old, handsome specimen of it any where. It is propagated by grafting on the common mountain ash. On account of the brilliant colour of the fruit, and the large size of the bunches in which it is produced, this species well deserves a place in collections.

¥ 31. P. MICROCA'RPA *Dec.* The small-fruited Service.

Identification. Dec. Prod., 2 p. 636.; Don's Mill., 2. p. 648.
Synonymes. Sórbus aucupària *a Michx. Fl. Bor. Amer.* 2. p. 291.; *S.* micrántha *Dum.Cours.* ed. 2. p. 464.; *S.* microcárpa *Ph. Fl. Amer. Sept.* 1. p. 341.
Engravings. Our *fig.*2090. in p. 1107.

Spec. Char., &c. Leaflets glabrous, acuminate, unequally incisely serrated; the teeth tipped with a bristle-like mucro. Petiole glabrous. Pome globose, scarlet. (*Dec. Prod.*) A large shrub or low tree. North America, Carolina to New York. Height 10 ft. to 12 ft. Introduced ?. Flowers white; May and June. Fruit scarlet, small; ripe ?.

According to Pursh, this species is very distinct from *P.* americàna; from which it is distinguished by the young branches being covered with a shining dark brown gloss, and by having small scarlet berries.—We have never seen it.

♆ 32. *P.* So'rbus *Gærtn.* The True Service.

Identification. Gærtn. Fruct., 2. p. 45. t. 87.; Dec. Prod., 2. p. 637.
Synonymes. Sórbus doméstica *Lin. Sp.* 684.; Pȳrus doméstica *Smith* in *Eng. Bot.* t. 350., *Wallr. Ann. Bot.* 145., *Don's Mill.* 2. p. 648.; the Whitty Pear Tree; Cormier, or Sorbier cultivé, *Fr.*; Speyerlingsbaum, or Sperberbaum, *Ger.*; Sorbo domestico, *Ital.*
Engravings. Eng. Bot., t. 350.; Gærtn. Fruct., 2. t. 87.; the plate in Arb. Brit., 1st edit., vol. vi.; and our *fig.* 792.

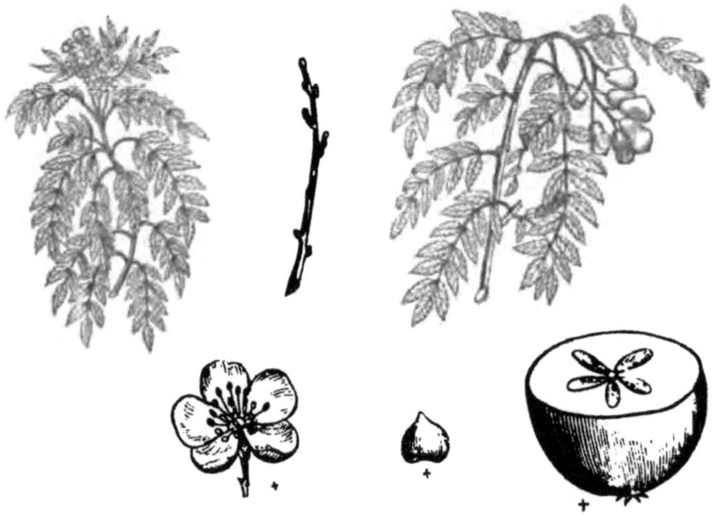

792. *P.* Sórbus.

Spec. Char., &c. Buds glabrous, glutinous, acuminate. Leaflets serrated, villose beneath, but becoming naked when old. Pome obovate, pear-shaped. (*Dec. Prod.*) A tree of the middle size. Europe, chiefly of the middle region; found also in some parts of Barbary, particularly in the neighbourhood of Algiers; and by some considered a native of Britain. Height 30 ft. to 60 ft. Flowers white; May. Fruit brown; October. Decaying leaves yellowish brown. Naked young wood grey, like that of the common mountain ash.

Varieties. In *Du Hamel* and the *Dictionnaire des Eaux et Forêts*, eight varieties of the true service are described; but in British gardens only the two following sorts are cultivated: —

♆ P. S. 2 *malifórmis* Lodd. Cat., la Corme-Pomme, *Fr.*, has apple-shaped fruit. Of this variety there are trees which bear abundantly in the Horticultural Society's Garden, and in the Hackney Arboretum.

♆ P. S. 3 *pyrifórmis* Lodd. Cat., la Corme-Poire, *Fr.*, has pear-shaped fruit; and of this, also, there are fruit-bearing trees in the places above referred to.

A tree, in foliage and general appearance, closely resembling the mountain ash; but attaining a larger size, and bearing much larger fruit, of a greenish brown colour when ripe. In France this tree attains the height of 50 or 60 feet: it requires two centuries before it reaches its full size; and lives to so great an age, that some specimens of it are believed to be upwards of 1000 years old. It grows with an erect trunk, which terminates in a large pyramidal head. This tree is readily known from the mountain ash, in winter, by its buds, which are smooth and green, instead of being downy and black; in the beginning of summer, by its leaflets being broader, downy above, and also beneath; and, in autumn, by its pear or apple shaped fruit, which is four or five times the size of that of *P.* aucupària, and of a dull greenish brown colour. It is said to be 30 years before it comes into a bearing state when it is raised from the seed; but, when scions from fruit-bearing trees are grafted on seedling plants, or on the mountain ash, they come into bearing in a few years as in the case of other fruit trees. (See *Gard. Mag.*, iv. p. 487.) The wood of the true service is the hardest and the heaviest of all the indigenous woods, of Europe. It weighs, when dry, no less than 72 lb. 2 oz. per cubic foot. It has a compact fine grain, a reddish tinge, and takes a very high polish; but it must not be employed until it is thoroughly seasoned, as otherwise it is apt to twist and split. It is much sought after in France, by millwrights, for making cogs to wheels, rollers, cylinders, blocks and pulleys, spindles and axles; and for all those parts of machines which are subject to much friction, and require great strength and durability. In France, it is preferred to all other kinds of wood for making the screws to wine-presses. In France, the fruit, when beginning to decay, is brought to table; though it is not highly prized, and is more frequently eaten by the poor than the rich. In Britain, the tree is chiefly to be recommended as one of ornament and rarity. A good, free, deep, dry soil, and a sheltered situation, are essential, wherever it is attempted to grow this tree in Britain. From the specimens in the neighbourhood of London, it does not appear to suffer from the climate, after it has been five or six years planted; but it is rather difficult to establish young plants. Seeds may be procured in abundance from France; and from them stocks may be raised on which the best fruit-bearing varieties may be grafted. The true service may also be grafted on the pear, the mountain ash, the hawthorn, and other allied species. The graft should be made close to the ground, or even under it, on the root; and care should be taken to retard the scion previously to grafting it, in order that the stock may be somewhat in advance of it. On the whole, the operation requires to be performed with the greatest care; because this is one of the most difficult of all non-resinous trees to graft successfully. The plants at Messrs. Loddiges's ripen fruit every year, from the seeds of which numerous young plants have been raised.

♀ 33. *P.* LANUGINO'SA *Dec.* The woolly-*leaved* Service Tree.

Identification. Dec. Prod., 2. p. 637.; Don's Mill., 2. p. 646.
Synonymes. *P.* hýbrida lanuginòsa *Hort.*; Sórbus lanuginòsa *Kit.* in Litt., and *Lodd. Cat.*
Engravings. The plate of this species in Arb. Brit., 1st edit., vol. vi.; and our *fig.* 793. from a tree in the Horticultural Society's Garden.

Spec. Char., *&c.* Buds woolly. Leaflets serrated, woolly beneath. Petiole woolly. Pome globose (*Dec. Prod.*) A fastigiate tree of the middle size. A hybrid, when and whence originated is uncertain. Height 20 ft. to 30 ft. Flowers white; May. Fruit small, like that of the mountain ash, but seldom coming to maturity.

The trees of this species in Loddiges's arboretum, and in the Horticultural Society's Garden, are very distinct from any other sort, and appear to be hybrids between *P.* pinnatífida, or perhaps *P.* Sórbus, and the common mountain ash. The general form of the tree is fastigiate, with numerous parallel, rigid, upright shoots. The flowers and fruit resemble those of the mountain ash, but are smaller: the former are frequently abortive; and the latter, when it is produced, is generally without seeds. It is a robust, hardy, vigorous-growing tree, which comes early into leaf, and is well deserving of a place in collections. This species, and all the others belonging to the section Sórbus, graft readily on the common hawthorn; and, as they make very handsome, small, round-headed trees, beautiful at every season of the year, common hedges might be grafted with them at regular distances, and the grafts would grow up, and become handsome standards.

793. *P.* lanuginósa.

Ÿ 34. *P.* SPU'RIA *Dec.* The spurious Service Tree.

Identification. Dec. Prod., 2. p. 637.; Don's Mill., 2. p. 649.
Synonymes. *P.* hýbrida *Mœnch Weiss.* t. 6. on the authority of *Willd. Sp.* 2. p. 1022.; Sórbus spùria *Pers. Syn.* 2. p. 39.; Méspilus sorbifòlia *Poir. Suppl.* 4. p. 72.; *P.* sorbifòlia *Bosc* and *Wats. Dend. Brit.* t. 53., but not of Smith; *P.* sambucifòlia *Cham.* and *Don's Mill.* 2. p. 648.
Engravings. Mœnch Weiss., t. 6., on the authority of Willd. Sp., 2. p. 1002.; Wats. Dend. Brit. t. 53.; Bot. Reg., t. 1196.; and our *fig.* 794.

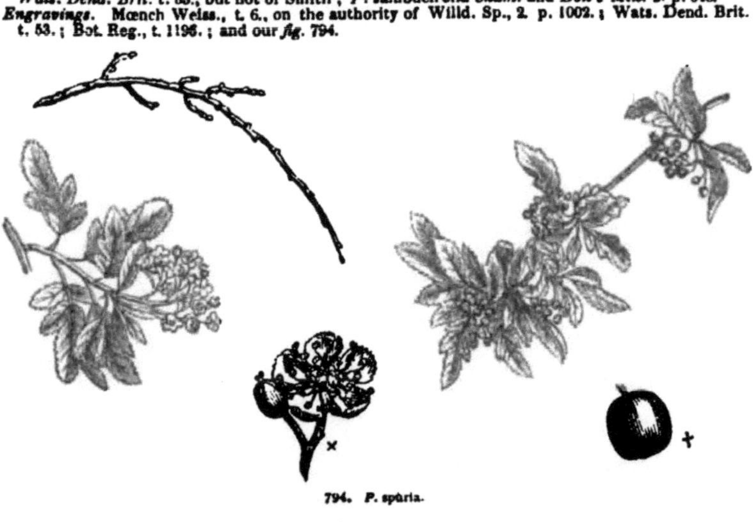

794. *P.* spùria.

Spec. Char., &c. Leaflets ovate, crenate; 3 pairs, with an odd one, which is longer than the others: all are hairy beneath. Petiole gland-bearing upon the upper side. Styles 5. Intermediate, and perhaps a hybrid, between *P.* aucupària and *P. a*rbutifòlia. (*Dec. Prod.*) A low tree, with rambling, slender, dark-coloured shoots, and fruit resembling that of *P. a*rbutifòlia. Height 10 ft. to 20 ft. In cultivation in 1800. Flowers white May. Fruit small, black; ripe in September. Decaying leaves dark purplish red. Naked young wood purplish.

Variety.

Ỵ P. *s.* 2 *péndula* Hort., *Só*rbus hýbrida péndula *Lodd. Cat.*, *P.* spùria sambucifòlia *Hort. Brit.* (the plate of this variety in *Arb. Brit.*, 1st edit. vol. vi.) has pendulous shoots, and is a very distinct and most interesting kind. There are fine low trees of it in the Horticultural Society's Garden; and, if grafted 10 or 12 feet high, instead of only 3 or 4 feet, as it is there, it would form one of the most beautiful of pendulous trees. It is prolific in flowers, and dark purple fruit; and the leaves die off of an intensely dark purplish red. Every hawthorn hedge might be adorned with this tree by grafting.

Both the species and variety are very desirable small trees for their leaves, their flowers, and their fruit; they are readily propagated by grafting on the common thorn or mountain ash, and require the same soil as that species.

Ỵ 35. *P.* FOLIOLO'SA *Wall.* The leafy Mountain Ash.

Identification. Wall. Cat. p. 677.; Pl. Asiat. Rar., 2. p. 81.; Don's Mill., 2. p. 648.
Engravings. Wall. Pl. Asiat. Rar., 2. t. 189.; and our *fig.* 795.

Spec. Char., &c. Leaves pinnate, with 7—8 pairs of elliptic-lanceolate, mucronate leaflets, which are serrated at the apex, pubescent beneath. Cymes branched, terminal, pubescent. (*Don's Mill.*) A deciduous tree. Nepal, on mountains. Height 15 ft. to 20 ft. Introduced?. Flowers white; June. Pome small, obovate roundish, red; ripe Nov. This very desirable and probably quite hardy species, we believe, has not yet been introduced, but it doubtless will soon be so.

795. *P.* foliolòsa.

P. *hircìna* Wall. Cat. p. 675., and *Don's Mill.* ii. p. 648., is a native of Nepal, with pinnate leaves, and numerous leaflets, rusty beneath; and with red fruit, about the size of that of the common mountain ash.

§ vii. *Adenórachis* Dec.

Sect. Char. Petals spreading, each with a claw, and a concave limb. Styles 2—5. Pome globose. Leaves simple, the midrib bearing glands on its upper surface (which is the character expressed in the sectional name). Flowers in branched corymbs. — Deciduous shrubs, natives of North America; growing to the height of 4 or 5 feet, and prolific in flowers, followed by red, dark purple, or black, fruit. They are all readily propagated by division, by suckers, or by grafting on the common hawthorn.

This section is so unlike the others in habit and general appearance, that it would be much more convenient to have it as a distinct genus; say, *Arònia*, as it was before that genus was united with *Pỳrus*.

■ 36. *P.* ARBUTIFO'LIA *L. fil.* The Arbutus-leaved Aronia.

Identification. Lin. fil. Suppl., 256.; Ph. Sept. 1. p. 339.; Dec. Prod., 2. p. 637.; Don's Mill., 2. p. 649.
Synonymes. Cratæ'gus pyrifòlia *Lam. Dict.* 1. p. 83.; *Arònia* pyrifòlia *Pers. Syn.* 2. p. 39.; ?Cratæ'gus serràta *Poir. Suppl.* 1. p. 292.; Méspilus arbutifòlia *Schmidt Arb.* t. 86., *Mill. Dict.* t. 109.
Engravings. Schmidt Arb., t. 86.; and our *fig.* 796.

796. *P.* arbutifòlia.

Spec. Char., &c. Leaves obovate, lanceolate, acute, crenate, tomentose beneath, especially when young, the midrib in each glandulous above. Calyx tomentose. Pome dark red or purple. (*Dec. Prod.*) A deciduous shrub. North America, from Canada to Carolina, in low copses and swamps, common. Height 4 ft. to 6 ft. Introduced in 1700. Flowers white; May. Fruit dark red or black; ripe in September. Decaying leaves intensely dark red, or purplish black.

Varieties.

■ P. a. 2 *intermèdia* Lindl. (Hort. Trans., vii. p. 229.; Don's Mill., ii. p. 649. and our *fig.* 798.) has the fruit globose and brown.

■ P. a. 3 *seròtina* Lindl. (Hort. Trans., l. c.; Don's Mill., l. c.) has the leaves shining above, and velvety beneath; and the fruit late, and party-coloured.

797. *P.* arbutifòlia pùmila.

798. *P.* arbutifòlia intermèdia.

■ P. a. 4 *pùmila*, *Méspilus* pùmila *Lodd. Cat.* (Krause, t. 86.; and our *figs.* 797. and 799.), appears to be different from the two preceding varieties. It is a low plant, seldom exceeding 1 ft. or 18 in. in height.

and rooting at the joints. The fruit is intermediate in colour between *P. arbutifòlia* and *P. melanocárpa*, being of a reddish black.

799. *P. a. pùmila.*

A very desirable shrub, frequent in collections, and known in the nurseries under the name of *Méspilus arbutifòlia*. It is prolific in flowers, which are produced in May, and which are followed by dark red or purple fruit, which, when not eaten by birds, will remain on the bushes till the following April or May, when the plant is again in flower. This species, whether as a bush, or grafted standard high on the common thorn, is highly ornamental in spring, when it is covered with its profusion of white flowers; in autumn, when its foliage assumes a deep red or purple; and in winter, after the leaves have dropped, when it is still enriched with its persistent fruit. It is propagated by layers, suckers, or seeds; but most frequently by suckers. There was, in 1835, a remarkably fine plant of this species, grafted standard high, in Knight's Exotic Nursery: it had attained the height of 10 or 12 feet; its branches hung down gracefully to the ground, not in one mass, but in varied tufts; and their appearance in autumn, when they were of an intensely purple red, was beyond expression interesting and beautiful.

37. *P. (A.)* MELANOCA'RPA *W.* The black-fruited Aronia.

Identification. Willd. Enum., 525.; Ph. Sept., 1. p. 339.; Dec. Prod., 2. p. 637.; Don's Mill., 2. p. 649.
Synonymes. *P. arbutifòlia β Willd. Sp.* 2. p. 1013.; *Arònia arbutifòlia Pers. Syn.* 2. p. 39.; *Méspilus capitàta Lodd.*; *M. floribúnda Lodd.*; *M. pùbens Lodd. Cat.* 1836.
Engravings. Schmidt Arb., t. 86.; Krause, t. 79.; and our *fig.* 800.

Spec. Char., &c. Leaves obovate-oblong, acuminate, serrated, glabrous beneath; the midrib glandulous above. Corymb more crowded than in *P. arbutifòlia*. Calyx glabrous. Pome black. (*Dec. Prod.*) A deciduous shrub. North America, in Canada, in bogs, and on the high mountains of Carolina and Virginia; and judging from the plants in the Horticultural Society's Garden, and in the arboretum of Messrs. Loddiges, nothing more than a variety of *P. arbutifòlia*. Height 4 ft. to 5 ft. Cultivated in 1700. Flowers white; May. Fruit large, black, resembling in taste those of *Vaccínium pennsylvánicum*; ripe in September.

800. *P. (a.) melanocárpa.*

Variety.

P. (a.) *m.* 2 *subpubéscens* Lindl. (Hort. Trans., vii. p. 232.; Don's Mill., ii. p. 649.), *P. m. xanthocárpa Hort.*, has the leaves, when young, tomentose beneath, but glabrous in the adult state.

P. (a.) melanocárpa or its variety, grafted standard high on the common hawthorn, forms a truly interesting pendulous, and at the same time picturesque, tree; and we can scarcely sufficiently recommend it for small shrubberies and suburban gardens. As its berries are not so greedily eaten by birds as those of most of the other *Rosàceæ*, in mild winters they remain on till the following summer, and mix beautifully with the flowers in June. It grafts readily on the common hawthorn; and it, and all the other species and varieties belonging to the section Adenórachis, might be introduced into our common hedges by any countryman who could graft, thus rendering them truly ornamental.

38. *P. (A.)* FLORIBU'NDA *Lindl.* The abundant-flowered Aronia.

Identification. Lindl. Hort. Trans., 7. p. 230.; Lindl. Bot. Reg., 1006.; Don's Mill., 2. p. 649.
Engravings. Lindl. Bot. Reg., t. 1006.; and our *fig.* 801.

Spec. Char., &c. Branches cinereous, reclinate. Leaves oblong-lanceolate, acute, on long petioles, tomentose beneath, as well as the calyxes. Fruit spherical. Corymbs many-flowered, and longer than the leaves. (*Don's Mill.*) A shrub resembling the preceding species, but with more pendulous branches. Originated in gardens. Height 3 ft. to 4 ft. Cultivated in 1815. Flowers white; May. Fruit black; ripe in September.

801. P. (a.) floribúnda.

The leaves die off of a purplish red; and the whole plant, from the time of its leafing till it becomes naked, is highly interesting and ornamental, more especially when grafted standard high.

39. *P.* (*A.*) DEPRE'SSA *Lindl.* The depressed Aronia.

Identification. Lindl. Hort. Trans., 7. p. 230.; Don's Mill., 2. p. 649.
Engraving. Our *fig.* 802. from a specimen in Dr. Lindley's herbarium.

802. P. (a.) depréssa

Spec. Char., &c. Stems humble, reclinate. Leaves oblong, obtuse, tomentose beneath, as well as the calyxes. Fruit pear-shaped. Corymbs length of the leaves. (*Don's Mill.*) A low shrub, with pendulous depressed branches. Of garden origin. Height 1 ft. to 2 ft. Cultivated in 1815. Flowers white; May. Fruit dark purple; ripe in September.

It is evidently a variety or modification of the preceding sort; and, from its profusion of flowers and fruit, and the purple hue of its foliage, it is highly ornamental.

40. *P.* (*A.*) PU'BENS *Lindl.* The downy-*branched* Aronia.

Identification. Lindl. Hort. Trans., 7. p. 232.; Don's Mill., 2. p. 649.
Engraving. Our *fig.* 803. from a living specimen in the Horticultural Society's Garden.

Spec. Char., &c. Stem erect. Branches pubescent. Leaves oblong or obovate, abruptly acuminated, smooth. Fruit spherical, and, as well as the calyxes, quite glabrous. Corymbs loose, and many-flowered. (*Don's Mill.*) A bushy shrub, with robust foliage, and shoots. Of garden origin. Height 4 ft. to 5 ft. Cultivated in 1810. Flowers white; May. Fruit round, large, dark purple; ripe in September.

This and the following kind have much of the robust foliage and habit of *P.* Chamæméspilus, and well deserve a place in collections.

803. P. (a.) pùbens.

41. *P.* (*A.*) GRANDIFO'LIA *Lindl.* The large-leaved Aronia.

Identification. Lindl. Hort. Trans., 7. p. 232.; Lindl. Bot. Reg., 1154.; Don's Mill., 2. p. 649.
Engravings. Bot. Reg., t. 1154.; and our *figs.* 804. and 805.

Spec. Char., &c. Stem erect, and, as well as the branches, smoothish. Leaves

804. P. (a.) grandifòlia.

oblong, or obovate, acute, glabrous. Fruit spherical, and, as well as the calyxes, glabrous. Corymbs few-flowered, coarctate. Fruit with a villous disk. (*Don's Mill.*) A bushy shrub. Of garden origin, probably a hybrid between *P. arbutifòlia* and *P. Chamæméspilus.* Height 4 ft. to 5 ft. Cultivated in 1810. Flowers white; May and June. Fruit dark purple; ripe Sept. Decaying leaves purple and yellow.

805. P. (a.) grandifòlia.

It bears a profusion of flowers, and dark purple fruit; and, on that account, and also on account of the purple tinge of its leaves, it is highly ornamental. Dr. Lindley considers it as the most valuable species of this division of *Pyrus* that has hitherto been described.

§ viii. *Chamæméspilus* Dec.

Sect. Char. Petals upright, conniving, concave. Styles 2. Pome ovate. Leaves simple, glandless. Flowers in a capitate corymb. (*Dec. Prod.*)

■ 42. *P.* Chamæme'spilus *Lindl.* The dwarf Medlar.

Identification. Lindl. in Lin. Soc. Trans., 13. p. 98.; Dec. Prod., 2. p. 637.; Don's Mill., 2. p. 649.
Synonymes. Cratæ'gus Chamæméspilus *Jacq. Austr.* t. 231.; Méspilus Chamæméspilus *Lin. Sp.* 685.; Sórbus Chamæméspilus *Crantz Austr.* 83. t. 1. f. 3.; the bastard Quince; niedriger Mispelbaum, *Ger.*; Camenespolo, *Ital.*
Engravings. Jacq. Austr., t. 231.; Crantz Austr., 83. t. 1. f. 3.; and our *fig.* 806.

Spec. Char., &c. Leaves ovate, serrated, glabrous; except bearing on the under surface, when young, down, which is deciduous. (*Dec. Prod.*) A stiff-branched shrub. Europe, in rough mountainous places. Height 5 ft. to 6 ft. Introduced in 1683. Flowers white, tinted with rose; May and June. Fruit round, orange-coloured, or red; ripe in September.

This species forms a compact bush, and flowers and fruits in the greatest abundance, and hence it merits to be much more extensively introduced into collections than it appears to have hitherto been. It grafts beautifully on the common hawthorn; and, indeed, whoever has a quickset hedge may have a collection of all the species of this genus.

806. P. Chamæméspilus.

Other Species of Pyrus. — P. *alnifòlia* Lindl. in Lin. Trans. xiii. p. 98. Leaves glabrous, roundish, feather-nerved, and rather glaucous beneath. Fruit black and sugary. North America, at Fort Mandon. — P. *tomentòsa* Dec. Prod. ii. p. 637.; Màlus tomentòsa *Dum. Cours.* ed. 2. v. p. 438. Allied to *P. baccàta*; but the flowers, as well as fruit, are unknown. Siberia.—P. *rubicúnda* Hoffmans. (*Verz.*, 1824, p. 192.; *Dec. Prod.*, ii. p. 637.) Leaves oval-acuminate. Fruit partly red and partly yellow. Native country unknown.

Genus XXI.

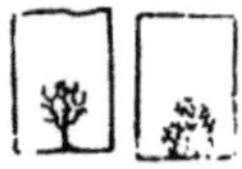

CYDO'NIA *Tourn.* The Quince Tree. *Lin. Syst.* Icosándria Di-Pentagýnia.

Identification. Tourn. Inst., 632. t. 435.; Dec. Prod., 2. p. 638.; Don's Mill., 2. p. 650.
Synonymes. Pȳrus sp. *Lin.*; Coignassier, *Fr.*; Quittenbaum, *Ger.*; Cotogno, *Ital.*
Derivation. From its native place, *Cydon*, in Candia.

Gen. Char. *Calyx* 5-cleft. *Petals* orbicular. *Stamens* erect. *Styles* 5. *Pome* closed, 5-celled; cells cartilaginous, many-seeded. *Seeds* covered with mucilaginous pulp. (*Don's Mill.*)

Leaves simple, alternate, stipulate, deciduous; serrated or entire. *Flowers* large, solitary, or few together in a kind of umbel. — Low deciduous trees or shrubs, natives of Europe and Asia; easily propagated by layers, or by grafting on the common thorn. Common soil, rather moist than dry.

† 1. *C.* vulga'ris *Pers.* The common Quince Tree.

Identification. Pers. Syn., 2. p. 40.; Dec. Prod., 2. p. 638.; Don's Mill., 2. p. 650.
Synonymes. Pȳrus Cydònia *Lin. Sp.* 687., *Jacq. Austr.* t. 342.; *C.* europæ'a *Sav. Alb. Tosc.* 1. p. 90.
Engravings. Jacq. Austr., t. 342.; the plate in Arb. Brit., 1st edit., vol. vi.; and our *fig.* 807.

807. *Cydonia vulgàris.*

Spec. Char., &c. Leaves ovate, obtuse at the base, entire, tomentose beneath. Calyx tomentose; its lobes serrulated, and a little leafy. Stamens in one row. (*Dec. Prod.*) A low tree. South of Europe, in rocky places and hedges; and by some considered indigenous in Britain. Height 15 ft. to 20 ft. Cultivated in 1573. Flowers white; May and June. Fruit large, of a fine orange yellow; ripe in November.

Varieties.

† C. *v.* 1 *pyrifórmis* Hort. — Fruit pear-shaped.
† C. *v.* 2 *malifórmis* Hort. — Fruit apple-shaped.

♀ C. *v.* 3 *lusitánica* Du Ham. — Broader leaves, and larger fruit, than the two preceding kinds; and, being of more vigorous growth, it is better adapted for being used as a stock for pears.

The quince is a low tree, with a crooked stem and tortuous rambling branches. It is of moderately rapid growth when young; attaining, in four or five years, the height of 6 or 8 feet; and, in ten or twelve years, the height of 15 feet: after which it continues to increase in width of head only; being very seldom found higher than 15 or 20 feet.

The quince prefers a moist free soil, near water, and a situation open, but sheltered.. In dry soil, neither the tree nor the fruit attains any size; and, in situations exposed to high winds, the fruit will not remain on the tree till ripe. Layers.

♀ 2. *C.* SINE'NSIS *Thouin*. The China Quince Tree.

Identification. Thouin Ann. Mus., 19. p. 145. t. 8. and 9.; Dec. Prod., 2. p. 638.; Don's Mill., 2. p. 650.
Synonyme. Pȳrus sinénsis *Poir. Suppl.* 4. p. 452.
Engravings. Ann. Mus., 19. t. 8. and 9.; the plate of this tree in Arb. Brit., 1st edit., vol. vi.; and our *fig.* 808.

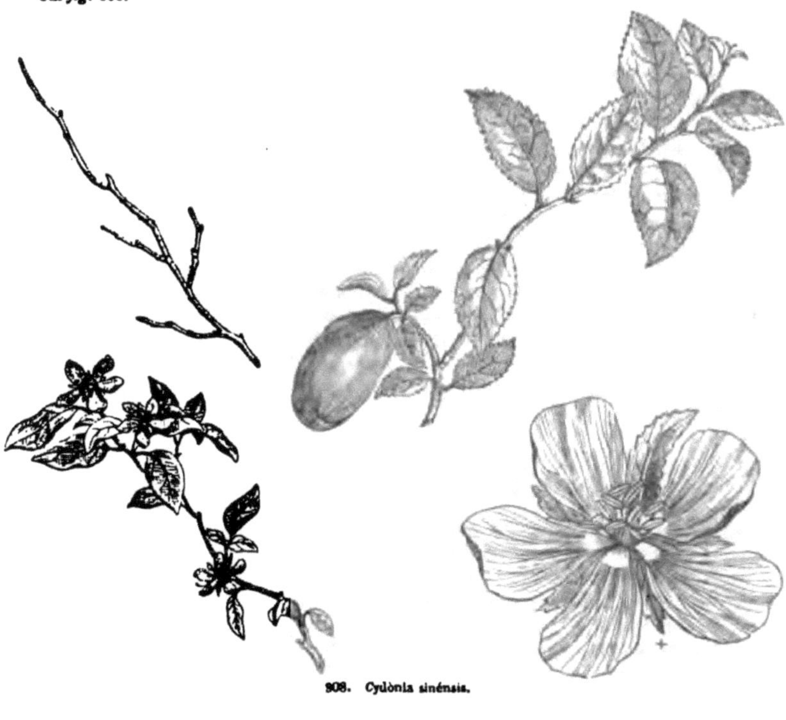

808. Cydònia sinénsis.

Spec. Char., &c. Leaves ovate, acuminated at both ends, acutely serrated, when young a little villose, and when adult glabrous. Stipules oblong linear, serrated; the teeth glanded. Flowers rosy, becoming red. Calyx glabrous; its lobes serrulated, and a little leafy. Stamens in one row. Fruit egg-shaped, large, hard, almost juiceless, and greenish. Seeds in each cell about 30, with many abortive. (*Dec. Prod.*) A very handsome low tree, very distinct in appearance from the common quince, from the shining surface of its leaves, and the regular serratures of their margins. China. Height in China 20 ft.; in England 10 ft. to 12 ft. Introduced in 1818. Flowers white or pale red; May and June. Fruit egg-shaped, greenish, hard, and nearly dry; ripe in October.

■ 3. *C.* JAPO'NICA *Pers.* The Japan Quince Tree.

Identification. Pers. Syn., 2. p. 40.; Dec. Prod., 2. p. 638.; Don's Mill., 2. p. 650.
Synonymes. ? Pýrus japónica *Thunb. Fl. Jap.* 207., and *Bot. Mag.* t. 692.; Chænomèles japónica *Lindl. Lin. Trans.* 13. p. 98.
Engravings. Bot. Mag., t. 692.; Morris Fl. Consp., t. 1.; our *fig.* 809.; and *fig.* 810. showing the fruit.

809. C. japónica.

810. C. japónica.

Spec. Char., &c. Leaves oval, somewhat cuneated, crenately serrated, glabrous upon both surfaces. Stipules kidney-shaped, and serrated. Flowers mostly 2—3 together, rarely solitary. Calyx glabrous; its lobes short, obtuse, entire. Stamens in two rows. (*Dec. Prod.*) A rambling deciduous shrub. Japan and China. Height 5 ft. to 6 ft. Introd. in 1815. Flowers scarlet; produced the greater part of the year. Fruit green, very fragrant, but not eatable; ripe in Oct.

Varieties.

■ C. *j.* 2 *flòre álbo* has cream-coloured, or very pale red, flowers, and forms a very distinct kind when in blossom.

■ C. *j.* 3 *fl. semi-plèno* has red flowers, somewhat semidouble. There are plants of this kind in the Kensington Nursery.

One of the most desirable deciduous shrubs in cultivation, whether as a bush in the open lawn, trained against a wall, or treated as an ornamental hedge plant. It has also been trained up with a single stem as a standard; and, in this character, its pendent branches and numerous flowers, give it a rich and striking appearance, especially in early spring. It is difficult to unite with its congeners by grafting; but, if it could be grafted standard high on the pear, the hawthorn, or even on the common quince, it would form a most delightful little tree. Readily propagated by layers or suckers, and it also grows by cuttings.

ORDER XXVII. CALYCANTHA'CEÆ.

ORD. CHAR. *Calyx* coloured, many-parted. *Segments* in many series, unequal. *Corolla* none. *Stamens* indefinite, in several series. *Anthers* adnate, extrorse. *Styles* exserted. *Carpels* 1-seeded, enclosed within the fleshy tube of the calyx. *Albumen* none. Agrees with *Rosàceæ* in fruit, and *Granàteæ* in leaves, but differs from both in the absence of petals (*G. Don.*)

Leaves simple, opposite, exstipulate, deciduous; feather-nerved, rough. *Flowers* axillary.—Deciduous shrubs, natives of Asia and America.

CALYCA'NTHUS. Stamens 12, deciduous.
CHIMONA'NTHUS. Stamens 5, persistent.

GENUS I.

CALYCA'NTHUS *Lindl.* THE CALYCANTHUS, *or* AMERICAN ALLSPICE. *Lin. Syst.* Icosándria Polygýnia.

Identification. Lindl. in Bot. Reg., t. 404.; Nees Nov. Act. Nat. Cur. Bonn., 11. p. 107.; Dec. Prod., 3. p. 2.; Don's Mill., 2. p. 652.

Synonymes. Calycánthi sp. *Lin.*, *Lam.*, *Willd.*; Büttnèria Du Ham. Arb., 1. p. 114., not of Lin.; Beurrèria Ehret Pict. t. 13.; Bastèria Adans. Fam. 2. p. 294.; Pompadoúra Buchoz; Calycanthe, *Fr.*; Gewürzstrauch (spice shrub), Kelch Blume, *Ger.*; Calicanto, *Ital.*
Derivation. From *kalux*, a calyx, and *anthos*, a flower; the calyx is coloured, and resembles a corolla. The name allspice was given to it by the inhabitants of Carolina, from the strong aromatic smell of the bark.

Gen. Char. *Lobes* of calyx disposed in many series, imbricate, lanceolate, all somewhat coriaceous and coloured. *Stamens* unequal, deciduous, 12 outer ones fertile. *Achenia* numerous. (*Don's Mill.*)

Leaves simple, alternate, stipulate, deciduous; entire, coriaceous. *Flowers* axillary, rising after the leaves, of a lurid purple colour, and sweet-scented, as well as the bark and leaves.

Deciduous shrubs, natives of North America; propagated, in England, by layers. DeCandolle states that the removal of the terminal leaf bud of a shoot causes the production of two new flower buds; and that by this practice a succession of flowers, during the whole summer, may be obtained when desirable. (*Dec. Prod.*)

1. C. FLO'RIDUS *L.* The flowery Calycanthus, *or Carolina Allspice.*

Identification. Lin. Sp., 718.; Nutt. Gen. Amer., 1. p. 312.; Dec. Prod., 3. p. 2.; Don's Mill., 2. p. 652.
Synonymes. C. stérilis *Walt. Car.* 151.; sweet-scented Shrub, *in Carolina*; common American Allspice; Calycanthe de la Caroline, *Fr.*; Carolinische Kelch Blume, *Ger.*; Pompadur, *Ital.*
Engravings. Bot. Mag., t. 503.; Du Ham. Arb., 1. t. 45.; Lam. Ill., t. 446. f. 1.; Guimp. Abb. Holz., t. 4.; our *fig.* 810.; and *fig.* 811. showing the fruit.

811. Calycánthus flóridus.

812. C. flóridus.

Spec. Char., &c. Wood of the trunk, and especially of the root, intensely camphor-scented. Branches spreading; branchlets tomentose. Leaves oval, tomentose beneath. Flowers mostly abortive. Fruit top-shaped. (*Dec. Prod.*) A dense orbiculate bush. Carolina, on the shaded banks of rivulets. Height 6 ft. to 8 ft. Introduced in 1726. Flowers dusky, purple, scented; May to August. Fruit brown, rarely seen in England. General aspect of the foliage, in summer, dark brownish green. Decaying leaves bright yellow. Naked young wood dark brown.

Varieties. DeCandolle gives two forms of this species: —

- *C. f.* 1 *oblóngus*, leaves oblong (*Ait. Hort. Kew.*, ed. 2., 3. p. 282.); and
- *C. f.* 2 *ovàtus*, leaves roundishly ovate (*Ait. Hort. Kew.*, ed. 2., 3. p. 282.).

The following varieties are in Loddiges's *Catalogue* for 1836; and plants of most of them are in their arboretum, and in that of the Horticultural Society: —

- *C. f.* 3 *asplenifòlius* has cut leaves.
- *C. f.* 4 *fèrax* has fertile flowers.
- *C. f.* 5 *glaúcus* has leaves somewhat glaucous.
- *C. f.* 6 *inodòrus* has flowers nearly scentless.
- *C. f.* 7 *longifòlius* has elongated leaves.
- *C. f.* 8 *variegàtus* has variegated leaves.

The flowers grow singly on short peduncles at the extremity of the branches; they have two series of narrow thick sepals, which spread open, and turn inward at the top, like those of the anemone or clematis. It thrives best in a light, rich, sandy soil, kept rather moist, and in a shady situation. The varieties differ very slightly from each other.

■ 2. C. (F.) GLAU'CUS *Willd.* The glaucous-*leaved* Calycanthus, *or fertile-flowered American Allspice.*

Identification. Willd. Enum., 559.; Pursh Fl. Sept. Amer., p. 357.; Dec. Prod., 3. p. 2.
Synonyme. C. fértilis *Walt. Car.* 151., *Lindl. Bot. Reg.* t. 404., *Guimp. Abb. Holz.* t. 5., *Don's Mill.* 2. p. 652.
Engravings. Bot. Reg., t. 404.; Guimp. Abb. Holz., t. 5.; and our *fig.* 813.

813. C. (f.) glaúcus.

Spec. Char., &c. Branches spreading. Leaves ovate-lanceolate, acuminate, glaucous beneath, pubescent. Flowers less odorous than those of C. flóridus. (*Dec. Prod.*) A dense orbiculate bush. Carolina, on mountains. Height 6 ft. to 8 ft. Introduced in 1726. Flowers lurid purple; May to August. Fruit, leaves, and young wood as in the preceding species.

Variety.

■ *C. g.* 2 *oblongifòlius* Nutt. Gen. Amer. i. p. 312., Dec. Prod. iii. p. 2. C. oblongifòlius *Hort. Brit.*—Leaves ovate-lanceolate, elongated. A native of North Carolina, on mountains. (*Dec. Prod.*, iii. p. 2.)

Closely resembling C. flóridus in general appearance; and requiring the same soil and culture. According to Pursh, the flowers are like those of C. flóridus; but their scent is not so agreeable, and is more faint. Whether there is much difference between this sort and C. f. 5 glaúcus, we have not had an opportunity of ascertaining; the plant in Messrs. Loddiges's arboretum not having flowered. We have therefore retained the description of this kind as a species, in deference to Pursh, DeCandolle, and G. Don, though we strongly suspect that they are identical.

■ 3. C. (F.) LÆVIGA'TUS *Willd.* The glabrous-*leaved* Calycanthus, *or American Allspice.*

Identification. Willd. Enum., 559.; Willd. Hort. Berol., t. 80.; Pursh Fl. Sept. Amer., p. 358.; Nutt. Gen. Amer., 1. p. 312.; Dec. Prod., 3. p. 2.; Don's Mill., 2. p. 652.
Synonymes. C. fèrax *Michx. Fl. Bor. Amer.* 1. p. 305.; C. pennsylvánicus *Lodd. Cat.*
Engravings. Bot. Reg., t. 481.; and our *fig.* 814.

814 C. (f.) lævigàtus.

Spec. Char., &c. Branches strictly upright. Leaves oblong or ovate, and gradually acuminated, slightly wrinkled; the upper surface rough to the touch, the under one glabrous and green. (*Dec. Prod.*) A dense orbiculate bush. Pennsylvania, Virginia and Carolina, on mountains. Height 3 ft. to 5 ft. Introduced in 1806, and resembling the two preceding sorts in appearance and culture, but with the leaves more pointed. Very probably the C. f. 4 fèrax of the preceding page.

GENUS II.

CHIMONA'NTHUS *Lindl.* THE CHIMONANTHUS, *or* WINTER-FLOWER. *Lin. Syst.* Icosándria Polygýnia.

Identification. Lindl. Bot. Reg., t. 404. and 451.; Dec. Prod., 3. p. 2.; Don's Mill., 2. p. 652.

Synonymes. Meràtia Nees Act. Soc. Nat. Bonn. 11. p. 107. ; Calycánthi sp. *Linn.*
Derivation. From *cheimōn*, winter, and *anthos*, a flower ; in allusion to the season of its flowering.

Gen. Char. *Lobes* of calyx imbricate, oval, obtuse ; outer ones in the form of bracteas ; inner ones larger, and appearing like a corolla. *Stamens* nearly equal, permanent, the five outer ones fertile, connate at the base. (*Don's Mill.*)
Leaves simple, alternate, stipulate, deciduous ; entire. *Flowers* rising before the leaves, in the axils of the leaves of the preceding year ; very sweet-scented ; yellowish, but purplish inside. Bark and leaves without scent.

A deciduous shrub, a native of Japan ; remarkable for the fragrance of its flowers, which are produced from December till March, even in the open garden, in the neighbourhood of London, and more especially if the plant is trained against a wall. The blossoms are produced singly, in the axils of the leaves, on the shoots of the preceding year, and also on spurs proceeding from the old wood. The soil, culture, &c., are the same as for Calycánthus.

■ 1. C. FRA'GRANS *Lindl.* The fragrant-*flowered* Chimonanthus.

Identification. Lindl. Bot. Reg., t. 404. 451. ; Dec. Prod., 3. p. 2. ; Don's Mill., 2. p. 652.
Synonymes. Calycánthus præ̀cox *Lin. Sp.* 1718., *Ait. Hort. Kew.* ed. 1. vol. 2. p. 220. t. 10., *Curt. Bot. Mag.* t. 466., *Lam. Ill.* t. 445. t. 2. ; Meràtia fràgrans *Nees Act. Soc. Nat. Bonn.* 11. p. 107. ; O'bai, or Ròbai, *Kæmpf. Amæn.* 879. ic. ; the Winter-Flower ; Calycante de Japon, *Fr.* ; Japanische Kelch Blume, *Ger.*
Engravings. Kæmpf. Amœn. ic. ; Ait. Hort. Kew., ed. 1. vol. ii. t. 10. ; Bot. Mag., t. 466. ; Lam. Ill., t. 445. t. 2. ; and our *fig.* 815.

Spec. Char., &c. Bark and leaves scentless. Flowers protruded before the leaves, solitary in the old axils of leaves, extremely odorous, yellowish, and purple within. Fruit flask-shaped, or thicker above the base, and in the upper part tapered into a cylindrical neck. (*Dec. Prod.*) A deciduous shrub. Japan. Height 6 ft. to 8 ft. as a bush, twice that height when trained against a wall. Introduced in 1776. Flowers yellowish and purple within ; December to March. Fruit flask-shaped. Decaying leaves greenish. Naked young wood light grey.

815. C. fràgrans.

816. C. fràgrans grandiflòrus.

Varieties.

■ *C. f.* 2 *grandiflòrus* Lindl. Bot. Reg. t. 451. ; and our *fig.* 816. — Flowers larger, and more spreading. Fruit oblong, tapered at the base. (*Dec. Prod.*)

■ *C. f.* 3 *lùteus* Hort. has the flowers yellow both inside and outside.

■ *C. f.* 4 *parviflòrus* Hort. — A distinct late-flowering variety. Plants in the Horticultural Society's Garden.

The flowers, which are produced in the greatest abundance from November till March (as the name, winter-flower, implies), and which are delightfully and refreshingly fragrant, scent the air to a considerable distance round the tree. It is grown in most choice gardens for its flowers ; a few of which are gathered daily, and placed in the drawingroom or boudoir, in the same manner as violets. The plant is generally propagated by layers ; but it frequently produces seeds, from which many plants have been raised. The variety C. f. grandiflòrus has the flowers rather less fragrant than the species, but they are much more ornamental. No garden whatever ought to be without this shrub. North of London, it deserves a wall as much as any fruit tree ; at least, judging from the measure of enjoyment which it is calculated to afford : and, south of London, it may be planted as a standard bush on the open lawn, or in the shrubbery.

ORDER XXVIII. *GRANATA'CEÆ.*

ORD. CHAR. *Calyx* 5—7-cleft, tube turbinate, limb tubular ; æstivation valvate. *Petals* 5—7. *Stamens* indefinite, free. *Style* filiform. *Stigma* capitate. *Fruit* large, spherical, divided horizontally into two compartments, lower part 3-celled ; upper part 5—9-celled. *Seeds* numerous, covered with pellucid baccate pulp. *Albumen* wanting. Differs from *Myrtàceæ* in the leaves being without dots. (*G. Don.*)

Leaves simple, opposite or alternate, exstipulate, deciduous ; lanceolate, entire. *Flowers* terminal, scarlet. — Shrubs or low trees, natives of Africa.

GENUS I.

PU'NICA Tourn. THE POMEGRANATE TREE. *Lin. Syst.* Icosándria Monogýnia.

Identification. Tourn. Inst., t. 401. ; Dec. Prod., 3. p. 3. ; Don's Mill., 2. p. 653.
Synonymes. The Carthaginian Apple ; Grenadier, *Fr.* ; Granate, *Ger.* ; Melograno, *Ital* ; Granados, *Span.*
Derivation. Pùnica is said, in the *Nouveau Du Hamel*, to be derived either from *puniceus*, scarlet, in allusion to the scarlet colour of the flowers ; or from the same word, or *Punicus*, both signifying " of Carthage ;" near which city, Pliny tells us, it was first found.

Gen. Char. Same as that of the order.

Leaves simple, opposite, sometimes whorled or alternate, exstipulate, deciduous ; oblong, entire. *Flowers* terminal, scarlet, with numerous stamens. — Low deciduous trees or shrubs, indigenous to Africa, and naturalised in the South of Europe.

♆ 1. *P.* GRANA'TUM *L.* The *common* Pomegranate Tree.

Identification. Lin. Sp., 676. ; Dec. Prod., 3. p. 3. ; Don's Mill., 2. p. 653.
Engravings. Bot Mag., t. 1832. ; and our *fig.* 817.

Spec. Char., &c. Stem arboreous. Leaf lanceolate. (*Dec. Prod.*) A deciduous tree. Mauritania, whence it may have migrated into the South Europe, where it is now perfectly indigenous. Height, in France and Italy, 15 ft. to 30 ft. ; in England generally trained against a wall, where it attains double that height when there is room. Introduced in 1548. Flowers scarlet ; June to September. Fruit globose, in warm seasons sometimes ripened in November.

Varieties.

♆ P. G. 1 *rùbrum* Dec. Prod. iii. p. 3. (Trew Ehret, t. 71. f. 1. ; Poit. et Turp. Arbr. Fr., 22. ; Schkuhr. Handb., t. 131. *b.* ; Sims Bot. Mag., t. 1832. ; and our *fig.* 817.) has the flowers red ; pulp of fruit reddish. Wild in Mauritania and the South of Europe, and enduring even the coldest winters. (*Dec. Prod.*, iii. p. 3.)

817 *P. Granàtum rùbrum.*

■ P. G. 2 *rùbrum flòre plèno* Trew Ehret t. 71. f. 2. has double red flowers. It is common in gardens, and is a little more impatient of cold than the preceding variety. (*Dec. Prod.*, iii. p. 4.)

■ P. G. 3 *albéscens* Dec. Prod. iii. p. 4., Andr. Bot. Rep. t. 16. — Petals white. Calyx slightly yellowish. Pulp of the fruit of a pale red. It is cultivated in gardens, and is rather more tender than *P. G.* rùbrum. (*Dec. Prod.*, iii. p. 4.)

■ P. G. 4 *albéscens flòre plèno* Dec. has double flowers, which are nearly white. It is cultivated in gardens, and is the tenderest of all the forms of the species. (*Dec. Prod.*, iii. p. 4.)

■ P. G. 5 *flàvum* Hort. has the flowers yellow, but is rare in gardens.

A tree, in magnitude and ligneous character, bearing considerable resemblance to the common hawthorn. In the South of France, and in Spain and Italy, it grows to the height of 18 or 20 feet; forming a very branchy twiggy tree, seldom found with a clear stem, unless it has been pruned up. In a wild state, about Marseilles, it forms a thorny bush; but, in the gardens about Nice and Genoa, it is a very handsome small tree, much admired both for its flowers and its fruit. In the South of Europe, the pomegranate is cultivated for its fruit; and, in some places, as a hedge plant. It is also grown as an ornamental tree; the stem being trained to the height of 6 or 8 feet, and the head afterwards allowed to spread, and droop down on every side. The single wild pomegranate will grow in almost any soil; but the double-flowered varieties, and the species when it is intended to bear fruit, require a rich free soil. Whether the object be flowers or fruit, the head of the tree should be thinned out, and so cut as to multiply, as much as possible, short slender shoots; on the points of which alone the flowers are produced. The plant is easily propagated by cuttings of the shoots or of the roots, by layers, or by grafting one sort on another. The double variety grafted on the single is more productive of flowers than when raised by cuttings or layers; and in good rich soil, judiciously supplied with water, it will continue producing its large scarlet flowers for four or five months. When the plant is raised from seeds, they ought to be sown immediately on being removed from the fruit; because they very soon lose their vital powers.

■ 2. *P.* (*G.*) NA'NA *L.* The dwarf Pomegranate.

Identification. Lin. Sp., 676.; Sims Bot. Mag., t. 634.; Dec. Prod., 3. p. 4.; Don's Mill., 2. p. 653.
Synonymes. *P.* americàna nàna *Tourn.*; *P.* Granàtum nànum *Pers.*
Engravings. Bot. Mag., t. 634.; Trew Ehret, t. 71. f. 3.; and our *fig.* 818.

818. *P.* (*G.*) nàna.

Spec. Char., &c. Stem shrubby. Leaf linear. (*Dec. Prod.*) A low deciduous bush; said to be a native of the Caribee Islands and of South America, about Demerara. Introduced in 1723. Flowers red; June to September.

Without doubt, only a variety of *P.* Granàtum.

ORDER XXIX. TAMARICA'CEÆ.

ORD. CHAR. *Calyx* 4—5-parted. *Petals* 4—5. *Stamens* 4—5 or 8—10, free or monadelphous. *Style* short. *Stigmas* 3. *Capsule* trigonal, 3-valved, 1-celled, many-seeded. *Placentas* 3 at base of cell, or along the middle of the valves. *Seeds* comose at apex. *Albumen* none. Near Frankeniàceæ in Thalamiflòræ.

Leaves simple, alternate, exstipulate, deciduous or sub-evergreen; linear-lanceolate. *Flowers* small. — Shrubs. Natives of Europe, Asia, and Africa; of easy culture in light sandy soil. The genera in British gardens are two, which are thus contradistinguished: —

Ta'marix. Stamens 4—5. Hairs on the seeds simple.
Myrica'ria. Stamens 10. Hairs on the seeds feathery.

Genus I.

TA'MARIX *Desv.* The Tamarisk. *Lin. Syst.* Pentándria Trigýnia.

Identification. Desv. Ann. Sc. Nat., 4. p. 348.; Dec. Prod., 3. p. 95.; Don's Mill., 2. p. 725.
Synonymes. The species of Támarix of authors that have 4 stamens and 5 stamens; Tamaris, *Fr.*; Tamarisken, *Ger.*; Tamarice, Ital.
Derivation. So called, according to some, from the plants growing on the banks of the river *Tamaras*, now Tambra, on the borders of the Pyrenees; or, according to others, from the Hebrew word *tamaris*, cleansing, on account of their branches being used for brooms.

Gen. Char. *Calyx* 4—5-parted. *Petals* 4—5. *Stamens* 4—5, alternating with the petals, almost free. *Ovarium* tapering to the apex. *Stigmas* 3, long, divaricate, glandular, and oblique at the apex. *Seeds* erect, inserted nearly at the base of the valves; tufted; tuft composed of numerous simple hairs arising from the apex. (*Don's Mill.*)

Leaves simple, alternate, exstipulate, deciduous or sub-evergreen; linear, stem-clasping, very small; adult ones diaphanous at the apex. *Flowers* in spikes, and usually disposed in panicles, small, red, seldom white.

Tall shrubs, natives of Europe, the North of Africa, and the West of Asia; sub-evergreen in British gardens; and highly valuable, as standing the sea breeze in situations where few other ligneous plants, and no other flowering shrubs, will grow. The whole plant is very bitter, and the young shoots were formerly employed as a tonic, and as a substitute for hops in brewing beer.

■ ■ 1. *T.* GA'LLICA *L.* The French Tamarisk.

Identification. Lin. Sp., 386.; Mill. Ic.; Dec. Prod., 3. p. 96.; Don's Mill., 2. p. 726.
Synonymes. *T.* narbonénsis *Lob. Ic.* 2. t. 218.; Tamaríscus gállicus *All.*; Tamaríscus pentándrus *Lam. Fl. Fr.*, not of Pall.; Mirice, *Ital.*
Engravings. N. Du Ham., vol. vii. t. 59.; and our *fig.* 819.

819. *T. gállica.*

Spec. Char., &c. Glabrous, glaucous. Leaves minute, clasping the stem or branch, adpressed, acute. Spikes of flowers lateral, somewhat panicled, slender, 5 times longer than broad. (*Dec. Prod.*) A sub-evergreen shrub, frequent in sandy places in the middle and South of Europe, and in the South of England. Height 5 ft. to 10 ft., sometimes twice that height. Flowers pinkish; May to October.

Varieties. In the *Linnæa*, 2. p. 267., 6 varieties are described, for which we refer to our first edition, as the plants are not in cultivation in Britain, and indeed appear to us not worth keeping distinct.

T. gállica prefers a deep, free, sandy soil; and will only attain a large size when it is in such a soil, and supplied with moisture from the proximity of some river, or other source of water. It is valuable as thriving on the sea shore, where few other shrubs will grow; as being sub-evergreen; and as flowering late in the season, and for several months together. It is readily propagated by cuttings, planted in autumn, in a sandy soil, with a northern exposure. In favourable situations in France and the South of Europe, it grows to the height of 15 or 20 feet; and there are instances, both in Britain and on the Continent, of its growing as high as 30 ft., and this, we suppose, has given rise to the alleged variety, *T.* g. arbòrea. In the South of Russia, and in Tartary, the species assumes a great variety of forms according to the soil and situation; the tops of the dwarf plants are there eaten by sheep, and the stems of the larger ones used as handles for whips.

Genus II.

MYRICA'RIA *Desv.* The Myricaria. *Lin. Syst.* Monadélphia Decándria.

Identification. Desv. Ann. Sc. Nat., 4. p. 349.; Dec. Prod., 3. p. 97.; Don's Mill., 2. p. 727.
Synonymes. The species of Támarix of authors that have monadelphous stamens.
Derivation. From *murikē*, the Greek name of the tamarisk, derived from *murō*, to flow; the species being generally found on the banks of running streams; or from the flowing of the sap as manna.

Gen. Char. *Calyx* 5-parted. *Petals* 5. *Stamens* 10, alternate ones shorter than the rest; filaments monadelphous from the base to about the middle. *Stigmas* 3, sessile, in a head. *Seeds* inserted in a line along the middle of the valves, tufted at one end; hairs of tuft feathery. (*Don's Mill.*)

Leaves simple, alternate, exstipulate, evergreen; linear or oblong, becoming gradually broader towards the base, sessile. *Flowers* in simple, solitary, terminal spikes.

Sub-evergreen shrubs, not growing to half the height of Támarix gállica, and readily distinguished from it by their longer and thicker leaves, placed at a greater distance from one another on the stem; and by their larger flowers, which have 10 stamens. Propagation and culture as in the preceding genus.

■ 1. M. germa'nica *Desv.* The German Myricaria, *or German Tamarisk.*

Identification. Desv. Ann. Sc. Nat., 4. p. 349.; Dec. Prod., 3. p. 97.; Don's Mill., 2. p. 727.
Synonymes. Támarix germánica *Lin. Sp.* 386.; Tamaríscus decándrus *Lam. Fl. Fr.*; Támarix decándra *Mœnch*; Tamariscus germánicus *Lob. Ic.* 2. t. 218.; Tamaris d'Allemagne, *Fr.*; Deutschen Tamarisken, *Ger.*; Tamarigia piccola, *Ital.*
Engravings. Mill. Ic., t. 262. f. 2.; and our *fig.* 820.

820 M. germánica.

Spec. Char., &c. Fruticulose, glabrous. Leaves linear-lanceolate, sessile. Spikes of flowers terminal, solitary. Bracteas longer than the pedicels. Capsules ascending. (*Dec. Prod.*) An evergreen shrub, with upright slender branches. Europe, in inundated sandy places, and the banks of rivers; and in Asia, on Caucasus, and the Himalayas. Height 6 ft. to 8 ft. Introduced in 1582. Flowers pinkish; June to September.

■ 2. M. dahu'rica *Dec.* The Dahurian Myricaria.

Identification. Dec. Prod., 3. p. 98.; Don's Mill., 3. p. 728.
Synonyme. Támarix dahúrica *Willd. Act. Berol.* No. 16.
Engraving. Our *fig.* 821. from a specimen in Dr. Lindley's herbarium.

Spec. Char., &c. Shrubby, glabrous. Leaves linear, almost oblong, sessile, slightly spreading. Spikes of flowers lateral, ovate-cylindrical, thick, blunt, with scales at the base. Bracteas extending as far as the flowers. (*Dec. Prod.*) Siberia, beyond the Baikal; and in Dahuria. Introduced in 1816, and closely resembling the preceding species.

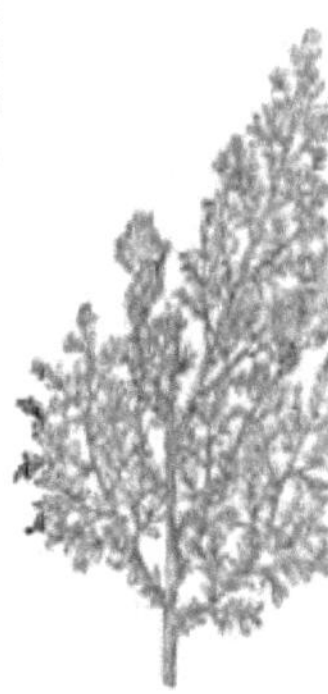
821. M. dahúrica.

Order XXX. PHILADELPHA'CEÆ.

Ord. Char. *Calyx* tube turbinate, limb 4—10-parted. *Petals* 4—10, æstivation convolutely imbricate. *Stamens* numerous. *Styles* distinct or combined. *Stigmas* numerous. *Capsule* half inferior, 10-celled, many-seeded.

Seeds subulate, heaped at the angles of the cells; each furnished with a loose membranous aril. *Albumen* fleshy. Differs from *Myrtàceæ* in the arillate albuminous seed, and in the toothed dotless leaves. (*Don's Mill.*)

Leaves simple, opposite, exstipulate, deciduous; toothed, serrated, or sometimes entire. *Flowers* large, white, showy.—Shrubs, natives of Europe, Asia, and America. Three genera are hardy in British gardens, which are thus contradistinguished:—

*P*HILADE'LPHUS. Calyx 4—5-parted. Petals 4—5. Stamens 20—40. Styles 4—5. Capsule 4—5-celled.

DEU'TZ*IA*. Calyx 5—6-cleft. Petals 5—6. Stamens 10—12. Filaments tricuspidate. Styles 3—4. Capsule 3—4-celled.

DECUMA'RIA. Calyx 7—10-toothed. Petals 7—10. Stamens 21—30. Style 1. Capsule 7—10-celled.

GENUS I.

*P*HILADE'LPHUS *L.* THE PHILADELPHUS, *or* MOCK ORANGE. *Lin. Syst.* Icosándria Monogýnia.

Identification. Lin. Gen., No. 614.; Dec. Prod., 3. p. 205.; Don's Mill., 2. p. 807.
Synonymes. *Syringa* Tourn. Inst. t. 389., not of Lin.; Philadelphus, *Fr.*; Pfeifenstrauch (Pipe Shrub), *Ger.*; Filadelpho, *Ital.*; Pipe Privet, *Gerard*; the Syringa of the gardens.
Derivation. Philadélphus is a name used by Athenæus for a tree which cannot now be identified: Bauhin applied it to this genus. Instead of the common trivial name Syringa, applied to this genus in gardens, as its English name, we have substituted its generic name, Philadelphus; *Syringa* being the generic name of the lilac.

Gen. Char. *Calyx* tube obovate, turbinate; limb 4—5-parted. *Petals* 4—5. *Stamens* 20—40, free, shorter than the petals. *Styles* 4—5, sometimes connected, and sometimes more or less distinct. *Stigmas* 4—5, oblong or linear, usually distinct, rarely joined. *Capsule* 4—5-celled, many-seeded. *Seeds* scobiform, enclosed in a membranous arillus, which is fringed at one end. (*Don's Mill.*)

Leaves simple, opposite, exstipulate, deciduous; oval, acuminate, serrated or entire. *Flowers* white, usually sweet-scented, disposed in corymbose cymes, or somewhat panicled on the extremities of the lateral shoots, rarely axillary and bracteate.

Deciduous shrubs, natives of Europe, North America, and Asia; cultivated for their very showy white flowers; most of which have a strong scent, resembling, at a distance, that of orange flowers, but, when near, disagreeably powerful. The species are in a state of utter confusion: there are probably only three: one a native of the South of Europe, or possibly of some other country; one of North America; and one, *P.* tomentòsus, of Nepal. All the kinds are of the easiest culture in any tolerably dry soil; and they are all propagated by layers, or by suckers or cuttings.

§ i. *Stems stiff and straight. Flowers in Racemes.*

▲ 1. *P.* CORONA'RIUS *L.* The garland Philadelphus, *or Mock Orange.*

Identification. Lin. Sp., 671.; Schrad. Diss.; Dec. Prod., 3. p. 205.; Don's Mill., 2. p. 807.
Synonymes. *Syringa* suavèolens *Mœnch Meth.* 678.; Wohlriechender, Pfeifenstrauch, *Ger.*; Fior angiolo, *Ital.*
Engravings. Bot. Mag., t. 391.; Schkuhr Handb., t. 121.; Lam. Ill., t. 420.; and our *fig.* 822.

Spec. Char., &c. Leaves ovate, acuminate, serrately denticulate, 3-nerved, rather glabrous, but hairy upon the veins beneath; inflorescence racemose. Flowers sweet-scented. Lobes of the calyx acuminate. Styles distinct almost from the base, not exceeding the stamens in height. (*Dec. Prod.*) A dense fastigiate bush. Native country uncertain; according to some, the South of Europe, but not common there. Height 10 ft. to 12 ft. Introduced in 1596. Flowers white, sweet-scented; May and June. Fruit brown; ripe in September. Decaying leaves yellowish green.

822. *P. coronàrius.*

Varieties. This species varies in having its leaves sometimes perfectly glabrous beneath, and sometimes slightly pubescent along the nerves; and, besides, as follows: —

- ■ P. *c.* 1 *vulgàris* Schkuhr Handb. t. 121., Lam. Ill. t. 420., Dec. Prod. iii. p. 205. — A shrub of about the height of a man. Leaves ovate-oblong, large, and rather distant.
- ■ P. *c.* 2 *nànus* Mill. Dict. 2.—A shrub, 2 ft. high; its branches and leaves crowded, and its flower-bearing branches incurved. It very seldom flowers.
- ■ P. *c.* 3 *flòre plèno* Lodd. Cat. is a dwarf plant, like the above, but with double flowers.
- ■ P. *c.* 4 *variegàtus* Lodd. Cat. has the leaves variegated with white or yellow, and is one of the few varieties of deciduous shrubs which preserve, through the summer, a tolerably healthy appearance with their variegation.

The shoots are clothed with a white bark, and interiorly they have a very large pith. The leaves are rough, and of a deep green above, though they are pale beneath. The flowers come out from the sides and ends of the branches, in loose bunches, before any of the other species of the genus. The flowers smell like those of the orange, and the leaves taste like the fruit of the cucumber. It will grow in almost any situation, whether open or shady; and it is easily propagated by division of the root, and by suckers, layers, or cuttings.

■ 2. *P.* (C.) INODO'RUS *L.* The scentless-*flowered* Philadelphus, *or Mock Orange.*

Identification. Lin. Sp., 671.; Pursh Sept., 1. p. 329.; Dec. Prod., 3. p. 206.
Synonymes. *Syringa* inodòra *Mœnch*; *P.* láxus in various English gardens; Sillndia senz' odore, *Ital.*
Engravings. Catesb. Car., 2. t. 84.; Bot. Mag., t. 1478.; and our *fig.* 823.

Spec. Char., &c. Leaves broad-ovate, acuminate, perfectly entire, 3-nerved, usually feather-nerved. Flowers singly, or in threes. Style, at the very tip, divided into 4 oblong stigmas. (*Dec. Prod.*) A large rambling shrub. Apparently of garden origin, or possibly from North America. Height 5 ft. to 10 ft. Cultivated in 1738. Flowers large, white, scentless; June and July. Fruit brown; ripe in September.

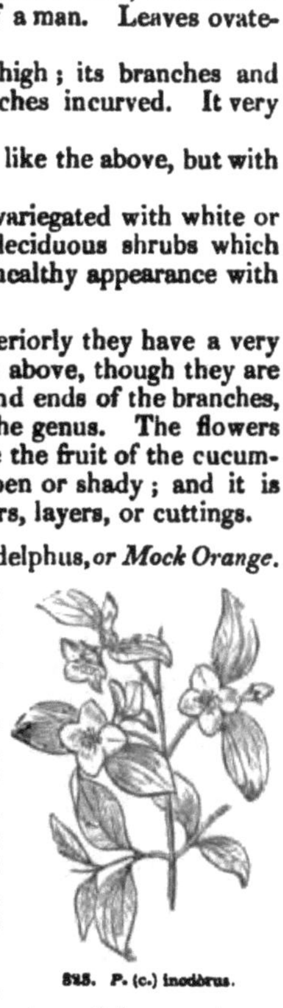

823. *P.* (c.) *inodòrus.*

Not altogether so hardy as *P.* coronàrius; though it appears to be only a variety of that species.

■ 3. *P.* (c.) ZEY'HERI Schrad. Zeyher's Philadelphus, *or Mock Orange.*

Identification. Schrad. Diss. Philad.; Dec. Prod., 3. p. 206.; Don's Mill., 2. p. 807.
Engravings. Schrad. Diss. Philad. ic. 1. and our *fig.* 824. from the Museum of the Jardin des Plantes.

Spec Char., &c. Not so tall as *P.* c. vulgàris. Leaves ovate, acuminate, serrately denticulate, rounded at the base, 3-nerved, hairy upon the veins beneath. Inflorescence some-

824. *P.* (c.) *Zeyheri.*

what racemose. Flowers fewer and larger than in *P.* c. vulgàris, and scentless. Lobes of the calyx long, acuminate. Style deeply 4-cleft. It differs from *P.* c. vulgàris, chiefly in its leaves being rounded at the base, and in its flowers being fewer, larger, and scentless. (*Dec. Prod.*) North America. Horticultural Society's Garden.

■ 4. *P.* VERRUCO'SUS *Schrad.* The warted Philadelphus, *or Mock Orange.*

Identification. Schrad. Diss. Philad.; Dec. Prod., 3. p. 205.; Don's Mill., 2. p. 807.
Synonyme. P. grandiflòrus *Lindl. Bot. Reg.* t. 570., *Lodd. Cat.* ed. 1836.
Engravings. Bot. Reg., t. 570.; and our *fig.* 825.

Spec. Char., &c. Leaves elliptic-ovate, acuminate, denticulate, pubescent with hairs beneath, and bearing beneath, upon the midrib and primary veins, warts at the base of the hairs. Similar warts are, also, on the peduncles, pedicels, and calyxes. Inflorescence racemose. Lobes of the calyx acuminate. Style, at the very tip, 4-cleft. (*Dec. Prod.*) A large vigorous-growing somewhat fastigiate shrub. North America, or possibly a garden production. Height 8 ft. to 10 ft. Cultivated in 1800, or before. Flowers white; May and June. Fruit brown; ripe in September.

825. P. verrucòsus.

Young shoots twice the thickness of those of *P.* coronàrius, and having a somewhat more fastigiate habit. When in flower, this sort and the two following make a splendid appearance; the plants, in fine seasons, being so entirely covered with bloom as scarcely to show the leaves.

■ 5. *P.* (v.) LATIFO'LIUS *Schrad.* The broad-leaved Philadelphus, *or Mock Orange.*

Identification. Schrad. Diss. Philad.; Dec. Prod., 3. p. 206.; Don's Mill., 2. p. 807.
Synonyme. P. pubéscens *Cels. Hort.*, *Lois. Herb. Amat.* t. 208.
Engravings. Lois. Herb. Amat., t. 208.; and our *fig.* 826.

Spec. Char., &c. Bark whitish. Leaves broad-ovate, acuminate, toothed, nerved with about 5 nerves, and pubescent with hairs beneath. Flowers in racemes. Lobes of the calyx acuminate. Style 4-cleft at the very tip. (*Dec Prod.*) A large rambling shrub, but still somewhat fastigiate, and crowded with branches. North America, or probably a garden production. Height 10 ft. Cultivated in 1815, or before. Flowers white; May and June. Fruit brown; ripe in September.

826. P. (v.) latifòlius.

Distinguishable by its bark being whitish; and by its leaves, especially those of the younger branches, being more broadly ovate; and by the hairs they bear not being based by warts. A tolerably distinct variety, splendid when in flower.

■ 6. *P.* (V.) FLORIBU'NDUS *Schrad.* The abundant-flowered Philadelphus, *or Mock Orange.*

Identification. Schrad. Diss. Philad.; Dec. Prod., 3. p. 205.; Don's Mill., 2. p. 807.
Engravings. Schrad. Diss. Philad., ic.; and our *fig.* 827. from a specimen in DeCandolle's herbarium.

827. *P. floribundus.*

Spec. Char., &c. Leaves ovate-oval, and with a long acuminate tip, serrately toothed, 3-nerved, pubescent with hairs beneath. Inflorescence subracemose. Flowers 5—7, showy, slightly scented. Lobes of the calyx long and acuminate. Style 4-cleft at the very tip. (*Dec. Prod.*) A shrub like the preceding. Of uncertain origin. Cultivated in 1815 or before.

■ 7. *P.* SPECIO'SUS *Schrad.* The showy-*flowered* Philadelphus, *or Mock Orange.*

828. *P. speciosus.*

Identification. Schrad. in Dec. Prod., 3. p. 206.; Don's Mill., 2. p. 807.; Bot. Reg., vol. 23.
Synonymes. *P.* grandiflòrus of German gardeners; *P.* grandiflòrus láxus of other gardeners.
Engravings Schrad. Diss. Philad., ic.; Bot. Reg., t. 2003.; and our *fig.* 828.

Spec. Char., &c. Leaves ovate, rarely oval-ovate, long acuminated, sharply serrate, toothed, clothed with hairy pubescence beneath. Flowers solitary or by threes. Style deeply 4-cleft, exceeding the stamens. Lobes of calyx with very long acumens, tube nearly terete. (*Don's Mill.*) A vigorous-growing shrub, with ascending branches gently bending at their extremities, and loaded with snow-white flowers of the largest size and scentless. Origin uncertain. Height 10 ft. to 14 ft. Cultivated in 1815 or before. Flowers white; June.

The handsomest species of the genus, and at present rare in British gardens. Horticultural Society's Garden.

■ 8. *P.* GORDONIA'NUS Lindl. Gordon's Philadelphus *or Mock Orange.*

Identification. Lindl. in Bot. Reg., 1839, No. 32.
Engravings. Bot. Reg., 1839, t. 32.; and our *fig.* 829.

829. *P. Gordonianus.*

Spec. Char., &c. Branches pendulous, scaly, the young shoots pubescent. Leaves ovate, acute, coarsely dentate, hairy beneath. Racemes terminal, compact, 5—9-flowered. Ovary half superior. Style 4-cleft. Calyx spreading from the fruit. (*Lindl.*) A hardy vigorous-growing shrub, having a weeping appearance in consequence of producing numerous slender side shoots. America, on the north-west coast, on the Columbia River, in woods. Height 8 ft. to 10 ft. Introduced in 1823. Flowers white, scentless, very hairy; end of July. Fruit large, smooth; ripe in October.

The leaves are bright green, rather small, ovate, pointed, 3-nerved at the base, and coarsely serrated. The flowers are large, pure white, and produced in great profusion. The species is readily known by its deeply serrated leaves, its nearly superior fruit, its broad spreading calyx, and the compact manner in which its flowers are arranged. It is the latest in flowering of all the species, and one of the most showy. It was not at all injured by the severe winter of 1837-8. It grows in any common soil, and is readily increased from seeds, or by cuttings of the half-ripened shoots in August. It was named in compliment to Mr. Robert Gordon, foreman of the arboretum, in the Hort. Soc. Garden, who has paid great attention to this genus.

§ ii. *Stems more slender, rambling, twiggy, and loose. Flowers solitary, or 2 or 3 together.*

■ 9. *P.* LA'XUS *Schrad.* The loose-*growing* Philadelphus, *or Mock Orange.*

Identification. Schrad. Diss. Philad.; Dec. Prod., 3. p. 206.; Don's Mill., 2. p. 807.
Synonymes. *P.* hùmilis *Hortul.*; *P.* pubéscens *Lodd. Cat.* edit. 1836.
Engravings. Schrad. Diss. Philad., ic.; Bot. Reg., 1839, t. 39.; and our *fig* 830. from a plant in the Horticultural Society's Garden.

Spec. Char., &c. Leaves oval-ovate and with a long acuminate tip, toothed, pubescent with hairs beneath. Flowers solitary, 2 or 3 together. Lobes of the calyx very long, acuminate. Style 4-cleft. Stigmas about level with the stamens. (*Dec. Prod.*) A low straggling shrub. North America. Height 4 ft. to 5 ft., but covering double that space upon the ground with its long slender deep brown shoots. Introduced about 1830. Flowers white, most commonly solitary and almost scentless.

830. *P.* láxus

The leaves are smaller than is usual in the genus, very sharp-pointed, with the toothing unusually sharp; the uppermost leaves become gradually narrow, till those immediately below the flowers are not unfrequently linear and entire. As this species leafs early, the young shoots are apt to be killed by frost, and when this takes place no flowers are produced that season, as it is from the ends of the lateral shoots that blossoms always appear in this genus. (*Bot. Reg.*)

■ 10. *P.* (L.) GRANDIFLO'RUS *Willd.* The large-flowered Philadelphus, *or Mock Orange.*

Identification. Willd. Enum., 1. p. 511.; Schrad. Diss. Philad.; Dec. Prod., 3. p. 206.
Synonymes. *P.* inodòrus *Hortul.*; *P.* làxus *Lodd. Cat.* edit. 1836.
Engravings. Guimp. Abb. Holz., t. 44.; Schrad. Diss. Philad., ic.; and our *fig.* 831.

Spec. Char., &c. Epidermis of the branches of a reddish brown colour. Leaves ovate, with a long acuminate tip, denticulate, 3-nerved, hairy upon the veins, and with groups of hairs in the axils of the veins. Flowers about 3 together, or solitary; scentless. Lobes of the calyx long, acuminate. Styles concrete into one, which extends beyond the stamens. Stigmas 4, linear. (*Dec. Prod.*) A vigorous-growing shrub. North America. Height 10 ft. to 12 ft. Introduced in 1811. Flowers white; June and July.

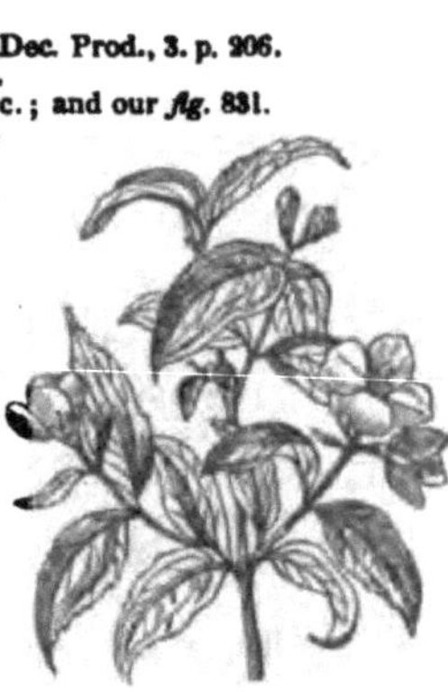
831. *P.* (l.) grandiflòrus.

We have given this description and figure from Schrader and Guimpel, because in 1837 there was a plant in the Hort. Soc. Garden which answered to it, and which differs from *P.* speciòsus in the leaves being nearly entire.

■ 11. *P.* HIRSU'TUS *Nutt.* The hairy-*leaved* Philadelphus, *or Mock Orange.*

Identification. Nutt. Gen. Am., 1. p. 301.; Dec. Prod., 3. p. 206.; Don's Mill., 2. p. 808.
Synonymes. *P.* villòsus *Lodd. Cat.*; *P.* grácilis *Lodd. Cat.*
Engravings. Wats. Dend. Brit., t. 47.; and our *figs.* 832, 833.

Spec. Char., &c. Leaves oblong-ovate, acute, dentate, 5-nerved, hairy on both surfaces, whitish on the under one. Flowers singly, or by threes. Styles concrete to the tip. Stigmas undivided. (*Dec. Prod.*) A straggling sarmentose shrub. America, in Tennessee. Height 3 ft. to 5 ft., but spreading several times that distance by its long shoots, which sometimes grow from 6 ft. to 10 ft. in a season. Introduced in 1820. Flowers white, scentless; middle of July.

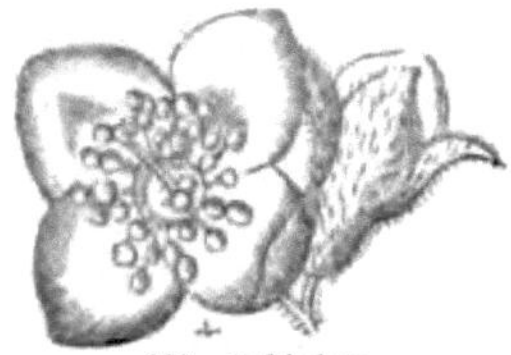

832. *P.* hirsùtus.

Very hardy, uninjured by the winter of 1837–8, and striking readily from half-ripened cuttings planted in sandy loam in shady situations and covered with a hand-glass. Grafted standard high, it would form a very ornamental object.

833. *P.* hirsùtus.

▲ 12. *P.* TOMENTO'SUS *Wall.* The woolly-*leaved* Philadelphus, *or Mock Orange.*

Identification. Wall. Cat., 3658.; Don's Mill., 2. p. 807.
Synonymes. *P.* nepalénsis *Lodd. Cat.* edit. 1836; ? *P.* triflòrus *Royle.*
Engravings. Royle Illust., t. 46. f. 1.; our *fig.* 834. from Royle; and *fig.* 835. from a specimen in the Linnæan herbarium.

834. *P.* tomentòsus.

Spec. Char., &c. Leaves ovate, acuminated, denticulated, tomentose beneath. Racemes terminal. Pedicels opposite. Lobes of calyx ovate, acute. (*Don's Mill.*) A rambling shrub. Nepal and Kamaon. Height 5 ft. to 6 ft. Introduced in 1822. Flowers white; July.

835. *P.* tomentòsus.

P. triflòrus, Royle observes, is, probably, only *P.* tomentòsus in a less advanced state. According to Mr. Gordon, *P.* triflòrus is very distinct from *P.* tomentòsus, and quite hardy.

P. mexicànus Schlecht. in Linnæa 13. 418., Plant. Hart. 61., and Bot. Reg. Chron. 1840, No. 70., was raised from seeds in the Horticultural Society's Garden in 1840, and is probably hardy.

GENUS II.

DEU'TZIA Thunb. THE DEUTZIA. *Lin. Syst.* Decándria Trigýnia.

Identification. Thunb. Nov. Gen., 19.; Jap. p. 10.; Juss. Gen., 431.; Don's Mill., 2. p. 808.
Synonymes. Philadélphus in part, Leptospérmum in part.
Derivation. Named in honour of *John Deutz*, a Dutch naturalist.

Gen. Char. *Calyx* tube campanulate, tomentose; limb 5—6-cleft. *Petals* 5—6, oblong. *Stamens* 10. *Filaments* tricuspidate. *Styles* 3—4, longer than the corolla. *Stigma* simple, club-shaped. *Capsule* globular, truncate, perforated, somewhat 3-cornered, scabrous; 3—4-valved, 3—4-celled. *Seeds* several in each cell. (*Don's Mill.*)

Leaves simple, opposite, exstipulate, deciduous; petiolate, ovate, acuminated, serrated, wrinkled, and veined; scabrous from stellate fascicles of down. *Flowers* white, in compound panicles; peduncles and pedicels tomentose and scabrous. — Much-branched shrubs, with the branches purplish and villous; natives of Asia; of the same culture as Philadélphus.

1. D. SCA'BRA. The scabrous Deutzia.

Identification. Lin. Syst., p. 425.; Thunb. Jap., t. 185.; Don's Mill., 2. p. 808.
Engravings. Thunb. Jap., t. 24.; Bot. Reg., t. 1718.; and our *fig.* 836.

Spec. Char., &c. Leaves ovate, acuminated, serrated, scabrous from stellate down. Flowers in compound panicles; peduncles and pedicels scabrous. Calycine lobes short and bluntish. (*Don's Mill.*) A sarmentose, showy, free-flowering shrub. Japan. Height 4 ft. to 6 ft. Introduced in 1822. Flowers white; May and June. Fruit brown; ripe in September. Decaying leaves yellowish brown. Naked young wood purplish brown.

836. Deützia scàbra.

2. D. (s.) CORYMBO'SA. The corymbose-*flowered* Deutzia.

Identification. R. Br. in Wall. Cat., 3652.; Don's Mill., 2. p. 808.
Synonymes. D. canéscens *Sieboldt*; Philadélphus corymbòsus *Wall.*
Engravings. Royle Illust., t. 46. fig. 2.; and our *fig.* 837.

Spec. Char., &c. Glabrous. Leaves ovate, acuminated, cuspidately serrated. Panicles corymbose, trichotomous. Panicle and outside of calyx dotted. Teeth of calyx short and rounded. Teeth of filaments, shorter than the anthers. (*Don's Mill.*) A sarmentose shrub. Nepal, on mountains. Height 3 ft. to 4 ft. Introd. in 1830. Flowers white, sweet-scented, produced in abundance; May and June. Fruit ?. Decaying leaves yellowish brown. Naked wood brown.

837. D. (s.) corymbòsa.

D. staminea R. Br., Philadélphus stamíneus *Wall.*, has entire, scabrous, lanceolate leaves, and white sweet-scented flowers. A native of Nepal, on high mountains; but not yet introduced.

D. Brunònia Wall., Leptospérmum scàbrum *Wall.*, has ovate leaves, and axillary white flowers. It is a native of Kamaon, but has not yet been introduced. All these are probably only varieties of one form.

GENUS III.

DECUMA'RIA *L.* THE DECUMARIA. *Lin. Syst.* Dodecándria Monogýnia.

Identification. Lin. Gen., No. 597.; Lam. Ill., t. 403.; Don's Mill., 2. p. 808.
Synonyme. Forsýthia Walt., not of Vahl.
Derivation. From *decuma*, a tenth; in reference to the prevailing number, in some of the parts of fructification, being ten. In DeCandolle's description of the genus, it is stated that the teeth and nerves of the calyx, the petals, the stigmas, and the cells of the capsule, are each usually ten.

Gen. Char. *Calyx* tube campanulate; limb 7—10-toothed. *Petals* oblong, equal in number to the teeth of the calyx, and alternating with them. *Stamens* thrice the number of the petals, disposed in one series. *Style* 1. *Capsule* ovoid, connate with the calyx. *Seeds* numerous, oblong. (*Don's Mill.*)

Leaves simple, opposite, exstipulate, deciduous; glabrous, entire or toothed at the apex. *Flowers* white, sweet-scented, disposed in terminal corymbs, sometimes diœcious. *Leaf buds* beset with short rufous pili.— Sarmentose shrubs, natives of North America.

They will grow in any dry soil, and are readily propagated by cuttings of the half-ripened wood.

⁂ 1. D. BA'RBARA *L.* The barbarous Decumaria.

Identification. Lin. Sp., No. 1668.; Pursh Sept., 1. p. 328.; Dec. Prod., 3. p. 206.
Synonymes. D. radicans *Mœnch Meth.* 17.; D. Forsythia Michx. Fl. Bor. Amer. 1. p. 282.; D. prostràta *Lodd. Cat.*
Engravings. N. Du Ham., 6. t. 20.; and our *figs.* 838. and 839.

838. D. bárbara.

839. D. bárbara.

Spec. Char., &c. Leaves ovate-oblong, acute at both ends, glabrous, entire or toothed at the tip. Buds hairy with short rufous hairs. (*Dec. Prod.*) A sarmentose shrub seldom seen in a vigorous state in British gardens. America, and Lower Carolina, in shady woods. Height 4 ft. to 5 ft.; against a wall 10 ft. to 12 ft. Introduced in 1785. Flowers white, in corymbose panicles, sweet-scented; July and August.

Variety.

⁂ *D. b.* 2 *sarmentòsa* Dec. Prod. iii. p. 206.; D. sarmentòsa *Bosc Act. Soc. Hist. Nat. Par.* i. p. 76. t. 13., *Pursh. Sept.* 1. p. 328.; Forsýthia scándens *Walt. Car.* 154. — Lower leaves rounded; upper leaves ovate-lanceolate. (*Dec. Prod.*) A native of moist shaded places in Virginia and Carolina, where, in company with Zízyphus volùbilis, it ascends the tallest cypress trees in the cedar swamps. Larger in all its parts than the species.

The flowers are only produced in favourable situations; and the plant seldom rises above 4½ or 5 feet, in the open air, in the climate of London.

ORDER XXXI. NITRARIA'CEÆ.

ORD. CHAR. *Calyx* 5-toothed, inferior, fleshy. *Petals* 5, inflexed; æstivation valvate. *Stamens* 15. *Anthers* innate. *Ovarium* superior, 3- or more celled. *Stigma* terminated by as many lines as there are cells in the ovarium. *Fruit* drupaceous, 1-seeded, opening by 3 or 6 valves. *Albumen* none. Differs from Ficöídeæ in æstivation of petals. (*G. Don.*)

Leaves simple, alternate or opposite, exstipulate, deciduous; small, generally in fascicles. *Flowers* small, greenish white. — Low shrubs, natives of Asia.

GENUS I.

NITRA'RIA *L.* THE NITRARIA. *Lin. Syst.* Dodecándria Monogýnia.

Identification. Lin. Gen., No. 602.; Lam. Ill., t. 403.; Gærtn. Fruct., 1. t. 58.; Dec. Prod., 3. p. 456.; Don's Mill., 3. p. 155.
Derivation. So named by Schober, from one of the species being discovered in certain nitre-works in Siberia, along with other saline plants.

Gen. Char. The same as that of the order.

Leaves simple, alternate or opposite, exstipulate, deciduous; oblong entire, small-fascicled. *Flowers* small, white, in panicled racemes.— Shrubs, with white bark; natives of Siberia.

Seldom rising more than 3 ft. in height; and, in British gardens, thriving best in a dry soil, composed partly of lime rubbish, which should be, about once a year, strewed with a thin coat of salt. Propagated by cuttings.

1. N. SCHO'BERI L. Schober's Nitraria.

Identification. Lin. Sp., 638.; Dec. Prod., 3. p. 456.; Don's Mill., 3. p. 155.

841 N. S. cáspica.

Spec. Char., &c. Leaves oblong, perfectly entire. Drupes ovate. (*Dec. Prod.*) A low bushy shrub, varying with branches spiny, and branches smooth. Russia, in the neighbourhood of salt lakes. Height 1 ft. to 3 ft. Introd. in 1788. Flowers white; May to August. Fruit blackish blue or red, rather larger than peas; rarely seen in England.

840. N. Schöberi sibírica.

Varieties.

- *N. S.* 1 *sibírica.* N. sibírica *Pall. Fl. Ross.* t 50. f. A., and our *fig.* 840.—Fruit of a blackish blue colour. Siberia. Hort. Soc. Garden.
- *N. S.* 2 *cáspica.* N. cáspica *Pall. Flora Ross.* t. 50. f. B., and our *fig.* 841.—Fruit red. Leaves longer. Young branches pubescent. Fruit larger, and much more acute. Hort. Soc. Garden.

N. tridentàta Desf., a native of Africa, is described in our first edition, but is rather tender in British gardens.

ORDER XXXII. GROSSULA'CEÆ.

ORD. CHAR. *Calyx* 4—5-parted. *Petals* 5. *Stamens* 4—5. *Anthers* introrse. *Ovarium* 1-celled, with two opposite parietal placenta. *Style* 2- 3-, or 4-cleft. *Fruit* succulent, crowned by the persistent calyx, many-seeded. *Seeds* arillate. *Albumen* horny. Differs from Cactàceæ in definite stamens, albuminous seed, calyx, corolla, and habit. (*G. Don.*)

Leaves simple, alternate, stipulate, deciduous; lobed or cut, plaited when folded in the bud. *Flowers* axillary, or terminal, greenish, whitish yellow or red.— Shrubs; natives of Europe, Asia, and North America; all included in the genus *Ribes.*

GENUS I.

RI'BES L. THE RIBES. *Lin. Syst.* Pentándria Monogýnia.

Identification. Lin. Gen., 281.; Dec. Prod., 3. p. 477.; Don's Mill., 3. p. 177.
Synonymes. Grossulària *Tourn.*; Chrysobótrya, Calobótrya, Coreósma, and *Ribes* Spach; Groseiller, *Fr.*; Johannisbeere, *Ger.*; Kruisbes, *Dutch*; Uva Spina, *Ital.*; Grosella, *Span.*
Derivation. The word *Ribes* is from the name of an acid plant mentioned by the Arabian physicians, which has been discovered to be the *Rhèum Ribes*; Grossulària is from the Latin *grossulus*, a little unripe fig.

Gen. Char. The same as that of the order.

Leaves simple, alternate, stipulate, deciduous; lobed or cut, plaited while in the bud. *Flowers* greenish white, yellow, or red; very rarely unisexual. There is one bractea at the base of each pedicel, which is cut more or less; and two much smaller ones, called bracteoles, under each ovarium. — Unarmed or spiny shrubs; natives of Europe, Asia, and North America; two of which (the common currant and gooseberry) are well known in British gardens, for their valuable fruits.

Many of the sorts here set down as species are, we have no doubt, only varieties; but, as we are not able to refer these to their aboriginal forms, we have followed the usual authorities, and more especially the nomenclature adopted in the Horticultural Society's Garden. All the species of *Rìbes* strike root readily from cuttings; and grow freely in any soil that is tolerably dry; but, as they are only ligneous in a subordinate degree, and are but of a temporary duration under any circumstances, they require to be grown in dug beds or borders, and are, therefore, more fitted for scientific collections or flower-borders, than for general shrubberies, undug arboretums, or lawns. The most showy species are *Rìbes* sanguíneum and aúreum, and their varieties. *R.* speciòsum has a singular fuchsia-like appearance when in blossom; and *R.* multiflòrum, though the flowers are greenish, is remarkably elegant, on account of the long many-flowered racemes in which they are disposed.

§ i. *Grossulàriæ* Ach. Rich. *Gooseberries.*

Synonymes. Groseiller à Maquereau, *Fr.*; Stachelbeere Strauch, *Ger.*; Kruisbes, *Dutch*; Uva Spina, *Ital.*; and Grosella, *Span.*

Sect. Char., &c. Stems, in most instances, prickly. Leaves plaited. Flowers in racemes; 1, 2, or 3, in a raceme. Calyx more or less bell-shaped. (*Dec. Prod.*, iii. p. 478.) — Shrubs with prickles; and with the leaves and fruit more or less resembling those of the common gooseberry.

A. *Flowers greenish white.*

■ 1. *R.* OXYACANTHÖI`DES *L.* The Hawthorn-*leaved* Gooseberry.

Identification. Pursh Sept., 2. p. 165.; Berlandier in Mém. Soc. Phys. Gen., 3. pars 2., not of Michaux.

Engravings. Mém. Soc. Phys. Gen., 3. pars 2. t. 1. f. 1.; Dill. Elth., t. 139. p. 166.; and our *fig.* 842.

Spec. Char., &c. Infra-axillary prickles larger, and mostly solitary; smaller prickles scattered here and there. Leaves glabrous, their lobes dentate, their petioles villous, and a little hispid. Peduncles short, bearing 1—2 flowers. Berry globose, glabrous, purplish blue. (*Dec. Prod.*) A prickly shrub. Canada, on rocks. Height 2 ft. to 3 ft. Introduced in 1705. Flowers greenish; April. Fruit small, red and green, or purplish blue; ripe in August; and agreeable to eat.

842. *R.* oxyacanthöides.

This shrub varies much in the number and colour of its prickles, and its more or less dense ramification and pubescence. The fruit resembles that of the common gooseberry. It is not common in British gardens, the *R.* oxyacanthöides of Michaux (*R.* lacústre *Poir.*) being different from it. Perhaps it is only one of the wild states of the common gooseberry; which varies so very much when in a state of culture, that it is reasonable to suppose that it will vary much also in a wild state.

2. *R.* SETO'SUM *Lindl.* The bristly Gooseberry.

Identification. Lindl. Bot. Reg.; Hook. Fl. Amer., 1. p. 230.; Don's Mill., 3. p. 177.
Engravings. Lindl. Bot. Reg., t. 1237.; and our *fig.* 843.

Spec. Char., &c. Branches beset with dense bristles. Prickles unequal, subulate. Leaves roundish, cordate at the base, pubescent, 3—5-lobed, deeply crenated. Peduncles 2-flowered, sometimes bracteate. Calyx tubularly campanulate, with the segments linear, obtuse, and spreading, twice the length of the petals, which are entire. Berries hispid. (*Don's Mill.*) A prickly shrub. North America, on the banks of the Saskatchawan. Height 4 ft. to 5 ft. Introduced in 1810. Flowers greenish; April and May. Fruit as in the preceding species.

843. R. setòsum.

3. *R.* TRIFLO'RUM *W.* The 3-flowered Gooseberry.

Identification. Willd. Enum., 1. p. 51.; Dec. Prod., 3. p. 479.; Don's Mill., 3. p. 177.
Synonymes. *R.* stamineum *Horn. Enum. Hort. Hafn.* p. 287.; *R* t. màjus *Hort.*
Engravings. Mém. Soc. Phys. Gen., 3. pars 2. t. 1. f. 4.; and our *fig.* 844.

Spec. Char., &c. Infra-axillary prickles solitary. Leaves glabrous, 3—5-lobed, incisely dentate. Peduncles bearing 1—3 flowers. Pedicels long. Bracteas membranaceous, sheathing. Calyx tubularly bell-shaped. Petals spathulately obcordate. Berries reddish, glabrous. (*Dec. Prod.*) A prickly shrub. North America. Height 3 ft. to 4 ft. Introduced in 1812. Flowers whitish; April and May. Fruit reddish, glabrous; ripe in July and August.

844. R. triflòrum.

Easily distinguished from *R.* Cynósbati by its smooth fruit, narrow flowers, and exserted stamens.

4. *R.* (T.) NI'VEUM *Lindl.* The snowy-*flowered Currant-like* Gooseberry.

Identification. Lindl. in Bot. Reg., t. 1692.
Engravings. Bot. Reg., t. 1692.; and our *fig.* 845.

Spec. Char., &c. Branches prickly, the prickles solitary, or in pairs, or in threes. Leaves glabrous, roundish, entire at the base, having in the outward part 3 blunt lobes that are crenately cut. Flowers about 2 together, on peduncles. Sepals reflexed. Stamens very prominent, conniving, hairy, longer than the style. (*Lindl.*) A prickly shrub. North America, on the north-west coast. Height 4 ft. to 5 ft. Introd. 1826. Flowers white, pendulous; April and May. Fruit deep rich purple, about the size of the black currant; ripe in July and August.

845. R. (t.) niveum.

The bush bears some similarity to *R.* triflòrum. The fruit resembles a small smooth gooseberry; "but its flavour is very different: it is entirely destitute of the flatness which is more or less perceptible in even the best gooseberries; in lieu of which it has a rich subacid, vinous, rather perfumed, flavour, which is extremely agreeable. The fruit is rather too acid to be eaten raw; but, when ripe, it makes delicious tarts, and would, probably, afford an excellent means of improving the common gooseberry by cross breeding." (*Lindl.*) *R.* níveum, apart from these considerations (which, however, will probably lead to its culture in the kitchen-garden), is, from its white pendulous flowers, a valuable addition to our ornamental hardy shrubs.

■ 5. *R.* (T.) CYNO'SBATI *L.* The Dog-Bramble Gooseberry.

Identification. Lin. Sp., 292.; Dec. Prod., 3. p. 479.; Don's Mill., 3. p. 178.
Synonyme. *R.* ? triflòrum var.
Engravings. Mém. Soc. Phys. Gen., 3. pars 2. t. 1. f. 3.; and our *fig.* 846.

Spec. Char., &c. Infra-axillary prickles 1—2. Leaves 3—4-lobed, softly pubescent. Peduncles bearing 2—3 flowers. Calyx campanulately cylindrical. Petals small, much shorter than the stigmas and stamens. Style simple, toward the middle hairy, rarely glabrous. Berry prickly. (*Dec. Prod.*) A prickly shrub. Canada, on mountains; and also Japan. Height 3 ft. to 4 ft. Introduced in 1759. Flowers whitish; April. Fruit reddish.

846. *R.* (t.) Cynósbati.

Varieties. There are two forms of this species: —

- ■ R. (*t.*) C. 1 *frúctu glàbro*, with whitish flowers and smooth fruit. Native of Hudson's Bay.
- ■ R. (*t.*) C. 2 *frúctu aculeàto*, with prickly branches and fruit, and flowers pubescent and purplish. Native of Lake Huron.

Hardly differs from *R.* divaricàtum, except in the broader tube of the corolla, and the shorter stamens.

■ 6. *R.* (T.) DIVARICA'TUM *Dougl.* The spreading-*branched* Gooseberry.

Identification. Dougl. in Bot. Reg., t. 1359.; Don's Mill., 3. p. 178.; Lodd. Cat., ed. 1836.
Synonymes. *R.* ? triflòrum var.; *R.* ? Grossulària var. triflòra subvar.
Engravings. Bot. Reg., t. 1359.; and our *fig.* 847.

Spec. Char., &c. Branches divaricate, bristly, at length naked. Spines 1—3 together, axillary, deflexed, large. Leaves roundish, 3-lobed, deeply toothed, nerved, glabrous. Peduncles 3-flowered, drooping. Calyx funnel-shaped; with the segments at length spreading, and twice the length of the tube. Style and stamens exserted. (*Don's Mill.*) A large prickly shrub, with ascending branches. North America, on the north-east coast, common on the banks of streams near Indian villages. Height 5 ft. to 7 ft. Introduced in 1826. Flowers white; April. Fruit black, smooth, spherical, agreeable to eat; ripe in July.

847. *R.* (t.) divaricàtum.

Nearly allied to *R.* triflòrum, of which, like *R.* Cynósbati and some of the following sorts, it is, probably, only a variety.

7. *R.* (T.) IRRI'GUUM *Dougl.* The well-watered Gooseberry.

Identification. Dougl. in Hort. Trans., 7. p. 516.; Don's Mill., 3. p. 178.
Synonyme. *R.* ? triflòrum var.
Engraving. Our *fig.* 848. from a plant in the Horticultural Society's Garden.

848. *R.* (t.) irriguum.

Spec. Char., &c. Prickles axillary, ternary. Leaves cordate, somewhat 5-lobed, toothed, ciliated, pilose on both surfaces, nerved. Peduncles 3-flowered, beset with glandular hairs. Calyx campanulate. Segments linear, about equal in length to the tube. Berries glabrous, spherical, half an inch in diameter, smooth, juicy, and well-flavoured. (*Don's Mill.*) A prickly shrub. America, on the north-west coast, on moist mountains and rocks, near springs and streams. Height 3 ft. to 4 ft. Introduced in 1820. Flowers white; April. Fruit reddish, glabrous; ripe in August.

8. *R.* HIRTE'LLUM *Michx.* The slightly hairy-*branched* Gooseberry.

849. *R.* hirtéllum.

Identification. Michx. Fl. Bor. Amer., 1. p. 111.; Dec. Prod., 3. p. 479.; Don's Mill., 3. p. 178.
Engraving. Our *fig.* 849. from a specimen in the Lambertian herbarium.

Spec. Char., &c. Spines infra-axillary. Branches sparingly hispid, with short hairs. Leaves small, cleft half-way down into 3 dentate lobes. Peduncles 1-flowered. Berries glabrous. (*Dec. Prod.*) A prickly shrub. Canada and Virginia, on rocky mountains. Height 3 ft. to 4 ft. Introduced in 1812. Flowers greenish white; April and May. Fruit red; ripe in August.

9. *R.* GRA'CILE *Michx.* The slender-*branched* Gooseberry.

Identification. Michx. Fl. Bor. Amer., 1. p. 111.; Pursh Fl. Amer. Sept.; Dec. Prod., 3. p. 479.
Engraving. Our *fig.* 850. from a specimen in the Lambertian herbarium.

850. *R.* grácile.

Spec. Char., &c. Infra-axillary spine very short. Petioles of leaves slender. Disks cut into acute lobes. Peduncles slender, upright, bearing about 2 flowers. Calyx glabrous, tubularly bell-shaped. Berries glabrous, purple or blue; of exquisite flavour. (*Dec. Prod.*) A prickly shrub. North America, on the mountains of Tennessee, and in mountainous meadows from New York to Virginia. Height 3 ft. to 4 ft. Introduced in 1812. Flowers whitish; April and May. Fruit purple or blue, high-flavoured; ripe in July and August.

10. *R.* ACICULA'RE *Smith.* The acicular-*spined* Gooseberry.

Identification. Smith in Rees's Cycl.; Don's Mill., 3. p. 178.
Synonyme. *R.* U'va-críspa *Sievers* in *Pall. Nord. Beytr.* 7. p. 274., ? *Pall. Fl. Ross.* 2. p. 37.
Engravings. Led. Fl. Ross. Alt. Ill., t. 230.; and our *fig.* 851.

851. *R.* aciculàre.

Spec. Char., &c. Very prickly. Prickles stipular, 3—5-parted. Leaves rather pubescent, nearly orbicular, 3—5-lobed. Lobes bluntish, deeply serrated. Peduncles usually 1-flowered, bracteolate in the middle. Calyx campanulate, smoothish. Berries bractless, and, as well as the styles, quite glabrous. Stem erect or procumbent. Petals white. Berries glabrous. (*Don's Mill.*) A prickly shrub. Siberia, on stony, rocky, mountainous places. Height 1 ft. to 2 ft. Introduced ? 1815. Flowers whitish; April and May. Fruit yellowish or purplish; grateful to the taste; ripe in July and August. Horticultural Society's Garden.

11. *R.* Grossula`ria *L.* The *common* Gooseberry.

Identification. Lin. Sp., p. 291.; Smith's Engl. Bot., t. 1292.; Don's Mill., 3. p. 179.
Synonymes. R. U`va-crispa *Œd. Fl. Dan.* 546.; Grossulària hirsùta *Mill. Dict.* No. 2.; *R.* U`va-crispa var. 5. sativa *Dec. Fl. Fr.* 4. p. 408.; Feaberry, *Cheshire and the North of England;* Feabes, *Norfolk*; Grozert *in Scotland*; Groseiller à Maquereau, *Fr.*; Griselle *in Piedmont*; gemeine Stachelbeere, *Ger.*; Uva Spina, *Ital.*
Derivation. U`va-crispa signifies the rough grape. Feaberry is a corruption of fever-berry, from the fruit being formerly, according to Gerard, considered a specific against fevers; Feabes, or Feapes, is an abbreviation of feaberry. Grozert is evidently taken from the French name. Groseiller à Maquereau is from the Latin name Grossulària, and the use made of the fruit as a sauce for mackerel. Stachelbeere signifies prickly berry; and Uva Spina, the prickly grape. Gooseberry is from gorse berry, from the prickliness of the bush resembling that of the gorse, or furze; or, more probably, from the use made of the fruit as a sauce to young, or green, geese.
Engravings. Eng. Bot., t. 1292.; and our *fig.* 852.

852. R. Grossulària.

Spec. Char., &c. Prickles 2 or 3 under each bud. Branches otherwise smooth, and spreading or erect. Pedicels 1—2-flowered. Leaves 3—5-lobed, rather villous. Bracteas close together. Calyx campanulate, with reflexed segments, which are shorter than the tube. Petals rounded at the apex, glabrous, but bearded in the throat. Style always beset with long down. (*Don's Mill.*) A prickly spreading shrub. Europe and Nepal, in woods and hedges. Height 2 ft. to 4 ft. Flowers greenish; April. Fruit commonly red, sometimes yellow or green; ripe in August.

Varieties.

853. R. G. Uva-crispa.

- R. *G.* 2 *U`va-crispa* Smith Engl. Fl. ii. p. 333.; *R.* U`va críspa *Lin. Sp.* 292., *Smith Engl. Bot.* t. 2057. (our *fig.* 852.); U`va-críspa *Fuch. Hist.* t. 187.; U`va spina *Math. Valgr.* 1. t. 151. f. 1.; *R.* U`va-críspa var. 1 sylvéstris *Berlandier*; has the berries smooth.
- R. *G.* 3 *spinosíssima* Berl. MSS. has the branches thickly beset with spines.
- R. *G.* 4 *reclinàta* Berl. MSS., *R.* reclinàtum *Lin. Sp.* 291., Grossulària reclinàta *Mill. Dict.* No. 1., has the branches rather prickly, and reclinate.
- R. *G.* 5 *Besseriàna Berl. MSS., R.* hýbridum *Besser Prim. Fl. Gall. Austr.* p. 186., has the branches prickly, and the fruit pubescent, intermixed with glandular bristles. Native of Cracow, in hedges.
- R. *G.* 6 *subinérmis* Berl. MSS. — Plant nearly glabrous. Bark smooth, brown. Prickles axillary. Flowers and leaves small. Native about Geneva. Perhaps a subvariety of *R.* G. reclinàta.
- R. *G.* 7 *macrocárpa* Dec. Prod. iii. p. 478. — Stigmas often longer than the petals. Flowers and berries large.
- R. *G.* 8 *bracteàta* Berl. MSS. — Berries clothed with 2—4—5 straight, coloured, nearly opposite, bracteas and bristles, resembling sepals, which fall off before the berry arrives at maturity. (*Don's Mill.*)
- R. *G.* 9 *himalayánus, R.* himalayànus *Royle,* was raised in the

Horticultural Society's Garden in 1838, and seems hardly different from the species. (*Gard. Mag.*, 1839, p. 4.)

Other Varieties. Till lately, botanists made even the rough and the smooth-fruited kinds of the cultivated gooseberry two distinct species, as may be seen by the synonymes to *R.* U'va-críspa above; though it was recorded by Withering, that seeds from the same fruit would produce both rough and smooth-fruited plants. If varieties were to be sought for among the sorts in cultivation, they would be found almost without number. The following selection of garden varieties has been made solely with reference to the habit of growth of the plants: —

The Red Champagne, or Ironmonger, has the branches erect and fastigiate, and will form a handsome bush, 6 or 7 feet high.

Horseman's Green Gage is a most vigorous-growing plant, with a spreading head, and will form a bush 10 ft. high.

The Red Rose is a vigorous-growing bush, with a pendulous head, but seldom rising higher than 3 ft., unless trained to a stake to some height before it is allowed to branch out.

B. *Flowers red.*

12. *R.* SPECIO'SUM *Pursh.* The showy-*flowered* Gooseberry.

Identification. Pursh Fl. Amer. Sept., 2. p. 731.; Dec. Prod., 3. p. 478.; Don's Mill, 3. p. 185.
Synonymes. *R.* stamíneum *Smith* in *Rees's Cycl.*, *Dec. Prod.* 3. p. 477.; ?*R.* fuchsioìdes Fl. Mex., ic. ined.; *R.* triacánthum *Menzies.*
Engravings. Sw. Fl.-Gard., 2d ser., t. 149.; and our *fig.* 854.

854. *R.* speciòsum.

Spec. Char., &c. Shrub prickly. Prickles infra-axillary, triple. Branches hispid. Leaves with petiole short, and disk wedge-shaped at the base, rounded at the outer end, indistinctly 3-lobed, incisely crenate, glabrous, and nerved. Peduncles longer than the leaves, and bearing 1—3 flowers. Pedicels and germens hairy with glanded hairs. Bracteas rounded or very obtuse. Flowers of a deep red. Calyx cylindrical, 4-parted; the lobes oblong, obtuse. Petals of the length of the lobes of the calyx. Stamens 4; in length double that of the calyx. Filaments red. Style as long as the stamens, simple, red. (*Dec. Prod.*) A very prickly-branched shrub, with a brownish red aspect. America, on the western coast, and in California. Height, in a wild state, 3 ft. to 4 ft.; in cultivation twice that height in rich deep soil. Introduced in 1829. Flowers deep red; May and June. Fruit red; ripe ?.

The shining leaves and large crimson glittering blossoms (resembling those of the fuchsia) of this species render it a most desirable acquisition to the flower-garden and shrubbery. The leaves, in favourable situations, are frequently retained during great part of the winter; so that it may almost be considered as an evergreen. It will grow by cuttings of the old or young wood, but not so readily as most other species; and, therefore, it is generally propagated by pegging down the shoots quite flat, and covering them with an inch of soil, as recommended for the propagation of the common plum for stocks. Plants of this species do not grow so rapidly as most others of the gooseberry sections; and their branches arch over and droop in such a manner, as not to display the flowers to advantage, unless the branches are raised at least to the level of the eye. For this reason, the plant ought either to be grown on elevated rockwork, or trained to an espalier or wall.

■ 13. *R.* MENZIE`SII Ph. Menzies's Gooseberry.

Identification. Pursh Sept., 2. App. p. 732.; Don's Mill., 3. p. 185.
Synonyme. *R.* ferox *Smith* in *Rees's Cycl.*
Engraving. Our *fig.* 855. from a specimen in the Lambertian herbarium.

Spec. Char., &c. Very prickly. Spines 3-partite. Leaves cordate, truncate at the base, 5-lobed, serrated, wrinkled from veins, clothed with pubescence beneath. Peduncles usually 1-flowered. Calyx cylindrically campanulate, deeply 5-parted, glandular. Stamens 5, enclosed. Style a little exserted. Germens and peduncles prickly. (*Don's Mill.*) A very prickly shrub. North California and at Port Trinidad. Height 4 ft. to 5 ft. Introduced in 1830. Flowers bright red or crimson, glandular, as showy as in the preceding species; May and June. Fruit red; ripe ?.

855. *R. Menziesii.*

R. *microphýllum* H. B. et Kunth is a native of the mountains of Mexico, at an elevation of 4200 ft., with the leaves small and nearly reniform, and the peduncles very short and 2-flowered. It grows to the height of from 4 ft. to 6 ft.

§ ii. *Botrycárpum* Dec.

Sect. Char. Fruit disposed in racemes; the plants having the prickles of the preceding section (Grossulària), and the racemose flowers of the following section (*Ribèsia*). (*Don's Mill.*, iii. p. 185.) Plants intermediate between gooseberries and currants.

■ 14. *R.* ORIENTA`LE *Poir.* The Eastern *Currant-like* Gooseberry.

Identification. Poir. Encycl. Suppl., 2. p. 856.; Desf. Arb., 2. p. 88.; Don's Mill., 3. p. 185.
Engraving. Our *fig.* 856. from a living specimen in the Birmingham Botanic Garden.

Spec. Char., &c. Plant rather prickly. Leaves 3—5-lobed, somewhat reniformly orbicular, cut, hairy; lobes rather deep, obtuse. Petioles hairy. Racemes erectish, few-flowered. Bracteas longer than the flowers. Style bifid at the apex. Flowers greenish yellow. Fruit lke those of the currant. (*Don's Mill.*) A vigorous-growing shrub. Syria. Height 4 ft. to 6 ft. Introduced in 1824. Flowers greenish yellow; April and May. Fruit red; ripe in September.

856. *R. orientàle.*

The plant in the Birmingham Botanic Garden does not agree altogether with the description, and may possibly be some other species.

■ 15. *R.* SAXA'TILE *Pall.* The rock *Currant-like* Gooseberry.

Identification. Pall. Nov. Act. Petr., 10. p. 726.; Don's Mill., 3. p. 185.
Synonyme. ?*R.* alpinum *Sievers* in *Pall. Nord. Beytr.* 7. p. 345.
Engravings. Led. Fl. Ros. Alt. Ill., t. 239.; and our *fig.* 857.

Spec. Char., &c. Prickles scattered. Leaves roundish-cuneiform, bluntly 3-lobed. Racemes erect. Bracteas linear, shorter than the pedicels. Calyx flat, scabrous. Sepals small, of a livid green colour. Flowers small, greenish purple. Petals spathulate. Berries smooth, globose, bractless, dark purple when mature, full of edible pulp, rarely so large as common currants, but like them. (*Don's Mill.*) A bushy shrub. Siberia. Height 4 ft. to 5 ft. Introduced in 1819. Flowers small, greenish purple; April and May. Fruit dark purple; ripe in August.

857. *R. saxàtile.*

■ 16. *R.* DIACA'NTHA *L. fil.* The twin-prickled *Currant-like* Gooseberry.

Identification. Lin. fil. Suppl., p. 187.; Dec. Prod., 3. p. 479.
Engravings. Schmidt Baum., t. 97.; and our *fig.* 858.

Spec. Char., &c. Stipular prickles twin. Leaves with a disk shorter than the petiole, and wedge-shaped, perfectly glabrous, and parted into 3 lobes which are dentate. Flowers upon long pedicels, in long upright racemes. Bracteas the length of the flowers. Sepals rounded, yellowish. Petals small, roundish. Berry ovate or globcse, red. (*Dec. Prod.*) A spiny shrub. Dahuria and Siberia, in rocky places. Height 4 ft. to 5 ft. Introduced in 1781. Flowers yellowish green; May and June. Fruit ovate, red; ripe in August.

858. R. Diacantha.

A very distinct sort, easily known by its cuneated leaves and yellowish flowers. In Messrs. Loddiges's collection there is a fastigiate-growing variety.

■ 17. *R.* LACU'STRE *Poir.* The lake-*side Currant-like* Gooseberry.

Identification. Poir. Encycl. Suppl., 2. p. 856.; Dec. Prod., 3. p. 478.; Don's Mill., 3. p. 178
Synonyme. ?*R. oxyacanthöides Michx. Flor. Bor. Amer.* 1. p. 111.
Engraving. Our *fig.* 859. from a plant in the Horticultural Society's Garden.

Spec. Char., &c. Infra-axillary prickles manifold; the stem hispid with minute prickles. Leaves lobed beyond the middle; glabrous beneath, rather pilose above. Petioles villous. Peduncles ? upright, ? reflexed, bearing 2—3 flowers upon hispid pedicels. Flowers small, yellowish green. Germen hispid. (*Dec. Prod.*) A very prickly shrub. Canada and Virginia, in moist places. Height 4 ft. to 5 ft. Introduced in 1812. Flowers small, yellowish; April and May. Fruit purplish black, about the size of the common black current; ripe in August.

859. R. lacustre.

Variety.

■ R. *l.* 2 *echinàtum*; *R.* echinàtum Dougl. MSS., and Arb. Brit. 1st edit. p. 992.; *R.* armàtum *Hort.*; has the stems prostrate, while those of the species are upright and rather slender.

The flowers are those of the currant, and the prickly stems those of the gooseberry. The fruit is about the size of black currants, in pendulous racemes, purplish black, shining, clothed with hairs, and unpleasant to the taste. The plant forms rather a spreading trailing bush, and is therefore more adapted for spreading over rockwork or stones, than for standing erect by itself. Horticultural Society's Garden.

§ iii. Ribèsia *Dec.* *Currants.*

Synonymes. Ribes sp. *Lin. and others*; Calobótrya, Coreosma, and *Rèbis* Spach; Groseilles en Grappes, or Groseillier commun, *Fr.*; Johannisbeere, *Ger.*; Bessenboom, *Dutch*; Ribes, *Ital.*

Sect. Char. Shrubs unarmed. Racemes, for the most part, many-flowered. Leaves plicate. Calyx campanulate or cylindrical. (*Don's Mill.*, iii. p. 185.) Shrubs, the branches of which are without prickles, and the leaves and fruit of which resemble those of the currant more than those of the gooseberry.

A. *Flowers greenish, or greenish yellow, or reddish; and Fruit, in a wild State, red.*

18. *R.* RUBRUM *L.* The *common* red Currant.

Identification. Lin. Sp., 290.; Dec. Prod., 3. p. 481.; Don's Mill., 3. p. 187.
Synonymes. *R. vulgàre N. Du Ham.*; Groseillier commun, *Fr.*; gemeine Johannisbeere, *Ger.*; Aalbessen Boom, *Dutch*; Ribes rosso, *Ital.*
Engravings. Smith Engl. Bot., t. 1289.; Krauss, t. 48.; and our *fig.* 860.

Spec. Char., &c. Leaves cordate, bluntly 3—5-lobed, pubescent beneath, when young, usually rather tomentose, glabrous above. Racemes drooping. Bracteas ovate, shorter than the pedicels. Calyx flatly campanulate, spreading. Sepals obtuse. Petals obcordate. Fruit quite glabrous. Flowers yellowish. (*Don's Mill.*) An upright shrub. Europe and Siberia, in woods; and throughout Canada to the mouth of the Mackenzie; in the North of England and in Scotland, in mountainous woods, and about the banks of rivers. Height 4 ft. to 6 ft. Flowers greenish yellow; April and May. Fruit red; ripe in July.

860. *R. rùbrum.*

Varieties. De Candolle gives the following forms of this species:—

- R. *r.* 1 *sylvéstre* Dec. Fl. Fr. iv. p. 406. — Leaves and berries small. Lobes of leaves short.
- R. *r.* 2 *horténse* Dec. l. c. *R.* rùbrum *Lois. Nouv. Dict.* iii. — Leaves large, sometimes variegated. Berries sweeter and larger than in var. 1. Cultivated in gardens.
- R. *r.* 3 *cárneum* Berl. MSS. ex Dec. Prod. iii. p. 481. *R.* rùbrum doméstĭcum 2 báccis cárneis *Wallr. Sched.* p. 106. — Leaves rather tomentose beneath. Sepals red. Cells of anthers distant. Berries pale red.
- R. *r.* 4 *variegàtum* Dec. Prod. iii. p. 481., Wallr. l. c., has the berries beautifully variegated; or, rather, distinctly striped with white and red. In cultivation in Austria, and well deserving of a place in every collection, from the beauty and singularity of its fruit.
- R. *r.* 5 *álbum* Desf. Cat. Bot. p. 164. — Berries white.
- R. *r.* 6 *fòliis lùteo variegàtis* Du Ham. has the leaves variegated with yellow, and the fruit red.
- R. *r.* 7 *fòliis álbo variegàtis* Du Ham. has the leaves variegated with white, and the fruit white.
- R. *r.* 8 *sibíricum* Oldaker. The Russian currant.—Of vigorous growth.

The propagation, culture, &c., of the currant, as a fruit shrub, will be found given at length in our *Encyclopædia of Gardening*, and in our *Suburban Horticulturist.*

19. *R.* (R.) ALPINUM *L.* The alpine red Currant.

Identification. Lin. Sp., 291.; Dec. Prod., 3. p. 480.; Don's Mill., 3. p. 186.
Synonyme. *R.* dioicum *Masters.*
Engravings. Schmidt Baum., t. 96.; and our *fig.* 861.

Spec. Char., &c. Leaves with 3—5 lobes, obtuse, hairy above, shining beneath. Racemes grouped. Bracteas lanceolate, inflated, sparingly glandulose, mostly larger than the flowers. Petals minute, as if in abortion. Anthers more or less sessile. Styles connate. Berries red. (*Dec. Prod.*) A spreading shrub. Alps of Europe and Si-

861. *R.* (r.) *alpinum.*

beria; and found in Britain, in woods, both in England and Scotland Height 3 ft. to 4 ft. Flowers greenish yellow; April and May. Fruit red; ripe in July.

Varieties. Berlandier has described two forms of the species, and Dr. Lindley has added a proper variety.

- R. (*r.*) *a.* 1 *stérile* Wallr. Sched. p. 108. *R.* dioícum *Mœnch Meth.* — Flowers many in a raceme, and densely disposed, flat, destitute of a germen, soon falling off. Anthers almost sessile, acute, bearing pollen.
- R. (*r.*) *a.* 2 *baccíferum* Wallr. Sched. p. 108. — Flowers few in a raceme, rather salver-shaped. Anthers upon obvious filaments, ? imperfect. Style bifid to a small extent. Germen obvious.
- R. (*r.*) *a.* 3 *pùmilum* Lindl. in Hort. Trans. vii. p. 244., and our *fig.* 862. — In every respect the same as the species, but not one third of the size, never exceeding 2 ft. in height, even when cultivated in gardens. The leaves are deeply cut, the flowers small, and the fruit seldom produced.

862. *R.* (r.) a. pùmilum.

- R. (*r.*) *a.* 4 *fòliis variegàtis* Hort. has variegated leaves. Horticultural Society's Garden.

20. R. (R.) PETRÆ'UM *Wulf.* The rock red Currant.

Identification. Wulf. in Jacq. Misc., 2. p. 36.; Don's Mill., 3. p. 187.
Synonymes. *R.* alpìnum *Delarb. Auvergn.* p. 166.; Ribes corallino, *Ital.*; the woolly-leaved Currant, the red Marshmallow-leaved Currant.
Engravings. Eng. Bot., t. 705.; and our *fig.* 863.

863. *R.* (r.) petræ'um.

Spec. Char., &c. Leaves acuminated, 3—5-lobed, rather cordate, deeply serrated, on long petioles, pilose above. Racemes erect, crowded, rather pubescent. Bracteas shorter than the pedicel. Sepals obtuse. Petals obcordate, small, white. Berries large, deep red, with an acid taste. Fruiting racemes pendulous. (*Don's Mill.*) A spreading shrub. Alps of Carinthia, Savoy, and on almost all the mountains of the continent of Europe. In England, it is found near Eggleston and Conscliffe, in the county of Durham; and in Scotswood Dean, Northumberland. Height 3 ft. to 4 ft. Flowers greenish yellow; May. Fruit red; ripe in July.

21. *R.* (R.) SPICA'TUM *Robs.* The spiked-*flowered* red, *or Tree*, Currant.

Identification. Robs. in Lin. Trans., 3. p. 240. t. 21.; Smith Engl. Bot., t. 1290.; Don's Mill., 3. p. 187.
Synonyme. The Tree Currant.
Engravings. Lin. Trans., 3. p. 240. t. 21.; Eng. Bot., t. 1290.; Berl., l. c., t. 2. f. 16.; and our *fig.* 864.

Spec. Char., &c. Leaves roundish-cordate, 3—5-lobed, covered with soft hairs above, and with tomentum beneath. Racemes erect.

864. *R.* (r.) spicàtum

Flowers more or less pedicellate. Bracteas obtuse, tomentose, much shorter than the pedicels. Sepals roundish-cuneated. Petals oblong. Styles bifid. Berries glabrous, globose, and in colour and taste resembling those of *R.* rùbrum. (*Don's Mill.*) An upright shrub. North of England, in woods near Richmond in Yorkshire, and Piersbridge and Gainford in Durham. Height 4 ft. to 6 ft. Flowers greenish yellow; April and May. Fruit red; ripe in July.

The tree currant affords a fruit rather smaller, and more acrid, than the common red currant; but by crossing and cultivation it might, no doubt, be greatly improved; and, from its comparatively tree-like habits, might be a more convenient fruit shrub in respect to the crops around it.

22. *R.* (R.) CARPA'THICUM *Kit.* The Carpathian red Currant.

Identification. Kit. in Schultes Œstr. Fl., 2. ed. 1. p. 432.; Don's Mill., 3. p. 187.
Synonyme. *R.* acérrimum *Rochel ex Ræm. et Schultes* 5. p. 493.
Engraving. Our *fig.* . in p. .

Spec. Char., &c. Stem erect. Leaves 5-lobed, cordate. Racemes pendulous, and, as well as the calyxes, pubescent. Petals flattish, smaller than the calyx. (*Don's Mill.*) Carpathian Mountains. Height 4 ft. Perhaps only a variety of *R.* rùbrum.

23. *R.* (R.) MULTIFLO'RUM *Kit.* The many-flowered red Currant.

Identification. Kit. in Rœm. et Schultes Syst., 5. p. 493., but not of H. B. et Kunth; Don's Mill. 3. p. 187.
Synonymes. *R.* spicàtum *Schultes Œstr. Fl.* ed. 1. p. 433.; *R.* vitifòlium *Hort.*
Engravings. Bot. Mag., t. 2368.; and our *fig.* 866.

Spec. Char., &c. Leaves 5-lobed, cordate, tomentose beneath. Racemes very long, pendulous, drooping. Bracteas shorter than the flowers. Petioles length of leaves. Petals wedge-shaped. Styles bifid, and sometimes distinctly trifid. (*Don's Mill.*) A spreading shrub, with vigorous branches and large leaves. Croatia. Height 4 ft. to 6 ft. Introduced in 1822. Flowers greenish yellow; April and May. Fruit small, red; ripe in July.

865. R. (r.) multiflòrum.

The long racemes of flowers, the vigorous growth of the shoots, the large leaves, and the luxuriant habit of the plant, altogether render this a very ornamental sort. From the luxuriance of the flowers and leaves, and of the plant generally, fruit is seldom produced; and, when it appears, it is generally of small size. On account of the gracefulness of the long drooping racemes of flowers, it well deserves a place in collections.

24. *R.* (R.) ALBINE'RVUM *Michx.* The white-nerved-*leaved* red Currant.

Identification. Michx. Fl. Bor. Amer., 1. p. 110.; Don's Mill., 3. p. 187.
Engraving. Our *fig.* 2092 in p. 1107.

Spec. Char., &c. Leaves short, petiolate, deeply and acutely lobed, smoothish, with whitish nerves. Racemes recurved. Flowers small. Berries red, glabrous. (*Don's Mill.*) A shrub. Canada and the Catskill Mountains, in the state of New York. Height 4 ft. Introduced ?. Flowers greenish yellow; April and May. Berries red; ripe in July.

25. *R.* ACUMINA'TUM *Wall.* The pointed-*leaved* Currant.

Identification. Wall. Cat.; Royle Illust., p. 225.; Don's Mill., 3. p. 187.
Engraving. Our *fig.* 866. from a specimen in the Linnæan herbarium.

866. R. (r.) acuminàtum.

Spec. Char., &c. Branches glabrous. Leaves glabrous above, but with a few scattered hairs beneath, 3—5-lobed; lobes acuminated, serrated. Racemes axillary, erect. Peduncles pubescent. Berries nodding. Calyx campanulate. Petals rounded at the apex. (*Don's Mill.*) A smooth shrub. Nepal, on Sirmore and Emodi. Height 4 ft. to 6 ft. Introduced ? in 1837. Flowers greenish yellow; April and May. Fruit red, about the size of that of the red currant; ripe in July.

★ 26. *R.* (R.) TRI'FIDUM *Michx.* The trifid-*calyxed* red Currant.

Identification. Michx. Fl. Bor. Amer., 1. p. 110.; Don's Mill., 3. p. 186.
Engraving. Our *fig.* 867. from a specimen in Sir W. J. Hooker's herbarium.

867. R. (r.) trifidum.

Spec. Char., &c. Leaves smooth, moderately lobed. Racemes loosely many-flowered, pubescent. Flowers small. Calycine segments rather trifid. Berries hairy, red. Lobes of leaves acutish. Racemes weak, nearly like those of *R.* rùbrum, but the flowers smaller. Petals purplish, spathulate, rounded at the apex. (*Don's Mill.*) A prostrate shrub. North America, near Quebec, and at Hudson's Bay. Height 1 ft. to 2 ft. Introduced in 1823. Flowers purplish; April and May.

B. *Flowers greenish yellow, sometimes with the Tips of the Sepals and Petals red. Fruit black.*

■ 27. *R.* NI'GRUM *L.* The black Currant.

Identification. Lin. Sp., 291.; Don's Mill., 3. p. 190.
Synonymes. *R.* ólidum *Mœnch Meth.* 683.; Capis and Poivrier, *Fr.*; schwartze Johannisbeere *Ger.*; Ribes nero, *Ital.*
Engravings. Engl. Bot., t. 1291.; and our *fig.* 868.

868. R. nigrum.

Spec. Char., &c. Leaves dotted from glands beneath, 3—5-lobed. Racemes loose. Bracteas minute, subulate or obtuse, much shorter than the pedicels. Petals oblong. Calyx campanulate, with reflexed segments. Flowers whitish, or yellowish green. Calyx often of a rich brownish red colour, or pink. Stamens sometimes more than 5, in which case there are fewer petals; so that when there are 10 stamens there are no petals. This change of petals into stamens is just the reverse of the process by which single flowers become double; and it is the only fact of the kind which has hitherto been observed. Stigmas bifid. Berries globose, black, glandular. (*Don's Mill.*) A shrub with smoothish branches, strong-smelling leaves.

Europe. Height 4 ft. Flowers yellowish green ; April and May. Fruit dark purple ; ripe in June and July.

Varieties.

- R. *n.* 2 *bácca flávida* Hort. — Supposed to be a hybrid between the black and white currants, and to have been originated near Bath, previously to 1827. The fruit is of a dingy greenish yellow ; but the plant has the habit and general appearance of *R.* nìgrum.
- R. *n.* 3 *bácca víridi* Hort. has the fruit green when ripe. This variety is common in Russia in a wild state. Hort. Soc. Garden.
- R. *n.* 4 *fòliis variegàtis* Vilm. — Leaves variegated with yellow streaks.

Garden Varieties. Six of these are enumerated in the Horticultural Society's *Fruit Catalogue* of 1831, the best of which are the black Naples and the large black. The fruit of the former variety is very large and handsome, more especially when the plant is grown in deep rich soil, and in a situation rather shady and moist.

The leaves, fruit, and the entire plant are powerfully diuretic. The treatment of the black currant, as a fruit tree, will be found in the *Encyclopædia of Gardening*, and in the *Suburban Horticulturist*.

28. *R.* (N.) TRI'STE *Pall.* The *sad-coloured, or dark-blossomed,* black Currant.

Identification. Pall. Nov. Act. Pet., 10. p. 378. ; Don's Mill., 3. p. 189.
Synonyme. *R.* althàicum *Lodd. Cat.*
Engraving. Our *fig.* 869. from a living specimen.

869. *R.* (n.) triste.

Spec. Char., &c. Leaves 5-lobed. Branches simple, twiggy, bearing leaves and racemes of flowers at the apex. Racemes pendulous, both when in flower and in fruit. Corollas flattish, of a dull brownish red on the outside, and yellowish inside. Petals revolute. Berries small, black, insipid. Root creeping. (*Don's Mill.*) A low shrub. Siberia, on the Mongol Mountains. Height 2 ft. to 3 ft. Introduced in 1820. Flowers brownish red and yellow ; April and May. Fruit black ; ripe in July. Differs from *R.* nìgrum only in the dull brown colour of the flowers.

29. *R.* (N.) FLO'RIDUM *L'Hérit.* The flowery black Currant.

Identification. L'Hérit. Stirp., 1. p. 4. ; Ph. Sept., 1. p. 164. ; Don's Mill., 3. p. 190.
Synonymes. *R.* nìgrum 2. *Lin. Sp.* 291. ; *R.* pennsylvánicum *Lam. Dict.* 3. p. 49. ; *R.* recurvàtum *Michx. Fl. Bor. Amer.* 1. p. 109., according to Torrey ; *Ribèsium* nìgrum, &c., *Dill. Elth.* 2. t. 244. f. 315.
Engravings. Schmidt Baum., t. 92. ; and our *fig.* 870.

870. *R.* (n.) flóridum.

Spec. Char., &c. Leaves full of resinous glands, 3 or 5-lobed, cordate, doubly serrated. Racemes pendulous, pubescent. Bracteas linear, longer than the pedicels. Calyx tubularly campanulate, glabrous : with the segments obtuse, and at length reflexed. Germens and black berries oval-globose, glabrous. This is in many respects nearly allied to *R.* nìgrum ; but its more copious and denser flowers, and especially their long bracteas, and more tubular calyxes, will always distinguish it : the solitary pedicel, too, at the base of the flowers, is wanting in this species. Petals oblong, rather erose at the apex ; greenish yellow. (*Don's Mill.*) A large shrub. Canada to Virginia, in hedges and woods. Height 4 ft. to 6 ft. Introduced in 1729. Flowers pale yellow ; April and May. Fruit black ; ripe in July.

Varieties.

R. (*n.*) *f.* 2 *grandiflòrum* Hort. *R.* rigens *Michx. Fl. Bor. Amer* i. p. 110., *Ph. Sept.* i. p. 136.—Flowers and racemes larger than those of the species.

R. (*n.*) *f.* 3 *parviflòrum* Hort. *R.* americànum *Mill.*, *R.* pennsylvánicum *Cels.*, *R.* campanulàtum *Hort.*—Flowers smaller, and the racemes shorter.

871. *R.* (*n.*) procúmbens.

30. *R.* (N.) PROCU'MBENS *Pall.* The procumbent black Currant.

Identification. Pall. Fl. Ross., 2. p. 35. t. 65.; Don's Mill., 3. p. 186.
Synonyme. *R.* polycárpon *Gmel. Syst. Veg.* p. 419.
Engravings. Pall. Fl. Ross., 2. p. 35. t. 65.; and our *fig.* 871.

Spec. Char., &c. Leaves bluntly lobed; lobes serrated, lateral ones a little cut. Racemes erect. Peduncles long, setaceous. Segments of the limb of the flower pubescent, acute, of a purplish colour. Anthers hardly rising from the calyx. Flowers flattish. Berries very grateful to the taste, rufescent when ripe. (*Don's Mill.*) A procumbent shrub. Siberia, in moist places. Height 1 ft. to 2 ft. Introduced in 1804. Flowers greenish yellow; May and June. Fruit brownish; ripe Aug.

31. *R.* (N.) PROSTRA'TUM *Lin.* The prostrate black Currant.

Identification. L'Hérit. Stirp., 1. p. 3. t. 2.; Don's Mill., 3. p. 186.
Synonymes. *R.* glandulòsum *Ait. Hort. Kew.* ed. 1. p. 279.; *R.* canadénse *Lodd.*
Engravings. Schmidt Baum., t. 95.; and our *fig.* 872.

872. *R.* (*n.*) prostratum.

Spec. Char., &c. Leaves deeply cordate, 5—7-lobed, glabrous. Lobes acutely cut, doubly serrate, naked on both surfaces. Racemes erect, loose, slender. Bracteas small, obtuse, much shorter than the pedicels, which are beset with glandular bristles. Calyx rotate. Germens and berries beset with glandular bristles. Berries large and black. (*Don's Mill.*) A prostrate shrub. Newfoundland, throughout Canada, and in the woods on the Rocky Mountains. Height 1 ft. to 3 ft. Introduced in 1812. Flowers greenish yellow; April and May. Fruit black; ripe in July.

Variety.

R. (*n.*) *p.* 2 *laxiflòrum.* *R.* affi'ne *Dougl. MSS.*, *R.* laxiflòrum *Pursh Fl. Amer. Sept.* ii. p. 731. — Racemes pubescent. Pedicels divaricate. A very distinct sort. North-west coast of America.

32. *R.* (N.) RESINO'SUM *Pursh.* The resinous black Currant.

Identification. Pursh Fl. Amer. Sept., 1. p. 163.; Don's Mill., 3. p. 186.
Synonymes. *R.* orientàle *Catros*; *R.* reclinàtum *Hort.*
Engravings. Bot. Mag., t. 1583.; Berl., 1. c., t. 2. f. 10.; and our *fig.* 873.

873. *R.* (*n.*) resinòsum.

Spec. Char., &c. All herbaceous parts of the shrub bear hairs tipped with resinous glands. Leaves 3—5-lobed, roundish. Racemes erect. Calyx flattish. Petals bluntly rhomboid. Bracteas linear, longer than the pedicels. Flowers greenish yellow. ? Berry hairy and black. Perhaps the flowers are diœcious. (*Don's Mill.*) A spreading shrub. North America, on the mountains. Height 3 ft. to 5 ft. Introduced in 1800. Flowers greenish yellow; April and May. Fruit black; ripe in July.

33. *R.* (N.) PUNCTA'TUM *Ruiz et Pav.* The dotted-*leaved* black Currant.

Identification. Ruiz et Pav. Fl. Per., 2. p. 12. t. 233. f. a.; Don's Mill., 3. p. 187.
Synonyme. *R.* glandulòsum *Ruiz et Pavon Fl. Per.* t. 233. f. 6., but not of Ait.; *Don's Mill.*, 3. p. 189., on the authority of Mr. Gordon, of the Horticultural Society.

Engravings. Lindl. Bot. Reg., t. 1278. and 1658.; and our *figs.* 874, 875.

874. *R.* (n.) punctàtum.

Spec. Char., &c. Leaves 3-lobed, serrated, beset with resinous glands beneath, as are also the bracteas. Racemes longer than the leaves, either drooping or erect. Bracteas cuneate-oblong, obtuse, at length reflexed. Calyx campanulate, yellowish. Berries oblong, hairy, black, and dotted. Petals small, yellow. (*Don's Mill.*) A sub-evergreen shrub. Chili, on hills. Height 3 ft. to 4 ft. Introduced in 1826. Flowers yellow: April and May. Fruit black; ripe in July.

875. *R.* (n.) punctàtum.

The leaves are shining, and of a yellowish green, and, when rubbed, have an agreeable odour. The short close bunches of rich yellow flowers are produced in the axils of the leaves. The plant throws up suckers from the roots: a circumstance which distinguishes it from almost every other species of the genus in British gardens. Dr. Lindley has given two figures of this species in the *Botanical Register*: one, t. 1658., of the wild plant, in which the spikes are pendulous, or nodding; and the other, t. 1278., of the cultivated plant, in which the spikes are erect. He observes that it is hardy enough to live in a dry border without protection, and that it is a rather pretty evergreen shrub. H. S.

876. *R.* heterótrichum.

■ 34. *R.* (N.) HETERO'TRICHUM *Meyer.* The variable-haired Currant.

Identification. Meyer in Led. Fl. Ros. Alt. Illus., 1. p. 270.; Don's Mill., 3. p. 187.
Engravings. Led. Fl. Ros. Alt. Illus., t. 235.; and our *fig.* 876.

Spec. Char., &c. Stem erectish. Leaves pubescent, bristly, and glandular, nearly orbicular, 3-lobed; lobes obtuse, toothed. Racemes erect. Pedicels equal in length to the bracteas. Calyx flat, pubescent. Berries puberulous, glandless, bractless. (*Don's Mill.*) An upright shrub. Altaia, on rocks, at the foot of the mountains. Height 2 ft. to 3 ft. Introduced in 1837. Flowers purple; April and May. Fruit like that of the red currant, but orange-coloured; ripe in August.

■ 35. *R.* (N.) BRACTEO'SUM *Dougl.* The bracteate Currant.

Identification. Hook. Fl. Bor. Amer., 1. p. 232.; Don's Mill., 3. p. 181.
Engraving. Our *fig.* 877. from a specimen in Sir W. J. Hooker's herbarium.

Spec. Char., &c. Leaves on long petioles, cordate, deeply 5—7-lobed; lobes acuminated, cut, doubly serrated, hispid above, but full of resinous dots beneath; racemes often terminal, at length reflexed. Pedicels erectly spreading, pubescent, exceeding the spathulate bracteas. Calyx rotate, glabrous. Petals minute, roundish. Germens and berries full of resinous dots. (*Don's Mill.*) A large shrub. North-west coast of America, at the confluence of the Columbia with the ocean. Height 5 ft. to 8 ft. Introduced ?. Flowers purplish yellow; April and May. Fruit about the size of the red currant, greenish, hairy.

A very remarkable and elegant shrub, with leaves

877. *R.* (n.) bracteòsum.

resembling those of the common sycamore, and nearly as large. It is much to be desired that it could be introduced into British gardens.

■ 36. *R.* VISCOSI'SSIMUM *Pursh.* The very clammy *black* Currant.

Identification. Pursh Fl. Amer. Sept., 1. p. 163.; Don's Mill., 3. p. 191.
Synonyme. Coreósma viscosíssima *Spach Ann. des Scien. Nat.* 1836.
Engravings. Hook. Fl. Bor. Amer., 1. p. 234. t. 74.; and our *fig.* 878.

878. *R.* viscosíssimum.

Spec. Char., &c. Leaves cordate, obtuse, 3—5-lobed, deeply crenated. Viscid and glandular pubescence. Glands on both surfaces. Racemes erect, corymbose. Bracteas linear-obovate, rather shorter than the pedicels, which are clothed with glandular hairs. Calyx tubularly campanulate, with erectly spreading obtuse segments. Germens and fruit ovate-oblong, clothed with viscid hairs. Berries oblong-ovate, black. Flowers large and white. *(Don's Mill.)* An upright shrub. North America, on the Rocky Mountains, towards the sources of the Columbia; also on the summits of the hills near the Spokan and Kettle Falls, at an elevation of 8000 ft. above the sea. Height 4 ft to 8 ft. Introduced in 1826. Flowers whitish or yellowish; April and May. Fruit black; ripe in July.

A very fine and remarkable species, somewhat difficult to keep.

■ 37. *R.* (N.) HUDSONIA'NUM *Richardson.* The Hudson's *Bay* black Currant.

Identification. Richards in Frankl. First Journ., ed. 2. append. p. 6.; Don's Mill., 3. p. 190.
Synonyme. *R.* petiolàre *Dougl. Hort. Trans.* 7. p. 514.
Engraving. Our *fig.* 879. in flower, and *fig.* 880. in fruit, from a specimen in Dr. Lindley's herbarium.

879. *R.* (n.) hudsoniànum.

Spec. Char., &c. Branches erect. Leaves 3-lobed, quite glabrous above, full of resinous dots beneath, and, as well as the petioles, villous. Germens dotted. Berries globose, glabrous. black. Racemes erect, pubescent. Bracteas short. Segments of the calyx, which is campanulate, spreading. Flowers small. Petals white. The fruit, and peculiar odour of the plants, are those of *R.* nigrum. (*Don's Mill.*) An erect shrub. North America, from Hudson's Bay to the Rocky Mountains, in the west, and as far north as lat. 57°, including the mountains of Columbia, about the Kettle Falls. Height 3 ft. to 4 ft. Introduced in ? 1820. Flowers whitish; April and May. Fruit black; ripe in July. Horticultural Society's Garden.

880. *R.* (n.) hudsoniànum.

■ 38. *R.* GLACIA'LE *Wall.* The icy *black* Currant.

Identification. Wall. Cat., No. 6833.; Don's Mill., 3. p. 189.
Engraving. Our *fig.* 881. from a specimen in the Linnæan herbarium.

Spec. Char., &c. Bracteas smooth. Leaves glabrous above, but with few scattered bristly hairs beneath, cordate at the base, 3—5-lobed at the apex;

lobes acute, serrated. Petioles long, serrated at the base. Racemes drooping. Calyx campanulate. Petals longer than the calyx. Flowers white. Berries black. (*Don's Mill.*) A shrub. Nepal, on Emodi and Gossainthan. Height 4 ft. to 6 ft. Introduced in 1823. Flowers white; April and May. Fruit black; ripe in July. Hort. Soc. Garden.

881. *R. glaciale.*

■ 39. *R.* INE`BRIANS *Lindl.* The intoxicating Currant.

Identification. Lindl. in Bot. Reg., t. 1471.; Don's Mill., 3. p. 193.
Engravings. Bot. Reg., t. 1471.; and our *fig.* 882.

Spec. Char., &c. Leaves roundish, deeply 3—5-lobed, and deeply toothed, truncate at the base, glandular on both surfaces. Petioles pubescent. Peduncles 3—5-flowered, pendulous. Flowers aggregate. Calyx tubular, glandular, with the segments recurved. Calyx greenish white, with the tube 4 lines long. Leaves smelling like those of *R.* flóridum. (*Don's Mill.*) An upright shrub. North America. Height 3 ft. to 4 ft. Introduced in 1827. Flowers greenish white; April. Fruit amber-coloured; ripe in July.

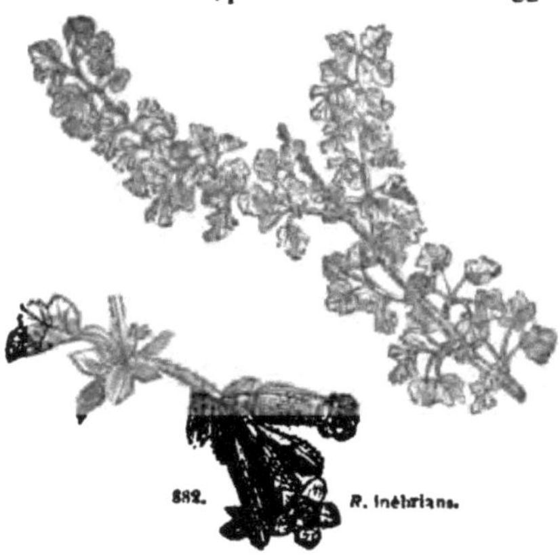

882. *R.* inébrians.

This species was received from Mr. Floy of New York, under the name of the intoxicating currant, but without any other account of its properties. The berries may probably possess some narcotic quality.

■ 40. *R.* CE`REUM *Dougl.* The waxy-*leaved* Currant.

Identification. Dougl. in Hort. Trans., 7. p. 512.; Don's Mill., 3. p. 190.
Engravings. Bot. Reg., t. 1263.; and our *fig.* 883.

Spec. Char., &c. Leaves small, cordate, lobed, serrated, clothed with glandular pubescence, glabrous, glaucous, full of white glands above. Racemes pendulous, rather capitate. Bracteas ovate, adpressed to the germens, which are glabrous. Flowers nearly sessile, cylindrical, rather angular. Calycine segments small, reflexed. (*Don's Mill.*) A low bush. North-west America, on the banks of the Columbia, and its southern tributary streams, from the Great Falls to the Rocky Mountains. Height 2 ft. to 3 ft. Introduced in 1827. Flowers whitish; April. Fruit amber; ripe in July.

883. *R.* céreum.

In its small foliage and few-flowered racemes, this species resembles the gooseberry tribe; but it has no thorns. The flowers are rather large and white, with a slight tinge of green, and are rather downy. White waxy dots like scales cove the upper surface of the leaf; whence the specific name.

C. *Flowers deep red. Fruit black.*

■ 41. *R.* SANGUI'NEUM *Pursh.* The bloody, *or red, flowered* Currant.

Identification. Pursh Fl. Amer. Sept., 1. p. 164.; Don's Mill., 3. p. 191.
Synonymes. *R.* malvàceum *Smith* in *Rees's Cycl.*; Calobótrya sanguínea *Spach.*
Engravings. Hort. Trans., 7. t. 13.; Bot. Reg., t. 1349.; and our *fig.* 884.

Spec. Char., &c. Leaves cordate, somewhat 5-lobed, serrated, veiny, smoothish above, but clothed with villous tomentum beneath. Racemes drooping, pubescent, twice the length of the leaves. Calyx tubularly campanulate, with oblong, obtuse, spreading segments, exceeding the petals, which are red, and quite entire. Bracteas obovate-spathulate. Berries turbinate, hairy. (*Don's Mill.*) A large branchy smooth shrub. North-west coast of America, in rocky situations, by the sides of streams. Height 4 ft. to 8 ft. Introduced in 1826. Flowers deep rose; March and April. Fruit purplish, with a glaucous bloom; ripe in August.

884. *R. sanguineum.*

Varieties.

■ R. *s.* 2 *glutinòsum.* *R.* glutinòsum *Benth. Hort. Trans.* 2d ser. vol. i. p. 476., *R.* angústum *Dougl. MS.* (Our *fig.* 885.)—The foliage is destitute of down, and slightly viscous. The racemes are rather larger than in the species, and the flowers are of a very pale rose colour. This variety comes into leaf a month before the species.

885. *R. s. glutinòsum.*

886. *R. s. malvàceum.*

■ R. *s.* 3 *malvàceum.* *R.* malvàceum *Benth.* l. c. (Our *fig.* 886.)—Leaves rough and hispid on the upper side, and clothed underneath with a whitish cottony down. The racemes of flowers are shorter and closer; and each flower is almost sessile on the common stalk. In colour, the flowers are rather darker than those of *R.* s. glutinòsum, and have more of a lilac tinge.

■ R. *s.* 4 *àtro-rùbens* Hort.—Flowers and racemes rather smaller, and of a much deeper and darker red, than those of the species. Horticultural Society's Garden.

By far the most ornamental species of the genus. It is easily propagated, and as hardy as the common black currant. It flowers profusely; and, coming into bloom early in the season, forms the most splendid bush to be seen in British shrubberies, from the middle or end of March to the beginning or middle of May. A great many seeds were sent over by Mr. Douglas, a number of which were distributed by the Horticultural Society; and the plants produced from them have varied in the colour of their flowers, from pale pink to deep red. The plants, also, seed freely in this country; and hence a number of varieties have been originated by nurserymen, independently of *R.* s. malvàceum and *R.* s. glutinòsum, which differ from the species, not only in the shades of colour of their flowers, but also in their leaves. The variety which has the darkest-coloured flowers is *R.* s. àtro-rùbens.

■ 42. *R.* A'TRO-PURPU'REUM *Meyer.* The dark-purple-*flowered* Currant.

887. R. atro-purpùreum.

Identification. Meyer in Led. Fl. Ross. Alt. Ill. t. 231.; Fl. Alt., l. p. 268.; Don's Mill., 3. p. 191.
Engravings. Led. Fl. Ross. Alt. Ill., t. 231.; our *fig.* 887. from a living plant in the Birmingham Botanic Garden.; and *fig.* 888. from Ledebour.

Spec. Char., *&c.* Stem erect. Leaves pubescent, nearly orbicular, cordate, 3—5-lobed; lobes acute, serrated. Racemes drooping. Pedicels exceeding the bracteas. Calyxes campanulate, ciliated. Berries glabrous, and bractless; dark purple, and the size of those of the common currant. (*Don's Mill.*) An upright shrub. Altaia, on mountains and subalpine places on the river Ursal; and also at the river Tscharysch. Height 4 ft. to 6 ft. Flowers deep purple; April and May. Fruit dark purple; ripe in July.

888. R. atro-purpùreum.

Varieties.

■ R. *a.* 1. — Flowers deep purple. Leaves rather pubescent beneath, but smooth and glabrous above, as well as the branches.

■ R. *a.* 2. — Leaves rather pubescent beneath, but hispid from bristles above, as well as the petioles and stems. Found near the river Volschoi Ulegumen.

■ R. *a.* 3. — Flowers paler. Leaves pubescent above, but most so below. Branches smooth.

§ iv. *Symphócalyx* Dec.

Derivation. From *sumphuō*, to grow together, and *kalux*; in reference to the sepals of the calyx of the species belonging to this section.

Sect. Char. The calyxes tubular, and yellow. The racemes many-flowered. Leaves compassing the bud. Unarmed shrubs. (*Dec. Prod.*, iii. p. 483.)

■ 43. *R.* AU'REUM *Pursh.* The golden-*flowered* Currant.

Identification. Pursh Fl. Amer. Sept., 1. p. 164.; Don's Mill., 3. p. 191.
Synonymes. *R.* palmàtum *Desf. C. Hort. Paris.*; Chrysobótrya revolùta *Spach.*
Engravings. Berl., l. c., t. 2. f. 23.; Bot. Reg., t. 125.; and our *fig.* 889.

Spec. Char., *&c.* Quite glabrous. Leaves 3-lobed; lobes divaricate, with a few deep teeth, shorter than the petioles, which are ciliated at the base. Calyxes tubular, longer than the pedicels. Tube slender. Segments oblong, obtuse. Petals linear, much shorter than the calycine segments. Bracteas linear, length of the pedicels. Style entire. Berries glabrous. Flowers golden yellow. Fruit yellow, seldom black, and of an exquisite flavour. (*Don's Mill.*) An upright branchy shrub, which before blowing has the appearance of a species of *Cratæ'gus.* North-west America, in light gravelly soils, from the Great Falls of the Columbia River to the mountains, and on the southern branches. Height 6 ft. to 8 ft. Introduced in 1812. Flowers yellow; April and May. Fruit yellow, seldom black; ripe in August.

889. R. aùreum.

Varieties.

■ R. *a.* 1 *præ'cox* Lindl. in Hort. Trans. vii. p. 242. *R.* frágrans *Lod.* (Bot. Cab., t. 1533.; and our *fig.* 890.) — Flowers earlier. Leaves cuneated at the base, pubescent beneath; lobes deeply serrated.

890. *R.* fràgrans.

Berries copious, earlier, turbinate. Racemes bracteate.

■ R. *a.* 2. *villòsum* Dec. Prod. iii. p. 483. *R.* longiflòrum *Fraser's Cat.* 1813.— Leaves rather villous.

■ R. *a.* 3 *serótinum* Lindl. l. c., and our *fig.* 891.— Flowers late. Leaves of various forms, smoothish beneath; lobes deeply serrated. Berries few, late, and round in shape. Racemes naked.

891. *R. a.* serótinum.

All the forms of this species are highly ornamental, from their fine, large, bright yellow flowers, which are produced in abundance; and their smooth, glossy, yellowish green leaves. The plants are, also, more truly ligneous, and of greater duration, than those of most other species of *Ribes*. Next to *R.* sanguíneum, and its varieties, they merit a place in every collection.

■ 44. *R.* (A.) TENUIFLO`RUM *Lindl.* The slender-flowered Currant.

Identification. Lindl. in Hort. Trans., 7. p. 242.; Bot. Reg., 1274.; Don's Mill., 3. p. 191.
Synonymes. *R.* aúreum *Colla Hort. Rip. Append.* 3. t. 1. f. A.; *R.* flàvum *Berl.* in *Dec. Prod.* 3. p. 483.; *R.* missouriénsis *Hort.*; Chrysobótrya Lindleyàna Spach.
Engravings. Bot. Reg., t. 1274.; and our *fig.* 892.

Spec. Char., &c. Unarmed, quite glabrous. Leaves roundish, 3-lobed, mealy; lobes bluntly toothed at the apex. Racemes pendulous, many-flowered. Calyx tubular, glabrous, longer than the pedicels, coloured. Petals quite entire, linear, one half shorter than the segments of the calyx, which are oblong and obtuse. Bracteas linear, length of the pedicels. Berries glabrous. (*Don's Mill.*) An upright branchy shrub. North America, on the rocky tracts of the Columbia, near the head waters of the Missouri. Height 6 ft. to 8 ft. Introduced in 1812. Flowers yellow; April and May. Fruit purple or yellow; ripe in August.

892. *R.* (*a.*) tenuiflòrum.

Varieties.

■ R. (*a.*) *t.* 1 *frúctu nìgro.*—Berries changing from yellow to red, and finally acquiring a deep blackish purple colour.

■ R. (*a.*) *t.* 2 *frúctu lùteo.*—Fruit yellow; always retaining the same colour.

In habit, this species is more erect than *R.* aúreum, and has the young wood more thinly clothed with leaves; its whole appearance is also paler, during the early part of the season. The flowers are not more than half the size of *R.* aúreum; and have entire, not notched, petals. The fruit is about the size of the red currant, of an agreeable flavour, but possessing little acidity.

■ 45. *R.* (A.) FLA`VUM *Coll.* The yellow-*flowered* Currant.

Identification. Coll. Hort. Ripul. Append., 3. p. 4. t. 1. f. β.; Don's Mill., 3. p. 191.
Synonymes. *R.* aúreum β sanguíneum *Lindl.* in *Hort. Trans.* 7. p. 242.; *R.* palmàtum *Desf. Hort. Par.*; *R.* aúreum *Ker Bot. Reg.* t. 125., but not of Pursh; Chrysobótrya intermèdia *Spach.*
Engravings. Coll. Hort. Ripul. Append., 3. p. 4. t. 1. f. 2.; and our *figs.* 893. and 894.

Spec. Char., &c. Unarmed, quite glabrous. Young leaves 3-lobed; adult

893. *R. flàvum.*

ones usually 5-lobed, deeply toothed, about equal in length to the ciliated petioles. Racemes short, 4—5-flowered. Calyx tubular, much longer than the pedicels. Tube slender. Segments rather spathulate, reflexed. Petals one half shorter than the calycine segments. Bracteas elliptic. Berries oblong, glabrous. Flowers yellow. (*Don's Mill.*) An upright shrub. North America. Height 6 ft. to 8 ft. Introduced in 1812. Flowers yellow; April and May. Fruit purple or yellow; ripe in August.

A very ornamental species, of vigorous growth, fine shining foliage, and of greater duration than many species of *Ribes*.

894. *R. flàvum.*

Order XXXIII. ESCALLONIA'CEÆ.

Ord. Char. *Calyx* 5-toothed. *Petals* 5, forming a tube by their cohesion, finally separating; æstivation imbricated. *Stamens* definite. *Disk* epigynous, surrounding the base of the style. *Ovarium* 2-celled, containing two large placentas in the axis. *Stigma* 2-lobed. *Capsule* crowned by the calyx and style, dehiscing at the base. *Seeds* numerous, minute. *Albumen* oily. The cohering petals, oily albumen, and situation of placentas separate this from Grossulariàceæ. (*G. Don.*)

Leaves simple, alternate, exstipulate, deciduous or evergreen; lanceolate serrated or entire. *Flowers*, terminal in spikes or racemes.—Shrubs, natives of North and South America, of which two genera are in British gardens, which are thus contradistinguished: —

I'tea. Stigma capitate. Capsule compressed.
Escallo'nia. Stigma peltate, 2-lobed. Capsule baccate.

Genus I.

I'TEA *L.* The Itea. *Lin. Syst.* Pentándria Monogýnia.

Identification. Lin. Gen., 275.; Dec. Prod., 4. p. 6.; Don's Mill., 3. p. 396.
Synonymes. Cedrèla *Lour.*; Diconángia *Michx.*
Derivation. *Itea* is the Greek name of the willow, which is given to this genus on account of the quick growth of the I'tea virginica.

Gen. Char. *Calyx* bell-shaped, with 5 teeth, persistent. *Petals* 5, their æstivation valvate. *Stamens* 5, shorter than the petals. Both petals and stamens inserted upon the tube of the calyx. Teeth of calyx, petals, and stamens, alternate with one another. *Ovary* not connate with the calyx. *Style*, at first, seemingly one; afterwards it parts into two portions: hence, there are rather 2 styles connate. *Stigmas* capitate, mostly divided by a furrow. *Carpels* two, connate into a capsule of 2 cells, that has 2 furrows, and parts from bottom to top. *Seeds* in two rows along the introflexed margins of the carpels. (*Dec. Prod.*)

Leaves simple, alternate, exstipulate, deciduous; lanceolate, toothed. *Flowers* small, white, in simple terminal racemes. — A shrub, native of North America.

■ 1. *I.* VIRGI'NICA *L.* The Virginian Itea.

Identification. Lin. Sp., 289.; Don's Mill., 3. p. 196.
Engravings. N. Du Ham., 6. t. 9.; Bot. Mag., t. 2409.; and our *fig.* 895.

Spec. Char., &c. Leaves lanceolate, acutely toothed. Racemes simple, terminal. (*Don's Mill.*) A deciduous shrub. Pennsylvania to Carolina. Height 3 ft. to 5 ft. Introduced in 1744. Flowers white; June to August. Carpels brown; ripe in October.

895. *I. virginica.*

It may be propagated by cuttings, but more readily by layers, suckers, or seeds, which are annually imported from America; and it thrives best in a sandy or peaty soil, kept moist. The plant, to be kept in vigour, should have the old wood frequently cut down to the ground. When grown in a situation that is rather moist, its flowers make a fine appearance late in the season, when there are few other shrubs in blossom.

GENUS II.

ESCALLO'N*IA* Mutis. THE ESCALLONIA. *Lin. Syst.* Pentándria Monogýnia.

Identification. Mutis in Lin. fil. Supp., t. 21.; Dec. Prod., 4. p. 2.; Don's Mill., 3. p. 192.
Synonyme. Stereóxylon *Ruiz et Pav. Fl. Per. Prod.* p. 38.
Derivation. From *Escallon*, the pupil and companion of Mutis, during his travels in New Spain.

Gen. Char., &c. *Calyx* tube semiglobose, adnate to the ovarium; limb 5-toothed or 5-lobed. *Petals* 5, arising from the calyx. *Stamens* 5; anthers ovate-oblong. *Stigma* peltate. *Style* filiform, permanent. *Capsule* baccate. *Seeds* numerous. (*Don's Mill.*)

Leaves simple, alternate, exstipulate, sub-evergreen; serrated or entire, full of resinous glands. *Flowers* terminal, bracteate, variously disposed, white or red.—Sub-evergreen shrubs, natives of South America, more especially of Chili. Propagated with the greatest ease by cuttings; and growing freely in any common soil.

■ ■ 1. E. RU'BRA *Pers.* The red-*flowered* Escallonia.

Identification. Pers. Ench., 2. p. 235.; Hook. Bot. Mag., t. 2890., and Don's Mill., 3. p. 193.
Synonyme. Stereóxylon rùbrum *Ruiz et Pav.*
Engravings. Ruiz et Pavon Fl. Per., 3. t. 236. f. b.; Bot. Mag. t. 2890.; and our *fig.* 896.

Spec. Char., &c. Shrubs smoothish; branches erect, when young clothed with glandular villi. Leaves obovate-oblong, acuminated, serrated, full of resinous dots beneath. Peduncles 2—7-flowered, bracteate. Lobes of calyx denticulated. Petals spathulate. (*Don's Mill.*) A sub-evergreen shrub. Chili, on the mountains of Colocolo, and in the fissures of rocks, and about Valparaiso. Height 3 ft. to 6 ft. Introduced in 1827. Flowers red; July to September.

896. E. rùbra.

Varieties. In the *Bot. Misc.*, iii. p. 252., three forms are recorded: —

■ ■ *E. r.* 1 *glabriúscula* Hook. et Arn., with glandular branches, leaves highly pubescent, and red flowers, which may be considered as the species.

E. r. 2 *albiflòra* Hook. et Arn., E. glandulòsa *Bot. Cab.* t. 291., with white flowers.

E. r. 3 *pubéscens* Hook. et Arn., with pubescent branches, and red flowers.

Very desirable shrubs for training against a wall.

2. E. MONTEVIDE'NSIS *Dec.* The Monte Video Escallonia.

Identification. Dec. Prod., 4. p. 4.; Don's Mill., 3. p. 194.
Synonyme. E. floribúnda var. β montevidénsis *Schlecht.* in *Linnæa* l. p. 543.; E. bífida *Link et Otto Abbild.* t. 23.
Engravings. Link et Otto Abbild., t. 23.; Bot. Reg., 1467.; and our *fig.* 897.

Spec. Char., &c. Shrub glabrous. Branches erect. Leaves oblong, cuneated at the base, acutish, finely serrated, full of resinous dots beneath. Panicle terminal, many-flowered, crowded, intermixed with foliaceous bracteas. Lobes of calyx acute, rather denticulated. Petals obovate, oblong. (*Don's Mill.*) A sub-evergreen shrub. Brazil, in many places, but especially on the sandy banks and pastures of the Uruguny. Height 6 ft. to 10 ft. Introduced in 1827. Flowers white, very like those of the hawthorn, with a style which becomes double the length of the fruit after flowering; July to September.

897. E. montevidénsis.

Variety.

E. m. 2 *floribúnda*, E. floribúnda *H. B. et Kunth*, is a native of New Granada, on the Andes, with white flowers, and shining leaves, which are clammy when young. A very distinct variety considered by some as a species.

This species forms a remarkably vigorous-growing bush, with long, flexible, rope-like shoots, and is very prolific in flowers. It is so hardy as to have stood through several winters, as a bush, in the open ground of the Kensington Nursery, though it was killed by the winter of 1837–8.

3. E. ILLINI'TA *Presl.* The varnished Escallonia.

Identification. Presl Reliq. Hænk. vol. ii. p. 49.; Don's Mill., 3. p. 193.
Engravings. Presl l. c., t. 59.; Bot. Reg., t. 1900.; and our *fig.* 898.

Spec. Char., &c. Quite glabrous. Branches spreading, anointed with resin. Leaves petiolate, obovate or oblong obtuse, crenulated, attenuated at the base, beset with glandular dots above, and clammy. Panicle terminal, many-flowered, leafy. Petals on long claws. Capsule turbinate, 5-nerved. (*Don's Mill.*) An evergreen shrub. Chili, at the streamlet of Los Lunes. Height 3 ft. to 6 ft. Introd. 1833. Flowers white; Aug. and Sept.

The whole plant emits a powerful odour, which to some persons resembles the smell of swine, and to others that of melilot or fenugreek. One of the hardiest species of the genus, and, like all the others, well deserving a place in collections.

898. E. illinita.

Other Species of Escallònia.—*E. resinòsa* Pers., Stereoxylon resinòsum *Ruiz et Pavon* (*Don's Mill.*, iii. p. 94.), is a native of Peru, on the cold parts of hills, which stood out at Kew for five years, till it was killed by the winter of 1837–8

E. pulverulénta Pers., Stereóxylon pulverulëntum *Ruiz et Pav.*, is a shrub, hairy in every part, with white flowers; growing to the height of 8 or 10 feet. It is a native of Chili; and plants of it were in the Horticultural Society's Garden from 1831 till 1837–8. Twenty other species are described in Don's *Miller*, iii. p. 193. to p. 195., all natives of South America, and probably as hardy as those above mentioned; but it does not appear that any of them have been introduced.

Order XXXIV. SAXIFRA'GEÆ,

Tribe HYDRA'NGEÆ.

Ord. Char. *Calyx* 4—5-parted. *Petals* 5, inserted between the lobes of the calyx. *Stamens* 5 or 10. *Disk* perigynous. *Ovarium* of 2 to 5 carpels. *Stigmas* sessile. *Fruit* 1—2-celled. *Seeds* numerous, minute. *Albumen* fleshy. Absence of stipules distinguishes this from *Rosàceæ* and *Cunoniàceæ*. (*G. Don.*)

Leaves simple, opposite, exstipulate, deciduous. *Flowers* in large corymbs, pink or white, often sterile.—Suffruticose shrubs, natives of North America and Asia. Easily propagated by cuttings, and growing freely in any soil that is rather moist.

Genus I.

HYDRA'NGEA *L.* The Hydrangea. *Lin. Syst.* Decándria Di-Trigýnia.

Identification. Lin. Gen., 557.; Dec. Prod., 4. p. 13.; Don's Mill., 3. p. 232.
Synonymes. Hydrángea, and Horténsia Juss.; Idrangea, *Ital.*
Derivation. From *hudōr*, water, and *aggos*, a vessel; with reference to some of the species which grow in water; or, as some suppose, from the capsule resembling a cup.

Gen. Char. *Flowers* generally deformed; but some of them hermaphrodite and fertile. *Calyx* tube hemispherical, 10-ribbed, rather truncate, adnate to the ovarium; limb permanent, 5-toothed. *Petals* 5, regular. *Stamens* 10. *Styles* 2, distinct. *Capsule* 2-celled, with introflexed valves, crowned by the teeth of the calyx and styles, flattish at the top, opening by a hole between the styles. *Seeds* numerous, reticulated. (*Don's Mill.*)

Leaves simple, opposite, exstipulate, deciduous; serrated or lobed. *Flowers* corymbose, pink, or yellowish white; the marginal ones sterile, and large, in consequence of the teeth of the calyx being dilated into broad, petal-like-coloured segments; the rest of the sterile flower partially abortive. — Shrubs. natives of North America and Asia.

A. *Species Natives of North America.*

899. H. arborescens.

■ 1. H. arbore'scens *L.* The arborescent Hydrangea.

Identification. Lin. Sp., p. 568.; Don's Mill., 3. p. 232.; Lodd. Cat., ed. 1836.
Synonymes. H. vulgáris *Michx. Fl. Bor. Amer.* 1. p. 268.; H. frutéscens *Mœnch Meth.* 1. p. 106.
Engravings. Bot. Mag., t. 437.; and our *fig.* 899.

Spec. Char., &c. Leaves ovate, rather cordate; superior ones lanceolate, coarsely toothed, pale and puberulous beneath. Corymbs flattish. Flowers nearly all fertile. Flower buds obtuse. Flowers white, small, having an agreeable odour. (*Don's Mill.*) A low shrub. Pennsylvania to Virginia. Height 4 ft. to 6 ft. Introduced in 1736. Flowers white, having an agreeable odour; July and August.

Variety.

H. *a.* 2 *discolor* Ser. in Dec. Prod. 4. p. 14.—Leaves almost white beneath from tomentum.

It prefers a moist soil, and is readily propagated by division of the roots.

2. H. (A.) CORDA'TA *Pursh.* The cordate-*leaved* Hydrangea.

Identification. Pursh Sept., 1. p. 309., exclusive of the synonyme of Michx.; Don's Mill., 3. p. 232.
Engravings. Wats. Dendr. Brit., t. 42.; and our *fig.* 900.

900. H. (a.) cordàta.

Spec. Char., &c. Leaves broadly ovate, acuminated, rather cordate at the base, coarsely toothed, glabrous beneath. Flowers all fertile, small, white, and sweet-scented. (*Don's Mill.*) A low shrub. Carolina, on mountains, and on the banks of the Missouri, above St. Louis. Height 6 ft. to 8 ft. Introd. in 1806. Flowers white; July, Aug.

Variety.

H. (*a.*) *c.* 2 *geórgica*, H. geórgica *Lodd. Cat.*, differs from the species in flowering a little later, and being rather more robust.

We agree with Torrey, in thinking this merely a variety of H. arboréscens.

3. H. NI'VEA *Michx.* The snowy-*leaved* Hydrangea.

Identification. Michx. Fl. Bor. Amer., 1. p. 268.; Don's Mill., 3. p. 232.
Synonyme. H. radiàta *Wal. Fl. Car.* 251., ex *Michx.*, but not of Smith.
Engravings. Wats. Dendr. Brit., t. 43.; and our *fig.* 901.

901. H. nìvea.

Spec. Char., &c. Leaves cordate, oval, acuminated, sharply toothed, clothed with white tomentum, or pubescence, beneath. Corymbs flattish. Sepals of sterile flowers entire. Flower buds depressed. Flowers white, rather large. (*Don's Mill.*) A low suffrutescent shrub. North America, on the Savannah River. Height in America 4 ft. to 6 ft.; in England 2 ft. to 3 ft. Introd. 1786. Flowers white; July and August.

Variety.

H. n. 2 *glabélla* Ser. in Dec. Prod. 4. p. 14.—Leaves nearly glabrous beneath. Flowers all fertile. This variety has, probably, originated in culture.

4. H. QUERCIFO'LIA *Bartram.* The Oak-leaved Hydrangea.

Identification. Bartram Trav., ed. Germ., p. 336. t. 7.; Don's Mill., 3. p. 233.
Synonyme. H. radiàta *Smith Icon. Pict.* 12., but not of Walt.
Engravings. Bot. Mag., t. 975.; and our *fig.* 902.

902. H. quercifòlia.

Spec. Char., &c. Leaves large, ovate, serrately lobed, and toothed, pilose beneath. Corymbs rather panicled, flattish. Sepals of sterile flowers entire. Flower buds depressed. Flowers white. Sterile, or outer, ones of the corymbs large. (*Don's Mill.*) A shrub. Florida. Height 4 ft. to 6 ft. in America; 2 ft. to 3 ft. in England. Introduced in 1803. Flowers white; June to September.

This is by far the most interesting of the North American hydrangeas, from its large, deeply lobed, and sinuated leaves; and its fine, large, nearly white corymbs of flowers, which are sterile, and appear from June till they are destroyed by frost. Culture as in the other species; but it is essential that the situation be sheltered, and the soil kept somewhat moist, otherwise the leaves are not perfectly developed, and the branches are apt to be broken off by high winds.

B. *Species Natives of Asia.*

5. H. HETEROMA'LLA *D. Don.* The diverse-haired-*leaved* Hydrangea.

Identification. D. Don Prod. Fl. Nep., p. 211.; Don's Mill., 3. p. 233.
Engraving. Our *fig.* 903. from a specimen in the Linnæan herbarium.

Spec. Char., &c. Leaves oval, acuminated, sharply serrated, tomentose beneath, 5 in. long, and nearly 3 in. broad. Corymbs supra-decompound, diffuse, pilose. Sepals of sterile flowers roundish oval, quite entire. Flowers white. (*Don's Mill.*) A shrub. Nepal, at Gossainthan. Height 4 ft. to 6 ft. Introduced in 1821. Flowers white; ? July, August.

A very vigorous-growing plant in its native country, and probably as hardy in British gardens as some of the North American species.

903. H. heteromálla.

6. H. ALTI'SSIMA *Wall.* The tallest Hydrangea.

Identification. Wal. Tent. Fl. Nep., 2. t. 50.; Don's Mill. 3. p. 233.
Engravings. Wall. l. c., t. 50.; and our *fig.* 904.

Spec. Char., &c. Leaves ovate, acuminated, serrated, smoothish. Corymbs flattish. Sterile flowers few, on pilose peduncles; alabastra, or fertile flowers, conical. (*Don's Mill.*) A rambling shrub, which, according to Dr. Royle, climbs lofty trees. Nepal, on mountains. Height ?. Introduced in 1839. Flowers white; ? July, August.

904. H. altíssima.

Other Species of Hydrángea.—H. Horténsia Sieb., H. horténsis *Smith*, a well-known ornament of gardens, is suffrutescent and hardy in the S. of England. Even in the climate of London it lives in sheltered situations in the open garden, because, though frequently killed to the ground, it always springs up again, and even flowers. — *H. vestìta* Wall., a native of Nepal, is probably as hardy as H. altíssima, and would be a most desirable introduction.

ORDER XXXV. UMBELLA'CEÆ.

ORD. CHAR. *Calyx* entire or toothed. *Petals* 5, entire, emarginate, or 2-lobed, each usually drawn out into a replicated or involuted point. *Stamens* 5. *Ovarium* 2-celled. *Styles* 2. *Fruit* of 2 separating pericarps, adhering by their faces to the carpophore. *Fruit* ribbed or winged. *Peri-*

carps 1-seeded. — Habit alone is sufficient to distinguish this order. (*D. Don.*)

Leaves simple, alternate, exstipulate, evergreen or sub-evergreen; quite entire. *Flowers* greenish yellow. — There are only one or two ligneous species hardy in British gardens, and these belong to the genus *B*upleùrum.

Genus I.

*B*UPLEU'RUM *Tourn.* The Bupleurum, *or Hare's Ear.*
Lin. Syst. Pentándria Digýnia.

Identification. Tourn. Inst., 309. t. 163.; Dec. Prod., 4. p. 127.; Don's Mill., 3. p. 296.
Synonymes. Tenòria and *B*uprèstis *Spreng. Syst.* 1. p. 880.; Bupliore, or Oreille de Lièvre, *Fr.*; Hasenöhrlien, *Ger.*
Derivation. From *bous*, an ox, and *pleuron*, a side; from the supposed quality of swelling cattle that feed on some of the species of the genus. The name of Hare's Ear, which is preserved in the French and German, has reference to the shape of the leaves.

Gen. Char. *Calyx* margin obsolete. *Petals* roundish, entire, strictly involute, with a broad retuse point. *Fruit* compressed from the sides. *Seed* teretely convex, flattish in front. (*Don's Mill.*)

Leaves as in the order. — Smooth shrubs, natives of Europe and Africa, and some of Asia. Only one hardy species is in cultivation in British gardens.

■ ■ 1. *B.* FRUTICO'SUM *L.* The shrubby Bupleurum, *or Hare's Ear.*

Identification. Lin. Sp., 343.; Don's Mill., 3. p. 301.; Webb Iter Hispan., p. 44.
Synonymes. Tenòria fruticòsa *Spreng.* in *Schultes Syst.* 6. p. 376.; *B*uprèstis fruticòsa *Spreng. Mag.*; Séseli æthiópicum *Bauh. Pin.* 161.; Séseli frùtex *Mor. Umb.* 16.
Engravings. Sibth. Fl. Græc., t. 263.; Wats. Dendr. Brit., t. 14.; and our *figs.* 905. and 906.

905. *B.* fruticòsum.

906. *B.* fruticòsum.

Spec. Char., &c. Shrubby, erect-branched. Leaves oblong, attenuated at the base, coriaceous, 1-nerved, quite entire, sessile. Leaves of involucre oblong. Ribs of fruit elevated, acute. Vittæ broad. Bark of branches purplish. Leaves of a sea-green colour. (*Don's Mill.*) A neat sub-evergreen glaucous shrub. Portugal, Spain, the South of France, about Nice, Corsica, Sicily, Mauritania, and Thessaly. Height 3 ft. to 4 ft. in a wild state; 6 ft. in British gardens. Introduced in 1596. Flowers yellow; July and August.

It is readily propagated by cuttings, is of free growth in any dry calcareous soil, and is particularly vigorous on the sea coast in Kent. The blue glaucous hue of its smooth shining foliage renders it a desirable addition to every collection. If planted in an open airy situation, in a deep soil, not moist, and allowed to extend itself on every side, it would soon form a large hemispherical bush, highly ornamental during winter from its evergreen foliage, and during summer from its bright yellow flowers.

B. *frutéscens* L. (Cav. Icon., ii. t. 106.; and our *fig.* . in p. .) has slender elongated branches, and linear-subulate, stiff, striated leaves. It is a native of Mauritania in Spain, and also at Tarragona.

B. *gibraltárica* Lam. Dict., *B.* arboréscens *Jacq.* (Ic. rar., ii. t. 351.; and our *fig.* 2094. in p. 1108.) grows to the height of 3 ft., and has fragrant flowers.

ORDER XXXVI. ARALIA'CEÆ.

ORD. CHAR. *Calyx* entire or toothed. *Petals* 5 or 10; æstivation valvate. *Stamens* same, or double the number of petals. *Anthers* peltate. *Ovarium* of 2 or more cells; cells 1-seeded. *Styles* numerous, usually distinct. *Berry* crowned by the limb of the calyx. *Albumen* fleshy. — Differs from the Umbellàceæ in inflorescence, numerous styles, and baccate, generally many-celled fruit.

Leaves simple or compound, alternate, stipulate, deciduous or evergreen; serrated or entire. *Flowers* small, greenish.

The genera belonging to this order, which contain ligneous plants, are Aràlia and *Hédera*, the former rather suffruticose than permanently woody: their characteristics are as under: —

ARA'LIA *L.* Petals 5. Stamens 5. Styles 5, expanded. Berry 5-celled.
HE'DERA Swartz. Petals 5—10. Stamens 5—10. Styles 5—10, conniving. Berry 5—10-celled.

GENUS I.

ARA'LIA *L.* THE ARALIA, *or ANGELICA TREE.* *Lin. Syst.* Pentándria Pentagýnia.

Identification. D. Don Prod. Fl. Nep., p. 185., in a note; Dec. Prod., 4. p. 257.; Don's Mill., 3. p. 388.
Synonymes. Aràlia sp. *Lin.*; Aràliæ véræ *Blum.*
Derivation. According to some, from *ara*, annoyance, the spines being very troublesome in its native country to travellers; but, according to others, a name of unknown meaning, under which one species was sent to Fagon, at Paris, from Quebec, in 1764, by one Sarrasin, a French physician.

Gen. Char. *Calyx* margin very short, entire or toothed. *Petals* 5, free, and expanded at the apex. *Stamens* 5. *Styles* 5, expanded, spreading divaricately. *Berry* 5-celled, usually torose. *Pyrenæ* chartaceous. (*Don's Mill.*)

Leaves compound, imparipinnate, alternate, stipulate, deciduous; large, rough. *Flowers* white, or greenish; in umbels, usually disposed in panicles. — Suffrutescent shrubs, with prickly branches and leaves, and with large pith. Natives of North America and Japan.

▲ 1. A. SPINO'SA *L.* The spiny Aralia, *or Angelica Tree.*

Identification. Lin. Sp., p. 392.; Don's Mill., 3. p. 389.; Lodd. Cat., ed. 1836.
Synonymes. Aralie, *Fr.* and *Ger.*; Angelica spinosa, *Ital.*; Spikenard, *N. Amer.*
Engravings. Schmidt Arb., t. 102. and t. 103.: Wats. Dend. Brit., t. 116.; and our *fig.* 907.

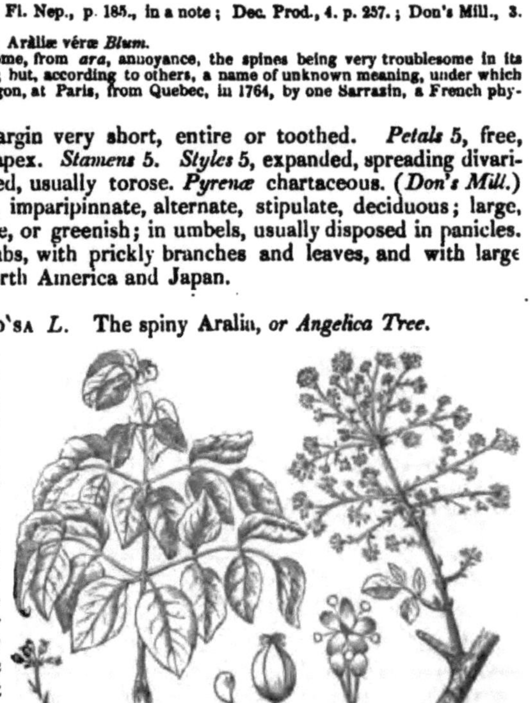

907. A. spinòsa.

Spec. Char., &c. Stem arboreous and prickly. Leaves doubly and trebly pinnate. Leaflets ovate, acuminated, and deeply serrated. Panicle much branched, beset with velvety stellate down. Umbels nu merous. Involucre small, of few leaves. Petals white and reflexed. Styles 5, divaricate, arched. Fruit 5-ribbed. (*Don's Mill.*) An erect suffrutescent plant, with the habit of a tree. Carolina and Virginia, in low, fertile, moist woods. Height 10 ft. to 12 ft. Introduced in 1688. Flowers greenish white; August and September.

An infusion of the fruit, in wine or spirit, is considered an effectual cure for the rheumatism. In British gardens, this species is propagated by cuttings of the roots; and, from its large doubly and trebly pinnate leaves, it forms a singularly ornamental plant, with a spreading, umbrella-like head, when standing singly on a lawn. After the plant flowers, the stem commonly dies down to the ground, like that of the raspberry, and, like it, is succeeded by suckers. Pursh "mentions a variety in which the petioles of the leaves are without prickles."

2. A. JAPO'NICA *Thunb.* The Japan Aralia.

Identification. Thunb. Jap., p. 128.; Blume Bijdr., p. 371.; Don's Mill., 3. p. 389.
Engraving. Our *fig.* 2091, in p. 1107.

Spec. Char., &c. Stem shrubby, unarmed. Leaves petiolate, 7-nerved, 7-lobed; lobes ovate, serrated at the apex. Panicles terminal. Peduncles umbelliferous. Leaves coriaceous, glabrous in the adult state, but when young woolly on both surfaces. (*Don's Mill.*) A suffrutescent erect shrub. Japan, near Nagasaki. Height 5 ft. to 6 ft. Introduced in 1838. Flowers white Fruit striated.

GENUS II.

HE'DERA *Swartz.* THE IVY. *Lin. Syst.* Pent-Decándria, and Pent-Decagýnia.

Identification. Swartz Fl. Ind. Occ., p. 581.; Dec. Prod., 4. p. 261.; Don's Mill., 3. p. 391.
Synonymes. Arália sect. Gymnópterum *Blum. Bijdr.* p. 871.; Hédera and Arália sp. *Lin.*; Lierre, *Fr.*; Ephen, *Ger.*; Edera, *Ital.*
Derivation. Various etymologies have been proposed for the word *Hédera*; but the most probable supposition appears to be, that it is derived from the Celtic word *hedira*, a cord. The English word Ivy is derived from the Celtic word, *iw*, green.

Gen. Char. Calyx margin elevated or toothed. *Petals* 5—10, not cohering at the apex in the form of a calyptra. *Stamens* 5—10. *Styles* 5—10, conniving, or joined in one. *Berry* 5—10-celled. (*Don's Mill.*)

Leaves simple, alternate, exstipulate, evergreen; lobed. *Flowers* umbellate or capitate. Fruit dark purple, or black.

Evergreen shrubs, climbing by the clasping roots produced by their stems, or creeping on the ground when without support. Natives of Europe and Asia.

1. *H. HE'LIX L.* The *common* Ivy.

Identification. Lin. Sp., 292.; Dec. Prod., 4. p. 261.; Don's Mill., 3. p. 391.
Derivation. Hélix is derived from *eileō*, to encompass, or turn round; in reference to the clasping stems, which, however, are not twining.

Spec. Char., &c. Stems climbing, throwing out roots from their sides to any object next which they may be placed. Leaves coriaceous, glabrous, shining, with 5 angular lobes; those on the old upright and rectangular branches, which form the tops of the plants, ovate, acute, quite entire. Umbels simple, pubescent. (*Don's Mill.*) A well-known evergreen climber and creeper. Europe and Britain, in woods. Stem 20 ft. to 60 ft. Flowers greenish yellow, or greenish; October and November. Fruit black; ripe in April.

Varieties. DeCandolle has enumerated three forms of this species which are independent of the varieties cultivated in British gardens: —

908. *H. H. vulgàris.*

L H. H. 1 *vulgàris* Dec. (Eng. Bot., t. 1267., and our *fig.* 908.) has the pedicels clothed with stellate down, and the fruit black. This is the commonest form of the ivy, throughout Europe, in a wild state; and there are varieties of it with white and yellow variegated leaves, in gardens.

L H. H. 2 *canariénsis* Dec.; *H.* canariénsis *Willd. Berol. Mag.* ii. p. 170. t. 5. f. 1.; the Irish Ivy, or Giant Ivy, of British gardens; has the pedicels scaly with pubescence. Floral leaves subcordate; those of the creeping branches 5-lobed and larger than those of the common ivy. Fruit ? red, or black. Canary Islands. Introduced in ? 1800, or before.

L H. H. ? 3 *chrysocárpa* Dec., *H.* poética *C. Bauh.*, H. chrysocárpos *Dalech.*, *H.* Dionýsias *J. Bauh.*, *H.* Hèlix *Wall.*, is a native of the North of India, with yellow fruit. It differs from the common ivy in its yellow fruit, and in being of more gigantic growth; in the leaves being more cuneated at the base; and in the pedicels being scaly. Hort. Society's Garden.

The Varieties in British Gardens, additional to the above, are: —

L H. H. 4 *fòliis argénteis* Lodd. Cat. *The Silver*-striped *Ivy.*

L H. H. 5 *fòliis aùreis* Lodd. Cat. *The Golden*-striped *Ivy.*

L H. H. 6 *digitàta* Lodd. Cat. *The palmate*, or Hand-shaped, *Ivy.*

L H. H. 7 *arboréscens* Lodd. Cat. *The arborescent*, or Tree, *Ivy.* — This variation is merely an extension of the flowering shoots, which are entire-leaved, and take an arborescent character; and, when a portion of them is cut off, and has rooted as a separate plant, it will sometimes produce an upright bush, which will retain its arborescent form for many years. Sooner or later, however, it resumes its native habit, and throws out rambling, or creeping, shoots, with 5-lobed leaves like the common ivy.

A variety with white berries is mentioned by Theophrastus, Pliny, Virgil, and Dioscorides.

The ivy will grow in any soil or situation, but thrives best when somewhat shaded. The common British variety, and its sub-varieties, are the best kinds for supporting themselves on walls, especially when young; at which period the giant ivy seldom throws out rootlets, though it does so subsequently.

ORDER XXXVII. *HAMAMELIDA'CEÆ.*

ORD CHAR. *Calyx* 4-lobed or repandly toothed. *Petals* 4, linear, rarely wanting; æstivation involutely valvate. *Stamens* 8, short, those opposite the petals barren. *Ovarium* half-inferior. *Styles* 2—3. *Capsule* 2-celled, 2-valved. *Ovules* bifid. *Albumen* horny. The flowers are sometimes diœcious, and sometimes polygamous. (*G. Don.*)

Leaves simple, alternate, bistipulate, deciduous; toothed or serrated. *Flowers* yellow or white. — Shrubs, deciduous; natives of North America.

HAMAME'LIS *L.* Calyx 4-lobed, furnished with 3—4 scales outside. Capsule coriaceous, 2-celled.

FOTHERGI'LLA L. Calyx campanulate, 5—7-toothed. Anthers in the form of a horseshoe. Capsule 2-lobed, 2-celled.

Genus I.

HAMAME'LIS L. The Hamamelis, *or Wych Hazel*.
Lin. Syst. Tetrándria Digýnia.

Identification. Lin. Gen., 169.; Dec. Prod., 4. p. 268.; Don's Mill., 3. p. 396.
Synonymes. Trilòpus *Mith. Act. Acad. Nat. Cur.* 8 *App.*; Hamamelide, *Ital.*
Derivation. Hamamèlis is a name by which Athenæus speaks of a tree which blossomed at the same time as the apple tree; the word being derived from *hama*, together with, and *mēlis*, an apple tree. The modern application seems to be from the *Hamamèlis* having its blossoms accompanying its fruits (*mēla*); both being on the tree at the same time.

Gen. Char. *Calyx* 4-lobed, adhering to the ovarium at the base, furnished with 2—3 scales on the outside. *Petals* 4, long, alternating with the teeth of the calyx. *Stamens* 4, alternating with the petals. *Ovarium* free at the apex. *Capsules* coriaceous, 2-celled, 2-valved. *Arils* 2 in each capsule. *Seed* oblong, shining. (*Don's Mill.*)

Leaves simple, alternate, bistipulate, deciduous; ovate or cuneated, feather-nerved, nearly entire. *Flowers* nearly sessile, disposed in clusters, in the axils of the leaves, girded by a 3-leaved involucrum. *Petals* yellow. — Shrubs or low trees, deciduous; natives of North America; interesting from producing their flowers in the autumn, which remain on during the winter.

■ ♀ 1. *H.* virgi'nica *L.* The Virginian Hamamelis, *or Wych Hazel*.

Identification. Dec. Prod., 4. p. 268.; Don's Mill., 3. p. 396.; Lodd. Cat., ed. 1836.
Synonymes. Hamamelie de Virginie, *Fr.*; Virginische Zaubernuss, *Ger.*; Pistacchio nera della Virginia, *Ital.*
Engravings. N. Du Ham., 7. t. 60.; Bot. Cab., t. 598.; and our *fig.* 909.

Spec. Char., &c. Leaves obovate, acutely toothed, with a small cordate recess at the base. (*Don's Mill.*) A deciduous shrub. Canada to Florida; in dry and stony situations, but frequently near water. Height 20 ft. to 30 ft. with a trunk 6 in. or more in diameter. Introduced in 1736. Flowers yellow; beginning of October to the end of February.

909. *H. virginica.*

Varieties.

■ ♀ H. *v.* 2 *parvifòlia* Nutt. — Leaves smaller, oblong ovate, and a more stunted habit than the species. Pennsylvania, on mountains. In British gardens, when planted in peat soil, this forms a very handsome little shrub; and is peculiarly valuable from being densely covered with fine yellow flowers throughout the winter.

■ ♀ H. *v.* 3 *macrophýlla. H.* macrophýlla *Pursh.* — Leaves nearly orbicular, cordate, coarsely and bluntly toothed, and scabrous from dots beneath. Western part of Georgia, and North Carolina, on the Katawba Mountains. Introduced in 1812, and flowers from May to November.

In British gardens, it has been but little cultivated, notwithstanding the singularity of its appearance in autumn and winter; when it is profusely covered with its fine rich yellow flowers, which begin to expand before the leaves of the previous summer drop off, and continue on the bush throughout the winter. After the petals drop off in spring, the persistent calyxes remain on till the leaves reappear in April or May. It will grow in any light free soil, kept rather moist; and it is propagated by layers and by seeds; which last, though rarely produced in Britain, are frequently sent to this country from America. They ought to be sown immediately on being received, as they are often two years before they come up.

Genus II.

FOTHERGI'LLA L. The Fothergilla. *Lin. Syst.* Icosándria Digýnia.

Identification. Lin. fil. Suppl., p. 42.; Dec. Prod., 4. p. 269.; Don's Mill., 3. p. 397.
Derivation. In memory of *John Fothergill*, M.D., an eminent physician and patron of botany, who introduced many new plants, and cultivated an excellent collection in his grounds, at Ham House, at Stratford-le-bow, in Essex. He was, besides, one of the most charitable men of his time.

Gen. Char. *Calyx* campanulate, adhering to the ovarium at the base, somewhat truncate, with 5—7 callous subrepand teeth. *Petals* wanting. *Stamens* about 25. *Styles* 2. *Capsule* adnate to the base of the calyx, 2-lobed, 2-celled, 1-seeded. *Seed* bony. (*Don's Mill.*)

Leaves simple, alternate, bistipulate, deciduous; feather-nerved, clothed with soft starry down. *Flowers* white, sweet-scented, sessile, anthers yellow; in terminal ovate spikes, having a solitary bractea under each flower; those bracteas at the base of the spike are trifid, and those at its apex are nearly entire.

Shrubs, deciduous, of which there is only one species, but several varieties. Natives of North America.

■ 1. F. alnifo'lia *L.* The Alder-leaved Fothergilla.

Identification. Lin. fil. Suppl., 257.; Dec. Prod., 4. p. 269.; Don's Mill., 3. p. 397.
Synonymes. F. Gárdeni Michx. Fl. Bor. Amer., 1. p. 313.; *Hamamèlis* monoica *Lin. ex Smith* in *Rees's Cycl.* vol. xvii.

Spec. Char., &c. See the generic character. The flowers, which are white and sweet-scented, appear before the leaves; the latter resembling those of the wych hazel. A low deciduous bush. North America, Virginia to Carolina, in shady woods on the sides of hills. Height 3 ft. to 6 ft. Introduced in 1765. Flowers white, sweet-scented; April and May.

Varieties. The following are very distinct:—

■ *F.* a. 1 *obtùsa* Sims Bot. Mag. t. 1341., Pursh Sept. 1. p. 335.; F. màjor *Bot. Cab.* t. 1520.; F. alnifòlia *Lin. fil. Supp.* 257.; and our *fig.* 910.; has obovate leaves, downy beneath.

■ *F.* a. 2 *acùta* Sims, Pursh Sept. 1. p. 335.; F. Gárdeni Jacq. Icon. rar. t. 100. (Bot. Cab., t. 1507.), has narrow leaves, nearly entire, white from down beneath.

■ *F.* a. 3 *màjor* Sims Bot. Mag. t. 1342., Pursh Sept. 1. p. 335. (Bot. Cab., t. 1520.; and our *fig.* 911.) has leaves ovate-oblong, somewhat cordate at the base, very black and serrated at the apex; when young, tomentose beneath.

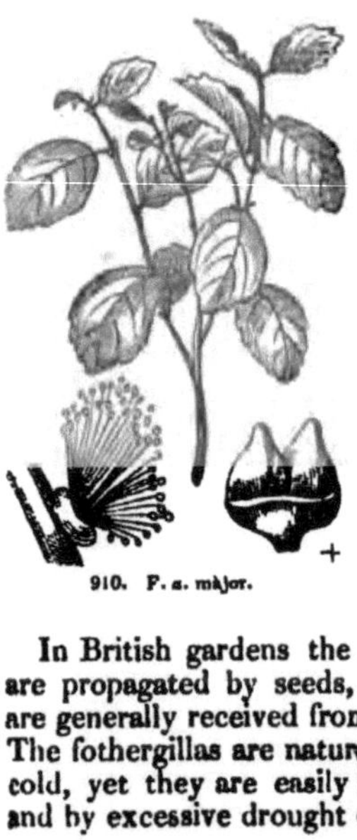

910. F. a. màjor.

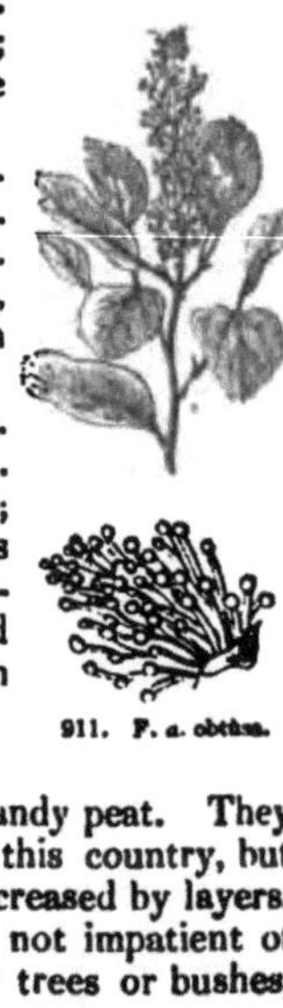

911. F. a. obtùsa.

In British gardens the fothergillas thrive best in moist sandy peat. They are propagated by seeds, which are sometimes ripened in this country, but are generally received from America. The varieties are increased by layers. The fothergillas are naturally somewhat tender, and though not impatient of cold, yet they are easily injured by the proximity of other trees or bushes, and by excessive drought or perpetual moisture.

ORDER XXXVIII. CORNA'CEÆ.

ORD. CHAR. *Calyx* 4-lobed. *Petals* 4; æstivation valvate. *Stamens* 4. *Style* filiform. *Stigma* simple. *Drupe* baccate, enclosing a 2-celled nut. *Seeds* solitary in the cells. *Albumen* fleshy. — Differs from Caprifoliàceæ tribe Sambùceæ, in the polypetalous corolla and drupaceous fruit. (*G. Don.*)

Leaves simple, opposite, rarely alternate, exstipulate, deciduous, or sub-evergreen; ovate or oval, entire. *Flowers* white or yellowish. — Shrubs or low trees; natives of Europe and Asia.

CO'RNUS *L.* Flowers in cymes. Stamens 4. Style 1. Pome baccate.

BENTHA'MIA Lindl. Flowers disposed in involucrated heads. Fruit constituted of many pomes grown together.

GENUS I.

CO'RNUS *L.* THE DOGWOOD. *Lin. Syst.* Tetrándria Monogýnia.

Identification. Tourn. Inst., 641. t. 410.; Dec. Prod., 4. p. 271.; Don's Mill., 3. p. 398.
Synonymes. Cornouiller, *Fr.*; Hartriegel, *Ger.*; Corniolo, *Ital.*
Derivation. From *cornu*, a horn; the wood being thought to be as hard and as durable as horn. Hartriegel signifies hard rail, or hard wood. The name of Dogwood is applied to this genus, because, as Parkinson says, in his *Paradisus*, the fruit of most of the species is not fit even for dogs; but it is more likely to have been given to it from the astringent properties of the bark and leaves, a decoction of which was formerly used as a wash for curing the mange, &c., in dogs.

Gen. Char. *Calyx* tube adhering to the ovarium; limb small, 4-toothed. *Petals* 4, oblong, sessile, valvate in æstivation. *Stamens* 4. *Style* 1. *Drupe* baccate, marked by the vestiges of the calyx, containing a 2-celled, rarely 3-celled nucleus. *Seeds* solitary, pendulous. (*Don's Mill.*)

Leaves simple, opposite, except in C. alternifòlia, exstipulate, deciduous; entire, feather-nerved. *Flowers* sometimes capitate and umbellate, involucrated; sometimes corymbose and panicled, without involucra. *Petals* white, rarely yellow. — Trees under the middle size, and shrubs, deciduous; natives of Europe, Asia, and North America.

Most of the species ripen their fruit in England; but they are usually propagated by suckers, or by layers or cuttings. The wood of all the species makes the very best charcoal. Common soil, and most of the species will thrive in the shade of other trees.

§ i. *Nudiflòræ* Dec.

Derivation. From *nudus*, naked, and *flos*, a flower; the inflorescence being without an involucre.

Sect. Char. Flowers corymbose or panicled, without an involucre.

A. *Leaves alternate.*

♣ ♈ 1. *C.* ALTERNIFO'LIA *L.* The alternate-leaved Dogwood.

Identification. Lin. fil. Suppl., p. 125.; L'Hérit. Corn., No. 11.; Don's Mill., 3. p. 398.
Synonyme. C. altérna *Marsh.*
Engravings. Guimp. Abb. Holz., t. 43.; Schmidt Baum., 2. t. 70.; and our *fig.* 912.

Spec. Char., &c. Leaves alternate, ovate, acute, hoary beneath. Corymbs depressed, spreading. Branches warted. Pomes purple, globose, about the size of a grain of pepper. Leaves on long petioles. Branches green or reddish brown. (*Don'*

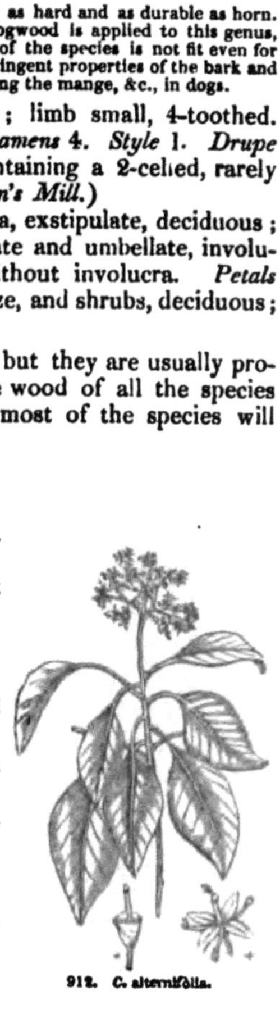

912. C. alternifòlia.

Mill.) A smooth deciduous shrub or low tree. North America, from Canada to Carolina, in shady woods on river banks. Height 15 ft. to 20 ft. Introduced in 1760. Flowers white; May to July. Fruit purple; ripe in October. Decaying leaves reddish yellow. Naked young wood greenish or reddish brown.

This species is easily known from every other, even at a distance, by the horizontal umbelliferous character assumed by the branches, which are also dichotomous, with clusters of leaves at the joints; and the general colour is that of a lively green. The leaves are generally alternate, but not unfrequently opposite.

B. *Leaves opposite.*

♣ 2. *C.* SANGUI'NEA *L.* The blood-*red-leaved, or common*, Dogwood.

Identification. Lin. Sp., p. 171.; Don's Mill., 3. p. 399.; Lodd. Cat., edit. 1836.
Synonymes. *C.* fœ'mina *Raii Syn.* 460.; Vírga sanguínea *Matth. Valgr.* 1. p. 236.; Female Cornel, Dogberry Tree, Hound Tree, Hound's-berry Tree, Prickwood, Gaten or Gatten Tree, Gater or Gatter Tree, Catteridge Tree, wild Cornel; Cornouiller sauvage, sanguin, or femelle, Puine or Bois punais, *Fr.*; rother Hartriegel, *Ger.*; Sanguinello, *Ital.*
Derivation. This species is called fœ'mina, and Female Cornel, because it bears fruit when very young; whereas Córnus más produces male blossoms only till the tree is 15 or 20 years old. Vírga sanguínea is literally the bloody twig, alluding to the colour of the shoots, though they are not nearly so red as those of Córnus álba. The names of Dogberry Tree, Hound Tree, &c., arise from the same source as Dogwood. (See above.) Prickwood alludes to the use of the wood for skewers; Gaten Tree is a corruption of *Gatr treow*, the Saxon name for this species; or, as some suppose, it is derived from *gayta*, the Spanish word for a pipe, the wood of this tree being more hollow, or full of pith, than that of *C.* más. Catteridge, and all the other somewhat similar names, are derived from Gaten. Chaucer calls the fruit Gaitres berries, evidently from the same origin. The French names of Puine, and Bois punais, bug-wood, are from the strong and unpleasant smell of the bark and leaves; and also because a decoction of them forms a wash to destroy bugs. Rother Hartriegel signifies red hard rail, or red hard wood.
Engravings. Eng. Bot., t. 249.; Fl. Dan., t. 481.; N. Du Ham., 2. t. 44.; and our *fig.* 913.

Spec. Char., &c. Bracteas straight. Leaves ovate, acute, smooth and green on both surfaces. Corymbs flat. Branches of a dark red when full grown. Leaves 2 to 3 in. long. Flowers greenish white, unpleasantly scented. Petals revolute at the sides. Fruit dark purple, and very bitter. (*Don's Mill.*) A large shrub. Europe and the North of Africa, in hedges and thickets, especially on a chalk and limestone soil; plentiful in Britain, in like situations; and also said to grow in North America, near the lakes of Canada and near New York; but it has probably been introduced there. Height 4 ft. to 15 ft. Flowers white; June. Fruit dark purple; ripe in August and September. Decaying leaves deep red. Naked young wood green.

913. C. sanguínea.

Varieties.

- ♣ C. *s.* 2 *Púrshii Don's Mill.* 3. p. 399.; *C.* sanguínea *Pursh, Schmidt Baum.* 2. t. 66.; has the flowers with yellow anthers, and the berries a dark brown. Lakes of Canada, and near New York; and only differs from the *C.* sanguínea of Europe in having the leaves pubescent, and in being of larger stature.
- ♣ C. *s.* 3 *fòliis variegàtis* Lodd. Cat. has the leaves variegated with white and yellow, and occasional streaks of red. A plant lately received into Messrs. Loddiges's collection, named *C.* candidíssima fol. var., appears, from the leaves, to be identical with this variety. *C.* candidíssima, in the same collection, appears from its leaves to be nothing more than *C.* sanguínea.

One of the commonest shrubs in old shrubberies; and easily known from all the other kinds of Córnus by the abundance of its dark purple fruit, and the intensely dark red of its leaves before they drop off in autumn. It is from this last circumstance, we suppose, that the specific name of sanguínea has been given to it, though it is much more obviously applicable to *C.* álba, on account of the redness of its shoots. *C.* purpùrea would be a much better name as contrasted with *C.* álba, both names applying to the fruit. The wood, which is hard, though not nearly so much so as that of *Córnus más*, was formerly used for mill-cogs, and for various purposes in rustic carpentry; and it still makes excellent skewers for butchers, toothpicks, and similar articles. The bark tastes like apples.

♠ 3. *C.* A'LBA *L.* The white-*fruited* Dogwood.

Identification. Lin. Mant., p. 40.; Don's Mill., 3. p. 399.
Synonymes. C. stolonifera *Michx. Fl. Bor. Amer.* 1. p. 109.; *C.* tatárica *Mill. Icon.* t. 104.
Engravings. Pall. Fl. Ross., 1. t. 34.; and our *fig.* 914.

Spec. Char., &c. Branches recurved. Branchlets glabrous. Leaves ovate, acute, pubescent, hoary beneath. Corymbs depressed. Branches of a fine red colour. Fruit white, or bluish white. (*Don's Mill.*) A large shrub. Siberia, at the rivers Oby and Irtysch, among bushes, &c.; North America, from Virginia to Canada, on the banks of rivers and lakes; and also in North California. Height 4 ft. to 10 ft. Introduced in 1741. Flowers white; May to July. Fruit white or bluish white; ripe in September. Decaying leaves red or reddish yellow. Naked young wood intensely red or coral colour.

914. C. álba.

Varieties.

♠ C. *a.* 2 *circinàta* Don's Mill. iii. p. 399., *C.* circinàta *Cham. et Schlecht.* in *Linnæa* iii. p. 139., has the berries of a lead colour. Throughout Canada, and from Lake Huron to lat. 69° N.

♠ C. *a.* 3 *sibírica* Lodd. Cat., ed. 1836, has the shoots of a fine orange red, covered with a delicate bloom. It makes a splendid appearance in the winter season.

Interesting in summer, from its fine large leaves and white flowers; in autumn, from its white fruit, which are about the size and colour of those of the mistletoe; and in the winter and spring, from the fine red of its young shoots.

♠ 4. *C.* (A.) STRI'CTA *Lam.* The straight-*branched* Dogwood.

Identification. Lam. Dict., 3. p. 116.; Don's Mill., 3. p. 399.
Synonymes. C. fastigiàta *Michx. Fl. Bor. Amer.* 1. p. 92.; *C.* sanguínea *Walt.*, but not of Lin.; *C.* cyanocárpos *Gmel. Syst. Veg.* 1. p. 267.; *C.* canadénsis *Hort. Par.*; *C.* cærùlea *Meerb. Icon.* 3., but not of Lam.
Engravings. Schmidt Baum., 2. t. 67. and our *figs.* 915, 916.

915. C. (a.) stricta.

Spec. Char., &c. Branches straight, fastigiate. Leaves ovate, acuminated, glabrous, green on both surfaces; when young, hardly pubescent beneath. Corymbs convex, somewhat panicled. Branches reddish brown. Anthers blue. Pomes globose, soft, blue on the outside, but white inside. (*Don's Mill.*) An upright shrub. North America, from Carolina to Canada, frequent on the banks of rivers; also in Mexico, between Tampico and Real del Monte. Height 6 ft. to 10 ft. rarely 20 ft. Introd. 1758. Flowers white; June and July. Fruit blue without and white within; ripe in October. Decaying leaves reddish green. Naked young wood green, or rusty green.

916. C. (a.) stricta.

Varieties.

■ C. (*a.*) *s.* 2 *asperifòlia*, *C.* asperifòlia *Lodd. Cat.* ed. 1836, if not identical with the species, differs from it but very slightly.

■ C. (*a.*) *s.* 3 *sempervìrens*, *C.* sempervìrens *Lodd. Cat.* ed. 1836, closely resembles the species, but differs from it in retaining its leaves throughout a part of the winter.

■ ♈ 5. *C.* (A.) PANICULA'TA *L'Hérit.* The panicled-*flowering* Dogwood.

Identification. L'Hérit. Corn., No. 10. t. 5. ; Don's Mill., 3. p. 398.
Synonymes. C. racemòsa *Lam. Dict.* 2. p. 116. ; *C.* fœ'mina *Mill. Dict.* No. 4. ; *C.* citrifòlia *Hort. Par.*
Engravings. Schmidt Baum., 2, t. 68. ; and our *fig.* 917.

917. C. (a.) paniculàta.

Spec. Char., &c. Branches erect. Leaves ovate, acuminated, glabrous, hoary beneath. Corymb thyrsoid. Ovarium silky. Branches pale purplish. Pomes roundish, depressed, watery, white, 3 lines in diameter. The dots on the under side of the leaves, which are only seen through a lens, bear bicuspidate short, adpressed hairs. Tube of calyx pubescent (*Don's Mill.*) A large shrub. Canada to Carolina, in swamps and near rivulets, among other bushes. Height 4 ft. to 6 ft. in America ; 20 ft. to 25 ft. in cultivation. Introduced in 1758. Flowers white ; July and August. Fruit white ; ripe in October. Decaying leaves reddish brown. Naked young wood purplish.

Varieties.

■ C. *p.* 2 *álbida* Ehrh. Beitr. iv. p. 16. — Leaves elliptic-lanceolate.

■ C. *p.* 3 *radiàta* Pursh Fl. Amer. Sept. i. p. 109. — Corymbs sterile, foliiferous.

■ 6. *C.* (A.) SERI'CEA *L'Hérit.* The silky Dogwood.

Identification. L'Hérit. Corn., No. 6. t. 2. ; Don's Mill., 3. p. 399.
Synonymes. C. lanuginòsa *Michx. Fl. Bor. Amer.* 1. p. 92. ; *C.* álba *Walt. Fl. Car.* 88., but not of Lin. ; *C.* cærùlea *Lam. Dict.* 2. p. 116. ; *C.* Amòmum *Du Roi Harbk.* 1. p. 165. ; *C.* rubiginòsa *Ehrh. Beitr.* 4. p. 15. ; *C.* ferruginea *Hort. Par.* ; *C.* candidissima *Mill.* ; C. cyanocárpos *Mœnch*, but not of Gmel.
Engravings. Schmidt Baum., 2. t. 64. ; and our *fig.* 918.

918. C. (a.) sericea.

Spec. Char., &c. Branches spreading. Branchlets woolly. Leaves ovate, acuminated, clothed with rusty pubescence beneath. Corymbs depressed, woolly. Pomes bright blue. Nut compressed. (*Don's Mill.*) A large shrub. Canada to Carolina, in swampy woods and on river banks. Height 5 ft. to 8 ft. Introduced in 1683. Flowers white; June and July. Fruit bright blue ; ripe in October. Decaying leaves rusty brown. Naked young wood brown and green.

Variety.

■ C. (*a.*) *s.* 2 *oblongifòlia* Dec. Prod. iv. p. 272., *C.* oblongifòlia *Rafin in Litt.*, has leaves oblong and glabrous above.

This sort is very distinct from the two preceding ones, and comes nearer, in general appearance, to *C.* álba than they do ; but it is a weaker plant, and smaller in all its parts than that species. The two preceding sorts, *C.* (a.)

stricta and *C.* (a.) paniculàta, have much narrower leaves, and a more compact fastigiate habit of growth, than any other species or variety of the genus. *C.* (a.) paniculàta is the handsomest of the three sorts for a small garden, as it is easily kept of a small size, and in a neat shape, and it flowers profusely.

♈ 7. *C.* (A.) CIRCINA'TA *L'Hérit.* The rounded-*leaved* Dogwood.

Identification. L'Hérit. Corn., p. 7. No. 8. t. 3.; Don's Mill., 3. p. 399.
Synonymes. C. tomentòsa *Michx. Fl. Bor. Amer.* 1. p. 91.; *C.* rugòsa *Lam. Dict.* 2. p. 115.; *C.* virginiàna *Hort. Par.*
Engravings. Schmidt Baum., 2. t. 69.; and our *fig.* 919.

Spec. Char., &c. Branches warted. Leaves broadly oval, acuminated, clothed with hoary tomentum beneath. Corymbs depressed, spreading. Branches slightly tinged with red. Leaves broad, waved on their edges. Flowers white, as in most of the species. Pomes globose, at first blue, but at length becoming white. (*Don's Mill.*) A large shrub. North America, from Canada to Virginia, on the banks of rivers; and probably of California. Height 5 ft. to 10 ft. Introduced in 1784. Flowers white; June and July. Fruit at first blue, and then turning white; ripe in October. Decaying leaves rusty brown. Naked young wood green, tinged with red.

919. C. (a.) circinàta.

Readily distinguished from all the other sorts, by its broader leaves, and its rough warted branches.

♈ 8. *C.* OBLO'NGA *Wall.* The oblong-*leaved* Dogwood.

Identification. Wall. in Roxb. Fl. Ind., 1. p. 432.; Don's Mill., 3. p. 398.
Synonyme. C. paniculàta *Hamilt. ex D. Don Prod. Fl. Nep.* p. 140.
Engraving. Our *fig.* 920. from a specimen in Dr. Lindley's herbarium.

Spec. Char., &c. Leaves oblong, acuminated, acute at the base, glaucous, and rather scabrous beneath, with many excavated glands along the axils of the ribs and nerves. Corymbs spreading, panicled. Young shoots clothed with short adpressed hair. Leaves 4 in. to 6 in. long, and 1 in. to 1½ in. broad. Petioles about an inch long. Flowers white or pale purplish, fragrant. Calyx clothed with adpressed silvery hairs, as well as the pedicels and petals. Ovarium 3-celled. Pome ovate-oblong. (*Don's Mill.*) A large shrub. Nepal, about Narainhetty, Katmandu, and the Valley of Dhoon. Height 10 ft. to 15 ft. Introduced in 1818. Flowers white or purplish, fragrant.

920. C. oblónga.

§ ii. *Involucràtæ* Dec.

Derivation. From *involucrum*, an involucre, with which the heads of flowers are severally surrounded.

Sect. Char. Flowers disposed in heads or umbels, surrounded by coloured involucres, which are usually composed of 4 leaves. (*Dec. Prod.*) Trees, with yellow umbelled flowers.

♈ 9. *C.* MA'S *L.* The male Dogwood, *the Cornel, or Cornelian Cherry Tree.*

Identification. Lin. Sp., 171.; Don's Mill., 3. p. 400.; Lodd. Cat., ed. 1836.
Synonymes. C. máscula *L'Hérit. Corn.* No. 4.; Long Cherry Tree; Cornella; Cornouiller mâle, Cornes, Corneilles, *Fr.*; Kornel Kirsche Hartriegel, *Ger.*; Corgnolo, *Ital.*
Derivation. The name of *más* has been applied to this species since the days of Theophrastus; in all probability, because young plants are barren for many years after they show flowers; these

flowers being furnished with stamens only. For an opposite reason, the name of *Córnus fœ'mina* was given to *C. sanguínea*. (See p. 502.) The name of Cornelian Cherry relates to the beautiful colour of the fruit, which resembles that of a cornelian.

Engravings. Black., t. 121.; the plate in Arb. Brit., 1st edit., vol. vi.; and our *figs.* 921. and 922.

921. C. más.

Spec. Char., &c. Branches smoothish. Leaves oval, acuminated, rather pubescent on both surfaces. Flowers protruded before the leaves. Umbels about equal in length to the 4-leaved involucre. Flowers yellow. Fruit elliptic, of a bright shining scarlet colour, the size and form of a small olive or acorn, very styptic in its immature state. (*Don's Mill.*) A large shrub or low tree. Europe, Britain excepted, and in the North of Asia, in hedges and among bushes. Height 12 ft. to 20 ft. Introduced in 1596. Flowers yellow; February to April. Fruit cornelian-coloured; ripe in October and November. Decaying leaves reddish green. Naked young wood brownish green.

Varieties.

- ♈ C. *m.* 2 *frúctu cèræ colòris* N. Du Ham. ii. p. 162. has the fruit of a wax colour.
- ♈ C. *m.* 3 *variegàtus* has the leaves edged with white or yellow.

922. Córnus más.

The wood has been, in all ages, celebrated for its hardness and durability; and it is at the same time tough and flexible. In a dry state, it weighs 69 lb. 5 oz. to the cubic foot. The small branches are said to make the most durable spokes for ladders; wooden forks for turning the grain on barn floors, and for making hay; hoops, butchers' skewers, and toothpicks. The wooden forks are made by selecting branches which divide into three near the extremity; and, after cutting the branch to a proper length, which is commonly about 5 or 6 feet, the bark is taken off, and the three branches which are to form the prongs are bent so as to form a triangle, like the wooden corn forks of England. In this state they are put into a hot oven, where they are kept till they are hardened, so as to retain the shape given to them. Similar hay and straw forks are made of the nettle tree in France, and of the willow in various parts of England, by the same procedure. The fruit, when thoroughly ripe, is somewhat sweet, and not disagreeable to eat; and, on the Continent, it is frequently used in confectionery, and for making marmalades. As an ornamental tree, the cornel is valuable, not only on account of its early flowering, and the fine display made by its ripe fruit, but because it is a low tree, never growing out of bounds, and one which, after it has attained the height of 10 or 12 feet, is of slow growth, and of very great duration. For these last reasons, it is particularly suitable for small suburban gardens, in which it will form a fit associate for small trees of *Cratæ'gus*, *Bérberis*, *Rhámnus*, *Euónymus*, *Hamamèlis*, &c. Seeds; but layers or suckers come much sooner into a flowering state. There are remarkably fine specimens of this tree in the old French gardens in the neighbourhood of Paris, and also in the old gardens of Germany.

♀ 10. *C.* FLO'RIDA *L.* The Florida Dogwood.

Identification. Lin. Sp., 1661.; Don's Mill., 3. p. 400.
Synonyme. Virginian Dogwood.
Engravings. Bot. Mag., t. 526.; Schmidt Baum., 2. t. 52.; and our *fig.* 923.

Spec. Char., &c. Branches shining. Leaves ovate, acuminated, pale beneath, beset with adpressed hairs on both surfaces. Flowers umbellate, protruded after the leaves. Leaves of involucre large, roundish, retuse, or nearly obcordate. Pomes ovate. Leaves of involucre white. Flowers greenish yellow, and very large. Pomes scarlet, about half the size of those of *C.* más; ripe in August. (*Don's Mill.*) A large shrub or low tree. Carolina to Canada, in woods; and on the banks of the Columbia, near its confluence with the sea. Height 20 ft. to 30 ft. Introduced in 1731. Flowers large, yellowish white; April and May. Fruit scarlet; ripe in August. Decaying leaves yellowish green. Naked young wood brownish green.

923. C. flórida.

Córnus flórida is universally allowed to be the handsomest species of the genus. It thrives best in a peat soil, which must be kept moist; and the situation should be sheltered, though the foliage of the plants must be fully exposed to the influence of the sun, otherwise they will not flower. Cuttings or layers, both of which readily strike root.

Córnus grándis Schlect. A small tree or shrub. Chico, in ravines. Introduced in 1838 by Hartweg, and probably hardy. "It has a beautiful foliage; the leaves being from 3 in. to 5 in. long, smooth and deep green above, hoary with down on the under side." The flowers are in small heads, and the fruit as large as a sloe, and purplish black, covered with bloom. (*Bot. Reg. Chron.*, 1839.)

C. officinàlis, a native of Japan, is figured by Sieboldt (t. 50.), and will probably prove hardy.

GENUS II.

BENTHA'MIA Lindl. THE BENTHAMIA. *Lin. Syst.* Tetrándria Monogýnia.

Identification. Lindl. in Bot. Reg., t. 1579.
Synonyme. Córnus sp. *Wall.*, *Dec.*, and *G. Don.*
Derivation. Named in honour of *George Bentham*, Esq., F.L.S., Secretary to the Horticultural Society; and nephew of the celebrated moralist and jurist, Jeremy Bentham.

Gen. Char. *Flowers* disposed in heads, each head attended by an involucre which consists of 4 petal-like parts, and resembles a corolla. *Calyx* with a minute 4-toothed limb. *Petals* 4, fleshy, wedge-shaped. *Stamens* 4. *Style* 1. *Fruit* constituted of many pomes grown together; endocarp in each pome with 2 cells. *Seeds* solitary and pendulous in each cell. (*Lindl.*)

Leaves simple, opposite, exstipulate, sub-evergreen : entire. *Flowers* large, white. — A large shrub or low tree. Himalayas. Culture as in Córnus.

♆ 1. B. FRAGI'FERA *Lindl.* The Strawberry-bearing Benthamia.

Identification. Hort. Trans., 2d series, 1. p. 458. ; Lindl. in Bot. Reg., t. 1579.
Synonymes. C. capitàta *Wall.* in Roxb. Fl. Ind. 1. p. 434., Don's Mill. 3. p. 399. ; Chung-wa, *in Nepal* ; Bhumowro, *in Serampore*.
Engravings. Bot. Reg., t. 1579. ; Hort. Trans., 2d series, 1. t. 17. ; and our *fig.* 924.

924. B. fragifera.

Spec. Char., &c. Branches spreading, smooth. Leaves lanceolate, acuminated at both ends, on short petioles, rather rough with small adpressed down. Flowers sessile, densely aggregate, forming a round head girded by a 4-leaved scabrous involucrum. (*Don's Mill.*) A large sub-evergreen shrub or low tree. Nepal, in Gossainthan. Height 10 ft. to 15. ft. Introduced in 1825. Flowers large, yellowish white; June. Fruit large, about the size of that of the common arbutus, reddish; ripe in October; yellowish white within, not unpleasant to eat Young shoots whitish green.

Rather tender in the climate of London, though it forms a fine evergreen in some parts of Cornwall : perhaps it might be rendered hardier by grafting it on Córnus sanguínea. Readily propagated either from seeds or cuttings, and of easy culture in loamy soil, kept moist.

Benthàmia japónica is figured by Sieboldt (t. 16.), and is probably hardy.

ORDER XXXIX. LORANTHA'CEÆ.

ORD. CHAR. *Calyx* calyculate, entire or lobed. *Petals* 4—8, distinct or cohering ; æstivation valvate. *Stamens* 4—8, opposite the petals, or more or less adnate to them. *Style* absent or present. *Stigma* capitate. *Berry* 1-celled, 1-seeded, crowned by the calyx. *Albumen* fleshy. *Flowers* generally unisexual. The habit, and the stamens being opposite the petals, distinguish this from Caprifoliàceæ and Cornàceæ. (*G. Don.*)
Leaves simple, opposite, exstipulate, evergreen; entire. *Flowers* small, whitish, or purplish.—Shrubs, parasitical or terrestrial. Europe and Japan.

The hardy ligneous plants are included in the genera *V*íscum, Loránthus, and *Aúcuba.*

VI'SCUM L. Flowers diœcious or monœcious. Margin of the calyx obsolete. Petals usually 4, connected at the base in the male flowers, but free in the female. Berry globular, viscid, of 1 cell.
LORA'NTHUS *L.* Flowers diœcious or hermaphrodite. Margin of calyx entire. Petals 5—6, linear, reflexed.
AU'CUBA Thunb. Flowers diœcious. Calyx 5-toothed. Petals 4, ovate lanceolate. Fruit fleshy, 1-seeded.

GENUS I.

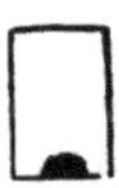

VI'SCUM L. THE MISTLETOE. *Lin. Syst.* Monœ'cia, or Diœ'cia, Tetrándria.

Identification. Tourn. Inst., p. 609. ; Dec. Prod., 4. p. 277. ; Don's Mill., 3. p. 402.

Synonymes. Misseldine, Gui, or Guy, *Fr.*; Mistl, or Missel, *Ger.*; Visco, or Vischio, *Ital.*; Legamodoga, *Span.*
Derivation. *Viscus*, or *viscum*, is the Latin for birdlime, which is made from the berries; and Mistletoe is by some supposed to be derived from *mist*, the German word for dung, or slimy dirt, and by others from *misteltа*, the Saxon name for the plant.

Gen. Char. *Calyx* a slight border in the male flowers, more evident in the female. *Corolla* in the male flowers gamopetalous, in 4 deep, ovate, acute, equal divisions; in the female flower of 4 ovate, equal, deciduous petals. *Anthers* in the female flowers none; in the male flower 4, compressed. *Ovarium* ovate. *Stigma* sessile. *Berry* globular. (*Don's Mill.*)

Leaves simple, opposite, rarely alternate, exstipulate, evergreen; undivided, entire, rigid. *Flowers* in fascicles or spikes, greenish. *Berries* white. —An evergreen shrub, parasitical on trees. Europe; in Britain, England.

£ 1. *V.* A'LBUM *L.* The white-*fruited, or common,* Mistletoe.

Identification. Lin. Sp., 1451.; Dec. Prod., 4. p. 277.; Don's Mill., 3. p. 403.
Engravings. N. Du Ham., 1. t. 115.; Eng. Bot., t. 1470.; Baxt. Brit. Pl., t. 40.; and our *fig.* 925., which exhibits a portion of a male plant, marked *m*, and of a female plant in fruit, marked *f*.

Spec. Char., &c. Stem much branched, forked; with sessile intermediate heads, of about 5 flowers. Branches terete. Leaves obovate-lanceolate, obtuse, nerveless. (*Don's Mill.*) A parasitical shrub, forming a confused tuft of branches with a yellowish green aspect. Europe and England, on trunks and branches of trees, most frequent on *Rosàceæ*. Height 2 ft. to 3 ft. Flowers greenish yellow; May. Berry white; ripe in December.

925. *V. álbum.*

The leaves vary considerably in different plants, as may be seen in *fig.* 926., which contains engravings of three different specimens. The durability of the plant is very great; for, when once established on a tree, it is seldom known to cease growing while the tree is in life; but, when it dies, or the branch on which it is rooted decays, or becomes diseased, the death of the mistletoe immediately follows. The trees on which the mistletoe grows belong to various natural orders; and, indeed, it would be difficult to say on what dicotyledonous trees it will not grow. In England, it is found on *Tiliàceæ*, *Aceràceæ*, *Rosàceæ*, *Cupulíferæ*, *Salicàceæ*, *Oleàceæ*, and, we believe, also on *Coníferæ*. It is found on the oak at Eastnor Castle (see *Gard. Mag.*, vol. xiii. p. 206.); and in the neighbourhood of Magdeburg we saw it growing in immense quantities on *Pìnus sylvéstris* in 1814. In France, it grows on trees of all the natural orders mentioned, but least frequently on the oak. It does not grow on the olive in France, though it abounds on the almond. In Spain, it grows on the olive; as it does in the neighbourhood of Jerusalem; and, in the latter locality, is found the variety with red fruit, which is perhaps a Loránthus.

The mistletoe is propagated by the berries being, by some means or other, made to adhere to the bark of a living tree. The common agency by which this is effected is supposed to be that of birds; and more especially of the missel thrush, which, after having satisfied itself by eating the berries, wipes off such of them as may adhere to the outer part of its beak, by rubbing it against the branch of the tree on which it has alighted; and some of the seeds are thus left sticking to the bark. If the bark should be smooth, and not much indurated, the seeds will germinate, and root into it the following spring; that is, supposing them to have been properly fecundated by the proximity of a male plant to the female one which produced them. The first indication of germination is the appearance of one or more radicles, like the sucker of a house fly, but larger; as at *h i*, in *fig.* 926., which are front views, and at *k l* in the same figure, which are side views, taken from mistletoe berries, which were stuck on the upright trunk of a cherry tree in our garden at Bayswater, in March, 1836, and germinated there, as they appeared on the 20th of May of the same year. When the white, viscous, pulpy matter of the mistletoe berry is removed, the kernel, or seed, appears of a greenish colour, and

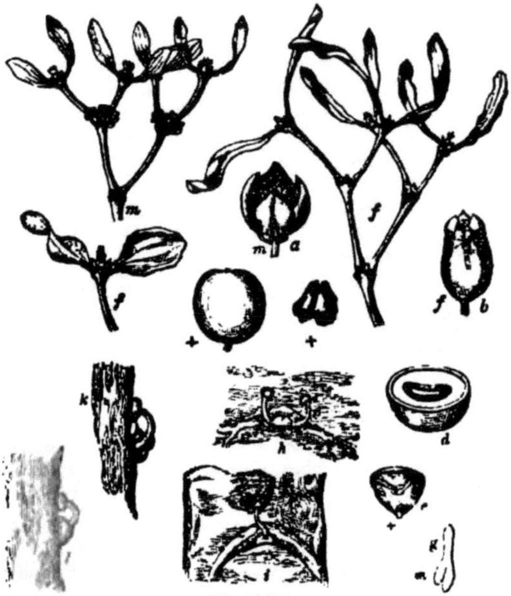

926. *V. album.*

flat; sometimes oval, at other times triangular, and at other times of various forms. In *fig.* 926., *a* is the male blossom magnified; *b*, the female blossom magnified; *d*, a berry cut through, transversely; *e*, a seed divided vertically, showing the two embryos; *g*, the embryo magnified; *h*, the two embryos, with the two radicles germinating; *i*, a single radicle; *k*, a side view, or section, of the two radicles; and *l*, a side view, or section, of the single radicle. Our mistletoes, at Bayswater, are now (Aug. 1840) from 4 in. to 8 in. in length, with 4 or 5 pairs of leaves, after having been four years sown. Mr. Moss, a nurseryman at Malvern, has succeeded in grafting the mistletoe standard high on young apple and pear trees, and also on poplars and willows. The grafts should be made in the first or second week in May; and they should never be lower than 5 ft. from the ground, or higher than 10 ft. Where the stock is not more than ½ in. in diameter, an incision is made in the bark, into which a scion of mistletoe, pared thin, is inserted, having a bud and a leaf at the upper end. In grafting longer pieces, a notch should be cut out of the stock; an incision made below the notch; and a shoulder left on the scion to rest on the notch, in the manner of crown grafting. In every case, there must be a joint on the lower extremity of the scion. The mistletoe may also be propagated by budding, taking care to have a heel of wood and a joint at the lower extremity of the bud. (See *Gard. Mag.*, vol. xiii. pp. 206. and 285.)

Genus II.

LORA'NTHUS *L.* The Loranthus. *Lin. Syst.* Pentahexándria Monogýnia.

Identification. Lin. Sp., 1672.; Dec. Prod., 4. p. 671.; Don's Mill., 3. p. 409.
Derivation. From *lorum*, a lash made of leather, and *anthos*, a flower; alluding to the long linear shape and leathery substance of the petals.

Gen. Char. *Flowers* diœcious or hermaphrodite. *Calyx* cup-shaped, adnate, with an entire border. *Petals* 5—6, linear, reflexed. *Stamens* inserted into the middle of the petals. *Filaments* short. *Anthers* globose. *Style* thickish. *Stigma* simple. *Berry* globose, 1-celled, 1-seeded. (*Don's Mill.*)

Leaves simple, opposite or nearly so, exstipulate, evergreen, entire, rigid. *Flowers* in spikes, axillary and terminal. — An evergreen shrub, parasitical on trees; native of Austria.

£ 1. L. EUROPÆ'US *L.* The European Loranthus.

Identification. Lin. Sp., 1672.; Jacq. Fl. Austr., t. 30.; Don's Mill., 3. p. 409.
Engravings. Jacq. l. c.; our *fig.* 927., and our *fig.* 928. of the natural size

927. L. europæ'us.

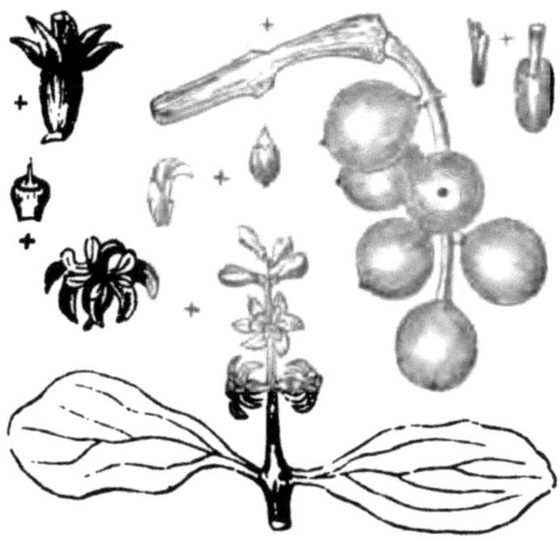

928. L. europæ'us.

Spec. Char., &c. Plant glabrous, much branched. Branches terete. Leaves opposite, petiolate, oval-oblong. Racemes terminal, simple. Flowers diœcious, of 6 petals. (*Don's Mill.*) An evergreen parasitical shrub, with the habit of *V*ìscum álbum. Austria, Hungary, Italy, and Upper Siberia, on oak and sweet chestnut trees. Height 2 ft. to 3 ft. Introduced in 1839. Flowers greenish; May. Berries oval, white or yellowish; ripe in December.

Berries of this plant were received from M. Charles Rauch of Vienna, by his brother, M. Francis Rauch, and sown on trees in the Horticultural Society's Garden, and also in our garden, at Bayswater, in January, 1839, though they have not yet vegetated.

GENUS III.

AU'CUBA Thunb. THE AUCUBA. *Lin. Syst.* Diœ'cia Tetrándria.

Identification. Thunb. Fl. Jap., p. 4.; Dec. Prod., 4. p. 274.; Don's Mill., 3. p. 433.
Synonymes. *Aúcuba* Kæmpf. Amœn. 5. p. 775.; Eùbasis *Salisb. Prod.* p. 68.
Derivation. *Aucuba* is the Japanese name.

Gen. Char. *Flowers* diœcious. *Calyx* closely adhering, with the margin a little elevated, and 4-toothed; teeth obtuse, very short. *Petals* 4, deciduous, alternating with the calycine teeth, inserted in the margin of the elevated fleshy 4-angled disk. *Stamens* 4. *Ovarium* cylindrical. *Style* very short, thick, terete. *Berry* fleshy, 1-seeded. (*Don's Mill.*)

Leaves simple, opposite, exstipulate, evergreen; petiolate, coriaceous. *Flowers* small, panicled.

An evergreen shrub or low tree; Japan; with dichotomous or verticillate branches, in the manner of those of Loránthus and *V*ìscum. The male blossom is unknown. Only the female state of this plant is in British gardens.

■ 1. *A.* JAPO'NICA Thunb. The Japan Aucuba.

Identification. Thunb. Fl. Jap., p. 64.; Dec. Prod., 4. p. 274; Don's Mill., 3. p. 433.
Synonymes. Eùbasis dichótomus *Salisb. Prod.* p. 68.; spotted-leaved Laurel, Japan Laurel.
Engravings. Thunb. Icon. Fl. Jap., t. 12. and 13.; Bot. Mag., t. 1197.; and our *fig.* 929.

Spec. Char., &c. Leaves ovate-lanceolate, acuminated, toothed, coriaceous, glabrous, shining, pale green; beautifully spotted with yellow, having the midrib rather prominent, the rest of the leaf reticulately veined. Petioles articulated with the branches, and dilated at the base. (*Don's Mill.*) An evergreen shrub. Japan. Height 6 ft. to 10 ft., and probably much higher in warm sheltered situations. Introduced in 1783. Flowers dark blood-coloured; May and July. Berries red; ripe in March; not yet seen in England.

929. *A. japónica.*

As hardy as, or hardier than, the common laurel; and, what is a very valuable property in England, it will endure coal smoke better than almost any other evergreen. It is readily propagated by cuttings; and grows freely in any soil tolerably dry, advancing steadily by shoots of from 6 in. to 9 in. long every season.

ORDER XL. CAPRIFOLIA'CEÆ.

ORD. CHAR. *Calyx* 5-lobed. *Corolla* monopetalous; tube short; limb 5-lobed; æstivation valvate. *Stamens* 5, adnate to the corolla, and alternating with its lobes. *Ovarium* 3-celled. *Style* exserted. *Stigmas* 3, distinct, or combined. *Berry* pulpy, rarely dry, crowned by the calyx, 1 or many celled. *Seeds* solitary, twin or numerous in the cells. *Albumen* fleshy. (*G. Don.*)

Leaves simple, or compound, generally exstipulate, deciduous, or evergreen. *Flowers* terminal, corymbose, or axillary. — Shrubs or low trees, natives of Europe, North America, and Asia, which may be arranged in two sections as under: —

Sect. I. SAMBU'CEÆ *Humb. et Kth.*

Sect. Char. Corolla monopetalous, regular, rotate, with 5 segments only connected a little at the base; rarely tubular. Style wanting. Stigmas 3, sessile.

SAMBU'CUS *Tourn.* Corolla rotate, urceolate. Berry roundish, pulpy, 1-celled, 3—4-seeded, hardly crowned.

VIBU'RNUM *L.* Corolla rotate, subcampanulate, and tubular. Berry 1-seeded, crowned by the calycine teeth. Low deciduous trees or shrubs; partly evergreen.

Sect. II. LONICE'REÆ Brown.

Sect. Char. Corolla monopetalous, more or less tubular, usually irregular Style filiform, crowned by three distinct or concrete stigmas.

DIERVI'LLA Tourn. Calyx bibracteate at the base, 5-parted. Corolla 3-cleft. Capsule 1-celled, many-seeded. A low deciduous shrub.

LONI'CERA Desf. Tube of calyx 5-toothed. Corolla 5-parted, usually irregular. Berry 3-celled. Cells few-seeded. Deciduous and evergreen shrubs, many of them twiners.

SYMPHORICA'RPOS *G. Don.* (Symphòria *Pers.*) Calyx 4—5-toothed. Corolla almost regular. Berry crowned by the calyx, 4-celled, two of them empty, and the other two containing 1 seed each Deciduous shrubs.

LEYCESTE'RIA Wall. Calyx 5-parted, irregular. Corolla 5-parted. Berry 5-celled, crowned by the calyx. Cells many-seeded. A rambling sub-evergreen shrub.

Sect. I. SAMBU'CEÆ.

GENUS I.

SAMBU'CUS *Tourn.* THE ELDER. *Lin. Syst.* Pentándria Trigýnia.

Identification. Tourn. Inst., 376.; Dec. Prod., 4. p. 321.; Don's Mill., 3. p. 436.
Synonyme. Phyteùma *Lour. Coch.* p. 138., but not of Lin.
Derivation. From *sambukē*, which the Latins have changed to *sambuca*, a musical instrument which is believed to have been made of elder wood.

Gen. Char., &c. *Calyx* small, but divided into 5 deep segments, permanent. *Corolla* rotate, urceolar, 5-lobed; lobes obtuse. *Stamens* 5, about the length of the corolla. *Filaments* awl-shaped. *Anthers* roundish, and heart-shaped. *Style* none. *Stigmas* 3, obtuse. *Berry* globular, pulpy, of 1 cell, containing 3—5 seeds, which are convex on the outside, and angular inside. (*Don's Mill.*)

Leaves compound, opposite, bistipulate, deciduous; stalked: leaflets toothed, pinnate, or jagged, often biglandular at the base. *Flowers* white or purplish, in terminal cymes, which are in some flat, and in others thyrsoid. *Berries* purplish, cathartic.—Those plants of the genus which have pinnate or jagged leaflets, are not true species, but only varieties; all the true species having only toothed leaflets.

Trees, low, deciduous, natives of Europe and North America; ornamental for their compound leaves, and large terminal cymes of flowers; which are succeeded by purplish, red, white, or green berries, from which a wine is made. All the species are of easy culture, in good soil, rather moist and loamy; and they are all readily propagated by cuttings.

A. *Leaves pinnate. Flowers cymose or corymbose.*

1. *S.* NI'GRA *L.* The *common, or black-fruited,* Elder.

Identification. Lin. Sp., 385.; Don's Mill., 3. p. 437.

930. Sambucus nigra.

Synonymes. Bourtry, or Bour Tree, Arntree, *Scotch*; Sureau, *Fr.*; Hollander, *Ger.*; Sambuco *Ital.*; Sauco, and Sabuco, *Span.*; Flaeder, *Swed.*; Hylde, *Dan.*

Engravings. N. Du Ham., 1. t. 55.; Eng. Bot., t. 476.; the plate of this species in Arb. Brit., 1st edit., vol. vi.; and our *figs.* 930. and 931.

931. *S. nigra.*

Spec. Char., &c. Arboreous. Leaves pinnate. Leaflets usually 5, smooth, deep green, ovate or oblong-oval, acuminated; the lower leaves sometimes trifoliolate. Cymes with 5 main branches. Branches, after a year's growth, clothed with smooth grey bark, and filled with a light spongy pith. Flowers cream-coloured, with a sweet but faint smell. Berries globular, purplish black. Stalks reddish. (*Don's Mill.*) A low tree. Europe, and part of Asia, in hedges, coppices, and woods; and plentiful in Britain, in like situations. Height 20 ft. to 30 ft. Flowers cream-coloured; June. Berries purplish black; ripe in September. Decaying leaves yellowish green. Naked young wood whitish green.

Varieties.

- ♀ S. *n.* 2 *viréscens* Dec. Prod. iv. p. 322.; *S.* viréscens *Desf. Arbr. Fr.* i. p. 348. — Fruit yellowish green.
- ♀ S. *n.* 3 *leucocárpa.* — Fruit white.
- ♀ S. *n.* 4 *laciniàta*; *S.* laciniàta *Mill. Dict.* No. 2. (Lob. Icon., 2. t. 164. f. 2.; and our *fig.* 932); the Parsley-leaved Elder; has the leaflets cut into fine segments. The handsomest of all the varieties.

932. *S. n. laciniàta.*

933. *S. n. fòliis argénteis.*

- ♀ S. *n.* 5 *rotundifòlia.* — Leaves trifoliolate. Leaflets petiolate, roundish, serrated. Corymbs few-flowered. Cultivated in the Chelsea Garden.
- ♀ S. *n.* 6 *monstròsa*, *S.* monstròsa *Hort.*, has the branches striped. Flowers of from 5—15 parts; and with from 5—15 stamens. Stigmas 5—12. Berries irregular.
- ♀ S. *n.* 7 *fòliis argénteis* (*fig.* 933.) has the leaves variegated with white, and forms a striking and lively-looking plant in a shrubbery.
- ♀ S. *n.* 8 *fòliis lùteis* has the leaves slightly variegated with yellow.

The elder is cultivated in some parts of Kent for its fruit, which is much in demand for making elder wine. The flowers and bark are much used by herbalists; and the wood of old trunks, being very hard, is used as a substitute for that of box and dogwood. The young shoots, having large pith, are made into pop-guns, and the pith is used by electricians in various experiments. The plant, both in Britain and on the Continent, is sometimes used for forming hedges, and also as a nurse plant for plantations exposed to the sea breeze. In the latter capacity, it has the great advantage of growing rapidly the first five or six years, and afterwards of being easily choked by the trees it has nursed up. The elder will not thrive except in a good soil, kept somewhat moist; and it will not flower and fruit abundantly, unless the situation be open, and fully exposed to the light and air. The plant roots so readily from cuttings and truncheons, that, where the soil is tolerably moist, a plantation or a hedge may be made at once, by the use of the latter, instead of employing rooted plants.

2. S. canade'nsis L. The Canadian Elder.

Identification. Lin. Sp., 385.; Don's Mill., 3. p. 436.
Engravings. Schmidt Baum., 2. t. 142.; and our *fig.* 934.

Spec. Char., &c. Frutescent. Leaves pinnate or sub-pinnate. Leaflets about 4 pairs, and an odd one; oblong, oval, stiffish, acuminated, more or less pubescent beneath, sometimes appendiculated at the base. Cymes of 5 main branches. Flowers almost scentless. Berries deep bluish black. (*Don's Mill.*) A shrub, nearly suffruticose. Canada to Carolina, in swamps and near hedges. Height 4 ft. to 6 ft. Introduced in 1761. Flowers white; July and August. Fruit bluish black; ripe in September.

934. *S. canadénsis.*

A bush, in foliage resembling the common elder, but it is less hardy, and, in Britain at least, never assumes any thing of a tree character.

B. *Leaves pinnate. Flowers panicled.*

3. S. racemo'sa L. The racemose-*flowered* Elder.

Identification. Lin. Sp., 386.; Don's Mill., 3. p. 438.
Synonymes. S. montána *Cam. Epit.* 976.; S. cervìni *Tabern.* 1029.; Sambuco montana, *Ital.*
Engravings. Jacq. Icon. Rar., 1. p. 59.; N. Du Ham., 1. t. 56.; and our *fig.* 935.

Spec. Char., &c. Shrubby. Leaves pinnate. Leaflets 5, membranous, oblong, acuminated, serrated, unequal at the base. Petioles glabrous. Panicle ovate. Leaves pale green, pretty smooth. (*Don's Mill.*) A low tree or large shrub. Middle and South of Europe and Siberia, on mountains. Height 10 ft. to 12 ft. Introduced in 1596. Flowers whitish green; April and May. Fruit scarlet; ripe in August.

Variety.

S. r. 2 *laciniàta* Koch in Dec. Fl. Fr. Supp. p. 3000.—Leaflets jagged. A native of the Palatinate of the Rhine.

935. *S. racemòsa.*

This tree has a splendid appearance when covered with its panicles of fine, large, scarlet fruit, which resemble miniature bunches of grapes of the most brilliant scarlet. Its large leaves, with their deeply serrated pinnæ, are also very ornamental. It grows as freely as the common elder, and deserves a place in every collection; though it is very seldom found, in British gardens, of such a size as to display its beauty. We should think it would succeed if budded on the common elder; and, as that species is abundant in many places, plants might be trained to a single stem, and budded with *S.* racemòsa standard high. It is very ornamental in the Paris gardens.

4. *S.* (R.) PU'BENS *Michx.* The downy Elder.

Identification. Michx. Fl. Bor. Amer., 1. p. 181.; Don's Mill., 3. p. 438.
Synonymes. *S.* racemòsa *Hook. Fl. Bor. Amer.* 1. p. 279., but not of Lin.; *S.* pubéscens *Lodd. Cat.*
Engravings. Our *fig.* 936. from a living plant in the Chelsea Botanic Garden.

936. *S.* (*r.*) pùbens.

Spec. Char., &c. Shrubby. Leaves pinnate. Leaflets 5, membranous, ovate-lanceolate, or oblong, acuminated, serrated, pubescent, but chiefly on the under side. Panicle thyrsoid. (*Don's Mill.*) A large shrub or low tree. Carolina to Canada, on the highest mountains. Height 6 ft. to 10 ft. sometimes 12 ft. Introd. 1812. Flowers whitish; April and May. Berries red; ripe in August.

Closely resembling *S.* racemòsa, of which it is probably only a variety. Sir W. J. Hooker mentions a variety with 7 leaflets, which may be designated S. (*r.*) *p.* 2 *heptaphýlla.*

GENUS II.

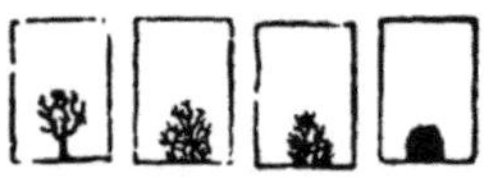

*V*IBU'RNUM *L.* THE VIBURNUM. *Lin. Syst.* Pentándria Trigýnia.

Identification. Lin. Gen., p. 370.; Dec. Prod., 4. p. 323.; Don's Mill., 3. p. 438.
Synonymes. O'pulus, *V*ibúrnum, and *T*inus, *Tourn. Inst.* p. 607. t. 376. and 377.; *V*ibúrnum and O'pulus, *Mœnch Meth.* p. 505.; Viorne, *Fr.*; Schneeball, *Ger.*; Viburno, *Ital.*
Derivation. According to Vaillant, the word *V*ibúrnum is derived from the Latin word *vieo*, to tie; on account of the pliability of the branches of some species. Viburna, in the plural, appears to have been applied by the ancients to any shrubs that were used for binding or tying.

Gen. Char. *Calyx* limb small, permanent. *Corolla* rotate, somewhat campanulate, or tubular, with a 5-lobed limb. *Stamens* 5, equal. *Stigmas* 3, sessile. *Berry* ovate or globose, 1-seeded from abortion, crowned by the calycine teeth. (*Don's Mill.*)

Leaves simple, opposite, stipulate, chiefly deciduous, but partly evergreen; petiolate. *Flowers* in terminal corymbs; usually white, but sometimes verging to a rose colour. Decaying leaves red and yellow.—Shrubs; natives of Europe, Asia, and North America; of easy culture and propagation, by seeds or layers, in any common soil.

§ i. *Tìnus* Tourn.

Synonymes. Lentago *Dec. Prod.* 4. p. 324.; *V*ibúrnum *Mœnch Meth.* p. 505.

Sect. Char., &c. Leaves quite entire, or toothed. Style almost wanting; stigmas 3, sessile.

1. *V.* *T*I'NUS *L.* The Laurustinus.

Identification. Lin. Sp., 383.; Dec. Prod., 4. p. 324.; Don's Mill., 3. p. 438.
Synonymes. *V.* lauriforme *Lam. Fl. Fr.* 3. p. 363.; *T*inus *Tourn. Inst.* p. 607. t. 377.; *T*inus *lauri*fòlia *Borkh.* in *Ræm. Arch.* 1. pt. 2. p. 20.; the Laurustine, wild Baie Tree, *Gerard*; Viorne, Laurier Tin, *Fr.*; Lorbeerartiger Schneeball, or Schwalkenstrauch, *Ger.*; Lagro salvatico, and Lauro Tino, *Ital.*
Derivation. Laurustinus is from *laurus*, a laurel, and *tinus*, the Latin name of the plant; the word *laurus* being added, by old authors, from the supposition that this shrub belonged to the same family as the *Laúrus nóbilis*, or sweet bay. Lorbeerartiger is laurel-like.
Engravings. N. Du Ham., 2. t. 37.; Bot. Mag., t. 38.; and our *fig.* 937.

Spec. Char., &c. Leaves ovate-oblong, quite entire, permanent; having the ramifications of the veins beneath, as well as the branchlets, furnished with glandular hairs. Corymbs flat. Flowers white, but rose-coloured before expansion, and sometimes afterwards for a little time. Berries dark blue. (*Don's Mill.*) A compact evergreen shrub. South of Europe, and North of Africa. Height 8 ft. to 10 ft. Introduced in 1596. Flowers white; December till March. Berries dark blue; ripe in June.

937. *Vibúrnum Tinus.*

Varieties.

- V. T. 2 *hírta* Ait. Hort. Kew. ii. p. 166.; *V. Tìnus Mill. Dict.* No. 4.; *V. lùcidum Mill., Pers.*, and *Schultes.* — Leaves oval-oblong, hairy beneath and on the margins. The flowers of this variety appear in autumn, and continue on the shrub all the winter. A native of Portugal and Spain, and the vicinity of Nice. Very distinct, from the comparative roundness of its leaves, and the hairiness both of the leaves and branches.
- V. T. 3 *lùcida* Ait. l. c. — Leaves ovate-oblong, glabrous on both surfaces, shining. The cymes, as well as the flowers and leaves, are larger than those of the common sort, and seldom appear till the spring. When the winters are sharp, the flowers are killed, and never open unless they are sheltered. This is quite a distinct variety, with fewer and more spreading branches than the common kind, and much larger leaves, which are shining. There is a subvariety of it with leaves more or less variegated with white. It is a native about Algiers, and on Mount Atlas.
- V. T. 4 *virgàta* Ait. l. c., Clus. Hist. No. iii. with a fig. — Leaves oblong-lanceolate, pilose on the margins, as well as on the under surface. It is a native of Italy, about Rome and Tivoli, &c.
- V. T. 5 *stricta* Hort. has a somewhat erect and fastigiate habit. Horticultural Society's Garden. There is also a variegated subvariety.

One of the most ornamental of evergreen shrubs, the foliage tufting in beautiful masses, and covered with a profusion of white flowers which commence expanding in November, and continue flowering till April or May. In British nurseries, it is frequently, for expedition's sake, increased by layers; but all the varieties are readily propagated by cuttings, taken off in autumn, and planted in a sandy soil, on a northern border. In two years, these cuttings will form saleable plants of the smallest size. The variety *V. T.* lùcida, being somewhat more difficult to strike than the others, is generally increased by layers, which are made in autumn, and root in a year.

§ ii. *Vibúrnum* Tourn.

Synonyme. Lentàgo *Dec. Prod.* 4. p. 424.

Sect. Char., &c. Leaves deciduous. All the flowers fertile, and equal in shape and size, except in *V.* lantanöìdes. Corolla rotate. Fruit oval.

♠ ♈ 2. *V.* Lenta'go *L.* The Lentago, *or pliant-branched*, Viburnum.

Identification. Lin. Sp., p. 384.; Dec. Prod., 4. p. 325.; Don's Mill., 3. p. 440.
Synonymes. Tree Viburnum, Canada Viburnum; Viorne à Rameaux pendans, Viorne luisante, *Fr.*; Birn-blätteriger Schneeball, *Ger.*; Canadische Schwalkenbeerstrauch, Schwalkenstrauch, *Hayne.*
Engravings. Wats. Dend. Brit., t. 21.; Schmidt Baum., 3. t. 176.; and our *fig.* 938.

Spec. Char., &c. Leaves broad-ovate, acuminated, sharply serrated, glabrous. Petioles with narrow curled margins. Corymbs terminal, sessile. Serratures of leaves hooked a little, and somewhat cartilaginous. (*Don's Mill.*) A robust shrub or low tree. New England to Carolina, among hedges and on the borders of woods; and found throughout Canada. Height 6 ft. to 10 ft. Introduced in 1761. Flowers white; July. Fruit black; ripe in September. Decaying leaves purple red and yellow Naked young wood yellowish and reddish green.

938 *V. Lentågo.*

In British gardens, this species forms, when pruned to a single stem, a handsome small tree, flowering freely and producing abundance of fruit, which is greedily eaten by birds. Propagated by layers, or by seeds.

3. *V.* (L.) PRUNIFO'LIUM *L.* The Plum-tree-leaved Viburnum.

Identification. Lin. Sp, 383.; Dec. Prod., 4. p. 325.; Don's Mill., 3. p. 440.
Synonyme. *V. Lentågo Du Roi.*
Engravings. Wats. Dend. Brit., t. 23.; N. Du Ham., 2. t. 38.; the plate in Arb. Brit., 1st edit., vol. vi.; and our *fig.* 939.

939. *V.* (L.) prunifolium.

Spec. Char., &c. Leaves roundish-obovate and oval, glabrous, rather membranous, crenately serrated, ending in a short acumen. Petioles marginate, glabrous. Cymes sessile. Berries ovate or roundish. (*Don's Mill.*) A large shrub or low tree. New England to Carolina, in hedges and fields; and also Canada, about Lake Huron. Height 8 ft. to 10 ft. Introduced in 1731. Flowers white; May and June. Fruit dark blue; ripe in September.

4. *V.* (L.) PYRIFO'LIUM *Poir.* The Pear-tree-leaved Viburnum.

Identification. Poir. Dict., 5. p. 658.; Dec. Prod., 4. p. 325.; Don's Mill., 3. p. 440.
Engravings. Dend. Brit., t. 22.; and our *figs.* 940. and 941.

940. *V.* (L.) pyrifolium.

Spec. Char., &c. Leaves ovate, acutish, glabrous, subserrated. Petioles smooth. Corymbs somewhat pedunculate. (*Don's Mill.*) A large shrub or low tree. Pennsylvania, New Jersey, &c., on the banks of rivers. Height 5 ft. to 8 ft. Introduced in 1812. Flowers white; May and June. Fruit ovate oblong, black; ripe in September.

941. *V.* (L.) pyrifolium.

Resembles the preceding species, but is not so straggling in its growth.

♠ ♆ 5. *V.* (L.) NU`DUM *L.* The naked-*corymbed* Viburnum.

Identification. Lin. Sp., 383.; Dec. Prod., 4. p. 325.; Don's Mill., 3. p. 440.
Synonyme. *V. pyrifòlium Poir.*
Engravings. Wats. Dend. Brit., t. 20.; Mill. Icon., 274.; and our *fig.* 942.

Spec. Char., &c. Leaves oval-oblong, angular at the base, bluntish, with revolute obsoletely crenulated margins, quite glabrous. Petioles beset with scale-like scurf or down. Corymbs pedunculate, not involucrate. (*Don's Mill.*) A large shrub or low tree. Canada to Georgia, in swamps, particularly on a sandy soil. Height 6 ft. to 10 ft. Introduced in 1752. Flowers whitish; May to June. Fruit globose, black or dark blue; ripe in September.

Variety.

♠ ♆ V. (*L.*) *n.* 2 *squamàtum*; *V.* squamàtum *Willd. Enum.* (Wats. Dend. Brit., t. 24.; and our *fig.* 943.); has the surface, midribs, and petioles of the leaves scaly (whence its name), and their margins crenate, subdentate. The peduncles and pedicels are also covered with minute ferruginous scales; and the leaves are smaller, and of a bluer green than those of *V.* nùdum.

943. *V.* (L.) n. squamàtum.

942. *V.* (L.) nùdum.

Sir W. J. Hooker says of this species, that he cannot satisfy himself of permanently distinguishing characters between it and *V.* Lentàgo and *V.* prunifòlium. We think all the four varieties of the same form.

♠ ♆ 6. *V.* CASSINOI`DES *L.* The Cassine-like Viburnum.

944. *V.* cassinoìdes.

Identification. Lin. Sp., p. 384.; Dec. Prod., 4. p. 326.; Don's Mill., 3. p. 440.; Lodd. Cat., ed. 1836.
Synonyme. *V.* punctàtum *Rafin.*
Engraving. Our *fig.* 944. from a specimen in the Lambertian herbarium.

Spec. Char., &c. Leaves ovate-lanceolate, acute at both ends, crenated, glabrous above, with subrevolute edges. Under side of leaves, as well as the petioles, which are keeled, and branches, which are tetragonal, covered with scurfy dots. Corymbs sessile. (*Don's Mill.*) A large shrub or low tree. New York to Carolina, in swamps. Height 3 ft. to 5 ft. in America; 10 ft. to 12 ft. in England. Flowers white; June and July. Fruit ovate, bluish black; ripe in September.

♠ ♆ 7. *V.* (C.) LÆVIGA`TUM *Willd.* The smooth Viburnum.

Identification. Willd. Sp., 1. p. 1491. Dec. Prod., 4. p. 326.; Don's Mill., 3. p. 440.
Synonymes. *V. cassinöìdes Du Roi Harbk.* 2. p. 486.; *V.* lanceolàtum *Hill, Hort. Kew.* t. 19.; *Vibúrnum* caroliniànum *Hort.*; *Cassìne* paràgua *Lin. Mant.* 220.; *Cassìne* corymbòsa *Mill. Icon.* t. 83. f. 1.; Thè americano, *Ital.*
Engravings. Mill. Icon., t. 83. f. 1.; and our *fig.* 945.

945. *V.* (c.) lævigàtum.

Spec. Char., &c. Leaves lanceolate, or oblong-lanceolate, smooth, remotely or unequally serrated, cuneated at the base, and quite entire, glabrous. Branches tetragonally 2-edged, and also glabrous. Corymbs sessile. (*Don's Mill.*) A large shrub or low tree. Virginia and Carolina, near the sea coast. Height 10 ft. to 14 ft. Introduced in 1724. Flowers white; June and July. Fruit black; ripe in September.

♠ ♆ 8. *V.* Lanta'na *L.* The Wayfaring Tree.

Identification. Lin. Sp., p. 384.; Dec. Prod., 4. p. 326.; Don's Mill., 3. p. 440.
Synonymes. *V.* tomentòsum *Lam. Fl. Fr.* 3. p. 363.; wild Guelder Rose, pliant-branched Mealy Tree; Viorne cotonneuse, Camara, Viorne commune, Coudre-moinsinne, Moncienne, *Fr.*; Schlingstrauch, wolliger Schneeball, or Schwalkenstrauch, *Ger.*; Lentaggine, *Ital.*
Engravings. Eng. Bot., t. 331.; Jacq. Austr., t. 341.; and our *fig.* 946.

Spec. Char., &c. Leaves cordate, rounded, finely serrated, veiny, clothed beneath, but more sparingly on the upper side, with starry mealy pubescence, like that on the branches, petioles, and peduncles. Under side of leaves and branches white from mealy down. Cymes pedunculate, broad, flat, of numerous crowded white flowers. Bracteas several, small, acute. (*Don's Mill.*) A large shrub or low tree, with copious, opposite, round, pliant, mealy branches. Europe and the West of Asia, in low woods and hedges, chiefly on calcareous soils. Height 12 ft. to 15 ft. Flowers white; May and June. Fruit compressed in an early state, red on the outer side, yellow, and finally black, with a little mealy astringent pulp; ripe in August and September. Decaying leaves of a fine deep red.

946. *V.* Lantàna.

Varieties.

- ♠ ♆ V. *L.* 2 *grandifòlia* Ait., *V.* L. latifòlia *Lodd. Cat.*, has leaves larger than those of the species, and, according to some, ought to constitute a separate species itself. Mr. Gordon thinks this variety the same as *V.* (L.) lantanöìdes.
- ♠ ♆ V. *L.* 3 *fòliis variegàtis* Lodd. Cat. has leaves variegated with white and yellow.

It grows rapidly when young, often producing shoots 5 or 6 feet long, from stools in coppice woods; but becoming stationary when it has attained the height of 12 or 15 feet, which it does in 5 or 6 years; and, when pruned to a single stem, forms a handsome durable small tree. In Germany, the shoots of one year are employed in basket-making, and for tying faggots and other packages; and those of two or three years old are used for tubes to tobacco-pipes. Plants may be raised from seeds, which should be laid up in a heap in the rotting-ground, like haws; for, if sown immediately after being gathered, they will not come up for 18 or 20 months.

♠ ♆ 9. *V.* (L.) lantanöì'des *Michx.* The Lantana-like Viburnum, *or American Wayfaring Tree.*

Identification. Michx. Fl. Bor. Amer., 1. p. 179.; Dec. Prod., 4. p. 326.; Don's Mill., 3. p. 441.
Synonymes. ? *V.* Lantàna β grandifòlia *Ait. Hort. Kew.* ed. 1., vol. 1. p. 392.; *V.* grandifòlium *Smith* in *Rees's Cycl.* No. 14.; *V.* Lantàna β canadénsis *Pers. Ench.* 1. p. 327.; Hobble Bush, *Amer.*
Engravings. Bot. Cab., t. 1070.; and our *fig.* 947.

Spec. Char., &c. Leaves roundish-cordate, abruptly acuminated, unequally serrated; serratures awnless. Branches, petioles, and nerves of leaves clothed with powdery tomentum. Corymbs terminal, almost sessile. Fruit ovate.

947. *V.* (L.) lantanöìdes.

(*Don's Miller.*) The outer flowers of the corymbs are abortive and radiant; a circumstance, as Sir W. J. Hooker observes, noticed by few botanists. A shrub or low tree, very like *V.* Lantàna, but of more humble growth, and the leaves are larger, and tomentose. Canada to Carolina, principally in the forests called Beech Woods, about Quebec and Lake Huron. Height 5 ft. to 10 ft. Introduced in 1820. Flowers white; June and July. Fruit first red, afterwards black. Hort. Soc. Garden.

■ 10. *V.* (L.) DAHU'RICUM *Pall.* The Dahurian Viburnum.

Identification. Pall. Fl. Ros., p. 52.; Dec. Prod., 4. p. 328.; Don's Mill., 3. p. 442.
Synonymes. Lonicera mongólica *Pall. Fl. Ros.* 1. t. 38. f. t. 58. f. F. G.; Córnus dàurica *Laxm.*
Engravings. Pall. Fl. Ros., t. 38., and t. 58. f. F. G.; Gmel. Sib., 3. t. 25.; and our *fig.* 948.

Spec. Char., &c. Leaves ovate, somewhat cordate at the base, crenately serrated, beset with stellate down, as well as the branchlets. Corymbs dichotomous, few-flowered. Corollas tubular, somewhat funnel-shaped, bluntly 5-toothed. Berry 5-seeded. (*Don's Mill.*) A shrub resembling *V.* Lantàna. Dahuria. Height 6 ft. to 8 ft. Introduced in 1785. Flowers yellowish white; June and July. Fruit at first red, but afterwards black and sweet; ripe in September. Horticultural Society's Garden.

A tolerably distinct variety; but, in our opinion, by no means entitled to be considered a species.

948. *V.* (L.) dahùricum.

■ 11. *V.* (?L.) COTINIFO'LIUM *D. Don.* The Cotinus-leaved Viburnum.

Identification. D. Don Prod. Fl. Nep., 1. p. 141.; Dec. Prod., 4. p. 327.; Don's Mill., 3. p. 441.
Synonyme. *V. Mullàha* Ham. in D. Don Prod. Fl. Nep. 1. p. 141.
Engravings. Bot. Reg., t. 1650.; and our *fig.* 949. from the plant in the Horticultural Society's Garden, and *fig.* 950. from the Bot. Reg.

949. *V.* (?L.) cotinifòlium.

Spec. Char., &c. Leaves roundish oval, quite entire, clothed with stellate tomentum on both surfaces, grey beneath, as well as the branches. Corymbs terminal, woolly. Flowers white. (*Don's Miller.*) A shrub. Himalayas, at the height of from 5000 ft. to 7000 ft., 30° N. lat. Height 6 ft. to 8 ft. Introduced in 1832, or before. Flowers white, tinted with pink; April and May.

950. *V.* (?L.) cotinifòlium.

In general appearance it closely resembles *V.* Lantàna; but the flowers are much larger, and more tinted with pink; and neither flat nor bell-shaped, but of a distinct obconical figure.

■ 12. *V.* DENTA'TUM *Lin.* The toothed-*leaved* Viburnum.

Identification. Lin. Sp., p. 384.; Dec. Prod., 4. p. 326.; Don's Mill., 3. p. 441.
Synonymes. *V.* dentàtum lùcidum *Ait. Hort. Kew.* 1. p. 372.; *V.* dentàtum glabéllum *Michx. Fl. Bor. Amer.* 1. p. 179.; Arrow-wood; Viorne dentée, *Fr.*
Engravings. Jacq. Hort. Vind., 1. t. 36.; Wats. Dend. Brit., t. 35.; and our *figs.* 951. and 952.

Spec. Char., &c. Partly glabrous. Leaves ovate, and nearly orbicular, plicate, coarsely and dentately serrated, with the nerves thick and feathered, glabrous on both surfaces. Cymes or corymbs pedunculate. Berries small, and nearly globose, of a dark blue colour, and crowned by the calyx. (*Don's Mill.*) A large shrub. New York to Carolina, in mountain woods; and also in Mexico. Height 4 ft. to 6 ft. Introduced in 1763. Flowers white; June and July. Fruit small, nearly globose, dark blue, and crowned by the calyx; not very frequently ripened in England.

951. *V.* dentàtum.

952. *V.* dentàtum.

Varieties. In the arboretum of Messrs. Loddiges, are plants named V. *d. pubéscens*, V. *d. fòliis variegàtis*, V. *acuminàtum*, V. *longifòlium*, and V. *montànum*, which are either varieties of, or identical with, this species.

■ 13. *V.* (D.) PUBE'SCENS *Pursh.* The downy Viburnum.

Identification. Pursh Fl. Amer. Sept., 1. p. 202.; Dec. Prod., 2 p. 326.; Don's Mill., 3. p. 441.
Synonymes. *V.* dentàtum β pubéscens *Ait. Hort. Kew.* 1. p. 168.; *V.* dentàtum semi-tomentòsum *Mich. Fl. Bor. Amer.* 1. p. 179.; *V.* tomentòsum *Rafin. Med. Rep.* 2. p. 860.; *V.* villòsum *Rafin.* in *Desf. Journ.* 1. p. 228.; *V.* Rafinesquiànum Schultes Syst. 6. p. 630.
Engraving. Our *fig.* 953. from a specimen in the British Museum.

Spec. Char., &c. Pubescent. Leaves ovate, acuminated, on short petioles, coarsely serrate-toothed, villous beneath, with the nerves feathered and prominent. Corymbs pedunculate. (*Don's Mill.*) A low shrub. Virginia and Carolina. Height 3 ft. Introduced in 1736. Flowers white; June and July. Fruit small ovate; smaller in every part than *V.* dentàtum. The fruit is dark blue, but sparingly produced; the leaves die off yellow and red.

953. *V.* (d.) pubéscens.

■ 14. *V.* NI'TIDUM *Ait.* The shining-*leaved* Viburnum.

Identification. Ait. Hort. Kew., 1. p. 371.; Pursh Fl. Amer. Sept., 1. p. 202.; Dec. Prod., 4. p. 326.; Don's Mill., 3. p. 440.
Engraving Our *fig.* 954. from a specimen in the Lambertian herbarium.

Spec. Char., &c. Quite glabrous. Leaves linear-lanceolate, shining above, obsoletely serrated or entire. Branches tetragonal. (*Don's Mill.*) A low shrub, with small leaves. Carolina and Georgia, in sandy barren woods. Height 2 ft. to 4 ft. Introduced in 1758. Flowers white; May and June. Fruit; ?. H. S.

Judging from the plant in the Horticultural Society's Garden, this is a very distinct species, and its smooth shining yellowish green leaves render it more ornamental than most of the others of this section.

954. *V.* (d.) nítidum.

§ iii. O'pulus Tourn.

Identification. Tourn. Inst., t. 376.; Dec. Prod., 4. p. 328.; Mœnch Meth., p. 605.

Sect. Char. Outer flowers of the corymbs radiant and sterile, much larger than the rest, which are fertile. Seed obcordate. (*Don's Mill.*) Leaves mostly 3-lobed, and deciduous.

■ 15. *V.* O'PULUS *L.* The Guelder Rose.

Identification. Lin. Sp., 384.; Dec. Prod., 4. p. 328.; Don's Mill., 3. p. 442.
Synonymes. *V.* lobàtum *Lam. Fl. Fr.* 3. p. 363.; O'pulus glandulòsus *Mœnch Meth.* p. 505.; O'pulus *Raii Syn.* 460.; Sambùcus aquática *Bauh. Pin.* 456.; Marsh Elder, Rose Elder Water Elder;

Viorne-Obier, l'Obier d'Europe, *Fr.*; Schwalkenbeer Strauch, Wasserholder, Schneeball, *Ger.*; Maggi, *Ital.*
Derivation. Altered from *Pópulus*, the poplar, from some supposed resemblance between the leaves of the plant and those of the poplar. The English name, Guelder Rose, is derived from Guelderland, where the double-flowered variety was first originated.
Engravings. Engl. Bot., t. 332.; N. Du Ham., 2. t. 39.; and our *fig.* 955.

Spec. Char., &c. Quite glabrous in every part. Leaves broad, 3-lobed, acuminated, unequally serrated, veiny. Petioles beset with glands towards the top, and several oblong leafy appendages lower down. Cymes pedunculate, white, with linear bracteas; with several of the marginal flowers dilated, flat, radiant, and without stamens or pistils. Seed compressed. (*Don's Mill.*) A shrub or low tree. Europe, and part of Asia, in moist hedges and swampy thickets; frequent in Britain, and also in Sweden, as far north as lat. 61°. Height 6 ft. to 12 ft. in a wild state, and higher in gardens. Flowers white; May and June. Fruit elliptical, crowned by the limb of the calyx, bright red, very juicy, but bitter and nauseous. Decaying leaves beautiful pink or crimson. Naked young wood smooth, green.

955. *V. O'pulus.*

Varieties.

- V. O. 2 *stérilis* Dec. Prod. iv. p. 328., Don's Mill. iii. p. 442. *V. O. ròseum Ræm. et Schult. Syst.* vi. p. 635.; the Snow-ball Tree, or Guelder Rose; Rose de Gueldres, Pellotte de Neige, Boule de Neige, Poire molle, *Fr.*; Schneeballe, *Ger.*—The specific name *ròseum* is applied on account of the form of the flowers, and not because of their colour. Layers, sometimes by suckers, or it might be grafted on the species.
- V. O. 3 *fòliis variegàtis* Lodd. Cat. ed. 1836 has the leaves variegated with white and yellow.
- V. O. 4 *nàna* Hort.—A very distinct little plant, scarcely 1 ft. in height. Horticultural Society's Garden.

The Guelder rose, in a wild state, is not remarkable for the beauty of its flowers; but its bright red berries, which ripen in September, and which, towards the middle of October, assume a beautiful pink, almost compensate for the inferiority of the species to the variety in point of flowers. The leaves of both die off of a fine red on the first approach of frost. The snow-ball tree, or the Guelder rose (*V. O.* 2 stérilis), is one of the most ornamental shrubs, or low trees, that can be planted in a pleasure-ground.

16. *V.* (*O.*) ACERIFO'LIUM *L.* The Maple-leaved Guelder Rose.

Identification. Lin. Sp., 383.; Pursh Sept., 1. p. 203.; Dec. Prod., 4. p. 327.; Don's Mill., 3. p. 442.
Engravings. Vent. Hort. Cels., t. 72.; Wats. Dend. Brit., t. 118.; and our *fig.* 956.

Spec. Char., &c. Branchlets and petioles pilose. Leaves ovate-cordate, usually 3-lobed, acuminated, sharply and loosely serrated, downy beneath. Petioles glandless, and, when young, stipulaceous at the base, and rather tomentose. Corymbs terminal, pedunculate, not radiant. (*Don's Mill.*) A low shrub. New England to Carolina, in rocky mountainous situations. Height 4 ft. to 6 ft. Introduced in 1736. Flowers white; May and June. Fruit black, oval, and compressed; ripe in September.

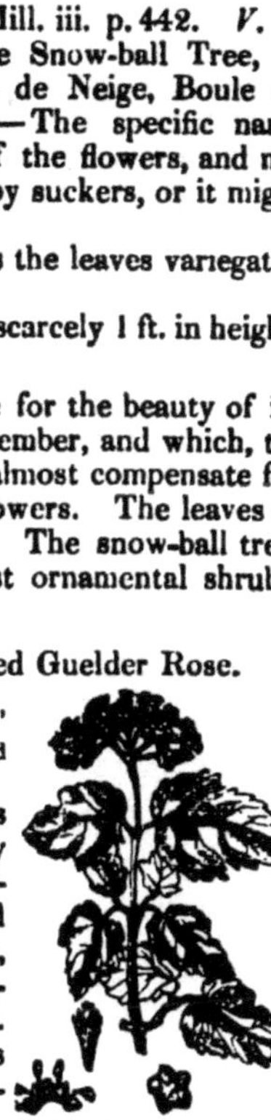

956. *V. O. acerifòlium.*

957. *V.* (*O.*) orientàle.

■ 17. *V.* (*O.*) ORIENTA'LE *Pall.* The Eastern Guelder Rose.

Identification. Pall. Ross., t. 58. f. H.; Dec. Prod., 4. p. 328.; Don's Mill., 3. p. 442.
Synonyme. O'pulus orientàlis fòlio amplíssimo tridentàto *Tourn. Cor.* p. 42.
Engravings. Pall. Fl. Ross., t. 58. f. H.; and our *fig.* 957.

Spec. Char., &c. Leaves 3-lobed, acuminated, coarsely and bluntly toothed. Petioles glandless, glabrous. Corymbs terminal, not radiant. Fruit oblong, compressed. Seed oval, furnished with two channels on both sides, as in *V.* Lantàna. (*Don's Mill.*) A low shrub. Georgia, in Asia Minor, in woods, on the mountains. Height 6 ft. to 10 ft. Introduced in 1827. Flowers white; July. Fruit; ?.

■ 18. *V.* (*O.*) OXYCO'CCOS *Pursh.* The Cranberry-*fruited* Guelder Rose.

Identification. Pursh Sept., 1. p. 203.; Dec. Prod., 4. p. 328.; Don's Mill., 3. p. 442.
Synonymes. *V.* opulöides *Mühl. Cat.* 32.; *V.* trilobum *Marsh. Arb.* p. 162.; *V.* O'pulus americàna *Ait. Hort. Kew.* 1. p. 373.
Engraving. Our *fig.* 958. from a specimen in the Lambertian herbarium.

958. *V.* (*O.*) Oxycóccos.

Spec. Char., &c. Leaves 3-lobed, acute behind, 3-nerved. Lobes divaricate, acuminated, coarsely and distantly serrated. Petioles glandular. Cymes radiant. (*Don's Mill.*) A large shrub or low tree. New York and New Jersey, on mountains and throughout Canada, to the arctic circle. Height 6 ft. to 12 ft. Flowers white; July. Fruit subglobose, red, of an agreeable acid, resembling that of cranberries, for which they are a very good substitute; ripe in September.

Varieties.

■ V. (O.) *O.* 2 *subintegrifòlius* Hook. Fl. Bor. Amer. i. p. 281., Don's Mill. iii. p. 442.—Leaves but little cut, very pubescent beneath. A native of the banks of the Columbia.

■ V. (O.) *O.* 3 *móllis*. *V.* mólle *Michx. Fl. Bor. Amer.* i. p. 180., *Don's Mill.* iii. p. 442.; *V. a*lnifòlium *Marsh Arb.* p. 162. (Our *fig.* 959.) — Leaves nearly orbicular, cordate, plicate, toothed. (*Don's Mill.*) Fruit oblong ovate, red.

959. *V.* (*O.*) O. mólle.

Very like *V.* O'pulus, and there can be no doubt but that it is only the American form of that species. The fruit is comparatively large, and not disagreeable to the taste. Were a great number of seedlings grown till they produced fruit, and then the plant producing the largest and best-flavoured fruit selected and propagated by extension, the cranberry-fruited guelder rose might be cultivated in our kitchen-gardens and orchards for the same purpose as the common cranberry. We have no doubt whatever that its fruit would be soon as much relished by the public as the cranberry; and, as the guelder rose is less difficult in regard to soil and situation than that plant, a crop of fruit might be depended on with greater certainty. At all events, this and similar experiments offer interesting and useful employment to the amateur who has nothing better to do.

■ 19. *V.* (*O.*) EDU'LE *Pursh.* The edible-*fruited* Guelder Rose.

Identification. Pursh Sept., 1. p. 203.; Dec. Prod., 4. p. 328.; Don's Mill., 3. p. 442.

Synonyme. V. O'pulus edùlis *Michx. Fl. Bor. Amer.* 1. p. 180.
Engraving. Our *fig.* 960. from a specimen in Dr. Lindley's herbarium.

Spec. Char., &c. Leaves 3-lobed, bluntish behind, and 3-nerved. Lobes very short, denticulately serrated; serratures acuminated. Petioles glandular. Outer flowers of corymb radiant. A smaller and more upright shrub than the preceding species. The berries of the same colour and size; but, when completely ripe, more agreeable to eat, and frequently employed as a substitute for cranberries. It does not seem to differ much from V. Oxycóccos, except in the broader base of the leaf. (*Don's Mill.*) Canada to New York, on the banks of rivers. Height 5 ft. to 10 ft. Introduced in 1812. Flowers white; July. Fruit as in the preceding species.

960. V. (O.) edùle.

Sect. II. Lonice'reæ.

Genus III.

DIERVI'LLA Tourn. The Diervilla. *Lin. Syst.* Pentándria Monogýnia.

Identification. Tourn. Act. Ac. Par., 1706. t. 7. f. 1.; Dec. Prod., 4. p. 330.
Synonymes. Lonicera sp. *L.*; Weigela Thunb. Fl. Jap. p. 6.; Weigel*ia* Pers. Ench. 1. p. 176.
Derivation. Named by Tournefort, in compliment to *M. Dierville*, a French surgeon, who was the first to introduce D. canadénsis into Europe.

Gen. Char. *Calyx* tube oblong, bibracteate at the base; limb 5-cleft. *Corolla* funnel-shaped, 3—5-cleft, spreading twice the length of the calyx. *Stamens* 5, somewhat exserted. *Stigma* capitate. *Capsule* oblong, acute, 1-celled. *Seeds* numerous, minute. (*Don's Mill.*)

Leaves simple, opposite, exstipulate, deciduous; ovate, acuminated, serrated. *Flowers* in axillary pedúncles, bibracteate, usually dichotomous. — Shrubs, deciduous. North America. Common soil, and suckers. Five species, natives of Japan, and figured by Sieboldt, are probably hardy, but they have not yet been introduced.

≡ 1. D. canade'nsis *Willd.* The Canadian Diervilla.

Identification. Willd. Enum., 1. p. 222.; Dec. Prod., 4. p. 330.; Don's Mill., 3. p. 444.
Synonymes. Lonicera Diervilla Lin. Mat. Med. p. 62.; D. Tournefórtii Michx. Fl. Bor. Amer. 1. p. 107.; D. hùmilis *Pers. Ench.* 1. p. 214.; D. lùtea *Pursh Sept.* 1. p. 162.; D. trífida *Mœnch Meth.* 492.; D. acadiénsis *Du Ham. Arb.* 1. t. 87.
Engravings. Bot. Mag., t. 1796.; Schmidt Baum., t. 116.; and our *figs.* 961. and 962.

Spec. Char., &c. Leaves on short petioles, ovate, acuminated, serrated, and, as well as the petioles, glabrous. Fruit a dry brown capsule. Root

961. D. canadensis.

creeping, throwing up suckers. (*Don's Mill.*) A bushy shrub. Carolina, New England, and Newfoundland, on rocks and the highest mountains. Height 3 ft. to 4 ft. Introduced in 1739. Flowers yellow; June and July. Fruit brown; ripe in September.

962. D. canadénsis.

There are a number of varieties of this species, differing in respect to the size of the flowers and of the leaves, but they are not worth keeping distinct.

Genus IV.

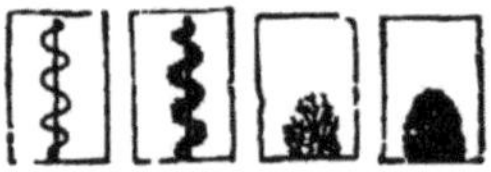

LONI'CERA Desf. The Lonicera, *or Honeysuckle.* *Lin. Syst.* Pentándria Monogýnia.

Identification. Desf. Fl. Atl., 1. p. 183.; Dec. Prod., 4. p. 330.; Don's Mill., 3. p. 444.
Synonymes. Lonicera sp. *Lin.*, and many authors; Caprifòlium and Xylósteum *Juss. Gen.* p. 212.; Xylósteum, Caprifòlium, Chamæcérasus, Periclýmenum *Tourn. Inst.* t. 378. and 379.; Caprifòlium and Lonicera Rœm. et Schult. Syst.; Lonícera and Xylósteum *Torrey Fl. Un. St.*; Chèvrefeuille, *Fr.*; Geissblätt, Honeigblume, and Lonicere, *Ger.*
Derivation. Named after *Adam Lonicer*, a German, who was born in 1528, and died in 1586. There was another Lonicer, John, who wrote comments on Dioscorides.

Gen. Char. *Calyx* tube 5-toothed. *Corolla* tubular, campanulate, or funnel-shaped, with a 5-cleft, usually irregular, limb. *Stamens* 5. *Style* filiform. *Stigma* capitate. *Berries* 3-celled. *Seeds* crustaceous. (*Don's Mill.*)

Leaves simple, opposite, stipulate, deciduous, or evergreen; sometimes connate, entire, occasionally runcinate in the same species. *Flowers* axillary, or capitate, variously disposed. — Shrubs, erect or twining; natives of Europe, the North of Africa, Asia, and America.

The greater number of the species and varieties are of easy culture in British gardens, in common garden soil; and they are all propagated by cuttings, or some of them more readily by layers. The flowers of some of the species are highly fragrant and ornamental; and that of the common European honeysuckle is supposed to have given rise to one of the most beautiful ornaments of Grecian architecture. "The honeysuckles offer an easy opportunity of improvement, by intermixing the fragrant and more vigorous with the yellow and the scarlet." (*Herb. Amaryll.* p. 363.) The genus Lonícera of Linnæus was separated by Rœmer and Schultes into the genera Lonícera and Caprifòlium; but they were reunited by DeCandolle, whose arrangement has been followed by Sir W. J. Hooker and G. Don, and is adopted by us on the present occasion. The distinctive characters of the sections are as follows:—

Caprifòlium. Plants twining. Flowers in capitate whorls.
Xylósteum. Plants twining or erect. Flowers axillary.

§ i. *Caprifòlium* Dec.

Identification. Dec. Fl. Fr., 4. p. 270.; Prod., 4. p. 331.
Synonymes. Caprifòlium *Juss. Gen.* 212.; Lonicera Torr. Fl. Un. St. 1. p. 242., but not of Schult.

Derivation. From *caper*, a goat, and *folium*, a leaf; in reference to the climbing habit of the species; or, as appears much more probable, because goats are fond of browsing on its leaves.

Sect. Char. Berries solitary, while young 3-celled, but when mature usually 1-celled, crowned by the tube of the calyx, which is permanent. Flowers disposed in capitate whorls. Twining shrubs, mostly deciduous; natives of Europe, the North of Africa, China, Nepal, and North America; all of easy culture, and tolerably hardy, but none of them of long duration.

A. *Flowers ringent.* — *Caprifòlium* Tourn. Inst., p. 608.

3 1. L. Pericly'menum *L.* The Woodbine, *or common Honeysuckle.*

Identification. Lin. Sp., p. 247.; Dec. Prod., 4. p. 331.; Don's Mill., 3. p. 445.

Synonymes. Periclýmenum *Ger. Emac.* p. 891.; Periclýmenum germánicum *Riv. Mon. Irr.* t. 122.; *P.* horténse *Gesn. Icon. Pict.* fasc. 1. 38. t. 7. f. 49.; Caprifòlium Periclýmenum *Ræm. et Schult.* 5. p. 262.; Caprifòlium sylváticum *Lam. Fl. Fr.* 3. p. 365.; Caprifòlium *Raii Syn.* p. 458.; Woodbind; Chèvrefeuille des Bois, *Fr.*; wildes gemeines Geissblätt, *Ger.*; gewoone Kamperfoelie, *Dutch*; Madre Selva, *Ital.* and *Span.*

Derivation. Periclýmenum, from *peri*, round about, and *kuliō*, to roll. Woodbine is a corruption of woodbind, and both allude to the habit of the common sort, of winding itself round every tree and shrub within its reach, and binding them together. In the time of Chaucer, the woodbine was considered as the emblem of true love, from this property. The name of Honeysuckle has reference to the fondness of children for this plant, who amuse themselves with drawing the trumpet-shaped corollas from the calyx, to suck the honey from the nectary. Chèvrefeuille and Geissblätt both signify literally, goat's leaf. The Spanish and Italian names, Madre Selva, wood mother, and the Dutch name Kamperfoelie, the champion mace, seem to have little relation to the plant.

Engravings. Engl. Bot., t. 800.; Schmidt Arb., t. 107.; and our *fig.* 963.

963. L. Periclýmenum.

Spec. Char., &c. Leaves all separate, deciduous, sometimes downy, glaucous beneath, ovate, obtuse, attenuated at the base; upper ones the smallest. Heads of flowers all terminal, ovate, imbricated. Flowers ringent. There are varieties of this species with either smooth, pubescent, or variegated leaves; and, when the plant grows by the sea side, they are occasionally more glaucous and rather succulent. Corollas externally deep red; or, in the earlier-flowering varieties, all over buff-coloured; in the maritime plant, smaller and greenish. Berries nearly globular, accompanied by permanent bracteas. (*Don's Mill.*) A twining deciduous shrub, which always turns from east to west. Europe; common in hedges, groves, and thickets; plentiful in Britain. Stem 15 ft. to 30 ft. Flowers rich yellow; June and July, and, in moist summers, also in August, and sometimes in September. Fruit deep red, bitter and nauseous; ripe in September.

Varieties.

964. L. P. serótinum.

3 *L.* P. 2 *serótinum* Ait. Hort. Kew., i. p. 378. Periclýmenum germánicum *Miller Dictionar.* No 4., (Schmidt, Oester. Baumz. t. 108.; and our *fig.* 964.) — Branches glabrous. Flowers late, and reddish. This, the late red honeysuckle, produces a greater number of flowers together than either the Italian (No. 3.) or Dutch honeysuckle, so that it makes a finer appearance than either of them during its period of flowering. Introduced in 1715.

3 *L.* P. 3 *bélgicum.* Periclýmenum germánicum *Mill. Dict.* No. 4.— Branches smooth, purplish. Leaves oblong-oval, of a lucid green above, but pale beneath, on long petioles. Flowers in terminal verticillate heads; each flower arising out of a scaly cover, reddish on

the outside, and yellowish within; of a very agreeable odour. This, which is commonly called the Dutch honeysuckle, may be trained with stems, and formed into heads; which the wild sort cannot, the branches being too weak and trailing for the purpose.

⅄ L. P. 4 *quercifòlium* Ait. Hort. Kew., the Oak-leaved Honeysuckle, has the leaves sinuated like those of an oak. Found wild in several parts of England. There is a subvariety of this, with the leaves slightly marked near the margin with yellow. The flowers are like those of the species.

All the varieties of the common honeysuckle are beautiful and fragrant; and, either trained against a wall, twining round a pole and over a parasol top, or climbing and rambling among bushes, form great ornaments to gardens. They are propagated by cuttings; but a large proportion of these do not succeed, owing to the tubular shoots admitting the wet during winter, and rotting the upper part of the cutting, that the more common mode of propagation is by layers. Both layers and cuttings are made in the autumn, as soon as the leaves have dropped; and they become sufficiently rooted in one year. (See *Encyc. of Gard.*, edit. 1835.)

⅄ 2. L. Caprifo'lium *L.* The Goat's-leaf, *or pale perfoliate*, Honeysuckle.

Identification. Lin. Sp., p. 246.; Dec. Prod., 4. p. 331.; Don's Mill., 3. p. 444.
Synonymes. Periclýmenum perfoliàtum *Ger. Emac.* p. 891.; Chèvrefeuille des jardins, *Fr.*; Durchwachsene, *Ger.*; Caprifolio, *Ital.*
Engravings. Eng. Bot., t. 799.; and our *fig.* 965.

Spec. Char., &c Leaves deciduous, obovate, acutish, glaucous; uppermost ones broader and connate. Flowers ringent, terminal, disposed in capitate whorls. Stems twining from left to right. Buds acute, glaucous. The lower leaves are distinct, and somewhat stalked; two or three of the upper pairs united; the uppermost of all forming a concave cup. Flowers in one or more axillary whorls, the uppermost whorl terminal; with a central bud, 6 in each whorl, highly fragrant, 2 in. long, with a blush-coloured tube. Berries each crowned by an almost entire calyx. (*Don's Mill.*) A deciduous twining shrub. Middle and South Europe, even to the river Tereck in Siberia, and on Mount Caucasus, in woods, hedges, and thickets; in England, it has been occasionally found in similar situations, in an apparently wild state Stem 15 ft. to 20 ft. Flowers blush-coloured; May and June. Fruit elliptical, tawny or orange-coloured; ripe September.

965. L. Caprifòlium.

⅄ 3. L. (*C.*) etru'sca *Santi.* The Etruscan Honeysuckle.

Identification. Santi Viagg., 1. p. 113. t. 1.; Dec. Prod., 4. p. 331.; Don's Mill., 3. p. 444.
Synonymes. L. etrúsca *Hort. Fl. Austr.* 1. p. 298.; Caprifòlium etrúscum *Ræm. et Schult. Syst.* 5. p. 261.; *Periclýmenum Gouan Hort.* p. 101.; *Caprifòlium itàlicum perfoliàtum præ'cox Tourn. Inst.* p. 608.; The Italian Honeysuckle; Mansorino, *Ital.*
Engravings. Santi Viagg., 1. p. 113. t. 1.; and our *fig.* 966.

Spec. Char., &c. Leaves deciduous, obovate, obtuse, pubescent; lower ones on short petioles, upper ones connately perfoliate, acute, glabrous. Flowers disposed in verticillate heads, with usually about three heads on the top of each branch; glabrous. (*Don's Mill.*) A deciduous twining shrub. South of France, Sicily, Valais, Carniola, and Dalmatia, on hills. Stem 15 ft. to 20 ft. Introduced ? 1700.

966. L. (C.) etrúsca.

Flowers purplish on the outside, yellow within, scented; May and June. Fruit yellow; ripe in August.

4. L. IMPLE'XA *Ait.* The interwoven, *or Minorca*, Honeysuckle.

Identification. Ait. Hort. Kew., 1. p. 231.; Dec. Prod., 4. p. 331.; Don's Mill., 3. p. 444.
Synonymes. Caprifòlium impléxum *Ræm. et Schult. Syst.* 5. p. 261.; Vincibosco sempreverde, *Ital.*
Engravings. Bot. Mag., t. 640.; and our *fig.* 967.

967. L. impléxa.

Spec. Char., &c. Quite glabrous. Leaves permanent, evergreen, glaucescent; lower ones oblong, distinct; middle ones perfoliate; uppermost ones connate, forming a hollow roundish cup. Flowers disposed in capitate whorls, ringent; purplish before they open, but becoming paler on the outside as they expand, white on the inside; but finally changing to yellow, as in the common woodbine. (*Don's Mill.*) A twining evergreen shrub. Balearic Islands, and Sicily. Stems 10 ft. to 15 ft. Introduced in 1772. Flowers purplish, finally changing to yellow; June to September. Fruit ?.

Variety.

L. i. 2 *balcárica* Viv. Camb. p. 84. Caprifòlium baleáricum *Dum. Cours. Bot. Cult.*, ed. 2. vol. iv. p. 358.; L. baleárica *Dec. Fl. Fr. Suppl.* 499.; L. Caprifòlium *Desf. Fl. Alt.* i. p. 183. — Lower leaves somewhat cordate; upper ones connate, obovate, glaucous beneath. Evergreen. Bark of branches violaceous, clothed with glaucous bloom. Flowers 4—6 in a head, large, cream-coloured, 15 in. to 18 in. long.

5. L. FLA'VA *Sims.* The yellow-*flowered* Honeysuckle.

Identification. Bot. Mag., t. 1318.; Dec. Prod., 4. p. 332.; Don's Mill., 3. p. 445.
Synonymes. Caprifòlium flàvum *Ell. Sketch.* 1. p. 271.; Caprifòlium Fràseri Pursh Sept. 1. p. 271.
Engravings. Bot. Mag., t. 1318.; and our *fig.* 968.

968. L. flàva.

Spec. Char., &c. Quite glabrous. Branches twining a little. Leaves ovate, sometimes glaucous beneath, with cartilaginous margins; upper leaves connately perfoliate. Flowers in terminal verticillate heads. Corollas rather ringent; with oblong obtuse lobes. Flowers bright yellow, but, as they fade, becoming orange-coloured; very fragrant. (*Don's Mill.*) A twining, deciduous shrub. Paris Mountains, in South Carolina; and the Catskill Mountains, New York. Stem 10 ft. to 12 ft. Introduced in 1810. Flowers bright yellow; June and July. Fruit ?.

A very desirable species, from the large size, rich yellow colour, and grateful fragrance of its flowers; but somewhat tender, and, even in the neighbourhood of London, requiring the protection of a wall.

969. L. (f.) pubéscens.

6. L. (F.) PUBE'SCENS *Sweet.* The pubescent Honeysuckle.

Identification. Sweet Hort. Brit., p. 194.; Dec. Prod., 4. p. 332.; Don's Mill., 3. p. 445.

Synonymes. Caprifòlium pubéscens *Goldie* in *Edin. Phil. Journ.* 1822, April, p. 323; L. hirsùta *Eaton Man. Bot. Ed.* 3. p. 341.; L. Góldii Spreng. Syst. 1. p. 758.
Engravings. Hook. Exot. Fl., t. 27.; Bot. Mag., t. 3103.; and our *fig.* 969.

Spec. Char., &c. Leaves broad-ovate-elliptic, on short petioles, pubescent and ciliated, glaucous beneath; upper ones connately perfoliate. Spikes or racemes composed of verticillate heads of flowers. Corollas beset with glandular pubescence. Flowers yellow. (*Don's Mill.*) This appears to hold the place in the more northern parts which L. flàva does in the south; of which, indeed, Dr. Torrey suspects it to be a variety. (*Hook. Fl. Bor. Amer.*, p. 282.) A deciduous twining shrub. North America, in Massachusetts, Vermont, New York, and Canada, in many places. Stems 10 ft. to 12 ft. Introduced in 1822. Flowers yellow; June and July. Fruit ?. It appears hardier than the preceding sort.

ꙃ 7. L. PARVIFLO'RA *Lam.* The small-flowered Honeysuckle.

Identification. Lam. Dict., 1. p. 728.; Dec. Prod., 4. p. 332.; Don's Mill., 3. p. 445.
Synonymes. Caprifòlium parviflòrum *Pursh Sept.* 1. p. 161.; Lonicera dioica *Lin. Syst. Veg.* ed. 13. p. 181.; L. mèdia *Murr. Nov. Comm. Gött.* 1776 p. 28. t. 3.; Caprifòlium bracteòsum *Michx. Fl. Bor. Amer.* 1. p. 105.; Caprifòlium dioícum *Ræm. et Schult. Syst.* 5. p. 260.; Caprifòlium glaúcum *Mœnch*; glaucous Honeysuckle; Chèvrefeuille dioique, *Fr.*; Meergrünes Geissblätt, *Ger.*; Middelboore Kamperfoelie, *Dutch*.
Engravings. Krauss, t. 27.; and our *figs.* 970. and 971.

970. L. parviflòra.

Spec. Char., &c. Quite glabrous. Leaves elliptic, sessile; lower ones somewhat connate; upper ones connately perfoliate, very glaucous beneath. Flowers disposed in verticillate heads. Corollas glabrous, with tubes gibbous at the base on one side. Filaments rather hairy. Flowers yellow, and smaller than in any of the foregoing species, but varying exceedingly in their colour; for there is a variety mentioned by Michaux in which they are purple. (*Don's Mill.*) A deciduous twining shrub. New England to Carolina, in rocky shady situations; frequent in Canada. Stem 10 ft. to 15 ft. Introduced in 1776. Flowers small, yellow; June and July. Fruit scarlet.

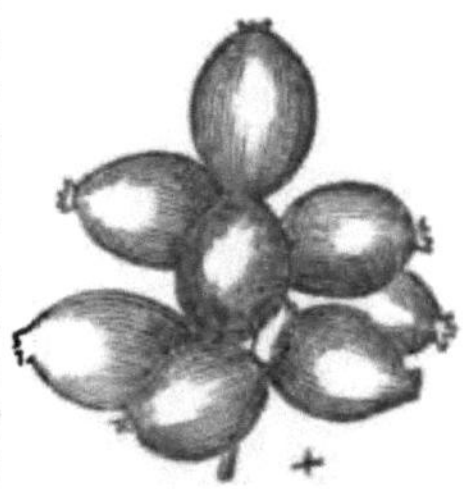

971. L. parviflòra.

ꙃ 8. L. (P.) DOUGLA'SII Dec. Douglas's Honeysuckle.

Identification. Dec. Prod., 4. p. 332.; Hook. Fl. Bor. Amer., 1. p. 282.; Don's Mill., 3. p. 445.
Synonyme. Caprifòlium Douglàsii Lindl. Hort. Trans. 7. p. 244.
Engraving. Our *fig.* 972. from a specimen in Dr. Lindley's herbarium.

Spec. Char., &c. Leaves oval, acute at both ends, petiolate, glabrous, ciliated, tomentose on the outside; upper ones connate. Flowers disposed in capitate whorls. Stigmas exserted. Stamens enclosed. Corollas pubescent, bilabiate, deep orange red. Leaves 4 in. to 6 in. long, deep green. (*Don's Mill.*) A deciduous twining shrub. Western coast of North America, on the banks of the Saskatchawan. Stems 10 ft. to 15 ft. Introd. 1824. Flowers deep orange yellow; July to September. Fruit ?.

972. L. (p.) Douglásii.

9. L. HISPIDULA *Dougl.* The bristly Honeysuckle.

Identification. Dougl. MSS.
Synonyme. Caprifolium hispidulum *Lindl. Bot. Reg.*
Engravings. Bot. Reg., t. 1761.; and our *figs.* 973. and 974.

973. L. hispidula.

974. L. hispidula.

Spec. Char., &c. Hispidly pilose. Umbels pedunculate. Corolla smooth, bilabiate, longer than the tube of the limb. Stamens exserted. Leaves petiolate, cordate-ovate, obtuse, glaucous beneath, sessile towards the summit. Stem slender. Flowers small. (*Lindl.*) A prostrate shrub. North-west America, in woods. Stem 2 ft. to 3 ft. Introduced in 1826. Flowers rose-coloured, nearly scentless; July and August. Fruit ?.

A very rare species, quite different from all the other honeysuckles. In common soil it can scarcely be kept alive; but in peat and loam it grows as readily as any other hardy American plant.

10. L. GRATA *Ait.* The pleasant, or *evergreen*, Honeysuckle.

Identification. Ait. Hort. Kew., 1. p. 231.; Dec. Prod. 4. p. 332.; Don's Mill., 3. p. 446.
Synonymes. Caprifolium grātum *Pursh Sept.* 1. p. 161.; L. virginiāna *Marsh Arb.* 136.; ? Periclymenum americānum *Mill. Dict.* No. 7.; Caprifolio sempreverde, *Ital.*
Engravings. Hort. Angl., p. 15. No. 10. t. 8.; and our *fig.* 975.

975. L. grāta.

Spec. Char., &c. Leaves permanent, obovate, rather mucronate, glaucous beneath, and reticulately veined, glabrous; upper ones connately perfoliate. Spikes composed of approximate whorls of flowers. Corollas ringent. Branches reddish brown. Flowers inclining to scarlet on the outside, according to Pursh. Corolla ringent, reddish on the outside, and yellow inside. Berries red. (*Don's Mill.*) A sub-evergeen twining shrub. Carolina to New York, on the mountains, rambling among rocks, in shady moist situations, but rare. Stems 15 ft. to 20 ft. Introd. 1730. Flowers yellow, red, and white; June or July to September, and sometimes till the commencement of frost. Fruit red; ripe in September.

The plant is of vigorous growth, with woody stems, and will live longer than most of the other species. It is inferior in vigour only to L. japónica, the Caprifòlium flexuòsum of the nurseries.

976. L. sempervirens.

B. *Limb of Corolla nearly equal.* — *Periclymenum* Tourn.

11. L. SEMPERVIRENS *Ait.* The evergreen *Trumpet* Honeysuckle.

Identification. Ait. Hort. Kew., 1. p. 230.; Dec. Prod., 4. p 332.; Don's Mill., 3. p. 446.
Synonymes. Caprifolium sempervirens *Michx. Fl. Bor. Amer.* 1. p. 105.; Periclymenum sempervirens *Mill. Dict.* No. 1.;

Alatérnus sempervirens Kœhl. ex Steud.; *Periclymenum virginiacum Riv. Mon.* 116.; Madre Selva de Virginia, *Ital.*
Engravings. Hort. Angl., t. 7.; Krauss, t. 1.; and our *fig.* 976.

Spec. Char., &c. Quite glabrous. Leaves persistent, sub-evergreen, obovate or ovate, glaucous beneath; upper ones connately perfoliate. Spikes nearly naked, composed of whorls of flowers. Tube of corolla ventricose on the upper side; limb nearly regular, with 5 roundish lobes. Branches brown. Leaves deep green above, 2 in. long and 1 in. broad. Whorls of flowers usually 3, at the top of each branch. Flowers of a beautiful scarlet outside, and yellow inside, about 1 in. long, inodorous. There are several varieties of this species, particularly one with an almost upright stem. (*Don's Mill.*) A sub-evergreen twining shrub. New York to Carolina, in dry stony woods. Stems 6 ft. to 10 ft. Introduced in 1656. Flowers scarlet; from May till August. Fruit reddish yellow; ripe in September.

Varieties.

977. L. s. màjor.

⚘ *L. s.* 2 *màjor* Ait. Curt. Bot. Mag. 1781. (Schmidt Baum. t. 104.; and our *fig.* 977.) — Leaves roundish, and flowers very large, and of a brilliant scarlet.

⚘ *L. s.* 3 *minor* Ait. Sims Bot. Mag. 1753. (Ker Bot. Reg. t. 556.; and our *fig.* 978.) L. connàta *Meerb. Icon.* t. 11.? — Leaves oblong, acute at both ends; upper ones obtuse, perfoliate. Flowers small, and scarlet both outside and inside.

⚘ *L. s.* 4 *Brównii Gordon.* — Flowers larger and brighter than those of the species. A very desirable variety.

978. L. s. minor.

The fine scarlet flowers of this species, and the length of time during which they are produced, render it a very desirable one; but it is somewhat tender, and rather capricious in regard to situation. It will not thrive in clayey or wet soil; neither in the smoke of cities, nor in a confined situation. It grows well in sand, but still better in sandy peat.

⚘ 12. L. CILIO'SA *Poir.* The ciliated-*leaved* Honeysuckle.

Identification. Poir. Suppl., 5. p. 612.; Dec. Prod., 4. p. 332.; Don's Mill., 3. p. 446.
Synonymes. Caprifòlium ciliòsum *Pursh Fl. Amer. Sept.* 1. p. 160.; L. ciliàta *Dietr. Lex. Suppl.* 4. p. 263.
Engraving. Our *fig.* 979. from a specimen in the Lambertian herbarium.

Spec. Char., &c. Upper part of the branches hairy on one side. Leaves coriaceous, reticulated, ovate, on short petioles, glaucous beneath, and ciliated on the margins; upper ones connately perfoliate. Spikes composed of approximate verticillate heads of nearly sessile flowers. Tube of corolla hairy, ventricose in the middle; limb nearly equal. Peduncles beset with glandular hairs. (*Don's Mill.*) A deciduous twining shrub. North America, on the banks of the Kooskoosky. Stem 6 ft. to 12 ft. Introduced in 1825. Flowers deep yellow; July and August. Fruit ?.

979. L. ciliòsa.

⚘ 13. L. OCCIDENTA'LIS *Hook.* The Western Honeysuckle.

Identification. Hook. Fl. Bor. Amer., 1. p. 282.; Don's Mill., 3. p. 446.
Synonymes. Caprifòlium occidentàle *Lindl. Bot. Reg.* t. 1457.; Caprifòlium ciliòsum *Douglas MSS.*
Engravings. Bot. Reg., t. 1457.; and our *fig.* 980.

Spec. Char., &c. Leaves oval, almost sessile, glabrous, ciliated, glaucous

980. L. occidentàlis.

beneath; upper ones connately perfoliate. Flowers disposed in verticillate heads. Corolla glabrous, with an elongated tube, which is gibbous above the base; the limb nearly equal. Stamens almost enclosed. (*Don's Mill.*) Branches and peduncles glabrous. A deciduous twining shrub. Fort Vancouver, on the Columbia. Stems 6 ft. to 12 ft. Introduced in 1824. Flowers large, orange red; June, July, and August. Fruit ?.

A great acquisition to our gardens; quite different from L. pubéscens, L. parviflòra, and L. Douglàsii; and, if the presence or absence of hairs in the corolla are to be depended on, it is also different from L. ciliòsa, which inhabits nearly the same country.

L. *pilòsa* Willd., Dec. Prod. iv. p. 233., Caprifòlium villòsum *H. B. et Kunth Nov. Gen. Amer.* iii. p. 427. t. 298. (and our *fig.* 981.), is a native of New Spain, in cold places, with purple flowers, not yet introduced.

981. L. pilòsa.

§ ii. *Xylósteum* Dec.

Identification. Dec. Prod., 4. p. 333.; Don's Mill., 3. p. 446.
Synonymes. Xylósteon *Juss. Gen.* 212.; Lonicera Rœm. et Schult. Syst. 5. p. 19.; Xylósteon and Chamæcérasus *Tourn. Inst.* p. 609.; Xylósteum and *Isika* Adans. Fam. 2. p. 501.; Cobœ'a Neck. Elem. No. 219.; the Fly Honeysuckle; Hackenkirsche, *Ger.*; Hondsbeziën or Hondskarsen, *Dutch.*
Derivation. From *xylon*, wood, and *osteon*, a bone; the wood of L. Xylósteum being as hard as bone.

Sect. Char., &c. Pedicels axillary, 2-flowered, bibracteate at the apex. Berries twin, distinct, or joined together more or less; 3-celled in the young state; rarely 2-celled in the adult state. The limb of the calyx is generally deciduous, therefore the fruit is usually not crowned. (*Don's Mill.*) Climbing or erect shrubs, deciduous, with leaves never connate. Of the easiest culture, and extremely hardy.

A. *Ovaries and Berries altogether distinct. Stems twining. Flowers irregular.*— Nintoòa *Dec. Prod.* iv. p. 33.

Derivation. *Nintoo*, or *Sintoo*, is the name of L. japónica in China.

⅋ 14. L. CONFU'SA *Dec.* The confused Honeysuckle.

Identification. Dec. Prod., 4. p. 333.; Don's Mill., 3. p. 446.
Synonymes. *Nintoòa* confùsa *Swt. Hort. Brit.* ed. 2.; Lonicera japónica *Andr. Bot. Rep.* t. 583.; Nintoo, Sintoo, *Kæmpf. Amæn.* 5. p. 785.; Caprifòlium japónicum *Loud. Hort. Brit.* 1st ed.
Engravings. Andr. Bot. Rep., t. 583.; Bot. Reg., t. 70.; and our *fig.* 982.

Spec. Char., &c. Branches twining, pubescent. Leaves ovate, acute, rounded at the base, downy on both surfaces, as well as the peduncles. Peduncles axillary, longer than the petioles, 2-flowered, opposite, disposed in something like a thyrse at the tops of the branches. Calycine segments ovate, and, as well as the corollas, pubescent. The flowers are snow-white at first, but gradually change to a golden yellow colour; hence it is called Suikadsara and Kinginqua, that is gold and silver flowers, by the Japanese. Corolla about an inch long, bilabiate. (*Don's Mill.*) A deciduous twining shrub. Japan, China, and the Himalayas. Stem 10 ft. to

982. L. confùsa.

15 ft. Introduced in 1805. Flowers silvery white, changing to gold colour; June and July. Fruit ?.

It is somewhat tender; nevertheless, it will grow and flower freely against an open wall in the neighbourhood of London; and the extraordinary fragrance of its flowers, which are produced in the greatest abundance, well entitles it to a place in every collection.

♃ 15. L. LONGIFLO'RA *Dec.* The long-flowered Honeysuckle.

Identification. Dec. Prod., 4. p. 333.; Don's Mill., 3. p. 447.
Synonymes. Caprifòlium longiflòrum *Sabine*; *Nintoòa* longiflòra *Swt. Hort. Brit.* ed. 2.; Caprifòlium japónicum *D. Don Prod. Fl. Nep.* 140.; Caprifòlium nepalénse *Loud. Hort. Brit.* 79.
Engravings. Bot. Reg., t. 1232.; and our *figs.* 983. and 984.

983. L. longiflòra.

984. L. longiflòra.

Spec. Char., &c. Glabrous in every part. Branches twining. Leaves petiolate, oblong-lanceolate, shining above, and pale beneath. Peduncles short, 2-flowered, about the length of the petioles. Tube of corolla very long and filiform; limb bilabiate. Flowers several inches long, at first snow-white, but finally changing to a golden yellow colour. (*Don's Mill.*) A deciduous twining shrub. China and Nepal. Stem 10 ft. to 15 ft. Introduced in 1826. Flowers snow white, changing to gold colour; July to September. Fruit ?.

A very showy species, but it is somewhat tender in British gardens.

♃ 16. L. JAPO'NICA *Thunb.* The Japan Honeysuckle.

Identification. Thunb. Fl. Jap., p. 89.; Don's Mill., 3. p. 447.
Synonymes. *Nintoòa* japónica *Swt. Hort. Brit.* ed. 2.; L. chinénsis *Hort. Kew.*; L. flexuòsa *Lodd. Bot. Cab.* 1037.; L. glabràta *Roxb.*; Caprifòlium chinénse *Loud. Hort. Brit.*; C. flexuòsum *Hort.*
Engravings. Dend. Brit., t. 117.; Bot. Cab., t. 1037.; Bot Reg. t. 712.; and our *figs.* 985. and 986.

985. L. japónica.

986. L. japónica.

Spec. Char., &c. Stems twining, flexuous, hairy. Branchlets opposite, very hairy, bearing 2 leaves and 2 sessile flowers at the base of each. Leaves about an inch long, petiolate, ovate, acutish, villous, pale beneath; uppermost ones the smallest. Corolla tubular, irregular, about an inch long, red and villous on the outside, and white inside, sweet-scented, equal in length to the stamens. (*Don's Mill.*) A twining shrub. China, Japan, and the Himalayas. Stems 15 ft. to 30 ft. Introduced in 1806. Flowers yellow and red; July to September.

Perhaps the most valuable species of the genus, next to the indigenous one. It is nearly evergreen, apparently as hardy as the common woodbine, and of far more robust habit of growth; and, probably, a much longer-lived plant. Its flowers, which are produced for several months together, are exceedingly fragrant; and, by pruning and watering, it may be kept in flower in the open garden from April to November, and in a conservatory throughout the year. No garden whatever, whether large or small, should be without this species. Intending purchasers of this species will find that plants in pots are much to be preferred, though they are one half dearer; because, if they are turned out into a large mass of prepared light rich soil, and placed against a wall, the ball being broken, and the roots spread carefully out in every direction, the shoots will cover several square yards of wall the first summer, and flower abundantly.

L. longifòlia Hort. and our *fig.* 987., of which there are plants in the London gardens, probably belongs to this section. The plants are apparently only half-hardy. Fruit blue.

987. L. longifòlia.

B. *Berries distinct, or usually connate together at the Base, and diverging at the Tip. Corolla hardly gibbous at the Base, or equal. Erect deciduous shrubs.* — *Chamæcérasi* Dec.

Derivation. The name signifies a kind of false cherry; the fruit of some of the species resemble cherries. (*Dec. Prod.* iv. p. 335.)

■ 17. L. TATA'RICA *Lin.* The Tartarian Honeysuckle.

Identification. Lin. Sp., 1. p. 247.; Dec. Prod., 4. p. 335.; Don's Mill., 3. p. 448.
Synonymes. Xylósteum cordàtum *Mœnch Meth.* p. 502.; X. tatáricum *Dum. Cours.*
Engravings. Pall. Fl. Ross., t. 36.; Jacq. Icon., t. 37.; Bot. Reg., t. 31.; and our *figs.* 988 and 989.

988. L. tatárica.

Spec. Char., &c. Quite glabrous, erect. Leaves cordate-ovate, hardly acute. Peduncles shorter than the leaves. Berries distinct when young, and nearly globose, but at length connate at the base. Flowers rose-coloured, short, somewhat gibbous at the base. Fruit black, with one of the berries usually abortive. Bracteas 2, linear-setaceous. Peduncles 2-flowered. (*Don's Mill.*) An upright shrub. Tartary. Height 4 ft. to 6 ft. Introduced in 1752. Flowers purplish; April and May. Berry bright red; ripe in August.

989. L. tatárica.

Varieties.

- ■ *L. t.* 2 *albiflòra* Dec. Prod. iii. p. 335. L. pyrenàica *Willd. Baumz.*, p. 181. — Flowers and fruit white.
- ■ *L. t.* 3 *rubriflòra* Dec. l. c. L. grandiflòrum *Lodd. Cat.*; L. sibírica *Hortul. ex Pers. Ench.* — Flowers and fruit red.
- ■ *L. t.* 4 *lùtea* Lodd. Cat. has yellowish flowers and yellow fruit.
- ■ *L. t.* 5 *latifòlia* Lodd. Cat. has broad leaves.

This is one of the most hardy of European shrubs, and one of the few which grow in the open gardens of Petersburg and Stockholm, without protection during winter. In British gardens, the plant is very common, and it is valued for its early leafing and flowering. It will grow in any soil, and almost in any situation, and is readily propagated by cuttings.

■ 18. L. (T.) NI'GRA *L.* The black-*fruited* Honeysuckle.

Identification. Lin. Sp., 247.; Dec. Prod., 4. p. 335.; Don's Mill., 3. p. 449.
Synonymes. Caprifòlium ròseum *Lam. Fl. Fr.* 3. p. 368.; Chamæcérasus nigra *Delarb. Fl. Auv.* ed. 2. p. 130.; Ciliegia salvatica, *Ital.*
Engravings. Jacq. Aust., t. 314.; Schmidt Baum., t. 110.; Gesn., fasc. 37. t. 8. f. 48.

Spec Char., &c. Erect. Leaves oval-oblong or elliptic, on short petioles, rather villous when young, but nearly glabrous in the adult state. Peduncles 2-flowered, elongated,

990. L. (t.) n. campaniflòra. 991.

shorter than the leaves. Corolla reddish, and pubescent on the outside, but whitish on the inside. Bracteas 4, under the ovaries; the two outer ones lanceolate, and the inner quadrifid. Berries black, globose, joined together at the side. (*Don's Mill.*) An erect shrub. Middle Europe, in subalpine woods, as in France, Switzerland, Austria, Silesia, Piedmont, &c. Height 3 ft. to 4 ft. Introduced in 1597. Flowers whitish; March to May. Fruit black; ripe in August.

Variety.

L. (*t.*) *n.* 2 *campaniflòra;* Xylósteum campaniflòrum *Lodd. Cab.* t. 1361., and our *figs.* 990, 991.; has the flowers bell-shaped.

19. L. (T.) CILIA'TA *Mühl.* The ciliated-*leaved* Honeysuckle.

Identification. Mühl. Cat., p. 22.; Dec Prod., 4. p. 335.; Don's Mill., 3. p. 448.
Synonymes. Xylósteum ciliàtum *Pursh Sept* 1. p. 161.; L. tatárica *Michx. Fl. Amer.* 1. p. 166. but not of Lin.; L. canadénsis *Ræm. et Schult. Syst.* 5. p. 260.
Engraving. Our *fig.* 992. from a living specimen.

Spec. Char., &c. Erect. Leaves ovate or oblong, cordate, thin, ciliated, villous beneath in the young state. Peduncles elongated. Bracteas 2, ovate, three times shorter than the ovaries, which are distinct. Corolla bluntly spurred at the base; with short, nearly equal lobes. Berries distinct, red, divaricate. Flowers white, with a tinge of red or yellow; tube ventricose above; limb with short acute segments; style protruded. (*Don's Mill.*) An erect shrub. Canada to Virginia, and throughout Canada, on mountains among rocks, in rich soils. Height 4 ft. to 6 ft. Introduced in 1824. Flowers reddish or yellowish white; June, July.

992. L (t.) ciliata.

20. L. PYRENA'ICA *L.* The Pyrenean Honeysuckle.

Identification. Lin. Sp., p. 248.; Dec. Prod., 4 p. 335.; Don's Mill., 3. p. 448.
Synonymes. Caprifòlium pyrenàicum *Lam. Fl. Fr.* 3. p. 366.; Xylósteum pyrenàicum *Tourn. Inst* 609.
Engraving. Our *fig* 993. from a specimen in the British Museum.

993. L. pyrenàica.

Spec. Char., &c. Glabrous, erect. Leaves obovate-lanceolate, acute, glaucous beneath. Peduncles 2-flowered, shorter than the leaves. Bracteas oblong-linear, foliaceous. Flowers almost regular. Berries globose, distinct. Corolla white, twice the size of that of L. Xylósteum, funnel-shaped: limb 5-cleft, flat; with equal, ovate, obtuse segments. (*Don's Mill.*) An erect shrub. Pyrenees, on calcareous rocks, in exposed situations. Height 4 ft. to 5 ft. Introduced in 1739. Flowers white; May. Berries white.

21. L. PUNI'CEA *Sims.* The crimson-*flowered* Honeysuckle.

Identification. Sims Bot. Mag., t. 2469.; Dec. Prod., 4. p. 335.; Don's Mill., 3. p. 448.
Synonyme. Symphoricárpos puniceus *Swt.*
Engravings. Bot Mag., t. 2469.; and our *fig.* 994.

Spec. Char, &c. Erect. Leaves ovate, subcordate at the base, of the same colour on both surfaces. Peduncles axillary, and almost terminal, 2-flowered, shorter than the leaves. Tube of corolla rather gibbous at the base; segments of corolla nearly equal, irregularly arranged, 3 one way and 2 another. Berries distinct ?. Leaves sometimes three in a whorl on the young shoots. (*Don's Mill.*) An erect shrub. Native country unknown. Height 2 ft.

994. L. punicea.

to 4 ft. Cultivated in 1822. Flowers deep red, scarlet, or crimson; April and May. Fruit ?.

■ 22. L. XYLO'STEUM *L.* The bony-wooded, *or upright, Fly* Honeysuckle.

Identification. Lin. Sp., 248.; Dec. Prod., 4. p. 335.; Don's Mill., 3. p. 448.
Synonymes. Caprifòlium dumetòrum *Lam. Fl. Fr.* 3. p. 367.; Xylósteum dumetòrum *Mœnch Meth.* p. 502.; Gisilostio, *Ital.*
Engravings. Eng. Bot., t. 916.; Fl. Græc., t. 223.; and our *fig.* 995.

Spec. Char., &c. Erect, downy. Leaves ovate, acute, petiolate, soft. Peduncles 2-flowered, shorter than the leaves. Bracteas hairy, double; the two outer ones lanceolate, spreading; inner a small concave scale under each germ. Berries oval, distinct, 1-celled, 6-seeded. Flowers small, cream-coloured, downy. Calyx of 5 obtuse lobes. Berries scarlet. (*Don's Mill.*) An upright shrub. Europe, to Caucasus, in thickets, hedges, and rocky places, and by the sides of woods. Height 8 ft. to 10 ft. Cultivated in 1596. Flowers cream-coloured; July. Fruit scarlet; ripe in September. Naked young wood greyish white.

995. L. Xylósteum

Varieties.

■ *L. X.* 2 *leucocárpum* Dec. Prod. iv. p. 335. has white berries.
■ *L. X.* 3 *xanthocárpum* Dec. l. c. has the berries yellow.
■ *L. X.* 4 *melanocárpum* Dec. l. c. has black berries.

Linnæus says that it makes excellent hedges in a dry soil; that the clear parts between the joints of the shoots are used in Sweden for tobacco-pipes; and that the wood, being extremely hard, makes teeth for rakes, &c., and yields only in beauty to that of L. tatárica for walking-sticks. It is one of the oldest and hardiest inhabitants of British shrubberies. In the English garden, or rather park, at Munich, it is planted in masses and groups, along with other masses and groups of Córnus álba, Sàlix vitellìna, and Vibúrnum O'pulus; and, in the winter time, the whitish-grey bark of its shoots contrasts finely with the red, yellow, or brown, bark of the shrubs mentioned.

■ 23. L. HI'SPIDA *Pall.* The hispid Honeysuckle.

Identification. Pall. ex Willd. MSS.; Led. Flor. Ross. Alt. Ill., t. 212.; Don's Mill., 3. p. 449.
Engravings. Led. l. c.; and our *fig.* 996.

Spec. Char., &c. Branches hispid. Leaves ovate, ciliated, petiolate, glabrous on both surfaces. Peduncles 2-flowered. Bracteas ovate-elliptic, exceeding the berries. (*Don's Mill.*) An upright shrub. Siberia, on the Altaian Mountains. Height 2 ft. to 3 ft. Introduced ?. Flowers greenish white, pendulous; May and June. Berries distinct, purple; ripe in August.

Branches opposite, glabrous or bristly, brownish. Leaves 1½ or 2 inches long, and 1 in. broad, glabrous on both surfaces, cordate at the base.

996. L. híspida.

■ 24. L. FLEXUO'SA *Thunb.* The flexible-*stemmed* Honeysuckle.

Identification. Thunb. in Lin. Trans., 2. p. 330., but not of Lodd., nor Ker; Don's Mill., 3. p. 449.
Synonymes. L. nigra *Thunb. Fl. Jap.* p. 89., but not of Lin.; L. brachýpoda *Dec. Prod.* 4. p. 335.

Engraving. Our *fig.* . in p. .

Spec. Char., &c. Erect, branched. Branches very villous at the apex. Leaves ovate-oblong, acute, on short petioles, glabrous; petioles villous; nerves of leaves puberulous. Flowers axillary, few, almost sessile. Berries globose, glabrous. Stems flexuous. Leaves about an inch long; upper ones the smallest. Peduncles hardly a line long. Berries distinct, ovate, acuminated, black. (*Don's Mill.*) An erect deciduous shrub. Japan. Height 4 ft. to 5 ft. Introduced in 1806. Flowers ?; June and July. Berries black; ripe ?.

C. *Berries either distinct or joined together. Corolla very gibbous at the Base. Erect bushy Shrubs.—Cuphánthæ* Dec.

Derivation. From *kuphos*, gibbous, and *anthos*, a flower; in reference to the flower being gibbous on one side at the base.

25. L. INVOLUCRA'TA *Banks.* The involucrated Honeysuckle.

Identification. Banks Herb. ex Spreng. Syst., 1. p. 759.; Dec. Prod., 4. p. 336.; Don's Mill., 3. p. 449.
Synonyme. Xylósteum involucràtum *Richards.* in *Frank. First Journ.* ed. 1. append. p. 6.
Engravings. Our *figs.* 997, 998, 999.

997, 998, 999. L. involucrata.

Spec. Char., &c. Erect. Branches acutely tetragonal. Leaves ovate or oval, petiolate, membranous, beset with appressed hairs beneath. Peduncles axillary, 2—3-flowered. Bracteas 4; two outer ovate, two inner broad, obcordate, at length widening, clothed with glandular pubescence. Corolla pubescent, gibbous at the base on the outside; yellowish, tinged with red. Style exserted. (*Don's Mill.*) An erect shrub. North-west America, between lat. 54° and 64° (but probably confined to the vicinity of the Saskatchawan); thence to the Rocky Mountains. Height 2 ft. to 3 ft. Introduced in 1824. Flowers yellowish, tinged with red; May. Fruit ?.

26. L. LEDEBOU'RII Eschsch. Ledebour's Honeysuckle.

Identification. Eschsch. Mém. Act. Soc. Petersb., 10. p. 284.; Hook. et Arn. in Beech Voy. pt. p. 145.; Don's Mill., 3. p. 449.
Engraving. Our *fig.* 1000. from a living specimen.

Spec. Char., &c. Erect. Branches elongated, acutely tetragonal. Leaves ovate or oblong, somewhat acuminated, stiff, pubescent, tomentose on the nerves. Peduncles axillary, 2—3-flowered. Bracteas 4; 2 outer ones ovate; 2 inner broad obcordate, pubescent, at length increasing in size. Corollas gibbous at the base on the outside. Berries distinct. (*Don's Mill.*) An erect deciduous shrub. California. Height 2 ft. to 3 ft. Introduced in 1838. Flowers yellow, tinged with red; June and July. Berries dark purple; ripe in September.

Very nearly allied to L. involucràta.

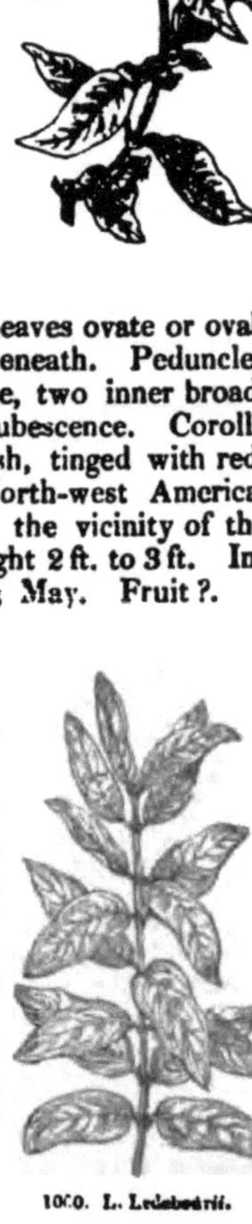

1000. L. Ledebourii.

D. *Berries two on each Peduncle, joined together in one, which is bi-umbilicate at the Apex. Erect, bushy, deciduous Shrubs.* — Isìkæ *Adans.*

Derivation. A name, the origin of which is unknown, employed by Adanson to designate this division of the genus.

27. L. ALPI'GENA *H.* The alpine Honeysuckle.

Identification. Lin. Sp., 248.; Dec. Prod., 4. p. 336.; Don's Mill., 3. p. 449.
Synonymes. Caprifòlium alpìnum *Lam. Fl. Fr.*; Caprifòlium alpìgenum *Gærtn. Fruct.* 1. p. 136. *Isìka* alpìgena *Börck.*; *Isìka* lùcida *Mœnch*; Xylósteum alpìgenum *Lodd. Cat.*; Chamæcérasus alpìgena *Delarb.*; Cherry Woodbine; Heckenkirsche, *Ger.*; Chamaeceraso, *Ital.*
Engravings. Jacq. Fl. Aust., t. 274.; N. Du Ham., 1. t. 16.; and our *figs.* 1001. and 1002.

1001, 1002. L. alpígena.

Spec. Char., &c. Erect. Leaves oval-lanceolate or elliptic, acute, glabrous or pubescent, on very short petioles, rather ciliated. Peduncles 2-flowered, shorter than the leaves. Corolla gibbous at the base, and greenish yellow tinged with red or purple. Berries red, and of the size and appearance of those of a cherry; whence it is called cherry woodbine by Johnson. Leaves large. (*Don's Mill.*) A large, upright, deciduous shrub. Middle and South of Europe, in subalpine places and mountains. Height 5 ft. to 8 ft. Introduced in 1596. Flowers greenish yellow, tinged with red; April and May. Fruit red; ripe in August.

Variety.

L. a. 2 *sibírica* Dec. Prod., iv. p. 336. L. sibírica *Vest* in *Rœm. et Schult. Syst.* 5. p. 259. — Lower leaves rather cordate. Peduncles thickened a little under the flowers. Like most other varieties of trees and shrubs, natives of the West of Europe, and also indigenous to Siberia, coming into leaf and flower a week, or more, earlier than the species.

28. L. (A.) MICROPHY'LLA *Willd.* The small-leaved Honeysuckle.

1003. L. (a.) microphylla.

Identification. Dec. Prod., 4. p. 336.; Don's Mill., 3. p. 450.
Synonymes. L. alpìgena *Sievers*; L. montàna, and L. mexicàna *Hort.*
Engravings. Led. Fl. Ros. Alt. Ill., t. 213.; and our *fig.* 1003.

Spec. Char., &c. Leaves elliptic, acute at both ends, glaucous beneath, rather villous on both surfaces, and sometimes rounded at the base. Peduncles 2-flowered, and shorter than the leaves. Corollas greenish yellow. Berries joined, of a reddish orange colour. The epidermis falls from the branches. (*Don's Mill.*) An erect shrub. Eastern Siberia. Height 3 ft. to 4 ft. Introduced in 1818. Flowers greenish yellow; April and May. Fruit reddish orange; ripe in August.

29. L. OBLONGIFO'LIA *Hook.* The oblong-leaved Honeysuckle.

Identification. Hook. Fl. Bor. Amer., 1. p. 284. t. 100.; Don's Mill., 3. p. 450.
Synonyme. Xylósteum oblongifòlium *Goldie* in *Edin. Phil. Journ.* 6. p. 232.
Engravings. Hook. Fl. Bor. Amer., 1. t. 100.; and our *fig.* 1004.

1004. L. oblongifolia.

Spec. Char., &c. Erect. Leaves oblong or oval, clothed with velvety pubescence beneath. Peduncles elongated, erect. Bracteas obsolete. Tube of corolla hairy, gibbous at the base on one side. Limb unequal, deeply 2-lipped; the upper lip 4-toothed, and the lower one nearly entire. Berries joined in one, which is bi-umbilicate at the top, bluish black in the dried state, and about the size of a pea. (*Don's Mill.*) An erect shrub. Island of Montreal, in the St. Lawrence, about Montreal; Lake Winnipeg; and the western parts of the state of New York. Height 3 ft. to 4 ft. Introduced in 1823. Flowers yellow; April and May. Fruit bluish black; ripe in August. Horticultural Society's Garden.

▪ 30. L. CÆRU'LEA *L.* The blue-*berried* Honeysuckle.

Identification. Lin. Sp., 349.; Dec. Prod., 4. p. 337.; Don's Mill., 3. p. 450.
Synonymes. L. villòsa *Mühl. Cat.* p. 22.; Xylósteon villòsum *Michx. Fl. Bor. Amer.* 1. p. 106.; X. Solònis Eaton Man. Bot. p. 518.; L. velutìna *Dec. Prod.* 4. p. 337.; L. altàica *Pall. Fl. Ross.* t. 37.; Xylósteum cærùleum canadénse *Lam. Dict.* 1. p. 731.; X. canadénse *Du Ham. Arb.* 2. p. 373.; Caprifòlium cærùleum *Lam. Fl. Fr.*, Chamæcérasus cærùlea *Delarb. Fl. Au.*; L. pyrenàica *Pall. Fl. Ross.* p. 58; L. Pallàsii Led. Fl. Ross. Alt. Ill. t. 131. Ciliegia alpina, *Ital.*
Engravings. Bot. Mag., t. 1965.; Pall. Fl. Ross., t. 37.; Led. Fl. Ross. Alt. Ill., t. 131.; and our *figs.* 1005. and 1006.

Spec. Char., &c. Erect. Leaves oval-oblong, ciliated, stiffish, densely clothed with pubescence while young. Peduncles short, 2-flowered, reflexed in the fructiferous state. Bracteas 2, subulate, longer than the ovaria. Tube of corolla glabrous, short, gibbous on one side at the base; lobes of limb short, nearly equal. Berries closely joined in one, which is bi-umbilicate at the apex. Flowers greenish yellow, tubular. Berries elliptic or globose, dark blue, and covered with a kind of bloom. Bark of young shoots purplish. There is no difference between the American and European plants of this species. (*Don's Mill.*) An erect shrub. Europe, and throughout the woody country of British North America, as far as lat. 66°; and of Siberia and Kamtschatka. Height 3 ft. to 5 ft. Introduced in 1629. Flowers greenish yellow; March and April. Fruit dark blue; ripe in August.

1005. L. cærùlea.

1006. L. cærùlea.

▪ 31. L. ORIENTA'LIS *Lam.* The Oriental Honeysuckle.

Identification. Lam. Dict., 1. p. 731.; Dec. Prod., 4. p. 337.; Don's Mill., 3. p. 450.
Synonymes. L. caucàsica *Pall. Fl. Ross.* 1. p. 57.; L. cærùlea *Güld. Itin.* 1. p. 423., Chamæcérasus orientàlis laurifòlia *Tourn. Cor.* p. 42.
Engraving. Our *fig.* 1007 from Tournefort's specimen in the British Museum.

Spec. Char., &c. Erect. Leaves on very shor petioles, ovate-lanceolate, acute, quite entire, smoothish. Peduncles 2-flowered, shorter than the leaves. Bracteas 2, setaceous. Berries joined in one, didymous and bi-umbilicate at the apex, 10-seeded. Leaves stiffish, veiny, larger than in L. cærùlea. Flowers greenish yellow. (*Don's Mill.*) An erect shrub. Iberia and Asia Minor, in woods. Height 3 ft. to 5 ft. Introduced in 1825. Flowers greenish yellow; April to June. Berries black or dark blue; ripe in September.

1007. L. orientàlis.

▪ 32. L. IBE'RICA *Bieb.* The Georgian Honeysuckle.

Identification. Bieb. Fl. Taur., and Suppl., 395.; Dec. Prod., 4. p. 337.; Don's Mill., 3. p. 450.
Synonyme. Xylósteon ibéricum *Bieb. Cent. Pl. Rar.* 1. t. 13. *ex Suppl.*, and *Lodd. Cat.* ed. 1836.
Engravings. Bieb. Cent. Fl. rar., 1. t. 13., ex Suppl.; and *figs.* 1008. 1009. from living specimens.

Spec. Char., &c. Erect. Leaves petiolate, cordate, roundish, tomentose or

1008. L. ibérica.

pubescent. Peduncles 2-flowered, shorter than the leaves. Bracteas oblong, ciliated. Berries joined together to the middle, globose. Corollas lucid, of the form of those of L. alpígena. Ovarium tomentose. Leaves like those of Cotoneáster vulgàris (*Don's Mill.*) An erect shrub. Georgia, about Teflis. Height 3 ft. to 4 ft. Introduced in 1824. Flowers greenish yellow; April and May. Fruit blood-coloured, sometimes pointed as in *fig.* 1008.; ripe in August.

A very neat little bush, which makes very good garden hedges.

1009. L. ibérica.

Genus V.

SYMPHORICA'RPOS *Dill.* The St. Peter's Wort. *Lin. Syst.* Pentándria Monogýnia.

Identification. Dill. Elth., p. 371.; Dec. Prod., 4. p. 338.; Don's Mill., 3. p. 451.
Synonymes. Symphoricárpa *Neck. Elem.* p. 220.; Symphòria *Pers. Ench.* 1. p. 214.; Anisánthus *Willd. Rel.*; Lonicera sp. *Lin.*
Derivation. From *sumphoreō*, to accumulate, and *karpos*, fruit; species bearing the fruit in groups. How it obtained the name of St. Peter's Wort we have not been able to ascertain.

Gen. Char. *Calyx* tube globose; limb small, 4—5-toothed. *Corolla* funnel-shaped, almost equally 4—5-lobed. *Stamens* 5, hardly exserted. *Stigmas* semiglobose. *Ovarium* adnate. *Berry* 4-celled. (*Don's Mill.*)

Leaves simple, opposite, exstipulate, deciduous; oval, quite entire. *Flowers* on short peduncles, axillary or many together, bibracteate, small, white or rose-coloured, on short pedicels.—Shrubs erect, bushy, oppositely branched; natives of Europe and North America; of the easiest culture in common garden soil; and readily increased by suckers, which they throw up in abundance.

1. S. vulga'ris *Michx.* The common St. Peter's Wort.

Identification. Michx. Fl. Bor. Amer., 1. p. 100.; Dec. Prod., 4. p. 339.; Don's Mill., 3. p. 451.
Synonymes. Lonicera Symphoricárpos *Lin. Sp.* 249.; S. parviflòra *Desf. Cat.*; Symphòria conglomeràta *Pers. Ench.* 1. p. 214.; Symphòria glomeràta *Pursh Sept.* p. 162.
Engravings. Schmidt Baum., t. 115.; and our *fig.* 1010.

Spec. Char., &c. Flowers disposed in axillary capitate clusters, composed of nearly sessile racemules. Corolla white. Berries red, size of hempseed; but, in America, according to Pursh, the flowers are small, red and yellow, and the berries purple. Branches brown, smooth. Leaves elliptic ovate, obtuse, glaucous, and pubescent beneath. The berries are numerous, and ripen in winter. (*Don's Mill.*) An erect bushy shrub. Virginia, Carolina, and Pennsylvania, in sandy dry fields. Height 3 ft. to 6 ft. Introduced in 1730. Flowers small, red and yellow; August and September. Fruit purple; ripe in December.

1010. Symphoricárpos vulgàris.

Variety.

■ *S. v.* 2 *fòliis variegàtis*, S. glomeràta fòliis variegàtis *Lodd. Cat.*, has the leaves finely variegated with green and yellow.

■ 2. S. MONTA'NUS *Humb. et Bonp.* The Mountain St. Peter's Wort.

Identification. Humb. et Bonp. Nov. Gen. et Spec., 3. p. 332.; Dec. Prod., vol. 4. p. 339.
Synonymes. Symphòria montàna *Spreng. Syst. Veg.* 1. p. 757.; S. glaucéscens *Don's Mill.* 3. p. 452.
Engravings. Maund's Botanist, 1. t. 20.; and our *fig.* 1011.

Spec. Char., &c. Leaves ovate, acute, slightly mucronate, rounded at the base, pubescent beneath. Flowers axillary, mostly solitary. A dense erect subevergreen shrub. Mexico, on mountains, 7000 to 8000 ft. of elevation. Height 5 ft. to 6 ft. Introduced in 1829. Flowers pinkish; August to October. Fruit globose white; ripe in December.

1011. S. montàna.

A very desirable shrub, perfectly hardy, and almost evergreen. It commences flowering in August, and does not cease till it is checked by frost. Layers in common soil.

■ 3. S. RACEMO'SUS *Michx.* The racemose-*flowered* St. Peter's Wort, *or Snowberry.*

Identification. Michx. Fl. Bor. Amer., 1. p. 107.; Dec. Prod., 4. p. 339.; Don's Mill., 3. p. 451.
Synonymes. Symphòria racemòsa *Pursh Sept.* 1. p. 162.; ?S. elongàta, and S. heterophýlla *Presl* in *Herb. Hænke*; S. leucocàrpa *Hort.*
Engravings. Bot. Mag., t. 2211.; Lodd. Bot. Cab., t. 230.; and our *fig.* 1012.

Spec. Char., &c. Flowers disposed in nearly terminal, loose, interrupted racemes, which are often leafy. Corolla densely bearded inside. Style and stamens enclosed. Leaves glaucous beneath. Corolla rose-coloured. Berries large, white. (*Don's Mill.*) A bushy shrub, with numerous ascending shoots. North America, on mountains, near Lake Mistassins, on the banks of the Missouri, and various other places. Height 4 ft. to 6 ft. Introduced in 1817. Flowers rose-coloured; July to September. Fruit large, white; ripening in October, and remaining on great part of the winter.

1012. S. racemòsus.

The S. elongàtus and S. heterophýllus *Presl* in *Herb. Hænke*, which were collected about Nootka Sound, do not differ from this species, in which the lower leaves are sometimes deeply sinuated. In small gardens, this shrub is rather troublesome, from the numerous suckers it throws up from the roots; but, as its flowers are much sought after by bees, and its berries are excellent food for game; that habit, when it is planted for these purposes, is found rather advantageous than otherwise. For single specimens in small gardens, it might be desirable to graft it on Lonícera Xylósteum, or some allied species of suitable habit. So grafted, standard high, it would form a very elegant little tree.

■ 4. S. OCCIDENTA'LIS *Richards.* The Western St. Peter's Wort.

Identification. Richards. and Frankl. 1st Journ., edit. 2., app. p. 6.; Hook. Fl. Bor Amer., 1. p. 285.
Synonyme. Wolf-berry, *Amer.*
Engraving. Our *fig.* 1013. from a specimen in Sir W. J. Hooker's herbarium.

Spec. Char., &c. Spikes dense, terminal and axillary, drooping. Corolla and seg-

1013. S. occidentàlis

ments densely bearded inside. Style and stamens a little exserted. (*Don's Mill.*) A dense shrub, bearing a close resemblance to S. racemòsus. British North America, in the woody country between lat. 54° and 64°. Height 4 ft. to 6 ft. Introduced ?. Flowers pinkish; July to September. Fruit white; October, and remaining on during the winter.

Distinguished from S. racemòsus by the larger, less glaucous, more rigid, and denser foliage, and by the flowers being arranged in dense drooping spikes, longer than in S. racemòsus, and by the prominent style and stamens.

Genus VI.

LEYCESTE'RIA Wall. The Leycesteria. *Lin. Syst.* Pentándria Monogýnia.

Identification. Wall. in Roxb. Fl. Ind., 2. p. 181.; Dec. Prod., 4. p. 338.; Don's Mill., 3. p. 451.
Derivation. Named by Dr. Wallich after his friend *William Leycester*, formerly chief judge of the principal native court under the Bengal Presidency; "who during a long series of years, and in various parts of Hindoostan, has pursued every branch of horticulture with a munificence, zeal, and success, which abundantly entitle him to that distinction."

Gen. Char. *Calyx* with an ovate tube, and an unequal, 5-parted, permanent limb. *Segments* unequal, small, linear, glandularly ciliated. *Corolla* funnel-shaped; having the tube gibbous above the base, and the limb campanulate, and divided into 5 ovate nearly equal lobes. *Stamens* 5. *Stigma* capitate. *Berry* roundish, 5-celled. (*Don's Mill.*)

Leaves simple, opposite, exstipulate, sub-evergreen; ovate-lanceolate, acuminated, petiolate, smooth, entire, membranous, glaucous, with an obtuse subcordate base. *Petioles* pilose. *Flowers* white, with a tinge of purple; disposed in whorls, forming short leafy drooping racemes, which terminate the branches and branchlets. *Bracteas* large, foliaceous, purplish, pubescent and ciliated, lanceolate, acuminated; generally 6 under each whorl of flowers. *Berries* deep purple, approaching to black, as large as a common-sized gooseberry. Shrub large, rambling, with elongated fistular branches, which rise from scaly buds. Native of Nepal.

This genus appears to be intermediate between Caprifoliàceæ and Rubiàceæ; but from the last it is distinguished by the want of stipules.

1014. Leycestèria formòsa.

■ 1. L. FORMO'SA *Wall.* The beautiful Leycesteria.

Identification. Wall. in Roxb. Fl. Ind., 2. p. 182.; Dec. Prod., 4. p. 338.; Don's Mill, 3. p. 451.
Synonyme. Hamèlia connàta *Pucrari MSS.*
Engravings. Plant. As. Rar., 2. t. 120.; and our *fig.* 1014.

Spec. Char., *&c.* As in Gen. Char. A large, rambling, sub-evergreen shrub. Nepal, on mountains; between 6000 ft. and 8000 ft. high, among forests of pine and oak. Height in England, against a wall, 6 ft. to 8 ft. Introduced in 1824. Flowers white, with a tinge of purple; August to October. Fruit purple; ripe in October.

Trained against a wall, this shrub has proved quite hardy, but in our cloudy atmosphere it has rather disappointed expectation in the colour of its bracteas, which are much less brilliant than they appear to be in the Himalayas. Cuttings or seeds, which are ripened freely, in common soil.

ORDER XLI. RUBIA'CEÆ.

ORD. CHAR. *Calyx* with a variable limb. *Corolla* monopetalous, with a variable limb, but generally 4—5-lobed; æstivation twisted or valvate. *Stamens* equal in number to the segments of the corolla, and more or less adnate to its tube. *Anthers* introrse. *Ovarium* 2- or many-celled, crowned by the limb of the calyx. *Style* 1. *Stigmas* 2. *Fruit* baccate or capsular. *Cells* 1—2- or many-seeded. *Albumen* horny and fleshy. (*G. Don.*)

Leaves simple, opposite, or 3 in a whorl, stipulate, deciduous. *Stipules* short, distinct, or a little combined. *Flowers* on peduncles, naked, rising from the axils of the leaves, or from the tops of the branches; heads globose, in consequence of the flowers being sessile, and seated on a sessile piliferous receptacle.

This order includes a great number of genera; but there is only one of these that contains any ligneous species truly hardy in British gardens.

GENUS I.

CEPHALA'NTHUS *L.* THE BUTTON-WOOD. *Lin. Syst.* Tetrándria Monogýnia.

Identification. Lin. Gen., No. 113.; Gærtn. Fruct., 2. t. 86. Lam. Ill., t. 59.; Juss. Mém. Mus., 6. p. 402.; Rich. Diss., with a fig.; Dec. Prod., 4. p. 538.; Don's Mill., 3. p. 610.; Lodd. Cat., ed. 1836.
Synonymes. Cephalante, *Fr.*; Knopflaum *Ger.*; Cefalanto, *Ital.*
Derivation. From *kephalē*, a head, and *anthos*, a flower; in allusion to the flowers being disposed in globular heads.

Gen. Char. &c. *Calyx* with an obversely pyramidal tube, and an angular 5-toothed limb. *Corolla* with a slender tube, and a 4-cleft limb; lobes erectish. *Stamens* 4, short, inserted in the upper part of the tube, hardly exserted. *Style* much exserted. *Stigma* capitate. *Fruit* inversely pyramidal, crowned by the limb of the calyx, 2—4-? celled, and separating into 2—4 parts; cells, or parts, 1-seeded, indehiscent, and sometimes empty by abortion. *Seeds* oblong, terminating in a little callous bladder. (*Don's Mill.*) —A shrub, with terete branches; native of North America.

Leaves and *Flowers* as in the order.

■ 1. C. OCCIDENTA'LIS *L.* The Western Button-wood.

Identification. Lin. Sp., 138.; Dec. Prod., 4. p. 538.; Don's Mill., 3. p. 610.
Synonymes. C. oppositifòlius *Mœnch Meth.* p. 487.; Swamp Globe Flower, *Amer.*
Engravings. Du Ham. Arb., 1. t. 54.; Schmidt Arb., 1. t. 45.; and our *figs.* 1015. and 1016.

Spec. Char., *&c.* Leaves opposite, or 3 in a whorl, ovate or oval, acuminated.

Peduncles much longer than the heads, usually by threes at the tops of the branches. Petioles reddish next the branches. Heads of flowers globular, size of a marble. Stipules deciduous. (*Don's*

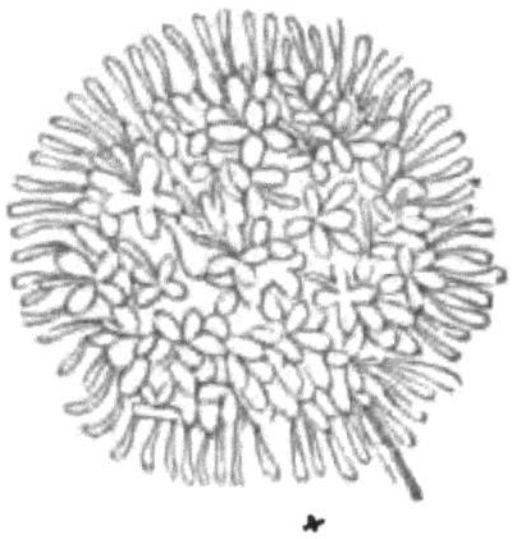

1015. Cephalánthus occidentàlis. 1016.

Mill.) A bushy shrub. Canada to Florida, in marshy places. Height 3 ft. to 6 ft. Introduced in 1735. Flowers yellowish white; July and August. Fruit brownish; ripe in October.

Variety.

▲ *C. o.* 2 *brachýpodus* Dec. Prod. iv. p. 539. — Leaves elliptic-oblong, 3 in a whorl, on short petioles. Petioles 3—4 lines long. There are varieties of this, with either glabrous or downy branches. North of Mexico, near Rio de la Trinidad and Bejar.

It will grow in common garden soil, but prefers peat kept moist; and is propagated chiefly by seeds, but will also grow by cuttings and layers. It is an interesting shrub, from its curious round heads of flowers, and from the lateness of the season at which these appear.

Order XLII. COMPO'SITÆ.

Ord. Char. *Calyx* limb membranous or wanting; or divided into bristles, paleæ, or hairs. *Corolla* 5-toothed or 5-lobed, tubular, ligulate, or bilabiate on the top of the ovarium. *Anthers* combined, rarely free. *Ovarium* 1-celled, 1-seeded. *Style* 1. *Stigmas* 2. *Fruit* an achenium, crowned by the limb of the calyx. *Albumen* none. Characterised by the cohesion of the anthers, and the arrangement of the flowers in involucrated heads on a common receptacle. (*G. Don.*)

Leaves simple, or compound, stipulate or exstipulate, deciduous or ever green. *Flowers* grouped in heads; those in each head so disposed, and so environed by an involucre composed of bracteas that corresponds to a calyx, as to seem to constitute but one flower.

The genera that include hardy ligneous species are mostly natives of Europe and North America: they are all of the easiest propagation and culture in any common garden soil, and are thus contradistinguished: —

Stæheli'na Lessing. Flowers bisexual. Receptacle with chaffy projections.

Ba'ccharis R. Br. Flowers diœcious, all tubular. Receptacle naked. Pappus pilose.

Iva L. Flowers monœcious, all tubular. Receptacle flat paleaceous. Achenia naked, but horned.

Santoli'na *L.* Receptacle furnished with somewhat flower-clasping paleæ. Achenia naked.

Artemi'sia Cass. Receptacle chaffless. Achenia naked. Heads discoid.

SENE'CIO *Lessing*. Receptacle naked. or alveolate. Styles penciled. Pappus pilose, caducous.

MUTI'SIA Cav. Receptacle naked. Achenia somewhat beaked. Pappus of many series, feathery.

With the exception of *Baccharis*, there is scarcely a plant belonging to the order Compósitæ which is truly ligneous, and at the same time hardy in British gardens, and sufficiently bulky for a general arboretum. Where an arboretum is planted on a lawn, and where it is not intended to cultivate the soil about the roots of the plants, there is not a single genus in this order, with the exception of that mentioned, which could with propriety be introduced. Even the common southernwood, if not planted in dug soil or on rockwork, would soon become stunted, and would ultimately die off. Nevertheless, in a technical enumeration of trees and shrubs, these species could not be omitted.

GENUS I.

STÆHELI'NA Lessing. THE STÆHELINA. *Lin. Syst.* Syngenèsia Æquàlis.

Identification. Lessing Synops. Gen. Compos., p. 5.; Ait. Hort. Kew., ed. 2., vol. 4. p. 512.
Synonyme. Stæheline, *Fr.* and *Ger.*
Derivation. So named in honour of *John Henry Stæhelin*, and his son *Benedict*, Swiss botanists and physicians.

Gen. Char., &c. Heads homogamous, equal-flowered. *Involucrum* cylindrical, the scales imbricated and adpressed. *Receptacle* flat, paleaceous; the paleæ narrow, persistent, hardly concrete at the base. *Corolla* 5-cleft, regular. *Filament* glabrous. *Anthers* appendiculate at top, bisetose at the base; the tails more or less bearded. *Style* bearded on the thickened part. *Stigmas* concrete at base, and free at apex, obtuse. *Fruit* oblong, areolate at apex. *Pappus* in one series, the hairs combined at the base into 4 or 6 bundles. (*G. Don.*)

Leaves simple, alternate, exstipulate, evergreen; linear, hoary or silky beneath *Flowers* in terminal spikes, usually naked. — Subshrubs, evergreen; South of Europe, of easy culture in dry soil, and propagated by cuttings or seeds.

ɑ 1. S. DU'BIA *L.* The doubtful, *or Rosemary-leaved*, Stæhelina.

Identification. Lin. Sp., 1176.; Less. Syn. Gen. Compos., p. 5.; Willd. Sp. Pl., 3. p. 1783.
Synonyme. S. rosmarinifòlia *Cass.*, according to Less. Syn. Gen. Compos., p. 5.
Engravings. Ger. Prov., p. 190. t. 6.; Lam. Ill., 666. f. 4.; and our *fig.* 1017.

Spec. Char., &c. Leaves sessile, linear, finely toothed, tomentose beneath. Inner bracteas of the involucre lanceolate, elongate. (*Willd.*) An evergreen undershrub. South of Europe. Height 2 ft. to 3 ft. Cultivated in 1640. Flowers purple, fragrant; June and July.

1017. Stæhelina dùbia.

GENUS II.

BA'CCHARIS *R. Br.* THE BACCHARIS, *or* PLOUGHMAN'S SPIKENARD. *Lin. Syst.* Syngenèsia Supérflua.

Identification. Less. Syn. Gen. Compos., p. 204.; Ait. Hort. Kew., ed. 2., vol. 5. p. 28.
Synonymes. Bacchante, *Fr.*; Baccharis, *Ger.*